Springer Geology

The book series Springer Geology comprises a broad portfolio of scientific books, aiming at researchers, students, and everyone interested in geology. The series includes peer-reviewed monographs, edited volumes, textbooks, and conference proceedings. It covers the entire research area of geology including, but not limited to, economic geology, mineral resources, historical geology, quantitative geology, structural geology, geomorphology, paleontology, and sedimentology.

Gad Mohamed El-Qady · Claudio Margottini
Editors

Sustainable Conservation of UNESCO and Other Heritage Sites Through Proactive Geosciences

Springer

Editors
Gad Mohamed El-Qady
National Research Institute
of Astronomy and Geophysics
Helwan, Cairo, Egypt

Claudio Margottini
Science and Technology Office
Embassy of Italy in Egypt
Cairo, Egypt

ISSN 2197-9545 ISSN 2197-9553 (electronic)
Springer Geology
ISBN 978-3-031-13812-6 ISBN 978-3-031-13810-2 (eBook)
https://doi.org/10.1007/978-3-031-13810-2

This Springer imprint is published by the registered company Springer Nature Switzerland AG
The registered company address is: Gewerbestrasse 11, 6330 Cham, Switzerland

Preface

Cultural heritage represents the legacy of humankind on planet Earth. It witnesses millennia of people's adaptation to their environment, as demonstrated in many monuments, sites and cultural landscapes. Such historical landmarks are subject to continuous changes and to the influence of modern growth and development. The impacts and alterations result from several internal and external conditions stemming from both natural and anthropogenic factors, such as rapid (i.e. earthquakes, landslides, floods, debris flows and slope movements) and slow onset (i.e. geological and geotechnical subsidence, soil and coastal erosion, sinkholes and hydro-geological conditions) including climate change. Nowadays, cultural heritage shows evidence of the impact of geohazards and weathering and calls for the need to rethink 'site' conservation and management plans. Consequently, geoscience discipline and affiliated empirical research studies and innovations in technology may need to bring a new paradigm for the preservation of cultural properties providing a resourceful platform for learning.

In the past decades, the shift in disciplines from working inward to opening up to inter-disciplinary ways of thinking draws special attention to the added value of merging Arts with Sciences among other disciplines for better management and preservation of cultural heritage. The 'New School' of thoughts is manifested by many showcases and projects on the mitigation of risks of geohazards whose aim is to maintain the integrity and authenticity of cultural and historical site heritage. The advanced scientific investigations involve a better understanding of the natural processes, coupled with conservation approaches mainly based on sustainable practices including the use of traditional knowledge in the recovery techniques and building local capacities to ensure effective conservation works with time.

In this context, this book titled 'Sustainable Conservation of UNESCO and Other Heritage Sites Through Proactive Geosciences' aims to share the work from various disciplines to share experiences and lessons learned on the management of heritage sites facing the risks of geohazards and other geo-environmental threats. The chapters included in this book will cover the innovation and advances in science and technology on mitigating the impacts of geohazards in heritage sites from all over the world. They highlight the existing diversity in management and preservation approaches under various types of geohazards, different countries and a variety of cultures.

These also include some work related to geohazards, climate change and weathering; satellite applications, monitoring and advanced modelling in geosciences; latest geophysical investigation; innovative mitigation projects, etc., in addition to *some* case studies on mitigation measures applied in various sites that are subject to different types of geohazards. Particular focus will be on the application of traditional knowledge in mitigation.

Gad Mohamed El-Qady
Professor of Geophysics and President
of NRIAG, Helwan, Cairo, Egypt

Claudio Margottini
Former Scientific Attachè at the Italian
Embassy in Egypt
Cairo, Egypt

Professor of Engineering Geology at
UNESCO Chair on Prevention and
Sustainable Management of
Geo-Hydrological Hazards
Florence University
Florence, Italy

Contents

Rock Mechanics, Rock Weathering and Slope Stability

Landslide Mitigation Measures for the Conservation of the Archaeological Site of Mata Ngarau, Orongo Village (Easter Island-Chile) ... 3
Claudio Margottini, Daniele Spizzichino, and Orlando Pandolfi

Geotechnical Design and Mitigation Measures for the Conservation of Akapana Pyramid in the Tiwanaku Archaeological Site (Bolivia) ... 23
Daniele Spizzichino, Julio Condori Amaru, Luca Lombardi, Jose Ignacio Gallego Revilla, Jose Antonio Fernando Merodo, Giorgio Vizzini, and Claudio Margottini

Multiscale Analysis of Geo-Hazards Affecting the Alhambra Cultural Heritage ... 47
José Antonio Fernandez-Merodo, Rosa Martin Mateos, Jose Miguel Azañon, Pablo Ezquerro, Juan Carlos García-Davalillo, Marta Bejar, Gerardo Herrera, Catuxa Novo, Emma Bee, Kyriacos Themistocleous, Giovanni Crosta, Paolo Frattini, Riccardo Castellanza, Gabriele Leoni, Daniele Spizzichino, and Claudio Margottini

Rock Mechanic Characterization and 3D Kinematic Analysis of Katskhi Pillar (Georgia) ... 63
Claudio Margottini, Giovanni Gigli, Daniele Spizzichino, Marco Camorani, Giovanni Fiorini, and Daniele Vicini

Petrological Study on the Roman Mortars from Kom El-Dikka Archaeological Site (Alexandria, Egypt) ... 81
Duygu Ergenç, Nevin Aly, Rafael Fort, Sayed Hemeda, and Mónica Alvarez de Buergo

Experimental Study of Different Electrokinetic Configurations for Desalination of a Brick Wall 89
Jorge Feijoo, Mónica Alvarez de Buergo, Rafael Fort, and Nevin Aly

Geotechnical Engineering

Investigation and Preservation of Historic Foundations 107
Guido Gottardi and Michela Marchi

Authenticity of Foundation and Proactive Conservation of the Central Tower of Bayon Temple Under a Changing Climate in Angkor, Cambodia 127
Yoshinori Iwasaki, Mitsuharu Fukuda, Mitsumasa Ishizuka, Robert McCarthy, Ichita Shimoda, Koichi Nakagawa, Tomofumi Koyama, Takeshi Nakagawa, and Vanna Ly

Slope Instability Induced by Climate Changes on the UNESCO Etruscan Necropolis of Monterozzi (Tarquinia, Italy) 147
Daniele Spizzichino, Gabriele Leoni, Paolo M. Guarino, Daniela Boldini, Saverio Mengoni, Ermanno Marino, Adele Cecchini, and Beatrice Casocavallo

Geophysics

Joint Application of Multiple-Geophysical Surveys for Archaeological Prospection 163
Amin Ibrahim, Khaled S. Gemail, Mahmoud M. Sensoy, Ali El-Khadragy, and Moamen Almaz

The Implementation of Shallow Geophysical Survey for Detection of Some Buried Archaeological Structures in Aswan City, Egypt 189
Abbas Mohamed Abbas, Raafat El-Shafie Fat-Helbary, Ahmed Hamed, Karrar Omar El-Faragawy, Ezzat M. El-Amin, and Gad Mohamed El-Qady

Ground Penetrating Radar for Investigating Painted Walls and Floors of Ancient Buildings 221
Massimiliano Pieraccini and Lapo Miccinesi

Laser Scanning and Virtual Reality

Developments in Digital Documentation of Cultural Heritage 239
Heinz Rüther, Roshan Bhurtha, Ralph Schroeder, and Bruce McDonald

Usage of Laser Scanning Systems to Document the Cultural and Historical Heritage 249
Abdel-Monem Sayed Mohamed, Gad Mohamed El-Qady, Abass Mohamed Abass, Abdel-Hamid Mohammed Elbshbeshy, and Ahmed Elhadi Sherif

HBIM Framework for Rehabilitation of Heritage Buildings 275
Mohamed Marzouk

Hydrogeology

Geophysics' Role in Investigating and Mitigating Groundwater Hazards on Archaeological Sites: Case Studies from "Sphinx-Giza, Kom Ombo Temple-Aswan and Hawara Pyramid-Fayoum" 287
Abbas Mohamed Abbas, Usama Massoud, Hany S. Mesbah, Ayman I. Taha, Mohamed Gamal Abdelmonem, and Gad Mohamed El-Qady

Strengthening the Conservation and Management of Lumbini, the Birthplace of Lord Buddha, World Heritage Property (Lumbini, Nepal) 313
Claudio Margottini, Daniele Spizzichino, Paolo Pagnin, and Luca Maria Puzzilli

Water Infiltration and Waterproofing of Susan-Ri Tomb (North Korea) 333
Claudio Margottini, Ugo Castellotti, Debi Ghoshal, and Rodolfo Lujan

Structural and Environmental Monitoring

Monitoring Cultural Heritage Sites Affected by Geohazards in Cyprus Using Earth Observation 359
Kyriacos Themistocleous

Innovative Structural Monitoring as Tool of Preservation and Valorisation of Monumental Architectures: The Case of Neptune Temple in Paestum (Salerno, Southern Italy) 379
Petti Luigi, Barone Fabrizio, Domenico Greco, and Gabriel Zuchtriegel

Structure Stability Analyses of Chapels Dedicated to Wives of Amun Using Non-Invasive Techniques, Case Study "Madinet Habu Temple—Luxor—Egypt" 393
Ayman Hamed, Nevin Aly, and Mathias Lang

Use of Integrated Regional Data and Local Low Impact Investigations for the Geo-Mechanical Characterisation of Rupestrian Cultural Heritage in Apulia (Southern Italy) 409
Ilenia Argentiero, Maria Dolores Fidelibus, Roberta Pellicani, and Giuseppe Spilotro

Geological Factors Controlling Evolution of Theban Tomb Stability, Luxor 429
Andrea Wolter, Martin Ziegler, Rachael Colldeweih, Andrea Loprieno-Gnirs, Rodrigo Alcaino-Olivares, and Matthew Perras

Seismic Hazard and Archeoseismology

Archeoseismology and the Lost Villages in Northern Syria, the Impact of Large Earthquakes on Cultural Heritage 445
Mustapha Meghraoui and Reda Sbeinati

Integrated Geoscience Investigations in Hittite Imperial Sites Affected by Earthquakes 463
Mahmut Göktuğ Drahor, Ökmen Sümer, Meriç Aziz Berge, Caner Öztürk, Atilla Ongar, Aygül Süel, and Andreas Schachner

Multi-scenario Physics-Based Seismic Hazard Assessment of Cultural Heritage Sites 501
Marco Fasan, Hany M. Hassan, Andrea Magrin, Franco Vaccari, Fabio Romanelli, and Mohamed ElGabry

Seismic Vulnerability Assessment of Historical Heritage Structures, Analysis and Intervention: Application in Skikda City, Northeast of Algeria 517
Hamidatou Mouloud, Lebdioui Saad, Hallal Nassim, Hugo Rodrigues, Assia Harbi, Ammouchi Nesrine, Karima Messaoudi, and Beldjoudi Hamoud

Seismic Assessment of a Cultural Heritage Minaret in Cairo 541
Hany M. Hassan, Mohamed A. Sayed, Marco Fasan, Fabio Romanelli, Claudio Amadio, Ayman Hamed, Mohamed ElGabry, and Islam Hamama

Remote Sensing

Notes on Artificial Intelligence and Big Earth Observation Data for the Study of the Human Past 569
Rosa Lasaponara, Xinyuan Wang, and Nicola Masini

The Potential of Satellite Interferometry for Geohazard Assessment in Cultural Heritage Sites 587
Federico Raspini, Silvia Bianchini, Davide Festa, Matteo Del Soldato, Pierluigi Confuorto, Pablo Ezquerro, and Nicola Casagli

Instability Processes and SAR Data Analysis in the Pompeii Archeological Park 597
Carla Iadanza, Gabriele Leoni, Daniele Spizzichino, Alessandro Trigila, Claudio Margottini, Massimo Osanna, Bruno de Nigris, Alberta Martellone, Mario Costantini, Elena Francioni, Francesco Trillo, and Federico Minati

Remote Sensing Applications for Cultural Heritage Sites Sustainability: Case Studies from Egypt 615
Abdelaziz Elfadaly, Mohamed A. R. Abouarab, Ayaat Shams Eldein, and Rosa Lasaponara

Rome Walls Satellite Monitoring and Protection by Prothego Methodology 641
Gabriele Leoni, Daniele Spizzichino, Marina Marcelli, and Cristina Carta

New Policies for Management and Conservation of Heritage Sites

The Project of Parco Archeologico Del Colosseo and the Italian Network of Archaeological Parks: From Satellite Monitoring to Conservation and Preventive Maintenance Policies 659
Alfonsina Russo, Irma Della Giovampaola, Daniele Spizzichino, Gabriele Leoni, Alessandro Coletta, and Maria Virelli

Adaptation, Traditions and Conservation: The Case of the Asante Traditional Buildings World Heritage Site in Ghana 679
Christopher Wetcher

Management of Cultural Assets in Sudan from the Perspective of Sustainable Development Goals 691
Naoyo H. Sekihiro

Improving the Georgian National Programme for the Preservation of the Rock-Cut Sites Through Interdisciplinary Geosciences 713
Nikoloz Antidze, Claudio Margottini, Tamar Meliva, Daniele Spizzichino, and Manana Vardzelashvili

Aswan Declaration

The Aswan Declaration on Sustainable Conservation of UNESCO and Other Heritages Sites Through Proactive Geosciences 725
Claudio Margottini and Gad Mohamed El-Qady

Rock Mechanics, Rock Weathering and Slope Stability

Landslide Mitigation Measures for the Conservation of the Archaeological Site of Mata Ngarau, Orongo Village (Easter Island-Chile)

Claudio Margottini, Daniele Spizzichino, and Orlando Pandolfi

Abstract The ceremonial village of *Orongo*, part of Rapa Nui National Park, is located in Easter Island, above a small crest looking down the inner wall of the crater of Rano Kau's volcano on one side (NE), and down 300 m cliff to the sea on the other (SW). Prominent historical remains of the village is the ceremonial altar of *Mata Ngarau*, where 7 small houses are radially arranged around a courtyard covered with a rock base relief engraving of *tangata manu* (birdman). Historical documentation and geomorphological analysis of the area have provided clear evidences of slope instability along the southern external flank of the caldera as well as the potential instability of rock masses constituting the *Mata Ngarau* altar. The weathering of surficial strata of the outcropping volcanic material has promoted rock sliding located in various part of the flank, till the limit of the crater edge, where the village is located. The back-slope evolution is now likely affecting even the stability of the individual rock blocks composing the ceremonial altar. An integrated approach to the global stability of the *Mata Ng*arau slope was then implemented, investigating the stability of entire cliff, the stability of upper part of the slope were the village is located and the stability of individual rock masses with petroglyph belonging to ceremonial altar. Rock mechanic data were collected on site and samples investigated in laboratory, to provide a reliable geotechnical model of the cliff. The archaeological top of the site was instead investigated with single station passive seismic analysis, to identify potential rock masses at risk of collapse. A Master Plan of mitigation measures was finally proposed to local authorities, by adopting low visual impact restoration works.

C. Margottini (✉)
former Scientific Attaché, Embassy of Italy in Egypt, 15, Abd El-Rahman Fahmy Str., Garden City, Il Cairo, Egypt
e-mail: claudio.margottini@gmail.com

D. Spizzichino
ISPRA, Geological Survey of Italy, Rome, Italy

O. Pandolfi
Studio Pandolfi, Piazza Duomo 11, Carrara, Italy

C. Margottini · D. Spizzichino
UNESCO Chair on Prevention and Sustainable Management of Geo-Hydrological Hazards, University of Florence, Florence, Italy

G. M. El-Qady and C. Margottini (eds.), *Sustainable Conservation of UNESCO and Other Heritage Sites Through Proactive Geosciences*, Springer Geology,
https://doi.org/10.1007/978-3-031-13810-2_1

Keywords Easter Island · Orongo · Mata Ngarua · Cultural Heritage · Rock slide · Passive seismic analysis

1 Geological and Geomorphological Setting

Easter Island (Rapa Nui) forms part of the Easter Line, a continuous latitudinal chain of volcanic seamounts and islands in the Pacific Sea (27° Lat. S). The island's roughly triangular shape (length: 24 km, surface: 166 km^2) derives from the merging of lava flows produced in the Upper Pleistocene by three main volcanoes, Rano Kau, Terevaka and Poike (Fig. 1). The Rano Kau volcano, where the village of Orongo is located, is made up of numerous basaltic lavas flows and has been reduced in size by faulting and marine erosion. Its crater (1.4 km wide) is a small caldera that collapsed after a late, large explosive phase, as attested by the presence of breccia deposits around the eastern rim of the crater (Fig. 2). This breccias comprises gabbroic blocks dispersed in a completely altered white matrix. A late aa-type lava flow of benmoreite composition extends to the east of the crater. The three small islets ("*motu*"), to the south of "Rano Kau" are composed of whitish and obsidian rhyolite.

The "Mamavai" is a parasite pyroclastic cone adjacent to a small dome of white rhyolite and spherulitic obsidian. Another exposure of obsidian occurs on the upper

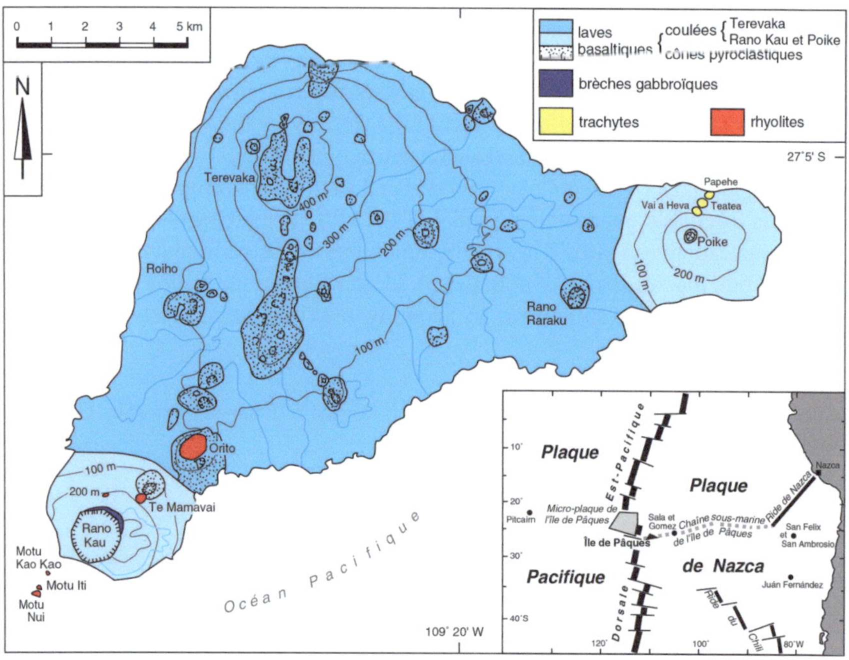

Fig. 1 Geological sketch map of Easter island (modified after Baker et al. 1974); inset: location of Easter Island in the eastern Pacific Ocean (after Hagen et al. 1990, in Déruelle et al. 2002)

Fig. 2 The Ranu Kau volcano and the Orongo village (red arrow) (background photo from https://www.windsurf.co.uk/featured/easter-island/2/)

slope of "Rano Kau", north of the crater. The *Orito* dome is probably related to Rano Kau; it essentially consists of whitish banded rhyolitic lava interlayered with (decimetre–to metre-thick) obsidian layers. Terevaka is the largest and highest (507 m a.s.l.) volcano of the island.

It consists essentially of basaltic lava that flowed mainly southwards and of more than 50 pyroclastic cones.

The Poike, which forms the eastern part of the island, is a basaltic volcano made up of numerous basaltic lava flows (up to 30 flows have been identified in the eastern cliff) with a small central crater. The volcano has been cut by faulting and is limited by vertical cliffs. Most of the outcropping materials are basalt flows (*hawaiite*), intermediate lavas, pyroclastic layers and hyaloclastic tuff. Smaller but no less important is the volcanic apparatus called "Rano Raraku", localized close to Poike and composed by layered deposits of coherent tuff probably erupted from a shallow submarine vent. Indeed all the Moai statues have been carved on its flanks.

Looking in detail at the cliff below the Orongo village, as reported before, the rocky material composing belongs to the Rano Kau volcano group (Fig. 3).

Vezzoli and Acocella (2009) well described the Rano Kau shield sequence, as follow. This is composed by 250-m-thick succession of gently dipping (7°–10°), thin (meter to decimeter in thickness) bedded lava flows (R1 in Fig. 3), consisting mainly of aphyric and microporphyritic (plagioclase, olivine, and clinopyroxene) tholeiitic to alkalic basalts. Outflow lavas of benmoreitic composition (sodic variety of trachyandesite) form a summit platform of extrusive domes and pahoehoe lava

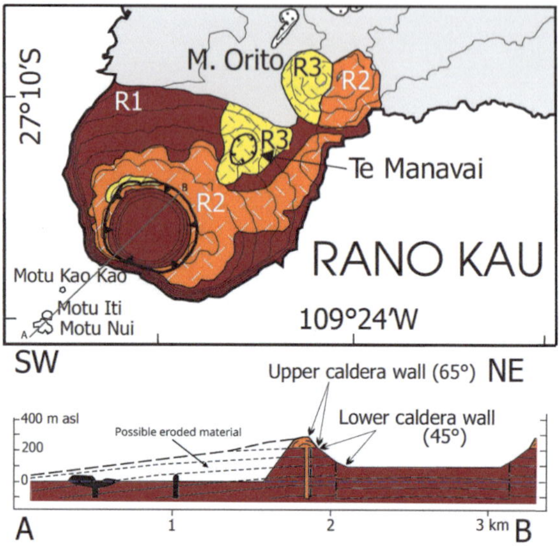

Fig. 3 Geological map and section of Rano Kau volcano. R1 is Shield lavas, transitional to alkalic basalt; R2 Caldera-outflow porphyritic lavas, peripheral small shield; benmoreite; R3 Lava domes, phreatomagmatic breccias, subvolcanic intrusions; rhyolite (excerpts after Vezzoli and Acocella 2009, modified)

flows with radial tumulus structures around the caldera rim (R2 in Fig. 3). Benmoreite is coarsely porphyritic, containing up to 20% plagioclase megacrysts (>10 mm) and minor clinopyroxene phenocrysts; the ground-mass is devitrified, and vesicular, with uniformly distributed rounded vesicles, several millimeters in size.

The final phase of the Rano Kau activity is represented by several monogenetic vents and subvolcanic intrusions (R3 in Fig. 3) of rhyolitic composition.

2 The Orongo Village

Orongo is a stone village and ceremonial centre at the south-western tip of the island, on a spectacular site over the crater edge of the ancient Rano Kau volcano (Fig. 4). Until the mid-nineteenth century 'Orongo' was the centre of the birdman cult, which hosted an annual race to bring the first "*Manutara*" (sooty tern) egg from the islet of Motu Nui to Orongo. The site has numerous petroglyphs, mainly of *tangata manu* mith(birdmen). The first half of the 53 stone masonry houses of ceremonial village's were investigated and restored in 1974 by American archaeologist William Mulloy. In 1976 Mulloy assisted by Chilean archaeologists Claudio Cristino and Patricia Vargas completed the restoration of the whole complex which was subsequently investigated by Cristino in 1985 and 1995. The dwellings are constructed of staked horizontal stone slabs and cantilevered stone roof covered with earth.

Orongo enjoys a dramatic location on the crater lip of Rano Kau at the point where a 250-m sea cliff converges with the inner wall of the crater of Rano Kau. Orongo now has world heritage status as part of the Rapa Nui National Park inscribed in

Fig. 4 Panoramic view of Orongo archaeological site and dwellings on the Rano Kau crater edge

1995 in the UNESCO WHL. Due to its vulnerability is also inscribed in the list of World Monuments Watch since 2000 by the World Monument Fund (WMF).

3 Recent Evolution of the Orongo Cliff

The marine slope of the caldera is presently affected by acceleration of slope instability processes, mainly rockslide and rock-falls.

Landslides are promoted by erosion and weathering processes (Sawada et al. 2001; Vouvé et al. 1994), produced by marine and atmosphere processes on the rocks forming the volcanic apparatus (sequences of lava flow and pyroclastic falls). Landslides phenomena affecting the Rano Kau caldera were collected and inventoried (starting from May 2003) especially along the slope flank located below the archaeological site (Fig. 5).

From a geomorphological point of view, the landslide can be classified as a rockslide, initially triggered as a rotational slide in the upper crown area, evolving into debris avalanche.

The oldest available information are from oral tradition by local inhabitants, who reported about a landslides just below the *Mata Ngarau* (Fig. 6). The morphology of the site seems to confirm such information. Similarly, the collapse at end '90 s of two Moai small statues, located just below the ceremonial altar, is very fresh in the memory of local people. The first real data is from a flight survey on May 2003, after two rock-slide on the cliff.

Two important rock-slides were also identified in the small edge between the caldera and the see, in 2007 and 2008. The crown area of 2007 event is located at approx. 100 m a.s.l. The mobilised material, according to a preliminary morphometric analysis, is about 5,000 m^3. After a field survey undertaken on 21st Jan. 2008, it has been recognized the reactivations of the landslide occurred in 2007. The new landslide was promoted mainly by the toe erosion (sea wave erosion) and by the

Fig. 5 Instability processes affecting the area immediately beneath the Orongo village

Fig. 6 The area below Mata Ngarua were a shallow landslide on weathered lava/debris/vegetation, possibly occurred in the past

slow but inexorable retrogressive activity along the slope. In this site, out of our area on investigation, there is the high risk that future collapse may endanger the lake inside the caldera.

In conclusion, the geomorphological investigation, made by multi-temporal analysis of aerial photos and direct survey, has highlighted a condition of instability, concerning both the scale of slope of the outer edge of the crater below the village of Orongo, both as regards the area immediately below the petroglyph's altar ceremonial. This instability process is actually in evolution as evidenced by the photos from 2003 until today.

As a main framework, it is possible to say that the lowest tholeiitic to alkalic basalts show mainly collapses related to weathering of surficial portion of the rock formation. The upper benmoreitic lava, exhibit lower strength then tholeiitic to alkalic basalts, with an important degree of weathering. Toppling and planar sliding possibly occurred in this formation. In the same benmoreitic lava, the weathered material may produce surficial sliding, also potentially affecting the stability of *Mata Ngarau* altar. Finally, the possibility that several engraved boulders of the ceremonial altar had moved toward the cliff edge was proposed by Hall (1995), one possibly by half meter. Also the stability of individual engraved boulders must be investigated.

4 Site Analysis and Rock Mechanic Proprieties

Geo-mechanical characteristics of the Orongo Village slope-forming rocks outcropping in the study area (mainly basaltic lava), have been reconstructed through geotechnical field techniques and laboratory tests (executed in Italy), on rock blocks (core samples and blocks) of the formation.

As previously mentioned, the material outcropping below the Orongo village is mainly benmoreitic lavas, a massive caldera-inward-flowing lava (Vezzoli and Acocella 2009), with moderate strength, for a thickness of about 40 m. The lowest part of the cliff is composed by 250-m-thick succession of gently dipping (7°–10°), thin (meter to decimeter in thickness) bedded lava flows, consisting mainly of aphyric and microporphyritic (plagioclase, olivine, and clinopyroxene) tholeiitic to alkalic basalts (Vezzoli and Acocella 2009).

From a geological point of view, the presence of many and different volcanic lava flows, is causing a vertical heterogeneity, with the alternance of levels showing different geomechanical parameters. Additionally, the most external part of the lava is affected by intensive weathering (Sawada et al. 2001), producing surficial bands and blocks with even reduced rock mechanic strength.

The geological profile of the cliff below *Mata Ngarua* was investigated with a direct survey by rock mechanic specialists, climbing on the slope (Fig. 7).

The log in the first 25–30 m below the *Mata Ngarua* ceremonial altar, within the benmoreitic lavas, is showing alternance of intact material lava with medium strength, exhibiting blocks with low strength, and weathered in the most surficial part. This is also reflecting the variability of field UCS, collected with L-type Schmidt Hammer.

The following Table 1 is reporting the UCS data from Schmidt Hammer for benmoreitic lava, just below the cerimonial altar.

In situ survey through Scan lines methodology were carried out, to define and reconstruct the main geo-mechanical parameters and indexes (e.g., RMR, GSI, Q system). Strength and deformation parameters were also collected starting from scientific and technical literature, local technical report and geo-structural analysis

Fig. 7 Field work on the Orongo cliff

Table 1 Uniaxial compressive strength on Benmoreite lava below the ceremonial altar

N. of test	Typology of material	Depth from surface	UCS from Schmidt Hammer (MPa)
1	Benmoreite Lava	About–24	48
1 bis	Weathered Benmoreite lava	About–25	14
2	Weathered Benmoreite Lava	About–21	14
3	Benmoreite Lava	About–18 m	35
4	Weathered Benmoreite Lava	About–13 m	13
5	Weathered Benmoreite Lava	About–7 m	15
6	Benmoreite Lava	Basement of Altar	13
7	Benmoreite Lava	Basement of Altar	13
8	Benmoreite Lava	Basement of Altar	17
9	Benmoreite Lava	Basement of Altar	15

(orientation and characteristics of discontinuities) (Barton 2006; Deere and Miller 1966).

Kinematic analysis is showing how toppling is the most favourable condition for collapse, but also planar slide and wedge failure are possible below the altar (Fig. 8).

Samples of local materials have been collected during the field work and investigated in rock mechanic laboratory in Italy (Table 2). Point Load Test for UCS determination were executed as well as and Brazilian test (Fig. 9) for the tensile strength definition. Natural density tests of weathered and intact lavas were also performed.

The parameter obtained from laboratory tests were compared with in-situ data, showing a general agreement. The following Table 2 is reporting the main collected data. This sample are surface intact material. Clearly, the weathered parts are showing less resistant values, as demonstrated by filed investigation, falling in the category of medium-to-weak rocks.

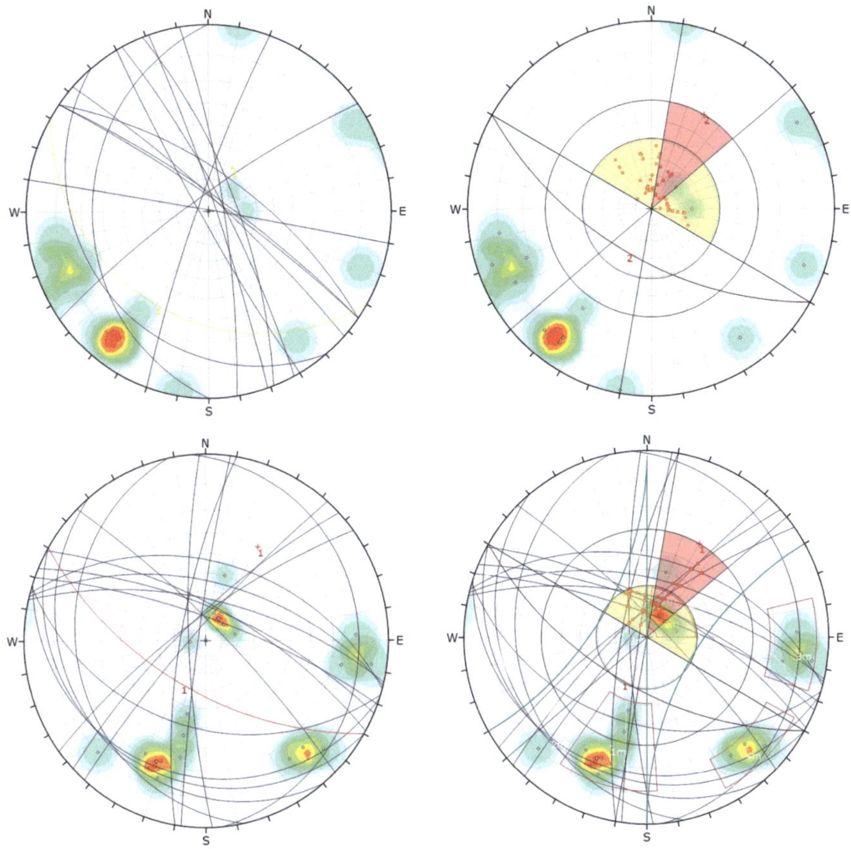

Fig. 8 StereoPlot Analysis along the S-West Flank of the caldera and major detected kinematism for toppling on the natural cliff (top) and altar (lower)

Table 2 Average geomechanical parameters for intact benmoreite lava

Lithology	γ KN/m^3	UCS P. Load (MPa)	Tensile strength σt (MPa)	Tilt test $\phi°$
Benmoreite lava	21.00	32,0	4,0	41

5 Geological Model and Slope Stability Analysis

The cliff below the ceremonial altar of Orongo is not homogeneous. A relevant vertical and horizontal variability has been detected, in terms of geological material and related geomechanical properties. This is the result of the many alternance of lava flows within the same formation (vertical variability), as well as weathering of most external part of cliff and presence of individual block with low strength (horizontal

Fig. 9 Brazilian test for the tensile strength determination

variability). As a consequence, it is difficult to reproduce such variabilities into a geotechnical model, to be used in stability analysis.

The stability of the cliff and the ceremonial altar was investigated making reference to:

(1) global stability of the entire slope;
(2) stability of upper part with benmoreitic lava and important weathering; debris and vegetation may also be involved:
(3) stability of individual engraved boulders composing the *Mata Ngarau* altar (see next paragraph).

In order to provide a set of data suitable to be used in a numerical model, mainly for the global stability of entire slope, reference was made to rock mass strength analysis using the generalized Hoek–Brown failure criterion by means of RocLab, a software by Rocscience©. This data can be used as input for the Shear Strength Reduction analysis, as in this case, to be implemented with the software Phase2, also by Rocscience©.

Having the entire slope constitute in the lower part by tholeiitic to alkalic basalt lavas (shield lava) and the upper part with the presence of benmoreitic lavas (sodic variety of trachyandesite), the following parameters have been obtained by RocLab, also displayed in Fig. 9 and Table 3. Using RocLab, the Hoek–Brown rock mass failure criterions were established, calculating the equivalent Mohr–Coulomb strength parameters (cohesion and friction angle).

Table 3 Adopted parameters for the global slope stability analysis

Rock type	Friction angle (peak) $\phi°$	Friction angle (residual) $\phi°$	Cohesion C (KPa)	Unit weight KN/m^3
Benmoreitic lava	52°	41°	350 kPa	20.0
Shield lava	52°	41°	1920 kPa	21.0

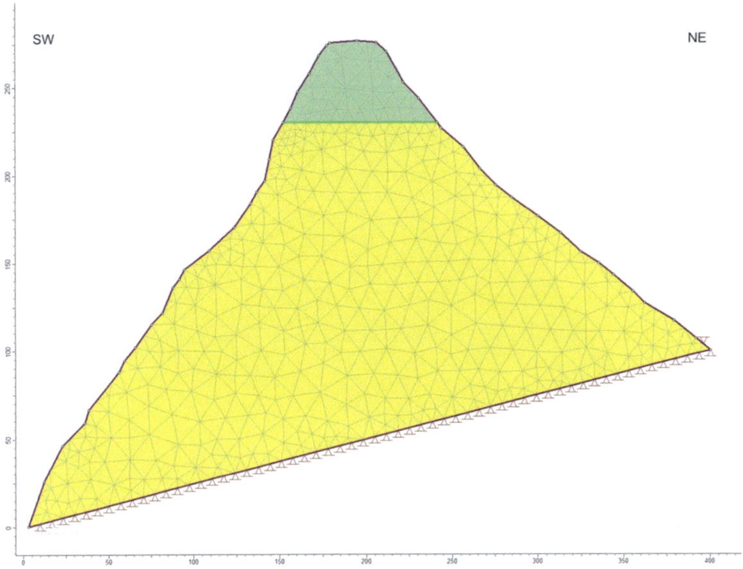

Fig. 10 The geotechnical model adopted for the global stability of the slope. Green is benmoreitic lavas and yellow is shield lava

The best-fit Mohr–Coulomb strength is determined in a failure envelope, depending from local principal and/or shear-normal stress. In the present case, the Mohr–Coulomb linearization was calculated considering the actual height of the slope. GSI was estimated between 55 and 65.

The above data have been used in a finite element stress analysis, such as Phase2 by Rocscience©, in order to investigate and assess the evolution of the cliff for various shear strength conditions (Fig. 10). The used approach is the Shear Strength Reduction (SSR) method, which allows to automatically perform a finite element slope stability analysis, and compute a critical strength reduction factor for the model. The critical strength reduction factor is almost equivalent to the "safety factor" of the slope (https://www.rocscience.com/).

The basic concept of the SSR method is actually quite simple. The strength parameters of a slope are reduced by a certain factor (SRF), and the finite element stress analysis is computed. This process is repeated for different values of Strength Reduction Factor (SRF), until the model becomes unstable (the analysis results do not converge). This determines the critical strength reduction factor (critical SRF), or safety factor, of the whole slope.

The Finite Element Shear Strength Reduction method is a robust alternative to limit-equilibrium slope stability methods. It is particularly beneficial in situations in which stress has a dominant influence on stability (Hamma et al. 2005).

According to the above data, but also in agreement with geomorphological and geotechnical survey, a global collapse of the entire slope is quite unlike to occur, with a Strength Reduction Factor, almost equivalent to Factor of Safety (FoS), of about

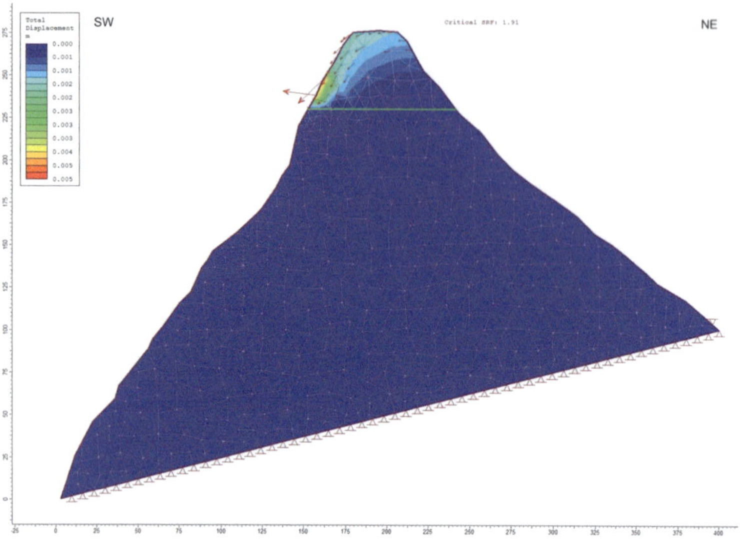

Fig. 11 Total displacement for the global Orongo slope, with SRF of 1.9 concentrated in the upper part of the cliff, below *Mata Ngarau*

1.9 and concentrated in the upper part of the slope (Fig. 11). To mobilize the entire slope a larger reduction of shear strength is required.

Looking in detail to the geology some tens of meters below *Mata Ngarau*, we need to add a shallow band of weathered lava, which can assume the following values, according to Mohr–Coulomb generalization of Hoek–Brown failure criterion (Roclab) and field data (see Table 4).

The related SRF from Phase2© is equal to 0.94, equivalent to the FoS of limit equilibrium model, somehow confirmed with FlacSlope, a finite-difference modelling code from Itasca©, which it has been implemented in order to detect the safety factors along the same topographic section, just below the ceremonial altar of *Mata Ngarua* (Fig. 11). In this case, in dry condition, the FoS value has been calculated equal to 1.05, close the potential unstable condition for the slope (Fig. 12). The minimum value for the Safety of Factor drops below unity (the simulation has been performed only in static condition) under the fully saturated slope condition. This assumption is in agreement with meteorological data showing important precipitant precipitation prolonged in many months, then favouring the infiltration (Table 5).

Table 4 Adopted parameters for weathered benmoreitic lava

Rock type	Friction angle (peak) ϕ°	Cohesion C (KPa)	Unit weight KN/m^3
Weathered benmoreitic lava	41°	81 kPa	16.0

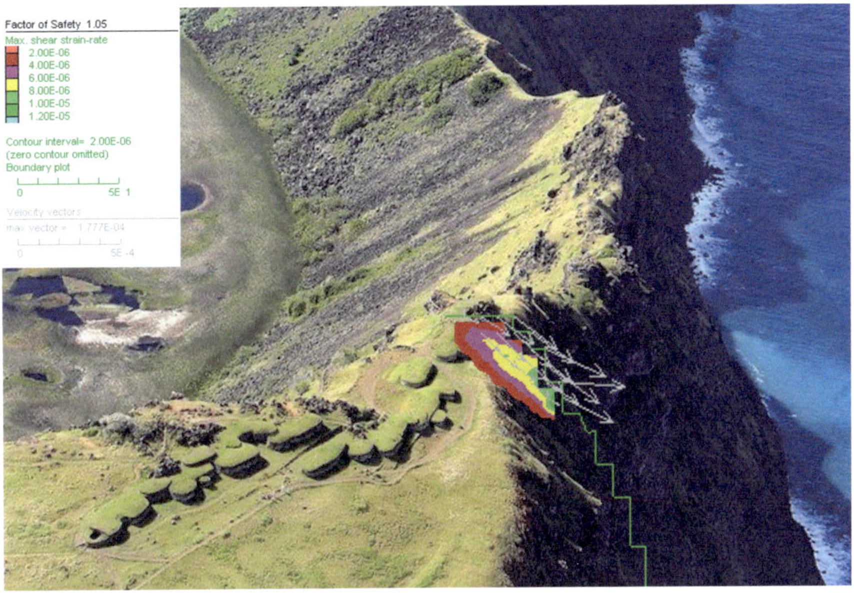

Fig. 12 The most critical surface for failure, according to a finite-difference modelling code from Flacslope by Itasca©

Table 5 Precipitation in Easter Island (https://climatologia.meteochile.gob.cl)

	Jan	Feb	Mar	Apr	May	Jun	Jul	Aug	Sep	Oct	Nov	Dec
1961–1990	72.8	84.8	95.6	120.7	152.9	106.3	105.1	93.8	86.8	68.0	74.0	86.4
Max 24 h	8.3	42.2	33.8	26.1	44.4	97.6	18.2	23.9	27.7	8.6	31.6	64.5

The occurred phenomena, mainly those of 2003, are suggesting that the weathering is surely more active in all the surface of slope. As a consequence, retrogressive phenomena may be possible in various part of the cliff and not only in the upper part. Thus, for the purpose of conservation of *Mata Ngarua* altar, the obtained results, concentrating the stress below the ceremonial altar, are more conservative and acting in favour of safety of the archaeological site.

6 Passive Seismic Analysis

After having investigated the slope stability of the whole cliff and the slope stability of the direct foundation of *Mata Ngarau* altar, the final issue is to understand the stability of individual engraved boulders composing altar. According to Hall (1995), several blocks had moved toward the cliff edge, in one case about half meter (Fig. 13).

Fig. 13 Recent photo of the cliff side reported by Hall (1995), as example of boulders possibly moving towards cliff face

The identification of unstable rock blocks is still unclear in terms of displacement, temporal occurrence and future evolution. Monitoring network can offer a great support but only during rupture phase or remarkable displacement, depending from the adopted instruments. It is also costly and depending from logistics and visual impact.

Considering the above constrains, it was decided to investigate the stability of engraved boulders by using the passive seismic monitoring and measuring micro-tremors and vibration (environmental noise) on cliffs (Tanaka et al. 2008).

The adopted method made use of a portable velocimeter (TrominoTM® by Micromed®), which is inexpensive and easy to operate.

This technique is measuring seismic passive environmental noise, from which, through vibrational analysis, it is possible to derive response spectra, Horizontal to Vertical Spectral Ratio (HSVR), particle motion, and other elaborations (Nakamura 1989).

Following the methodology proposed by Tanaka et al. (2008), the vibration that has been measured is analysed in the different rock blocks, to evaluate their stability and connection to bedrock in terms of response spectra, HSVR and particle motion.

Fig. 14 The *Mata Ngarau* ceremonial altar of Orongo in Easter Island and the individual monitored blocks

Data from Mata Ngarau altar were finally compared with e records from a stable area outside the site, in free field.

The investigation site is reported in the following Fig. 14.

The powers spectra graphs, which describes the distribution of power into frequency components composing that signal, is showing that in five blocks (2, 5, 8, 11 and 13) concentrates the energy in the frequencies 7–8 Hz. In the free field graph the energy is almost completely at lower frequencies (Fig. 15).

The H/V graph, is reporting the ratio between the Fourier amplitude spectra of the horizontal (H) to vertical (V) components of ambient noise vibrations recorded at one single station.

The method was relaunched and divulgated by Nakamura (1989; 2001) as a fast tool to measure the local seismic amplification. Consensus has not been reached on this point, although it is widely recognized that HVSR is capable of providing a reliable estimate of the main resonance frequencies of subsoil, which is nevertheless a crucial information for the seismic engineer. In the present case study may provide an useful tool discriminating blocks responding similarly and possibly connected among them or to bedrock from isolated blocks, moving independently.

According to the monitoring, boulders n. 5, 11, 13 and 14 are showing a well recognisable peak at 7–8 Hz while in others such peak is not very evident (Fig. 16). In free field is no reported.

Finally, the particle motion analysis is showing the vibration of horizontal components. The blocks n. 5, 11, 13 and 15 are showing a vibration higher than others, even if the differences are not very high (Fig. 17).

According to present elaboration is possible to notice that the boulders n. 5, 11 and 13 behaviour are atypical in any elaboration. Since they are on the edge of the

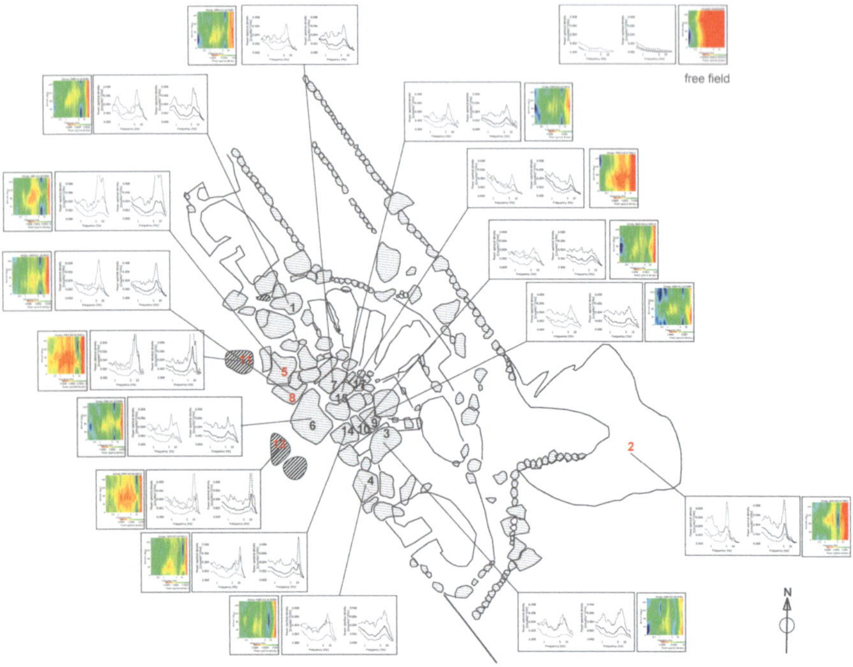

Fig. 15 Power spectra density of the investigated blocks. In red are atypical signals. N. 2 is also free field. The sea is on West

cliff, this is suggesting that they are disconnected by the plateaux/bedrock where all the other boulders are resting and need a special attention (Fig. 18).

The limited differences among all the blocks is not suggesting an immediate action but surely a precautionary stabilisation. As a confirmation it is possible to say that the monitoring was performed in 2009 and, after 10 years, they did not.

Clearly, what above said is valid if the morphological conditions are not evolving from present state, for instance as a consequence of shallow debris/vegetation landslides that may occur in the cliff, below the blocks.

7 Discussion

The conservation of ceremonial altar of *Mata Ngarua*, Orongo (Easter Island) is potentially affected by collapse of individual engraved boulder or shallow slide of weathered lava/debris/vegetation just below the altar. The shallow landslide is likely also involving the most external rock blocks of the archaeological site.

A global instability affecting the entire slope is considered of low probability, according to the adopted numerical model and available data.

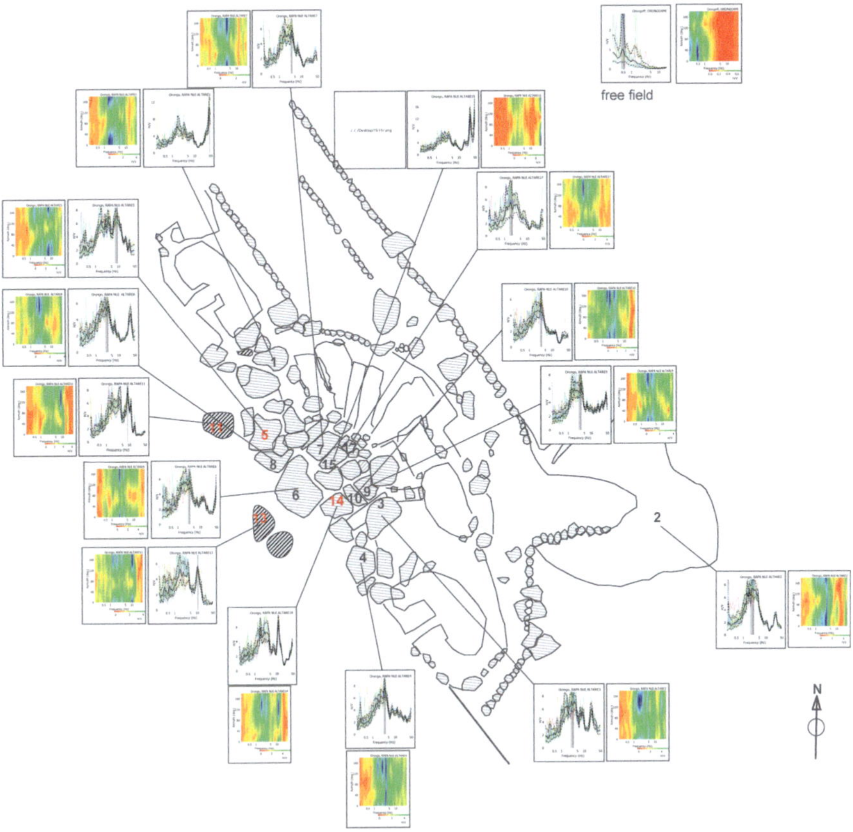

Fig. 16 Horizontal to Vertical Spectral Ratio (HSVR) of the investigated blocks. In red are atypical signals. N. 2 is also free field. The sea is on West

In order to reduce the possibility of triggering new slope instability phenomena along the Rano Kau cliff beneath the Orongo village, there is the need to consider a general master plan of mtitgation measure considering: proper waterproofing of ground surface to avoid infiltration and pore pressure excess; strengthening the cliff blocks with anchors and grouting; consolidation of individual unstable blocks on the cliff and iron net with green re-forestation beneath the altar (nature-based-solution).

As far as the endangered blocks from the ceremonial altar, after having executed the above measure for consolidation of the cliff, it will be possible to realize special support anchored in depth, avoiding them to rotate down in the slope. A temporary support during the consolidation of the cliff is also considered fundamental.

All the activities should consider the minimum visual impact and the adoption of nature-based-solutions (Margottini and Spizzichino 2021). Also the involvement of local manpower will be essential to ensure periodical maintenance.

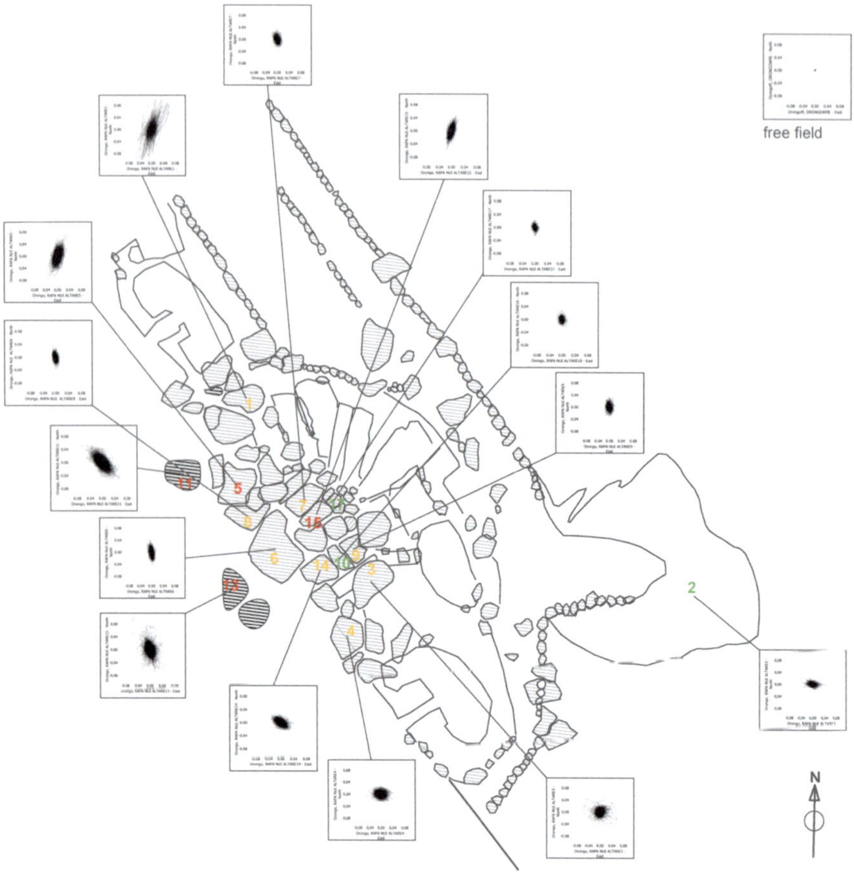

Fig. 17 Particle motion of the investigated blocks. In red are atypical signals. N. 2 is also free field. The sea is on West

8 Conclusion and Follow Up

The present paper reports the main output concerning landslide risk assessment affecting the Orongo village, part of Rapa Nui National Park (Easter Island). The historical remains are composed by numerous petroglyphs that represent the ancient ceremonial of the birdman cult.

From geological and geomorphological point of view the area is located on the southern part of the island on the top of the external border of the Rano Kau caldera, constituted mainly by lava flows and basaltic layers.

The spectacular location of the village, threat its stability condition due to recent landslides (mainly surficial) affecting rock slope flank (reactivation of several phenomena below the ceremonial sacred area). Detailed analysis of landslide activity was carried out through a 2D-GIS modelling coupled with in situ and laboratory test.

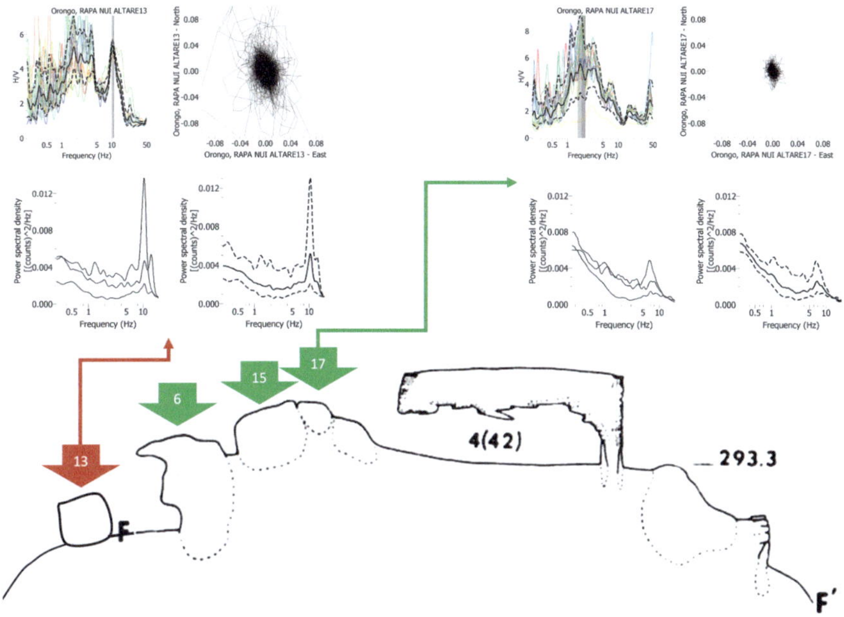

Fig. 18 A topographical profile after Mulloy (1975) with superimposed the block n. 13, considered more susceptible to slide. In the upper part are Horizontal to Vertical Spectral Ratio (HSVR), horizontal particle motion and power spectra density for two selected blocks, n. 13 unstable and n. 17 stable

Geotechnical stability model were implemented in order to define present slope stability conditions and possible future evolution of the slope. The main output of the stress strain analysis and limit equilibrium method implemented, suggest general stability condition (in static condition), of the whole area.

Passive seismic analysis through HVSR and particle motion methodologies were carried out in order to verify stability conditions of each single block constituting the ceremonial altar. At the moment the main outputs of the analysis suggest that no block is completely disjointed and in conditions of incipient collapse. Even if the monitoring of the whole area must still be promoted and continued.

A preliminary general Master Plan of mitigation measures was discussed by the adoption of low visual impact restoration works such as passive bars, nails and bolts coupled with steel mesh and drainage system in order to reduce erosion and loss of cohesion. This to provide useful tools for local authorities in implementing preventive landslide risk mitigation measures for the protection of the cultural heritage in Orongo.

Acknowledgements The authors are grateful to all the native communities on the island, and to the council of the elderly for their support during the survey and campaign activities.

References

Baker PE, Buckley F, Holland JG (1974) Petrology and geochemistry of Easter Island. Contrib Mineral Petrol 44:85–100

Barton N (2006) Rock quality, seismic velocity, attenuation and anisotropy. CRC Press

Deere DU, Miller RP (1966) Engineering classification and index properties for intact rock. Technical Report n. AFWL-TR-65–115. Air Force Weapons Laboratory, Kirkland Air Base, New Mexico

Déruelle B, Figueroa AO, Joron J, Schilling DM, Silva PC, Hervé AF, Demaiffe D (2002) Le volcanisme de l'île de Pâques (Chili). Géologie de la France 2:53–67

Hagen RA, Baker NA, Naar DF, Hey RN (1990) A SeaMARC II survey of recent submarine volcanism near Easter Island. Marine Geophys Res 12:297–315. Pisa University

Hall N (1995) Observation and preliminary assessment of the stability of the Mata Ngarau area, Orongo Ceremonial Village, Rapa Nui (Easter Island). Internal report for the Centro de Restauracion y Conservation, Chile and the World Monuments Fund

Hammah RE, Yacoub TE, Corkum BC, Curran JH (2005) The shear strength reduction method for the generalized hoek-brown criterion. American Rock Mechanic Association

Margottini C, Spizzichino D (2021) Traditional knowledge and local expertise in landslide risk mitigation of world heritages sites. In: Sassa K, Matjaž M, Sassa S, Bobrowsky PT, Takara K, Dang K (Eds.) Understanding and reducing landslide disaster risk, vol 1. Sendai Landslide Partnerships and Kyoto Landslide Commitment. Springer Verlag

Mulloy W (1975) Investigation and restoration of the ceremonial centre of Orongo, Easter Island. Bulletin n. 4Easter Island Committee, International Fund for Monuments Inc.

Nakamura Y (1989) A method for dynamic characteristics estimation of subsurface using microtremor on the ground surface. Quart Rep Railway Technol Res Instit 30:25–33

Sawada M, Koezuka T, Kohdzuma Y, Inoue S, Bahamondez M (2001) In-situ weathering tests of conservation materials applied to volcanic tuff samples from Ahu Tongariki, Easter Island. In: Stevenson CM, Lee G, Morin FJ (eds) Pacific 2000. Proceedings of the fifth international conference on Easter Island and the Pacific. Hawai'I preparatory Academy, Kamuela, Hawai'I

Tanaka H, Fujisawa K, Asai K (2008) Identifying unstable rock blocks by measuring micro-tremors and vibration on cliffs. Adv Geosci 14:165–171

Vezzoli L, Acecella V (2009) Easter Island, SE Pacific: an end-member type of hotspot volcanism. Geol Soc Am Bull

Vouvé J, Aurouze J, Lacazedieu G, Malaurent Ph., Vidal P., Vouvé F. (1994). Deterioration phenomena of lavas abd tuffs under subtropical and temperate conditions: field and laboratory study using multisequential methodology. In E. Charola (Editor), "Lavas and volcanic tuff". Proceedings of the ICCROM International meeting. Easter Island, Chile, October 25–31, 1990.

https://climatologia.meteochile.gob.cl. Accessed 15 Jan 2021

Geotechnical Design and Mitigation Measures for the Conservation of Akapana Pyramid in the Tiwanaku Archaeological Site (Bolivia)

Daniele Spizzichino, Julio Condori Amaru, Luca Lombardi, Jose Ignacio Gallego Revilla, Jose Antonio Fernando Merodo, Giorgio Vizzini, and Claudio Margottini

Abstract Tiwanaku is an Archaeological World Heritage Site, located at an elevation of 3,885 m.a.s.l., south of Lake Titicaca (La Paz), in the high plateau region of Bolivia. The present paper reports the actions carried out in the frame of the project: "*Preservacion y conservacion de Tiwanaku y la piramide de Akapana, Bolivia*". The consultancy was requested by UNESCO Office in Quito, following previous activities conducted since 2010 and directly by the CIAAAT. During the first mission in the archaeological site of Tiwanaku (from 9 to 16 May 2016), an interdisciplinary team, carried out the preliminary diagnosis concerning the geotechnical processes affecting the Akapana pyramid. After that, another filed mission was carried out in order to implement a general master plan of actions aimed at the conservation and risk reduction of the whole pyramid. For this scope during 2017 two different field survey were carried out. Main scope of the activities was: (i) prepare and develop a long term executive full proposal conservation project for the Pyramid; (ii) implement a Terrestrial Laser Scanning (TLS) survey in order to produce a very detail topographical setting of the whole pyramid; (iii) collect core drilling samples in the Akapana

D. Spizzichino (✉) · G. Vizzini
ISPRA- Geological Survey of Italy, via V. Brancati 48, Rome, Italy
e-mail: daniele.spizzichino@isprambiente.it

J. C. Amaru
CIAAAT, Tiwanaku, Bolivia

L. Lombardi
UNIFI, Firenze, Italy

J. I. G. Revilla
Museo de Segovia, Segovia, Spain

J. A. F. Merodo
IGME, Madrid, Spain

C. Margottini
MAECI, Italian Embassy, il Cairo, Egypt

D. Spizzichino · L. Lombardi · C. Margottini
UNESCO Chair, Florence, Italy

G. M. El-Qady and C. Margottini (eds.), *Sustainable Conservation of UNESCO and Other Heritage Sites Through Proactive Geosciences*, Springer Geology,
https://doi.org/10.1007/978-3-031-13810-2_2

pyramid for laboratory analysis; (iv) perform stability analysis 3D Model; (v) identify mitigation measures for a test site. After the summer 2017 field mission, the main topographical outcomes of the TLS survey were elaborated, incorporated and adopted for the implementation a pilot project of mitigation measures. Final design has been implemented along the east flank of the pyramid, and support was given to. local authorities and workers during all the mitigation measures implementation.

Keywords Akapana · Cultural heritage · Erosion · Mitigation · Risk

1 Site Location and Archaeological Setting

Tiwanaku is an Archaeological World Heritage Site, located at an elevation of 3,885 m.a.s.l., south of Lake Titicaca, in the high plateau region of the Plurinational State of Bolivia. It is in the province of Ingavi, department of La Paz, 70 km far from the city of La Paz. An area of 700 hectares has been calculated for the ancient urban settlement, and 40 hectares for the ceremonial core, where main examples of monumental architecture are located (see Fig. 1).

An updated archaeological prospection using UAV was carried out, in order to support the definition of the real extension and limits of each archeological zone and the associated regulations to be applied in terms of land use, preservation of the archaeological heritage, and cultural landscape.

Fig. 1 Location of Tiwanaku (Bolivia)

1.1 Archaeological Setting of the Area

The archaeological research refers to several periods of occupation, to different styles and techniques of construction, and to different uses given to the built-up area. In synthesis, a Village Period [Período Aldeano] is mentioned for the area, at around 1200 B.C., with crops of potatoes, oca [a root crop], and quinoa, related with the cultural complexes known as Chiripa and Wankarani (Ponce 1981; Kolata 1993); a Formative Period [Período Formativo] of the Tiwanaku culture in a society with communal government, between 100 B.C. and 250 A.D., which is transformed into a regional state between 250 and 500 A.D.; becoming a pan-regional theocratic state between 500 and 800 A.D. (Januseck 2004).

In the Formative period, the sites were centred on a sunken patio and raised platforms structures that possibly derive into pyramidal constructions (i.e., Akapana) across next periods, and with large inhabited compounds around them for the people, who took care of the ceremonial buildings and organized its maintenance and the exercises conducted there.

Later, the orientation was changed to an east–west axis and a central plaza was formed. At first, the early structure was modest, but with time construction projects were executed to produce the massive architectural monument.

New transformations of the previous structures took place in Akapana, Kalasasaya, and other parts of the city (Fig. 2). A large moat of rectangular shape was built, delimited by large clay platforms that mark its internal perimeter. In its interior the space is reconstructed to configure a ceremonial core, gradually eliminating the neighbourhoods that were arranged next to the main public buildings (Janusek 2004; Gallego-Revilla and Pérez-González 2018). Residential structures identified around the ceremonial buildings are adobe constructions, smaller, rectangular, located in relationship with streets. These compounds were moved outside, expanding the inhabited area of the city to about 750 hectares, and establishing a clearer formal division between private spaces and those dedicated to social, political and religious activities (Fig. 2).

It is important to know that historically, continuous works of reconstruction, maintenance, and repairs were carried out in the monuments. (Januseck 2004:487, Vranich 2006). Also, the drainage systems had to be cleaned and repaired, and in the case of Akapana several events of repair and reconstruction work have been identified (Vranich 2009), as well as for the sunken plazas. The importance of the religious ideology in the development of the Tiwanaku State and its socio-political interaction is also noted. It is believed to have achieved the integration of different societies, settled in environments with different resources, which maintained their own identities, practices, and authorities. Tiwanaku had a population characterized by the hierarchy of its social structure, with authorities and specialists, and the surpluses necessary for the construction of the monumental public architecture, based on clay-made platforms, covered mainly by red sandstone (*arenisca roja*) or andesite blocks, some of them massive.

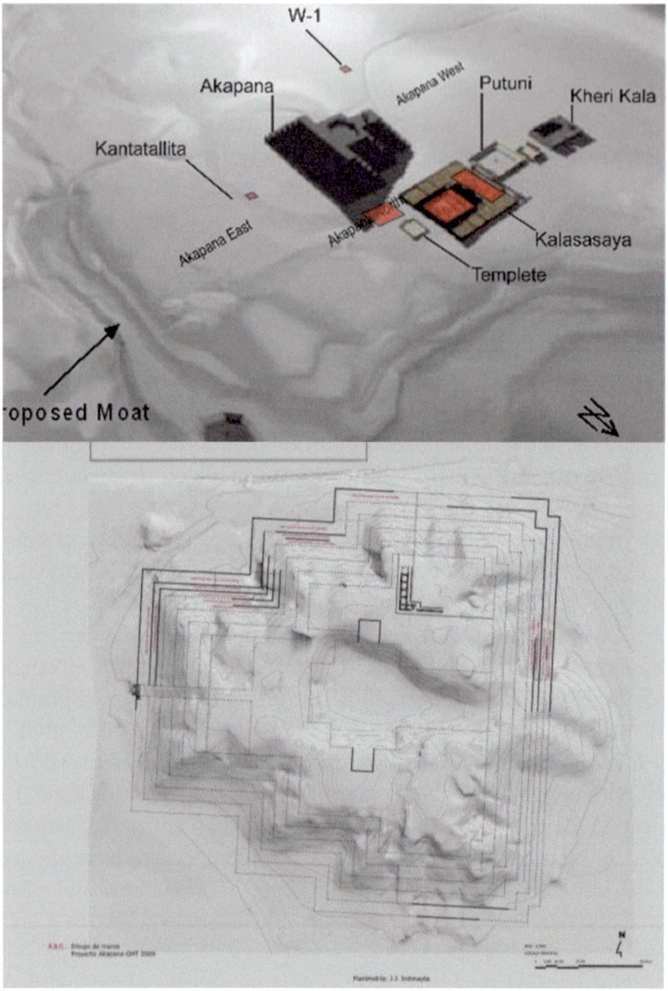

Fig. 2 Main archaeological buildings (up), Akapana pyramid detail (bottom)

The Akapana pyramid, the biggest pre Inca structure in the *Andean* world, is a particular symbol of the Tiwanaku archaeological culture. The seven stepped pyramid earth mound was originally covered and sustained by stones that were later used as rock quarry from the ending moment of Tiwanaku culture up to recent days (Fig. 2). Its stones have been reused into other historic buildings, as Puma Punku Inka reconstructions (Vranich 2006), church of San Pedro and Spanish governor's structures in Tiahuanaco village, as well as for the ballast of the current railway track. This has turned the inner earth mound very exposed, fragile and susceptible to soil erosion and weathering.

Table 1 Optimum measure references of Akapana Pyramid

Platform number	Area (m^2)	Perimeter (m)
First	32.12	774
Second	29.12	750
Third	27.62	726
Forth	25.48	702
Fifth	23.41	678
Sixth	21.41	654
Seven	19.48	630
Total sum	178.63	4.914

Based on what we know from different excavations, as well as our own evaluation works prior to the intervention, the dimensions of the building in its finished stage would correspond approximately to that described in Table 1 (Gallego-Revilla and Pérez-González 2018). In this sense, the clay core that composes it would involve a total of about 339.158 m^3. From the specific weight of this material, we can infer that the mass of the core of the structure, at its final moment, would reach 6.441,4 tons. In the case of the red sandstone that composes the facades, we would be talking about 18.108 m^3, which according to its specific weight, the clay volume would have to add a weight of 471,08 tons. In any case, a very remarkable effort for a culture that did not know the wheel.

2 Instability Processes Threatening the Pyramid

The Akapana pyramid is a great mound of earth, composed of clay, gravel, and sandy materials from the river and lake deposits of the valley, with a large hole at the summit, structured as seven raised platforms building, with at least one stairway and evidences of topping religious buildings and monoliths (see Fig. 3).

At the present, the pyramid is suffering extensive degradation due to phenomena such as:

- Rainy erosion and washout;
- Superficial mass movement;
- Solifluction
- Wetting-induced deformations;
- Creeping;
- Softening and swelling

The main geomorphological process (Spizzichino et al. 2019) affecting the Akapana pyramid induced by climate (e.g., rainy erosion, washout and snow melt) are reported in the following flow chart schema (Fig. 4).

Fig. 3 Akapana pyramid top view

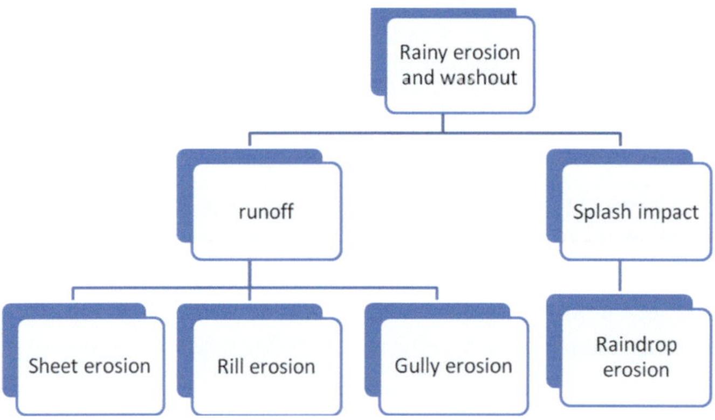

Fig. 4 General flow chart of erosion process affecting the Akapana pyramid in Tiwanaku

The majority of the above-mentioned phenomena are triggered by the uncontrolled flow of the rainwater and snowmelt (slow or rapid), along depressions (natural or manmade) but which, also, infiltrates and deforms the materials of the deposits and seems to be affecting the stability of the internal structures.

Figure 6 is reporting the major geomorphological features and processes affecting the pyramid of Akapana collected after new filed survey campaign carried out on May 2016 and confirmed by the April 2017 and 2018 last mission.

Starting from this it is has been possible to note: (i) the large toe deposit originated by the erosion/dismantling of the earth mound; (ii) some potential superficial mass movement and creeping (iii) the evidence of tilted walls, suggesting internal active deformation.

With reference to the erosion of the mound, there is a clear evidence (rill erosion and small valleys on the mound) that a large part of the original structure has been affected by erosion and deposited at the base or foot of the pyramid; moreover, it is not possible to exclude that the stepped design in the shape of the "*chakana*" or Andean cross could have played some roles in such erosion.

In the Fig. 6 is also possible to notice that the erosion and dismantled processes are particularly evident (in terms of magnitude and intensity) in all the flanks of the pyramid in which the terraced steps are absent (south west side, south side and south east side).

3 Geophysical and Geotechnical Caractherisation of the Akapana Pyramid

In order to understand the homogeneity of the earth mound, a geophysical investigation, was required by UNESCO (Vella et al. 2017). Electric tomography was then developed on June 2010 (Fig. 5).

According to these data no important contrast of resistivity was identified, suggesting that no major lithological inhomogeneity is occurring inside the mound. In addition, four samples of natural soil, constituting the Akapana pyramid, were collected during the April 2017 field mission.

The main objective collecting samples was to define (through a proper laboratory tests) the main physic and mechanical proprieties of the soil. In order to reduce the disturbance (always affecting the samples of natural soil) the collection was performed directly on the pyramid (S1, S2, S3 and S4).

In addition, three more samples (S5, S6 and S7) were investigated consisting of a compound (Argamasa) made up of variable percentages of silt, sand and gravel, coming from the site, mixed with straw and Penka.

The scope was to verify the physical–mechanical characteristics also of the materials to be used for the future conservations actions and mitigation measures.

The laboratory test program, provide the following analysis:

- Unit weight of solid;
- Granulometry (sieving and hydrometric).
- *Atterberg* limits and indices.
- Permeability.
- Unconfined compression tests (ELL).
- Direct shear test (on reconstituted sample) (Table 2)

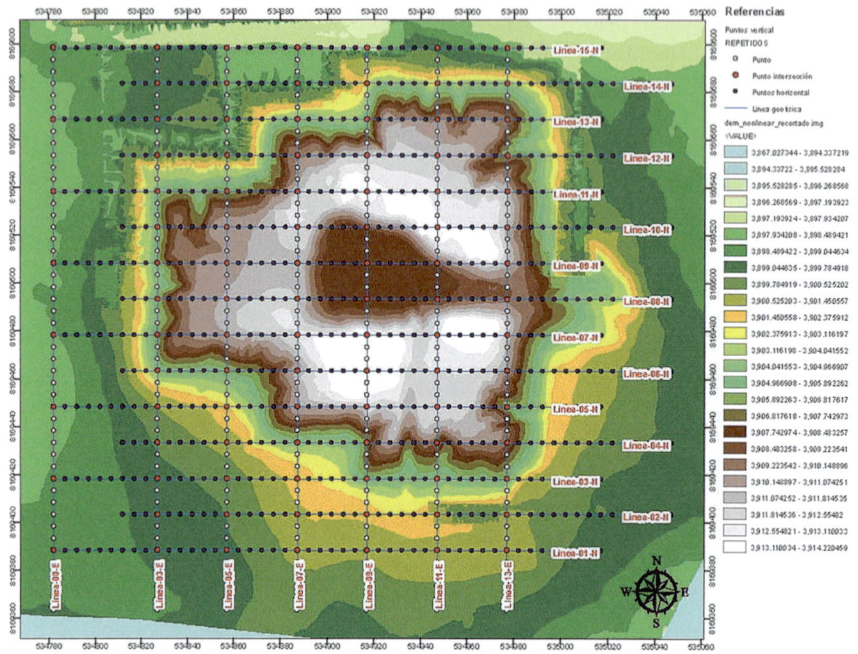

Fig. 5 Electrical tomography (Lasaponara and Masini 2014)

Akapana Pyramid is constituted by a homogeneous backfill, mainly silt and clay in equal percentages, while the Arganasas compound has variable percentages depending on the intended use ("Enlucidos" for the plaster, "Load-bearing or structural" for the load-bearing structures and "Impermeables" for the areas to be waterproofed).

The plot of grain size distribution is reported in Fig. 7.

The determination of Atterberg limits allowed classifying the samples as: "inorganic clay medium and low plasticity". The relative's values are reported in the next Table 3.

3.1 *Unconfined Compression Test on Reconstructed Samples of Argamasas*

Free expansion permeability and compression tests (ELL) were carried out on the *Argamasas* samples. The permeability tests showed values of the permeability coefficient (K) between 1*10–6 cm / s and 1*10–7 cm / s; the values relating to simple compression, performed on reconstituted specimens, show values of 300 and 500 kPa (Fig. 8).

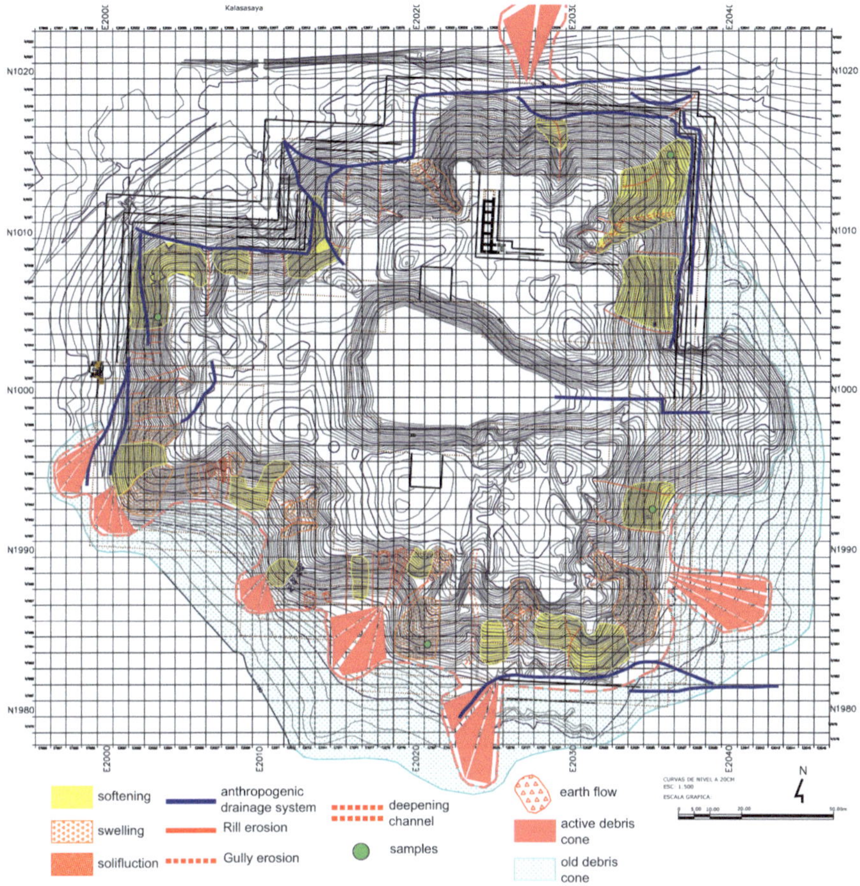

Fig. 6 Local detail geomorphological and active processes map

Table 2 Grain size distribution

Samples	Gravel	Sand	Silt	Clay
%				
1	0	7	44	49
2	0	3	60	37
3	1	1	58	41
4	2	16	46	36
5	3	41	32	24
6	16	12	38	34
7	27	13	36	24

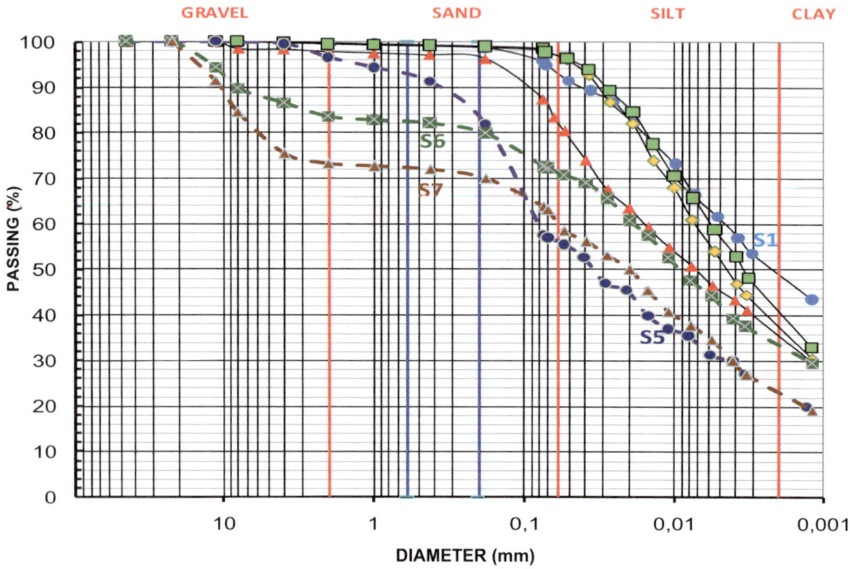

Fig. 7 Cumulative grain size distribution

Table 3 Atterberg limits and indices

Samples	WL	LP	IP	A
1	50	23	26	0,54
2	39	21	18	0,48
3	44	22	21	0,52
4	36	19	17	0,47
5	29	28	11	0,46
6	38	20	17	0,5
7	40	21	18	0,75

3.2 Direct Shear Test on Reconstructed Samples

Two direct shear tests were performed on reconstructed (disturbed) samples in order to obtain a preliminary estimation of strength parameters of the material.

The material is characterized by an average unit weight equal to 15.7 kN/m^3, a Cohesion C equal to 9 kPa and an average friction angle (ϕ) equal to 24° (see Fig. 9).

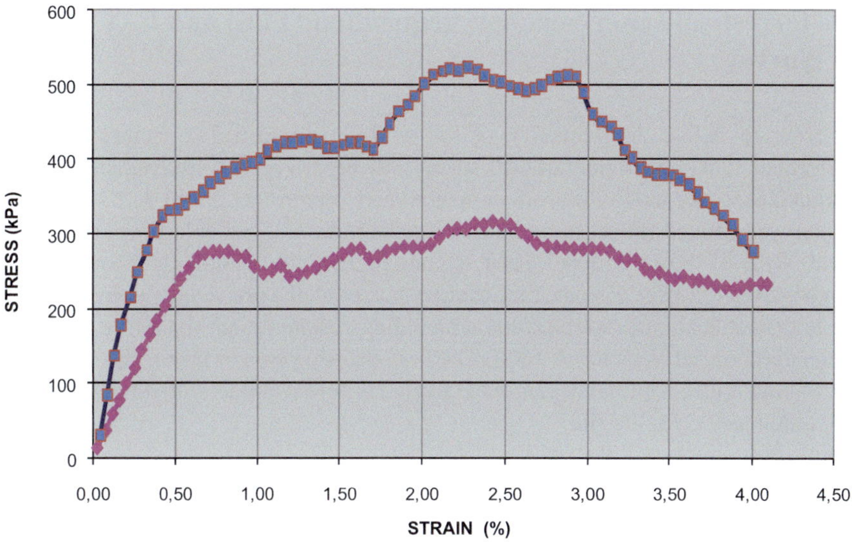

Fig. 8 Unconfined compression tests plot

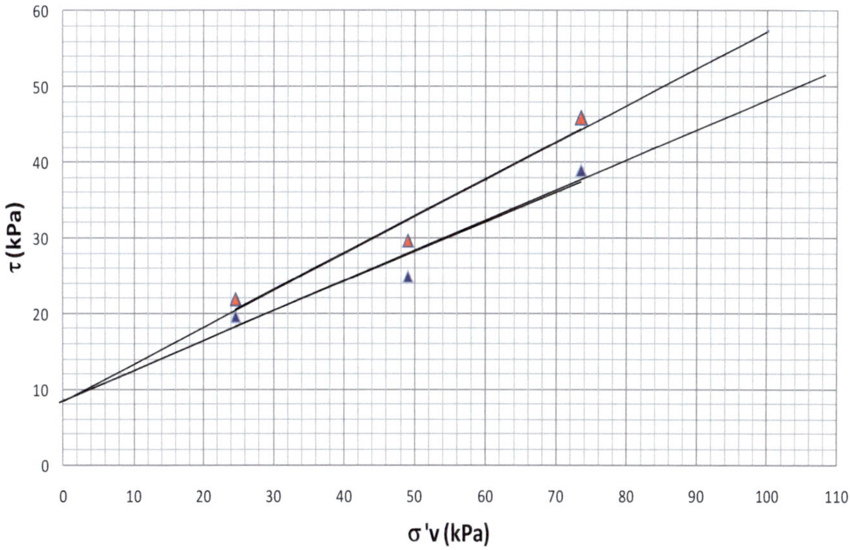

Fig. 9 Direct shear tests plot

4 Terrestrial Laser Scanner Acquisition (TLS) and UAV Survey

In order to define the geometry of the entire pyramid a TLS acquisition was performed. This new topographical survey has been used for dimensioning of the mitigations works and for future documentation (Cothren et al. 2008). A 3D computer model was created (Figs. 10 and 11). Two methods are available for the development of the 3D surface model, laser scanning and digital photogrammetry. During investigations and discussions, the accuracy potential, the practicality and the cost of the both two techniques were explored including considerations regarding the use of Unmanned Aerial Vehicles (UAVs). The final decision was in favour the combined use of both. Indeed, the two techniques covered different targets with different spatial resolution useful for the site.

4.1 *Laser Scanning Survey–April 2017 and 2018*

An extensive laser scan survey was completed by the University of Florence team during the all field missions (2017–2018).

Laser scanning for the generation of surface models acquires 3D surface information by determining the xyz co-ordinates of large numbers of surface points, referred to as a point cloud. The scanner can only measure and record surfaces in the field-of- view of the instrument and it is therefore necessary to move the instrument to multiple positions to cover an entire object. The scans acquired in this way, must overlap each other sequentially to allow the combination of all acquired scans into a

Fig. 10 Detail of laser scanner acquisition—April 2017 field mission

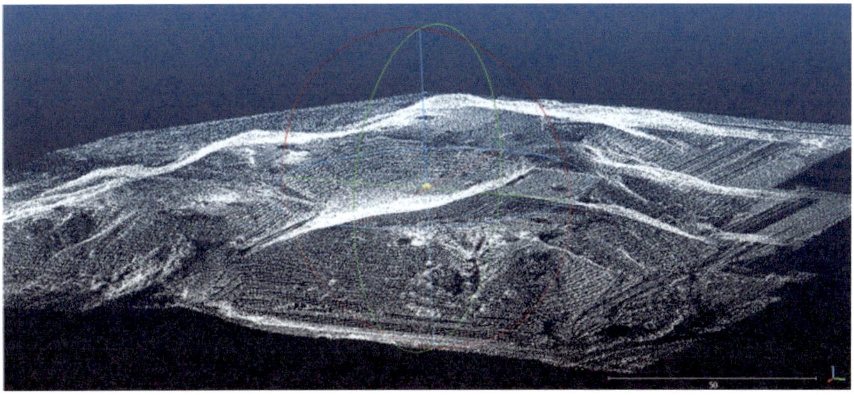

Fig. 11 Akapana 3d view from TLS acquisition—April 2017 field mission

single point cloud. In subsequent processing the individual points are connected to form a triangulated mesh which can be further processed to create a full surface.

Finally, photos can be draped over the surface to create a photo- realistic appearance. The interval between points of the point cloud, referred to as resolution, can be chosen by the operator of the laser scanner depending on the required detail and the complexity of the surface. For the Pyramid survey, the scanner was set up positioned around the entire structures with different acquisition at intervals of not more than 20 m (Fig. 12).

The total number of scans in the Pyramid was 25, with an average point interval of approximately 3 cm. Given this a point cloud of 62 billion of points was generated. The relative accuracy of points, i.e., the accuracy of neighbouring points is estimated to be in the order of a cm or better, while whereas the absolute accuracy, i.e., the accuracy of points over the entire pyramid is the one or two decimetre range. The final model (example in Fig. 11) can be viewed in 3D viewing and processing software.

The 3D model as well as and the GIS are referenced to the official Bolivian Reference system co-ordinates (WGS84 UTM 19S). The 3D model of the pyramid also had to be made available in 2D formats for processing and display in a variety of CAD and analysis software applications. For this purpose, sections and 2d topographical map were generated. For the investigated area (East sector) a very detail mesh, topographic map and cross section were carried out (Fig. 12).

Fig. 12 Derivate products from TLS acquisition (3Dshaded relief; 2Dtopographical map; 1D section along the cannel)—April 2017 field mission

5 Mitigation Measures Proposed and Adopted for the Control of Erosion Processes Affecting the Akapana Pyramid

Different part of the Akapana pyramid are affected by erosion processes (both linear/channelized and areal/superficial) in the last and recent time as results of meteo climatic external forces (e.g., rain, temperature, snow melt).

Following the main result and outcomes of the mission survey, jointly with local authority (UNESCO, CIAAT), it was decided to carried out a first Pilot project of consolidation works (in order to reduce and control erosion process) along the most eroded channel located in the eastern flank of the pyramid (Fig. 13).

For this particular morphological process affecting the pyramid, the most suitable mitigation measure is to control the deepening of the small channels due to accelerated linear erosion. Surely, one of the main problems is the uncontrolled flow of the rain water coupled with the snowmelt (Ortloff and Janusek 2016).

Fig. 13 Eroded channel (red circle) location. Pilot test n°1

In order to reduce the above free flow and to avoid linear and superficial erosion along the slope of the pyramid, the drainage system (able to collect and evacuate the water) must be properly designed, implemented and coupled with the consolidation.

This can be achieved with the diffused use along the selected channel of:

- Implementation of a system of small earthen/rock-and-wood channels at the upper part of the catchments, which would start from the top to collect the surface water;
- Final collection and extraction of water outside the area using pipes;
- Small gabions (filled with local stone) for the control of the deepening eroded channels;

Possible alternatives to the use of gabions may include:

- drywall (always realized with local stone);
- Tiwanaku style walls;
- small check dams (stone and woods);

5.1 Limits and Constrains

Any adopted solution must be clearly identifiable as new construction and never confused with ancient works. The use of gabions should be preferred because of their flexibility, cheapness and simplicity of production and installation. Due to possible limit and constrains and after several meeting with UNESCO consultant and CIAAT advisor the final adopted measures were the use of small gabions (filled with local stone- sand stone –*Arenisca*).

In order to reduce and mitigate the impact of the measures (sustainable mitigations) and to recreate the original shape of the Akapana Pyramid it was also decided that any adopted solutions must be located following the ancient terraced wall path (rocky staircase), (see Fig. 24).

6 Project Pilot Design Implementations

6.1 General Schema

The general above-mentioned layout of mitigation works was implemented following the logical sequences reported below:

Cleaning of vegetation. Before any kind of restoration works a detailed cleaning must be conducted along the whole area and in the closest surrounding. Bush, grass, garbage and vegetation must to be cut and evacuated by the area (see Fig. 14). The main target of this action will be the preparation of the whole area for the next works.

Archaeological excavation. In order to verify and reconstruct the original archaeological level of rocky steps of the pyramid, a small archaeological excavation must be conducted under the supervision of the CIAAT staff (see Fig. 15). The main target of this action must be also the exact location and discover of the ancient rocky steps original level.

Fig. 14 Sheet un-channelized erosion (solifluction, softening and swelling) and uncontrolled vegetation affecting the Akapana pyramid

Fig. 15 Archaeological excavation implemented during August—September 2017 survey

The archaeological excavation was carried out with the support and in accordance with local authority (CIAAT).

Gabions foundation, positioning and filling. After the reconstruction and location of the original rocky step level, the earth base floor foundation along the channel must be created and regularized in order to locate de gabions. Before the location of the empty gabions, a plane layer must be prepared in order to create a homogeneous and horizontal level to install the gabions before the fulfilling. A small gradient will have to be realized, providing a slight slope (3°–5°) towards the inside of the pyramid, with the objective of evacuating water through the drain pipe. The empty gabions mesh must be located (starting from the bottom and the back side) and constructed for each level. The gabions must be filled with stones of adequate mass, geometry and mechanical properties (average dimension is an average volume for single block between 0,001 and 0,03 cubic meters).

In the context of Akapana both massive limestone and/or river boulder should be adopted. Each gabion must be realized with a specific frontal pocket.

The frontal part of the gabions. (the above-mentioned pocket) must be fulfilled with local sandstone ("*Arenisca*") for the first two layer (starting from the bottom) and alternatively with local mortar (earth from the surroundings of the pyramid) for the last two level (see Fig. 16).

The final option for the location was decided directly with local authority (CIAAT) on site.

Drainage systems. Behind each level of gabions, a drainage pipe (ϕ 130 mm) must be located, with the creation of the correct gradient (2%) in order to be sure to evacuate the excess of water pressure in the back side of the gabions (see Fig. 17).

Geotextile. The back side of the gabions and the drainage pipes must be protected by geotextile in order to ensure long time durability and functionality (see Fig. 18).

Fig. 16 Example of local stone (Arenisca) to be used for the external part of the gabions

Fig. 17 Drainage pipe located in the back side of the gabion

Earth filling. Between each level of gabions, the original earth profile must be reconstructed using local material and earth. In order to reproduce the original shape and slope (at the end of gabions installation), the original slope gradient between each level must be realized. Original earth at the toe of the slope must be collected, mixed with "*Penka*" and reused (see Fig. 19).

Upper small channel. In the upper part of the channel a small ditch, useful to collect and evacuate superficial water, must be realized with a small excavation with a regular trapezoid shape. The small channel should be finished and stabilized with local stone.

Fig. 18 Geotextile installation on the back side of the gabion

Fig. 19 Penka preparation. The same technique must be adopted fort the slope between each gabion's level

Final reshaping. The original slope gradient and shape must be reconstructed between each level of gabions (average slope gradient around 15°). At the end of the work the general slope of the pyramid (from the bottom to the top) must be stabilised. In order to guarantee durability, stability and efficiency, the technique of local earth mixed with "*Penka*", must be adopted.

Topographical monitoring. Before and during all the mitigation works, regular and constant control of topographical marker point will assure the respect of geometric and topographic level of the original design. Any topographical modification must be considered (and possibly avoided). Any possible displacement and deformation during the works along the channel must be taken under control and monitored. A topographical total station tool must be used for all the works period in order to guarantee correct position and tilt of gabions and to measure slope gradient after final reshaping. The monitoring activities must be implemented at least after six months by the end of the work.

7 Mitigation Mesures Implementation Along the Selected Pilot Site

During the last field missions a strong support has been provided to local authority and CIAAAT workers. In cooperation with local technicians (e.g., archaeologist, architect) the design of mitigation measures was fitted to the real executive situation. The different levels at which positioning the gabions were defined (Fig. 20), the earth platform preparation was controlled and the new slope profile were measured.

The final equilibrium profile was designed after the modification determined by the earth platform built by local worked and not foreseen in the original design. Starting from the main results of the archaeological filed survey the types, geometry and position of the first gabions was decided. The final foundation level was defined directly in the field, with the support of topographical measures from local CIAAAT archaeologist. The operation of excavation started immediately after. In the same

Fig. 20 The final position of gabions level and earth platform were decided jointly with local authority

Fig. 21 Training day and capacity building with Maccaferri® company and local worker

mission support was also provided for the gabion's installation and filling procedures (Fig. 21).

With the support and help of Maccaferri® Company a training day with capacity building of the local worker was implemented before start gabions fulfilling and installation (Fig. 21).

In order to reduce the visual impact of the structures (following the design suggestions and constrains) the front of each gabions was filled with local sandstone ("*Arenisca*") (for more detail see Fig. 22). After the capacity building and training session, final location and number of platform and gabions levels were defined jointly with CIAAAT support. Also, the best solution for the drainage system were discussed and adopted (see Figs. 23 and 24).

The rendering of the final design of the pilot site is reported in Fig. 25.

8 Conclusions

The implementation of measures in the pilot site as well as research and works are not yet finished and, at the moment, only the first (upper part) and the last (lower part) two gabions wall have been installed.

The CIAAAT is preparing the phase 2 in order to complete the pilot and define other two channels to replicate de design able to reduce the erosion processes.

To confirm the fact that the consolidation of the terraces steps, beside any restoration interest, is certainly one of the main mitigation measures in order to reduce intensity and speed of the morphological processes affecting the site.

Fig. 22 Front view of the gabions adopted to reduce the visual impact

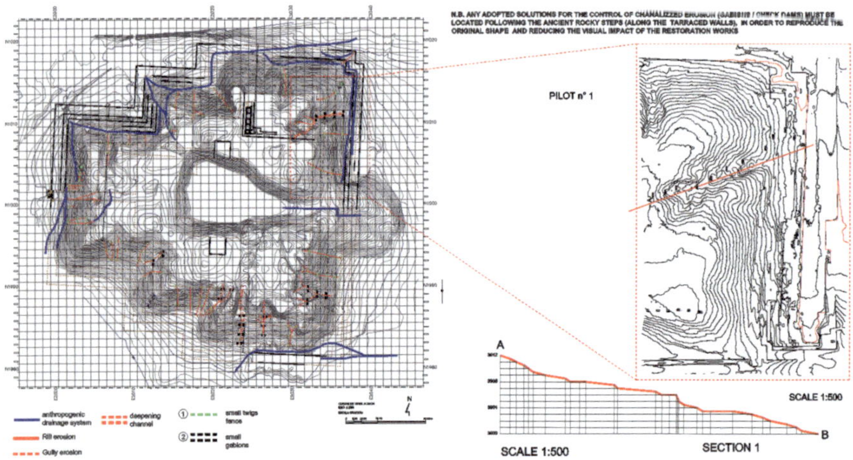

Fig. 23 Test area selected for the project pilot 1 (design project; June 2017)

Fig. 24 General schema of the preliminary location of the gabions following original ancient terraced wall path (rocky staircase)

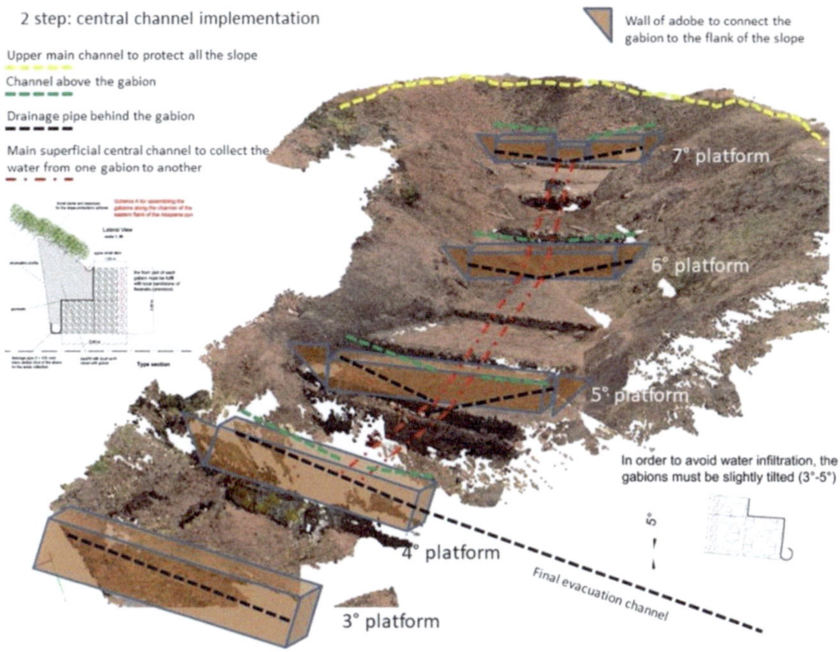

Fig. 25 Final rendering of the pilot design along the channel

Acknowledgements The research and works described in this paper were funded by Japan Funds in Trust and managed by UNESCO Office in Quito and representation in Bolivia, Colombia, Ecuador and Venezuela 's international experts' team, dispatched for attending "Pilot Project of Conservation of Akapana Pyramid", in Tiwanaku (Bolivia), 2015 – 2018. The authors are very grateful to all the CIAAAT workers and technicians, as well as to all the Quito UNESCO office staff. Special thanks go to Alcira Sandoval Ruiz. The work is also dedicated in memory of David Ayllon, good friend and best driver prematurely disappeared.

References

Cothren J, Goodmaster C, Barnes A, Ernenwein E, Vranich A, Fredrick Limp W, Payne A (2008) Fusion of three-dimensional data at Tiwanaku: an approach to spatial data integration. In: Proceedings of the computer applications and quantitative methods in archaeology
Gallego-Revilla JI, Pérez-González ME (2018) Tiwanaku, entre el cielo y la tierra. UNESCO, Paris
Gallego-Revilla JI, Spizzichino D, Dealaveris I (2018) Conservation Plan of Tiwanaku. CIAAAT/UNESCO, Bolivia
Janusek JW (2003) The changing face of Tiwanaku residential life: state and social identity in an Andean city. In: Kolata AL (ed) Tiwanaku and its hinterland: archaeology and paleoecology of an Andean civilization, 2. Smithsonian Institution Press, Washington, DC, pp 264–295
Janusek JW (2004) Identity and power in the Ancient Andes. Tiwanaku Cities Through Time. Routledge, New York
Kolata AL (1993) The Tiwanaku: portrait of an Andean civilization. Blackwell, Cambridge, MA
Lasaponara R, Masini N (2014) Beyond modern landscape features: new insights in the archaeological area of Tiwanaku in Bolivia from satellite data. Int J Appl Earth Obs Geoinf 26:464–471. http://dx.doi.org/10.1016/j.jag.2013.09.00
Latrubesse EM, Bakerb PA, Argolloc J (2009) Geomorphology of natural hazards and human-induced disasters in Bolivia. Developments in Earth Surface Processes 13:181–194
Ortloff CR (2016) New discoveries and perspectives on water management at 300 bc–ad 1100 Tiwanaku's urban center (Bolivia). Open Access investigation. https://www.academia.edu/30336746/new_discoveries_and_perspectives_on_water_management_at_300_bc-ad_1100_tiwanakus_urban_center_bolivia_hydrology_current_research_3_1_open_access_2016
Ortloff C, Janusek JW (2016) Hydrologic engineering of the Tiwanaku. In: Encyclopaedia of the history of science, technology, and medicine in non-western Cultures. Springer
Ponce Sanginés C (1981) Tiwanaku: Espacio, Tiempo y Cultura 4th edn. Editorial Los Amigos del Libro, La Paz
Spizzichino D, Fernández JA, Margottini C, Condori J, Gallego JI (2019) Hydro geological hazard and mitigation measures for the conservation of Akapana Pyramid in the Tiwanaku archaeological site (Bolivia). In: VV.AA. (2019). Monitoraggio e manutenzione nelle aree archeologiche. Cambiamenti climatici—Dissesto idrogeologico–Degrado chimico ambientales. Roma
Vella MA, Thiesson J, Rivera Casanovas C, Argollo J, Guédron S, Brisset E, Biévre G, Sánchez C, Guerin R, Ortuño T, Núñez Regueiro P (2017) The Guaquira-Tiwanaku Project (Bolivia): a multidisciplinary approach of Ancients Societies/Environment interactions. In: International congress on archaeological prospection, 2017. Poster
Vranich A (2006) The construction and reconstruction of ritual space at Tiwanaku, Bolivia (A.D. 500–1000). J Field Archaeol 31(2):121–136
Vranich A (2009) The development of the ritual core of Tiwanaku. In: Young-Sanchez M (ed.) Tiwanaku: papers from the 2005 mayer center symposium at the Denver Art Museum (pp. 11–34). Denver Art Museum, Denver, CO

Multiscale Analysis of Geo-Hazards Affecting the Alhambra Cultural Heritage

José Antonio Fernandez-Merodo, Rosa Martin Mateos, Jose Miguel Azañon, Pablo Ezquerro, Juan Carlos García-Davalillo, Marta Bejar, Gerardo Herrera, Catuxa Novo, Emma Bee, Kyriacos Themistocleous, Giovanni Crosta, Paolo Frattini, Riccardo Castellanza, Gabriele Leoni, Daniele Spizzichino, and Claudio Margottini

Abstract This paper presents a multiscale methodology to perform successful geo-hazards assessment in the context of Cultural Heritage. The methodology, defined in the European PROTHEGO project, has been applied to the Alhambra case study. Alhambra is an important Word Heritage site located in Andalusia, Spain. The site is prone to suffer flood, earthquake and landslide phenomena. The proposed multiscale methodology leans on new remote monitoring and modelling technics that not entail aesthetic and functional impacts on the site. For large-scale monitoring, satellite remote sensing technology enables to detect and characterize spatiotemporal ground and structure deformation as a whole, with up to millimeter precision. For very local phenomena analysis, small-scale monitoring based on Terrestrial Laser Scanner and airborne drone digital photogrammetry surveys enables to detect erosion processes that affect the slopes of the citadel. Furthermore, numerical modelling enables geo-mechanical instabilities to be quantified at different scales. Large-scale modelling

J. A. Fernandez-Merodo (✉) · R. M. Mateos · P. Ezquerro · J. C. García-Davalillo · M. Bejar · G. Herrera
Geological Survey of Spain, Ríos Rosas 23, 28003 Madrid, Spain
e-mail: jose.fernandez@igme.es

J. M. Azañon
University of Granada, Av. de La Fuente Nueva S/N, 18071 Granada, Spain

C. Novo
Patronato de La Alhambra Y Generalife, C\ Real de La Alhambra S/N, 18009 Granada, Spain

E. Bee
British Geological Survey, Nottingham NG12 5GG, UK

K. Themistocleous
Cyprus University of Technology, 30 Archbishop Kyprianos St, 3036 Limassol, Cyprus

G. Crosta · P. Frattini · R. Castellanza
University of Milano-Bicocca, Piazza Dell'Ateneo Nuovo 1, 20126 Milano, Italy

G. Leoni · D. Spizzichino · C. Margottini
Geological Survey of Italy, Via Vitaliano Brancati 60, 00144 Roma, Italy

G. M. El-Qady and C. Margottini (eds.), *Sustainable Conservation of UNESCO and Other Heritage Sites Through Proactive Geosciences*, Springer Geology,
https://doi.org/10.1007/978-3-031-13810-2_3

shows that the most dangerous slope is located at the San Pedro cliff site. Small-scale stability analysis in the San Pedro cliff shows that a slope failure, triggered by a possible earthquake scenario, could reach the wall foundations of the Alhambra citadel located 22 m behind it. Finally, very detailed small-scale modelling, using very precise digital elevation models, enables to track local instabilities and erosion phenomena. Such a multiscale, interdisciplinary approach is the most effective way to identify, assess and monitor risks, strengthening disaster preparedness at heritage property.

Keywords Cultural heritage · Geo-hazards · Remote sensing · Numerical modelling · Space-borne radar interferometry · Slope stability · Alhambra

1 Introduction

Natural hazards, worsened by the effects of climate change, are consistently putting Natural and Cultural Heritage (NCH) under pressure, with the frequency and magnitude of disasters increasing over time. These hazards menace our social and cultural assets, risk public safety, and impact local economies linked to tourism. The analysis of exposure and vulnerability of NCH to natural hazards is unfortunately almost totally neglected, despite being fundamental for risk assessment and management. This imposes new challenges for NCH management and an urgent need for innovative conservation and safeguarding approaches.

Here, a new multiscale methodology based on new remote monitoring and modelling technics that not entail aesthetic and functional impacts on the NCH sites is proposed to overcome the aforementioned shortcomings. The multiscale methodology leans on a downscaling approach, from regional to detailed local scale, that identifies geology, analysis geo-hazards, and quantifies the risk through advanced monitoring and mechanical modelling. The methodology is validated in Alhambra CH that has been affected by numerous dramatic natural disasters throughout history. It is worth noting that stability and erosion phenomena located in the San Pedro cliff has been recognized as a major concern by the site managers.

The property of the Alhambra, Generalife and Albayzín is located in Granada, Andalusia, Spain. This Cultural Heritage, listed as World Heritage Site by UNESCO in 1984, encompasses Moorish and Christian palaces, fortresses, houses and gardens from the 13th to sixteenth centuries. Vestige and testimony of the resplendent Moorish culture in the Iberian Peninsula and Europe, it is certainly one of the most important monumental complex in Spain. Exceeding 2,6 million visitors in 2016, it is the economical lung of Granada city.

In this case study, the downscaling approach and the regional, large, local, detailed local-scales are defined in Fig. 1. The **Regional-scale** covers the Granada Province (with 12.531 km^2 it represents 2,48% of Spain surface). The **large-scale**, or municipality scale, surrounds the Cultural Heritage buffer zone with 12 km^2 (approximately 4,7 × 2,7 km). The **local-scale** encompasses the San Pedro cliff with 0,01 km^2

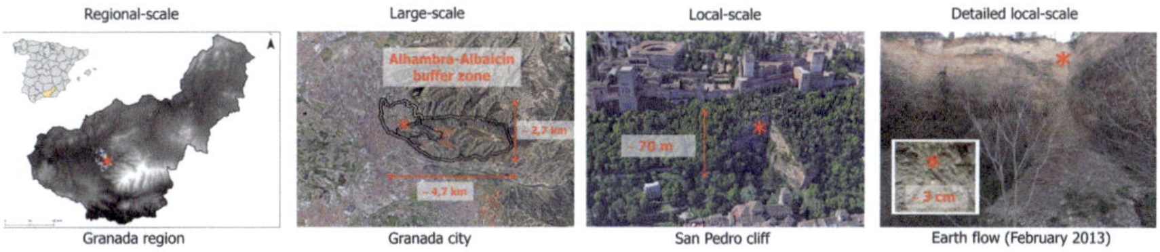

Fig. 1 Downscaling approach. Regional, large, local, detailed local-scales definitions (Red star: location of San Pedro cliff)

(approximately 100 × 100 m). From the engineering point of view, this scale is important to analyze global failure risk, quantify slope stability and evaluate the citadel integrity. A **detailed local-scale**, with very high space resolution (up to cm precision) is needed to study erosion phenomena and catch the earth flow event occurred in February 2013 (see last picture in Fig. 1).

2 Geology

Alhambra geology can be described using the defined downscaling approach from regional-scale to detailed local-scale (Fig. 2).

At regional-scale, Granada and its metropolitan area are located in the central sector of the Betic Cordillera and in the contact between de External and Internal zones of the range. The city lies on the eastern edge of the Granada Basin where the foothills rise into the adjacent Sierra Nevada (3,482 m). The river Genil flows through the center of the basin and it is joined by smaller tributaries, outstanding the river Darro that divides the old town into two parts: The Alhambra and the Albayzín. The basin is bordered by several faulted, Late Miocene, Pliocene and Quaternary-age alluvial fans, as a result of the rapid uplift and erosion of Sierra Nevada (Braga et al. 1990). They are predominately coarse-grained deposits (conglomerates) which prograde to the center of the Basin.

Large-scale geology can be described crossing the 1:50.000 scale geological map MAGNA 50, sheet 1009 (Lupiani and Soria 1988) with the defined Alhambra buffer zone, it appears that the Alhambra CH rests on the following materials: (1) Red clays,

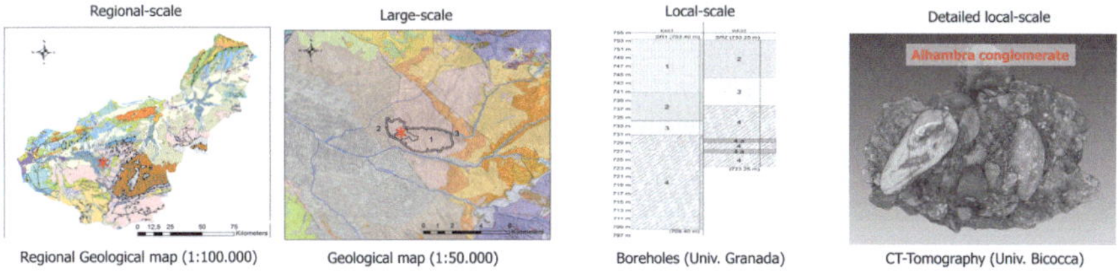

Fig. 2 Alhambra geology (Red star: location of San Pedro cliff)

sands and conglomerates. Alluvial fan materials. Upper Pliocene–Lower Pleistocene age. (2) Vertic soils. Holocene age. And (3) Undifferentiated alluvial fan materials. Holocene age. Granada is built on alluvial fan materials and specifically on the named Alhambra Formation. The Alhambra Formation is cropping out with a NNW-SSE trend along the eastern border of the Granada Basin covering a wide area of the city. It is an alluvial fan deposit, up to 200 m in thick, of lower Pliocene age ($\approx$5My). They are very coarse-grained, poorly-sorted and well-rounded conglomerates with metamorphic lithoclasts. They are cemented by caliche and very small gold plates are present in the matrix of the conglomerates, which comes in turn from erosion of an older Miocene conglomerate. The gold has been historically panned in the rivers of Granada (Martín 2000).

At local-scale, in the San Pedro Cliff, at the foot of the Alhambra, the Alhambra formation outcrops. Borehole surveys (Justo et al. 2008) show that the materials are sub-horizontal-bending conglomerates with numerous faults and joints, which play a relevant role in the stability of the cliff. In fact, the western part of the escarpment corresponds to a fault-scarp with recent activity.

A very detailed local-scale description can be obtained using CT-Tomography. This advanced technic highlights three kind of coarse materials: (i) weak cemented matrix, (ii) strong cemented matrix, and (iii) boulders.

3 Geo-Hazards

The particular geological, geomorphic and tectonic setting of Granada determine several natural hazards in the region. Granada has historically been the scene of numerous dramatic natural disasters that have threatened the population and construction throughout history. Geo-hazards can also be analyzed using the defined downscaling approach from regional-scale to detailed local-scale (Fig. 3).

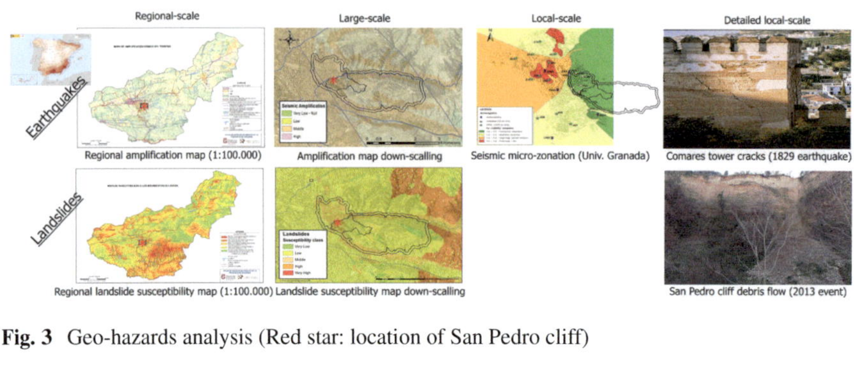

Fig. 3 Geo-hazards analysis (Red star: location of San Pedro cliff)

3.1 Earthquakes

Earthquakes have recurred over the centuries, as Granada is located on a seismically active region. In fact, Granada is at the maximum seismic hazardous area of Spain. The historical record of earthquakes in Granada reveals at least nine large events with the epicenters close to the city. One of the most destructive was the 1431 earthquake, with an epicenter intensity of IX (MSK scale) and an estimated magnitude (Mw) of 6.5. This earthquake destroyed many Nasrid palaces in the surroundings of the Alhambra fortress. In 1884, the Andalusian earthquake took place, with the epicenter located in the locality of Arenas del Rey (53 km from Granada). With an epicenter intensity of IX (MSK scale) and an estimated magnitude (Mw) of 6.8, the Andalusian earthquake caused great damage in a large area of the Granada province. Sanz de Galdeano and Alfaro (2004) postulate that many faults could potentially generate earthquakes with magnitudes as great as Mw 6.0 in the surroundings of Granada.

Earthquake hazard assessment at regional-scale (1:100.000) was performed by the Geological Survey of Spain (Ferrer 2007) showing middle seismic amplification in the Alhambra buffer zone. Local-scale assessment and seismic microzonation in Granada city were also performed by the University of Granada (Feriche 2012). Some of these results are shown in Fig. 3.

3.2 Landslides

The slope dynamics in the Granada Basin is very intense. Numerous landslides, rock and earth falls, solifluctions and debris flows have been well documented.

for a long time. The inventory developed by Chacón et al. (2007) reports almost 300 landslides affecting a 28 km^2 area. The last regional rainfalls triggering large amount of landslides occurred in 1996–1997 (Irigaray et al. 2000). Recently, the rainy winters of 2010–2013 caused some failures in the surroundings of the Alhambra. That is the case of the Jesús del Valle landslide, a complex landslide which affected an area of 50.000 m^2 and distorted the course of the river Darro in January 2010 (Ferreira et al. 2015). In the San Pedro Cliff, numerous earth-falls and debris flows have been documented. One of the latest took place in February 2013 after a heavy period of rain (Fig. 1). In general, the retreat of the escarpment is very slow, except in its western part where the presence of traction joints parallel to the slope suggest a faster retreat by the slope dynamics. At present, the base of the cliff is covered by a debris cone that protects it from the river Darro undercutting.

Landslide hazard assessment has been performed by the Geological Survey of Spain (Ferrer 2007) at regional-scale (1:200.000) showing middle susceptibility class in the slopes located inside the Alhambra buffer zone. Space distribution of lithological units, elevations, slope angles and tectonic units factors have been taken into account in the analysis.

3.3 *Floods*

In Granada, the Mediterranean climate alternates periods of droughts and heavy rains; flooding is very frequent. Most floods in Granada are related to the ephemeral behavior of the river Darro and they have been historically documented for a long time. The historical data on floods in Granada include documents from the beginning of the twelfth century. Nevertheless, the more precise information is related to 59 years of floods recorded between 1478 and 2010 (Páez 1996). The Darro´s flood in 1478 caused severe damage in the Albayzín and numerous fatalities. On March 1660, another flood generated a large earth-fall in the San Pedro cliff. From the historical data, the increasing incidence of floods with urban development shows the highest rate in the late twentieth century (Chacón et al. 2012). In fact, the river was channeled under the city at the end of the XIX century to avoid the periodical flash floods.

A first hazard flooding assessment of the Cultural Heritage can be performed by crossing the official national cartography of floodplains (SNCZI 2011) with the defined Alhambra buffer zone. It appears that the only affected area is the course of the Darro river that has a narrow morphology.

Ongoing local studies aim to perform more precise flood hazard and risk assessments.

4 Monitoring of Landslides and Earth Flows

Monitoring aims to control land deformation that could affect the studied Cultural Heritage. Those deformations are a reliable indicator of the risks associated to it (subsidence, landslides, sinkholes, etc.). Due to its proximity to the Alhambra walls (22,8 m actually), evolution of San Pedro cliff has been observed for a long time. It is possible to follow the local erosion process from historical paintings and from historical pictures. It is worth mentioning the monitoring work done by the Department of Geodynamics of the University of Granada in the last 10 years. From all the observations, the erosion rate has been estimated to 1,85 cm/year. Land deformation monitoring can also be aboard through the proposed downscaling methodology.

4.1 *Regional-Scale and Large-Scale Monitoring*

New space technologies based on satellite images (radar and optical) are now capable of monitoring, with a high spatial/temporal resolution, land use, surface deformation and physical parameters (e.g., temperature, humidity, vegetation) as well as anthropogenic effects. Here, three independent sets of SAR images acquired by three different satellites have been used: the ENVISAT (C-band) and the Sentinel-1A satellites (C-band) belonging to the European Space Agency (ESA), and the Cosmo-skyMed constellation (X-band) owned by the Italian Space Agency. Table 1 summarizes their main features and characteristics. The processed areas covered by the three satellites are depicted in Fig. 4. Under ideal conditions (large datasets, high Persistent Scatterer PS density, minimum atmospheric effects, etc.) sub-millimeter accuracy can be obtained (Ferretti et al. 2007) although millimeter accuracy is usually obtained under more realistic conditions.

Due to the spatial resolution 3 × 3 m and higher PS density, Cosmo-SkyMED processing is very suited for urban areas (Fig. 4). The Alhambra buffer zone seems to be stable; no wide deformation has been monitored inside. An example of measured

Table 1 Main characteristics of ENVISAT, Cosmo-SkyMed and Sentinel-1A data as well as the processing methodology applied for each dataset

Parametres/satellite	ENVISAT (ESA)	Cosmo-SkyMED (ASI)	Sentinel-1A (ESA)
Wavelength	C-BAND/5.6 cm	X-BAND/3.1 cm	C-BAND/ 5.6 cm
Temporal span	May 2003–Dec 2009	May 2011–May 2017	May 2015–April 2016
Geometry of acquisition	Ascending	Descending	Descending
Incidence angle	23°	26.57°	38°
N° Images	23	71	25
N° of Interferograms used	138	216	24
Original spatial resolution (approx.) (m)	4 × 20	3 × 3	30 × 30
Vlos std dev (mm/year)	1	0.8	13.6
Processed area (km)	35 × 45	8 × 10	37 × 48
Methodology of processing	PSIG Cousins (Devanthèry et al. 2014)	CPT (Blanco-Sánchez et al. 2008)	Direct integration approach (Barra et al., 2016)

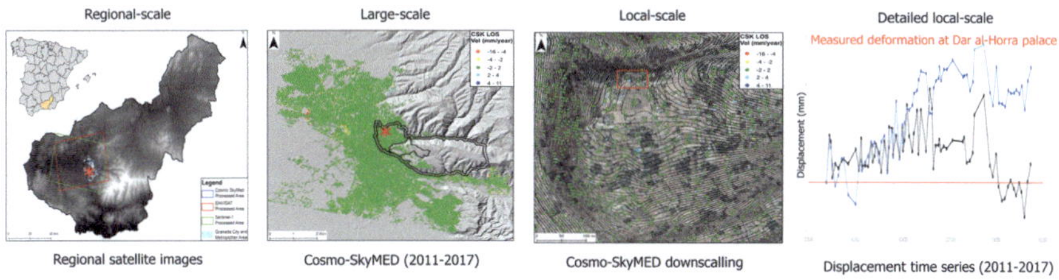

Fig. 4 Regional-scale and large-scale monitoring. Cosmo-SkyMED processing. (Red star and box: location of Dar al-Horra palace)

time-series deformation at Dar-Al-Horra palace (period May 2011 – December 2017) is showed in Fig. 4. It has to be mentioned that no PS has been detected near the San Pedro cliff.

4.2 Detailed Local-Scale Monitoring

Advanced remote monitoring systems that not entail aesthetic and functional impacts on the site, as Terrestrial Laser Scanner TLS and UAV (unmanned aerial vehicle) Photogrammetry, can be used to quantify erosion phenomena at local-scale.

Two TLS campaigns using Leica ScanStation C10 apparatus have been undertaken in the San Pedro cliff, the first one in July 2017 during the summer season and the second one in February 2018 during the winter season. The TLS apparatus was installed in the bell tower of San Pedro church located just in front the cliff 30 m away. Distance between the two date generated 3D point clouds is showed in Fig. 5. A small area of material (zone 1) 0.6 m deep in the apex of the dihedral is missing. This erosion process coincides with an earthflow occurred the 6 January 2018.

Two UAV operations have also been programmed in the same dates. The UAV operations consist on an automatic overhead flight and three manual oblique flights along the cliff at different heights. The drones, INSPIRE 1 Pro and Phantom 4 Pro, were also controlled from the bell tower of San Pedro church. 452 pictures and 432 were taken during the summer and winter campaigns respectively. Photogrammetry processing were made over these pictures using Agysoft PhotoScan 1.2.6 64 bits, Pix4Dmapper 4.1.10 64 bits and ArcGIS Desktop 10.5 software. An important earth flow event can be detected in the apse of the cliff who acts as a corridor. This event occurred the 6 January 2018 after an important snow event, it is the same as the one detected using TLS. After cleaning the vegetation, the two generated high resolution point clouds (26 million points) were quantitatively compared by a 3D cloud distance

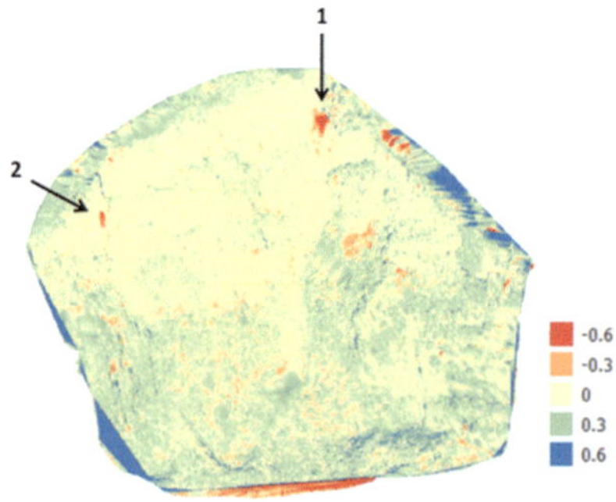

Fig. 5 Detailed local-scale monitoring. TLS monitoring results in San Pedro cliff. Left: generated 3D model in February 2018. Right: point cloud distance (in m) between July 2017 and February 2018 data acquisitions

Fig. 6 Detailed local-scale monitoring. UAV photogrammetry monitoring results in San Pedro cliff. Left: generated 3D model in February 2018. Right: point cloud distance (in m) between July 2017 and February 2018 data acquisitions

computation using M3C2 plugging (Lague et al. 2013) in PointCloud software. The results of the computed cloud distance are shown in Fig. 6. It can be seen that eroded material comes from the top of the cliff, just in the right side of the apex, zone 1. Using the recorded 0.60 m depth, the volume of the detached material can be approximated around 2 m^3.

5 Modelling of Landslides and Earth Flows

Even if no slow land deformation has been detected by the large and local-scale monitoring, stability analysis of the whole Cultural Heritage can be very useful to identify where landslide phenomena are prone to happen and to give a valuable hazard assessment based on prediction. Moreover, in the San Pedro cliff specific case, a local stability analysis can give a very useful picture of the stability conditions of the site. For this purpose, a downscaling modelling approach has also been undertaken, 3D stability analysis has been performed at different scale using the Finite Element code GeHoMadrid (Fernández-Merodo 2001). Safety Factor (SF), that expresses how strong a system is by means of a ratio between the soil's actual shear strength and the reduced shear strength at failure, is computed using the Shear-Strength Reduction technique and the numerical algorithms described in (Fernández-Merodo et al. 2014) and (Bru et al. 2017). A Mohr–Coulomb failure criteria has been used for the mechanical behaviour and the average representative parameters for the all in one equivalent and representative material are: dry density $\gamma = 2100$ Kg/m^3, friction angle $\phi = 40°$ and cohesion c = 90 kPa. More details of the modelling strategy and material characterization can be found in (Fernández-Merodo et al. 2018).

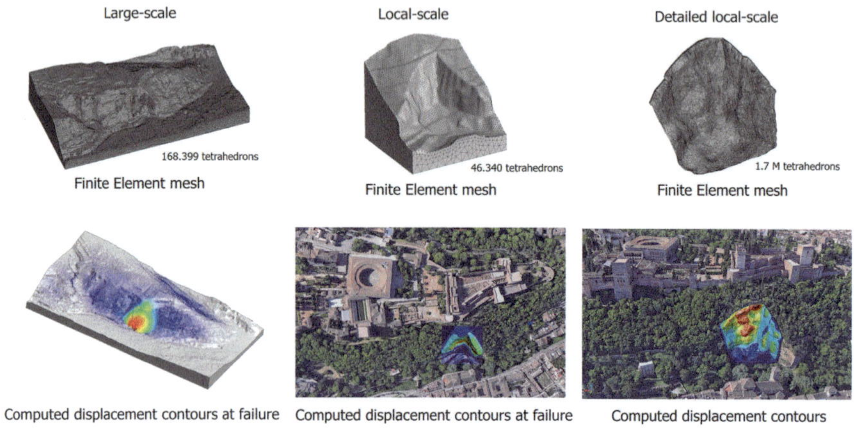

Fig. 7 Landslide and earth flow modelling. Stability analysis

5.1 *Large scale modelling*

3D finite element mesh has been generated over the Al Sabika hill (size 800 m × 400 m). It consists of 168.399 quadratic tetrahedral elements and 3.5 million nodes with a characteristic element size equal to 3 m matching the used digital terrain model (DTM) resolution from LIDAR data taken out in 2000. The stability analysis indicates that the most critical slope in the studied area is the San Pedro cliff with an associated safety factor SF around 2. Figure 7 depicts the failure mechanisms by plotting displacement contours contours at failure.

5.2 *Local scale modelling*

A more precise local-scale stability analysis has been performed in the San Pedro cliff. In this case, the 3D finite element mesh has been generated over a smaller box (size 100 m × 100 m). It consists of 46.340 quadratic tetrahedral elements and 70.889 nodes with a characteristic element size equal to 1 m matching a more precise topographic map done in 2007. The computed safety factor SF is in this case more precise equal to 1.5. Failure mechanism is depicted in Fig. 7 plotting displacement contours at failure. Plastic deformation develops along a shear surface, the mobilized mass slides along this shear surface with a typical rigid-body motion. It can be observed that the computed failure attains the wall foundations of the Alhambra citadel located 22 m behind the Cliff.

5.3 Detailed Local Scale Modelling

The global slope failure mechanism obtained in the previous local-scale modelling is not representative of the local superficial erosion and scarp retreat observed and recorded by the local-scale monitoring (see pictures of the 2013 and 2018 earth flow events depicted in Fig. 1 and 6). A new numerical analysis is made using the point clouds generated by the TLS scanning for the superficial topography. Two dates are selected from the different data acquisitions: (i) 18/04/2007 and (ii) 14/10/2014 such a way that both dates enclose the important flow slide occurred in February 2013, obtaining a topography pre-event and a topography post-event. The point clouds are filtrated with a 0.1 m distance resolution. Volumes are generated from these surfaces by a 5 m extrusion on a perpendicular direction. It is expected that erosion phenomena appears on the surface always bellow this 5 m distance. The bottom surface is fixed and only gravity load is considered. This analysis aims to reproduce the mechanical behaviour of the cliff "skin". The generated finite element meshes are quite huge reaching about 1,7 million tetrahedrons. Due to speed computation and hardware memory limitations, only linear computation is performed. Even though this simple numerical analysis is limited, no safety factor is computed, it serves to identify areas susceptible to be affected by erosion phenomena. Red colour in last column of Fig. 7 represents maximum computed displacements in a static linear analysis. Computed red areas coincide with cantilever material that can not be directly identify in the point cloud obtained by LIDAR neither in the classical plane representation used in the previous large and local-scale analysis. It is suspected that these areas located in the upper part of the cliff are susceptible to be affected by detachment mechanisms. Further stability analysis at detailed local scale will confirm this previous suspicion.

6 Discussion

Validation of the proposed multi-scale methodology for geo-hazards assessment in the Alhambra case study can be accomplished by cross comparison of the new generated results at regional, large and local scales investigation, monitoring and modelling with traditional assessments and recorded past events.

A preliminary geo-hazard assessment and the detection of CH site in danger from a geological point of view has been achieved by crossing the exact geo-localization of the CH buffer zone and existent regional geological hazard maps and datasets. It has been shown that Alhambra is located in a moderate seismic zone with middle seismic amplification, in a zone with middle landslide susceptibility class. On the other hand, seismic, and landslide stigmata of past events (Fig. 3) are evident and confirm those outcomes. The use of more detailed and available local-scale hazard maps and studies, as for instance the seismic micro-zonation study in Granada city, increases the precision of the required geo-hazard assessments.

Large scale-monitoring performed with different Interferometric Synthetic Aperture Radar (InSAR) techniques using different ENVISAT, COSMO-SKYMED and SENTINEL satellite datasets shows no important deformation in the Alhambra CH for the analyzed period 2003–2017. No active geo-hazard has been detected. It has to be mentioned that no PS, namely no deformation data, has been recorded near the San Pedro cliff. INSAR large-scale techniques do not give any information in this place due to six main reasons: (i) The cliff is nearly vertical and in some local areas there are cantilevered rocks. Satellite images are not suited for this kind of geometry. (ii) The size extension of the cliff is small, 70 m width $\times$ 70 m high, it does not cover a large area. INSAR monitoring could be only applied using high resolution satellites as the Cosmo-skyMed constellation (X-band) (3 $\times$ 3 m resolution). (iii) The orientation of the cliff is north, this direction is the worst one to be tracked by the line of sight (LOS) of the satellites. (iv) The surrounding area is totally covered by a leafy forest; dense vegetation cannot be penetrated by satellite using the X-band. (v) Erosion process cannot be tracked using INSAR methodology based on PS persistent scatters, the eroded material does not give points as continuous reflectors. vi) INSAR monitoring can only track slow deformation process and is not suited for sudden events (brittle failure). Therefore, large-scale monitoring has important limitations. This limitation can be beat through small-scale monitoring.

Performed small-scale monitoring in the San Pedro Cliff (2017–2018 period) has shown that TLS and UAV photogrammetry technics are able to catch the small material detachment and earth flow occurred the 6 January 2018. Those techniques are therefore suited to quantify erosion phenomena.

Large-scale and local-scale monitoring seem to be, in this case, parallel and complementary activities that must be performed at the same time.

As no deformation has been recorded for the analyzed period 2003–2017 by the large-scale monitoring, it makes no sense to propose a time-dependent stress–strain numerical model catching the observed deformation pattern, nevertheless a stability analysis of the whole Cultural Heritage is a very useful tool to identify where landslide phenomena are prone to happen. The large-scale scale stability analysis indicates that the most dangerous slope is located at the San Pedro cliff site coinciding with previous prognosis and assessments performed by the local CH managers. A more detailed (local-scale) stability analysis in the San Pedro cliff shows that the actual safety factor SF is around 1.5 and that a global failure could reach the wall foundations of the Alhambra citadel located 22 m behind it, endangering the CH site integrity. With the available geotechnical information it seems that global failure can only be possible with an earthquake scenario. Nevertheless, advanced numerical modelling must go on taking into account more precise geophysical setups (materials, faults, pore-pressure, etc.).

Moreover, the computed global failure does not represent the observed small material detachment and erosion phenomena occurred the 6 January 2018. This limitation can be overcome through a detailed local-scale modelling increasing the spatial resolution up to cm size. Preliminary stability analysis, using very precise 3D topographies, show that important cantilevered material located in the upper part of the cliff can be affected by detachment mechanisms. Further numerical analysis must be completed on clusters and super-computers to circumvent the high-consuming time drawback due to the size of high-resolution meshes (1.7 million tetrahedrons). It is expected that these new analyses could reproduce the observed and monitored small material detachments and erosion phenomena.

In this case too, large-scale and detailed local scale modelling should rather be complementary, one approach should not suppress the other.

7 Conclusions

The proposed integrated and simultaneous multi-scale investigation, monitoring and modelling is the most effective way to perform a successful geo-hazard assessment. It has been showed that Alhambra CH is susceptible to flooding, earthquakes and landslide phenomena.

Satellite remote sensing technology is a powerful tool to do large-scale monitoring, to detect and control spatiotemporal ground and structure deformation as a whole, with up to millimeter precision. Small-scale monitoring based on Terrestrial Laser Scanner and airborne drone digital photogrammetry surveys enables to detect and quantify erosion processes that affect the slopes of the citadel.

Numerical modelling enables the quantification of geo-mechanical instabilities at different scales. Large-scale modelling describes and predicts global failure mechanisms. Detailed local-scale modelling describes and predicts local failure and erosion phenomena.

Such a multiscale, interdisciplinary approach is the most effective way to identify, assess and monitor risks, strengthening disaster preparedness at heritage property.

Acknowledgements The research leading to these results has been supported by the PROTHEGO Project (Protection of European Cultural Heritage from Geo-hazards), funded through the framework of the Joint Programming Initiative on Cultural Heritage and Global Change (JPICH), and under ERA-NET Plus and the Seventh Framework Program (FP7) of the European Commission.

References

Azañón JM, Azor JL, De Justo Alpañés J, Martín Rosales W, Mateos RM, Pérez-Peña V (2007) Deslizamientos e inundaciones cuaternarias en la cuenca vertiente del río Darro: la génesis del Tajo de San Pedro (La Alambra, Granada). In: Resúmenes XII Reunión Nacional de Cuaternario. Ávila, pp 13–14

Barra A, Monserrat O, Mazzanti P, Esposito C, Crosetto M, Scarascia Mugnozza G (2016) First insights on the potential of Sentinel-1 for landslides detection. Geomatics. Nat Hazards Risk 1–10

Blanco-Sánchez J, Mallorquí J, Duque S, Monells D (2008) The coherent pixels technique (CPT): an advanced DInSAR technique for nonlinear deformation monitoring. Pure Appl Geophys 165:1167–1193
Braga JC, Martín JM, Alcalá B (1990) Coral reefs in coarse-terrigenous sedimentary environments (upper Tortonian, southern Spain). Sediment Geol 66(1–2):135–150
Bru G, Fernández-Merodo JA, García-Davalillo JC, Herrera G, Fernández J (2017) Site scale modelling of slow-moving landslides, a 3D viscoplastic finite element modelling approach. Landslides https://doi.org/10.1007/s10346-017-0867-y
Chacón J, Irigaray C, Fernández T (2007) Movimientos de Ladera. In: Ferrer, M. (Coordinator) Atlas Provincial de Riesgos Naturales en Granada: Diputación Granada-IGME. Spain. pp. 45–82, Maps 1:200,000/400,000, 1 CD
Chacón F, Irigaray C, El Hamdouni R, Valverde-Palaciuos I, Valverde-Espinosa I, Calvo F, Jiménez-Perálvarez J, Chacón E, Fernández P, Garrido J, Lamas F (2012) Engineering and environmental geology of granada and its metropolitan area (Spain). Environ Eng Geosci 3:217–260
Devanthéry N, Crosetto M, Monserrat O, Cuevas-González M, Crippa B (2014) An approach to persistent scatterer interferometry. Remote Sens 6(7):6662–6679
Feriche M (2012) Elaboración de escenarios de daños sísmicos en la ciudad de Granada. Universidad de Granada, PhD
Fernández-Merodo JA (2001) Une approche à la modélisation des glissements et des effondrements de terrains: Initiation et propagation. PhD, École Centrale Paris
Fernández-Merodo JA, García-Davalillo JC, Herrera G, Mira P, Pastor M (2014) 2D viscoplastic finite element modelling of slow landslides: the portalet case study (Spain). Landslides 11:29–42
Fernandez-Merodo JA, Mateos R, Azañon JM, Ezquerro P, García-Davalillo JC, Lorenzo C, Hernandez M, Bejar M, Herrera G, Castellanza R (2018) PROTHEGO Deliverable D.06.01: PROTHEGO PILOTS: THE ALHAMBRA. Local scale investigation, monitoring and advanced modelling of the geo-hazards affecting the Alhambra world heritage case study sit*e* version 1.3. JPI-CH Heritage Plus PROTHEGO project, Open Report. Date 11/05/2018. 66 Pages. http://www.prothego.eu/
Ferreira T, Mateos RM, Roldán FJ (2015) Los deslizamientos de la Cuenca baja del río Darro (Granada, España). Geogaceta 57:103–106
Ferrer M (2007) Atlas de riesgos naturales en la provincia de Granada. Diputación de Granada & Instituto Geológico y Minero de España, Spain. ISBN: 978-84-7807-438-9
Ferretti A, Savio G, Barzaghi R, Borghi A, Musazzi S, Novali F, Prati C, Rocca F (2007) Submillimeter accuracy of InSAR time series: experimental validation. IEEE Trans Geosci Remote Sens 45(5):1142–1153
Irigaray C, Lamas F, El Hamdouni R, Fernández T, Chacón J (2000) The importance of precipitation and the susceptibility of the slopes for the triggering of landslides along the roads. Nat Hazards 21(1):65–81
Justo JL, Azanón JM, Azor A, Saura J, Durand P, Villalobos M, Morales A, Justo E (2008) Neotectonics and slope stabilization at the Alhambra, Granada, Spain. Eng Geol 100:101–119
Lague D, Brodu N, Leroux J (2013) Accurate 3D comparison of complex topography with terrestrial laser scanner: application to the Rangitikei canyon (N-Z). ISPRS J Photogram Remote Sens 80:10–26
Lupiani E, Soria J (1988) Mapa Geológico y Memoria de la Hoja nº 1009 (Granada). Mapa Geológico de España E. 1:50.000 (MAGNA). Segunda Serie, Primera edición. IGME, 73 pp. Depósito legal: M-30.451–1988
Martín JM (2000) Geología e historia del oro de Granada. Boletín Geológico Minero 111(2–3):47–60
Mateos RM, Ezquerro P, Luque-Espinar JA, Béjar-Pizarro M, Notti D, Azañón JM, Montserrat O, Herrera G, Fernández-Chacón F, Peinado T, Galve JP, Pérez-Peña V, Fernández-Merodo JA, Jiménez J (2017) Multiband PSInSAR and long-period monitoring of land subsidence in a strategic detrital aquifer (Vega de Granada, SE Spain): an approach to support management decisions. J Hydrol 553:71–87

Páez J (1996) El Clima de Al-Andalus. In: Chacón J, Rosúa JL (eds) I Conferencia Internacional de Sierra Nevada, vol 5. University of Granada. Granada, Spain, pp 9–21
Sanz de Galdeano C, Alfaro P (2004) Tectonic significance of the present relief of the Betic Cordillera. Geomorphology 63(3–4):175–190
Sistema Nacional de Cartografía de Zonas Inundables SNCZI (2011) http://www.mapama.gob.es/es/agua/temas/gestion-de-los-riesgos-de-inundacion/snczi/ Accessed 16 May 2018

Rock Mechanic Characterization and 3D Kinematic Analysis of Katskhi Pillar (Georgia)

Claudio Margottini, Giovanni Gigli, Daniele Spizzichino, Marco Camorani, Giovanni Fiorini, and Daniele Vicini

Abstract Katskhi pillar is a natural limestone monolith located close to Katskhi village in western Georgian region of Imereti, near the town of Chiatura. It is approximately 40 m high, and overlooks the small river valley of Katskhura, a right affluent of the Q'virila river. On top of the pillar a small church is located, together with the house of the unique monk guardian taking care of the site. The pillar is affected by slope instability in its rocky components and his long time of standing is not clear. A 3D laser scanner (TLS), integrated with drone survey, has been conducted in order to identify the geometry and slope conditions of the pillar. Point clouds have been also used for automatic identification of rock mass structural setting, integrated with traditional manual survey. Geological, geomorphological and geotechnical data have been collected to define the main strength and deformation parameters, integrated with passive seismic surveys, to achieve information about resonant frequency of the site. Finally, the integration of 3D point clouds with rock mechanic parameters, allowed the realization of an advanced kinematic analysis for the whole pillar and identifying most potential unstable areas. Some preliminary proposals for monitoring and mitigation have been provided.

Keywords Limestone pillar · Laser scanner · Rock mechanic · Kinematic analysis

---

C. Margottini (✉)
former Scientific Attaché, Embassy of Italy in Egypt, 15, Abd El-Rahman Fahmy Str., Garden City, Il Cairo, Egypt
e-mail: claudio.margottini@gmail.com

C. Margottini · G. Gigli · D. Spizzichino
UNESCO Chair on Prevention and Sustainable Management of Geo-Hydrological Hazards, University of Florence, Florence, Italy

D. Spizzichino
Geological Survey of Italy (ISPRA), Rome, Italy

M. Camorani · G. Fiorini · D. Vicini
Gruppo Servizi Topografici, Reggio Emilia, Italy

G. M. El-Qady and C. Margottini (eds.), *Sustainable Conservation of UNESCO and Other Heritage Sites Through Proactive Geosciences*, Springer Geology,
https://doi.org/10.1007/978-3-031-13810-2_4

1 The Katskhi Pillar and the Monastery

The Katskhi pillar (Margottini et al. 2014a, b) is a natural limestone monolith located close to Katskhi village in western Georgian region of Imereti, near the town of Chiatura. It is approximately 35 m high, and overlooks the small river valley of Katskhura, a right affluent of the Q'virila river (Fig. 1). The rock column, with visible church ruins on its top surface of around 150 m^2, has been venerated by locals as the "Pillar of Life", symbolizing the True Cross, and has become surrounded by legends. It remained unclimbed by researchers and unsurveyed until 1944 and was more systematically studied from 1999 till now. These studies revealed the early medieval hermitage, dating from the 9th or tenth century. A Georgian inscription dated to the thirteenth century suggests that the hermitage was still extant at that time. Religious activity associated with the pillar started to revive in the 1990's and the monastery building had been restored within the framework of a state-funded program by 2009. The Katskhi pillar complex, in its current state, consists of a church, a crypt (burial vault), three hermit cells, a wine cellar, and a curtain wall on the uneven top surface of the column. At the base of the pillar are the newly built church of Simeon Stylites and ruins of an old wall and belfry. The church of St. Maximus the Confessor is located at the south-easternmost corner of the top surface of the Katskhi pillar. A small simple hall church design with the dimensions of 4.5 $\times$ 3.5 m., it is a modern restoration of the ruined medieval church built of stone.

Beneath and south of the church is an elongated rectangular crypt with the dimensions of 2.0 $\times$ 1.0 m., which had served as a burial vault. Digs at the ruined wine

Fig. 1 Panoramic view of Katskhi Pillar Monastery

cellar revealed eight large vessels known in Georgia as k'vevri. Also of note is a rectangular cellar grotto with the entrance and two skylights—on the vertical surface of the rock, some 10-m below the top. At the very base of the pillar there is a cross in relief, exhibiting parallels with similar early medieval depictions found elsewhere in Georgia.

In historical records, the Katskhi pillar is first mentioned by the 18th-century Georgian scholar Prince Vakhushti, who reports in his Geographic Description of the Kingdom of Georgia: "There is a rock within the ravine standing like a pillar, considerably high. There is a small church on the top of the rock, but nobody is able to ascend it; nor know they how to do that." No other written accounts of monastic life or ascents survive. A number of local legends surround the pillar. One of them has it that the top of the rock was connected by a long iron chain to the dome of the Katskhi church, located at a distance of around 1.5 km from the pillar (Gagoshidze 2010).

In July 1944 a group led by the mountaineer Alexander Japaridze and the writer Levan Gotua made the first documented ascent of the Katskhi pillar. Vakhtang Tsintsadze, an architecture specialist with the group, reported in his 1946 paper that the ruins found on top of the rock were remains of two churches, dating from the fifth and sixth centuries and associated with a stylite practice, a form of Christian asceticism (Gagoshidze 2010). The topographic surveys are reported in the following Fig. 2.

Since 1999, the Katskhi pillar has become the subject of more systematic research.

Based on further studies and archaeological digs conducted in 2006, Giorgi Gagoshidze, an art historian with the Georgian National Museum, re-dated the structures to the 9th or tenth century. He concluded that this complex was composed of a monastery church and cells for hermits. Discovery of the remnants of a wine cellar also undermined the idea of extreme ascetism flourishing on the pillar. In 2007, a small limestone plate with the asomtavruli Georgian inscriptions was found, dated to the thirteenth century and revealing the name of a certain "George", responsible for the construction of three hermit cells. The inscription also makes mention of the "Pillar of Life", echoing the popular tradition of veneration of the rock as a symbol of the True Cross.

Religious activity started to revive in 1995, with the arrival of the monk Maxim, a native of Chiatura. Between 2005 and 2009, the monastery building on the top of the pillar was restored with the support of the National Agency for Cultural Heritage Preservation of Georgia (Fig. 2). The rock is now accessible through an iron ladder running from its base to the top (Fig. 3).

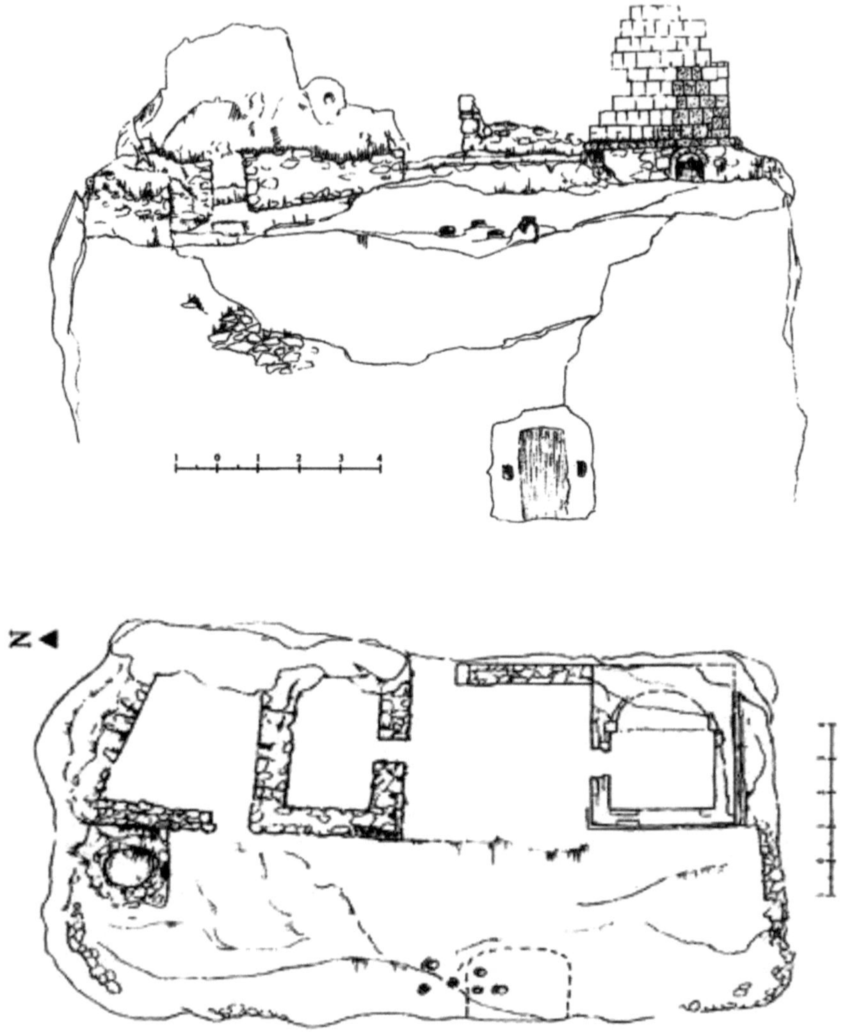

Fig. 2 Front view (top) and map (low) of Katski pillar in the first topographic survey from (Tsintsadze 1946; Gagoshidze 2010)

2 Site Investigations

2.1 Geomatic Survey

A proper geomatic survey was considered of fundamental relevance for the geological study in Katskhi. The surveying was aimed at the three-dimensional (3D) reconstruction of the pillar, collection of data for structural analysis and evaluation of future

Fig. 3 Ruins of the monastery before restoration works (Gagoshidze 2010)

potential slope stability of the site. Given the morphology and peculiarity of Katskhi, the survey was carried out with 3 different types of instrumentation:

- Leica HDS 7000 laser scanner for surveying accessible areas of the monolith. 26 high definition scans (6 mm/10 m) were performed. The union of the point clouds was performed with the Leica Cyclone© program with the "cloud to cloud" technique. The average alignment error did not exceed 3 mm. Also in Cyclone© the model obtained with the laser scanner was combined with that obtained from the survey with drone;
- Phantom 3 drone with 12 MP/2, 7 K video camera for the detection of portions of walls not visible with the laser scanner and for the realization of the video. Through the use of the Photoscan© software from the union of the photographs taken (1250), the sequence of which was appropriately programmed, the point cloud was obtained which in turn was combined with that of the laser scanner, as shown above. The precision achieved in this activity was found to be sub centimeter;
- GPS Leica 1200, reference + rover, in RTK mode for the local georeferencing of the laser scanner survey stations and that of the targets used for framing the flight with drone.

The result is reported in the following Fig. 4, compared with the real view.

2.2 *Rock Mechanic Characterization*

Geo-mechanical characteristics of the pillar have been reconstructed through geotechnical field techniques and in situ laboratory tests, on rock blocks. A rock mass classification and failure criteria have been implemented (Hoek 2007). More

Fig. 4 3 Point clouds final model (left) and real image (right)

in detail, the following activities have been carried out during 2012 field survey: implementation through scan lines techniques in order to define and reconstruct the main geo-mechanical parameters and indexes (e.g., RMR, GSI, Q system); strength and deformation parameters are collected starting from scientific and technical literature as well as from local technical report; geo-structural analysis (orientation and characteristics of discontinuities); Schmidt-hammer test on joint surfaces and intact rock block for in-situ analysis of UCS; sampling of blocks in the field for the laboratory tests; laboratory Point Load test to provide UCS data from sampled blocks and compare them with in-situ data (ISRM 1978, 1981).

The preliminary site engineering geology investigation for reconstruction of the strength of slope forming materials and geo-structural characteristics of the joints, have provided the following results. Uniaxial Compressive strength is about 153 MPa with laboratory point load, and an average density of 22,6 KN/m^3. Basic friction angle with Titl Test has been estimated in 46° (Table 1).

The field survey allowed to differentiate the rock mass of the pillar in three different GSI zones. They are reported in the following Fig. 5.

Table 1 Synthesis of preliminary geotechnical parameters for the rock material, in Katskhi limestone formation

Natural unit weight (KN/m^3)	JRC	UCS from Schmidt-Hammer test (MPa)	ϕ° Tilt test	UCS from Point Load Test (MPa)
22,6	6–8	50–70	46°	153

Fig. 5 Geological strength index (GSI) zoning

2.3 *Structural Analysis*

Laser scanning data have been managed in Coltop3D©, to identify dip and dip/direction of the identified slope faces. Coltop3D© is a software that performs structural analysis by using digital elevation model (DEM) and 3D point clouds acquired with terrestrial laser scanners. A colour representation merging slope aspect and slope angle is used in order to obtain a unique code of colour for each orientation of a local slope (Fig. 6). Coltop3D© extends the idea of having a unique colour for both dip and dip direction by adapting a computer graphics classical Hue Saturation Value (HSV) wheel to a lower or upper Schmidt-Lambert projected stereo-net.

The collected data, coupled with data from manual survey, allow a semi-automatic identification of dip/dip direction surfaces, to be transferred to a stereographic projection program such as Dips©, for the analysis and presentation of orientation based data (Fig. 7).

According to the 517.000 detected planes through Coltop3D, three major sets have been identified. They are 86/278 and 90/187 (sub-vertical, perpendicular each other) and a sub-horizontal plan 10/113.

2.4 *Types of Rock Slope Failures*

Rock slope failures occurring at Katskhi pillar can be classified into one of the following categories listed below depending on the type and degree of structural control and mechanical properties of rock masses.

Fig. 6 Global structural analysis performed with Coltop3D software

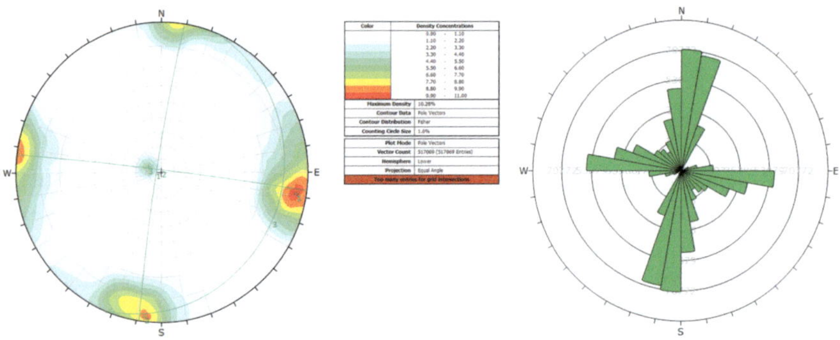

Fig. 7 Stereographic projection in Dips © of data collected in Coltop3D©

2.4.1 Small Medium Rock Fall and Wedge Failure

The whole pillar is affected (mainly in the lower/middle portion) by diffused small size rockfall and wedge, size <1 m^3 (Fig. 8).

These phenomena, although are not directly correlated to the overall stability of the pillar, can be very dangerous for pilgrims and visitors.

2.4.2 Small/medium Rock Fall and Wedge Failure

In the middle portion of the column small block sliding phenomena have been identified and mapped, block size from 1 to 2 m^3 (Fig. 9).

Fig. 8 Small size rockfall in the lower portion of the pillar

Fig. 9 Rock slide failure detected along the slope face

Also, in that situation there is no a real hazard for the stability of the entire column but there is a high risk for tourists, pilgrims and the community of monks and hermits of the site.

2.4.3 Medium Rock Fall and Wedge Failure

In the upper part of the pillar some large unstable rock masses have been detected and mapped (Fig. 10). Due to the dimensions of the unstable rock masses also the edge and the boundary of the monastery could be involved.

This situation is the most dangerous involving at the same time both safety of tourists and Monastery edge.

Fig. 10 Unstable rock masses in the upper portion of the pillar

3 Kinematic Analysis

The Katskhi pillar has been investigated with respect to both shallow mass movements as well as overall global stability.

Shallow mass movements are strongly depending from orientation of discontinuities with respect to slope face and friction angle along the joints (Goodman 1976). This can be evaluated by verifying graphically well determined geometrical conditions. The approach is well known as kinematic analysis which is a method used to analyze the potential for the various modes of rock slope failures (plane, wedge, toppling failures), that occur due to the presence of unfavorably oriented discontinuities.

Kinematic analysis is based on Markland's test which is described in Hoek and Bray (1981).

Traditional elaborations are developed in limited portion of the cliff and computed with the support of dedicated softwares, as in the case of Dips©. The result is shown in the Fig. 11. for the Eastern and Norther cliff. Major mechanisms for instability are respectevely flexural topplig and planar sliding.

In recent period, rock faces with rugged shape can be investigated by inspecting the discontinuity surfaces exposed on the slope. Such 3D approach requires the extraction of clusters of points belonging to the same discontinuity plane from the point cloud; subsequently, a spatial analysis for the quantitative description of discontinuities within the rock mass has to be performed.

The proposed approach is described in detail in Gigli and Casagli (2011) and is based on the definition of least squares fitting planes on clusters of points extracted by moving a sampling cube on the point cloud. If the associated standard deviation is below a defined threshold, the cluster is considered valid. By applying geometric criteria, it is possible to join all the clusters lying on the same surface; in this way discontinuity planes can be reconstructed, and rock mass geometrical properties are calculated.

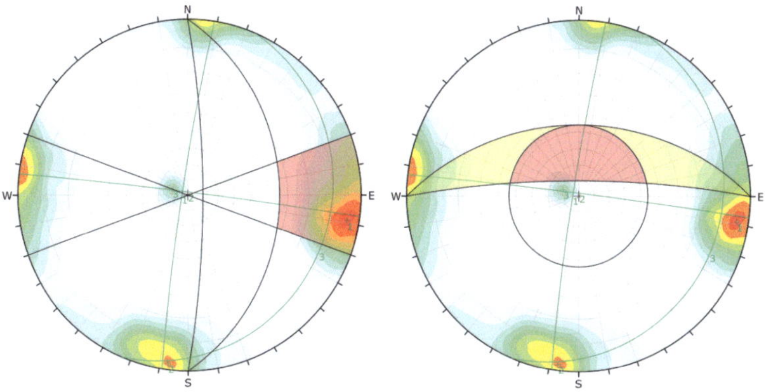

Fig. 11 Example of Kinematic analysis for Eastern and Northern cliff. Major mechanisms for instability are respectively flexural toppling (left) and wedge failure (right)

Finally, for the identification of the potential rockfall source areas, a spatial kinematic analysis has been performed by using discontinuity orientation data extracted from the point cloud by Diana and the high resolution DEM obtained from the TLS survey.

This kind of analysis is able to establish where a particular instability mechanism is kinematically feasible, given the geometry of the slope and the orientation of discontinuities (Goodman and Bray 1976; Hoek and Bray 1981; Matheson 1983; Hudson and Harrison 1997). The main instability mechanisms investigated with this approach are: plane failure; wedge failure; block toppling; flexural toppling, free fall.

The elaborations on the Katskhi pillar considered all the above potential mechanisms of rock collapse and has been developed on the 3D model of the pillar. This is a relevant advancement with respect to the traditional approach, generally restricted to specific part of the cliff. The result is in Fig. 12 where it is possible to notice that most potential slope instability are planar failure and wedge failure.

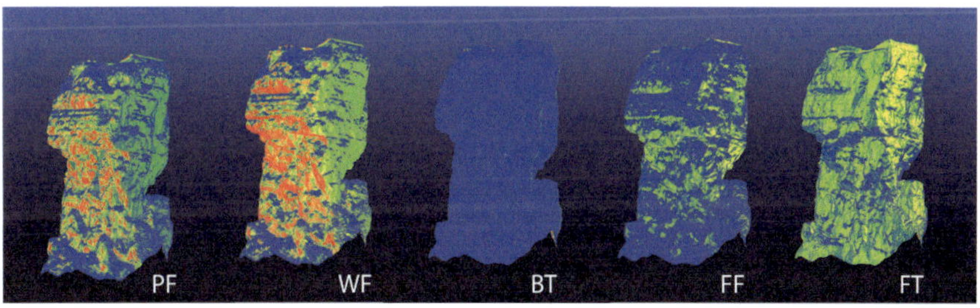

Fig. 12 Potential rock mass instabilities in the Katskhi pillar. PF stands for planar failure, WF wedge failure, BT direct toppling, FF free fall, FT flexural toppling

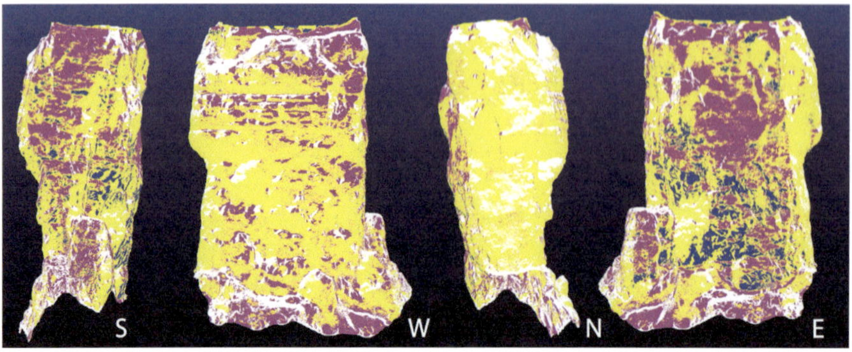

Fig. 13 Worst mechanism for slope instability in the various areas of the Katskhi pillar. Red stands for planar sliding, yellow wedge failure, green direct toppling, violet flexural toppling, blue free fall

Generalizing for the whole surfaces, the most endangered mechanisms for collapse are in the following Fig. 13. According to the results, the worst detected mechanisms are mainly wedge failure and flexural toppling.

4 Passive Seismic Analysis

Single station measurements of environmental seismic noise, were conducted around the base of the pillar and at the top of the pillar in order to assess frequency peak and possible double resonance effect. The traces recorded at the base of the pillar pointed out at least 3 frequency peaks with a range for the first between 1.0 and 1.5 Hz and two minors at 3 and 5 Hz: only the trace n. 3 is unclear. Concerning the other two traces collected on the top of the pillar, there is a clear frequency peak between 1 and 1.5 Hz, like the bottom traces, as well as the other minor peaks (Fig. 14).

Finally analysing the direction of the vibratory ground motion, it is possible to state that most of the directionality of the 1–1.5 Hz peak is in the range 40–140° that means around EW (Fig. 15).

As a consequence, it is possible to formulate the hypothesis that there is a higher focalisation of the signal around 1.5 Hz in the EW direction, as revealed also from the particle motion analysis (Fig. 16).

5 Overall Stability

A very crude investigation on the overall stability problem can be carried out by considering a two-dimensional cross-section cutting the pillar in the direction of the max slope.

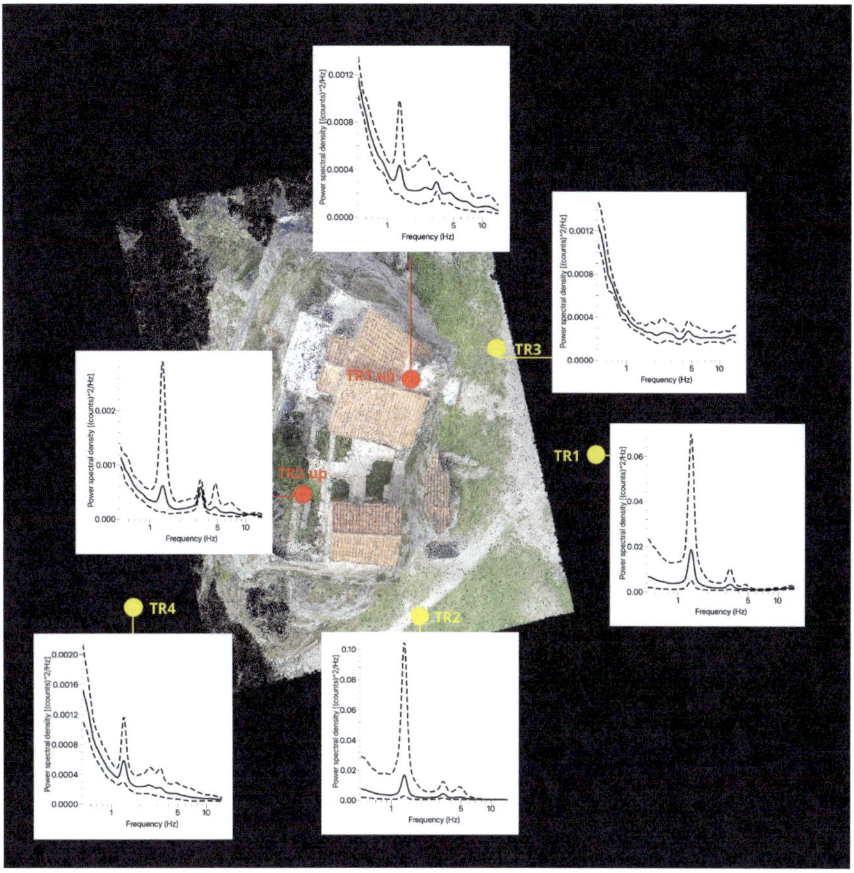

Fig. 14 Power spectral density for the registered signals

The results can only be used to obtain a general and preliminary indication of the behaviour of the pillar. A stress/strain analysis was then implemented through a finite element model by means of the software Phase2© by Rocscience.

Input parameters were collected according to the following conditions:

- the pillar have been devided in three parts, according to GSI index;
- the rock mass was considered to be isotropic;
- friction angle (ø = 46°) is the basic value determined from tilt test and and c = cohesion from the Hoek and Brown curve in RockLab1.0©, where ø = friction angle; the curve is calculated for the sigci, GSI, mi and D, corresponding to the different secotr of the pillar;
- Hoek and Brown parameters have been calculate from RockLab1.0©;
- Dilatation angle was estimated as $\partial = \frac{1}{3}\phi$;
- Tensile strength as $\sigma t = \sigma ci * \frac{s}{mb}$;

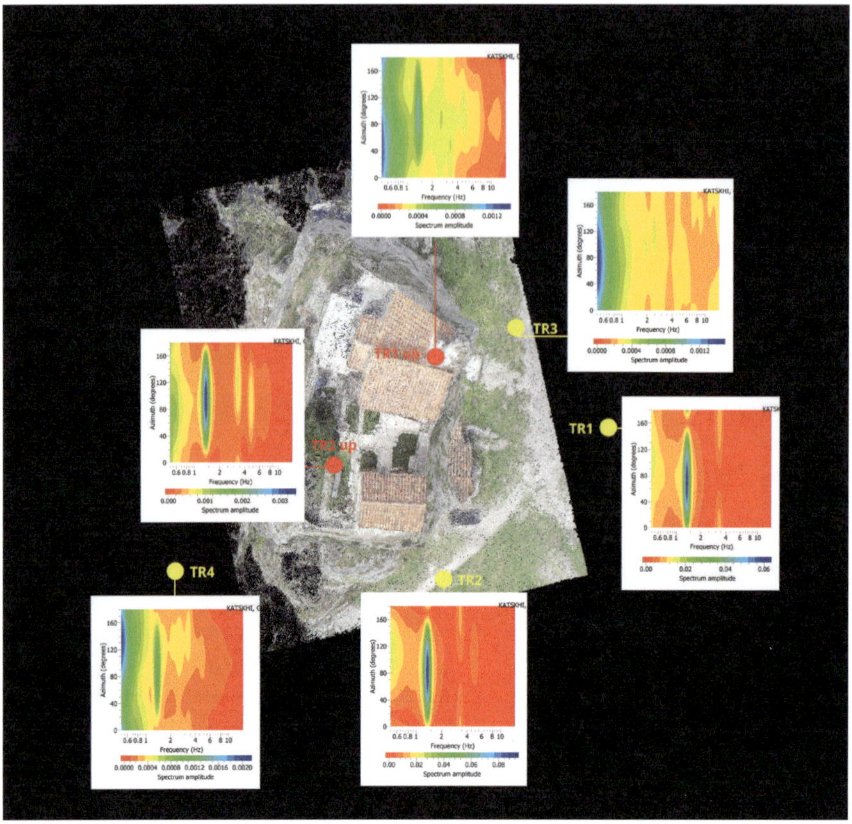

Fig. 15 Focalisation of spectral amplitude in different directions

- Young modulus (Er) is calculated from Em in RockLab1.0© according to the relationship $Em = \lambda e * Er$, where $\lambda e = 0,0231(RQD) - 1,32$.

The following Fig. 17 is synthesizing the various parameters adopted in the analysis and describing the geotechnical section used in phase2©.

The analysis reveal that the pillar is naturally deforming towards West, in the direction of the downslope. The rate is not relevant and of the order of 1 mm (Fig. 18), reaching the value of 3 mm in case of horizontal seismic acceleration equal to PGA = 0,18 g, according to Giardini et al. (2018).

The following Fig. 19 shows contour plots for principal stress $\sigma 1$, underlining the concentration on the lowest part of the column, then suggesting a high attention to such area.

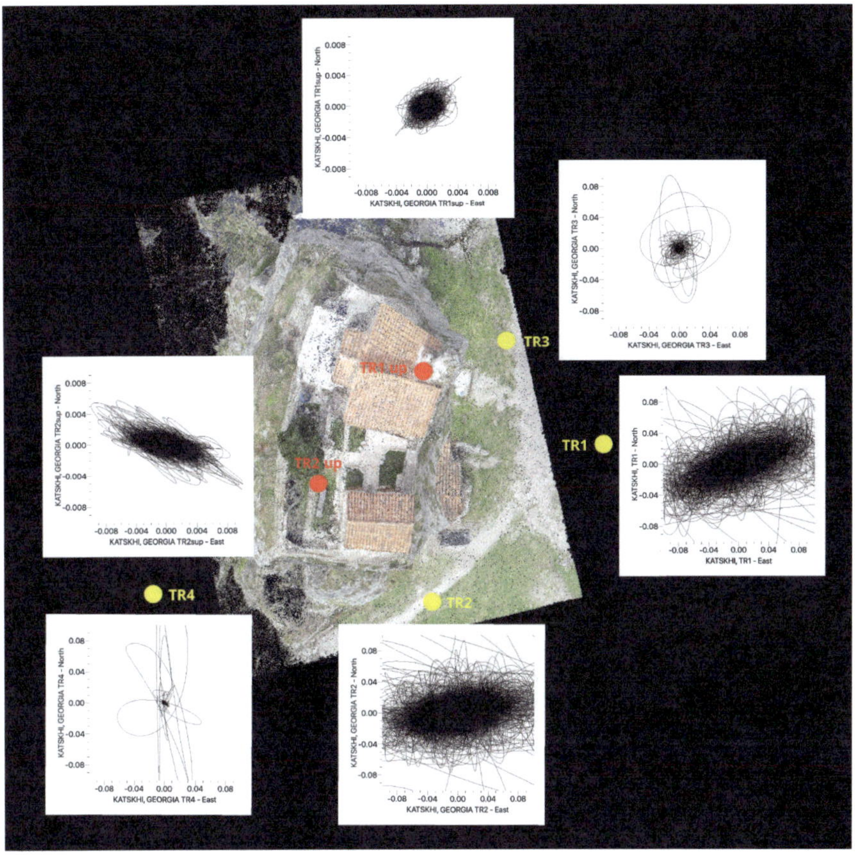

Fig. 16 Particle motion in the range 1–4 Hz. The directivity of the peak at 1–1.5 Hz is quite evident

6 Discussion and Conclusion

According to the present investigation, the Katskhi pillar is a limestone monolith standing 35 m above the flank of the valley. The origin of the pillar is not visible, most likely being the consequence of erosional phenomena.

The pillar is composed by medium-hard limestone, not presenting major discontinuities. Three set of discontinuities have been identified on the site, mainly affecting the stability of the shallower part of the pillar. This can be confirmed by the long life of the pillar, dated back at least to 9th or tenth century as a monastery church and cells for hermits.

Passive seismic analysis revealed that the vibratory ground motion has time histories highly focalised around 1.5 Hz, and particle motion in the EW direction.

The E-W direction is exactly the same downslope direction where also the 2D deformation analysis is most likely to occur.

Material Name	Color	Initial Element Loading	Unit Weight (MN/m3)	Elastic Type	Young's Modulus (MPa)	Poisson's Ratio	Failure Criterion	Material Type	Tensile Strength (MPa)	Tensile Strength (residual) (MPa)	Dilation Angle (deg)	Friction Angle (peak) (deg)	Friction Angle (residual) (deg)	Cohesion (peak) (MPa)	Cohesion (residual) (MPa)	Piezo Line	Ru
upper limestone		Field Stress and Body Force	0.023	Isotropic	41000	0.3	Mohr Coulomb	Plastic	0.91	0	15	46	44	4.82	0	None	0
middle limestone		Field Stress and Body Force	0.023	Isotropic	41000	0.3	Mohr Coulomb	Plastic	0.59	0	15	46	44	3.6	0	None	0
lower limestone		Field Stress and Body Force	0.023	Isotropic	41000	0.3	Mohr Coulomb	Plastic	0.38	0	15	46	44	2.69	0	None	0
bedrock		Field Stress and Body Force	0.023	Isotropic	41000	0.3	Mohr Coulomb	Plastic	0.91	0	15	46	44	4.82	0	None	0

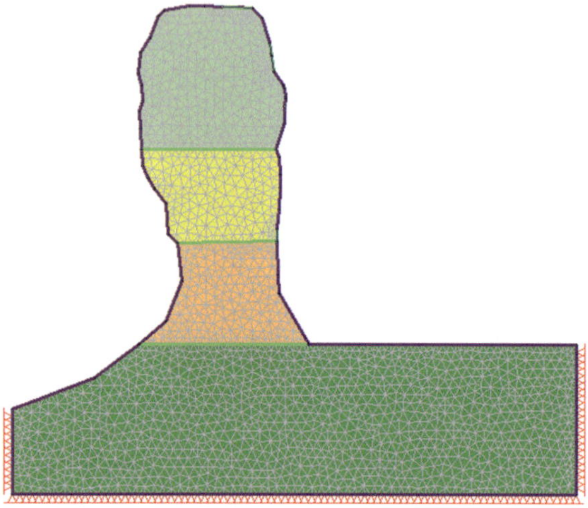

Fig. 17 Geotechnical section and adopted geomechanical parameters in Phase2©

As a conclusion, the pillar is relatively safe with respect to the overall stability, but it is presenting shallow wedge failure and flexural toppling among the many kinematic possibilites.

In order to reduce possible and localised block collapse (from small to medium size) a preliminary cleaning of the North, South and Est flank should be implemented. At the same time, the fixing (through bolts and passive anchors) of unstable medium-sized plates should also be provided.

An in situ monitoring system (wirless modality with one tiltmeter and a couple of crack gauges coupled with temperature and air humidity sensors) should be implemented in order to control possible deformations and opening of cracks at whole pillar scale.

Finally, the focalisation of the seismic motion in E-W direction may amplify the vibratory ground motion in occasion of strong earthquakes.

This is suggesting the need of further investigations coupling seismic hazard analysis with rock mass stability.

The future investigation should useful to address possible low impact mitigation measure (long term strategies) to adopt for the safety of the structure, monk's community and sustainable tourism exploitation.

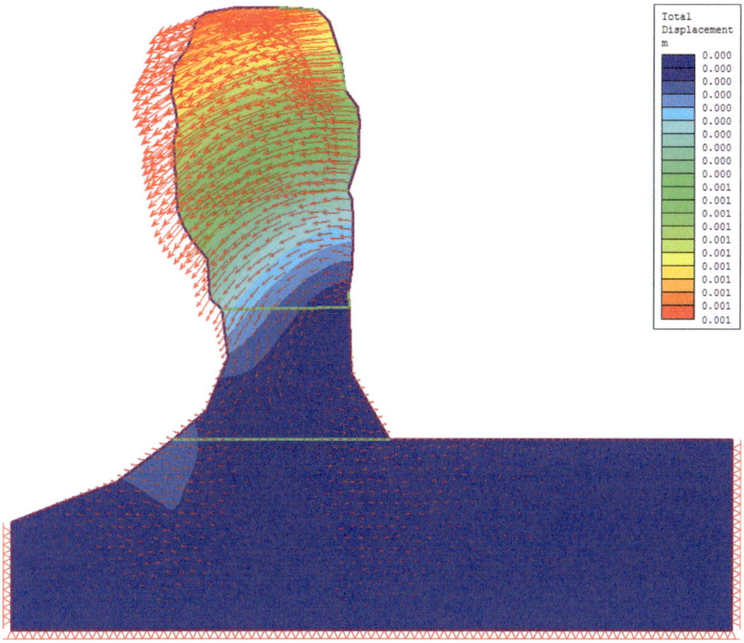

Fig. 18 2D deformation analysis of Katskhi pillar, in static condition

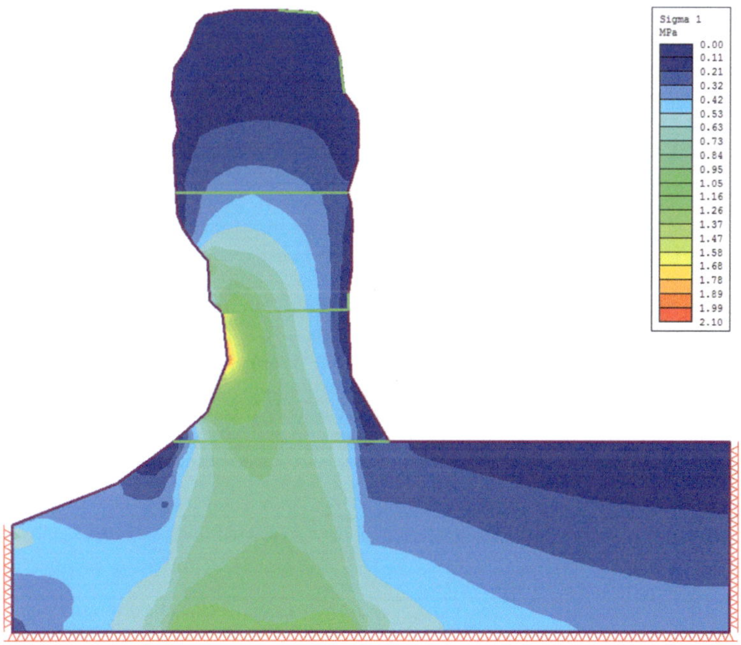

Fig. 19 Contour plots for principal stress $\sigma 1$

Acknowledgements The authors are grateful to all the staff members of NACHPG and especially to Director Nika Antidze for believing in the recovery and the valorization of rupestrian sites in Georgia, and for the support in the overall development of the activities. This work was carried out in the scientific framework of the scientific collaboration between the Geological Survey of Italy (ISPRA), the UNESCO Chair on Prevention and sustainable management of geo-hydrological hazards of the University of Florence (Italy) and Gruppo Servizi Topografici in Reggio Emilia (Italy).

A special thanks to the Embassy of Italy in Georgia for having promoted the topographic survey in Katskhi.

References

Gagoshidze G (2010) კაცხის სვეტი ("Katskhi Pillar"). Academia No. 1: 55–68. ISSN 1512–0899

Giardini D, Danciu L, Erdik M et al (2018) Seismic hazard map of the Middle East. Bull Earthquake Eng 16:3567–3570. https://doi.org/10.1007/s10518-018-0347-3

Gigli G, Casagli N (2011) Semi-automatic extraction of rock mass structural data from high resolution LIDAR point clouds. Int J Rock Mech Min Sci 48(2):187–198

Goodman ER (1976) Methods of geological engineering in discontinuities rocks. West Publishing Company

Goodman RE, Bray JW (1976) Toppling of rock slopes. In: Proceedings of the Special Conference on Rock Engineering for Foundations & Slopes (vol. 2, pp 201–234). ASCE Boulder, Colorado

Hoek E, Bray JW (1981) Rock slope engineering, 3rd edn. The Institute of Mining and Metallurgy, London, England, p 358

Hoek E (2007) Practical rock engineering, 2007 edn. https://www.rocscience.com

Hudson JA, Harrison JP (1997) Engineering rock mechanics. Pergamon ed

ISRM (1978) International society for rock mechanics commission on standardization of laboratory and field tests. Suggested methods for the quantitative description of discontinuities in rock masses. Int J Rock Mech Min Sci Geomech Abstr 15:319–368

ISRM (1981) International society for rock mechanics and rock engineering. Suggested Methods for the rock characterization, testing and monitoring. Pergamon Press, Oxford, p 211

Margottini C, Spizzichino D, Sonnessa A, Puzzilli LM (2014a) Natural hazard affecting the Katskhi Pillar Monastery (Georgia). In: Lollino et al. (eds) Engineering geology for society and territory—Vol. 8—preservation of cultural heritage. Proceeding of IAEG XII CONGRESS. Springer Verlag, Torino

Margottini C, Puzzilli LM, Sonnessa A, Spizzichino D (2014b) Instability processes affecting the Katskhi pillar monastery (Georgia). In: Sassa K, Canuti P, Yin Y (eds) Landslide science for a safer geoenvironment. Targeted Landslides. Springer-Verlag, ISBN: 978–3–319–04995–3

Matheson GD (1983) Rock stability assessment in preliminary site investigations–graphical methods. Transport and Road Research Laboratory Report, p 1039

Tsintsadze V (1946) Katskhis Sveti, Sakartvelos SSR Mgtsnierebata Akademiis Moambe, vol VII:8 (Tbilisi, 1946), p 557

Petrological Study on the Roman Mortars from Kom El-Dikka Archaeological Site (Alexandria, Egypt)

Duygu Ergenç, Nevin Aly, Rafael Fort, Sayed Hemeda, and Mónica Alvarez de Buergo

Abstract Two types of Roman mortars were collected from the buildings of the *Thermal Baths*, *Villa of birds* and *Cisterns* located at Kom el-Dikka archaeological site (Alexandria, Egypt). It is believed that these mortars date back to the first or second century A.D. Petrographical, and chemical characterization of these mortars were performed, to differentiate the various construction phases of Kom el-Dikka archaeological site. Results showed that the analysed samples are lime mortars, with different types of aggregates. The samples from the Cisterns, which showed a high superficial strength, also have a different isotopic ratio ($\delta^{13}C$ and $\delta^{18}O$) than the rest of the studied mortar samples. The analysis of soluble salts in the mortar samples was conducted. It reveals a main content on sulfates, nitrates and chlorides, of which provenance probably accounts for atmospheric pollution and marine aerosol, as well as possibly, previous restoration interventions.

D. Ergenç · R. Fort · M. A. de Buergo
Institute of Geosciences (CSIC-UCM), Doctor Severo Ochoa 7, 28040 Madrid, Spain
e-mail: duyguerg@ucm.es

R. Fort
e-mail: rafael.fort@csic.es

M. A. de Buergo
e-mail: monica.alvarez@csic.es

D. Ergenç
Civil Engineering Department, Middle East Technical University, Ankara, Turkey

N. Aly (✉)
Faculty of Petroleum and Mining Engineering, Science and Engineering Mathematics Department, Suez University, Suez, Egypt
e-mail: nevin.aly@suezuniv.edu.eg

S. Hemeda
Faculty of Archaeology, Archaeological Conservation Department, Cairo University, Giza, Egypt
e-mail: sayed.hemeda@cu.edu.eg

School of Basic and Applied Science- BAS, Egypt-Japan University of Science and Technology (EJUST), Alexandria, Egypt

G. M. El-Qady and C. Margottini (eds.), *Sustainable Conservation of UNESCO and Other Heritage Sites Through Proactive Geosciences*, Springer Geology,
https://doi.org/10.1007/978-3-031-13810-2_5

Keywords Kom el-Dikka · Mortar · Lime · Petrography · Soluble salts · δ13C · δ18O

1 Introduction

Historic mortars are composite materials consisting of a combination of non/hydraulic binding materials, aggregates and additives (Moropoulou et al. 2000, 2005). The identification of their mineralogical and chemical composition is very useful in the recognition of the damage causes of the host building (Middendorf et al. 2005).

Non-hydraulic lime binder was used in most of the Roman buildings (Hamey and Hamey 1990; Siddall 2011). The lime mortar hydraulicity has been enhanced by highly reactive silica and alumina rich aggregates and/or additives leading to high bonding capacity and more pozzolanic reactions (Rodríguez-Navarro 2004; Mertens et al. 2009). The hydraulic and mechanical behaviour of lime mortars were improved by using these pozzolanic materials (Jonaitis et al. 2019). Roman masons widely used volcanic pyroclasts, ceramic bits and dust as pozzolanic admixtures when natural pozzolans were lacking (Stefanidou 2016). To make the mortar hydraulic, either hydraulic lime or non-hydraulic (pure) lime was used and the above-mentioned pozzolanic reactions were provided by aggregates or additives. Reactive clayey sands, chert, slates and silica rich additives could provide pozzolanicity (Ergenç and Fort 2019).

One of the threats facing most lime-based mortars is their susceptibility to salt damage (Rossi-Manaresi and Tucci 1991) because of the high water evaporation rate they have due to their porous structure (Stefanidou and Papayianni 2006). The random pore system and the pore interconnectivity of the lime-based mortars provide them a high open porosity (Papayianni et al. 2013). Thisresults in easy entrance of saline solutions penetrating into the masonryleading to seriously compromising the stability of the structure (Papayianni et al. 2013).

This study aimed to investigate and characterize the mortars employed in different buildings at Kom el-Dikka archaeological site focusing understanding the reasons of the growth of various soluble salts on the studied samples.

2 Study Area

Alexandria extends about 20 miles along the Mediterranean coast of Egypt. Alexander the Great founded it around 331 BC (Abdelhady 2014). Kom El-Dikka archaeological site seems like a whole city (Fig. 1), since it has an amphitheatre, a theatre auditorium, a school, some baths and cisterns (Theodore 2001; Hemeda 2013). The site is the only historic example allows researchers to study the urban fabric of the ancient Alexandria city in a wider urban context. Nowadays, the site

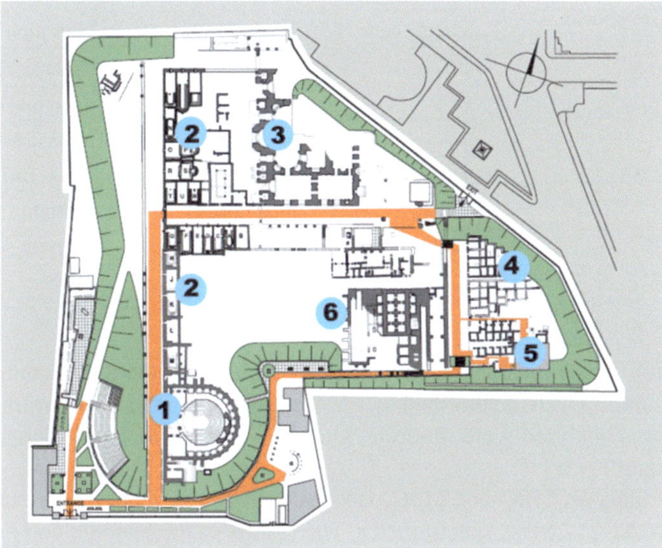

Fig. 1 Map of Kom el Dikka archaeological site (*source* Centrum Archeologii srodziemnomorskiej UW: Alexandria: Kom el-Dikka). 1. Theatre, 2. Auditoria 3. Baths 4. Domestic Quarter 5. Villa of the Birds 6. Cistern

embedded in the city, located nearby to Qaitbay citadel, towards the Southwest part of Alexandria.

3 Material and Methods

Joint mortar samples were taken from the *Thermal baths*, *Villa of birds* and *Cistern* structures, while from this last one, a rendering mortar was also sampled. The collected samples dated back to the period from the 2nd to the sixth century CE (Fort et al. 2021). The *Cistern* and *Baths* mortars showed a dark color and they were used to bond red brick blocks, while the *Villa of birds* mortar was pale in color and used to bond limestone (see Table 1).

4 Analytical Methods

The following analytical techniques were used in this study:

- Polarizing optical microscope Olympus BX51 polarized optical microscope (POM) with an attached Olympus DP12 digital camera was used for examining the mineralogical and petrographical features of the thin sections.

Table 1 Description of samples

Sample code	Type	Location
1	Joint mortar	Thermal Baths
2	Joint mortar	Thermal Baths
3	Joint mortar	Villa Birds
4	Joint mortar	Villa Birds
5	Joint mortar	Cistern
6	Joint mortar	Cistern
7	Rendering mortar	Cistern

- Ion Chromatography for the analysis of soluble salts: the anion content in the mortar samples were determined by a Metrohm 761 Compact ion chromatograph. The anions analysed were: fluoride, chloride, nitrite, nitrate, phosphate, sulfate and oxalate.
- Isotope analysis: δ18O and δ13C. Isotopic analyses of mortars were carried out with a MAT-252 Mass spectrometer. The results of the isotopic composition of carbon and oxygen were performed with the notation with respect to the Vienna Belemnite Standard Pee Dee (VPDB). The standard deviations were $\delta 13C = 0.03$ ‰; $\delta 18O = 0.06$ ‰.

For ion chromatography and isotope analyses, whole samples were ground for the analyses and binder-rich fragments of the samples were tested where applicable.

5 Results

5.1 *Petrographic Characterization*

The petrographic investigation of the studied mortars displayed different characteristics according to their hosting buildings, which correspond more likely to the different construction phases of the site. Samples collected from the Thermal Baths are lime mortars with a microcrystalline appearance (< crystal size 4 μm) with intercrystalline *vug* type porosity, and occasionally fissures in the binder (a petrographic estimation of porosity >15%). Two types of aggregates were identified, irregular ceramic fragments of ~1 to 0.1 mm size, in which quartz and quartzite could be identified (Fig. 2a). The second type of aggregates is limestone (mainly fossiliferous and oolitic) fragments seen larger than 4 mm (Fig. 2b). Other less frequently observed aggregates are poly-crystalline quartz (<1 mm), and some feldspars (potassium feldspar and plagioclase).

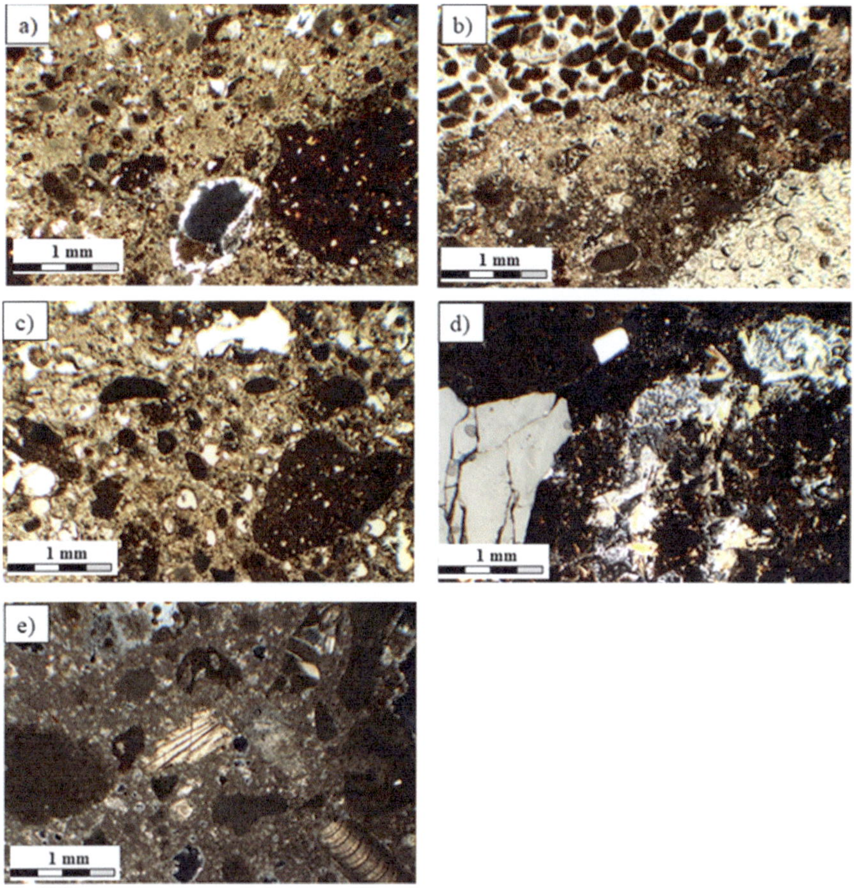

Fig. 2 Photomicrographs **a**, **b** *Thermal baths* joint mortar, **c** *Villa Birds* joint mortar, **d** *Cistern* joint mortar, **e** *Cistern* rendering mortar

In mortars of Thermal Baths, the binder: aggregate ratios ranged from 1:2 to 1:4. Both dissolution processes, resulting in an increase of the porosity, andrecrystallization processes leading to an increase of crystals size in the binder and therefore a reduction in the porosity were observed.

The binder of the mortars used in the *Villa of birds* presented a similar appearance to that of the mortar of the *Baths*. The aggregates of this mortar were mainly ceramic fragments, existing also lime lumps. Aggregates were smaller than those of the *Baths*´s mortars, together with quartz and some feldspar grains, smaller than 0.25 mm. They showed sub-rounded to angular shapes (Fig. 2c). The binder: aggregate ratio ranged from 1:3 to 1:4.

The samples from the *Cistern* were lime mortars, too. The carbonated binder consisted of microcrystalline calcite. It had an inter-crystalline porosity (>15%) in form of *vug*-type pores and fissures. The aggregates were fragments of fossiliferous

Table 2 Anion content in the mortar samples (ppm)

Sample	Fluoride	Chloride	Nitrate	Sulfate	Oxalate	Total
1 (Baths)	0.06	98.94	78.96	734.75	0.79	913.49
2 (Baths)	0.25	313.43	127.40	48.72	0.00	489.81
3 (Villa Birds)	0.31	90.47	60.61	1338.10	0.00	1489.49
4 (Villa Birds)	0.38	115.09	131.36	90.84	1.22	338.88
5 (Cistern)	0.21	727.03	924.85	577.64	0.00	2229.74
6 (Cistern)	0.27	725.09	475.64	39.40	0.00	1240.40

and oolitic limestone, and monocrystalline quartz (>2 mm). The presence of lime lumps was abundant. The binder: aggregate ratio ranged from 1:3 to 1:4. Drusy-type calcite crystals grew inside some pores. Filling of pores (secondary crystallization in the border) and shrinkage cracks in the binder were observed (Fig. 2d). The sample of the rendering mortar of the *Cistern* was composed of porous microcrystalline lime binder with aggregates consisting of fossil fragments and lime lumps (Fig. 2e). There was *vug*-type porosity within small drusy-type calcite crystal grown inside it.

Aggregate shapes, mainly quartz and feldspar, were irregular and surface textures were angular to sub-angular. This implies that the sands were transported by fluvial water currents.

5.2 *Ion Chromatography*

The results show that sulfates were abundant in some mortars such as Villa of Birds, which accounts for the inclusion of gypsum to the mortar (Table 2). Samples from the *Cisterns* showed higher amount of chlorides and nitrates than the others, which indicates a salt source related to rising damp. Neither nitrite nor phosphate was detected.

5.3 *Isotope Analysis*

The content of δ13C and δ18O in mortars showed values between −12.85 and −5.03 for δ13C and from −12.11 to 3.87 for δ18O (Table 3). The joint mortars from *Villa of birds* were those that have a lighter content of isotopes along with the rendering mortar of the *Cistern*. Mortars from the joints of the *Baths* and the *Cistern* showed a behaviour that moves away from those obtained in *Villa of birds*. The joint mortar of the *Cistern* had positive values of δ18O close to 3.8, and those of the *Thermal baths* between - 4 and 2. Both types had δ^{13}C values between −9,5 and −5,4.

The different isotopic composition of *Villa of bird* mortars with those of *Baths* and *Cistern* joints may be due to the presence of limestone aggregates, or to an

Table 3 Results of isotope analysis

	$\delta^{13}C$ (VPDB)	$\delta^{18}O$ (VPDB)
Baths joint mortar-1	−9,27	−3,96
Baths joint mortar-2	−5,36	1,96
Villa of birds-3	−5,04	−7,26
Villa of birds-4	−12,85	−12,11
Cistern joint-5	−8,7	3,82
Cistern joint-6	−9,45	3,87
Cistern rendering-7	−9,4	−11,67

uncompleted calcination of the limestone used to produce lime, which can increase the ratio of δ13C (Usdowski and Hoefs 1993; Dotsika et al. 2009). Another reason can be the presence of salt phases and ceramic dust, and even greater processes of dissolution and reprecipitation of calcium carbonate (Dotsika et al. 2018).

6 Conclusion

In this study petrographic and elemental description of Kom EL Dikka mortars are presented. Deterioration level of the mortar samples is discussed considering salt contents. Results showed that in the site, similar Roman construction technologies were used as those mentioned by Vitruvius.

Collected mortars confirmed to be lime mortars; fossiliferous limestone and ceramic were used as aggregates. This shows the selection among the available local raw materials.

Aggregate selection was based on where the mortars would be applied; humid or dry environment. Mortars from *Villa of Birds* have different characteristics, which implies different construction phase. Alternatively, this could be due to the weathering induced by salt presence.

The elaborate mason workmanship can be ruled out for the elaboration of mortars due to the high number of lime lumps due to the lack of a regular calcination temperature.

The high salt content has a high effect on dissolution–recrystallization processes. Therefore, classification based on the isotopic data according to the original composition is turned out to be complicated.

Acknowledgements This research has been financed by CLIMORTEC project (BIA2014-53911-R) of Spanish Ministry of Science, Innovation and Universities and TOP Heritage-CM (P2018/NMT4372) of Community of Madrid. Special acknowledgments to the professional support of the Interdisciplinary Thematic Platform from CSIC Open Heritage: Research and Society (PTI-PAIS). and to an I-COOP cooperation project (COOPB20379) between most of the co-authors of this paper (Egypt and Spain), and funded by CSIC (2019–2020).

References

Abdelhady K (2014) Factors affecting the conservation and regeneration of the urban fabric of old cities: case study of Old Alexandria. Sustain City IX 2:1175. https://doi.org/10.2495/SC140992

Dotsika E, Kyropoulou D, Christaras V, Diamantopoulos G (2018) 13C and 18O stable isotope analysis applied to detect technological variations and weathering processes of ancient lime and hydraulic mortars. Geosciences 8:339. https://doi.org/10.3390/geosciences8090339

Dotsika E, Psomiadis D, Poutoukis D, Raco BP (2009) GamaletsosIsotopic analysis for degradation diagnosis of calcite matrix in mortar. Anal Bioanal Chem 395:2227–2234. https://doi.org/10.1007/s00216-009-3135-8

Ergenç D, Fort R (2019) Multi-technical characterization of Roman mortars from Complutum, Spain. Measurement 147https://doi.org/10.1016/j.measurement.2019.106876

Fort R, Ergenç D, Aly N, de Buergo MA, Hemeda S (2021) Implications of new mineral phases in the isotopic composition of Roman lime mortars at the Kom el-Dikka archaeological site in Egypt. Constr Build Mater 268. https://doi.org/10.1016/j.conbuildmat.2020.121085

Hamey LA, Hamey JA (1990) The roman engineers. Published in cooperation with Cambridge University Press. Lerner Publications Co. ISBN 10: 0521225116

Hemeda S (2013) Laser induced breakdown spectroscopy and other analytical techniques applied on construction materials at Kom El-Dikka, Alexanderia, Egypt. Mediterr Archaeol Archaeom 13:103–119

Jonaitis B, Antonovic V, Sneideris A, Boris R, Zavalis R (2019) Analysis of physical and mechanical properties of the mortar in the historic retaining wall of the Gediminas castle hill (Vilnius, Lithuania). Materials 12:8.https://doi.org/10.3390/ma12010008

Mertens G, Snellings R, Van Balen K, Bicer-Simsir B, Verlooy P, Elsen J (2009) Pozzolanic reactions of common natural zeolites with lime and parameters affecting their reactivity. Cem Concr Res 39:233–240. https://doi.org/10.1016/j.cemconres.2008.11.008

Middendorf B, Hughes JJ, Callebaut K, Baronio G, Papayanni I (2005) Investigative methods for the characterisation of historic mortars. Part 1: mineralogical characterisation. Mater Struct 38

Moropoulou A, Bakolas A, Bisbikou K (2000) Investigation of the technology of historic mortars. J Cult Herit 1:45–58. https://doi.org/10.1016/S1296-2074(99)00118-1

Moropoulou A, Bakolas A, Moundoulas P, Aggelakopoulou E, Anagnostopoulou S (2005) Strength development and lime reaction in mortars for repairing historic masonries. Cement Concr Compos 27:289–294. https://doi.org/10.1016/j.cemconcomp.2004.02.017

Papayianni I, Stefanidou M, Pachta V, Konopisi S (2013) Content and topography of salts in historic mortars. In: Proceedings of 3rd historic mortars conference, 11–14 September 2013, Glasgow, Scotland. https://doi.org/10.1007/978-3-319-91606-4_9

Rodríguez-Navarro C (2004) Binders in historical buildings: traditional lime in conservation. SEM 09:91–112

Rossi-Manaresi R, Tucci A (1991) Pore structure and the disruptive or cementing effect of salt crystallization in various types of stone. Stud Conserv 36:53–58

Siddall R (2011) From kitchen to bathhouse: the use of waste ceramics as pozzolanic additives in Roman mortars. In: Ringbom Å, Hohlfelder RL (eds) Building Roma aeterna: current research on roman mortar and concrete, vol 128. The Finnish Society of Sciences and Letters, pp 152–168

Stefanidou M, Papayianni I (2006) Salt accumulation in historic and repair mortars. In: Fort R, de Buergo MA, Gomez-Heras M, Vazquez-Calvo C (eds) Heritage, weathering and conservation. Taylor & Francis, pp 269–272

Stefanidou M (2016) Use of natural pozzolans with lime for producing repair mortars. Environ Earth Sci 75:758. https://doi.org/10.1007/s12665-016-5444-5

Theodore V (2001) Alexandria, city of the western mind. Free Press, New York

Usdowski E, Hoefs J (1993) Oxygen isotope exchange between carbonic acid, bicarbonate, carbonate, and water: a re-examination of the data of McCrea (1950) and an expression for the overall partitioning of oxygen isotopes between the carbonate species and water. Geochim Cosmochim Acta 57(15):3815–3818. https://doi.org/10.1016/0016-7037(93)90159-T

Experimental Study of Different Electrokinetic Configurations for Desalination of a Brick Wall

Jorge Feijoo, Mónica Alvarez de Buergo, Rafael Fort, and Nevin Aly

Abstract Salt weathering is recognized as one of the main deterioration mechanisms that affect cultural heritage. The damage caused by this decay mechanism can range from simple aesthetic damage, when salts crystallize on the surface as efflorescence, to cause the loss of material, when salts crystalize beneath the material surface as sub- and crypto-efflorescence. To reduce the risk associated with salt-induced deterioration, the use of electrokinetic treatments, as desalination applied on-site, is undoubtedly the one that has allowed obtaining high extraction percentages in a short period of time. This study evaluates the efficacy of a new double electrode system at the anode (DA setup), to overcome the results achieved with the traditional electrokinetic configuration, using only one electrode (SA setup), to desalinate a brick masonry wall, located in the ancient stables of the eighteenth century Bernstorff Palace (Gentofte-Denmark), which is affected by salt-induced decay. The obtained results show that the DA setup allows maintaining higher moisture content in poultices located at the anode than using SA setup. This fact allows a proper and lasting current flow, which enhances the removal efficacy of all the anions present in the wall.

Keywords Salt-induced decay · Brick wall · Nitrates · Sulfates · Electrokinetic technique · Desalination · Double anode system · Anodic configuration

J. Feijoo · M. A. de Buergo · R. Fort
Institute of Geosciences - IGEO (CSIC-UCM), C/Doctor Severo Ochoa, 7, 28040 Madrid, Spain
e-mail: j.feijoo@csic.es; jfeijoo@cud.uvigo.es

J. Feijoo
Defense University Centre, Spanish Naval Academy, Plaza de España S/N, 36920 Marín, Spain

N. Aly (✉)
Faculty of Petroleum and Mining Engineering, Department of Science and Engineering Mathematics, Suez University, Suez 43512, Egypt
e-mail: nevin.aly@suezuniv.edu.eg

G. M. El-Qady and C. Margottini (eds.), *Sustainable Conservation of UNESCO and Other Heritage Sites Through Proactive Geosciences*, Springer Geology,
https://doi.org/10.1007/978-3-031-13810-2_6

1 Introduction

Salt weathering is considered as one of the most widespread decay mechanisms that affect porous materials used in building constructions (ornamental rocks, mortars, bricks…). The damage caused by this agent can cause material loss at the surface, which in the case of cultural heritage can be considered as an historical and artistic damage of inestimable value(Aly and Hamed 2020; Rovella et al. 2020). For this reason, there are numerous studies focused on: (1) modelling the physical damage caused by salts, i.e. crystallization pressure (Rossi-Manaresi and Tucci 1991; Scherer 2004), hydration pressure (Flatt 2002; Tsui et al. 2003) and the pressure caused by the different coefficients of thermal expansion between the porous material and salt (Cooke and Smalley 1968; Lubelli et al. 2004), (2) analysing environmental conditions that affect crystallization-dissolution and hydration cycles of each salt (Arnold 1981; Charola 2000; Charola et al. 2006; Aly et al. 2015), and (3) developing and optimizing treatments, both for desalinitation by poultices (Vergès-Belmin and Siedel 2005; Lubelli and van Hees 2010; Pel et al. 2010), immersion baths (Unruh 2007), sacrificial mortars (Feijoo et al. 2021; Ergenç et al. 2020; Husillos-Rodríguez et al. 2018), surfactants (Lubelli et al. 2010; Gupta et al. 2012; Rivas et al. 2017), electrochemical techniques (Ottosen and Rörig-Dalgaard 2009; Feijoo et al. 2015, 2017a), or to reduce the accessible porosity by the application of consolidation treatments (de Rosario et al. 2017; Feijoo et al. 2017b, 2020a), in order to diminish the damage caused by this agent.

The main strategy currently used to reduce the damage caused by salts is focused on decreasing the content of salts present in the material (i.e. desalination techniques). The electrokinetic technique is always shows a great efficiency, in terms of salt content removed versus time both at laboratory scale and in pilot tests (Ottosen and Christensen 2012; Feijoo et al. 2013, 2017d). The higher desalination rate achieved by this technique is due to (1) the faster migration speed experienced by ions under the presence of an electric field (as stated in Paz-García et al. (2013), for a monovalent ion and at 298 °K, the ionic mobility is approximately 40 times higher per applied volt than the diffusion coefficient of the same anion), (2) the lesser influence exerted by the material's pore structure when the ions must be mobilized to the outside, which allow achieving similar extraction percentages regardless of the material to be treated (Feijoo et al. 2015) and (3) the higher penetration depth of the treatment. By means of electrokinetic techniques, it is possible to force the mobilization of all those ions that are found in areas of the material through which the electric field will circulate. Therefore, this technique allows to overcome the depth limitations that others have, such as the application of poultices, which hardly desalinate above 4 cm in depth (Rivas et al. 2017, Sawdy-Heritage et al. 2008).

The physical principle by which electrokinetic techniques function is relatively simple: when an electric DC field is applied, through a porous material with high ion content, the ions transport the current and are forced to migrate towards the electrode of opposite polarity. Consequently, there is a decrease in the ionic content in the material (which is reflected by an increase in the electrical resistance that opposes against the current flow), and an increase in the ionic content in the vicinity of the electrodes. During this process, some mechanisms take place: (1) the electrolysis of water at the surface of the electrodes, which causes that extreme pH values are reached in these areas, and (2) movement of water towards one of the electrodes, depending on the surface charge of the porous material. In most porous materials this movement of water (by electro-osmosis) occurs from the anode towards the cathode, causing a differential drying rate between the materials located at each electrode (Ottosen et al. 2008; Bertolini et al. 2009; Franzoni 2014). This fact has an influence on the proper current flow between anodes and cathodes due to the loss of water that hinders the contact between the electrode and the poultice, and also reduces the amount of free ions that can transport the current.

In order to improve the operation of the technique, an electrokinetic casing that works with two electrodes connected in series, within the same compartment, has recently been developed at laboratory scale (Feijoo et al. 2018). This new approach allows modulating the net amount of H^+ ions generated on the surface of the electrode closest to the porous material to be treated, in order to (1) maintain the pH at the desired value, and (2) humidify the materials located inside the casing, which are used to retain the ions removed and improve the contact between the electrode and the material to be treated. This humidifying process guarantees the correct transport of the current for longer.

The present paper assess the desalination results achieved from a pilot test plant made on a brick wall, comparing the results achieved using electrokinetic casings with a double electrode system at the anodes with those achieved with only one electrode.

2 Material and Methods

2.1 Description of the Building

The locality selected for the test plant was the stable of the Bernstorff Palace built in the middle of the eighteenth century. This stable, built with Danish bricks bonded with lime mortar and coated with plaster, is situated in Gentofte, municipality located at 9 km from Copenhagen (Denmark).

The desalination test was performed on the inner face of the west wall (Fig. 1a), which showed an advanced state of alteration caused by salts: efflorescence, granular disintegration of the mortar, scaling and roundness of the brick's corners, among other factors. External facade is partially covered with cement layer, as part of a previous desalination process (Fig. 1b). However, this cement layer has not been removed so it is possible that it is currently acting as a source of salts.

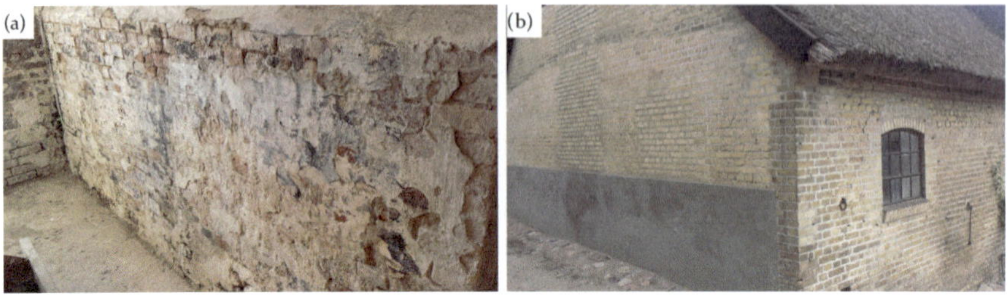

Fig. 1 Brick wall on which the pilot test was made (inner part: **a**, outer part: **b**)

2.2 *Electrokinetic Desalination Setup*

The test was carried out from 02.06.2014 to 07.08.2014 (i.e. 66 days). For this pilot test, 12 electrokinetic casings (6 anodes and 6 cathodes) were used, which were distributed following a staggered distribution along the wall in two rows (Fig. 2).

The electrokinetic casings used in this study (50 cm long) are similar to those used in previous interventions (Ottosen et al. 2012) but with some necessary modifications, in order to allow them hosting one or two electrodes depending on the case (Feijoo et al. 2017c). The configuration with two electrodes at the anode connected in series is named double-anode (DA) and with only one electrode single-anode (SA).

Each casing is composed of two main pieces (Fig. 3): (1) A plastic frame whose mission is similar to the electrokinetic reactors used in the laboratory, i.e. to house all the materials necessary to carry out the desalination treatment, such as electrodes (in this case inert titanium-MMO mesh electrodes) and poultices, and (2) a movable plastic bottom located inside the plastic frame but behind all the previous elements. This piece must provide the necessary pressure, as the treatment progresses and the poultices lose volume, through the drying process, in order to guarantee the contact between the poultice and the wall.

The poultices used were (Fig. 3): At the cathode a poultice made of kaolin with a water content of 37.9% (named as P–C). At the anode, two different poultices were used in the DA configuration, the furthest was a poultice made of cellulose with a water content of 83% approximately (named as P-A1), while the closest made of calcium carbonate: kaolin in a weight ratio 2:1 with a water content of 30%. In the SA configuration, the casing was filled only with the poultice P-A2.

The electric field established between the electrodes E1 and E2, allows regulating the net amount of H^+ generated at each electrode surface, as shown in the following equations that model the behavior of the DA configuration:

$$\text{At electrode E1}: \Delta\left[H^+\right] = Q_1/Fw = I_1 t/Fw \tag{1}$$

$$\text{At electrode E2}: \Delta\left[H^+\right] = Q_2/Fw = (I_2 - I_1)\, t/Fw \tag{2}$$

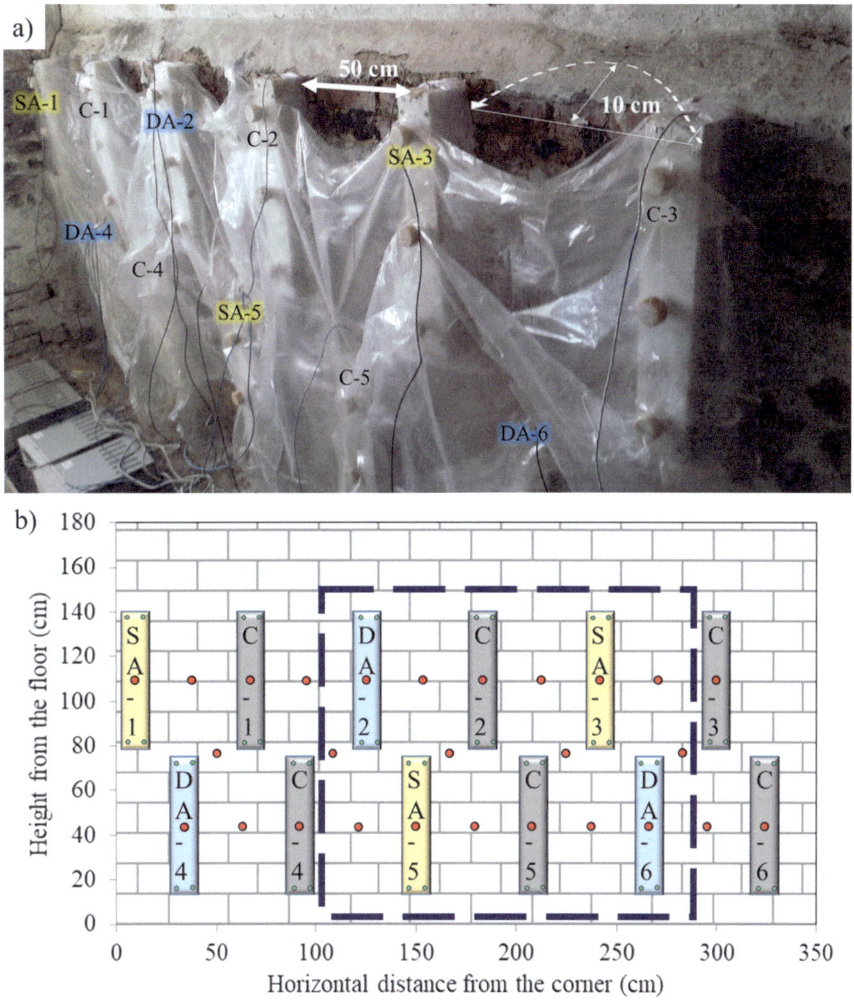

Fig. 2. **a** Photograph of the pilot test conducted on the west wall of the stable, showing the staggered distribution of the electrokinetic casings, the distance between them and the maximum expected depth of electric field penetration, and **b** Scheme with the distribution of the casings and sampling points made on the brick wall before (green circles made in the 4 corners of each casing to fasten it to the wall with screws), and after (red circles) the treatment. Also, the area used to compare the effectiveness of each type of anode is highlighted with a blue box in dashed line.

where Q is the circulating charge in C; F the Faraday's constant (96,485 C/mol), w the volume of the compartment, I the circulating current in A and t the time elapsed in s.

Therefore it is possible to regulate the amount of H^+ and OH^- ions generated on the electrode E2 surface to buffer the pH to the desired value and generate water.

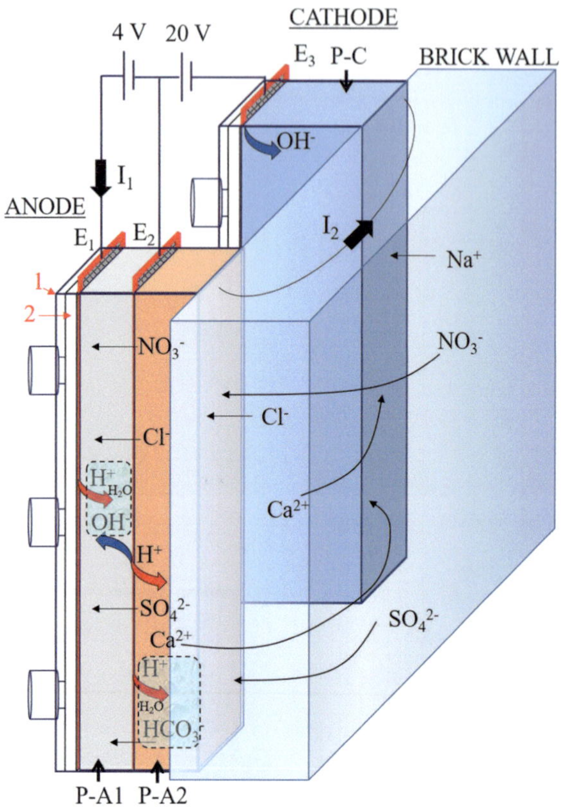

Fig. 3 Diagram of one anodic casing, with the double anode configuration (electrodes E1 and E2), and one cathodic casing where the electrode E3 is hosted. Also, the different poultices used in each compartment (Anode: P-A1 and P-A2; cathode: P–C) and the two main parts that make up each casing (the plastic frame (1) and the movable plastic bottom (2)) are shown

$$H_3O^+ + OH^- \rightarrow H_2O.$$

For example, $[H^+] = [OH^-] \rightarrow pH = 7$ if $I_1 = I_2$.

Once the electrokinetic casings were fastened and the titanium electrodes were connected to the respective power supplies (2 power supplies for the DA configuration and only 1 for the SA one), a constant voltage of 20 V was applied between electrodes E2 and E3, and in the case of DA configuration a constant voltage of 4 V was applied between electrodes E1 and E2. In order to ensure that all the potentials supplied by the sources are correct, all cathodes were short-circuited and grounded.

In addition, an automatic irrigation system was installed to moisten the brick wall with water during the treatment. This system provided water through drippers for 20 min every 3 h.

2.3 Desalination Treatment Monitoring and Effectiveness

Before carrying out the treatment, the plaster coating was removed and the entire wall was brushed. After that, and in order to identify the main anions present in the brick wall, forty-eight powder samples were collected, during the mounting of the electrode casings, by drilling (from the surface up to 10 cm in depth) in each of its four corners. The samples were dried, weighed and stirred in 25 mL of ultrapure water during 4 h. After filtration, the anion content was quantified with ion-chromatography (IC), using an IC Dionex ICS-1100. With these data, a mapping of each anion was performed using kriging as a statistical interpolation method. In order to know whether the anion content can involve a risk for the masonry wall or not, the only accessible threshold values established by the Austrian standard ÖNORM B 3355–1 (2006) was used. This standard establishes, specifically for brick masonries in general, three categories depending on the concentration of each anion. Concentrations above the upper limit involve a risk for the material and for this reason, it is necessary to carry out an active desalination intervention. Below the lower limit, no damage is expected, and between both limits, it is advisable to carry out a specific study. For the main anions, these limits are: Cl^- (0.10–0.03 wt.%), NO_3^-(0.15–0.05 wt.%), and SO_4^{2-} (0.25–0.10 wt.%).

During the treatment, measurements of intensity and resistivity between the electrodes connected to the same power supply were taken using a digital multimeter. The evolution of these parameters gives an idea about the progress of the treatment. Their subsequent analysis, along with other parameters, can indicate whether the increase in resistance is caused by a decrease in the ion content or by a loss of water in any of the materials used (i.e. poultices or brick wall). In addition, the loss of water in the poultices was analyzed indirectly by inserting, from the upper part of each electrokinetic casing, a plastic rod inside each poultice to assess their resistance against penetration.

At the end of the treatment, three samples per poultice were taken from each casing at different heights (from top to bottom). Each sample was used to measure the water content (following the gravimetric method), conductivity (adjusting the conductivity values following the expression proposed by Unruh (2007)), pH, and the anion content by IC (following the same procedure commented above).

Finally, in order to quantify the anion content that remains in the brick wall, a new in-depth sampling was carried out once the wall was already dry. Drilling powder samples were collected in different areas: on every area where each casing was placed, between casings, and between rows (Fig. 2b red circles). The efficacy of the desalination treatment was calculated comparing the ratio between the concentrations of each ion before and after the treatment, taking in both cases, as a data point, the concentration that exists just underneath the midpoint of each casing.

In all cases, to compare the results obtained with each anodic configuration, this study compares those casings that have similar boundary conditions, in terms of the row in which is placed (i.e. similar height) and the number of casings that surrounds it. For this reason, the following anodic casings are compared: SA-3 versus DA-2 and SA-5 versus DA-6.

3 Results and Discussion

3.1 Anion Content Before the Treatment

The anion content measured in the drilling samples taken from the corners of each electrokinetic casing (Fig. 2b, green dots) showed that the main anions present in the brick wall were sulfates (with an average value around 5.52 wt.% along the entire wall) and nitrates (1.46 wt.% on average) respectively. In both cases the average concentration of both anions exceeded the upper limit established by the Austrian standard ÖNORM, which justifies that an active salt removal is necessary.

The distribution of both anions in height showed that the highest concentration of nitrates was achieved above 1 m high (Table 1), especially in the area where the casings SA-3 and C-3 were located (with concentration values ranging from 2.47 to 7.79 wt.%). In the case of sulfates, the highest concentrations were achieved 1 m high from the ground, with an average value around 7.37 wt.%.

The presence of both ions, as stated in (Arnold 1981; Grossi and Esbert 1994; Charola 2000; Gómez-Heras et al. 2004), seems to be related to (1) leaching processes of some construction materials such as the plaster coating, (2) residues of animal droppings and (3) the capillary rise of groundwater from areas where fertilizers are usually used.

Table 1 Average of nitrate, sulfate and chloride content (in wt.%) analysed in the brick wall samples taken underneath the corners of each electrokinetic casing (initial content) and just underneath the midpoint of each casing (final content). Also, the efficacy achieved after the desalination treatment is shown

Row	Anode	Nitrate			Sulfate			Chloride		
		Before	After	Efficacy	Before	After	Efficacy	Before	After	Efficacy
Upper	SA-1	2.348	0.372	84.16	6.080	0.999	83.57	0.031	0.002	93.55
	DA-2	1.247	0.023	98.16	6.056	1.028	83.03	0.019	0.014	26.30
	SA-3	2.467	0.057	97.69	7.197	0.837	88.37	0.022	0.003	86.36
Lower	DA-4	0.049	0.013	73.47	0.507	0.222	56.21	0.013	0.002	84.62
	SA-5	0.112	0.003	97.32	1.217	0.068	94.41	0.004	0.011	−175.00
	DA-6	0.315	0.088	72.06	1.987	0.036	98.19	0.007	0.014	−100.00
	Cathode	Before	After	Efficacy	Before	After	Efficacy	Before	After	Efficacy
Upper	C-1	0.776	0.073	90.59	12.862	0.388	96.98	0.014	0.010	28.57
	C-2	1.688	0.020	98.82	9.516	0.281	97.05	0.014	0.015	−7.14
	C-3	7.785	0.095	98.78	2.503	0.604	75.87	0.110	0.016	85.45
Lower	C-4	0.025	0.004	84.00	0.104	0.055	47.12	0.006	0.012	80.00
	C-5	0.177	0.000	100.00	0.708	0.047	93.36	0.006	0.009	−50.00
	C-6	0.574	0.109	81.01	17.465	1.194	93.16	0.009	0.003	66.67

As shown in Table 1, the presence of chloride was very low, being in practically all cases below the risk level established by the Austrian standard.

3.2 Resistivity Measurements

The evolution of the resistivity showed a similar trend (Fig. 4): low and similar values at the beginning that increase during treatment. However, as the treatment progresses, there was a distancing between the resistivity measurements recorded in the casings with SA and DAS configuration. From day 20 approximately, the resistivity values reached in casings with SA configuration were much higher than those with DAS configuration, regardless of the row in which they were located. This different behavior was related to the different drying rate that occurs in the poultices located in the different type of anodic configurations. While in the SA configuration the anode poultice dried fast, hindering, in consequence, the current flow, with the DA configuration this drying process was slowed down by the generation inside the casing of water molecules. This fact was corroborated by inserting plastic rods inside each electrokinetic casing to check the degree of humidity of each poultice. The poultices hosted in the SA configuration hardened much earlier than those hosted in the DA configuration.

Summarizing, in the DA casings the values of current intensity were greater for longer than those reached in the SA casings, in virtually the entire pilot test. However, on day 66, the difference between the intensity values was reduced (moment in which it was decided to stop the test).

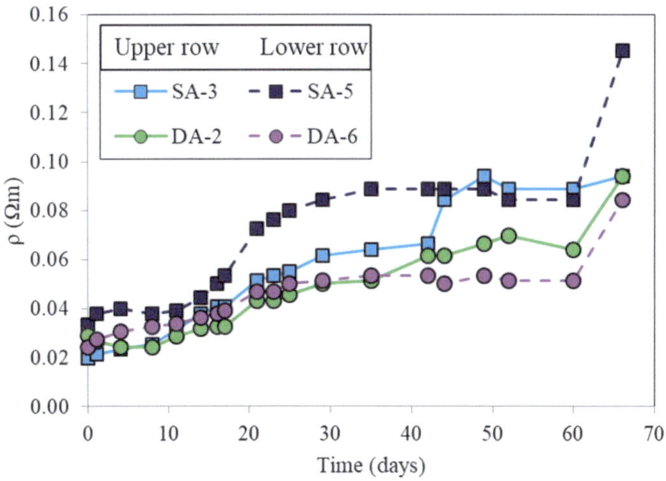

Fig. 4 Resistivity measurements registered during the treatment in the anodes with the same boundary conditions (SA-3 vs DA-2 and SA-5 vs DA-6)

3.3 Analysis of the Poultices

3.3.1 Water Content

The differences observed between the final water contents reached in the anodic P-A2 poultices, after 66 days of treatment, were very low between both configurations (around 21% of water in both cases). However, this does not mean that during the previous days the water content present in the DA configuration was not higher than that reached in this poultice hosted in the SA configuration. As indicated above, the periodic checks made by inserting a plastic rod clearly showed that the poultice P-A2 hosted in the SA configuration dried faster than that hosted in the DA configuration. Therefore, this result only indicates that at the end of the treatment, the differences in the water content between P-A2 poultices were reduced. The application of constant voltage, instead of constant current, in the power supplies did not allow to generate enough net amount of OH- ions in the vicinity of the electrode E2 to maintain the degree of humidity inside this P-A2 poultice for a longer time. For this reason, on day 66 the intensity values were similar for all the electrokinetic casings.

Regarding the poultices P-A1 used in the DA configuration, although they have lost about 50% of their initial water content, they still maintain a sufficient water content (approximately 40 wt.%) for maintaining an electric field between the electrodes E1 and E2.

Finally, with respect to the cathodic poultices (P–C), the weight loss was much lower than that registered by the anodic poultices (these poultices only loss around 15% of the initial water content). This result could be related to the establishment of an electroosmotic process which has slowed down this drying process.

3.3.2 pH and Conductivity

At the end of the treatment, the pH values reached by each poultice were different depending on the electrode that they hosted (Table 2). In the anode poultice P-A2, that hosted the electrode E2, the pH values were similar regardless of the type of anode (around 7). This fact indicates that the pH buffering system of both configurations worked perfectly, as occurs in previous studies (Ottosen et al. 2012; Feijoo et al. 2017a, 2018), preventing the brick wall to be exposed to acidic pH values that can cause chemical alteration processes.

In the anode poultice P-A1 (used in the DA configuration), in which the electrode E1 is hosted, acidic pH values around 3 were reached. This fact is related to the proper operation of the double electrode system since in this configuration the true anode of the system is the electrode E1. This electrode is hosted in a medium that has a zero buffering capacity of the pH (i.e. cellulose), favoring in consequence the acidification of the medium around the electrode, due to the generation of H^+ groups at the electrode surface; it also favors the generation of OH^- groups on the surface of the electrode E2 that is in the front. The migration of the H^+ groups, from the

Table 2 pH and adjusted conductivity values (Kad, in μS · L/g · cm) for each poultice used to fill each casing

Anode	pH		Kad	
	P-A1	P-A2	P-A1	P-A2
SA-1	–	7.05	–	0.027
DA-2	4.10	7.30	0.068	0.043
SA-3	–	7.54	–	0.027
DA-4	1.76	7.46	0.107	0.007
SA-5	–	7.49	–	0.012
DA-6	3.15	7.26	0.072	0.036

Cathode	pH	Kad
	P–C	P–C
C-1	11.34	0.019
C-2	11.30	0.012
C-3	11.51	0.014
C-4	11.47	0.010
C-5	11.36	0.009
C-6	11.40	0.015

electrode E1 towards the vicinity of the electrode E2, favors the generation of water molecules.

In all the cathodic casings, the pH values were similar and high (around 11.30). These high pH values show that in future on-site applications a solution similar to that used in the anode with a double electrode configuration could be applied to buffer the pH reached in the closest areas to the material to be treated, as has been demonstrated at laboratory scale in Feijoo et al. (2020b).

Regarding conductivity, the highest values were reached with the DA configuration, especially in the poultice P-A1. This result is indicative of: (1) the ionic content in the poultices hosted in the DA casings was higher than in the other casings. This result could be related to the greater intensity that flowed in these anodes. (2) A large part of the anions that were initially retained by the poultice P-A2 was subsequently forced to migrate to the poultice P-A1. This fact favors that the storage capacity of the poultice P-A2, which is smaller by the pore size of the kaolinite than that of the poultice P-A1 made of cellulose (Bourgès and Vergès-Belmin 2008; Pel et al. 2010), is maintained for a longer time.

Using the SA configuration, the conductivity values were similar and higher to those obtained in the cathodes, with the exception of those achieved with the SA-5 casing, where the value was similar to those reached at the cathodic casings. The lower conductivity achieved in the poultices hosted in the SA casings is related to the lower current flow, caused in large part by the faster loss of water, which made it difficult to mobilize a greater quantity of anions within the SA casings.

3.3.3 Anion Content

The chromatography results showed that in general: (1) nitrate has been removed to a greater extent than sulfate. This is due to the greater mobility of nitrate, which causes that the current was transported to a greater extent by this anion, similar result to that achieved in previous studies (Feijoo et al. 2017a). (2) The amount of each anion retained inside the casings was higher for those casings located in the upper row than for those located in the lower row. This fact is related to the initial distribution of salts in height (Fig. 5 and Table 1).

Regarding the two anode configurations assessed in this study, for anodes that had the same boundary conditions, the configuration that allowed extracting a greater amount of nitrate and sulfate was the DA configuration (Fig. 5), which agrees with the conductivity measurements discussed above.

This different behavior is clearly related to the double anode configuration and with the ability that this system provides to keep the poultices moist for longer. For the same ionic content in the medium, the higher the humidity, the higher the intensity value that circulates through the material, and the greater the amount of ions mobilized inside the casings (it must be taken into account that the ions are the responsible for the transport of the current).

At the cathode casings, the nitrate content was in general very low (between 0.007 and 0.04 wt.%), being in this case the sulfate the main anion retained by the P–C poultices. In both cases the extraction is caused by the typical process associated to desalination with poultices: advection and diffusion processes (Pel et al. 2010).

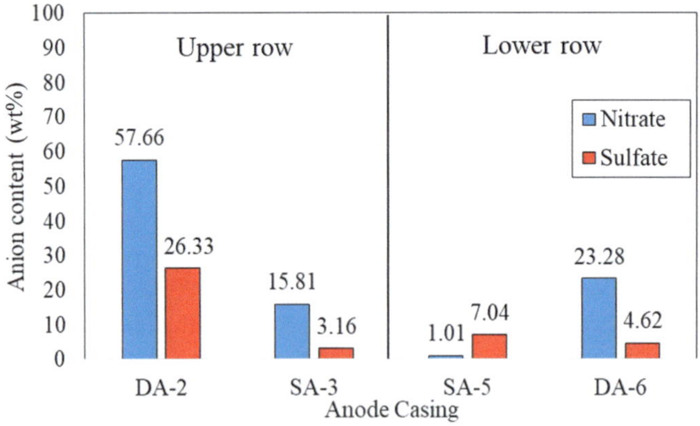

Fig. 5 Concentrations of NO_3^- (blue) and SO_4^{2-} (red) retained in the poultices that fill the anodes with the same boundary conditions (DA-2 vs SA-3 and SA-5 vs DA-6)

3.4 Analysis of the Brick Wall. Final Ion Content

At the end of the treatment, the anion content present in the entire wall was considerably reduced, reaching efficacy percentages in most cases that exceeded 80%, even 90% (Table 1). These percentages allow classifying this technique as a treatment of high or very high efficacy, according to the classification established by a desalination European project (Desalination Project E 2006).

In general, the highest efficiency percentages were reached in those wall areas where the cathodic casings were located. This is due to the fact that the anions, under the DC field, are forced to migrate from these areas towards those closest to the anodes, where their extraction takes place.

With respect to the two anodic configurations, and comparing those casings with similar boundary conditions, it is appreciated that although in almost all cases the greatest reductions were achieved with the DA configuration, these results cannot be considered as data to establish what the best configuration is. This is because the distribution of salts is not homogeneous along the brick wall. For example, in the particular case of nitrate, the efficiency percentages were higher in the upper part of the wall than in the lower part, although the final contents were lower in the lower part of the wall. This is related to the initial distribution of this anion, which was higher above 1 m high. In the case of sulfate, the differences between rows were smaller due to the fact that its initial distribution was more or less homogeneous up to 1 m high.

In spite of the high reductions achieved, there is still in the brick wall some areas with anion concentrations that exceed the upper limit value established by the Austrian standard ÖNORM (0.15 wt.% for nitrate and 0.25 wt.% for sulfate): in the case of nitrate, it would be the area surrounding the SA-1casing, while in the case of sulfate, it would be the entire upper area of the wall and the lower area located in the vicinity of the C-6 casing. Therefore, it would be advisable that in the future a second intervention of the wall would be carried out again to remove completely the anions presents.

4 Conclusions

From the obtained results, the following conclusions can be established:

- The use of the electrokinetic treatment allows reaching high efficiency percentages of reducing salt content (nitrate and sulfate) throughout the entire wall (higher than 80% in both cases), removing after the treatment around 206 g nitrate and 106 g sulfate.
- The resistivity measurements registered during the treatment, together with the analysis of the ionic content retained in the poultices and the resistance that these oppose to the penetration of a plastic rod, showed that the use of two electrodes at the anode (DA configuration) improves the wetting process of the anodic poultice

P-A2, which is in contact with the brick wall. This fact improves the current flow between the DA casings with those casings of different polarities that surround it (i.e. cathodes). In consequence, with this DA configuration, more ions can be mobilized from the wall towards the poultices.
- The use of the DA configuration allows buffering the pH by means of the electric field established between the electrodes located in the anode. This fact allows reducing the amount of the poultice made of calcium carbonate: kaolin and introducing in the casing a poultice made of cellulose, which increases the ion storage capacity of the casing.

Acknowledgements This research was funded by the TOP Heritage program (P2018/NMT-4372) of the Community of Madrid. J. Feijoo's work is supported by the Ministerio de Ciencia Innovación y Universidades, Spanish Government, through a Juan de la Cierva grant. Special acknowledgments to the professional support of the Interdisciplinary Thematic Platform from CSIC Open Heritage: Research and Society (PTI-PAIS) and an I-COOP 2018 cooperation project (COOPB20379) between the coauthors of this paper (Egypt and Spain), funded by CSIC.

References

Aly N, Hamed A (2020) The impact of salt crystallization on the building stones of AL-Azhar mosque from historic Cairo–Egypt. Int J Conserv Sci 11(4):895–904

Aly N, Gomez-Heras M, Hamed A, Álvarez de Buergo M, Soliman F (2015) The influence of temperature in a capillary imbibition salt weathering simulation test on mokattam limestone. Mater Construcc 65(317):e044. https://doi.org/10.3989/mc.2015.00514

Arnold A (1981) Nature and reactions of saline minerals in walls. In: The conservation of stone, II: preprints of the contributions to the International Symposium, Bologna, 27–30 October 1981. Part A, Deterioration. Part B, Treatment

Bertolini L, Coppola L, Gastaldi M, Redaelli E (2009) Electroosmotic transport in porous construction materials and dehumidification of masonry. Constr Build Mater. https://doi.org/10.1016/j.conbuildmat.2007.12.013

Bourgès A, Vergès-Belmin V (2008) Comparison and optimization of five desalination systems on the inner walls of Saint Philibert Church in Dijon, France. In: Salt weathering on buildings and stone sculptures, 22–24 October 2008, The National Museum Copenhagen, Denmark

Charola AE, Pühringer J, Steiger M (2006) Gypsum: a review of its role in the deterioration of building materials. Environ Geol 52:339–352https://doi.org/10.1007/s00254-006-0566-9

Charola AE (2000) Salts in the deterioration of porous materials: an overview. J Am Inst Conserv. https://doi.org/10.2307/3179977

Cooke RU, Smalley IJ (1968) Salt weathering in deserts. Nature 220:1226–1227

de Rosario I, Rivas T, Buceta G et al (2017) Surfactant-synthesized consolidants applied to a granitic medieval necropolis in NW Spain. Laboratory and in situ effectiveness evaluation. Int J Archit Herit 11:1166–1176. https://doi.org/10.1080/15583058.2017.1354097

Ergenç D, Feijoo J, Fort R, Alvarez de Buergo M (2020) Effects of potassium ferrocyanide used for desalination on lime composite performances in different curing regimes. Constr Build Mater 259:120409

Desalination Project E (2006) Assessment of desalination mortars and poultices for historic masonry

Feijoo J, Ergenç D, Fort R et al (2021) Addition of ferrocyanide-based compounds to repairing joint lime mortars as a protective method for porous building materials against sodium chloride damage. Mater Struct 54(1):14
Feijoo J, Fort R, Gomez-Villalba LS et al (2020a) Electroprecipitation of magnesium and calcium compounds for weathering protection of ornamental rocks. Cryst Growth Des 20(4):2337–2355
Feijoo J, Rivas T, Nóvoa XR, Ottosen LM (2020b) New double electrode system for the electrochemical desalination of building stones. Int J Archit Herit 14(5):678–693
Feijoo J, Nóvoa XR, Rivas T, Ottosen LM (2018) Enhancing the efficiency of electrochemical desalination of stones: a proton pump approach. Mater Struct Constr 51https://doi.org/10.1617/s11527-018-1224-x, https://doi.org/10.1080/15583058.2018.1561962
Feijoo J, Matyščák O, Ottosen LM et al (2017a) Electrokinetic desalination of protruded areas of stone avoiding the direct contact with electrodes. Mater Struct 50:82. https://doi.org/10.1617/s11527-016-0946-x
Feijoo J, Ottosen LM, Nóvoa XR et al (2017b) An improved electrokinetic method to consolidate porous materials. Mater Struct 50:186. https://doi.org/10.1617/s11527-017-1063-1
Feijoo J, Rivas T, Nóvoa XR et al (2017c) Electrokinetic method and device for extracting ions from a porous structure (ES2617971)
Feijoo J, Rivas T, Nóvoa XR et al (2017d) In situ desalination of a granitic column by the electrokinetic method. Int J Archit Herit 12:1–12. https://doi.org/10.1080/15583058.2017.1370509
Feijoo J, Ottosen LM, Pozo-Antonio JS (2015) Influence of the properties of granite and sandstone in the desalination process by electrokinetic technique. Electrochim Acta 181:280–287. https://doi.org/10.1016/j.electacta.2015.06.006
Feijoo J, Nóvoa XR, Rivas T et al (2013) Granite desalination using electromigration. Influence of type of granite and saline contaminant. J Cult Herit 14:365–376. https://doi.org/10.1016/j.culher.2012.09.004
Flatt RJ (2002) Salt damage in porous materials: how high supersaturations are generated. J Cryst Growth. https://doi.org/10.1016/S0022-0248(02)01429-X
Franzoni E (2014) Rising damp removal from historical masonries: a still open challenge. Constr Build Mater 54:123–136
Gómez-Heras M, Benavente D, Álvarez de Buergo M, Fort R (2004) Soluble salt minerals from pigeon droppings as potential contributors to the decay of stone based cultural heritage. Eur J Mineral. https://doi.org/10.1127/0935-1221/2004/0016-0505
Grossi CM, Esbert RM (1994) Las sales solubles en el deterioro de rocas monumentales. Revisión bibliográfica. Mater Construcción. https://doi.org/10.3989/mc.1994.v44.i235.579
Gupta S, Terheiden K, Pel L, Sawdy A (2012) Influence of ferrocyanide inhibitors on the transport and crystallization processes of sodium chloride in porous building materials. Cryst Growth Des. https://doi.org/10.1021/cg3002288
Husillos-Rodríguez N, Carmona-Quiroga PM, Martínez-Ramírez S et al (2018) Sacrificial mortars for surface desalination. Constr Build Mater. https://doi.org/10.1016/j.conbuildmat.2018.04.029
Lubelli B, Nijland TG, van Hees RPJ, Hacquebord A (2010) Effect of mixed in crystallization inhibitor on resistance of lime–cement mortar against NaCl crystallization. Constr Build Mater 24:2466–2472https://doi.org/10.1016/j.conbuildmat.2010.06.010
Lubelli B, van Hees RPJ (2010) Desalination of masonry structures: fine tuning of pore size distribution of poultices to substrate properties. J Cult Herit 11:10–18. https://doi.org/10.1016/j.culher.2009.03.005
Lubelli B, Van Hees RPJ, Groot CJWP (2004) The role of sea salts in the occurrence of different damage mechanisms and decay patterns on brick masonry. In: Construction and building materials
ÖNOR.M.B. 3355-1 (2006) Trockenlegung von feuchtemMauerwerk — Teil 1: Bauwerksdiagnose und Planungsgrundlagen Berlin. ASI Austrian Standards Institute ÖsterreichischesNormungsinstitut (Herausgeber) Deutschland, Bundesrepublik, BeuthVerlag. 1:3355

Ottosen LM, Christensen IV (2012) Electrokinetic desalination of sandstones for NaCl removal—test of different clay poultices at the electrodes. Electrochim Acta 86:192–202. https://doi.org/10.1016/j.electacta.2012.06.005

Ottosen LM, Christensen IV, Rörig-dalgaard I (2012) Electrochemical desalination of salt infected limestone masonry of a historic warehouse. Struct Faults Repair, Edinburgh Proc

Ottosen LM, Christensen IV, Rorig-Dalgård I et al (2008) Utilization of electromigration in civil and environmental engineering–processes, transport rates and matrix changes. J Environ Sci Health A Tox Hazard Subst Environ Eng 43:795–809. https://doi.org/10.1080/10934520801973949

Ottosen LM, Rörig-Dalgaard I (2009) Desalination of a brick by application of an electric DC field. Mater Struct 42:961–971. https://doi.org/10.1617/s11527-008-9435-1

Paz-García JM, Johannesson B, Ottosen LM, Ribeiro AB, Rodríguez-Maroto JM (2013) Simulation-based analysis of the differences in the removal rate of chlorides nitrates and sulfates by electrokinetic desalination treatments. Electrochim Acta 89:436–444

Pel L, Sawdy A, Voronina V (2010) Physical principles and efficiency of salt extraction by poulticing. J Cult Herit 11:59–67. https://doi.org/10.1016/j.culher.2009.03.007

Rivas T, Feijoo J, de Rosario I, Taboada J (2017) Use of ferrocyanides on granite desalination by immersion and poultice-based methods. Int J Archit Herit 11:588–606. https://doi.org/10.1080/15583058.2016.1277282

Rossi-Manaresi R, Tucci A (1991) Pore structure and the disruptive or cementing effect of salt crystallization in various types of stone. Stud Conserv. https://doi.org/10.2307/1506452

Rovella N, Aly N, Comite V, Randazzo L, Fermo P, Barca D, Alvarez de Buergo M, La Russa M (2020) The environmental impact of air pollution on the built heritage of historic Cairo (Egypt). Sci Total Environ. https://doi.org/10.1016/j.scitotenv.2020.142905

Sawdy-Heritage AM, Heritage A, Pel L (2008) A review of salt transport in porous media: assessment methods and salt reduction treatments. In: Salt weathering on buildings and stone sculptures (SWBSS), 22–24 October 2008, Copenhagen, Denmark

Scherer GW (2004) Stress from crystallization of salt. Cem Concr Res 34:1613–1624. https://doi.org/10.1016/j.cemconres.2003.12.034

Tsui N, Flatt RJ, Scherer GW (2003) Crystallization damage by sodium sulfate. J Cult Herithttps://doi.org/10.1016/S1296-2074(03)00022-0

Unruh J (2007) A revised endpoint for ceramics desalination at the archaeological site of gordion, Turkey. Stud Conserv. https://doi.org/10.2307/1506839

Vergès-Belmin V, Siedel H (2005) Desalination of masonries and monumental sculptures by poulticing: a review. Restor Build Monum an Int J = Bauinstandsetz und Baudenkmalpfl eine Int Zeitschrift. https://doi.org/10.1515/rbm-2005-6000

Geotechnical Engineering

Investigation and Preservation of Historic Foundations

Guido Gottardi and Michela Marchi

Abstract Italian experiences provide a long list of successful engineering solutions for the preservation of historic sites and, in particular, of tall structures like medieval and bell towers. Among them, Pisa Tower and Frari Bell Tower in Venice can demonstrate that innovative approaches and methodologies for the strengthening of historic foundations can be devised without any direct intervention on the structures and with the main aim of preserving their full integrity. Through purposely implemented and extensive real-time monitoring systems, it has been possible to minimize the impact of stabilization measures on the existing structures, including the historic foundations. The technologies used in the case studies described in this paper have proved to be especially gradual and flexible. Together with a number of other well described and successful examples of preservation activities currently available in the literature, such case studies will be hopefully used as a source of inspiration for future strengthening interventions aimed at preserving the integrity and the authenticity of historic foundations.

Keywords Preservation · Heritage structures · Historic foundations · Geotechnical engineering · Towers

1 Introduction

The noticeable contribution of Geotechnical Engineering to the preservation of monuments and historic sites has progressively grown over the last 50 years, together with the scientific and technologic advancement of the discipline. Indeed, it can substantially contribute to the understanding of soil-structure interaction aspects, since it is the discipline that more than any other investigates the characteristics and the causes of soil displacements and it is therefore the best suited to find engineering solutions to any possible problem. In 1981, the International Society of Soil

G. Gottardi (✉) · M. Marchi
Department of Civil, Chemical, Environmental and Materials Engineering (DICAM), University of Bologna, Viale del Risorgimento 2, Bologna, Italy
e-mail: guido.gottardi2@unibo.it

G. M. El-Qady and C. Margottini (eds.), *Sustainable Conservation of UNESCO and Other Heritage Sites Through Proactive Geosciences*, Springer Geology,
https://doi.org/10.1007/978-3-031-13810-2_7

Mechanics and Geotechnical Engineering (ISSMGE) dedicated one of its technical committee (at present TC301) to specifically work on the topic of preservation of monuments and historic sites. The approach to preservation has radically changed since then (Flora 2013).

More recently, the basic principles for the conservation of the foundations of heritage structures and historic sites have been introduced in the international Standard ISO13822-2010: "*Assessment of existing structures*". In the *Annex I* the document states the concept of «*Authenticity and Integrity of Foundations of Heritage Structures*», defining the foundations as integral part of the whole structural system and of the original design concept of the structure itself. As a consequence, the foundations and the surrounding soil have become authentic elements of a historic building, to be preserved as much as the elevation structure. Therefore, a monument, its foundation and the supporting ground should be considered as parts of a unique complex system, that any possible intervention should respect. Preservation of authenticity and integrity of such systems usually require interdisciplinary competences, focused on finding out the reasons of possible problematic issues and then committed to the design of minimal interventions, with incremental approaches based on real-time monitoring and removable measures.

In this framework, Italian experiences provide a long list of successful engineering solutions for the preservation of historic sites and, in particular, of tall structures like medieval and bell towers. Among them, the stabilization of the Leaning Tower of Pisa is, without any doubt, the worldwide best known example that embodies the conservation principles described above.

The Pisa Tower, with its extraordinary inclination, has a strikingly long history of stabilization proposals. For decades, experts have faced the issue of its conservation, finally ending up with a successful intervention by under excavation. Such a solution was conceived only after a careful analysis of monitoring data collected in a sufficiently long timespan, together with extensive analyses and experimental tests (Burland et al. 2003). The Tower had showed a significant differential settlement since the time of its construction, which began on 9 August 1173. Geotechnical investigations ascribed the initial inclination of the Tower to local stratigraphic inhomogeneities in the subsoil beneath the south side of the tower foundation (Fig. 1). Nonetheless, engineers have found difficulties for a long time in interpreting the phenomena that were still producing the continuous progressive inclination of the Tower. Eventually, a combined analysis of piezometric monitoring and precision levelling enabled to identify a seasonal differential fluctuation of the water table in the upper layer of sandy and clayey silts, which was producing a "*shallow-seated motion of the tower foundation*", characterized by a rotation of the foundation base (Burland et al. 2009).

The full understanding of the mechanisms that were causing such long-lasting stability problem to the Tower was of crucial importance for the final choice and subsequent design of the stabilization measure. Indeed, the investigations showed that even a small decrease of the inclination would have substantially improved the situation of the tower, especially with respect to the stability of equilibrium issue, potentially more threatening than a bearing capacity failure mechanism. A

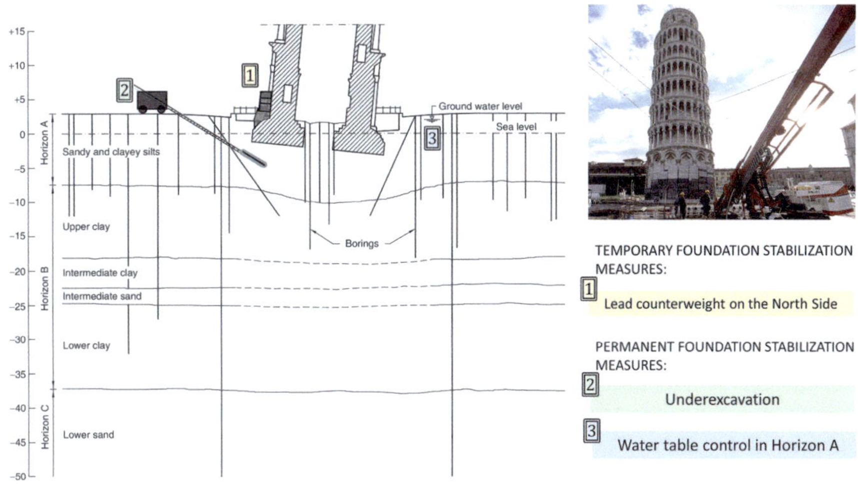

Fig. 1 Soil profile of the Pisa Tower (modified from Lancellotta (2008)) and schematic representation of the main temporary and permanent stabilization measures carried out from the second half of 1993 to 2002, as reported in Burland et al. (2009)

soil extraction intervention was therefore designed and then implemented by the International Committee for the Safeguard of the Tower of Pisa, appointed in 1990. In the preliminary design stage, the response of the Tower to the intervention was carefully analysed by means of numerical models and small-scale physical models. Then, a field trial confirmed the effectiveness of the devised intervention (Leoni et al. 2018).

The main phases of the intervention are summarized in Fig. 1. Before soil extraction, a temporary stabilization measure was carried out. In 1993, lead ingots, for a total weight of about 6 MN, were placed on the north edge of the Tower base; subsequently, in 1995, the load was increased to 9 MN. The underexcavation was finally implemented in two stages during year 1999–2000. It involved the removal of small soil volumes from underneath the elevated part of the foundation. At the end of the intervention the tower recovered part of its tilt, approximately 1800 arcsec. In 2002, after the stabilization intervention, a new drainage system for the control of water table fluctuations was also introduced. This was aimed at permanently mitigate additional movements, thus ensuring long lasting effectiveness of the overall intervention.

The stabilization measure carried out on the Pisa Tower has become one of the best known and brilliant examples of respectful approach to preservation, and, as such, its peculiarities are described in numerous volumes and papers (e.g. Burland et al. 2009, 2020; Leoni et al. 2018; Viggiani 2019).

2 Historic Towers: A Distinctive Italian Heritage

Historic towers are distinctive features of the Italian heritage, mainly typical of Northern and Central Italy. Since Middle Age, bell towers and civic towers started to populate little towns, like San Gimignano (Tuscany), and major urban areas, like Bologna and Venice. The number of towers could become quite impressive with respect to the city size at the time of their construction. Unfortunately, most of them collapsed or were demolished over the centuries. Nowadays, only a little part of these delicate structures is still standing and their conservation is clearly a priority for local governments.

The geotechnical perspective in the investigation process of these structures is fundamental. It starts from a deep understanding of the design concept of their foundations, in order to identify the soil-structure interaction mechanisms, both in static and seismic conditions (Lancellotta and Sabia 2015; Flora et al. 2018; Marchi et al. 2011). As a consequence of local stratigraphic conditions, water table depth and availability of materials and economic resources, a variety of structural configurations were adopted in different sites. As an example, in Bologna, where the shallowest soil layers are characterized by medium stiff clay and the water table is few meters deep, the foundations were built adopting relatively simple techniques. On the contrary, more sophisticated structures and procedures are typical of more complex environments, as for the foundations of major historic buildings in Venice, where very soft soil and water table at the ground surface were predominant. More details on these case studies are provided in the following sections.

2.1 Construction Methodology of Medieval Tower Foundations in Bologna

Bologna is known as “La Turrita”, which means City of Towers. The origin of this name can be found in the remarkable number of towers and tower houses that populated Bologna in the Middle Age, probably more than 75. The construction of towers in Bologna started at the beginning of the second half of XI century; at present, only 22 towers are still standing (Roversi 1989). Among them, Garisenda and Asinelli Towers—often simply referred as the Two Towers—are the best conserved and, standing one close to the other right in the heart of the city centre, definitely the most well-known and a symbol of the city itself (Fig. 2). In addition, the Garisenda tower is also famous because since its construction it has been dangerously leaning toward its east side, as much as the Pisa Tower.

The construction method of historic foundations in Bologna typically followed the steps described below. Depending on the properties of the local soil deposits and on the depth of the founding level, the excavation faces could be vertical or slightly inclined. In case a support were needed, wooden boards would have been placed all around the sides of excavations. The bottom of the hole was kept dry during

Fig. 2 The medieval Two Towers in Bologna: Garisenda (in the front) and Asinelli (in the background)

digging operations. The main body of the foundation was made of a cast-in-place lime concrete block, composed of sand, gravel, fragments of bricks and slaked lime. It could be executed directly on the virgin soil or on wooden piles, despite only limited evidence of their use has been found so far. When contemplated, pile installation was carried out by means of hammers. Piles were short, their length rarely exceeding 3 m, and the diameter varied between 15 and 25 cm; the head of the piles was left up over the ground level and thus included into the following cast. If adopted, the wooden boards all around the excavation area were removed soon after casting. Once the foundation was completed, the tower basement was built using squared selenite blocks (Bergonzoni 1989). These operations, and the subsequent construction of the masonry walls of the tower could last five to ten years (Gozzadini 1875).

Studies for the preservation of Garisenda and Asinelli towers started at the beginning of the last century and are still on-going. The analysis of historic documents (Marchi et al. 2019) suggests that both of them were constructed around the end of the XI century—beginning of the XII century. At the end of its construction, the Garisenda tower was 60 m high, but already in 1353 the tower was shortened to 48 m because of its risky leaning. On the contrary, the Asinelli tower, which was originally 60 m high, was subsequently raised to its present remarkable height of 97.20 m.

The foundation structures of Garisenda and Asinelli towers were investigated through vertical and inclined boreholes, carried out at different stages of investigation, starting from the first campaign on the Asinelli foundation in 1973, to the most recent in 2016, which completed the data set on both foundations. The geometry of the foundation blocks of the two towers, built as just described, is clearly displayed from the relevant cross sections of Fig. 3, while Fig. 4 shows a picture of cores extracted from one of the two inclined boreholes executed on the Asinelli foundation during the 2016 campaign (dashed line on Section A-A in Fig. 3). The Asinelli foundation is a block made of lime and well-graded cobbles, 6.5 m deep. The foundation width is 10.45 m wide. The block is not homogeneous, but divided in two parts: the lower one (2.90 m thick) is compact, with rare voids, the upper (1.80 m thick) is weaker, it can be easily scratched and is very powdery when touched. The geometry of the Garisenda foundation was first investigated in 2000, using inclined boreholes. The vertical borings carried out in 2016 confirmed the findings of the previous investigations.

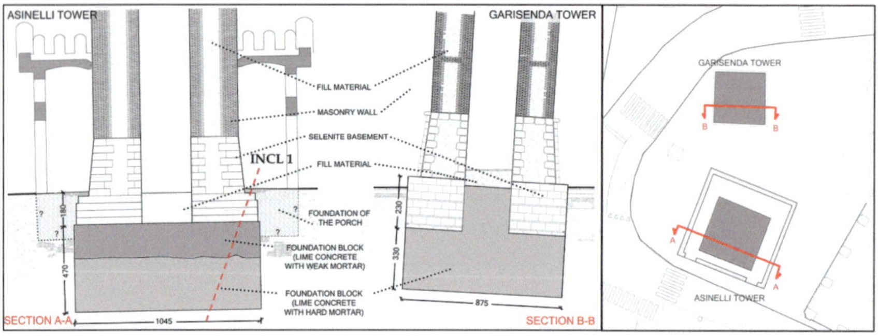

Fig. 3 Schematic cross sections of the two tower foundations (see the plan view of the Ravegnana Square in the centre of Bologna on the right), with the trace of the inclined borehole INCL1 (dashed red line) on the Asinelli foundation and shown in Fig. 4 (modified from Marchi et al. (2019))

Fig. 4 Picture of cores extracted from the upper 5 m of one of the two inclined boreholes executed on the Asinelli foundation (INCL 1, dashed line on Section A-A, in Fig. 3) (modified from Marchi et al. (2019))

The foundation block is a massive rectangular prism, with a squared base, 8.75 m wide and 3.30 m high. The outer walls are vertical and the foundation level is 5.6 m deep from the ground surface.

The soil profile beneath the Two Towers was mainly deduced from observation of core samples, from interpretation of CPTU tests via Robertson's (2009) Soil Behaviour Type chart and from laboratory classification tests (Marchi et al. 2019). The soil deposit basically consists of a floodplain succession characterized by an alternation of silty-clay and clayey-silt with medium–high plasticity (PI from 13 to 36). Field investigations and laboratory tests show a remarkable heterogeneity of the mechanical properties of the deposits, in spite of their apparent lithological uniformity. The sedimentological study of the core material (Bruno et al. 2020) has provided a robust scientific framework for the development of the local geotechnical model. Inhomogeneities were found to be limited around the Asinelli foundation, while they were more pronounced around the Garisenda. Beyond this lack of uniformity in the subsoil, which could have caused the Garisenda initial tilt, other factors that could affect the current behaviour of Garisenda foundation are still under investigation.

2.2 *Bell Tower Foundations in Venice*

Historical documents report that Venice used to have more than one hundred bell towers (Fig. 5): a few were demolished, others collapsed for various reasons (like differential settlements, lightning, fires, etc.). To date, 85 are still standing, thus playing their role of distinctive architectural feature of the Venetian skyline (Lionello 2011).

Scarce availability of materials, water table at the ground level and very low bearing capacity of the immediate subsoil imposed severe constraints to the building foundations in Venice, producing a typical and recurrent structural form which has proved to be extremely successful and, in fact, it has remained substantially unchanged until the beginning of the last century. Two main foundation types were

Fig. 5 Panoramic view of Venice in the XVIII century, after Friedrich Bernard Werner, Augsburg

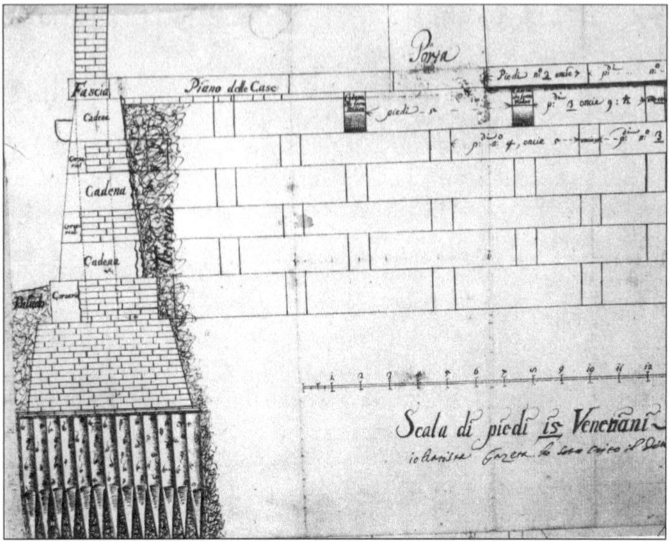

Fig. 6 Sketch of a Venetian wooden piled foundation (from Zuccolo 1975)

used (Fig. 6): shallow masonry foundations (for low-rise buildings, not bordered by canals) and wooden piled foundations (for major buildings and for walls bordering the canals) (Zuccolo 1975). Due to the presence of shallow soft silty clays, the major buildings of the historic centre in Venice could easily develop settlements up to several tens of centimetres, producing potentially severe stability issues in case of relevant differential settlements.

Such circumstance is at the base of the following well-documented case study, concerning the damaging interaction between the Frari bell tower and the adjacent Basilica.

3 The Monitoring-Driven Intervention on the Foundations of the Frari Bell Tower in Venice

3.1 The Frari Bell Tower: A Story of Maintenance Interventions

The Basilica of *Santa Maria Gloriosa dei Frari* (Fig. 7) is one of the largest and most splendid churches in Venice, located right at the heart of the city. Historical archives have told us that the Franciscans had been initially granted a land to build a church in 1250, but the formerly designed smaller building was not completed until 1338.

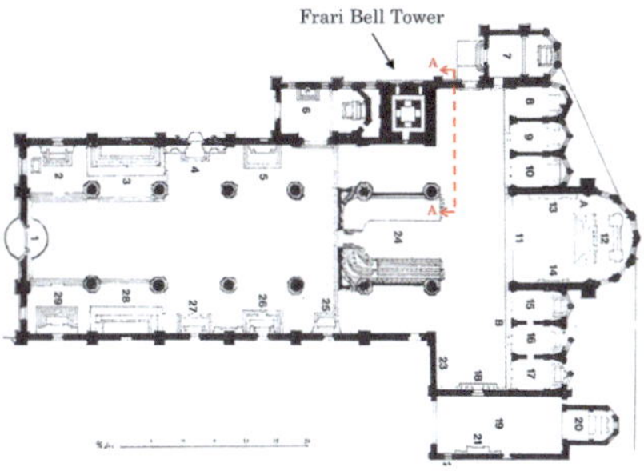

Fig. 7 Left: a view of the Basilica of Santa Maria Gloriosa dei Frari and of its bell tower. Right: plan of the Basilica with the bell tower at the south-east corner of transept and left aisle and the adjacent St Peter's chapel

Works soon started again on its much larger replacement, the current church, which took over a century to be completed.

The Frari bell tower, the second tallest in the city after San Marco, was built between 1361 and 1396. The bell tower square structure is 9.5 m wide at the ground level, 65 m tall and weighs about 57 MN. The internal ramp staircase up to the belfry is supported by a double structure of thick brick masonry. It was originally conceived as a fully independent structure, but during the reconstruction of the wider Basilica, the bell tower was included into the masonry walls, at the south-east corner of transept and left aisle (see plan of Fig. 7). The close connection between the bell tower and the Basilica is at the origin of the subsequent problems that have always affected their structures. In 1432 the St Peter's chapel was added adjacent to the Basilica and to the bell tower, even if structurally independent. The first documented signs of deterioration of the structures are dated back to the end of sixteenth century. Between the end of nineteenth century and the first decade of twentieth century, three main interventions were carried out (Lionello 2008). In fact, the differential settlement of the bell tower foundation, which had developed over the centuries, had produced major damages both to the St Peter's chapel and to the vaults of the left aisle of the church.

Then, interventions on the masonry walls and at the level of the foundations were periodically carried out. In the first years of the twentieth century the tower showed a differential settlement of about 0.40 m with respect to the Basilica and an out of plumb toward south-east of 0.765 m at a height of 42.5 m (resulting in an inclination of about 1°).

In 1904, following the universal concern after the sudden collapse of the world-wide famous St Mark bell tower, an extensive investigation was carried out on several Venetian slender structures considered at stake. In the Frari case, the surveys revealed inadequate foundations with respect to the bulk of the bell tower. For this reason, a strengthening intervention on the bell tower raft foundations was carried out, consisting of widening its base, starting from the south side (toward which the tower was leaning) and according to the traditional Venetian soil strengthening technique. Such raft foundation enlargement had been most probably designed to be extended to all the other sides; as a matter of fact, however, it was never completed (Lionello 2008).

3.2 Investigations and Monitoring Before the Intervention

Within an extensive survey of the general safety conditions of Venice bell towers planned at the beginning of the 1990's by the local Office of the Italian Ministry for Cultural Heritage and Activities, a rather detailed diagnostic investigation of the bell tower structures was carried out, including photogrammetric and crack-pattern survey, endoscopies, single and double flat-jack tests on the masonry walls, sonic tests on steel ties, monitoring of the main cracks by means of extensometers and installation of clinometers for the detection of rotations of the bell tower, as well as preliminary geotechnical investigations of the foundation soil. In 2000, a further survey aimed at detecting the actual movements of the bell tower indicated that the bell tower was tilting in the opposite direction with respect to the "historical" trend, essentially moving back toward its vertical position and against the church structures. This was most probably the consequence of the 1904 foundation widening.

Flat-jack testing technique was used to measure the existing state of stress on the masonry structures of the bell tower and of the adjacent Basilica. Very high values of compressive stress were recorded at the top of the column sustaining the propped arch: 1.76 MPa on the outer side and 3.20 and 3.04 MPa on the inner side. A further analysis carried out on the wall of the Basilica adjacent to the bell tower indicated the presence of a thrust line going from the bell tower to the structures of the Basilica, with stress values ranging from 0.56 to 0.95 MPa (Fig. 8a).

A rather extensive and diversified geotechnical investigation was preliminarily carried out in May 2003 (Fig. 9), consisting of 4 piezocone tests, 2 vertical and 5 inclined continuous coring boreholes (Fig. 10) and 4 continuous borings into the foundation block, together with the extraction of several undisturbed soil and foundation block samples for the subsequent execution of laboratory tests, which enabled the stratigraphy, the subsoil properties and the geometry of the foundation block to be defined with some detail (Gottardi et al. 2009). A section of the peculiar bell tower foundation structure along the SE-NW direction is provided in Fig. 11. As typical in Venice (Marchi et al. 2006), the Frari bell tower foundation is made up of Istrian limestone squared blocks and short timber compaction piles (1.702.20 m long), with an interposed 0.40–0.50 m thick larch boarding. The soil profile under the tower, shown in Fig. 11, consists of: Unit A, between ground level and a depth of about 3.2 m: anthropic fill; Unit B, between 3.2 m and about 6.7 m: dark grey, soft, silty clay, with occasional organic material, normally consolidated or slightly overconsolidated, with organic inclusions and shells.

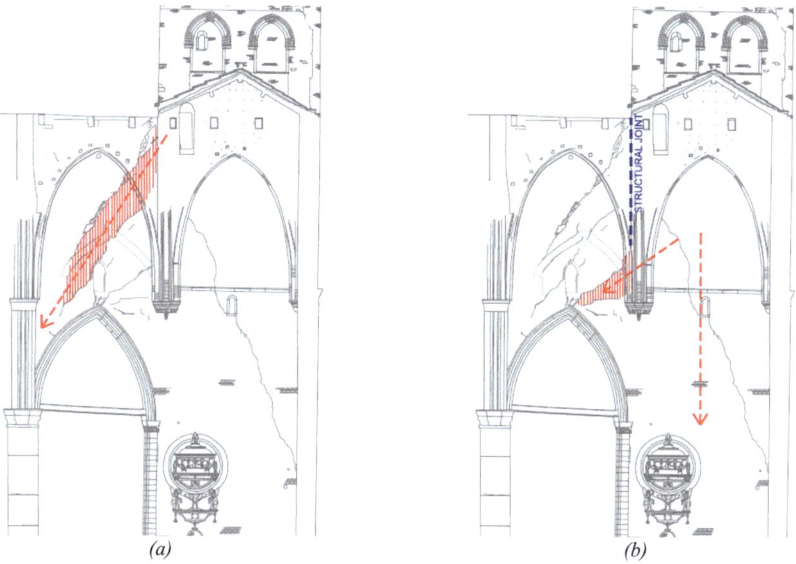

Fig. 8 Direction of the thrust between the bell tower and the Basilica along the section A-A of Fig. 7: **a** before and **b** after the structural joint execution

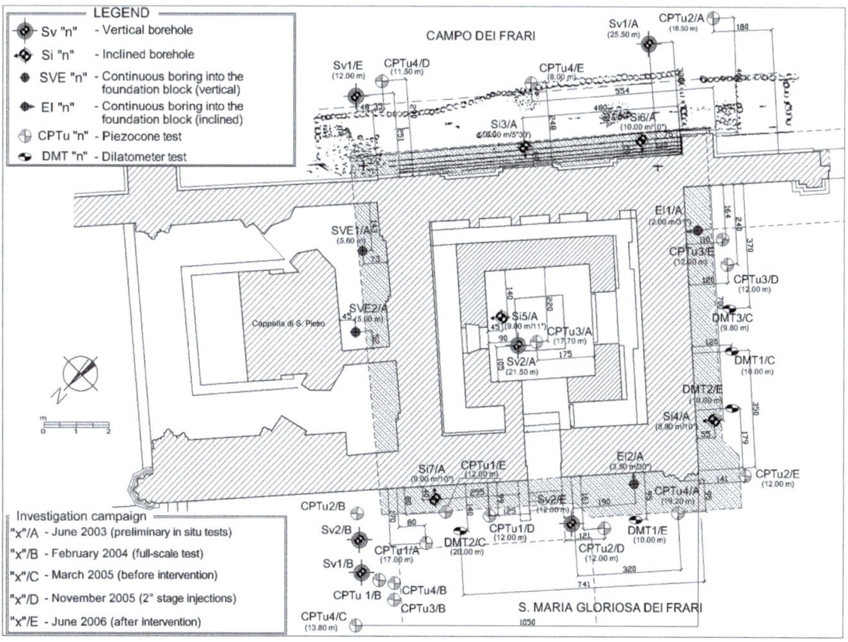

Fig. 9 Plan view of the Frari bell tower (St. Peter's chapel on the left) with the location of the several geotechnical in-situ investigations carried out at various stages of the intervention, as from legend (Gottardi et al. 2013)

Fig. 10 Inclined borehole carried out from inside the Basilica to investigate the foundation block of the Frari bell tower as well as the subsoil below it

Unit C, between 6.7 and 14 m: grey medium-fine sand, not plastic, from dense to very dense; between 14 m and the maximum investigated depth: alternation of soft clayey silt and medium-fine dense sand.

A careful evaluation of settlement trend with time excluded that the current movements of the tower could be entirely ascribed to secondary settlements in confined conditions. Hence, the reasons of the continuous foundation problems were mainly attributed to a slow lateral plastic flow under high stress gradients within the soft silty clay layer and to the possible progressive decay of timber piles.

A comprehensive monitoring system was then installed to analyse the deformation behaviour and the structural conditions of the bell tower and of the adjacent portion of the Basilica during all the phases of the strengthening interventions, consisting of: crack-gauges and long-base extensometers installed on the main cracks of the

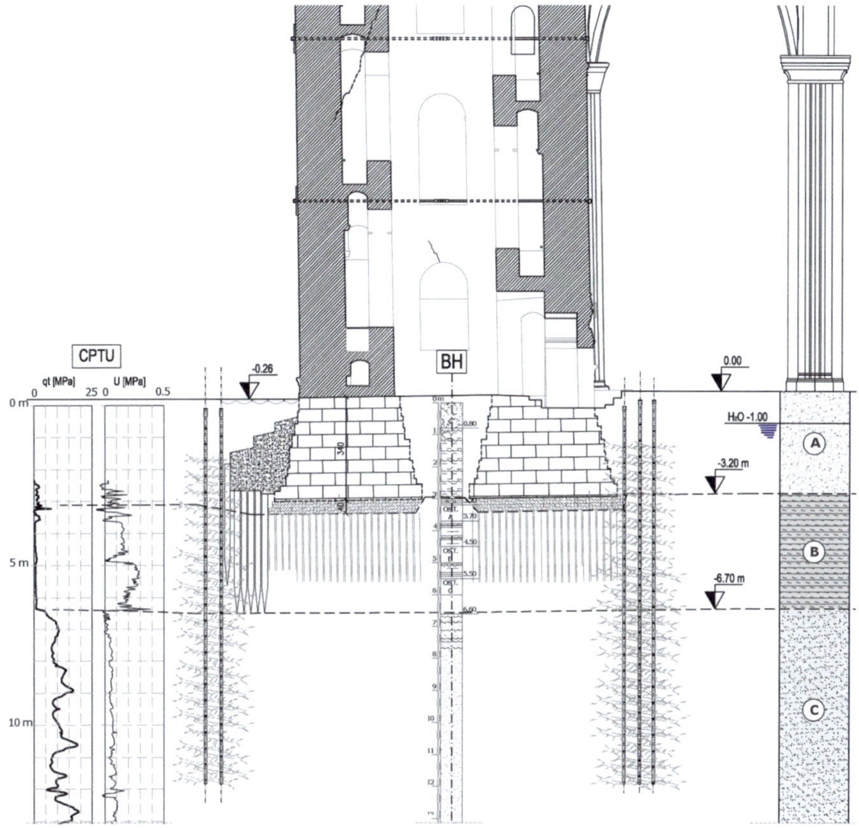

Fig. 11 Schematic section of the tower foundation and of the relevant subsoil along SE-NW direction, with the in-situ test logs, the strengthening intervention of 1904 (on the left-hand side of the foundation) and the Tubes à Manchettes (TAMs) of the new fracture grouting intervention

masonry walls; strain-gauges to measure the deformation of the steel cable installed in the bell-tower; thermal gauges to measure the air temperature outside as well as inside the masonry, at different distances from the outer wall; geotechnical instrumentation, including electrical piezometers, multi-base extensometers and biaxial inclinometers; a direct pendulum equipped with automatic telecoordinometer, for the measurement of the absolute horizontal movements of the top of the tower. All the instruments were connected to an automatic data acquisition system that enabled to follow in real time the effect of the works on the structures, thus enabling to introduce possible suitable modifications to the intervention design. In addition, it was considered of vital importance to measure as often as possible the vertical movements of the bell tower and of the adjacent portion of the Basilica. A high-precision and accurate manual levelling system, with several measuring points, was thus installed and periodical surveys were carried out and intensified during the most significant phases of the works.

3.3 The Multi-phase Strengthening Intervention

The slow but continuous differential settlement of the bell tower was cause of major concern for the present and future stability, not only of the bell tower but, above all, of the structurally connected Basilica. In fact, from the results of the diagnostic investigations and of the following numerical simulations (Lionello 2008), it clearly emerged that the interacting structures of the bell tower and the Basilica could not bear further differential settlements without serious consequences. It followed the need of a strengthening intervention at the level of the foundations, aimed at reducing the differential settlements of the bell tower.

A rather innovative strengthening intervention of soil fracturing (Fig. 11), also known as fracture grouting, was eventually devised in order to improve the mechanical characteristics of the soft silty clay.

Once the aim of improving the stability of the soil-foundation system had been achieved, a new solution was then required to reduce the damaging interaction between the masonry structures, activated by the foundation settlements. A structural joint between the bell tower and the Basilica was finally executed to increase the system deformability (Fig. 8b). In order to guarantee the safety of the whole Basilica and of the bell tower, a gradual strengthening intervention was designed, with a strict and constant monitoring control throughout.

A preliminary provisional intervention (Phase 1) was carried out in order to increase the safety level of the mentioned column of the Basilica. A steel cable was thus positioned connecting to the bell tower structure the stone ashlars just above the capital of the column, the closest to the bell tower, at a height of 14.40 m, aimed at supporting part of the horizontal thrust acting on the column. Two strain-gauges were installed on the steel cable and the relevant tension constantly monitored during the whole intervention period.

The principles on which the main soil strengthening intervention (Phase 2: soil fracturing) was based were: making compatible the remaining settlements of the complex Basilica-bell tower; preserving as much as possible the original foundation structures and the relevant stress distribution; avoiding to introduce a rigid foundation system for the bell tower; enabling a flexible and modular intervention, in constant agreement with the outcome of the real time monitoring.

The fracture grouting intervention (Fig. 11) consisted of the installation of special injection pipes in the foundation soil, fitted with equally spaced valves at different depths (typically known by the French acronym TAMs). Each valve could be selectively injected by means of a double packer device.

The slow-rate injection of suitable cement and bentonite mixtures was carefully repeated at incremental stages, in order to obtain progressive increments of the mechanical characteristics. The final outcome was expected to be a reinforced soil, made up of the original material and of an indented web of thin layers of injected grout. In order to evaluate the feasibility of the soil fracturing intervention (Marchi et al. 2014) and to calibrate the relevant design parameters (injection pressures, injection rate, grout mixture, etc.), a full-scale test site was preliminarily carried out on the northern corner of the bell tower, inside the Basilica. It demonstrated the effectiveness of the devised strategy.

The actual intervention was then carried out by means of eighty-eight 12 m-deep sleeve steel pipes installed all around the perimeter of the bell tower, along two or three alignment rows, according to local geometric constraints. A cement-bentonite grout was injected from each valve in three main separate cycles. A fourth cycle was introduced to enhance the intervention on the clay layer. Injection pressures varied between about 200–500 kPa in the first cycle to about 400–1200 kPa and more in the final cycle. During the intervention the flow rate was kept low and fixed at about 6 l/min. Constant grout volumes of 20 l/valve were injected each cycle in clay, whereas in sand variable volumes between 14 and 20 l/valve. Injection started in Spring 2005 and ended at the beginning of Summer 2006. At the end of the intervention, a total of about 100 m^3 of grout had been injected.

Finally, a structural joint (Phase 3) was created between the bell tower and the walls of the Basilica, in order to tackle the existing interaction between the two structures (Fig. 8b), according to a suitably conceived preservation approach aimed at reducing the current force distribution (Lionello 2008). The structural joint was created inserting a gap between the bell tower and the adjacent church. This would allow the relative movements and prevent or reduce the stress transmission to the column of the Basilica. The structural joint was eventually carried out at the beginning of June 2008 and, to preserve the Basilica architecture, it was placed only above the vaults where, on the other hand, mechanical interaction was greater. Its effect was a marked change of direction of the thrust applied by the bell tower to the adjacent column (see Fig. 8). The execution of the structural joint was very slow and lasted about 6 months. During this period a detailed analysis of the information obtained by the monitoring system enabled to carry out the different steps of the intervention with a continuous check of the structural response, thus avoiding to induce damages to the bell tower and to the supporting structures of the Basilica.

3.4 *Investigations and Monitoring During and After the Intervention*

Investigations and monitoring were carried out during and after the intervention, both on the foundation soil and on the structures of the bell tower.

First of all, the effect of the soil fracturing intervention on the soil was carefully controlled by means of geotechnical instrumentation installed throughout the area of the intervention. As designed, the injections led to the formation of evident and diffuse cement lenses in the soil, as found in undisturbed continuous core sampling carried out at the end of the intervention. Specific additional in situ geotechnical investigations (piezocone and dilatometer tests) were planned and carried out to assess their effect on the mechanical properties of soil. The global effect of the injections could be assessed by comparing the results of in situ tests performed at different

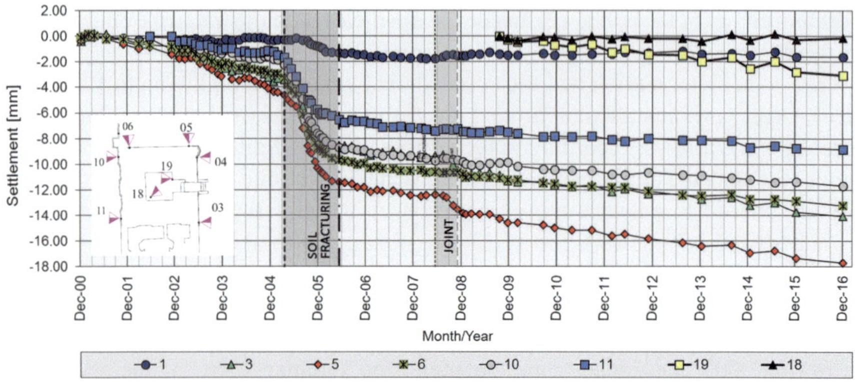

Fig. 12 Settlements of several benchmarks from precision levelling (courtesy of FOART srl, Parma), up to 27/12/2016 (Modified from Gottardi et al. (2017))

stages of the intervention, from which the soil strength and stiffness improvement appears clearly noticeable. However, more comprehensive information for the analysis of the structure behaviour during the soil fracturing works have come from the direct pendulum records and precision levelling surveys (Fig. 12). The analysis of the collected data showed a more pronounced vertical movement of the bell tower west and south corners during the soil fracturing intervention and a significant reduction immediately after, also with respect to the earlier rates, consistently with design expectations. Such observation is confirmed by the differential settlement trend along the main alignments and, above all, between the critical column of the Basilica and the adjacent bell tower corner: 1.01 mm/year before, 5.04 mm/year during and 0.27 mm/year after the intervention (before the joint execution).

At mentioned, structural monitoring was also carried out in order to follow with special care the deformation behaviour during the execution of the structural joint. New crack-gauges as well as new long-base extensometers were installed. Furthermore, in order to check the thrust modification between the bell tower and the Basilica, special flat-jacks devices were also installed. A significant decrease of the state of stress in the upper part of the wall between the bell tower and the Basilica could be observed, a clear experimental confirmation of the assumptions on the thrust reduction applied by the bell tower. In particular, the positive effects of the structural joint can be synthetized with a substantial reduction of the load acting on the column of the Basilica (about -17%) and with a restoration of the original conditions when the two structures were statically more independent (Gottardi et al. 2015). Despite such substantial modification of the load distribution, which has inevitably produced a new increase of the differential settlement between the bell tower corner and the column (up to about 0.6 mm/year from January 2009 to November 2012), all the monitoring devices in operation have consistently shown that the subsequent, rather uniform, average settlement of the bell tower has proved to be only about 0.3 mm/year in the 5 years from February 2012 to December 2016. When compared to the rate of more than 1 mm/year before the soil fracturing intervention, it appears that its overall effectiveness can be confirmed.

Finally, thanks to the current reduced static interaction, the bell tower can accommodate further movements without inducing excessive stresses on the adjacent Basilica. The separation of the two structures has proved to be especially positive also in relation to seismic actions, if after the shakes of the Emilia earthquake in May 2012 no specific consequences have been recorded in the isolated structures, whilst new cracks have appeared in the nearby St. Peter's Chapel, still interested by structural links. Careful monitoring of settlement rates should continue in order to keep under control possible future trends and to fully understand the ongoing mechanisms.

4 Concluding Remarks

Three thoroughly documented case studies of well-known historic towers have been mentioned and briefly described. They have been presented from a geotechnical perspective, with a distinctive focus on their foundations and on the soil-structure interaction issues. The account of the extremely accurate strengthening interventions carried out in the last decades is the story of a most complex and diversified process, in which expert engineers have played a fundamental role in full respect of the whole structure integrity and authenticity and in close collaboration with other disciplines.

Potential hazards for the stability and preservation of historic sites and monuments often reside on their foundations and on the subsoil around them. It is therefore of great importance to implement a careful investigation programme with the preliminary vital aim of understanding the structural concept at the base of the construction and the effect produced by the circumstances which occurred along their often long history. Subsequently, an extensive monitoring system has to be designed and made operative as early as possible, in order to learn about the current behaviour and the possible response to future events.

The Pisa Tower and the Frari Bell Tower case studies, together with their successful strengthening interventions, clearly demonstrate the importance of such accurate investigation and monitoring activities, which have enabled to conceive effective and low-impact stabilization measures. In both cases, innovative criteria and methodologies for the strengthening of historic foundations were devised, without any direct intervention on the relevant structures. Through a diversified and extensive real-time monitoring system, purposely implemented, it was possible to focus on the intervention areas, calibrating and minimizing its impact on the existing structures. The technologies used have proved to be especially gradual and flexible. Such case studies can be hopefully used as a source of inspiration for future strengthening interventions aimed at preserving the integrity and the authenticity of historic foundations.

References

Bergonzoni F (1989) Pietra su pietra verso il cielo: tecniche, tempi e costi di costruzione, in le torri di Bologna. Roversi G (eds) Grafis Ed, Bologna, Italy, pp 29–48 (in Italian)

Bruno L, Marchi M, Bertolini I, Gottardi G, Amorosi A (2020) Climate control on stacked paleosols in the Pleistocene of the Po Basin (northern Italy). J Quat Sci 35(4):559–571

Burland JB, Jamiolkowski M, Viggiani C (2003) The stabilization of the leaning Tower of Pisa. Soils Found 43(5):63–80

Burland JB (2004) The leaning tower of Pisa revisited. In: Proceedings of fifth international conference on case histories in geotechnical engineering, 13–17 April 2004, New York, NY, pp 1–13

Burland JB, Jamiolkowski MB, Viggiani C (2009) Leaning tower of Pisa: behaviour after stabilization operations. Int J Geoeng Case Hist 1(3):156–169

Burland JB, Jamiolkowski MB, Squeglia N, Viggiani C (2020) The tower of Pisa-history, construction and geotechnical stabilization. CRC Press, 62 p. ISBN 9780367469047

Flora A (2013) General report du TC 301 monuments, historic sites and case histories. In: Proceedings of the 18th international conference on soil mechanics and geotechnical engineering, 2–6 September 2013, Paris, France, pp 3087–3094

Flora A, Lancellotta R, Sabia D, Viggiani C (2018) A first insight into towers' behavior: geotechnical and structural mechanisms, leaning instability, long-term behavior. In: Lancellotta R, Flora A, Viggiani C (eds) Geotechnics and heritage: historic towers. Taylor & Francis Group, London, pp 5–13. ISBN 978 1 138 03272 9

Gottardi G, Cavallari L, Marchi M (2009) Soil fracturing of soft silty clays for the reinforcement of a bell tower foundation. In: Proceedings of 2nd international workshop on geotechnics of soft soils-focus on ground improvement, 3–5 September 2008. Glasgow, Scotland. Taylor & Francis Group, London, UK, pp 31–41

Gottardi G, Lionello A, Marchi M, Rossi PP (2013) Preservation and monitoring of the Frari Bell Tower in Venice. In: Proceedings of 2nd international symposium on geotechnical engineering for the preservation of monuments and historic sites (IS TC301), 30–31 May 2013. Napoli, Italy. Taylor & Francis Group, London, UK, pp 411–424

Gottardi G, Lionello A, Marchi M, Rossi PP (2015) Monitoring-driven design of a multiphase intervention for the preservation of the Frari bell tower in Venice. Rivista Italiana Di Geotecnica 49(1):45–64

Gottardi G, Marchi M, Lionello A, Rossi C (2017) Preservation of historic towers in Venice: the instructive monitoring-driven intervention on the foundations of the Frari bell tower. In: Lancellotta R, Flora A, Viggiani C (eds) Geotechnics and heritage: historic towers. Taylor & Francis Group, London, pp 73–97. ISBN 978 1 138 03272 9

Gozzadini G (1875) Delle torri gentilizie di Bologna, e delle famiglie alle quali prima appartennero. Zanichelli Editore, Bologna, Italy, 772 p. (in Italian).

ISO 13822:2010 (2010) Bases for design of structures—assessment of existing structures. International Organization for Standardization. 44 p

Leoni M, Squeglia N, Viggiani C (2018) Tower of Pisa: Lessons learned by observation and analysis. In: Lancellotta R, Flora A, Viggiani C (eds) Geotechnics and heritage: historic towers. Taylor & Francis Group, London, pp 15–35. ISBN 978 1 138 03272 9

Lancellotta R (2008) Geotechnical engineering. CRC Press, 520 p. eBook ISBN 978 04 29080 22 7

Lancellotta R, Sabia D (2015) Identification technique for soil-structure analysis of the Ghirlandina Tower. Int J Archit Herit 9(4):391–407

Lionello A (ed) (2008) Il campanile di Santa Maria Gloriosa dei Frari in Venezia–conoscenza, consolidamento e restauro. Electa, Milano, 183 p. (in Italian).

Lionello A (ed) (2011) Tecniche costruttive, dissesti e consolidamenti dei campanili di Venezia. Corbo & Fiore Publishers, Venezia. 191 p. (in Italian). ISBN-10: 8870861260

Marchi M, Gottardi G, Lionello A (2006) Sulle fondazioni dei campanili di Venezia. V Convegno Nazionale dei Ricercatori di Ingegneria Geotecnica-Fondazioni Superficiali e Profonde, Bari, 15–16 September 2006. Hevelius Ed., Benevento, Italy, pp 177–192. (in Italian)
Marchi M, Butterfield R, Gottardi G, Lancellotta R (2011) Stability and strength analysis of leaning towers. Geotechnique 61(12):1069–1079
Marchi M, Gottardi G, Soga K (2014) Fracturing pressure in clay. J Geotech Geoenviron Eng (ASCE) 140(2):04013008(1–9)
Marchi M, Bertolini I, Gottardi, G, Amorosi, A, Bruno, L (2019) From geological and historical data to the geotechnical model of the Two Towers in Bologna (Italy). In: Proceedings of the XVII European conference on soil mechanics and geotechnical engineering-geotechnical engineering, foundation of the future, 1–6 September 2019, Reykjavík, Iceland, pp 1–8
Roversi G (1989) Le torri di Bologna. Quando e perché sorsero, come vennero costruite, chi le innalzò, come scomparvero, quali esistono ancora. Grafis Ed., Bologna, Italy, 321 p. (in Italian)
Viggiani C (2019) Senza neanche toccarla-La stabilizzazione Della Torre di Pisa. Hevelius Edizioni, Benevento, Italy. 152 p. (in Italian). ISBN 978 88 86977 96 8
Zuccolo G (1975) Il restauro statico nell'architettura di Venezia. Istituto Veneto di Scienza Lettere ed Arti, Venezia. 208 p. (in Italian)

Authenticity of Foundation and Proactive Conservation of the Central Tower of Bayon Temple Under a Changing Climate in Angkor, Cambodia

Yoshinori Iwasaki, Mitsuharu Fukuda, Mitsumasa Ishizuka, Robert McCarthy, Ichita Shimoda, Koichi Nakagawa, Tomofumi Koyama, Takeshi Nakagawa, and Vanna Ly

Abstract Soil and foundation are recognized as important factors in general building construction but not as heritage. They have easily changed or modified without discussion of the special characteristics of the authenticity. The present paper introduces the authenticity of the foundation of heritage structure and the character defining elements of the authenticity of the foundation are discussed. Heavy rain sometimes has caused failures of soil structures in Angkor. The slope of Baphuon temple had failed in 1943 and the embankment of the west moat of Angkor Wat failed in 1997 because of heavy rains. The temple structures in Angkor stand on sandy fill and often with steep slopes, which failed due to heavy rain. The tall central tower of Bayon stands upon thick manmade fill by a direct shallow foundation without special supporting elements. The sandy soil used for the platform mound was found to have very strong strength in dry conditions, but weakened in wet conditions. The global warming could potentially bring heavy rain with long duration and is anticipated to

Y. Iwasaki (✉)
Geo Research Institute, 2-1-2, Otemae, Chuo-Ku, Osaka 540-0008, Japan
e-mail: yoshi-iw@geor.or.jp; dec19yoshi1+torino@gmail.com

M. Fukuda
Taisei Geotech, 1174-10, Nishimachi, Kurume, Fukuoka-Pref. 830-0038, Japan

M. Ishizuka · R. McCarthy
JASA, #56, Gr6, Phum Tropeang Ses, Khum Kokchork, Srok, Siem Reap, Cambodia

I. Shimoda
University of Tsukuba, 1-1-1, Tennodai, Tsukuba-City, Japan

K. Nakagawa
Osaka City University, 3-3-138, Sugimoto 3-Chome, Sumiyoshi, Osaka, Japan

T. Koyama
Kansai University, 7-1, Hakubai-Cho, Takatsuki 569-1098, Japan

T. Nakagawa
Waseda University, 3-4-1, Ohkubo, Shinjyuku, Tokyo 169-8555, Japan

V. Ly
Cambodia National Committee for World Heritage, Phnom Penh, Cambodia

G. M. El-Qady and C. Margottini (eds.), *Sustainable Conservation of UNESCO and Other Heritage Sites Through Proactive Geosciences*, Springer Geology,
https://doi.org/10.1007/978-3-031-13810-2_8

weakening the soils which might result in damages and failures of heritage structures. For the sustainable conservation, the preventive measures, and observational method are being prepared to avoid any damage and failure by heavy and long rain in Angkor.

Keywords Authenticity of foundation · Bayon in Angkor · Global warming

1 Introduction

Geotechnical engineering plays one of the important roles to safeguarding cultural heritages and has made such key contribution as in restoration work for the inclined Pisa Tower as well as many other historic monuments. In the past, foundations were generally considered as simply nothing but an element to support the heritage structures and not considered as a part of the heritage. However, the recent trend of conservation of heritage indicates that a foundation system has begun to be considered as one of the basic components of the heritage structure. In 2010, the international standard of ISO 13822 on assessment of structural safety was renewed and added as an Annex "Heritage Structures" and stressed the importance of the foundation ISO 13822: (2010).

The new concept of "authenticity of foundation" is discussed as well as the character defining elements of authenticity that should be evaluated and restored.

2 Authenticity

2.1 Authenticity of Cultural Heritage Subsection

Authenticity was defined in the Venice Charter of 1964 as heritage composed from original material, original position, original design as well as original procedure (Venice Charter 1964). The concept behind the Venice Charter is called "Anastylosis (Greek)," which means "take column back to original position." Anastylosis implies that original stone columns spread over a historical ruin shall be rebuilt in their original positions.

The principle of Anastylosis was developed based on conservation of stone structure in Europe and does not give any heritage values to such repaired materials as often seen in wooden structures in Japan.

Later in 1994, the concept of authenticity was expanded by the Nara Document on Authenticity to cover various methods characterized by the region to which the heritage belongs. Region specific methods that were developed in some area are also accepted as the characteristic of authenticity (Nara Doc 1994).

Character-defining elements are defined as historic materials, forms, locations, spatial configurations, morphology, concept and details, structural design, uses, and cultural associations that contribute to the heritage value of a structure that shall be retained in order to preserve its heritage value.

2.2 *Authenticity of Soils and Foundation*

In 2005, ISO 13822: (2010) (Bases for design of structures—Assessment of existing structures) was reviewed for renewal. ISCARSAH (International Scientific Committee on the Analysis and Restoration of Structures of Architectural Heritage under ICOMOS) had proposed to include heritage structures in the standard and worked together for five years to develop such a standard. ISO 13822 has been updated in 2010 and added as an Annex-I (informative) Heritage Structure, which has expanded the heritage structure to include foundation as a part of the structures.

Annex I clearly state in paragraph of I.5.3 Authenticity of foundation that "From the point of view of conservation, foundations are not different from the rest of the structure and should be assessed and rehabilitated taking into consideration their heritage value. This involves the requirement to identify their authenticity and character-defining elements.

3 Bayon Temple in Angkor Thom

Bayon, as shown in Fig. 1, is the central temple of Angkor Thom constructed around in late 12th to early thirteenth century. Figure 2 shows section and plan view of the Bayon temple. Japanese Government of Safeguarding Angkor Team (JSA) performed an archaeological trenching study at the north-eastern corner as "Long trench" as shown in Fig. 2 (Narita et al. 2000).

4 Trenched Foundation of Bayon Temple

The result of the long trench has revealed the following facts as shown in Fig. 3 (Narita et al. 2000).

1. The surface of the first terrace was covered with sandstone and laterite block, beneath which a densely compacted sandy filled layer was identified.
2. Compacted sandy filled layer continued with an additional laterite block layer to the level of the original ground surface.
3. It was identified that the compacted sandy filled layer continues from the original ground level to 2–3 m in thickness.
4. The compacted sandy filled layer below the original surface was found to continue horizontally to about 10 m outside of the outer gallery plinth.

Fig. 1 Bayon temple

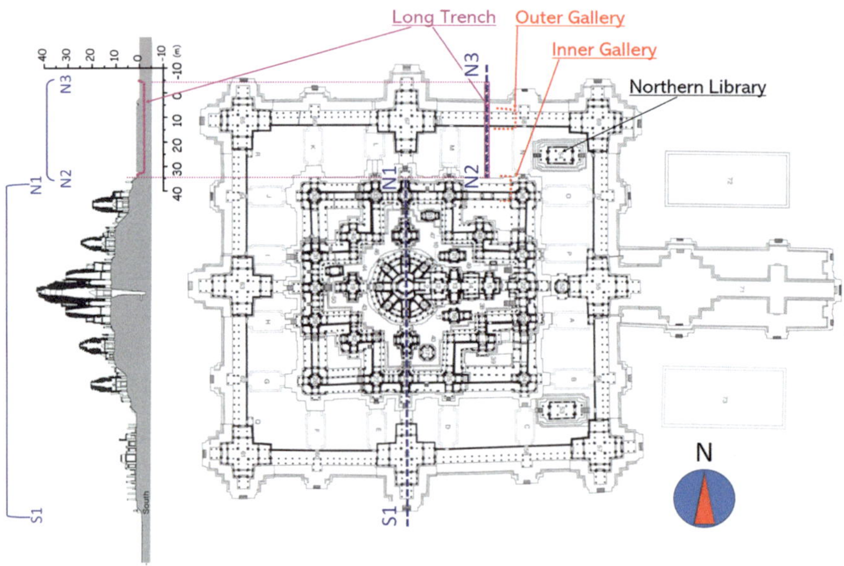

Fig. 2 Section and plan of Bayon

Based upon the trenched foundation of very well compacted sandy fill was constructed a three stepped terrace, where masonry structures with four faces and a central tower of about 42 m in height from the ground has been standing over 700 years (see Fig. 4).

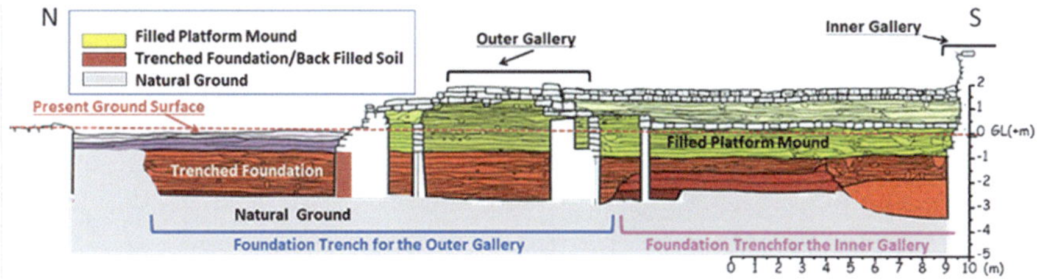

Fig. 3 Long trench of archaeological study

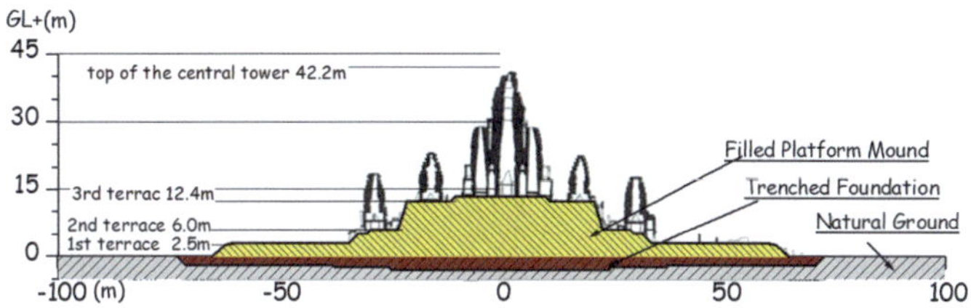

Fig. 4 Trenched foundation with filled high mound of Bayon, N-S Section

5 Foundation Structure of Bayon

5.1 Direct Shallow Foundation

A base stone is placed beneath the bottom of the central tower as shown in Fig. 5. JSA carried out archaeological trench excavation along the inside of the base stone and geotechnical hand auger sounding beneath the stone to determine if any special base structure was installed to support the heavy central tower (Shimoda et al. 2008).

Horizontal hand auger tests were carried out at 5 points as shown in Fig. 6 and has resulted in no supporting stones, but only very dense sandy fills beneath the base stone support.

5.2 Platform Mound with Vertical Shaft at the Center

EFEO, a France sponsored organization, in 1933 dug out the center of the base of the main central tower below the pavement and found a Buddha statue. It was recorded that the vertical shaft had been backfilled. Geotechnical boring was performed at the backfilled vertical shaft and at the original manmade filled mound as shown in Fig. 7.

The back filled soil for BY09 was found in a very loose state of SPT, N-values $N < 4$ (BYV2009, Fig. 7). Another boring (BYV2010) at the top terrace shows the

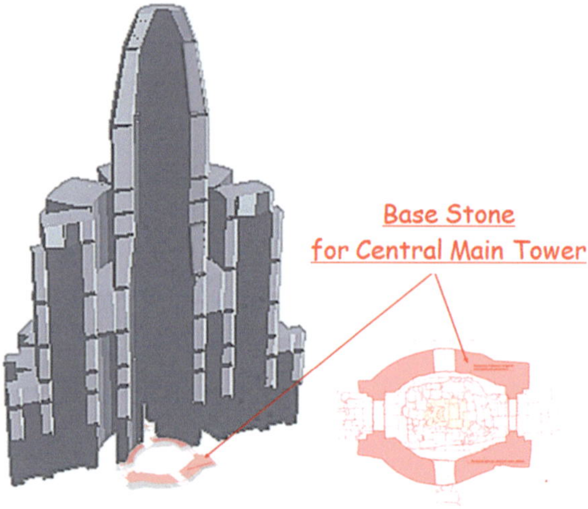

Fig. 5 Base stone for central tower Bayon

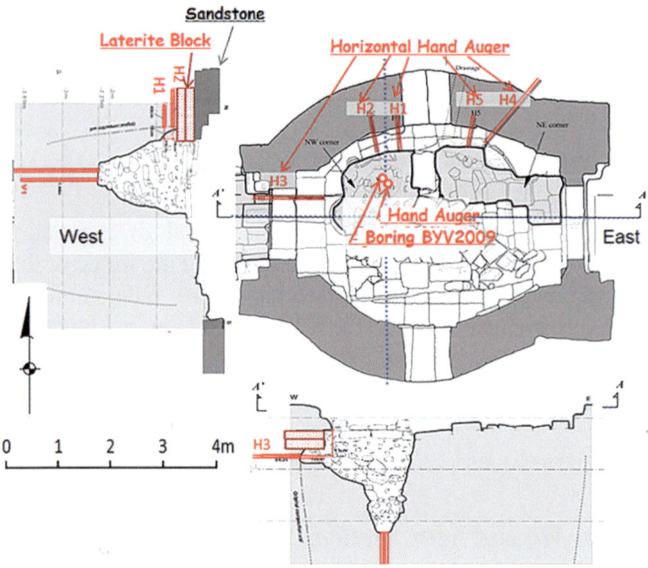

Fig. 6 Hand auger soundings beneath the base stone

sandy fill lower than GL-6 m of N = 100–150, which is a very large value compared to the expected values of 20–40 for common filled sandy soil (Iwasaki and Fukuda 2009; Iwasaki et al. 2014).

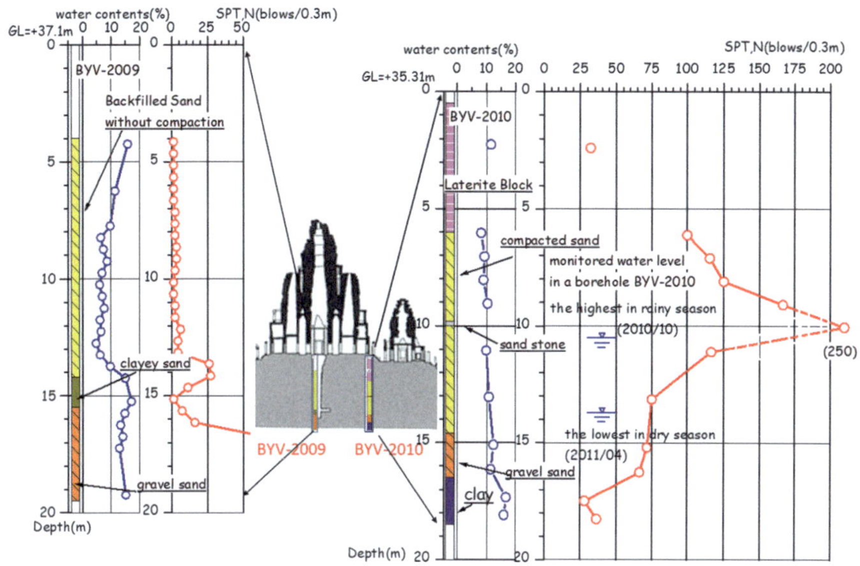

Fig. 7 Boring BYV2009 at the backfilled shaft and Boring BYV2010 at the original mound

5.3 Characteristics of Stiff Sand Fill of the Platform Mound

The obtained SPT, N-values are plotted against water contents of the sampled soils for both borings and shown in Fig. 8. No relationship is found for BYV2009 of the backfilled soil; however, the decrease of water is found to result in the increase of the strength for boring BYV2010 (Nakagawa et al. 2016).

Figure 9 shows the response of the dense sandy fill when it was submerged in water.

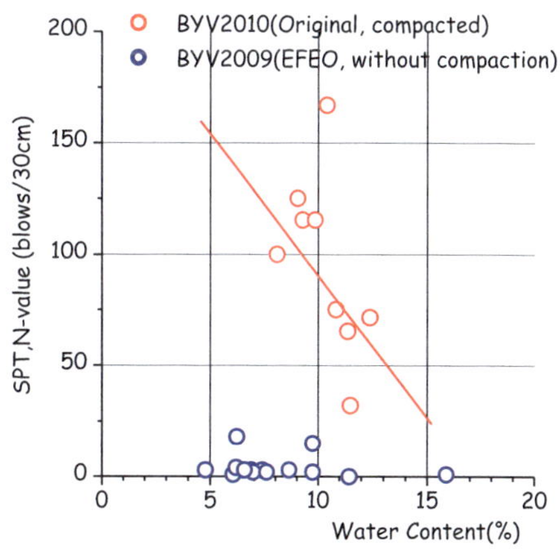

Fig. 8 Relationship between SPT, N-values for B09 (BY2009) and B10 (BY2010)

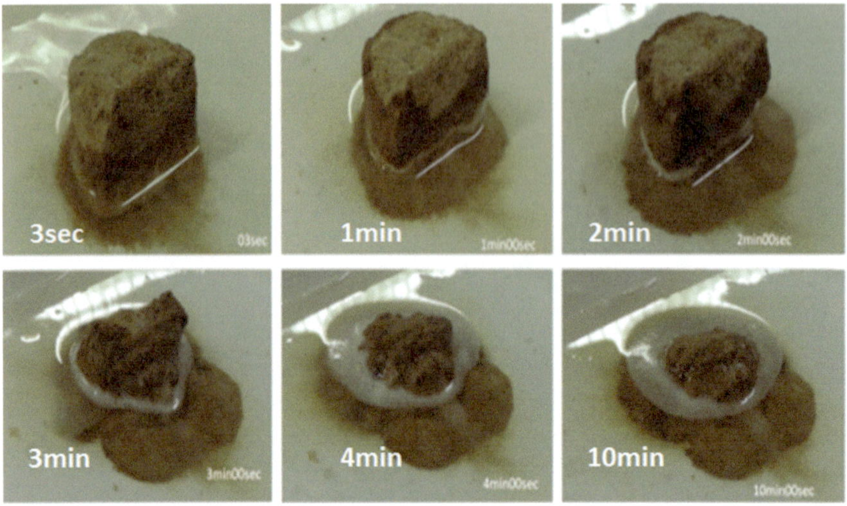

Fig. 9 Collapse of stiff sand under water within 10 min

Sampled soil looks like soft sandstone, however, it collapsed under water within 10 min as shown in Fig. 9.

The grainsize distributions of the sampled soil by the boring BYV2010 are shown in red color in Fig. 10. The soil is clayey sand and the entire samples of the filled soil show the same distribution, which implies very uniform filled soil.

X-ray diffraction analysis was applied to the fine particle of the filled sand shown in Fig. 11. In addition to quartz sand, Halloysite (Kaolinite group) was detected as the clay component.

Micrograph of the section of the sample is shown in Fig. 12, where the round shape of quartz is seen filled with clay material.

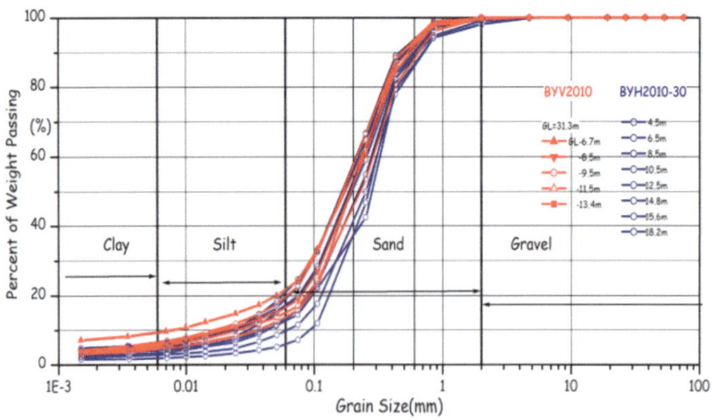

Fig. 10 Grain size distribution of BY2010 and BH

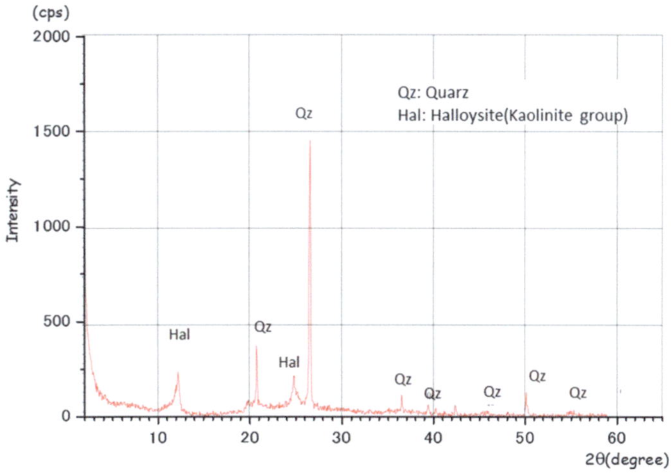

Fig. 11 X-ray diffraction for fine particle of the filled soil

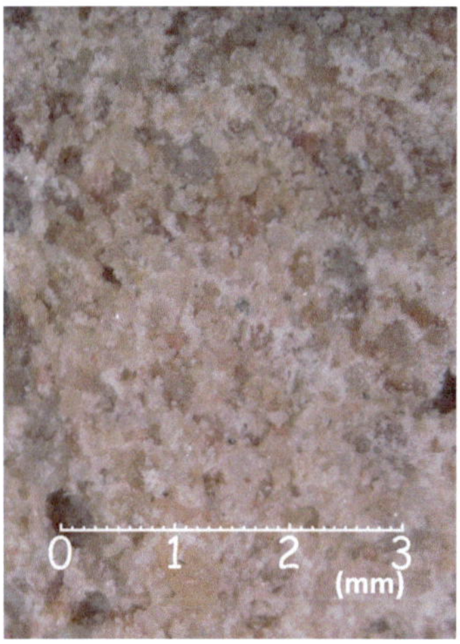

Fig. 12 Micrograph of clayey sand of the filled soil

5.4 Increase of Strength by Decreasing Moisture

A series of laboratory tests were performed to see how much strength changes due to the decrease of the moisture contents. More than 25 samples in containers were prepared with water content of 15%, which almost creates a 100% saturated condition. The samples were placed outdoor and the water evaporates from the sample and the water content decreases day by day. A Yamanaka cone penetration test mechanism tested these samples as shown in Fig. 13. The test results in Fig. 14 shows

Fig. 13 Yamanaka cone test

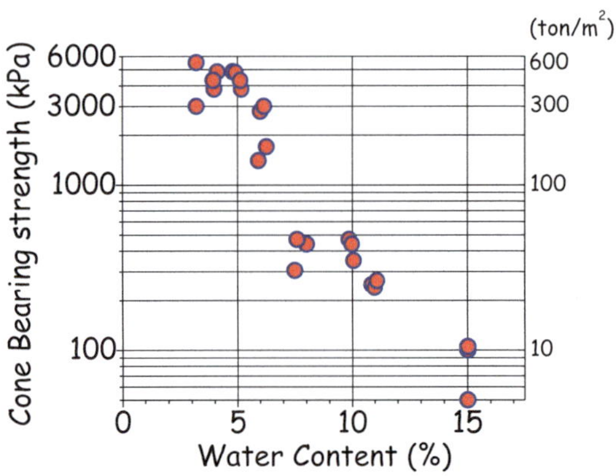

Fig. 14 Increase of cone bearing strength

clearly an increase of strength more than 50 times due to the decrease of the water contents.

5.5 Circular Laterite Structure Embedded in the Platform Mound

Laterite blocks were found underneath the pave stone at BY10 to the depth of 6 m from the surface. To investigate the shape and size of the laterite blocks, several additional horizontal and inclined borings were performed in N-S section as shown

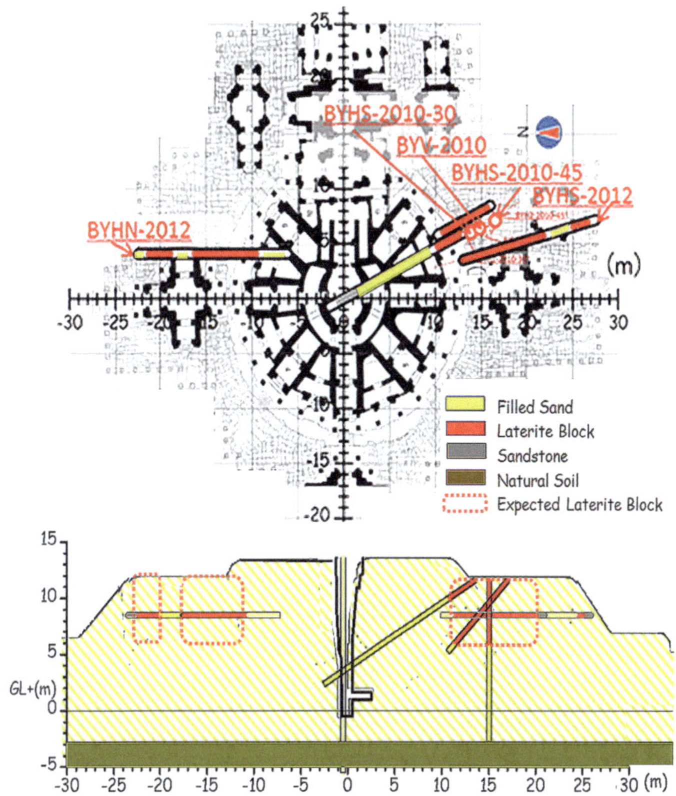

Fig. 15 Boring at foundation platform mound of Central Tower, Bayon

in Fig. 15. The inclined boring was set to cut the vertical shaft at the center of the mound.

The distribution of the laterite blocks in boring logs is shown in red in Fig. 16. Electric survey was performed to study the distribution characteristics of the platform mound (Nakagawa et al. 2016). The result is shown in Fig. 17. The electric resistance of 500–1500 Ω-m corresponds to the laterite block zone. The central zone of low resistivity corresponds to high moisture loose backfilled soil. Figure 15 shows the distribution of the electric resistivity along the edge of the masonry stone structures on the platform mound. The expected plan view of the laterite feature is also shown in Fig. 16. The strong laterite structure surrounding the central tower is expected to support the construction of the central tower from foundation failure (see Fig. 18).

The expected structure of the platform mound of the central tower of the Bayon Temple is shown in Fig. 19.

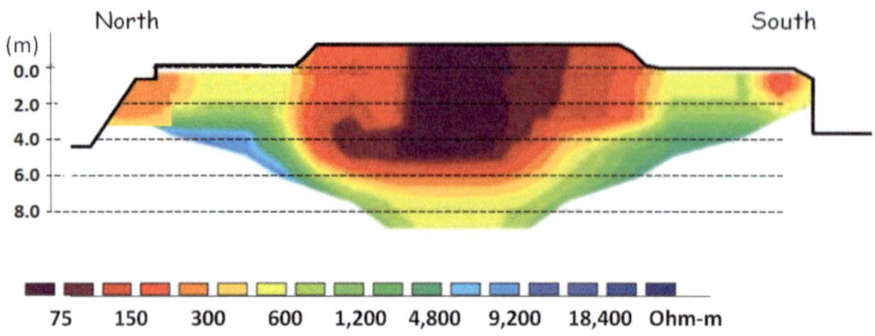

Fig. 16 Electric Survey in NS section

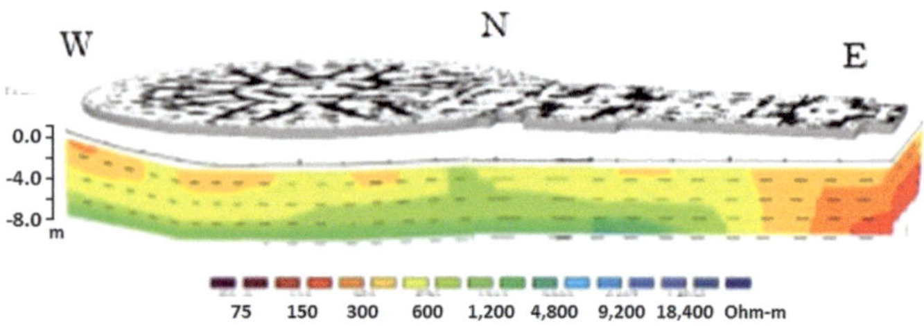

Fig. 17 Result of electric survey along the foundation of masonry structures

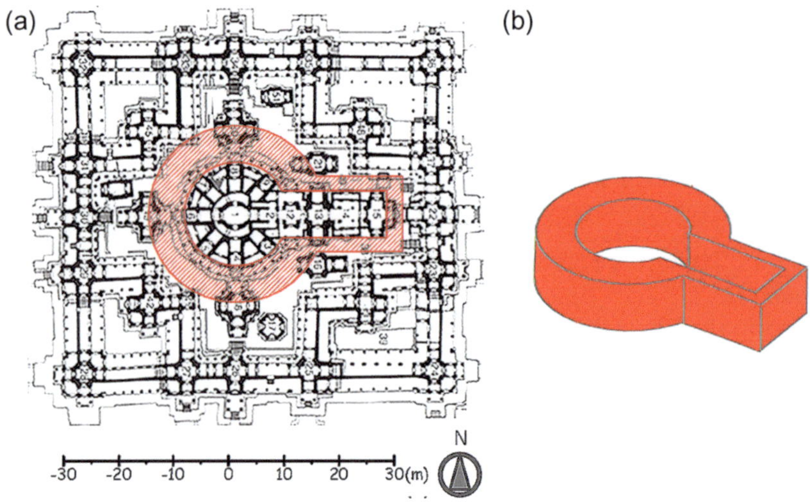

Fig. 18 Expected Laterite Body within the platform foundation. **a** plan view of the expected position of laterite block, **b** 3D image of the block

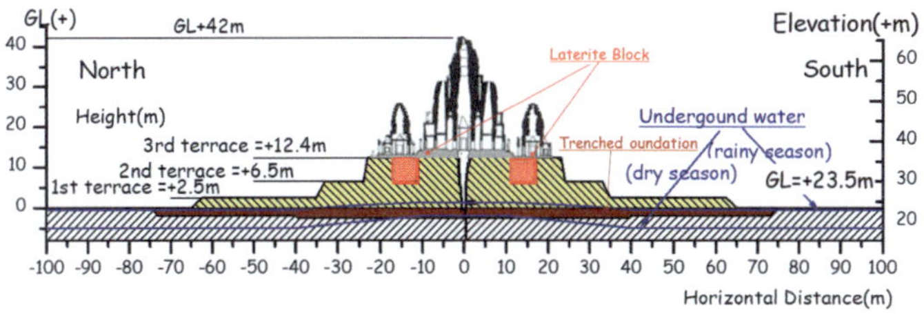

Fig. 19 Expected foundation structure of the mound of the Main Tower, Bayon

5.6 *Authenticity of the Foundation of Bayon Temple, Angkor Thom*

The study of soil and foundation structure of Bayon temple have revealed the following special characteristics as the character defining elements of the authenticity of Bayon Temple in Angkor Thom.

1. Trenched foundation extended outside of the temple.
2. The uniform sandy soil for filled mound that is characterized 10–20% of fine contents of kaolin sand.
3. The well compacted sandy fill shows very high strength under dry condition but collapses in water.
4. Laterite structure embedded in the platform mound just outside of the base of the main tower.
5. Direct shallow foundation of the masonry stone tower of Bayon.

The above-mentioned character defining elements of the authenticity are identified for Bayon Temple and are unique and may be applicable to most of the heritage structures including Angkor Wat, where the high platform mound of 25 m in height with the tall main tower of 60 m in height from the ground surface.

6 Anticipated Geotechnical Risk in Angkor

There are various factors of to cause damages and failures of cultural heritages. Among such causes, heavy rain and strong wind are major actions to have caused foundation and upper structure in the past. Heavy rain, which shall be associated with global warming, is anticipated to cause damage and failure for soils and foundations in Angkor.

Fig. 20 Slope failure of the Baphuon in 1943 (1963, EFEO) (Pascal 2016)

6.1 Heavy Rain

Heavy rains have caused damages and failures of the temples and other structures in Angkor. Recent records of these failures are as follows (see Fig. 20),

1943 Baphuon Temple: slope failure of northern east slope
1952 West Causeway, Angkor Wat: failure of stone retaining wall
1997 Moat of Angkor Wat: failure of embankment

Baphuon temple was constructed in about AD1060 and is located near Bayon in Angkor Thom shown in Fig. 21 from east front and the vertical section in Fig. 22. The temple mound was constructed by sandy fill, which was found having the same characteristics of grain size distribution and estimated to have the same strong strength under dry condition but weaken under wet condition as Bayon. The mound was very high at 30 m from the ground with very steep slopes of 45–60°. It is amazing fact such steep and high mound of manmade fill of natural sandy soil has been standing for 900 years (see Fig. 23).

6.2 Change of the Water Contents by Rain

When rain falls and infiltrates into the mound, the water content within the mound will be increased. Monitoring the change of the moisture in the soil mound was performed in the platform mound at Bayon temple (Koyama et al. 2016).

Moisture sensors (Fig. 24) had been installed at several depths at GL-0.5, -0.75, and -1.0 m and the monitored results are shown in Fig. 25 with rain fall results.

Fig. 21 Baphuon temple

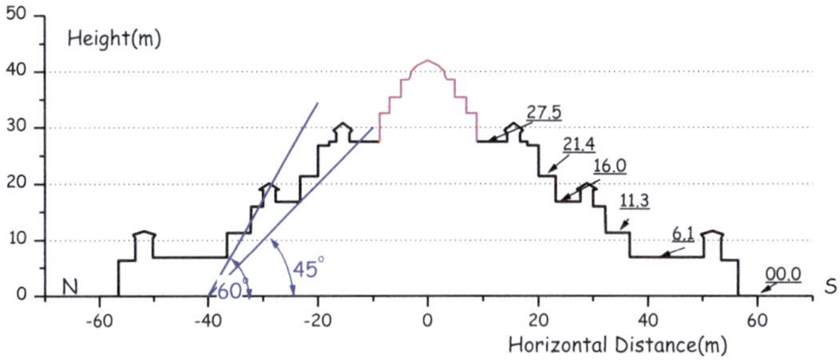

Fig. 22 Section of Baphuon

The senor at the shallowest responses first when rain fall begins. The deeper sensor responses with some delay by infiltration from the surface to the depth at which the sensor was installed. The seepage rate from the surface in the mound was found to be about 1 m/h.

When the rain stops, evaporation starts and the moisture begins to decrease.

In the past, a climate pattern of a squall shows, very heavy but short time rain fall. The maximum depth was limited to a rather shallow depth. However, the global warming potentially could change the rain pattern from the squall to heavy and long rain, which might bring an increase of the water contents resulting in failure of the soil mounds.

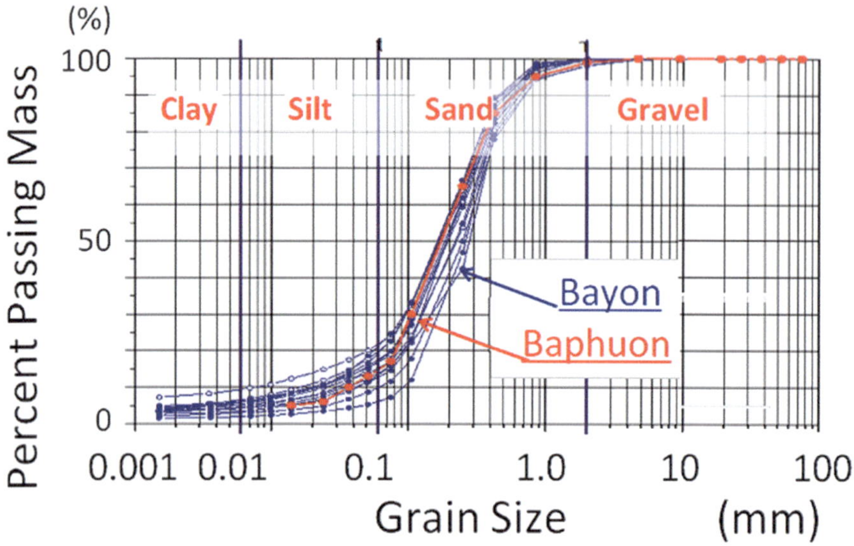

Fig. 23 Grain size distribution of mound of Baphuon and Bayon

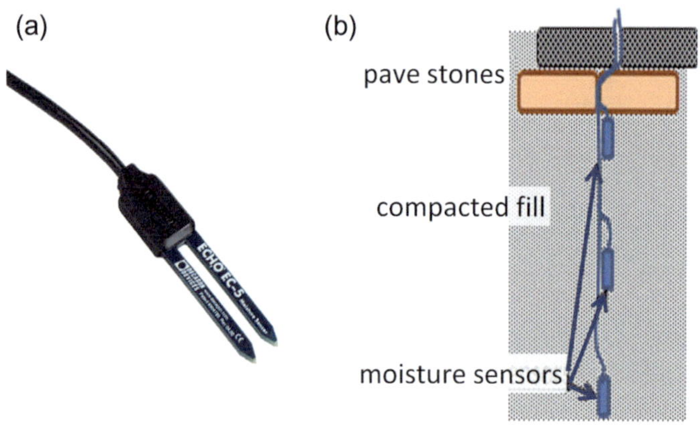

Fig. 24 **a** Moisture sensor and **b** Installation in the ground

6.3 Countermeasures with Preventive Monitoring

Global warming is potentially anticipated to induce the deeper seepage into filled soil and results in failure due to a decrease of bearing capacity. In general, the upper structures are required to be dismantled for treating foundation. However, the improvement of the foundation of Bayon is possible by establishing anti-seepage system into the ground and pavement. Among various countermeasures, the preventive/protective monitoring is selected as the best based upon the principles of the

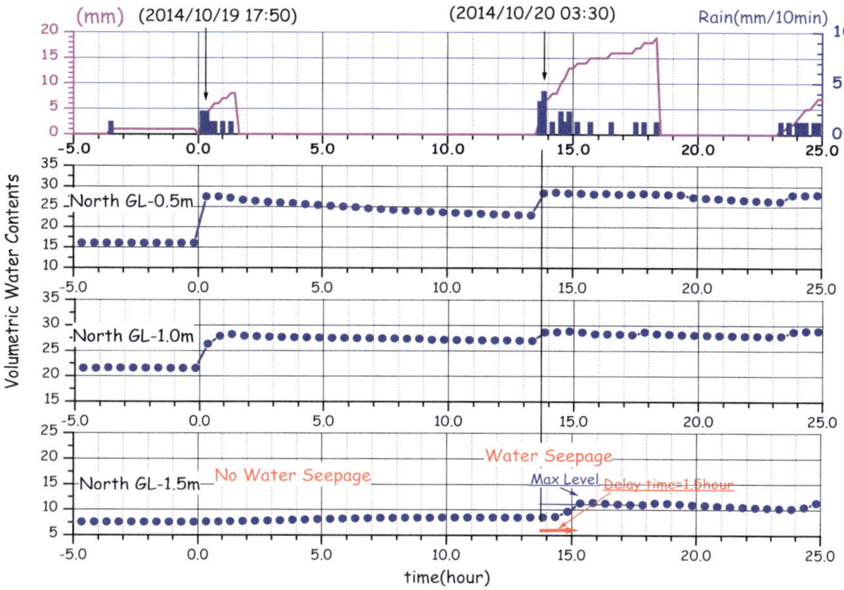

Fig. 25 Moisture monitoring to rainfall at the mound at Bayon

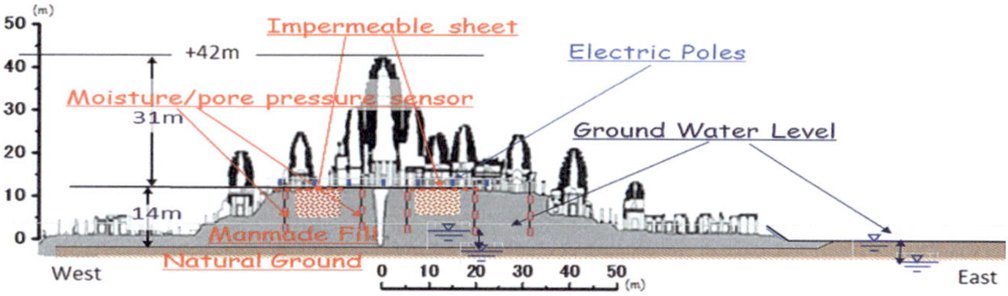

Fig. 26 Observational system of countermeasures against global warming

conservation, ① minimum intervention, ② removable measures, and ③ incremental approach. Impervious sheets of fabric will be placed beneath the pavement stones with monitoring and verifying the effects by moisture and pore water pressure as well as 3D electric survey to identify the seepage of rainfall into the filled soil of the foundation (see Fig. 26).

7 Conclusions

Following the intergovernmental conference for Safeguarding and Development of Angkor by UNESCO, which was held in Tokyo in 1993, Japanese Government Team for Safeguarding Angkor was organized consisting of various specialists including archaeology, geotechnology, architecture, and others.

Geotechnical study of soils and foundations of Bayon and related heritage structures in the past 25 years has resulted in the new prospective in the deeper understanding of the authenticity of the foundations of temple structures in Angkor.

The character defining elements of the authenticity of the foundation in Bayon temple are summarized as follows,

1. Trenched foundation extended outside of the temple plinth.
2. The uniform sandy soil for filled mound that is characterized 10–20% of fine contents of kaolin sand.
3. The well compacted sandy fill shows very high strength under dry condition but collapses in water.
4. Laterite structure embedded in the platform mound just outside of the base of the main tower.
5. Direct shallow foundation of the masonry stone tower of Bayon.

Heavy rains have caused damage and failures of heritage structures due to the weakening filled soils in the foundation mounds of structures or the embankment itself.

One of the potential threats of the global warming to the heritage structures in Angkor is the increase of moisture in the filled soil that could result in failure of the heritage due to heavy and long rain. One of the best countermeasures to protect the Central Tower, the symbol of the Bayon Temple, is to install anti-seepage system into the foundation mound. This preventive measure and monitoring system are to be checked and modified by observational method.

Geotechnical engineering has been contributing to the safeguarding of heritage structures around in the world. Without the geotechnical specialist, the restoration of the leaning Tower of Pisa would never have been realized.

However, we, geotechnical engineers, only concentrate on the technical aspects of the restoration and rarely if ever consider the authenticity of the heritage value of the foundation. Geotechnical engineer, in general, does not know the ISO 13822 that describes the concept of the authenticity of foundation of cultural heritage protection and preservation.

Geotechnical engineer must expand the boundary of the geotechnical engineering themselves to relate to the world of general cultural heritage.

The present conclusions are based upon cooperation between archaeology, architects, and geotechnical engineering. In archaeology, soils are described as clay, silt, sand, and gravel with its colors. No tests on the sampled soil are performed. In soil geotechnical engineering, soils are classified not only by name but through grain size distribution.

It was surprising to find the almost similar distribution curves for the filled soil in Bayon, Baphuon, and other sites in the Angkor Thom.

Since without the laboratory test of grain size, accompanying soil mechanics, and physical characteristics, the character defining elements of the authenticity of the filled soil has never be been adequately clarified.

References

ISO (2010) https://www.iso.org/standard/46556.html

Venice Charter (1964) https://www.icomos.org›charters›venice-e

Nara Doc (1994) https://www.icomos.org›charters›nara-e

Narita T, Nishimoto S, ,Shimizu N, Akazawa Y (2001) Trench excavation of outer gallery, Bayon. annual report on the technical survey of Angkor monument 2000, pp 3–18, 317

Shimoda I, Yamamoto N, Iwasaki Y, Fukuda M (2009) Excavation survey of the central tower chamber. Annual technical report on the survey of Angkor monument 2008, JASA, Tokyo, pp 67–88

Iwasaki Y, Fukuda M (2009) FY2009 report-geoengineering/environment unit. Annual technical report on the survey of Angkor monument 2008, JASA, Tokyo, pp 323–356

Iwasaki Y, Fukuda M, Haraguchu T, Kitamura A, Ide Y, Tokunaga T, Mogi K (2014) Structural of platform mound of central tower based upon boring information. Annual technical report on the survey of Angkor monument 2012–2013, JASA, Tokyo, pp 93–113

Nakagawa K, Iwasaki Y, Ishizuka M (2016) Electric survey for foundation mound of central tower, Bayon. Annual technical report on the survey of Angkor monument 2014–2015, JASA, Tokyo, pp 128–131

Pascal R (2016) Le Baphuon. Memories archaeology 27, EFEO, Paris, p 240 (EFEO/CAM12559-2)

Koyama T, Yamada S, Iwasaki Y, Fukuda M, Shimoda I, Ishizuka M (2016) Installation of moisture sensor. Annual technical report on the survey of Angkor monument 2014–2015, JASA, Tokyo, pp 132–137

Slope Instability Induced by Climate Changes on the UNESCO Etruscan Necropolis of Monterozzi (Tarquinia, Italy)

Daniele Spizzichino, Gabriele Leoni, Paolo M. Guarino, Daniela Boldini, Saverio Mengoni, Ermanno Marino, Adele Cecchini, and Beatrice Casocavallo

Abstract The Necropolis of Monterozzi (Tarquinia, Central Italy), with hundreds of painted rock cut tombs of VII-II centuries B.C., is a unique evidence of the Etruscan vanished culture. During year 2017, in the northern side of the necropolis, near the edge of the hill where the site stands, several tombs began to be affected by wall cracks. After the request of intervention from the local Soprintendenza, ISPRA (Geological Survey of Italy Dept.) and University of Bologna (DICAM) started to analyse the stability of the most threaten tombs. First step was the understanding of the climatic, geologic, geomorphologic and geotechnical setting, to verify the overall stability conditions and the landslide hazard possibly induced by climate changes. In fact, year 2017 registered minimum rainfall of last 40 years and high temperatures. Some tombs were equipped with a crack gauge system to monitor the deformation pattern. The biogenic calcarenite (Macco formation) in which the tombs are carved out was tested in the laboratory to define its physical–mechanical properties. Finally, a 3D model was generated from a detail topographic survey, combined with more precise surveys (Terrestrial Laser Scan and UAV modelling), to carry out stability

D. Spizzichino (✉) · G. Leoni · P. M. Guarino
ISPRA, Geological Survey of Italy, via V. Brancati 48, 00144 Rome, Italy
e-mail: daniele.spizzichino@isprambiente.it

D. Boldini
Department of Chemical Engineering Materials Environment, Sapienza University of Rome, Rome, Italy

S. Mengoni
Department of Civil, Chemical, Environmental and Materials Engineering, University of Bologna, Bologna, Italy

E. Marino
Stage srl, Marcianise, Italy

A. Cecchini
Associazione Amici Tombe dipinte di Tarquinia, Tarquinia, Italy

B. Casocavallo
Soprintendenza Archeologia, Belle Arti e Paesaggio per l'area metropolitana di Roma, la provincia di Viterbo e l'Etruria meridionale, Rome, Italy

G. M. El-Qady and C. Margottini (eds.), *Sustainable Conservation of UNESCO and Other Heritage Sites Through Proactive Geosciences*, Springer Geology,
https://doi.org/10.1007/978-3-031-13810-2_9

analysis of both slope and tombs. The study achieved the preliminary identification of low impact mitigation and conservation measures for the safeguard of the site against climate changes related slope instability.

Keywords UNESCO · Etruscan necropolis · Slope stability · Climate change · Macco calcarenite

1 Introduction

This research is the product of two different agreements. The first one (starting since December 2017), between the *Soprintendenza Archeologica, belle Arti e Paesaggio per l'area metropolitana di Roma, la provincia di Viterbo e l'Etruria Meridionale* and the Italian Institute for Environment Protection and Research (ISPRA, Geological Survey of Italy), was aimed at the analysis and assessment of stability conditions of the Etruscan necropolises distributed in the entire territory of competence. The second agreement, between ISPRA and University of Bologna (DICAM), was focused on the evaluation of stability conditions of the underground tombs belonging to the Monterozzi Etruscan Necropolis and in particular to the tombs placed at the edge of the plateau set on the *Formazione del Macco*, the most exposed hypogea of the Monterozzi site. They are: the *Tomba della Porta di Bronzo* (Tomb of the Bronze Gate), the *Tomba delle Iscrizioni* (Tomb of the Inscriptions) and the *Tomba dei Vasi Dipinti* (Tomb of the Painted Vases), see Fig. 3.

In general, the contact between the bedrock clays, characterized by a mainly ductile/plastic behaviour, and the outcropping lithology (by selective erosion) of the Macco formation, with a rigid/fragile response, determines a high degree of fracturing and a high permeability, thus promoting alteration and mechanical breakdown at different scales (from the material to rock-mass levels). At microscale the karst dissolution of calcarenite is the most active morphological process. In this context, new fractures were observed starting from 2017, a particularly dry and hot year.

The main targets of the cooperative research were:

- to evaluate the stability conditions of the tombs and possible instability triggering mechanisms through a multidisciplinary approach (e.g. archaeological, geological, topographical, hydrological, geotechnical);
- to install a monitoring system to control any potential paroxysmal phenomena affecting the tombs;
- to characterize (for the first time) the Macco formation from a mechanical point of view;
- to carry out numerical analyses, preliminarily in 2D conditions, to assess the stress–strain and stability conditions of the tombs and the slope;
- to define a prototype approach exportable in other similar situations;
- to implement a mitigation measures master plan both in short- and long-term situation.

2 The Necropolis of Monterozzi

The main necropolis in Tarquinia spreads over 6 km on the Monterozzi hill, a Pleistocene marine terrace between the sea and the inner Civita hill, where the town arose. Some canyons divide the hill into four sectors, that are named: *Secondi Archi*, *Arcatelle*, *Primi Archi* and *Calvario*. The necropolis counts more than 6000 tombs, few of which containing wall paintings and only 3% of these shows figurative art. The wall painting practise spanned from late 8th to third century B.C., thus covering the entire Etruscan period. This huge heritage, representing the 80% of the total production known so far, makes Tarquinia the outstanding site of Etruscan wall paintings, therefore of the ancient world (Cecchini 2017). The painted tombs showed the richness of the owner, allowing to understand the way of life of aristocracy that through the paintings celebrated itself (Adinolfi et al. 2019; Marzullo 2016).

2.1 *Tomba dei Vasi Dipinti*

The *Tomba dei Vasi Dipinti*, dated to 510 B.C., is located along the cliff east of cemetery and is a one chamber square tomb (3.16 m wide and 1.95 m high) with a double pitched ceiling and a short access corridor (*dromos*) weakly damaged, with a small vestibule at the edge (Fig. 1). The paintings show a manifold banqueting scene with some *kylikes* (small food tables) and a *kylikeion* (main table) with one crater and two amphorae; below there are two more *kylikes*. On the center wall a man and a woman are stretched out on a *kline* (bed used during meals and for relaxing) enriched with drapes and cushions, some servants standing on the side. The gesture of the man, caressing his partner's chin, makes the scene familiar. On the side walls there are musicians and dancers cheering the banquet, that seems to be held in a grove.

2.2 *Tomba delle Iscrizioni*

The *Tomba delle Iscrizioni*, located on the northern slope in the *Secondi Archi* sector, is a one chamber tomb with a double pitched ceiling and a short *dromos*. On the center wall a door is painted, that refers to the border between the living and the dead's world; on both sides' athletic competitions and sacrifice scenes in honour of the deceased are shown. On the right wall two bearded men stand facing each other, while on the right a *komos* (ritual procession) scene is painted where five *comasti* (dancers, singers and drinkers), all identified by inscriptions, proceed to the left. The inscriptions, providing information on different families, refer to a consortium membership. The tomb is dated to 250 B.C. and shows Northern Ionian influences.

Fig. 1 Entrance (left) and interior (right) of the *Tomba dei Vasi Dipinti*

2.3 Tomba della Porta di Bronzo

The *Tomba della Porta di Bronzo*, dated to 510-500 B.C., is carved out along the slope of the *Secondi Archi* sector. Even this is a one chamber tomb with a double pitched ceiling and a door painted on the center wall.

2.4 Preservation of the Three Tombs Over the years

These three tombs are on the same rocky cliff, the so-called *Ripa di Tarquinia*, and they were discovered, respectively, in 1867, 1827 and 1873. Their location, given the difficult access, always made them less safeguarded.

If the wall paintings had been detached by the *Istituto Centrale del Restauro*, as it occurred with many others during the mid-decades of the last century, today they would have been better preserved by the devasting action of *tombaroli* (wall painting thieves), that in 60s–70s of the last century definitely damaged this part of the Etruscan heritage. The *Tomba dei Vasi Dipinti* was almost destroyed, with saw marks nowadays still clearly visible on the walls. In 1971, the *Tomba delle Iscrizioni*, already damaged and hard to access due to the steep descent leading to its entrance, was also violated, even if to a less extent than the *Tomba dei Vasi Dipinti*. The *Tomba delle Iscrizioni* was discovered in 1872 by August Kestner, together with the *Barone* and *Bighe* tombs; in 1881 several cement grouting consolidation works were executed, which are still clearly visible. In 1971 the local archaeological authority closed the tomb with a brick wall after the devastation made by *tombaroli*, and reopened it in 1985 with a new security door. The painting detachment, extrema ratio, nowadays is no longer practiced, but it is necessary to preserve and safeguard the paintings by controlling all phenomena related to the stability of the rock-mass, such as the root penetration, the action of insects and animals, the deep plowing and finally, and above all, the geological instability, which made tombs' already precarious conditions worse.

After an overview of the geological and morphological setting of the site, the following sections describe the research carried out for the *Tomba dei Vasi Dipinti*, due to its historical legacy, exposure and vulnerability.

3 Geological and Geomorphological Setting

The Necropolis of Monterozzi is located on the top of a long flat hill, NW–SE elongated, originated as a marine terrace carved in a hard calcarenitic layer, which overlays a clay bedrock. The upper terrace is a monocline, fairly leaning to southwest, composed by alternating thick layers of fossiliferous calcarenite and thinner layers of bioclastic cemented sands (Mid-Upper Pliocene, *Formazione del Macco* Auct., #14 in Fig. 2). The *Formazione del Macco*, up to 40 m thick, overlays, by erosional unconformity, a marine sequence of Pliocene grey clay and sandy silty clay, up to 100 m thick (#10 in Fig. 2). The regional uplift, during Pleistocene volcanic phases, caused the fragmentation of calcarenitic layers along directions longitudinal and transversal to the north-eastern scarp. The three most threatened tombs, investigated in the present paper, are reported in Fig. 3.

Along this scarp many falls and topples of calcareous blocks are due to the fractures caused by the lithological rigid/ductile boundary and by the neotectonic activity. Inside the terrace there are several horizontal discontinuity planes at the bed of calcarenitic layers, where sandy levels are intercalated. These levels are more prone to erosion because of cement's dissolution caused by the abundance of groundwater, that infiltrates via high primary and secondary permeability. Most likely these weakness points and natural caves were used by Etruscan to make easier the tombs excavation (Margottini & Spizzichino, 2021); one of these caves is still present along the cliff, about 5 m below the *Tomba dei Vasi Dipinti* (Fig. 3).

New fractures developed in some tombs during 2017, the drier over the last 40 years (Fig. 4).

4 Topographical Survey

In order to define morphological setting of the investigated area, including the *Tomba dei Vasi Dipinti* hypogeum (located in the upper part of the slope) and a second cave

LEGEND

3) Alluvial deposits (Holocene - Pleistocene)
9) Coastal sand deposits of marine and marginal facies (Lower Pleistocene – Upper Pliocene)
10) Coastal clay deposits of marine facies (Pliocene)
14) Calcarenite and bioclastic limestone (*Formazione del Macco* Auct., Pliocene)
18) Arenaceous and arenaceo-pelitic flysch (Upper Miocene)
20) Marl and pelitic flysch (*Flysch Tolfetano* Auct., Oligocene – Upper Cretaceous)
22) Clay with gypsum (Upper Miocene)
43) Lithic tuffs (Middle Pleistocene)

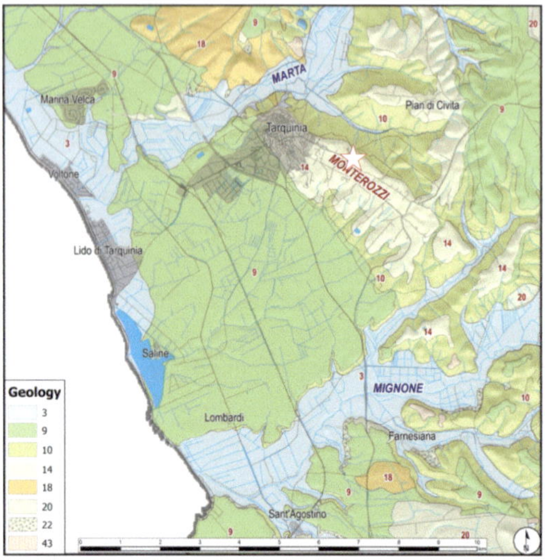

Fig. 2 Geographical location and geological setting

Fig. 3 Location of the three investigated tombs (upper) and geomorphological setting of the *Tomba dei Vasi Dipinti* (lower)

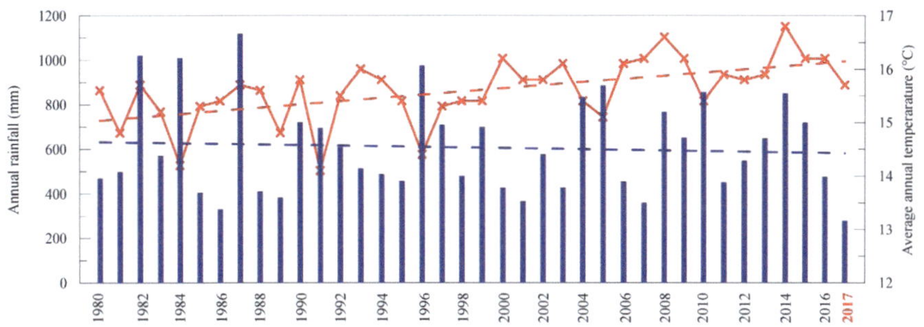

Fig. 4 Annual rainfall and average annual temperature over the last 40 years

(located at a lower elevation close to the tomb), an integration of several survey techniques was carried out, according to the following phases.

First of all, a topographic and geodetic placement was created, by means: (i) the construction of an open polygonal consisting of 6 vertices detected by Total Station mod. Leica TS60, (ii) the survey of the vertices using precision GNSS instrumentation mod. Leica GS08 in NRTK mode, in order to geolocate the measurements performed in the UTM33-WGS84 Coordinate Reference System and (iii) the topographic positioning and survey of 12 Ground Control Points (GCPs) located within the cave, by Total Station (Fig. 5a).

Afterwards, a 3D survey with TLS (Terrestrial Laser Scanner) of the external areas and of the tomb was carried out, using the Laser Scanner mod. Stonex X300 (Fig. 5b).

(a) (b)

Fig. 5 TLS survey inside (**a**) and outside the *Tomba dei Vasi Dipinti* (**b**)

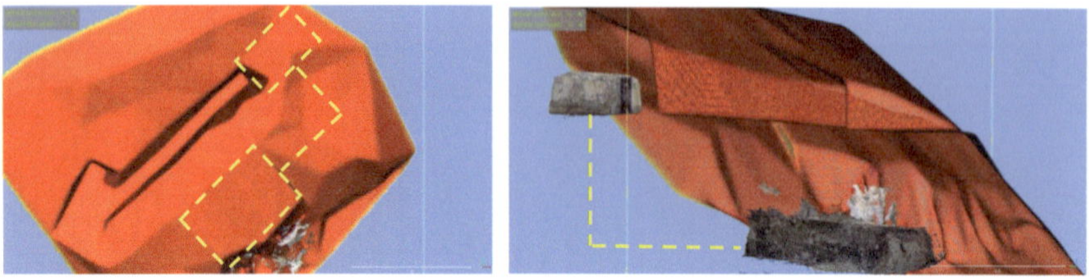

Fig. 6 3D geometric relationships between caves (i.e. the tomb itself and the lower cave) and topographic surface

Lastly, a close-range photogrammetric survey (STM—Structure from Motion) of the *Tomba dei Vasi Dipinti* and of the lower cave was carried out by means of a Fujifilm X-T1 camera, with the aid of the GCPs positioned inside the cave and framed in the new topographic network.

The data processing was performed by means of different software platforms, starting from the analysis and management software of topographic and geodetic networks, passing through laser scanning and photogrammetric processing software, up to the use of open source platforms to integrate spatial data from different software and analyze them jointly to obtain summary plots. In particular, CloudCompare© software was used, in order to define the 3D geometric relationships between caves and surface (DTM) (Fig. 6), and to extract 2D mapping. From the integrated 3D model, consisting of both the DTM and the caves, the cross-sections considered to be the most significant (i.e. those passing through the tomb) were exported in vector format, useful for the subsequent geotechnical modelling.

Additionally, precision frontal orthophotos of the internal walls of the *Tomba dei Vasi Dipinti* were obtained by means of digital photogrammetry procedures, with a resolution of 5 mm, useful to perform degradation analysis with a centimeter detail (Fig. 7).

5 In Situ Monitoring

Following the first surveys (December 2017 to December 2018), a manual deformometric monitoring system was also implemented through the installation of optical crack gauges and manual reading sights with digital calliper (Fig. 8) on the main fractures present in the three tombs.

The main objective of the monitoring system, whose quarterly readings are still in progress, is to verify potential deformation trends and to control their evolution over time for conservation purposes.

After the first zero lecture (1st February 2018), measurements have shown an almost constant trend highlighting the general stability of the complex (tomb plus the rock slope). Only some small oscillations due to thermal and seasonal effects can be appreciated (Fig. 9).

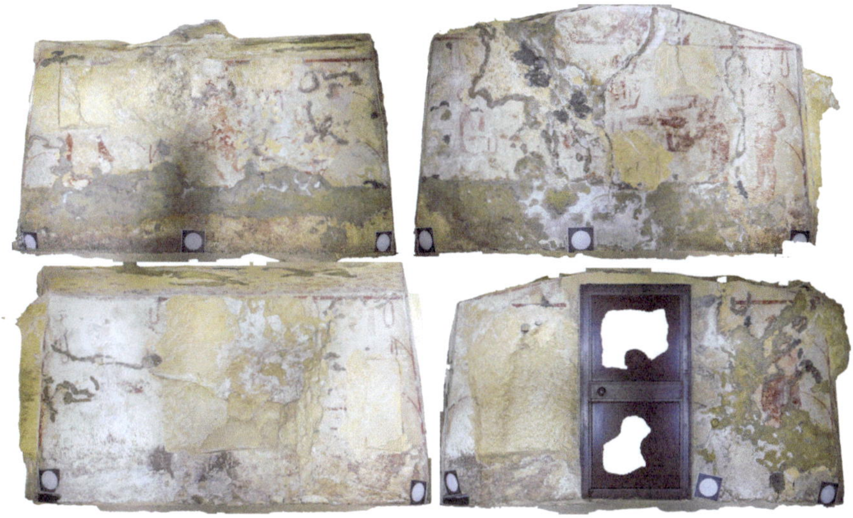

Fig. 7 Digital photogrammetry application: frontal orthophotos of the internal walls of the *Tomba dei Vasi Dipinti*

Fig. 8 Manual deformometric monitoring

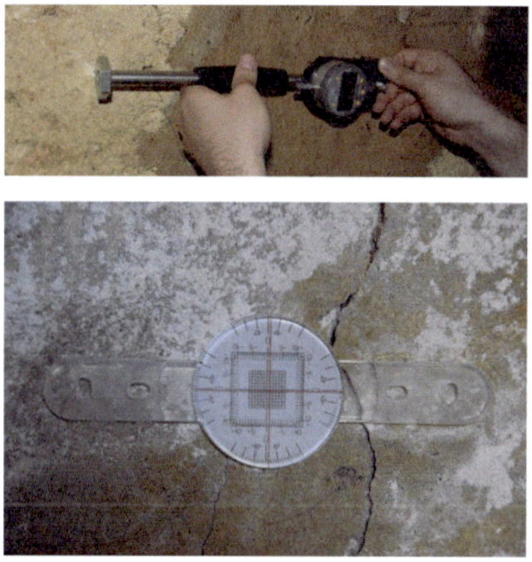

Fig. 9 Deformation measurements inside the *Tomba dei Vasi Dipinti* and *Tomba delle Iscrizioni*

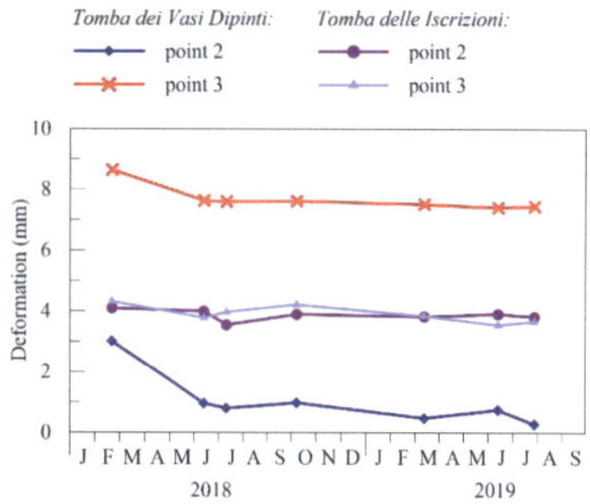

6 Characterisation of the Rock Material and Rock-Mass

The quantitative assessment of the stability conditions of the tomb required the definition of the mechanical properties of the calcarenite rock material and of the entire rock-mass (including the discontinuities).

For the characterisation of the rock material, 8 blocks (Fig. 10) were collected along the plateau of the archaeological site and were brought to the LAGIRN laboratory of the University of Bologna. They were found on the ground surface without the need of any drilling or excavation and are probably the final product of previous instability phenomena (and as such the collected material should be considered as representative of the most altered portion of the Macco formation). More specifically, blocks numbered as 6, 7 and 8 were recovered just outside the *Tomba dei Vasi Dipinti*.

Field investigations were instead carried out to collect information on the rock-mass at the scale of the site. Essentially, they consisted in the survey of the characteristics of the discontinuities along selected scanlines, including their spacing, orientation (measured by a compass) and roughness (by the Barton comb).

6.1 Laboratory Testing

A number of laboratory tests were performed for the evaluation of the physical and mechanical properties of the Macco calcarenite. A total number of 52 cylindrical specimens were prepared and, after the determination of the unit weight of volume and porosity, they were tested to obtain information about the shear (by uniaxial and triaxial tests) and tensile (by Brasilian tests) strength of the material in dry and saturated conditions.

Fig. 10 Example of rock block sampled near the *Tomba dei Vasi Dipinti*

Figures 11 and 12 show respectively the values of the uniaxial compressive strength and the tensile strength of all specimens. Data indicate a marked influence of porosity (the higher the porosity the lower the strength) and saturation conditions, especially for the first strength quantity. In fact, average value of the uniaxial compressive strength is 10.96 MPa in dry conditions and 2.98 MPa in saturated ones. For the tensile strength, the average values are respectively equal to 0.94 MPa and 0.67 MPa.

Limiting the analysis to failure points obtained in uniaxial and triaxial compressive tests performed on dry specimens cored from blocks 6, 7 and 8, the linear Mohr–Coulomb strength envelope provides the following values: cohesion 791 kPa and friction angle 36.6° (see Fig. 13).

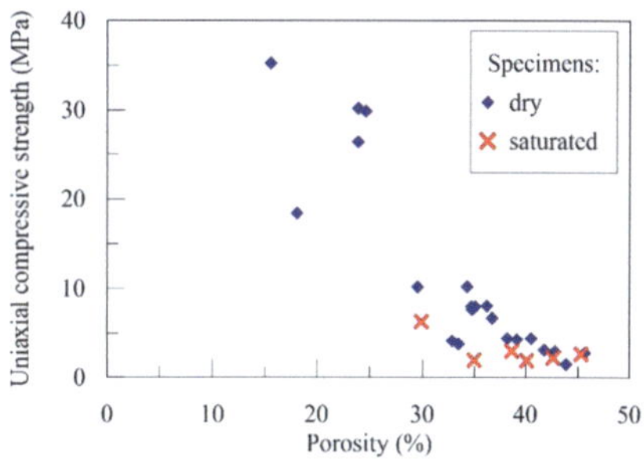

Fig. 11 Values of the uniaxial compressive strength against porosity for dry and saturated specimens

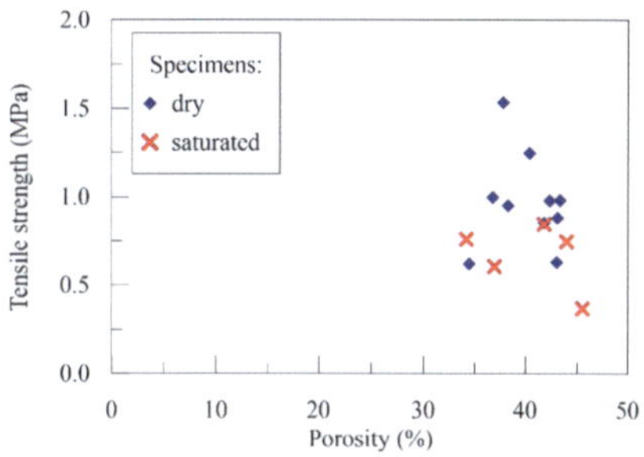

Fig. 12 Values of the tensile strength against porosity for dry and saturated specimens

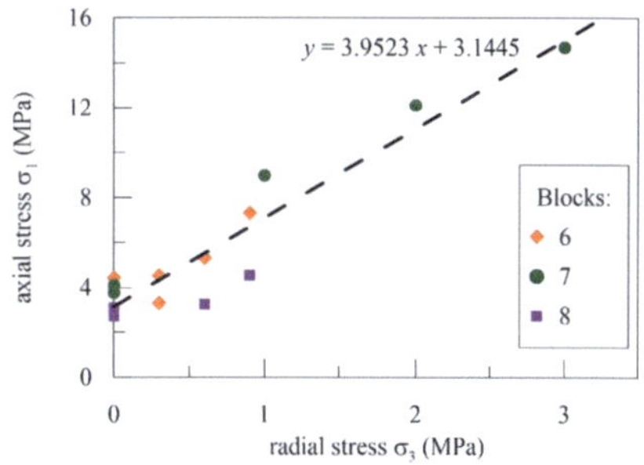

Fig. 13 Failure points of dry specimens cored from blocks 6, 7 and 8 and linear strength envelope

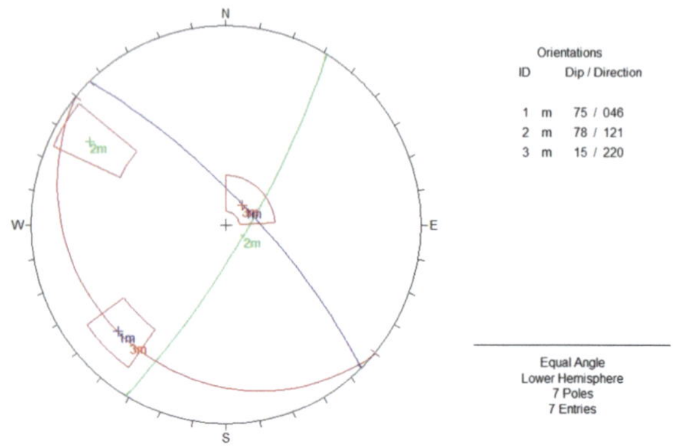

Fig. 14 Orientation of the three main families of discontinuities

6.2 *In Situ Survey*

The in-situ survey allowed to identify three main sets of discontinuities (Fig. 14).

The first one, almost parallel to the slope (orientation 045°/60° in terms of dip direction and dip angles) where the *Tomba dei Vasi Dipinti* is located, has average orientation of 046°/75°, elevated spacing (from 30–40 cm to 1–2 m), extension and persistency, variable roughness (two JRC values of 7 and 18 were measured) and opening (from 3–4 mm to values larger than 10 cm) and practically no filling.

Along these discontinuities most of collapses affecting the slope occur, due to sliding or toppling failure mechanisms. Also, the new detected fracture intersecting the tomb belongs to this set.

The second set is perpendicular to the slope, is characterised by millimetric aperture and limited extension and persistency.

Finally, the third set is given by the layering strata of the Macco formation.

All data were considered for the identification of the GSI index, averagely equal to 64, indicating a rock-mass of overall good quality.

7 Stability Analyses

Given the complex geometry of the *Tomba dei Vasi Dipinti* in relation to the slope, stability conditions were investigated performing a set of numerical analyses with the Finite Element code Plaxis®.

More specifically, only 2D plane strain analyses were carried out at this stage, considering a section perpendicular to the slope strike and passing through the tomb (Fig. 15). As such, this section includes also the fracture intersecting the cave schematised with its real dip (75°).

The model, summarised in Fig. 15, was set up adopting for the rock material an elasto-plastic law with Mohr–Coulomb strength parameters in dry conditions and elastic parameters $E = 1.5$ GPa, $\nu = 0.3$, The fracture was simulated by an interface

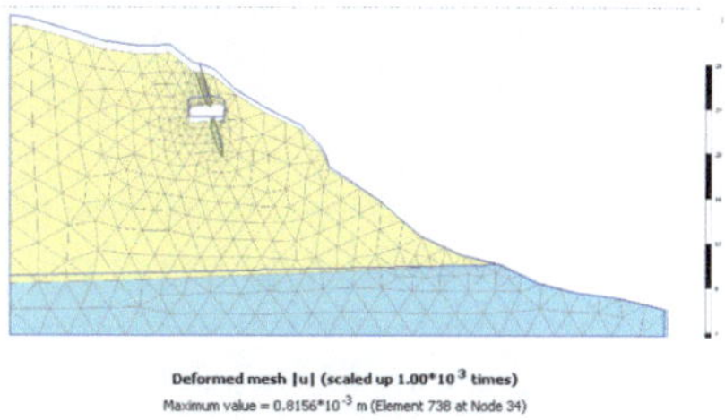

Fig. 15 Finite Element discretisation and deformed mesh at the end of the simulation

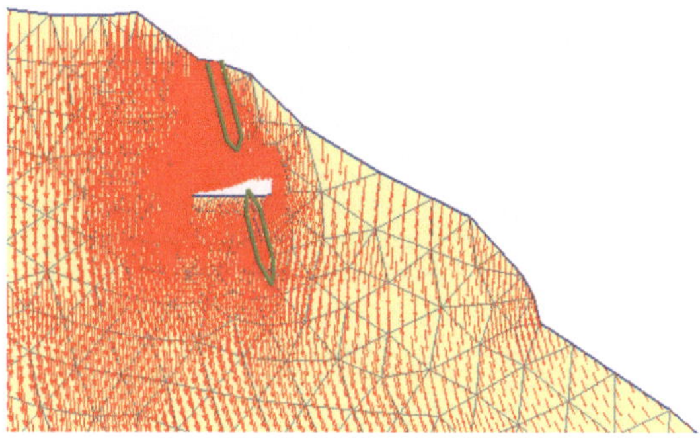

Fig. 16 Distribution of total displacement vectors around the tomb at the end of the simulation

with $c = 0$ and $\varphi = 50°$. For the clayey substratum, in absence of direct investigations, an elastic model was considered with $E = 100$ MPa, $\nu = 0.3$

The distribution of total displacement vectors shown in Fig. 16 was obtained imposing a progressive reduction of cohesion to 1/3 of its original value to simulate, consistently with the experimental evidence (see Fig. 11), the effect of rainfall in saturating the rock material. The numerical results indicate a very limited deformation level and an overall stability of the tomb and the cliff.

8 Conclusive Remarks and Mitigation Strategies

The present study represents a prototype model potentially exportable to other cases from a theoretical and methodological point of view. The physical and mechanical characterization of the Macco calcarenite was carried out for the first time and the resulting strength properties could be adopted in the future both in the archaeological field (excavation campaigns) and for the implementation of specific consolidation works. The study promotes the extensive use of topographical and geomatics survey techniques, even at low cost and of expeditious type.

The manual monitoring data and the numerical simulations indicate the general stability condition of the *Tomba dei Vasi Dipinti* in the current conditions. The main short-term mitigation proposals, besides the monitoring continuation (with less frequency), are: the implementation of water collection and drainage system from the top of the cliff to the foot of the slope; the adoption of techniques for controlling surface erosion (e.g. geotextile or turf) and for protecting the underground structures; the regulation of the tillage in particular on the edge of the plateau. Finally, in the long-term, it would be advisable to promote the consolidation of the external edge of the Macco, through the use of passive dowels hided in the slope with a specific technique already adopted in similar archeological contexts (e.g. Spizzichino et al. 2016; Boldini et al. 2018), and the installation of a deep drainage system at the contact between calcarenite and clays in order to reduce the increase of pore pressure and control water table fluctuations.

References

Adinolfi G, Carmagnola R, Cataldi M (2019) Istantanee dal passato. Pittura etrusca a Tarquinia. Larth, Quaderni dell'Associazione "Amici delle tombe dipinte di Tarquinia"

Boldini D, Guido GL, Margottini C, Spizzichino D (2018) Stability analysis of a large-volume block in the historical rock-cut city of Vardzia (Georgia). Rock Mech Rock Eng 51:341–349

Cecchini A (2017) Le tombe tarquiniesi riprodotte nelle copie Della collezione Morani: conservazione e restauri. L'Etruria di Alessandro Morani, Firenze, pp 179–185

Marzullo M (2016) Spazi Sepolti e Dimensioni Dipinte nelle tombe etrusche di Tarquinia. Tarchna Suppl 7. Ledizioni, MilanoClaudio Margottini, Daniele Spizzichino. Weak rocks in the Mediterranean region and surroundings: threats and mitigation strategies for selected rock-cut heritage sites. Eng Geol 297:2022. 106511. ISSN 0013-7952, https://doi.org/10.1016/j.enggeo.2021.106511

Spizzichino D, Margottini C, Chiessi V, Boldini D (2016) Assessment of the stability conditions of a large-volume sandstone block in the northern sector of the Siq of Petra. In: landslides and engineered slopes. Experience, theory and practice. Proceedings of the 12th International Symposium on Landslides (Napoli, Italy, 12–19 June 2016). CRC Press, pp 1851–1858

Geophysics

Joint Application of Multiple-Geophysical Surveys for Archaeological Prospection

Amin Ibrahim, Khaled S. Gemail, Mahmoud M. Sensoy, Ali El-Khadragy, and Moamen Almaz

Abstract Nowadays, the integration of Electrical Resistivity Tomography (ERT) and magnetic methods has been widely used in archaeological prospections for producing high resolution images of multidimensional targets. This integration was applied for the first time in Tell Dibgou, Northeastern Nile Delta (Egypt) to image the Islamic architectures which has been proved by the findings of mud-bricks, pottery and glass in the highly landscapes. The ERT and magnetic surveys were conducted in three integrated steps. At first, a Wenner Beta (WB) resistivity profile was measured perpendicular to the axis of the hill. Then, the Total Magnetic Field (TMF) survey was carried out on a grid covered the highest part of the area. Finally, the expected targets in the form of walls and closed rooms were imaged by high resolution 2D and 3D ERT surveys. The correlation between ERT (2D/3D) and magnetic results was satisfactory for imaging the unseen archaeological structures as indicated from the excavations during and after the geophysical surveys. Therefore, the joint application of ERT and magnetic techniques for near surface prospecting represents a very useful tool for multidimensional archaeological investigations and can provide a quantitative contribution to describe the spatial distribution of unseen objects. The interpretation of the 2D and 3D resistivity imaging provided information on existence of linear resistivity anomalies corresponding to buried walls and some small archaeological remains detached from the main walls. Additionally, the Tilt Angle and Euler Deconvolution techniques have been used to the magnetic data in a comparative manner with ERT results to determine the positions and depths of buried structures.

A. Ibrahim · K. S. Gemail (✉)
Environmental Geophysics Lab (ZEGL), Faculty of Science, Geology Department, Zagazig University, Zagazig 44519, Egypt
e-mail: khaledgemail@zu.edu.eg

M. M. Sensoy
Faculty of Science, Geology Department, Assiut University, Assiut, Egypt

A. El-Khadragy · M. Almaz
Faulty of Science, Geology Department, Zagazig University, Zagazig, Egypt

M. Almaz
Department of Geosciences, University of Arkansas, Fayetteville, AR 72701, USA

G. M. El-Qady and C. Margottini (eds.), *Sustainable Conservation of UNESCO and Other Heritage Sites Through Proactive Geosciences*, Springer Geology,
https://doi.org/10.1007/978-3-031-13810-2_10

The results constitute an encouraging approach using joint application of ERT and magnetic methods for further archaeological prospection in other parts of Tell Dibgou or elsewhere.

Keywords Electrical resistivity tomography (ERT) · Total magnetic field (TMF) · Multidimensional structures · 3D resistivity imaging · Archaeological investigations · Tell Dibgou

1 Introduction

Imaging of archaeological ruins of multilayered archaeological sites requires a detailed knowledge about the nature of unseen structures and their subsurface boundaries. The site of multilayered occupation periods is a subcircular, nucleated settlement composed of mud soil and stone-based or bricks structures remaining from different archaeological era, and therefore it presents a complex and challenging archaeological setting (Berge and Drahor 2011). In such sites, the integration surveys among geophysical methods is very helpful in defining zones of interest and providing detailed information for future excavations. Therefore, performing this investigation requires an understanding of the utilized geophysical methods and the physical properties of the buried archaeological structures (Schmidt 2009). Tell Dibgou site (Fig. 1) in the northeastern Nile Delta (Egypt) is a good example of multilayered archaeological settlements which include the Saito-Persian period in the lowlands, and the Ptolemaic, Roman, Byzantine periods caped by fired mud bricks constructions from the Muslim Era in highly terrains (Brissaud and Desbordes 2014). In such cases, the combined use of multiple geophysical methods and field techniques is a powerful approach for detecting the vertical and horizontal boundaries of the buried structures (Salvatore et al. 2019; Ibrahim et al. 2021). In addition, the prospecting of near-surface archaeological remains by single geophysical method can be difficult because of low signal to noise ratios (S/N) and the impact of near surface heterogeneities. In the last few years, the application of multiple geophysical techniques in archaeological investigations has acquired an increasing interest, because of their particularity of being non-destructive and time-consuming methods (Patella and Hesse 1999; Simms and Albertson 2000; Bevan and Roosevelt 2003). Therefore, the understanding of the site- specific conditions and the physical property contrasts between the buried features and the hosting soil is a critical aspect in choosing the suitable geophysical method. Practically, the combination between different geophysical surveys can offer detailed information on the dimensions, depth and nature of the buried archaeological targets where each geophysical survey responds to different physical property (Cardarelli et al. 2008; Arato et al. 2015; Gaber et al. 2021).

Nowadays, DC resistivity and magnetic methods are widely applied for archaeological aims, where the DC resistivity (2D/3D surveys) can be employed to describe the shape, nature, and the depth of near surface structures (Papadopoulos et al. 2006; Drahor et al. 2008; Cardarelli and Di Filippo 2009; Berge and Drahor 2011;

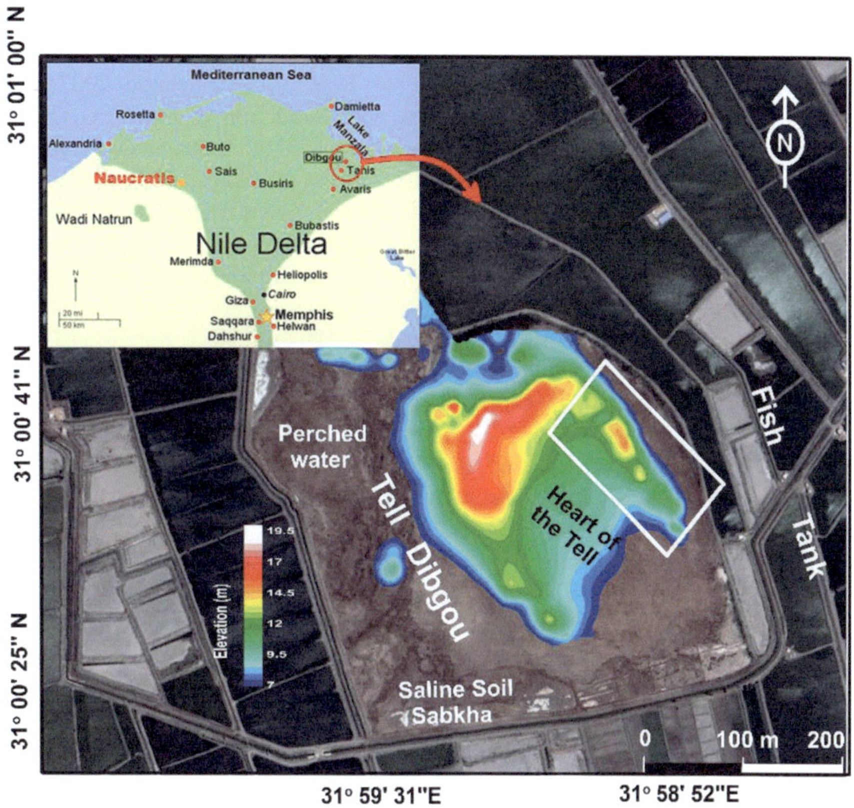

Fig. 1 Google earth image shows the location of Tell Dibgou in the northeastern Nile Delta, superimposed by color image of topography. The surveyed area is highlighted by white rectangle in the highly terrainous eastern side of the Tell

Papadopoulos and Sarris 2011; Tsokas et al. 2013; Papadopoulos et al. 2014; Gemail 2015; Bernardes et al. 2017; Gemail et al. 2017a; Deiana et al. 2018; Ortega-Ramírez et al. 2020; Gaber et al. 2021), while the magnetic method is applied to detect the buried objects characterized by a magnetic susceptibility contrast in respect of the surrounding ground or with a remnant magnetization (Chianese et al. 2004). The use of electrical methods in archaeology predates the use of magnetic method. However, the magnetic method is the most frequently used tool in archaeological prospection because of the speed and the high resolution in mapping large areas.

Magnetic survey is widely used in archaeological prospecting for detecting structures such as pottery, fired brick walls, fire pits, pathways, tombs, and numerous archaeological structures. These features are detected and mapped due to having more magnetic properties than the hosting materials (Patella 1991; Linford and Canti 2001; Gaffney et al. 2004; Geoffrey 2007). The magnetic survey measure the variation in the earth magnetic field including the effect of the buried features with high magnetic properties. The basics of magnetic surveys and interpretation of the measurements, have been clearly described by several authors (e.g., Weymouth and Huggins 1985; Blakely and Simpson 1986; Blakely 1995; Schmidt 2007; Fedi et al. 2017; Abdallatif et al. 2019; Colombero et al. 2020; Eppelbaum 2021). One of the significant advantages in magnetic data is the fact that the magnetic anomalies contain edge information of buried causative sources (Karaaslan 2020) where, the edge detection plays an essential role in interpreting the magnetic data to delineate the archaeological boundaries and structures. Therefore, several edge detection techniques from magnetic data have been extensively applied to describe the edge and lineament of buried archaeological structures (Cheyney et al. 2011; Sertcelik, and Kafadar 2012; Yuan et al. 2014; Zuo and Hu 2015; Wu et al. 2017; Karaaslan 2020).

In recent years, the development in the automated data acquisition systems and inversion modeling techniques of ERT has greatly expanded the practical applicability of resistivity imaging in archaeological prospection. Consequently, the application of 2D resistivity tomography, then a 3D resistivity survey is able to outline the accurate dimension of the archaeological structures The representation of the 3D inversion processes (depth and horizontal slices) enhances the detection and definition of the buried targets where the presence of an anomalous volume caused by the target could be highlighted (Arato et al. 2015; Ibrahim e al. 2021; Gaber et al. 2021).

In the present work, 2D/3D electrical resistivity tomography (ERT) and magnetic surveys were used in combination in the eastern border of Tell Dibgou in the northeastern Nile Delta (Egypt), to identify the buried archaeological structures and to report the efficiency of each survey according to the nature of the archaeological resource and the local environment. This work attempts to image the top most archaeological layer dated back to the last centuries of Islamic Era.

2 Geological and Archaeological Setting

Tell Dibgou has a remarkable landscape, with low land at the heart. An elevated rim of approximately 20 m forms an enclosure that opens to the southeast (Fig. 2). From the geological point of view, Tell Dibgou is occupied by Quaternary deltaic sediments (Abu Al Izz 1977) and is classified as a wet Sabkha (Elwan et al. 1983). Additionally, the deltaic clayey sediments dominating the area are strongly affected by salinization and salt efflorescence in the form of thin crusts and rather thin pans of nearly 5 cm thick (Fig. 2a). The distribution of salts and gypsum deposits (Sabkha soils) are controlled by the moisture regime due to the oscillation of shallow water table in the site (Fig. 2b). This geological information is worth mentioning here because it has a strong relation to the geoelectric resistivity characteristics of the area.

Nile Delta contains a lot of ancient archaeological settlements specifically in the northeastern part. In the archaeological point of view, Sharkia Governorate, Egypt is one of the most important regions. Because of its geographical location; Sharkia was Egypt's eastern gate and first defense line throughout the millennia. The area contains

Fig. 2 3D topographic map of Tell Dibgou with the probable enclosure of the heart of Tell zooming on the eastern ramp of the present surveyed area. The photos show the main feature in the area including, **a** Sabkha with dark brown color, thin crust and efflorescence of salts, **b** the concentration of Sabkha in the low land at the vicinity of fish farms, **c** Water tank excavated on the mound attributable to the Byzantine period built of baked bricks, **d** scattered fired brick and slag debris covered the surveyed area, **e** Double walls joined to each other in the eastern side of the Tell

important archeological sites such as Tell El-Dab'a (Avaris), Qantir (Piramesse), Tell San El-Hagar (Tannis) and Tell Dibgou which represent the most important periods of civilization.

The documentation for Dibgou is very rare, only few Arabic sources discussed the earlier history of the city, in the last centuries of its existence and the terminal phase of its development. Dibgou is considered as multi-layered site where the city was established and grew continuously throughout the Greek manner periods, Roman and Byzantine as indicated by the presence of water tanks (Fig. 2c). The city was flourished at a time when the neighboring town of Tanis lived its last hours therefore, it probably has many architectural elements taken from the ruins of Tanis ancient capital to cover its material requirements (Brissaud and Desbordes 2014). In addition, examination of the ceramic fragments, by the archaeologists, indicated that the site was regularly occupied during the Third Intermediate Period, Saito-Persian period, the Ptolemaic, Roman and Byzantine history of the city ended in the early centuries of the Muslim Era.

Tell Dibgou was considered a production center for precious fabrics and Luxurious textiles of linen and silk, which was the core of activity and inserted in a large urban network until the end of the twelfth century AD. Several Arabic manuscripts mentioned that Dibgou was famous for manufacturing very fine fabrics from linen and silk named “Dabiqi”, derived from the name of the city (Brissaud and Desbordes 2014). It is therefore likely that the site still contains the remains used for textile production (workshops of weaving and dyeing) and trade in these tissues (shops and warehouses), famous throughout the Arab Middle East.

3 Site Description and Field Surveys

Unfortunately, the ground conditions of Tell Dibgou are typically uneven and covered by a dense scatter of fired brick and slag debris (Fig. 2d). Before 2014, there was no any planned archaeological excavation carried out at the ancient settlement of Tell Dibgou prior to conducting geophysical survey except some random digging by non-specialists and/or robber actions. Accordingly, it is required to find some means other than excavation for assessing the buried remains. Based on the field observations and the nature of the minor excavated parts, DC resistivity and magnetic surveys were carried out for such archaeo-geophysical setting through gradual steps. The geophysical field work was executed from December 2013 to February 2015.

According to the soil conditions, field observations and the analysis of the topographic terrains of Tell Dibgou, a site of particular interest in the eastern margin was chosen to carry out the geophysical surveys (Fig. 3). The choice of the eastern mound based on the recommendation of the archaeologists of Mission of Archaeology of Tell Dibgou (MATD) who worked in this area during the geophysical field survey and their plan of excavations in the forthcoming season.

The selected site is a small mound of an area ~300 m^2 and located at the elevated rim of the eastern enclosure which surrounds the heart of the Tell. This site is characterized by a presence of a minor excavation (robber actions), and showing a part of fired brick wall dating from Islamic periods (Fig. 2e).

The geophysical survey was organized and scheduled in the following steps. Firstly, electrical resistivity prospecting was carried out through measuring two ERT profiles of Wenner Beta (WB) and Dipole–Dipole (DD) arrays with length 78 and 30 m, respectively (Fig. 3). In practice, WB is commonly used to image the different archeological prospection (e.g., Papadopoulos et al. 2006). It is characterized by a high signal to noise ratio (Dahlin and Zhou 2004; Gemail 2017b and Abu Salem et al. 2021). Because WB is a special case of DD, it is anticipated to detect the vertical fired bricks walls within the investigated site. In addition, WB provides a good vertical resolution, which means that the depth to the base of the archaeological layer can be well detected. Because the maximum information is extracted when survey lines are oriented perpendicular to the strike of geological or man-made linear features, the WB profile was built to have the direction of 80° NE perpendicularly to the principal direction of the neighboring excavated walls (Fig. 3). The profile setup

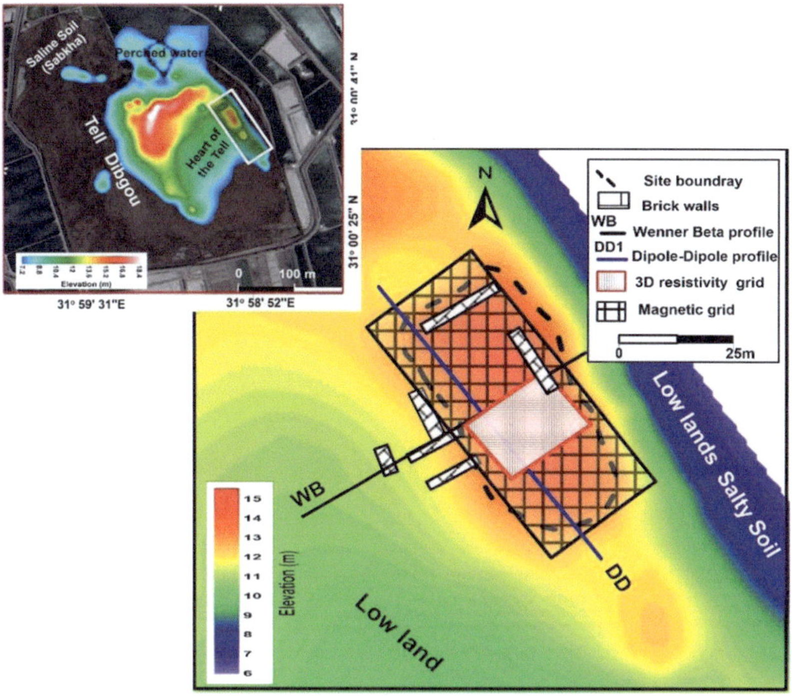

Fig. 3 Location map of the geophysical survey layout conducted at the eastern boundary of Tell Dibgou site. The brick walls are the man-made structures extending from neighbor excavation area into the surveyed site (not to scale)

involved 40 electrodes with minimum and maximum spacing (*a*) 2 m and 22 m, respectively. Accordingly, the maximum data level (*n*) and subsequently the total number of measurements along this profile are 11 and 242, respectively. The electrode number 15 was located directly above the exposed wall (Fig. 3). Further, a profile of DD was set up perpendicular to the direction of the neighboring WB profile over an interesting anomalous zone (Fig. 3). The potential and current dipole length was fixed to 1 m and the inter-dipole separation varies from 1 to 7 m, (i.e., 7 data level, *n*). A total number of 346 data points of apparent resistivity were measured using n = 1 to n = 7.

In the second step, a magnetic survey in the form of total magnetic intensity (TMI) was performed to cover the whole selected site. The magnetic profiles were 1 m apart and the measurement step was about 0.5 m, i.e., slightly less than the sensor height. The spatial resolution of the sampling was sufficient for the detection of the archaeological structures down to a size of 1–0.5 m (Fig. 4a). As illustrated in Fig. 4b, the measurements were started from the bottom right-hand corner of the site with zigzag traverse pattern (bidirectional survey). In order to minimize as far as possible one source of noise represented by the top soil, the measurements were carried out with the sensor detached from the console and connected to 1.8 m

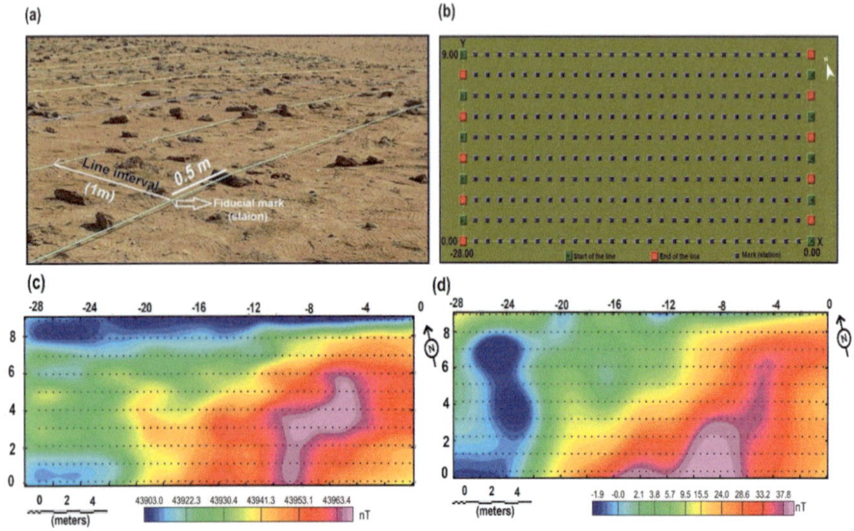

Fig. 4 The total field magnetic datasets and field setup **a** planning of the surveyed site for magnetic measurements using 1 m profile distance and 0.5 m station interval **b** Data after transferring from the G-856AX console to the computer through MagMap2000 software, **c** Total field magnetic raw data, and **d** magnetic data after applying correction and noise elimination

staff. To ensure valid readings of the earth's magnetic field, several readings were taken in succession almost at every station without changing any position. Valid registered readings were only considered when the successive values were within the difference of 1nT. Magnetic data were acquired with a Geometrics model G-856AX Memory-Mag Proton Precession Magnetometer with precision 0.1nT.

A base station was established, at an easily accessible position of approximately 10 m to the north of the investigated site, to monitor the temporal variation in the earth's magnetic field. The few preliminary magnetic measurements near the base station showed that there were no large magnetic variations. Also, it was reoccupied at the beginning and end of every day, as well as every 10–15 min interval to calculate and detect any diurnal variations. The intensity of the total magnetic field for the base station was ranged between 43,916.8 and 43,959.8 nT during the magnetic survey of this site. In order to reduce the remaining effect of the temporal variation, three tie lines were carried out crossing the survey lines.

The processing of magnetic measurements was initiated by cleaning the data from the abnormal high and/or low values (spikes). Accordingly, a pre-processing step of despiking involved removing the drop outs (zero readings) and non-zero anomalous readings of a short duration. Also, the contribution of the main magnetic field was removed from the magnetic measurements by carrying out IGRF correction. Then, the magnetic data was subjected to sequential processing steps to remove the systematic error and noise which make the interpretation more difficult. The sources of systematic error include diurnal variation of the earth's magnetic field, traverse stripping and staggering which were removed by diurnal, heading (zero mean traverse) and de-staggering corrections, respectively. The remaining non-coherent noise was eliminated by using conventional leveling (tie and regular survey lines). Finally, a low-pass Butterworth filter was used to remove the high wavenumber components with large amplitude (attributed to surface noise) before applying residualization and derivative-based approaches. Figure 4c, d displays the raw and processed magnetic maps, respectively.

Finally, a detailed 3D electrical resistivity tomography was carried out at an area selected from the results of magnetic and 2D resistivity measurements (Fig. 3). In this survey, the pole-pole (PP) array was employed to cover a grid of 7 × 7 with 1 m spacing. The implemented electrode spacing was chosen based on the expected depth that was deduced from the result of 2D resistivity imaging and the priori archaeological information. In general, PP array is popular for 3D resistivity survey where it provides a greater penetration depth and simplified data interpretation as well as faster field operations (Loke and Barker 1996a; Gemail et al. 2004; Nyquist and Roth 2005; Nemtsova et al. 2019). Therefore, the inverted 3D boundaries of local archaeological structures will make the pole-pole possible to determine the area of interest for further detailed excavations (Nemtsova et al. 2019). Furthermore, PP array is extremely sensitive to vertical structures and can be used to locate anomalies in two dimensions and to estimate the size and shape of buried structures. Due to the reciprocity, it is only necessary to measure the potential at the electrodes with higher index number than the current electrode (Loke and Barker 1996b; Gemail et al. 2004).

For a grid of 7 by 7 electrodes, a survey of measuring the complete data set (Fig. 5a) would have 1176 data points. To reduce this large number of measurements without seriously degrading the quality of the model obtained, "crossed-diagonal survey" method (Fig. 5b) was applied (Loke and Barker 1996a, b). This measurement scheme is an effective and speed subset of the exhaustive set of PP array (Nyquist and Roth 2005). In which, the total number of data points, in the arrangement of a grid 7 by 7 with 1 m spacing (Fig. 5c), was reduced to 476 which is about one-third of that required by a complete data set survey.

The measured 2D and 3D ERT data sets were inverted using Res2Dinv and Res3Dinv programs, respectively to construct 2D/3D images of the expect walls based on the true resistivity distributions. The inversion algorithms in both the programs have been developed by Loke and Braker (1995, 1996a, b) based on smoothness-constrained Gauss–Newton least-squares technique to create a 2D/3D model of the buried structures from the observed resistivity data sets. During the inversion procedures, both default (conventional) smoothness-constrained and blocky (robust) inversion were employed (Fig. 6). The default inversion produced a model of gradational and smooth variation in the resistivity values and tends to smear out the archaeological boundaries (Fig. 6a). Hence, the interpreted resistivity distribution is blurred and does not reflect the real shape of the anomalous body. Contrarily, the model produced by the robust inversion has sharper and straighter boundaries and the distribution of the isolines is closer to the real shape (Fig. 6b). A cut-off

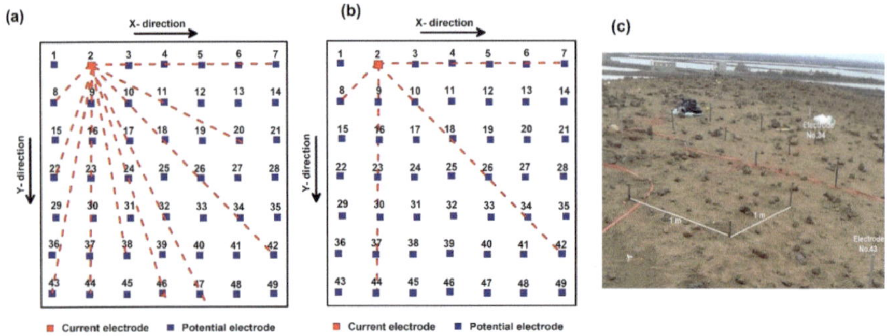

Fig. 5 Arrangement of potential and current electrodes in the 3D resistivity survey using 7 × 7 grid **a** the complete data set survey, and **b** crossed-diagonal data set survey, and c) electrode setup during field survey

factor of 0.05 was employed in the robust inversion to greatly reduce the effect of data points with a misfit between the measured and calculated apparent resistivity values of more than 5%. In addition, the preliminary excavations, topography, and field observations were included to constrain the inversion process.

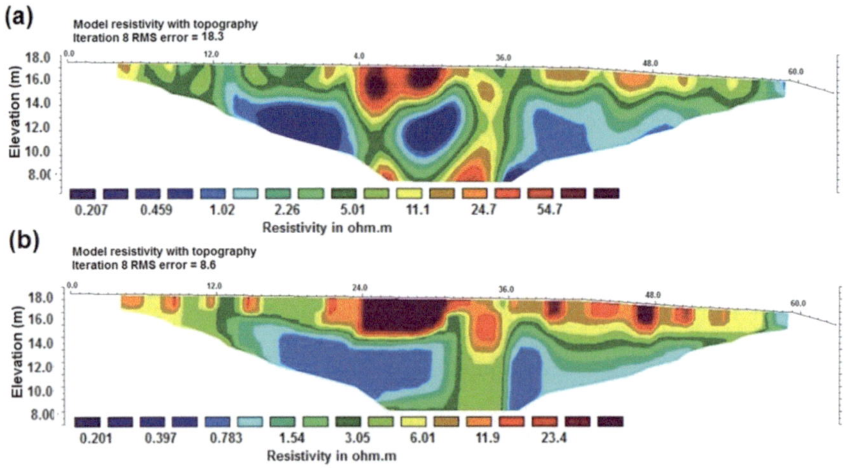

Fig. 6 Example of inversion results produced by RES2DINV software using: **a** the conventional least-squares smoothness-constrained method, and **b** the robust inversion method

4 Results and Discussion

4.1 2D ERT Imaging of the Archaeological Layer

The inverted ERT section shows the true resistivity distributions along the WB profile. The tentative interpretation of the constructed model is shown in Fig. 7. The inspection of the inverted WB section indicates a double layer model with a sharp boundary separating between the Islamic settlements in upper resistive zone where the fired bricks walls are dominant and the lower conductive zone of salty clay at a depth of approximately 2.75 m as marked by white dashed line in Fig. 7. According to resistivity distributions, it is possible to interpret high resistivity anomalies with sharp vertical boundaries in the upper zone as fired bricks walls with different orientations. In areas of compacted materials, the pore volume decreases which in turn decreases the amount of electrolytic conductivity within the pore space, resulting in an increase in the resistivity value. It can be noticed that a general decrease in the resistivity values with depth which is attributed to an increase in the moisture content of the deep salty soil materials because of drainage processes from the closed fishing tanks in the lowlands.

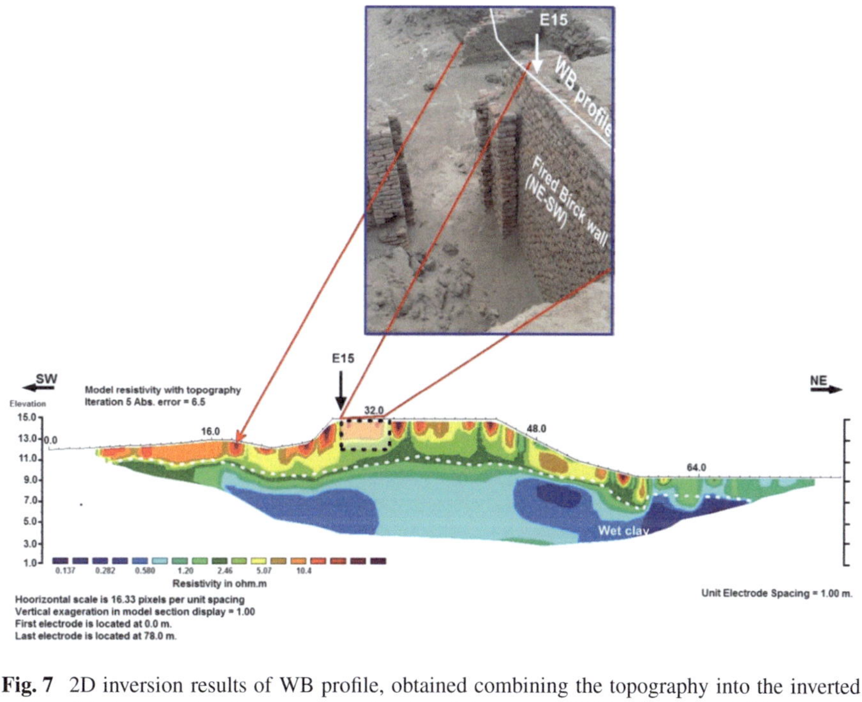

Fig. 7 2D inversion results of WB profile, obtained combining the topography into the inverted section. Black rectangle refers to excavated part carried out by MATD in September 2014 and white dashed line represents the boundary between the archaeological zone and the underneath saline clay sediments

From the field observation, the electrode (15) was located directly above a partially exposed wall unearthed by weathering and random digging (Fig. 7). A systematic excavation was carried out in September 2014 at the investigated site after measuring the WB profile (Fig. 7). The excavation depicts an ancient room with walls constructed from fired mud bricks. The length, thickness and height of the wall are 3 m, 0.55 m and 2.75 m, respectively. There is a good match between the interpreted walls, denoted by the black dashed box, and the results of excavations (Fig. 7). Most of the resistive anomalies are concentrated in the upper layer.

The inversion result of WB profile shows that the RMS% is relatively small (6.5%) and therefore the results can be reliable. This confirms that the WB configuration has a good signal-to-noise ratio for highly resistive topsoil conditions (Drahor et al. 2008). Therefore, the obtained results can be used to highlight the applicability of ERT surveys using Wenner array for a rough assessment of the variation of the archaeological features with depth. Nevertheless, the relatively utilized large spacing (2 m) has a negative impact on the spatial resolution of thinner walls. In addition, the broader Wenner spacing produced relatively smooth anomalies and could not provide accurate estimation for the dimensions of the interpreted walls. To overcome this problem, the DD profile was carried out with 1 m electrode spacing to image the closed fired brick walls in perpendicular direction with WB profile. It can be noticed that the dipole–dipole array gives high resolution and best image where the DD section of Fig. 8 displays a clear picture for three walls to maximum depth of 2.5 m with high resolution. This advantage of DD array could be noticed by mapping the vertical walls as sharp resistive blocks (7–15 Ω.m) due to its high sensitivity where DD to horizontal changes in the resistivity within the upper archaeological layer. Therefore, the comparison between the DD and WB sections (Figs. 7 and 8) gives a good idea about the diminution and extension of fired brick walls in the low resistivity soil background. Although DD produces sharp boundaries between archaeological structures and background deposits, nonetheless a shadow zone of high resistivity appeared below the targets (Fig. 8), which shows that the deep resolution of DD is not as good as WB array (Gemail et al. 2017b) due to the DD array having high noise level at deeper data levels.

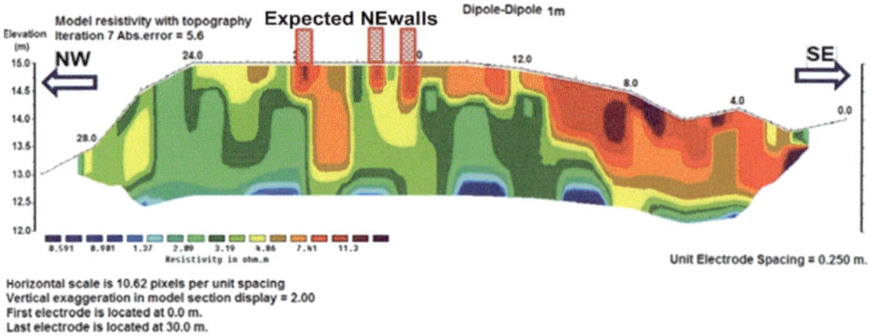

Fig. 8 Inverted section and possible interpretation of profile DD using 1 m electrode spacing

4.2 Archaeological Interpretation of Magnetic Data

4.2.1 Total Magnetic Field Map

A 2D plane view of the total magnetic intensity map was prepared by incorporating the surface elevation to the magnetic data (Fig. 9). From this presentation it can be noticed that the highest magnetic intensity values are located at a topographically high land (13–15 m) where the fired-mud bricks of Islamic architectures are covered the surveyed area. Beyond this distance, local positive and several negative anomalies (green to blue color) of limited dimensions and random distribution is observed which could be related to the scattered fragments of fired bricks.

The visual inspection of the final total magnetic intensity map (Fig. 9a) reveals numerous anomalies with a maximum range of variation fall within approximately 57 nT (from −6.86 to 50.03 nT). Positive anomalies (A, B and C; Fig. 9a) are marked by pink color and characterized by high magnetic intensity value, reaching 50 nT, and could be related to magnetic bodies of shallow depth. On the other hand, the magnetic

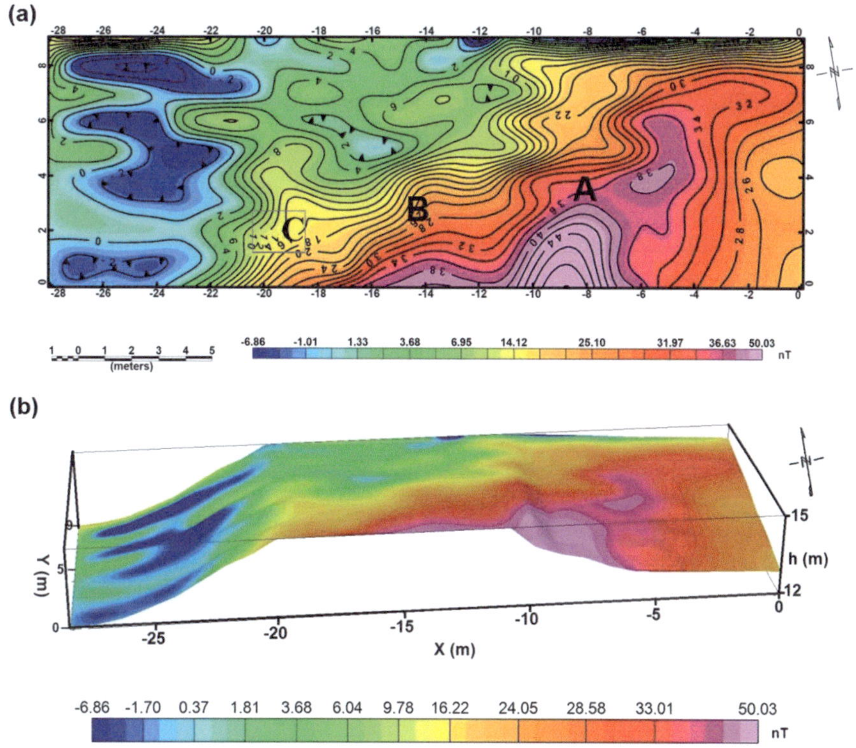

Fig. 9 The corrected magnetic map, **a** Total magnetic field intensity contour map, characters A through C refer to magnetic anomalies, **b** 2D plane view of magnetic image after incorporating surface elevation to the magnetic data

anomaly of uniform amplitude and falls in a medium range of magnetic field value (25 nT) might be interpreted as archaeological wall of greater depth. The map also shows linear magnetic anomaly of high gradient occupying the top right portion of the map and extending E-W (Fig. 9a). It seems to be associated with the presence of contact or boundary between baked brick wall and clay sediments. A 2D plane view of the total magnetic intensity map was prepared by incorporating the surface elevation to the magnetic data (Fig. 9b). It can be noticed that the highest magnetic intensity values are located at a topographically high land (13–15 m) where the fired-mud bricks of Islamic architectures are covered the surveyed area. Conversely, local positive and several negative anomalies (green to blue color) of limited dimensions and random distribution are observed in the low elevation area (Fig. 9b).

4.2.2 Isolation and Enhancement of Magnetic Anomalies

In the present work, wavelength filter (high-pass), continuation filter (downward continuation), and horizontal and vertical derivatives, were applied to isolate and enhance achaeo-magnetic anomalies. Additionally, the tilt angle derivative (TAD) and Euler deconvolution (ED) techniques were employed to delineate the edges of buried walls and estimate the depth to the archaeological structures.

High-pass filter

The high-pass filter was applied with wavelength cut off 4 m. The high pass filtered map (Fig. 10a) shows the distribution of magnetic field after removing the regional effect. It characterized by dominance of various alternating local positive and negative anomalies of different sizes, shapes, and orientations. The anomalies of relatively high amplitude and sharp gradients (A, B, and C; Fig. 10a) reflect different causative magnetic sources of shallow depth seated origin.

The high-pass filter was applied with a cut off wavelength of 4 m to highlight the local magnetic anomalies and suppress the regional magnetic field. The 2D plan view of the high-pass filtered map (Fig. 10a) shows a large number of elongated positive and semicircular negative anomalies distributed throughout the investigated area with amplitude range from −13.64 to 13.2 nT. The anomalies of relatively high amplitude and sharp gradients reflect magnetic sources of shallow depth seated origin (less than 1 m). Rectilinear positive anomalies of trends NNE-SSW (A, B, and C; Fig. 10a) resulted from the extension of partially excavated baked brick walls of ancient chambers. In addition, semicircular negative anomalies were interpreted as clay sediments filled the ancient chambers which have relatively low magnetization with respect to the circumference of fired brick walls.

Derivative techniques

The main advantage of the derivative-based techniques is the ability to resolve composite anomalies into their individual constituents and removing the regional magnetic gradient for better defining shallower anomalies (Nabighian 1984; Salem et al. 2007; Ekinci 2016).

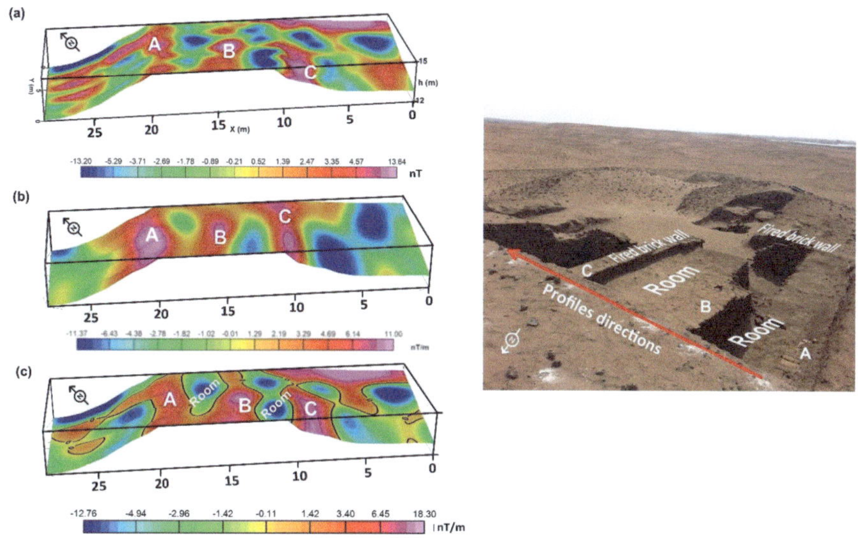

Fig. 10 Anomaly isolation and enhancement using **a** high pass filter, **b** horizontal derivative and **c** second vertical derivative. The photo shows the first stage of excavations in September 2014. The viewpoint from the northeastern view

The Horizontal Derivative (HD) of the magnetic data in X direction was calculated as shown in Fig. 9b to enhance the short wavelength component and suppress the impact of the deeper geomagnetic sources, which are normally not prospective archaeological targets. Some discrete and local linear anomalies, such as those located in central part of the map, were clearly outlined. Most of these emphasized anomalies have a trend of NNE-SSW direction. The well-defined walls in Fig. 10b indicate that the horizontal derivative (HD) can detect the shallow archaeological ruins and clearly identify the edges of the baked brick walls. Consequently, the features striking mostly perpendicular to the direction of the applied derivative (X-direction) are appeared with sharp boundaries, while the features in the other directions almost vanished.

The Vertical Derivative (VD) aims to increase the resolving power of the shallow features at the expense of the broad or regional magnetic effect. Thus, it gives a sharper image than the map of the total field intensity. Figure 10c displays the vertical derivative of the total magnetic field. By comparing the vertical and horizontal derivative maps (Fig. 10b, c, respectively), it can be noticed that the derivative in Z direction gives a more detailed picture about the geometrical shape of the buried construction. The linear positive anomalies of fired brick walls and the circular negative anomalies of ancient rooms are more readily apparent. The localized anomalies are recognized in various places specifically at the central high land part. Moreover, the zero second vertical derivative contours (Fig. 10c) approximate the plan view of the outline margins of buried fired brick walls of ancient chambers (Vacquier et al. 1951;

Hinze et al. 2013; Arsoy 2014; Mehanee et al. 2021). However, the second derivative of magnetic anomalies is typically useful only to increase the perceptibility of residual anomalies but limited in quantifying the source of the anomaly.

Downward continuation

Downward continuation is utilized to accentuate features at a particular depth, below the acquisition level (Cooper 2004). The crude depth to the buried fired brick walls is achieved by successively downward-continuing the observed magnetic anomaly at depth levels 0.1, 0.2, 0.3, 0.4 and 0.5 m which correspond to the expected depths of the buried archeological features. Figure 11 shows 3D visualization of the successively downward-continuing magnetic anomalies.

At depth level of 0.1 m, the continued field (Fig. 11) shows interference between the adjacent anomalies which have a gradational and smooth variation (the magnetic

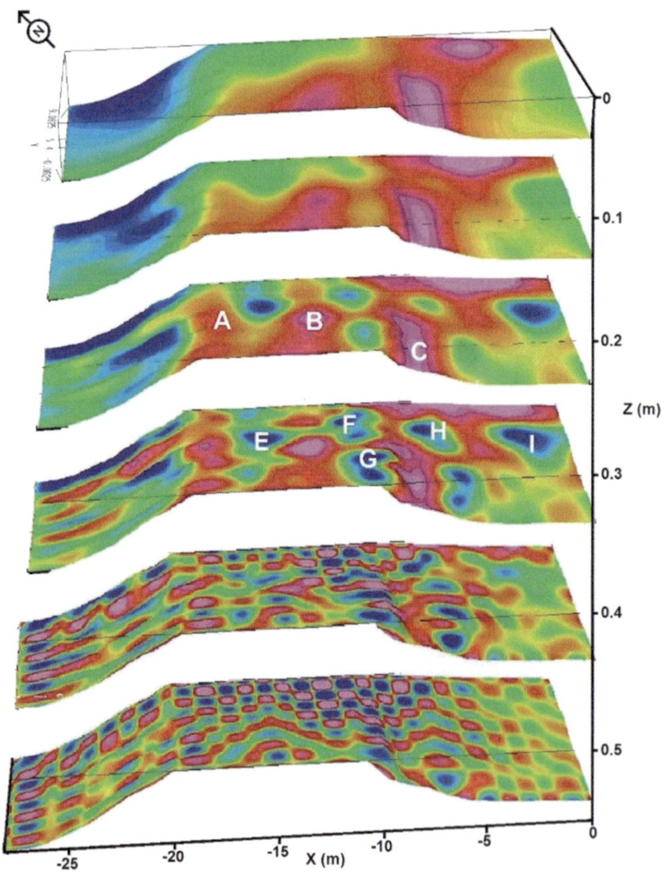

Fig. 11 Downward continuation results at different depth level, the first map is the total magnetic field. A, B, and C labels are NNE walls while, E, F, H, G, and I are the closed rooms

anomalies were flattened out). Accordingly, the archaeological boundaries tend to be smeared out and the distribution of anomalies is blurred and does not reflect the real shape of the anomalous body. By increasing the depth level of continuation to 0.2 m (Fig. 11), the positive magnetic anomalies started to be separated. Several small negative anomalies with circular geometry are slightly appeared and can be detected.

At a depth level of 0.3 m (Fig. 11), the symmetrical rectilinear positive anomalies (A, B, and C) are will remarkable and appeared with sharp edges. These pronounced anomalies could be interpreted as the top of the buried archaeological features which have a higher magnetization than the surrounding materials. Also, negative anomalies (E, F, G, H, and I) are clearly evident at the central and top right portions of the map which indicate the rooms of a construction dated back to the ancient Islamic settlers. The downward continuation at this level reflects the real shape of the archaeological features.

The anomalies of one geometric outlines, even positive or negative, are divided into two or the three anomalies at continuation levels 0.4 and 0.5 m (Fig. 11). The horizontal slice no. 6 (at depth 40 cm) shows oscillating signatures between negative and positive anomalies and serves no any interpretational purpose. This indicates that the continuation slightly exceeds the top of the buried archeological features (Cooper 2004). Accordingly, the maximum approximate depth to the top of archaeological walls is 0.4 m.

At a depth level of 0.3 m (Fig. 11), the symmetrical rectilinear positive anomalies (A, B, C and D) are will remarkable and appeared with sharp edges. These pronounced anomalies could be interpreted as the top of the buried archaeological features which have a higher magnetization than the surrounding materials. Also, negative anomalies (E, F, G, H and I) are clearly evident at the central and top right portions of the map which indicate the rooms of a construction dated back to the ancient Islamic settlers. The downward continuation at this level reflects the real shape of the archaeological features.

4.2.3 Edge Detection and Depth Estimation

Both tilt angle derivative and standard 3D Euler deconvolution were used to quantitatively interpret the total magnetic field data (Wang et al. 2015; Oliveira et al. 2017). The output of these methods basically aimed to produce maps showing the locations and depths of the archaeological features of the magnetic anomalies (Aziz et al. 2013).

Euler deconvolution

Euler deconvolution is both a boundary finder (contact mapper) and a depth estimator (Ravat 1996; Reid et al. 1990; Marson and Klingele 1993; Stavrev 1997; Hsu 2002; Ravat et al. 2002; Cooper and Cowan 2003, Cooper 2014; Oliveira et al. 2017). It is commonly used as a first step in the interpretation of magnetic data.

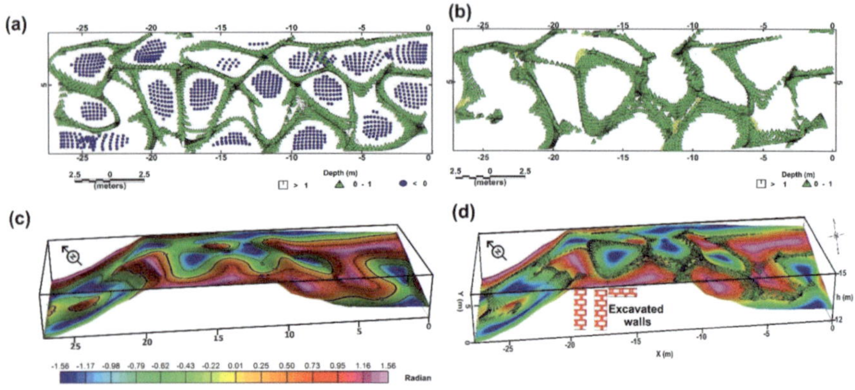

Fig. 12 **a** and **b** Euler depth solution map shows the cluster of source depth locations positioned at the boundaries of magnetic anomaly which is probably related to walls of ancient construction. **c** TAD image displaying only the 0 contours radian facilitating the recognition of the extent of anomalous sources edges assuming vertical contact and **d** 2D plane view of Euler deconvolution map

Figure 12a shows the 3D Euler deconvolution technique applied to the first vertical derivative data, which calculated from the total magnetic field, using structure index (SI) = 0. It can be clearly noticed that the solutions are closed to each other along the axes of the archaeological features whereas, the higher SI resulted in much more scattered and mis-located solutions. However, utilizing small SI value generated large number of spurious solutions. The plot was cleaned up by rejecting Euler solutions with depth and horizontal uncertainties of greater than 10% and 40%, respectively. Figure 12b shows the retained solutions which clustered near the edges of the fired brick walls of an ancient Islamic construction.

The Euler solutions are well clustered around the edges of the linear and circular magnetic anomalies. The depth solutions range from less than 1 m to more than 2 m (Fig. 12b). The map shows that, the archaeological fired brick walls are aligned in NNE-SSW and ESW-WNE directions. The given topographic evidence expressed on this site and the availability of 0.5 m resolution digital elevation data surveyed by (MATD) integrated with the results of ED can also provide useful insights. The integration indicates that the interesting clustering lineation of the ED solutions coincide with the highest points on the site (elevation ranges between 13 and 15 m). The interesting anomalies occur just to the center and east of the site (at local coordinate 0–20 m) indicating the presence of the archaeological ruins of fired brick walls and buried rooms, while the interruption in the clustering lies on the west side of the eastern hill of low land.

Tilt angle derivative

The location and the depth of the boundaries of the pre-mentioned archaeological structures can be detected through the analysis of TAD images (Miller and Singh 1994; Verduzco et al. 2004; Oruc and Keskinsezer 2008). The source edge location of a vertical side such as fired brick walls aligned with zero contour value of tilt derivative (Oruc 2010). Tilt derivative image (Fig. 12c) shows the positive contour value above the source bodies and negative away from the source, while the edges are marked by zero contour value. The depth to the edge of the magnetized sources can be estimated through measuring the half distance between $\pm\pi/4$ contours (± 0.8 radians) or the distance between zero and $\pm\pi/4$ radians (Salem et al. 2007). Accordingly, the TAD indicated that the depth to the anomalous sources ranges from 0.3 to 1 m with an average of 0.6 m (Fig. 12c).

The superposition of TAD and ED results (Fig. 12d) displays that the source locations obtained from the ED (Fig. 12b) are compatible with the zero contours of the TAD image (Fig. 12c). Consequently, ED and TAD have agreed adequately for detecting the horizontal locations and edges of the magnetized source bodies are well correlated. Furthermore, the depths computed from ED are also in a good agreement with that obtained from the TAD. The TAD almost confirmed the results of the ED under the assumption that the edges of anomalous sources are caused by vertical contacts.

The identified locations of double fired brick walls demarcated by field observation and archaeological excavation are superimposed on both clustering of ED solutions and zero contours of TAD. The results derived from both techniques provided more additional information regarding walls which are yet to be marked and not identified from field mapping study. In this work the sensitivity and the effectiveness of ED and TAD techniques in the interpretation of archaeo-magnetic data was verified to guide the subsequent excavation, where the main goal is to point out the shape and depth of the main archaeological features.

4.3 Three-Dimensional Resistivity Imaging

The 3D resistivity imaging was carried out using pole-pole array with electrode spacing 1 m to obtain a clear picture about the possible buried room in the surveyed site where the 2D ERT and magnetic techniques are failed to reconstruct the archaeological architecture in three dimensions. However, the priory 2D ERT results were considered during the elaboration of the tomographic slices. Accordingly, the specified intervals are 0.3, 0.64, 1.04, 1.5, 2.02, 2.63, 3.24, 4.12, and 5.04 to image the building of Islamic settlers within the first layer as illustrated in Fig. 13. Interestingly, the horizontal slices display more details about the exact depth and locations of the buried walls in 3D view. In the first two slices (depth range 0–0.64 m), the high resistivity values (>10 Ω.m) are laterally extensive and probably results from the effect of the damaged fired bricks paving mostly the area and accumulated around the standing walls which confirmed later by the excavation (Fig. 13d).

Accordingly, four perpendicular linear resistive anomalies with respect to the host can be clearly identified. The analysis of these anomalous patterns in the slices shows the presence of room structure with a strong linear development beginning from 0.64 m in depth and characterized by high resistivity intensity in comparison

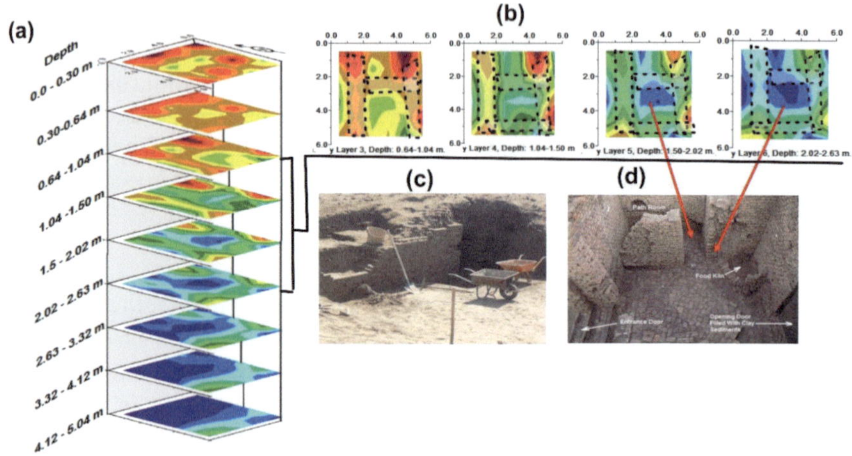

Fig. 13 The horizontal slices of the inverted 3D model of pole-pole array and the results of excavation in September 2015, **a** Resistivity distributions at different depths, **b** The PP slices related to depth range 0.64–2.63 m (extracted from 3D resistivity inversion model) and its probable interpretation **c** and **d** Photos show a house from outside before the excavations and inside after finishing excavations

to the host environment (clay-rich soil) in the middle part of measured grid. From the analysis of the PP tomographic slices (from third through sixth) which have depth range 0.64–2.63 m (Fig. 13b), two resistive anomalies parallel to the y axis can be clearly observed. On the other hand, another two distorted anomalies with relatively low resistivity appeared parallel to x-axis, which offer a smaller resistance to the current flow. The regularity in the shape of the resistive anomalies allowed confirming the strong relationship with built structures such as good-preserved wall. The resistivity and shape of these anomalies are mainly based on the nature of building materials, the orientation of buried walls with surveyed grid and position of current and potential electrodes. The importance of 3D survey compared with 2D profile could be noticed clear in the vertical sections in both x and y directions (Fig. 14a, b respectively) where the buried walls is appeared as well-defined vertical contacts in x-sections and distorted in y-sections. Finally, the high resistivity zone is replaced by resistivity values of less than 2 Ω.m at horizontal slices No. 6, 7 and 9 (Fig. 13) to display the end of Islamists settlement structures which are replace by wet clay soil.

5 Summary and Conclusion

The use of multiple geophysical surveys in the archaeological investigations allows understanding the variability of physical properties of the buried targets and the possibility of complementary of the applied techniques to construct models of multi-dimensional context At the eastern side of Tell Dibgou (Eastern Nile Delta, Egypt),

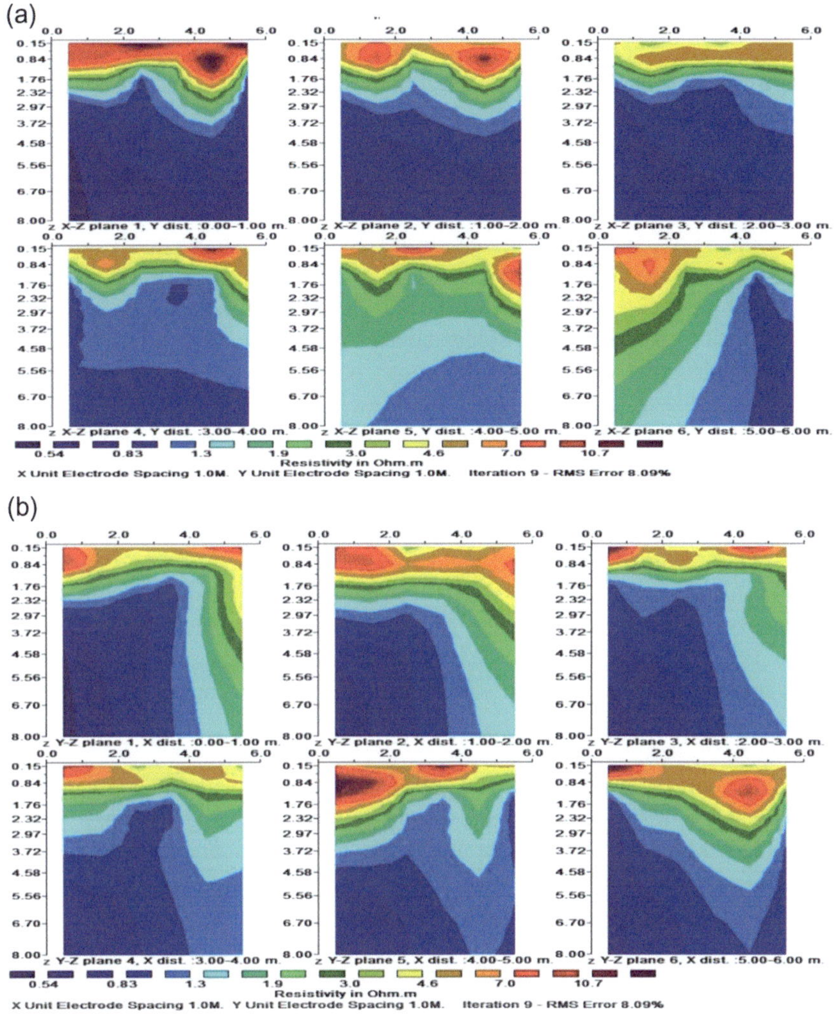

Fig. 14 The vertical slices of the inverted 3D model of pole-pole array along, **a** X direction **b** along Y direction

the DC resistivity imaging (2D/3D), and the magnetic surveys were used in combination to study the shallow depth of still unexplored Islamic archaeological settlement. Accordingly, better understanding the dual magnetic and electrical response of the archaeological materials entails a detailed comparison between the results of the DC resistivity imaging and the magnetic surveys. The interesting hidden archaeological features in the surveyed site appeared in all inverted sections and magnetic maps, but with different degrees of clarity and spatial resolution.

Both 2D resistivity imaging and magnetic method were successful in detecting the horizontal locations and the depth to the archaeological walls. Conversely, both types of surveys were not able to provide accurate detection of the dimensions of the interpreted walls. The 2D resistivity imaging failed to show the lateral discrimination between the archaeological remains and the surrounding. This could be related to the wet environments which tend to flatten out the resistivity anomalies and lessen the resistivity contrast between the archaeological and natural deposits. Additionally, the interference between the collapsed bricks and pottery shards around the walls makes the contrast in the electrical resistivity between the archaeological remains and the surrounding is inadequate. Likewise, the archaeological features of fired brick walls appeared broader in magnetic maps. Despite the magnetic survey was carried out using a proper sampling interval (0.5 m), it failed to discriminate between the adjacent walls separated by a distance smaller than twice the depth. However, the applied magnetic derivative-based techniques have been created successful models for estimating the depths of the buried structures in consistence with the results of the ERT method. Accordingly, for precise position and depth estimation in archeological investigations, the magnetic derivative-based techniques should be used first and then followed by the ERT measurements.

Fortunately, 3D ERT model successfully reconstructed the shape, location, depth, and vertical extent of the archaeological structure with sufficient resolution. The small inner walls parallel to the Y axis, which divide the building into two rooms, can now be identified more clearly. Furthermore, the 3D resistivity imaging was able to detect the dimension of archaeological walls more precisely (less than 1 m) which reveals a good agreement with the subsequent excavations (real dimension is approximately 60 cm). This indicates the superiority of full three-dimensional images in archaeological prospection even in complex areas.

It can be concluded that, the joint application of electrical and magnetic measurements seems to be the best approach for the investigation of the eastern hill of Tell Dibgou and other sites with similar conditions. It has been confirmed that geophysical surveying techniques are useful tools for obtaining valuable information for archaeologists to carry out subsequent excavations.

Acknowledgements The authors would like to express their sincere thanks to Dr. Metwally Salama, the Director of Antiquities of San El-Hagar, and Kafr Sakr and the archaeologists for their cordial support during field works. In addition, they appreciate the Egyptian Supreme Council of Antiquities for permitting the collection of data. They gratefully acknowledge the French Mission of Archaeology of Tell Dibgou (MATD) and the director, Prof. Philippe Brissaud, for their valuable information and guidance when selecting areas for the survey and during data acquisition.

References

Abdallatif T, Odah HH, El Emam AE, Mohsen A (2019) Geomagnetism exploration of the egyptian archaeology: thirty-years of success and challenges. In: El-Qady G, Metwaly M (eds) Archaeogeophysics. Natural science in archaeology. Springer, Cham. https://doi.org/10.1007/978-3-319-78861-6_7

Abu Al Izz MS (1977) Landforms of Egypt. The American University Press, 283p

Abu Salem H, Gemail KS, Nosair AM (2021) A multidisciplinary approach for delineating wastewater flow paths in shallow groundwater aquifers: a case study in the southeastern part of the Nile Delta, Egypt. J Contam Hydrol 2021(236):103701
Arato A, Piro S, Sambuelli L (2015) 3D inversion of ERT data on an archaeological site using GPR reflection and 3D inverted magnetic data as a priori information. Near Sur Geophys 13:545–556. https://doi.org/10.3997/1873-0604.2015046
Arsoy MÖ (2014) Edge detection of archaeomagnetic data: a study from the city of Pisidia antiocheia, Turkey. Archaeol Prospect 21:277–291. https://doi.org/10.1002/arp.1492
Aziz AM, Sauck WA, Shendi E-AH, Rashed MA, ElMaksoud MA (2013) Application of analytic signal and Euler deconvolution in archaeo-magnetic prospection for buried ruins at the ancient city of Pelusium, NW Sinai, Egypt: a case study. Surv Geophys 34(4):395–411
Berge MA, Drahor MG (2011) Electrical resistivity tomography investigations of multilayered archaeological settlements: part I–modelling. Archaeol Prospect 18:159–171. https://doi.org/10.1002/arp.414
Bernardes P, Alves M, Pereira B, Madeira J, Martins M, Fontes L (2017) Visualization of ERT data for archaeological purposes. In: Sablatnig R, Štular B (eds) EUROGRAPHICS workshop on graphics and cultural heritage, GCH 2017, Eurographics, The Eurographics Association. https://doi.org/10.2312/gch.20171294. ISBN 978-3-03868-037-6
Bevan BW, Roosevelt AC (2003) Geophysical exploration of Guajará a prehistoric earth mound in Brazil. J Geoarchaeol 18(3):287–331
Blakely RJ (1995) Potential theory in gravity and magnetic applications. Cambridge University Press, Cambridge
Blakely RJ, Simpson RW (1986) Approximating edges of source bodies from magnetic or gravity anomalies. Geophysics 51:1494–1498
Brissaud A, Desbordes C (2014) Activity report—2014 campaign, MATD, mission of archaeology at tell Dibgou, No 28, 23P. http://www.telldibgou.fr/index.php/rapports-d-activite/2019. Accessed 1 Mar 2020
Cardarelli E, Fischanger F, Piro S (2008) Integrated geophysical survey to detect buried structures for archaeological prospecting. A case-history at Sabine Necropolis (Rome, Italy). Near Surf Geophys 6(1):15–20
Cardarelli E, Di Filippo G (2009) Integrated geophysical methods for the characterization of an archaeological site (Massenzio Basilica–Roman forum, Rome. Italy). J Appl Geophys 68(4):508–521
Cheyney S, Hill I, Linford N (2011) Advantages to using the pseudogravity transformation to aid edge detection of total field archaeomagnetic datasets. Archaeol Prospect 18:81–93. https://doi.org/10.1002/arp.408
Chianese D, D'Emilio MG, Di Salvia S, Lapenna V, Ragosta M, Rizzo E (2004) Magnetic mapping, ground penetrating radar surveys and magnetic susceptibility measurements for the study of the archaeological site of Serra di Vaglio (Southern Italy). J Archaeol Sci 31(5):633–643
Colombero C, Elia D, Meirano V, Sambuelli L (2020) Magnetic and radar surveys at Locri Epizephyrii: a comparison between expectations from geophysical prospecting and actual archaeological findings. J Cult Herit 42:147–157.https://doi.org/10.1016/j.culher.2019.06.012. ISSN 1296-2074
Cooper GRJ (2004) The stable downward continuation of potential data. J Explor Geophys 35:260–265
Cooper GRJ (2014) Determining the location, depth and dip of contacts from aeromagnetic data. Geophysics 79(3):35–41
Cooper GRJ, Cowan DR (2003) Applications of fractional calculus to potential field data. J Explor Geophys 34:51–56
Dahlin T, Zhou B (2004) A numerical comparison of 2D resistivity imaging with 10 electrode arrays. J Geophys Prospect 52:379–398

Deiana R, Bonetto J, Mazzariol A (2018) Integrated electrical resistivity tomography and ground penetrating radar measurements applied to tomb detection. Surv Geophys 39(6):1081–1105. https://doi.org/10.1007/s10712-018-9495-x

Drahor MG, Göktürkler G, Berge MA, Kurtulmu TÖ, Tuna N (2008) 3D resistivity imaging from an archaeological site in south-western Anatolia: a case study. Near Surf Geophys 5(3):195–201

Ekinci YL (2016) MATLAB based algorithm to estimate depths of isolated thin dike like sources using higher order horizontal derivatives of magnetic anomalies. Springerplus 5:1384. https://doi.org/10.1186/s40064-016-3030-7

Elwan AA, Harga MA, El Kadi HA, El Demerdash S (1983) Preliminary studies on the soil of North Sinai Peninsula on aerial photo interpretation. Egypt J Soil Sci 23(1):37p

Eppelbaum LV (2021) Review of processing and interpretation of self-potential anomalies: transfer of methodologies developed in magnetic prospecting. Geosciences 2021(11):194. https://doi.org/10.3390/geosciences11050194

Fedi M, Cella F, Florio G, Manna ML, Paoletti V (2017) Geomagnetometry for archaeology. In: Masini N, Soldovieri F (eds) Sensing the past. Geotechnologies and the environment, vol 16. Springer, Cham. https://doi.org/10.1007/978-3-319-50518-3_10

Gaber A, Gemail KS, Kamel A, Atia HM, Ibrahim A (2021) Integration of 2D/3D ground penetrating radar and electrical resistivity tomography surveys as enhanced imaging of archaeological ruins: a case study in San El-Hager (Tanis) site, northeastern Nile Delta, Egypt. Archaeol Prospect 28:251–267https://doi.org/10.1002/arp.1810

Gaffney V, Patterson H, Piro S, Goodman D, Nishimura Y (2004) Multimethodological approach to study and characterize Forum Novum (Vescovio, Central Italy). Archaeol Prospect 11:201–212

Gemail K (2015) Application of 2D resistivity profiling for mapping and interpretation of geology in a till aquitard near luck Lake, southern Saskatchewan, Canada. Environ Earth Sci 73:923–935. https://doi.org/10.1007/s12665-014-3441-0

Gemail K, Samir A, Oelsner C, Mousa SE, Ibrahim S (2004) Study of saltwater intrusion using 1D, 2D and 3D resistivity surveys in the coastal depressions at the eastern part of Matruh area, Egypt. Near Surf Geophys 2(2):103–109. https://doi.org/10.3997/1873-0604.2004007

Gemail KS, Atwa M, Elarky M, Zamzam S (2017a) Imaging of wastewater percolation in heterogeneous soil using electrical resistivity tomography (ERT): a case study at east of tenth of Ramadan City, Egypt. Environ Earth Sci 76(11):666. https://doi.org/10.1007/s12665-017-7013-y

Gemail KS, El Alfy M, Ghoneim MF, Shishtawy AM, El-Bary MA (2017b) Comparison of DRASTIC and DC resistivity modeling for assessing aquifer vulnerability in the central Nile Delta, Egypt. Environ Earth Sci 76:350

Geoffrey J (2007) Geophysical survey of the mansard baptist church cemetery (34 ms 407): an historic cemetery on Camp Gruber, Oklahoma. Report to the LOPEZGARCIA Group, Dallas, TX, from Archaeo-Physics, LLC, Report of Investigation, No 115. Minneapolis, MN

Hinze WJ, von Frese RRB, Saad AH (2013) Gravity and magnetic exploration, principles, practices and applications. Published in the United States of America by Cambridge University Press, New York, p 512p

Hsu SK (2002) Imaging magnetic sources using Euler's equation. J Geophys Prospect 50:15–25

Ibrahim A, Gemail KS, Abdelrahman K, Al-Otaibi N, Ibrahim E, Saada SA (2021) Multi-scale geophysical methodologies applied to image archaeological ruins at various depths in highly terraneous sites. Remote Sens 13:2055. https://doi.org/10.3390/rs13112055

Karaaslan H (2020) Edge detection for the buried archaeological structures with the geophysical image processing method in the Alabanda Ancient Cistern in Turkey. Archaeol Prospect 27:275–284https://doi.org/10.1002/arp.1771

Linford NT, Canti MG (2001) Geophysical evidence for fires in antiquity: preliminary results from an experimental study. Paper given at the EGS XXIV general assembly in The Hague, April 1999. J Archaeol Prospect 8:211–225

Loke MH, Barker RD (1995) Least-squares deconvolution of apparent resistivity pseudosections. Geophysics 60:1682–1690

Loke MH, Barker RD (1996a) Rapid least-squares deconvolution of apparent resistivity pseudo-sections using a quasi-Newton method. J. Geophys Prospect 44:131–152
Loke MH, Barker RD (1996b) Practical techniques for 3D resistivity surveys and data inversion. J Geophys Prospect 44:499–523
Marson I, Klingele EE (1993) Advantages of using the vertical gradient of gravity for 3-D interpretation. J Geophys 58:1588–1595
Mehanee S, Essa KS, Diab ZE (2021) Magnetic data interpretation using a new R-parameter imaging method with application to mineral exploration. Nat Resour Res 30:77–95 (2021). https://doi.org/10.1007/s11053-020-09690-8
Miller HG, Singh V (1994) Potential field tilt—a new concept for location of potential field sources. J Appl Geophys 32:213–217
Nabighian MN (1984) Toward a three-dimensional automatic interpretation of potential field data via generalized Hilbert transforms fundamental relations. Geophysics 49:780–786
Nemtsova O, Zhurbin I, Zlobina A (2019) Vector analysis of pole–pole array for determining the 3D boundary of object. Near Surf Geophys 17:563–575. https://doi.org/10.1002/nsg.12065
Nyquist JE, Roth MJS (2005). Improved 3D pole-dipole resistivity surveys using radial measurement pairs. Geophys Res Lett 32(21). https://doi.org/10.1029/2005GL024153
Oliveira SP, Ferreira FJF, de Souza J (2017) EdgeDetectPFI: an algorithm for automatic edge detection in potential field anomaly images–application to dike-like magnetic structures. Comput Geosci 103:80–91
Ortega-Ramírez J, Bano M, Cordero-Arce MT, Villa Alvarado LA, Chavez FC (2020) Application of non-invasive geophysical methods (GPR and ERT) to locate the ancient foundations of the first cathedral of Puebla, Mexico. A case study. J Appl Geophys 174:103958. https://doi.org/10.1016/j.jappgeo.2020.103958
Oruc B (2010) Edge detection and depth estimation using a tilt angle map from gravity gradient data of the Kozakli-Central Anatolia region, Turkey. J Pure Appl Geophys 176(10):1259–1272
Oruc B, Keskinsezer A (2008) Structural setting of the northeastern Biga Peninsula (Turkey) from tilt derivatives of gravity gradient tensors and magnitude of horizontal gravity components. Pure Appl Geophys 165:1913–1927
Papadopoulos N, Sarris A (2011) Integrated geophysical survey to characterize the subsurface properties below and around the area of Saint Andreas church (Loutraki, Greece). In: Proceedings of the 14th international congress "Cultural Heritage and New Technologies", pp 643–652. ISBN 978-3-200-02112-9
Papadopoulos NG, Sarris A, Parkinson WA, Gyucha A, Yerkes RW, Duffy PR, Tsourlos P (2014) Electrical resistivity tomography for the modelling of cultural deposits and geomophological landscapes at Neolithic Sites: a case study from Southeastern Hungary. J Archaeol Prospect 21(3):169–183
Papadopoulos N, Tsourlos PI, Tsokas GN, Sarris A (2006) 2D and 3D resistivity imaging in archaeological site investigation. Archaeol Prospect 13:163–181https://doi.org/10.1002/arp.276
Patella D (1991) I principi della magnetometria e della gravimetria. Possibilità di app.licazione in archeologia. In: Proceedings of Geofisica per lper lica per, Quaderni delllrni dell per le Tecnologie App.licate ai Beni Culturali, vol 1, pp 71–81
Patella D, Hesse DA (1999) Electric, magnetic and electromagnetic methods applied to cultural heritage. J Appl Geophys 41:135–311
Ravat D (1996) Analysis of the Euler method and its applicability in environmental investigations. J Environ Eng Geophys 1:229–238
Ravat D, Kirkham K, Hildenbrand TG (2002) A source-depth separation filter: using the Euler method on the derivatives of total intensity magnetic anomaly data. Lead Edge 21:360–365
Reid AB, Allsop JM, Granser H, Milett AJ, Somerton I (1990) Magnetic interpretation in three dimensions using Euler deconvolution. Geophysics 55:80–91
Salem A, Williams S, Fairhead JD, Ravat D, Smith R (2007) Tilt depth method: a simple depth estimation method using first-order magnetic derivatives. Lead Edge 26:1502–1505

Salvatore P, Enrico P, Daniela Z, Melda K (2019) Multimethodological approach to investigate urban and suburban archaeological sites. In: Innovation in near-surface geophysics. Elsevier BV, Amsterdam, The Netherlands, pp 461–504
Schmidt A (2007) Archaeology, magnetic methods. In Gubbins D, Herrero-Bervera E (eds) encyclopedia of geomagnetism and paleomagnetism. Encyclopedia of earth sciences series Heidelberg. Springer, New York, pp 23–31. (8) (PDF) Archaeology, magnetic methods. https://www.researchgate.net/publication/228666190_Archaeology_magnetic_methods. Accessed 26 Jun 2021
Schmidt A (2009) Electrical and magnetic methods in archaeological prospection. In Campana S, Piro S (eds) Seeing the unseen. Geophysics and landscape archaeology. Taylor & Francis Group, London, pp 67–81
Sertcelik I, Kafadar O (2012) Application of edge detection to potential field data using eigenvalue analysis of structure tensor. J Appl Geophys 84:86–94. https://doi.org/10.1016/j.jappgeo.2012.06.005
Simms JE, Albertson PE (2000) Multidiscipline investigation to locate the Kentucky Shipwreck. J Geoarchaeol 15(5):441–468
Stavrev PY (1997) Euler deconvolution using differential similarity transformations of gravity or magnetic anomalies. J Geophys Prospect 45:207–246
Tsokas G, Diamanti N, Tsourlos P, Vargemezis G, Stampolidis A, Raptis K (2013) Geophysical prospection at the Hamza Bey (Alkazar) monument, Thessaloniki, Greece. J Mediterr Archaeol Archaeom 13(1):9–20
Vacquier V, Steenland NC, Henderson RG, Zietz I (1951) Interpretation of aeromagnetic maps. Geological Soc Am Memoir 47
Verduzco B, Fairhead JD, Green CM, Mackenzie C (2004) New insights into magnetic derivatives. Lead Edge 22:116–119
Wang J, Meng X, Li F (2015) Improved curvature gravity gradient tensor with principal component analysis and its application in edge detection of gravity data. J Appl Geophys 118:106–114
Weymouth JW, Huggins R (1985) Geophysical surveying of archeological site in archeological geology. Edited by Rapp G, Gifford JA, pp 191–235
Wu H, Li L, Xing C, Zhang S (2017) A new method of edge detection based on the total horizontal derivative and the modulus of full tensor gravity gradient. J Appl Geophys 139:239–245. https://doi.org/10.1016/j.jappgeo.2017.02.026
Yuan Y, Huang D, Yu Q, Lu P (2014) Edge detection of potential field data with improved structure tensor methods. J Appl Geophys 108:35–42. https://doi.org/10.1016/j.jappgeo.2014.06.013
Zuo B, Hu X (2015) Edge detection of gravity field using eigenvalue analysis of gravity gradient tensor. J Appl Geophys 114:263–270

The Implementation of Shallow Geophysical Survey for Detection of Some Buried Archaeological Structures in Aswan City, Egypt

Abbas Mohamed Abbas, Raafat El-Shafie Fat-Helbary, Ahmed Hamed, Karrar Omar El-Faragawy, Ezzat M. El-Amin, and Gad Mohamed El-Qady

Abstract Aswan is one of Egypt's most attractive cities, with numerous of historical monuments. The current work in the Aghakhan archaeological site, is primarily focused on the acquisition, processing, and interpretation of the obtained geophysical data, for delineating any possible buried archaeological relics or tombs. For achieving this goal, Ground Penetrating Radar (GPR) and Multichannel Analysis of Surface Wave (MASW) have been applied with minimal time, effort, and expense. First of all, GPR measurements were conducted over the archaeological sites with 26 profiles running from south to north and 16 profiles extend from west to east direction with profiles interval 2 m. This study provides a broad overview of the subsurface archaeological features. In additions, GPR data in the form of time-slice maps reveal several notable anomalies buried in Aghakhan site. Sequentially, to validate the results, the Multichannel Analysis of Surface Wave (MASW) technique is applied; eight profiles were conducted in the same location of the GPR's profiles. Indeed, the integration and comparison between the GPR's and MASW's profiles is commonly utilized to complete the image, and it revealed successfully a range of probable archaeological features and structures buried at a depth of around 2–3 m, which may be interpreted as tombs and ancient structure walls.

Keywords Aghakhan · GPR · MASW · Archaeological features · Time-slice maps

1 Introduction

The ancient Egyptian monuments distribute all across Egypt, from the Nile Delta in the north to Aswan in the south; some have been discovered, while others have yet to be discovered. (Hemeda 2018). This means that, the Egyptian land still contains

A. M. Abbas (✉) · R. E.-S. Fat-Helbary · A. Hamed · E. M. El-Amin · G. M. El-Qady
National Research Institute of Astronomy and Geophysics, Helwan, Egypt
e-mail: abbas@nriag.sci.eg

K. O. El-Faragawy
Faculty of Science, Aswan University, Aswan, Egypt

G. M. El-Qady and C. Margottini (eds.), *Sustainable Conservation of UNESCO and Other Heritage Sites Through Proactive Geosciences*, Springer Geology,
https://doi.org/10.1007/978-3-031-13810-2_11

more and more archaeological sites, which vary in their sizes, shapes, composition, and historical importance. We shall obtain additional information about the growth phases of sciences and civilization by unearthing ancient Egyptian archaeological relics, and these new locations will be added to Egypt's tourism map.

Different shallow geophysical methods are utilized, in the field of archaeological exploration in Egypt to find out information for excavation planning. Many studies have been conducted in this topic, including the use of ground penetrating radar (GPR) and magnetic surveys near the Karnak Temple's eastern gate to discover any concealed expansions of the temple. (Abbas et al. 2005). In addition, the GPR method was employed separately in the Isis Temple to map buried archaeological artefacts (Shaaban et al. 2003). Different geophysical approaches have also been used, for archaeological exploration and conservation of some ancient sites on the western bank of Luxor (Alwasif 2006). These studies are mostly focused on the physical differences between the buried archaeological artefacts and their burial medium (e.g., electrical conductivity, wave propagation velocity).

Various archaeological investigations have been carried out at the west of Aswan, essentially contiguities to the Nile for more than a hundred years (Storemyr 2010). One of the earliest investigations was carried out in 1890 as part of J. de Morgan and his collaborator's survey of the area between Kom Ombo and Aswan (Bloxam et al. 2007). Junker excavated two multi-period cemeteries; the other excavation is the Coptic monastery constructed on a Ptolemaic temple in the same location. Then different excavations used archaeological and geological survey were taken place at Gebel Gulab and Gebel Tingar, sponsored by the British–Norwegian–Egyptian mission headed by Elizabeth Bloxam (Shawarby et al. 2009). The survey region covered the entire south of Gharb Aswan village (West Aswan), however, it was mostly focused on the southern half of it (Storemyr 2010). Nowadays, German and Swiss institutes are involved in archaeological work at Gharb Aswan village, their main long-standing work is at Elephantine Island and within modern Aswan city.

Aswan governorate is one of Egypt's richest towns in terms of ancient relics like as temples and tombs; some of these relics have been excavated, but too many are yet to be discovered. In the current study, two geophysical techniques have been applied; the Ground Penetrating Radar (GPR) which is a common technique in archaeological prospection (El-Qady et al. 2005a; Papadopoulos et al. 2009). Traditionally, Shallow seismic methods have demonstrated capabilities in in-situ geotechnical evaluation and site effects estimation (Othman 2005; Fat-Helbary et al. 2019a, b), consequently, the Multi-Channel Analysis of Surface Wave (MASW), represents one of these valuable seismic techniques which not only applied in the geotechnical engineering field (Hamed, 2019), but also, in the archaeological prospection trend, for detecting any hidden low velocity objects beneath the earth surface (El-Qady et al. 2005b; Kamai 2015; Anbazhagan et al. 2018 and Mohamed et al. 2019). Indeed, both of these geophysical approaches have been used to identify possible anomalies; that may be interpreted as buried ancient structures in Aswan city, west of the Nile River; the location is located to the south of Aghakhan tomb. (Fig. 1).

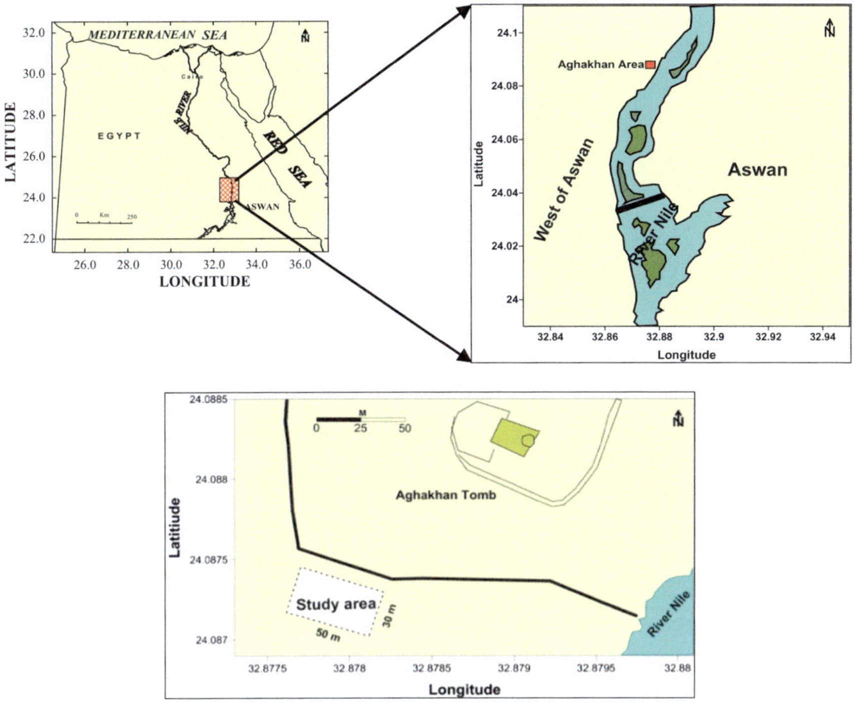

Fig. 1 Location map of the Aghakhan Study area in Aswan

2 Historical Background

Between around 3100 BC and 332 BC, Egyptian history may be split into eight periods, each divided into thirty dynasties (Sitek 2003). The city of Memphis is said to have been founded by King Mena, who unified Egypt at the beginning of the Early Dynastic Period (Baines and Malek 1992; Black and Norton 1993; Kinnaer 2003). Many Egyptologists equate Menes to Narmer or his successor Aha, the first and second Pharaohs of the 1th dynasty.

The study area is located to the south of the Aghakhan Tomb. Since it was unexcavated, it has not been evaluated to identify its era. on the other hand, The Directorate antiquities of Aswan and Nubia dates the discovered ornamated graves (Figs. 2 and 3), to the Greco-Roman era (323 BC 395 AD). This time is the period after Alexander the Great's death, when his successors continue his restoration strategy in Egypt. (Black and Norton 1993). They sponsored the construction of new temples all over the country. Ptolemy II Philopator initiated the restoration of the 26th Dynasty temple of Isis on the island of Philae; after that, Ptolemy XII's daughter Cleopatra became embroiled in the power battle between the Romans Octavianus (Augustus) and Antonius, but she selected the wrong side. Egypt became a Roman province once her and Antonius' ships were destroyed at the Battle of Actium, and Cleopatra

Fig. 2 Tomb's owner's feet wearing a gold sandal

committed suicide. The policy of constructing temples in Egypt was followed by the Roman emperors as well. (Kinnaer 2003). When an Egyptian-Italian archaeological team uncovered a rock-cut tomb dating from the late Pharaonic Greco-Roman period, in the Aghakhan Mausoleum region on the west bank of the Nile in Aswan city, they emphasized the earlier description of the discovered tombs. They also found numerous treasures from the Greco-Roman Period including a considerable collection of cartonnage, such as one of the tomb's owner's feet wearing a gold sandal (Fig. 2), while another is painted white. Another cartonnage is topped with a painting of the sun in addition to two gold painted funerary masks and two statues, among which one is in good condition preserving the Ba-bird, representing the soul of the deceased, still presenting all the details of the decoration. The mission also found many amphorae and offering vases, as well as a funerary structure containing two mummies, likely of a mother and her child, still covered by painted cartonnage (Fig. 3). The tomb consists of a stairway (Fig. 4) partly flanked by sculpted blocks leading to the funerary chambers. The entrance was sealed by a stone wall found in its original place over the stairway. The mission dating the discovered tombs from the sixth century BC to the fourth century AD (Majed 2019).

3 Geological Setting

The four geomorphic characteristics present in the Aswan area are; the Aswan Hills, the Nile River Valley, the Nubian Plain, and the Sin El-Kaddab Plateau (Fig. 5). Inside the Aswan Hills, which are defined by rugged terrain, the Precambrian basement complex is visible. The basement is flanked by clastic sedimentary layers of moderately slopping sandstone, shale, and ferruginous sandstone. Structurally, the Aswan Hills are a broad north-to-northwest trending arch with local faults and folds superimposed on it. The Nile River Valley runs along the western edge of the Aswan

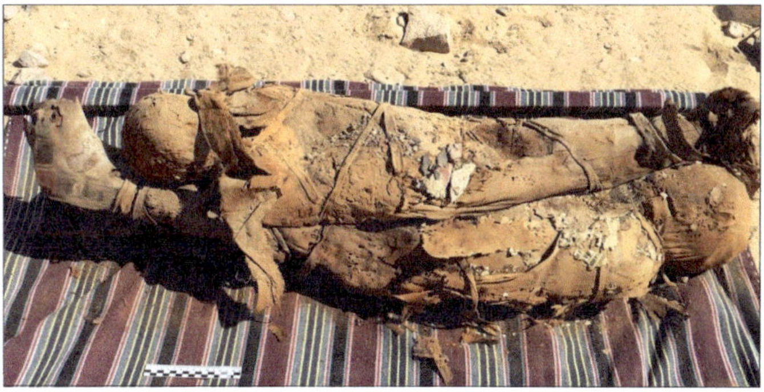

Fig. 3 The two mummies have been discovered in the area

Fig. 4 The entrance of the tomb with stairway

Hills. The Nubia Formation sandstone and the Precambrian basement complex in the Aswan region have been cut down to reveal the present narrow valley and steep-sided bedrock canyon, which is a remnant of an ancient canyon's upper section. (Khedr et al. 2010). The Nubia Formation or Kalabsha Plain is inhabited with many Wadies (valleys), such Wadi Kurkur and Wadi Kalabsha. Notably Isolated outliers of the limestone that makes up the Sin El-Kaddab Plateau may be seen to the west. The Nubia Formation is devoid of Quaternary deposits except along the escarpment of the Sin El-Kaddab Plateau. (Issawi 1969; El Shazly 1977).

The research site is predominantly characterized by unconformable sandstone layers that lie above Pre-Cambrian hard igneous and metamorphic rock. El-Naggar (1970) categorized the sandstone section in the Aswan area as Abu Aggag sandstone

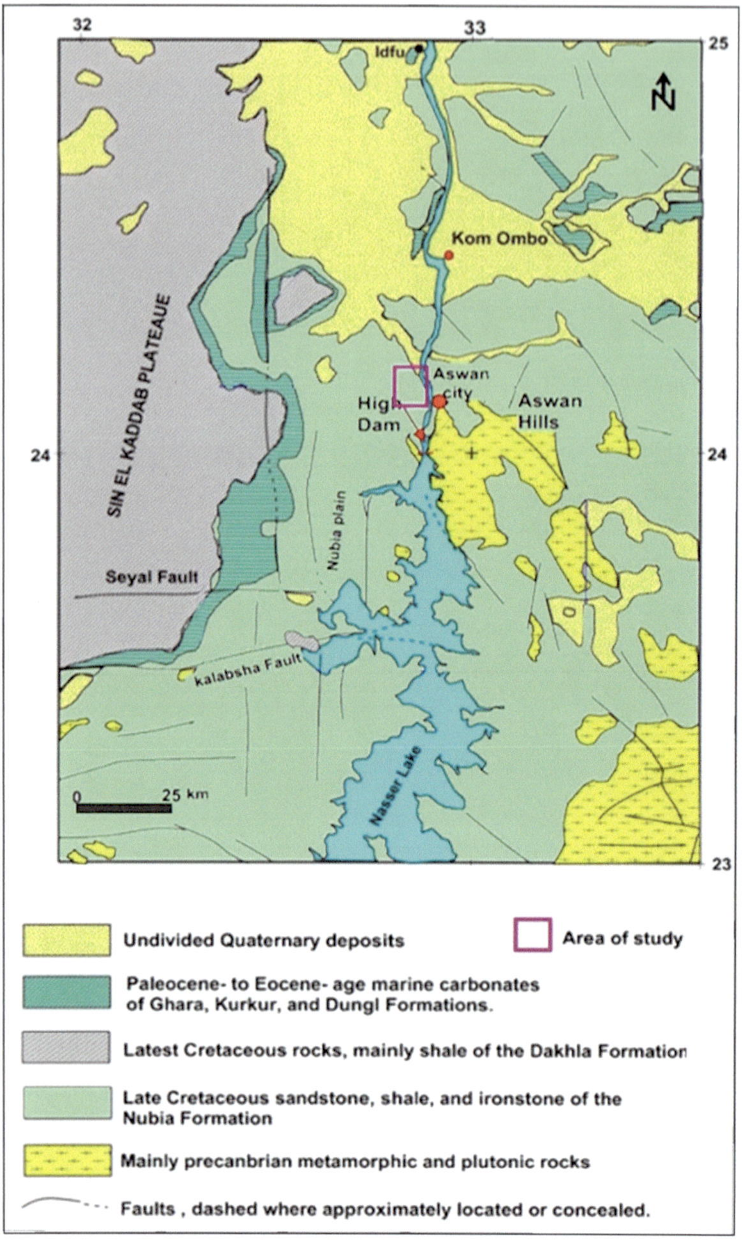

Fig. 5 Location and geological map of the study area, modified after Egyptian Geological Survey and Mining Authority (EGSMA) (1981)

at the bottom, Timsah claystone with iron ore layers in the middle, and Umm Barmil at the top. In this context El Naggar replaced the lower, middle, and upper units of Attia (1955) by these three units. Recent studies indicate that the Abu Aggag sandstone includes Cambrian fossils at base (Araba Formation) overlain by Devonian Bifungites Sandstone, followed by the Carboniferous Gelf Formation, which mainly made up of fine to medium grained laminated and cross laminated sandstone with lamellar joints filled by ferruginous material and cavernous structures (Issawi and Osman 1993, 1996; Issawi 2002, 2005a, b; Osman et al. 2002, 2005), Finally, fine to medium grained sandstones from the upper cretaceous Taref sandstone Formation cap this sequence. (Fig. 6).

4 GPR Data Acquisition, Processing and Interpretation

Usually, Ground-Penetrating Radar data are collected along closely spaced transects within a grid, each of which consists of many thousands of radar waves, that have been reflected from interfaces in the ground, the pulse of radiation propagates into the subsurface and is partially reflected from the contact surface of materials with different electrical properties. It is an active approach, that delivers electromagnetic pulses into the ground subsurface from surface antennas, and then measures the time between sending the pulses and receiving them back at the surface. Radar travel times are measured in nanoseconds, which are billionths of a second (Conyers 2009). Individual reflected waves (called waveforms) received from inside the earth are then digitized to create a reflection trace, which is a succession of waves reflected back to one surface location. A two-dimensional vertical profile is provided when numerous traces are placed next to each other sequentially along the transect where the antenna was moved. Thousands of reflection traces in many profiles within a grid can then be analyzed, to establish both two- and three-dimensional images, enabling the interpreter to determine the spatial location of all imaged subsurface features (Kraus 1950; Rojansky 1979).

The GPR survey was conducted using a GSSI Subsurface Interface Radar System-3000 equipped with 200 MHz antenna, which provide the required depth penetration (~4 m). The trace length was set at 150 ns; 40 traces (radar scans) were gathered per meter, in addition to the sample per scan value is 512 sample/scan. Due to the conical spreading of the transmitted radar waves and the attenuation of the energy as it passes through the ground, later arrival on a reflection trace will always have lower amplitudes than nearer arrivals. Gain control is often used during acquisition to retrieve this lower amplitude information. (Maijala 1992; Shih and Doolittle 1984; Sternberg and McGill 1995).

Forty-two profiles in zigzag pattern were conducted at this site on area 50 × 30 meters (Fig. 5), Twenty-six of them (G1 to G26) were oriented approximately north–south, with a length of 30 and 2 m intervals between profiles, another sixteen (G53 to G68) were aligned east–west, with a length of 50 m and a 2 m interval between profiles.

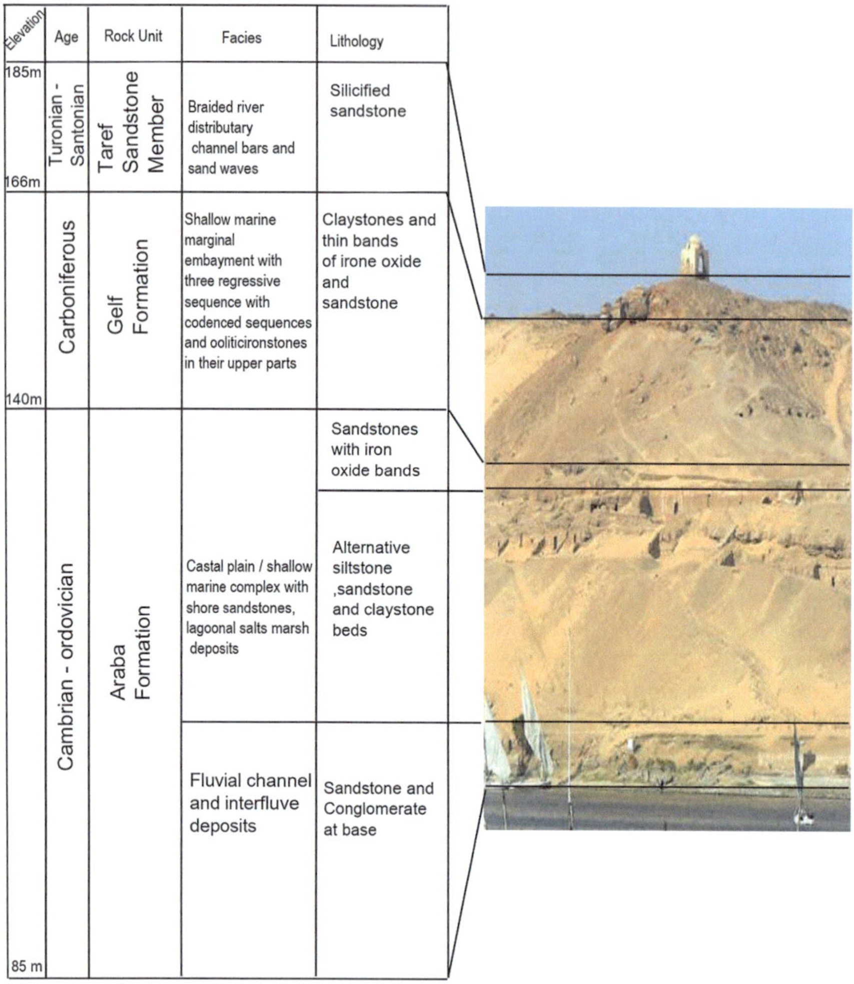

Fig. 6 Stratigraphy and facies sequences of the western bank of the Nile at Abu El-Hawa section, one kilometer from Aghakhan site, (Personal contact with Prof. Kamal Abu El Magd- Aswan University)

The overall goal of data processing is to create a cross-section that can be interpreted in order to find hidden targets. Therefore, the obtained data were processed in 1-D and 2-D scales (Abbas et al. 2005). The GPR reflected data seen immediately after acquisition in the field, it sometimes has 'noise'–unwanted reflections that make it difficult to analyze. These raw field data are also usually not displayed with corrected depth or horizontal scales. The raw data must be processed in order to remove noise and adjust the horizontal and vertical scale before being interpreted. (Conyers and Goodman 1997).

REFLEXW software version 6.0.5 was used to process the whole GPR data set. (Sandmeier 2001). A variety of post-processing procedures were used at this stage; static correction, background removal, gain energy decay, running average, in addition to profile length adjustment. In archaeological investigation, we commence by implementing static correction (zero-time adjustment) to the raw data (Fig. 8) to eliminate undesired reflections collected in the top portion due to the separation between the antenna and the ground surface. (Annan 2003), it determines the value of shifted time upwardly for the traces in the radar section (Fig. 9). The second step of data processing is the applying of background Removal, whereas GPR data are often contaminated by clutter. The clutter mainly consists of the GPR system noise, ground bounce, soil roughness scattering and reflection signals from external anomalies. The clutter mostly appears as nearly horizontal and periodic ringing. Clutter reduction is thus one of the most critical difficulties, as clutter occasionally fully obscures events, especially those that are deeper or weaker. A simple background removal (subtraction of an average trace) filter may readily reduce the GPR system-based coherent noise ringing (Fig. 10). Thereafter, to transform the radar data from time-depth profiles to depth profiles, the velocity of the rocks in the investigated site should be estimated, using the relative dielectric permittivity (RDP) of this area which is represented by sandstone (~4) (Table 1) wherein the velocity of light (0.3 m/ns) and V is the velocity of radar waves. The estimated velocity is 0.15 m/ns according to Eq. (1). Also, the hyperbolic shape technique has also been used to define the radar wave's velocity. On the radar profiles, many hyperbolas originating from prominent tombs in the research area were identified at known depths and used in the velocity calculation with the REFLEXWIN software.

$$\sqrt{RDP} = \frac{c}{V} \tag{1}$$

Band-pass filter applied to remove unwanted background noise resulting from electronic ringing of the antenna and other cultural noises in the study site, where the lower cut-off was 10 MHz, the lower plateau was 120 MHz, the upper plateau was 230 MHz and the upper cut-off was 350 MHz, in order to enhance the response from strong reflectors (Fig. 11).

The fundamental purpose of most GPR surveys in archaeology is to identify the size, shape, depth and location of the buried Pharaonic remains (Booth et al. 2008; Conyers 2012). The most straightforward way to accomplish it, is by identifying and correlating important reflection within two-dimensional reflection profiles. These reflections can then be correlated from profile to profile through a grid. The running average is then used to get the mean of the number of traces. This filter horizontally smooths data and highlights flat-lying reflectors while suppressing dipping reflections. (Fig. 12). The length and number of traces for each radar profile are adjusted in this processing phase. Trace increment and profile length are the two parameters. In the current example, the trace increment on the sections is 0.01, and the profile length is 50 m (Fig. 13).

Table 1 Dielectric constants and propagation velocities of common earth materials (Cardimona 2002)

Material	Dielectric constant (−)	Propagation velocity (m/ns)
Air	1	0.30
Ice (Frozen soil)	4	0.15
Granite	9	0.10
Limestone	6	0.12
Sandstone	4	0.15
Dry sand	4 to 6	0.12 to 0.15
Wet sand	30	0.055
Dry clay	8	0.11
Wet clay	33	0.052
Asphalt	3 to 6	0.12 to 0.17
Concrete	9 to 12	0.087 to 0.10
Water	81	0.033
Metal	∞	0

The GPR profiles can be subdivided into two visually distinct zones: an upper radar zone it is thickness about 2–3 m. From the direct field view and from the nearest outcrop to these sites, this zone represents loose sediments and clay, while the upper surface of second zone appeared at depths from 2 to 3 m, it is composed of clayey sandstone. Excavated tombs near the investigated region were used to interpret GPR data. During the field survey, it has been noticed that the opened tombs occupy the depths from 2 to 3 m below ground surface (Fig. 14). These tombs are of rectangular shapes and may be connected to other tombs in the same direction (Fig. 15).

5 GPR Profiles Measured in X-direction

This group of profiles takes N-S direction with a total length 30 m for each profile (Fig. 6). Many hyperbolas appear obviously at distances (1–23 m), (17–21–26 m), (5–19–24 m) and (2 m) at profile G1–G2–G3–G4 respectively, with comparable depths 2 to 3 m. The presence of buried archaeological features is indicated by these anomalies.

6 GPR Profiles Measured in Y-direction

In fact, this group runs from East to West, each profile extending 50 m in length. (Fig. 7). The anomalies in profiles G53–G54–G55 and G56 at offsets of (4–7–14–22 m), (19–34 m), (29 m), and (41 m), respectively, with average depth 2–3 m could

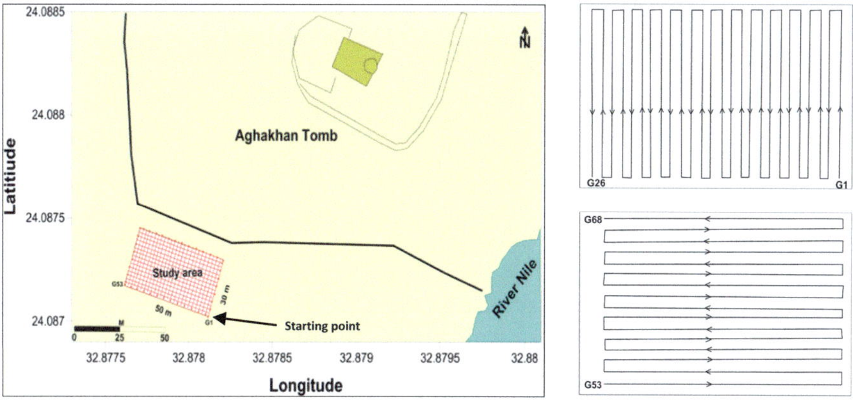

Fig. 7 The location map of GPR profiles in the study area of Aghakhan site (left), The GPR profiles measured in zigzag pattern: 26 profiles in X-direction (upper right) and 16 profiles in Y-direction (lower right)

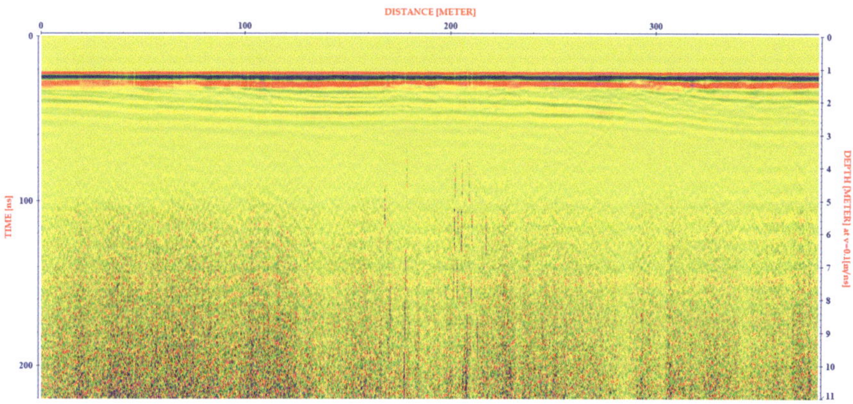

Fig. 8 GPR profile as a raw data before applying static correction

be interpreted as buried archaeological features or considered as doorways leading to underground burial chambers or tombs, as evidenced by the previously discovered tombs (Fig. 14 and Fig. 16).

7 GPR Time-Slice Maps

Viewing amplitude changes in a series of horizontal time slices within the ground is analogous to studying geological and archaeological changes in equal time depth

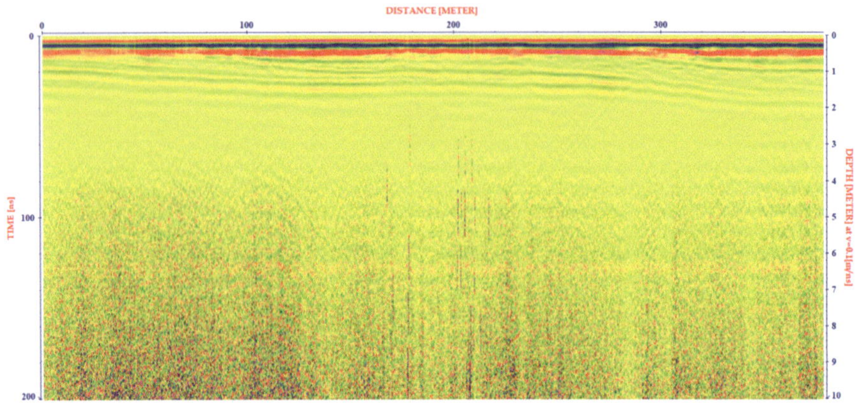

Fig. 9 GPR profile after applying move start time static correction

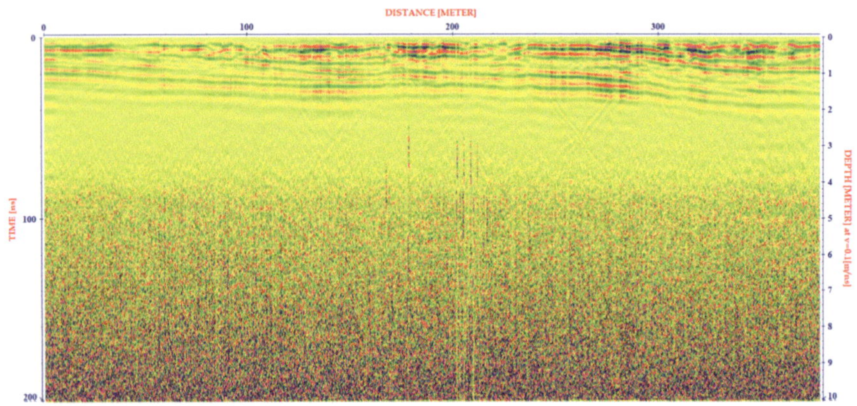

Fig. 10 GPR profile after applying background removal filter

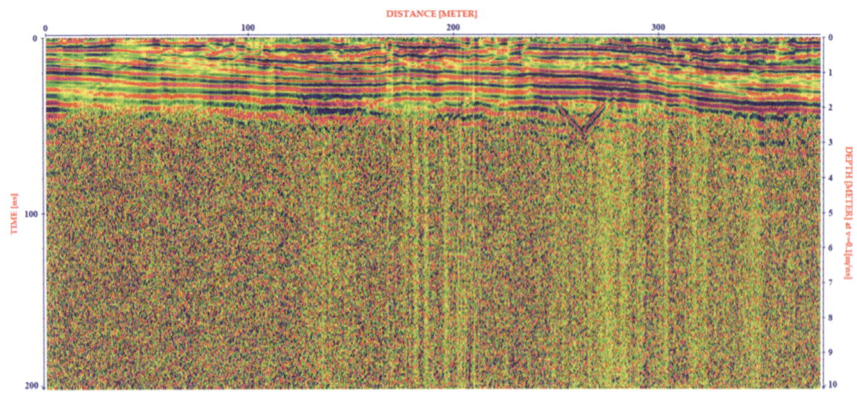

Fig. 11 GPR profile after applying band pass filter

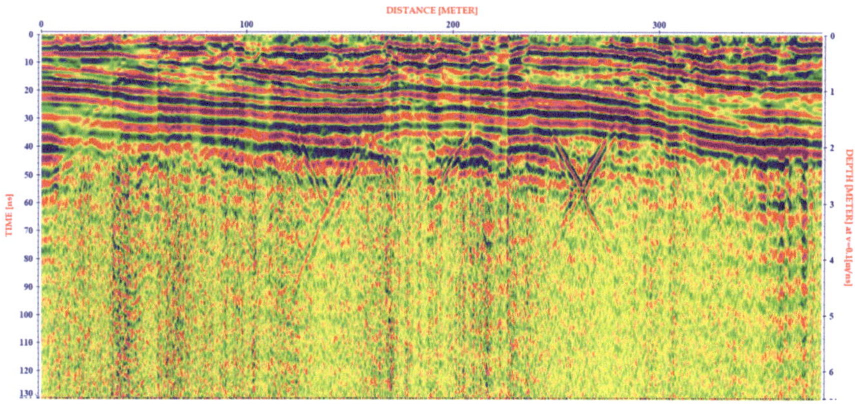

Fig. 12 GPR profile after applying running average

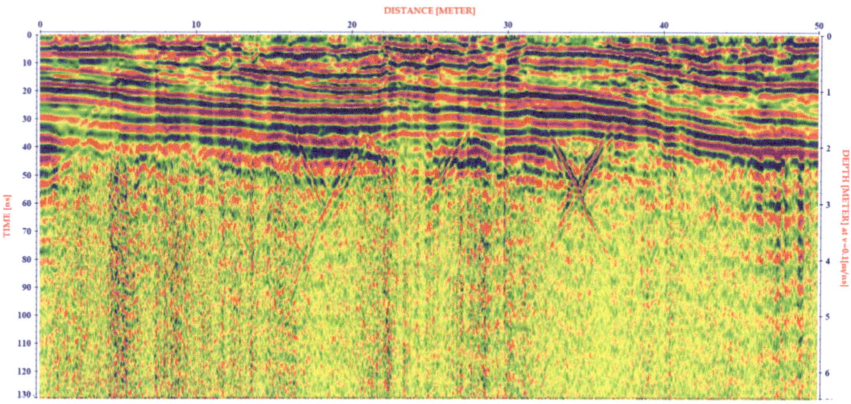

Fig. 13 GPR profile after applying length correction (trace interpol 3D file)

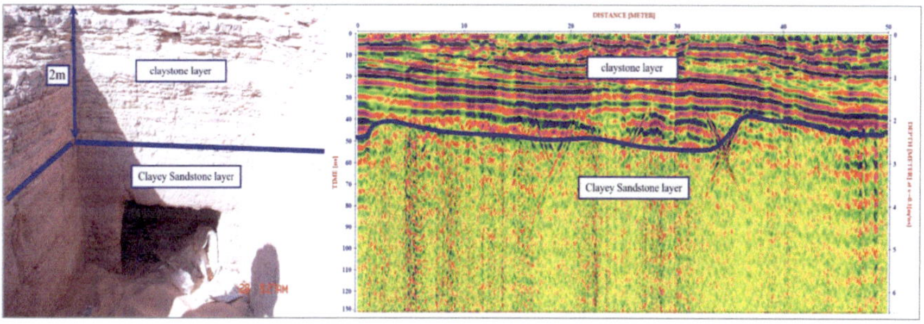

Fig. 14 Comparison between one of the opened tombs (left) in the study area and the radargram for it (right) showing two distinct sediment zones

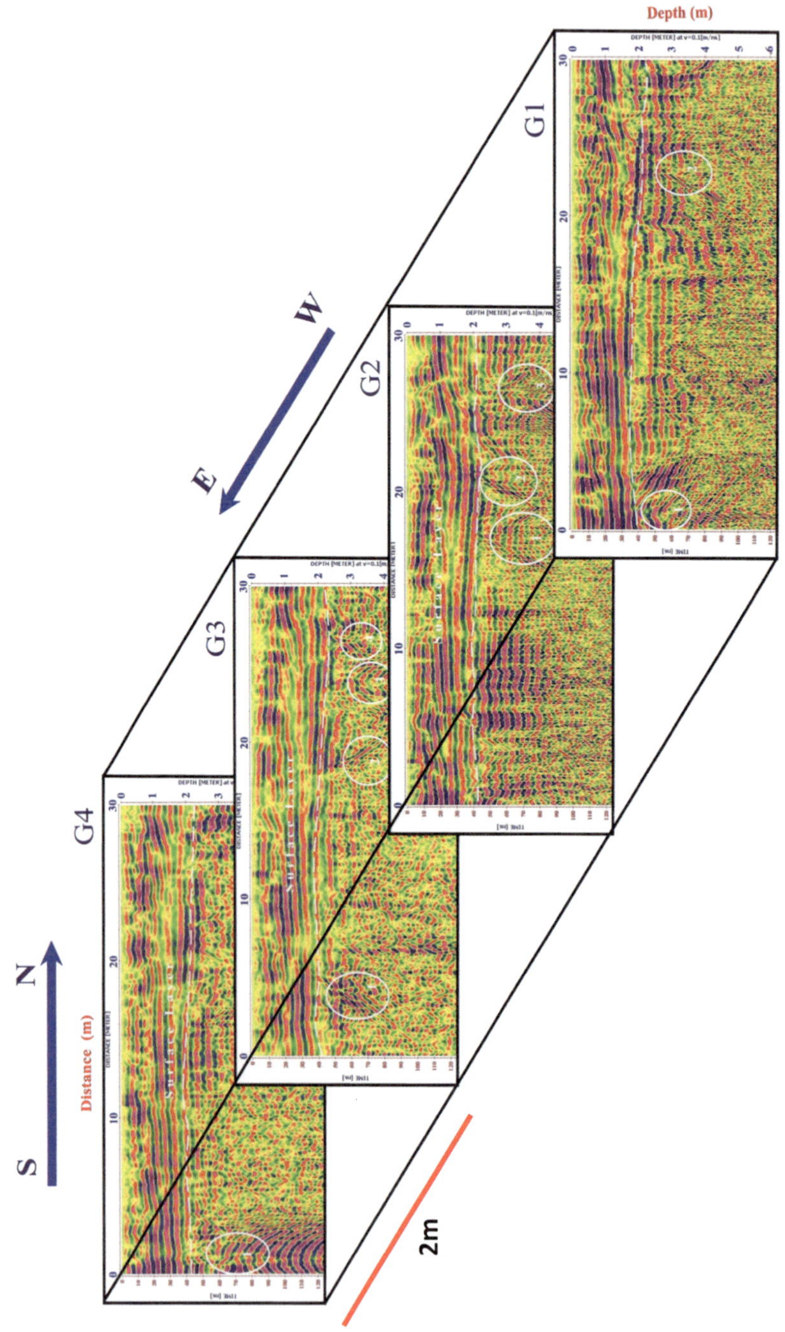

Fig. 15 Two-dimensional GPR crossesction running in X-direction

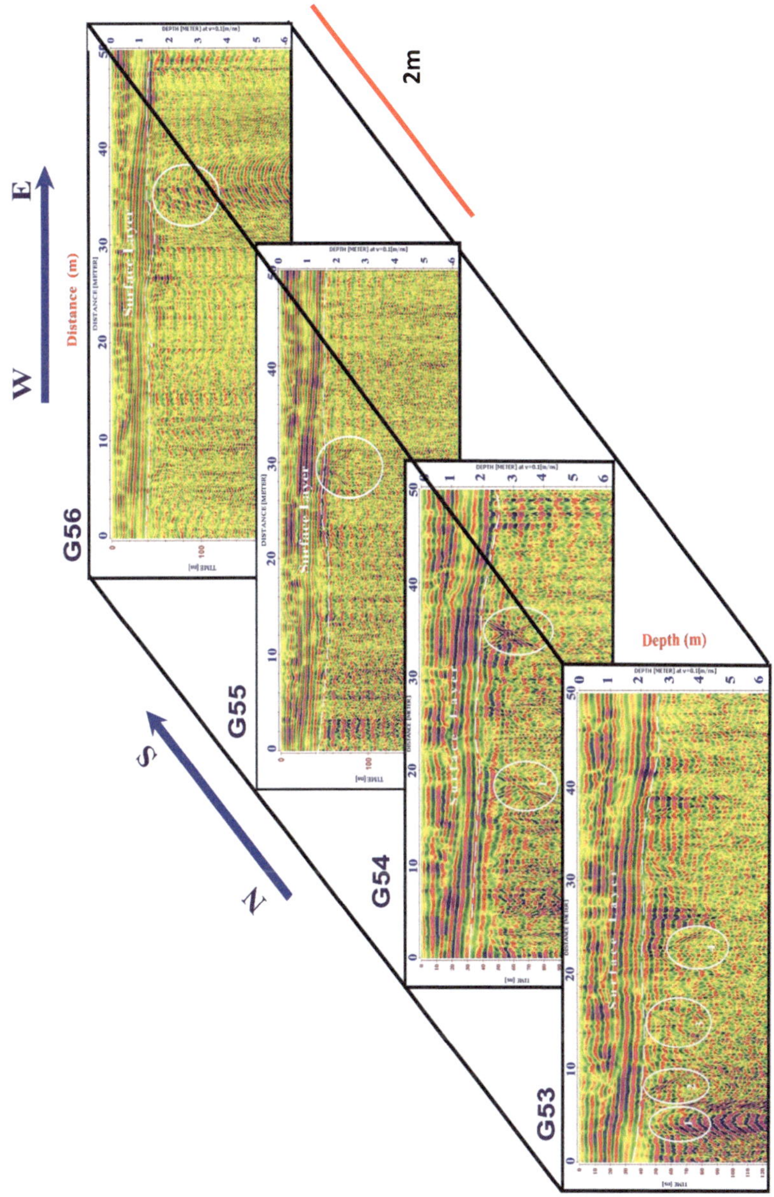

Fig. 16 Two-dimensional GPR crossesction running in Y-direction

layers (Arnold et al. 1997; Goodman et al. 1995; Malagodi et al. 1996; Milligan and Atkin 1993; Conyers and Goodman 1997). The possibility to map this site in three dimensions using amplitude slice maps enables for an interpretation of the archaeological remains found on the location. Because it reflects the consistency of the subsurface units, the fluctuation in the amplitude of the reflected waves is crucial. Therefore, the higher the contrasting velocity at a buried interface, the greater the amplitude of the reflected wave (Forte et al. 2014). The change in amplitude could be related to important archaeological features and stratigraphy, the location of higher or lower amplitudes at specific depths can be utilized to reconstruct the subsurface in three dimensions. Areas of low amplitude waves indicate uniform matrix materials or soils, while those of high amplitude denote area of high subsurface contrast such as buried archaeological features, voids or important stratigraphic changes (Conyers and Goodman 1997). For the study area, shallow GPR time slices maps have been generated. Warm colors (red, orange) correspond to relatively high reflections, whilst cool colors (blue, green) correlate to comparatively low reflections. In the current time slicing maps (Fig. 17a and b), walls of the construction typically appear as linear and polygon reflections, which can be used to define structure extents. Hence, walls and columns generally produce relatively large amplitude reflection, as a result of constructing velocities between walls and surrounding materials.

8 MASW Data Acquisition, Processing and Interpretation

A seismic survey can help evaluate the geotechnical properties of a subsurface material by detecting surface wave dispersion. The most practical technique to determine the elastic characteristics of shallow subsurface material is to use the Multi-Channel Analysis of Surface Wave (MASW) method. Sinkholes (ex. Pharaonic tombs) filled with loose sands have a lower shear wave velocity than surrounding material (Anbazhagan et al. 2018; Mohamed et al. 2019); as a result, differential velocities can contribute in identification of those buried tombs.

Traditionally, this technique uses multiple numbers of receivers deployed in a linear style of equal receiver interval with each receiver attaching to an individual recording channel as seen in Fig. 18 (Dobrin and Savit 1988). One record consists of more than twelve traces of seismic wave have been stacked at different distances from the source. Continues acquisition of multichannel surface wave data along linear pattern resulted in the generation of S-wave velocity profile (1-D S-wave velocity model, i.e. Vs vs. depth) at the center of geophone spread (Park et al. 1998a, b; 1999a, b; 2007). Because data are gathered in the standard CMP format, phase velocities of ground roll can be obtained from each shot gather so that the numerous 1-D S-wave profiles along a survey line can be produced. A two-dimensional (2-D) vertical S-wave velocity cross-section can be created by integrating the 1-D velocity models along the surveyed profile into 2-D profile. This cross-section contains information about horizontal and vertical continuity and physical characteristics of materials at different depths. It has lately shown great promise in detecting shallow cavities and

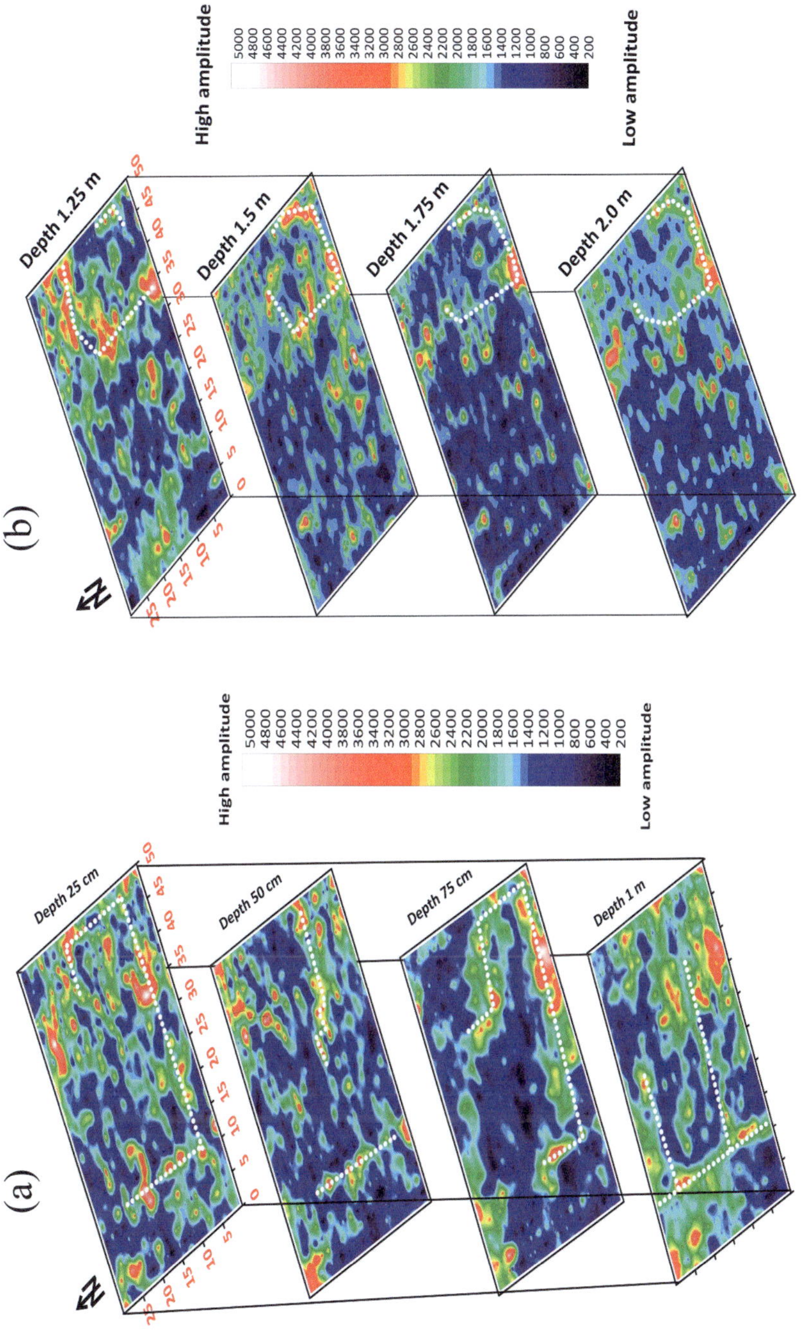

Fig. 17 Shallow time slices for the sudy area; **a** from 25 cm to 1 m depth, **b** from1.25 m to 2.0 m

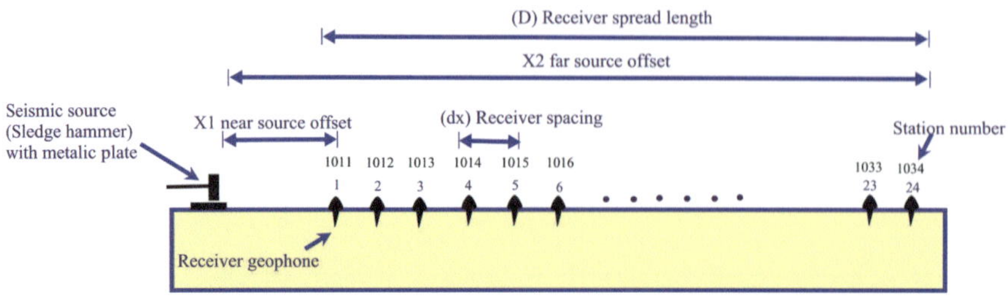

Fig. 18 Schematic of MASW field survey

tunnels (Park et al. 1998a), mapping the bedrock surface (Xia et al. 1998; Miller and Xia 1999).

Notable variations in S-wave velocity are anticipated at boundaries between several materials. The main aim of this study was to identify, and map buried different strata and any anomalous zones (Anbazhagan et al. 2018), it could be interpreted as archaeological features like tombs or buried walls and pillars at western bank of Nile River (Nolan et al. 2011). MASW profiles have been conducted along eights profile in Aghakhan site in order to locate the unexcavated monuments at target depth 2–3 m below ground surface.

For data acquisition 48-channel signal enhancement seismograph has been used in the present study. Geophones are used to convert the seismic energy into a measurable electrical voltage, In MASW survey Low frequency geophones of 4.5 Hz are always recommended (Foti et al. 2018).

A seismic source is usually used to produce a large enough signal into the ground to ensure sufficient depth penetration and high enough resolution to image the subsurface, a fairly heavy sledgehammer will be a good choice at the most of soil site. Metallic plate is used to help the source impact point to become less intrusive into soil (Asabere et al. 2016).

In comparison to body-wave survey techniques such as reflection or refraction, the surface wave seismic method always has a far greater tolerance in the selection of optimum field parameters. The essential reason for this tolerance is that the surface waves have the strongest power through all other kinds of seismic waves, ensuring the highest signal-to-noise ratio (S/N). Selection of some key parameters has to be addressed before actual surveying take place (Ivanov et al. 2000). This issue was briefly summarized in Table 2. Eight seismic profiles were conducted along the same GPR profiles location for verifying the results (Fig. 19). The profiles have a total length of 292 m, with four of them (M1 to M4) representing the first group that runs N-S and the second group (M53 to M56) running E-W and perpendicular to the first group.

Table 2 Field parameters of MASW profiles in Aghakhan site

Field parameters	Profiles							
	M1	M2	M3	M4	M53	M54	M55	M56
Near source offset (×1) (m)	4	4	4	4	4	4	4	4
Far source offset (×2) (m)	28	28	28	28	52	52	52	52
Geophone type (Hz)	4.5	4.5	4.5	4.5	4.5	4.5	4.5	4.5
Length of receiver spread (D) (m)	24	24	24	24	48	48	48	48
Receiver spacing (dx) (m)	1	1	1	1	1	1	1	1
Seismic source	Sledgehammer 20Ib with metallic plate							
Total recording Time (T) (ms)	1000	1000	1000	1000	1000	1000	1000	1000
Sampling interval (ms)	0.5	0.5	0.5	0.5	0.5	0.5	0.5	0.5
Low cut filter	No	No	No	No	No	No	No	No

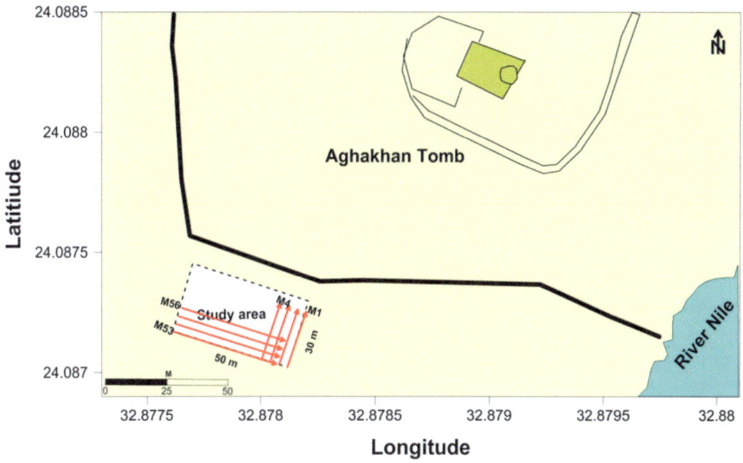

Fig. 19 Location map of MASW profiles in the study area of Aghakhan site

9 Steps for Data Processing

1. Convert data from SEG-2 to the KGS format. Most of engineering seismographs today have adopted the SEG-2 format as standard. The Kansas Geological Survey (KGS) adopted a format called (KGS) modified SEG-Y or simply (KGS) format, so seismic data downloaded from a seismograph need to be convert into KGS format before being used with the SurfSeis program (KGS 2006).
2. Apply field geometry to data based on the location of the source and geophones. this option imports the generated output file (KGS), and then you could specify the desired settings in the graphical field setup box, such as geophone spacing, source receiver spacing, and source receiver movement direction.

3. Appling F-K filter; it is a 2-D (time-offset) filter that removes all seismic events (Fig. 20), whose linear velocities fall into a specified range (Yilmaz 1987). To specify a range of the filtering velocity, draw two lines on the displayed seismic data.
4. Picking of the dispersion curve; this step is the most critical because it has the greatest influence on the confidence in the final Vs output. In other words, the Vs output will have, at best, as much confidence as the dispersion curve provided to the inversion step. For each shot, surface wave phase velocities of the fundamental mode were manually picked with frequency from 10 to 45 Hz with an interval 1 Hz. Corresponding signal –to –noise ratio (S/N) is displayed, this ratio is calculated from the background overtone data if exist, and therefore can be used as an indicator to judge the quality of the new data (Fig. 21).
5. Inversion of dispersion curves; the inversion uses the dispersion curve as the only empirical data with no reference to the original seismic record. Inversion step includes invert each dispersion curve individually, and the results from the inversion analysis basically contain shear–wave velocity (Vs) information in 1-D depth and surface location format (Fig. 22).
6. Using of Surfer V.18. (Golden software package) for contouring the obtained shear wave seismic data.

The MASW data were contoured and displayed in a 2D-shape to reveal lateral and vertical variations in the underlying strata, as well as any anomalies in seismic wave velocity that may be caused by buried archaeological objects. (Fig. 23).

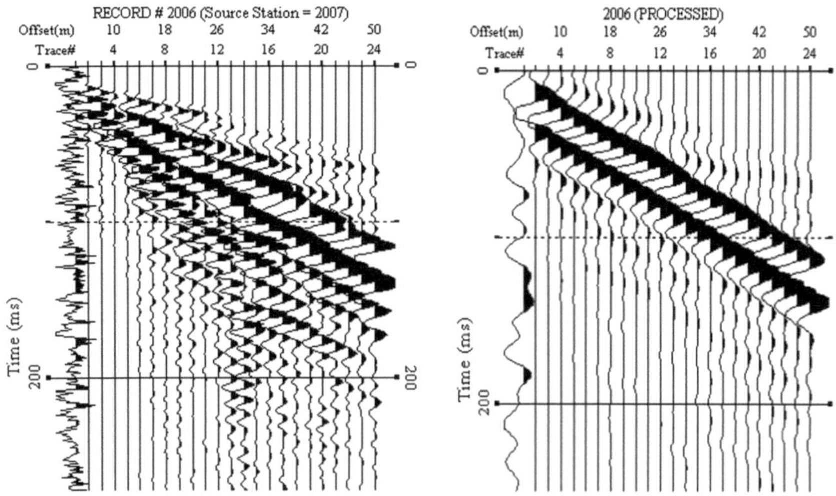

Fig. 20 A raw shot gather (left); noise-filtered shot gather by applying F-K filter (right) for profile M54

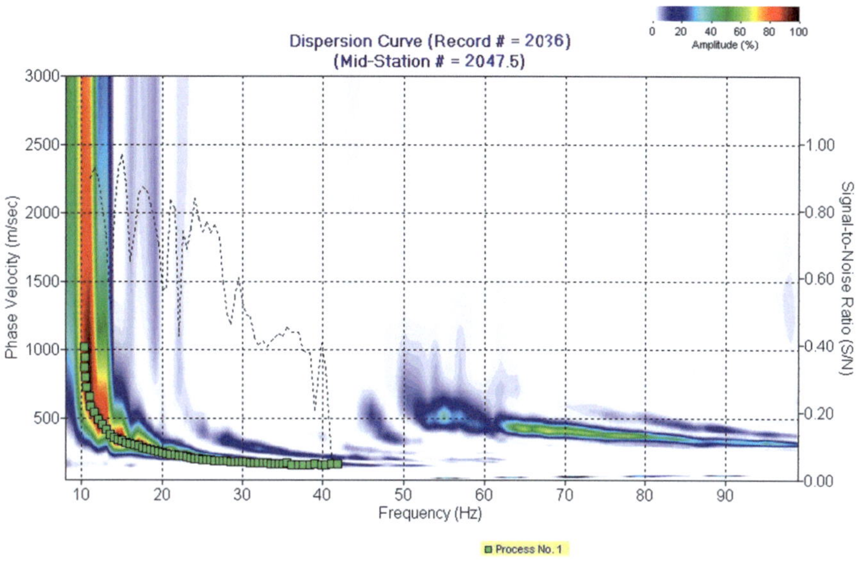

Fig. 21 Dispersion curve showing the approximate phase velocity and reference frequency for profile M54

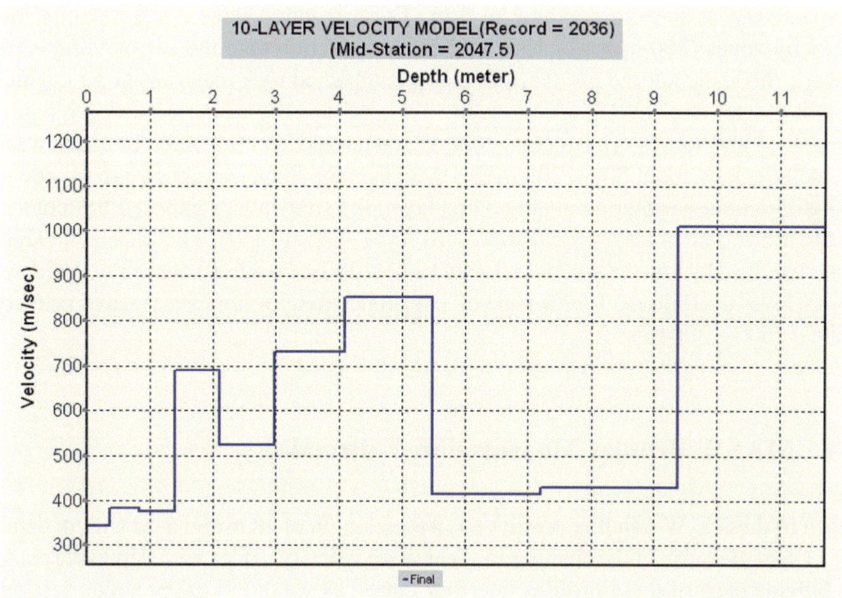

Fig. 22 The measured 1-D shear wave velocity model (phase velocity vs depth) for profile M54

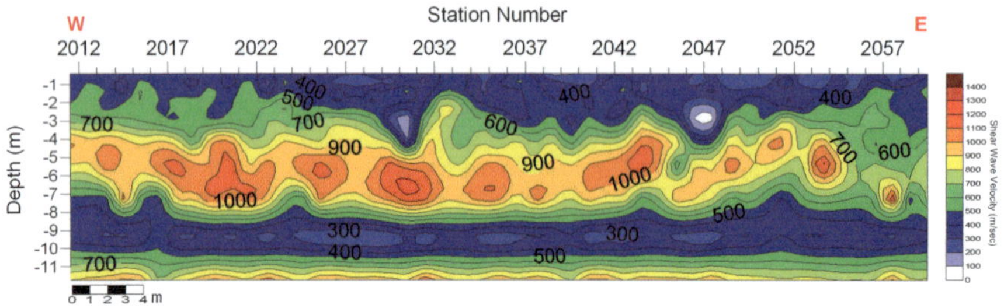

Fig. 23 The constructed 2-D shear wave velocity cross-section for profile M54

10 MASW Profiles Measured in X-direction

These profiles almost run north–south, with a total length of 24 m and a maximum depth of 11.5 m for each profile. (Fig. 24). Obviously, according to the variations in shear wave velocity, each profile can be classified vertically into four zones. The first zone, with a moderate shear wave velocity of 300 to 500 m/sec and an average thickness of 3 m, could be interpreted as claystone based on the rock exposure in the study area. (Fig. 14 left). Features of particular interest along profile M1, M2, M3 and M4 are located beneath stations 5012–5013, 6027–6033, 7017–7030 and 8013 respectively, at average depths 2 to 3 m. These features show a significantly low velocity zones (200 m/sec or less) in 2-D cross-section than the surrounding earth layers, this indicates the existence of a non-compacted rock materials or air cavities (Anbazhagan et al. 2018). As a result, it's possible they're buried archaeological chambers and tombs. The second zone characterized by high value of shear wave velocity values about 500–1200 m/sec, with average thickness about 3–5 m, due to the differential weathering acted on this layer, the study area's exposed lithological sequence reveals a clayey sandstone hard layer. (Fig. 14 left). The third zone/layer characterized by low shear wave velocity values (300–500 m/sec), interpreted as claystone layer overlain the fourth zone of sandstone strata of high shear wave velocity values (500–1300 m/sec).

11 MASW Profiles Measured in Y-direction

Each of these E-W trending profiles has a total length of 48 m and a maximum depth of 11.5 m. Because of the changes in shear wave velocity values with depth, we could subdivide each of these profiles into four zones, as we did in the previous section. (Fig. 25). The first zone, interpreted as claystone layer, is characterized by modest shear wave velocity values (300–500 m/sec) and average thickness of around 3 m, with increasing thickness at some locations throughout the profiles. Anomalies of very low shear wave velocity values observed beneath stations 1014–1019–1033, 2030–2047, 3039 and 4047 for profiles M53, M54, M55 and M56 respectively, these

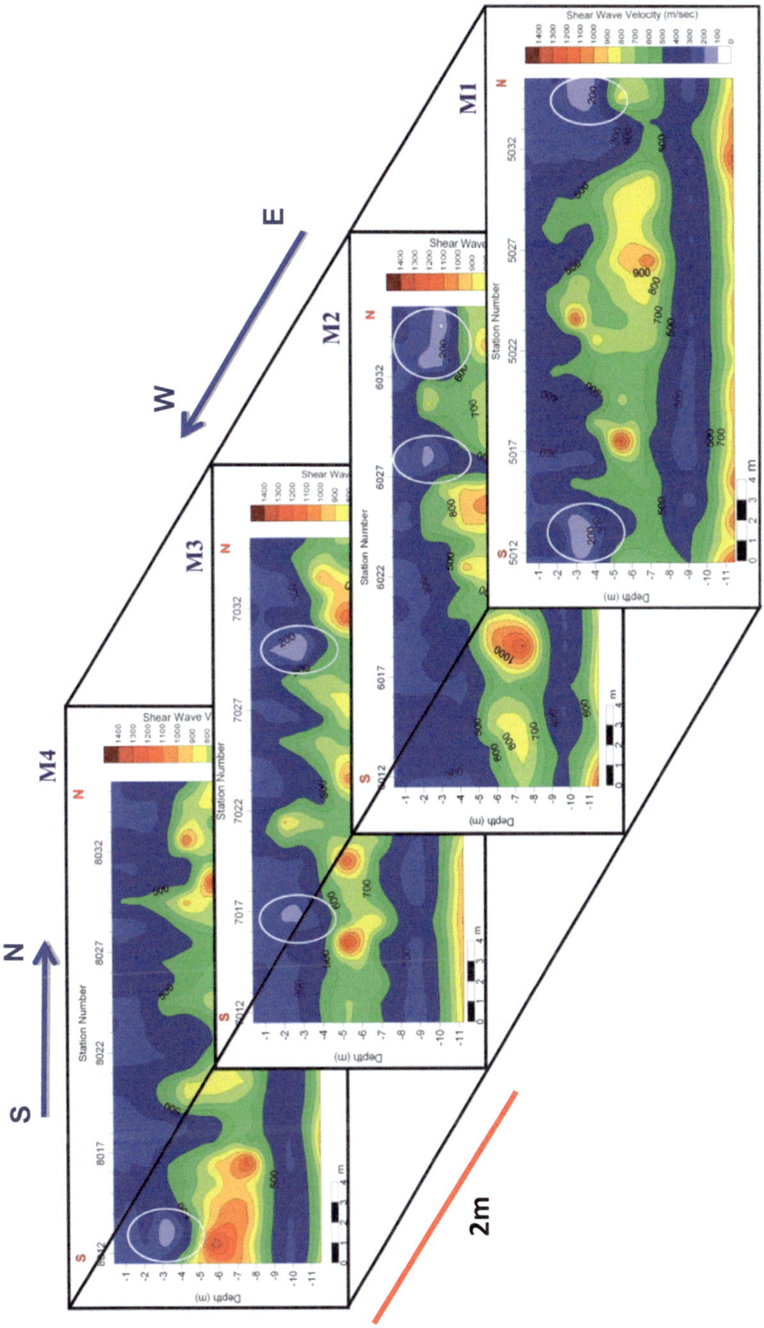

Fig. 24 Two-dimensional MASW crossesction in N-S direction

anomalies could be interpreted as buried tombs (Sloan et al. 2013; Anbazhagan et al. 2018). The second zone appears at depth 1.5–3 m with Vs values about 300–500 m/sec, and average thickness 7 m. The third zone appears at an average depth about 8 m with low shear wave velocity values (300–500 m/sec) and average thickness 3 m and classified as claystone layer. The fourth zone (sandstone layer) appears at depth 10.5 m, with high values of shear velocity (500–800 m/sec) due to the compaction effect of the above layers.

12 GPR's and MASW's Cross-Sections Comparison

In terms of position and depth of the inferred cavities, the results of the two geophysical techniques show good agreement (tombs). Figure 26 shows the anomalies appeared in the two obtained cross-sections at a distance range of 3–7–13–22.5 m with depth range from 2 to 3 m. Figure 27 illustrates the anomalies that appeared in the two obtained 2-D cross-sections at a depth of 2.5 to 3 m, and with comparable distances 19–35 m from the starting point of the two cross-sections. High reflectivity electromagnetic waves and low shear wave velocity anomalies could be interpreted as buried ancient tombs.

13 Discussion and Conclusions

Egypt is recognized for its profound civilization roots that stretch back thousands of years in human history. This civilization left behind a large number of antiquities. The ancient Egyptians regarded the west bank of the Nile River to be the region of death. So, we encouraged discovering tombs and other archaeological relics in Aghakhan archaeological site by utilizing two geophysical tools. GPR is a powerful method for identifying buried archaeological buildings in a non-destructive manner, and it acts as a guide for verifying the MASW results. In the study region several near-surface cavities (tombs) were identified, the comprehensive survey in the area which containing exposures of claystone and sandstone reveals zones of high GPR wave reflectivity and low seismic S-wave velocity. Consequently, these anomalous zones, which are found at depth between 2 and 3 m below ground surface, could be interpreted as buried archeological tombs. This perception is based on the excavated tombs which locate at the same depth from the ground surface in the study region. In addition to this, the constructed time-slice maps from the GPR data displays linear and polygon structures characterized by high reflection of electromagnetic waves. Therefore, those shapes may be interpreted as walls of ancient structures in the study area.

We can conclude that; several significant anomalies of archaeological potentiality were defined and represent a very promising for archaeological investigation and tourism industry. Also, there is a very clear similarity between GPR's and MASW's

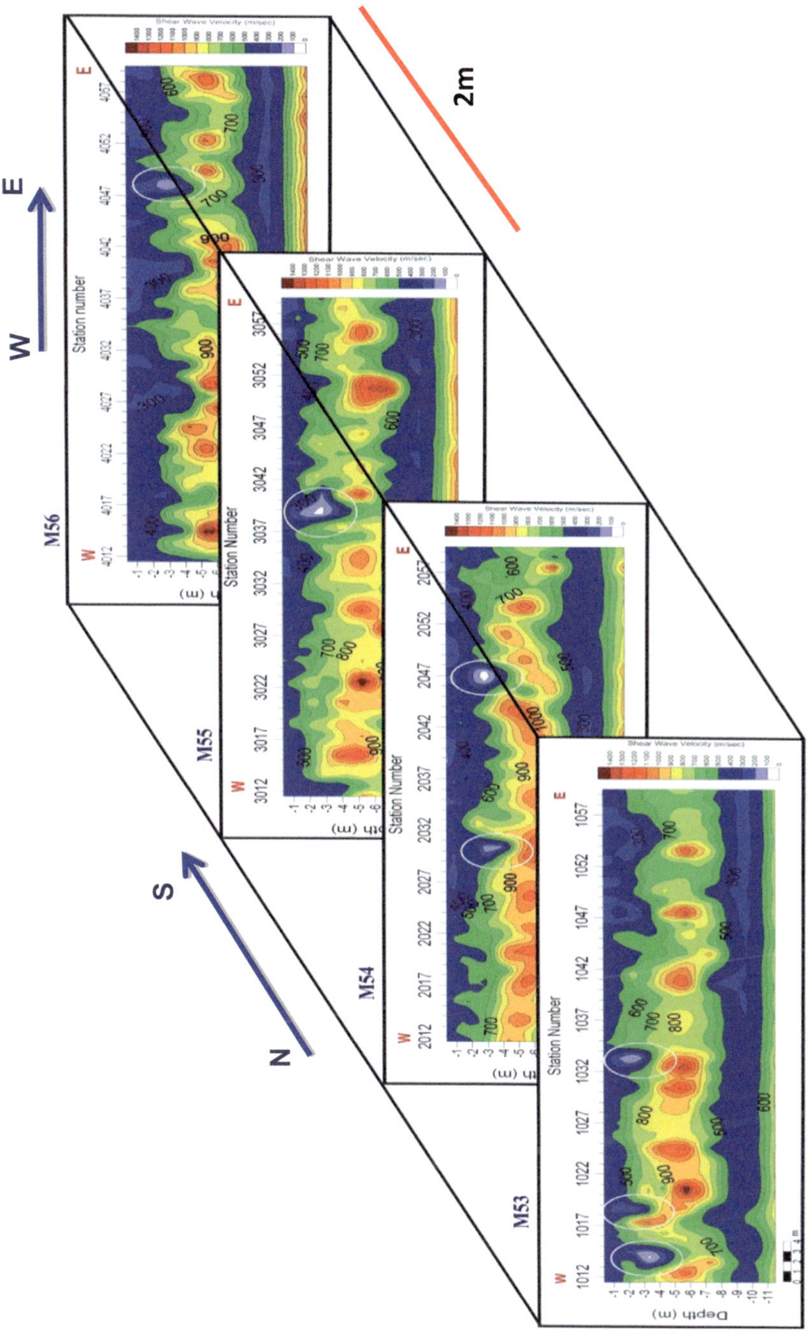

Fig. 25 Two-dimensional MASW crosssection in E-W direction

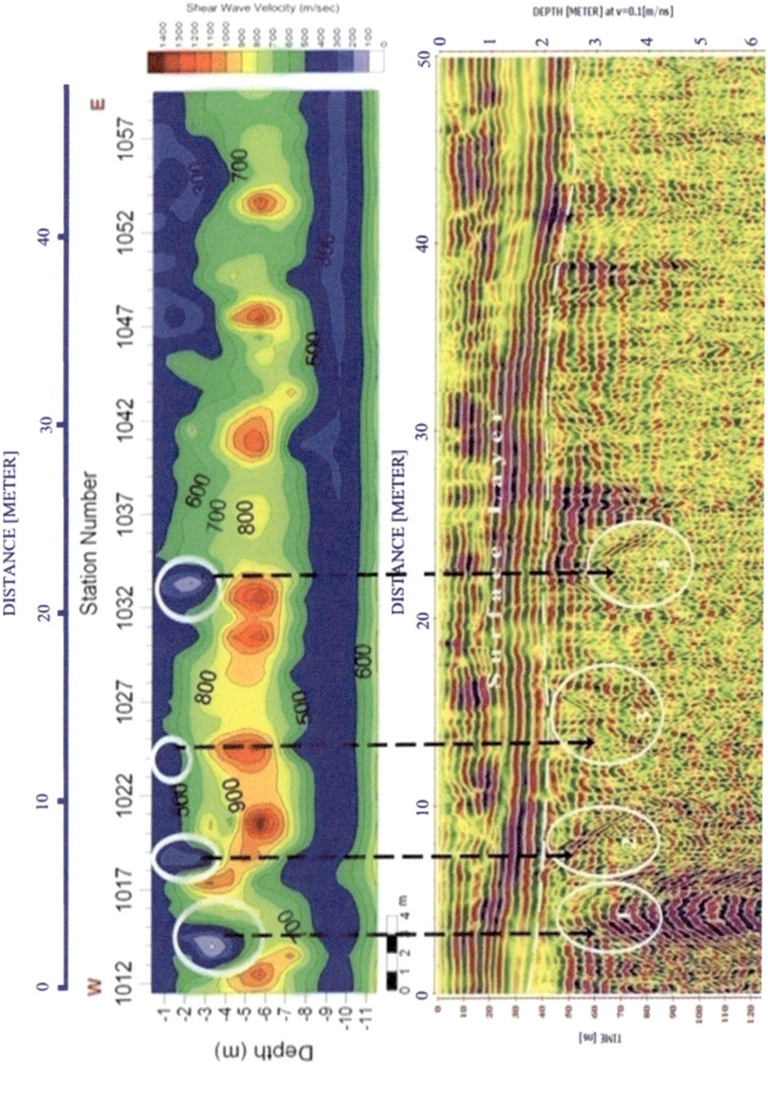

Fig. 26 Comparison showing good agreement in results from two different geophysical methods: 2-D sections of MASW (profile M53) and GPR (profile G53)

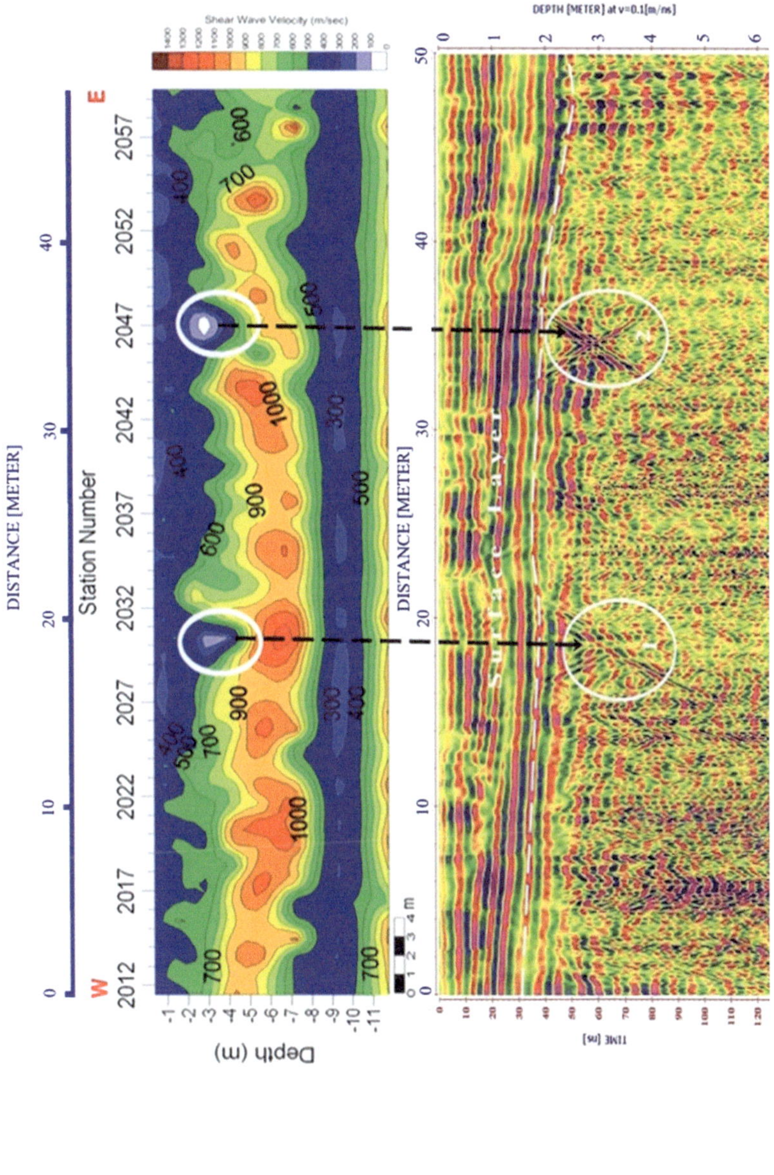

Fig. 27 Comparison showing good agreement in results from two different geophysical methods: 2-D sections of MASW (profile M54) and GPR (profile G54)

results, through the common anomalies that associated between them. Besides its fundamental use in geotechnical investigations, the MASW approach emphasizes its capability to outline the subsurface anomalies of archaeological potentiality.

Acknowledgements This study has been carried out as part of a successful collaboration between the National Research Institute of Astronomy and Geophysics and the Egyptian Ministry of State of Antiquities. Many thanks for all individuals who contribute to the collection of the data used in this work. The authors greatly appreciate the valuable comments from the unknown reviewers which helped for enhancing the final shape of the research.

References

Abbas AM, Abdallatif TF, Shaaban FA, Salem A, Sub M (2005) Archaeological investigation of the Eastern extension of the Karnak temple using ground penetrating radar and magnetic tools. Geoarchaeol Int J 20(5):537–554

Abd El Rahman MM (1991) Rock materials competence assessed by seismic measurements with emphasis on soil competence scales and their applications in some urban areas in Yemen. In: Proceedings of the 9th Annual Meeting on EGS. vol 9, pp 206-228

Alwasif MA (2006) Application of geophysics for archaeological prospection and conservation of some archaeological sites in the western bank of Luxor, Egypt. M. Sc. Thesis, Faculty of Science, Ain-Shams University

Anbazhagan P, Rohit D, Prabhakaran A, Vidyaranya B (2018) Identification of karstic features in lateritic soil by an integrated geophysical approach. Pure Appl Geophys 175(2018):4515–4536

Annan AP (2003) Ground penetrating radar: principles, procedures, & applications. Sensors & Software Inc., Technical Paper

Arnold JE, Ambos EL, Larson DO (1997) Geophysical surveys of stratigaphicaaly complex island california sites: new implications for household archaeology. Antiquity 71:157–168

Asabere P, Coe JT, Mahvelati S (2016) The effects of base plate material on dispersion curves acquired during multichannel analysis of surface wave (MASW) testing. In: Proceedings of SAGEEP. pp 464–468

Attia MI (1955) Topography, geology and iron ore deposits of East Aswan. Geology Survey, Egypt, Cairo

Baines J, Malek J (1992) Atlas of ancient Egypt. Andromeda, Oxford, p 240p

Black AC, Norton WW (1993) Blue guide Egypt. Bedford, London, p 762p

Bloxam EG, Heldal T, Storemyr P (2007) Characterisation of complex quarry landscapes: an example from the West Bank quarries, Aswan, QuarryScapes report, Deliverable 4. p 275. www.quarryscapes.no

Booth A, Linford NT, Clark RA, Murray A (2008) Threedimensional, multi-offset ground-penetrating radar imaging of archaeological targets. Archaeol Prospect 15:93–112

BS 8004 (1986) Code of practice for foundations

Conyers LB, Goodman D (1997) GPR an introduction for archaeologists: Walnut Creek. Alta Mira Press, CA, p 232p

Conyers LB, Ernenwein EG, Bedal L (2002) Ground-penetrating radar (GPR) mapping as a method for planning excavation strategies, Petra, Jordan, E-tiquity, n. 1. http://e-tiquity.saa.org/%7Eetiquity/title1.html

Conyers LB (2009) Ground-penetrating radar for landscape archaeology: method and applications. In: Campana S, Piro S (eds) Seeing the unseen: geophysics and landscape archaeology. CRC Press/Balkema, Leiden, pp 245–255

Conyers LB (2012) Interpreting ground-penetrating radar for archaeology. Left Coast Press, Walnut Creek, CA

Cardimona S (2002) Subsurface investigation using ground penetrating radar. In: 2nd International conference on the application of geophysical and NDT methodologies to transportation facilities and infrastructure. Los Angeles, California

Davis A, Taylor SD (1979) Dynamic elastic moduli logging of foundation materials. In: Ardus DA (ed) Proceeding of International conference on offshore site investigation. Graham and Trotman Limited, London

Dobrin MB, Savit CH (1988) Introduction to geophysical prospecting, 4th edn. McGraw-Hill, Inc., p 867

Dumbleton MJ, West G (1976) A guide to site investigation procedure for tunnels, DoE/DTp. Transport and Road Research Laboratory Report LR740, p 24

Egyptian Geological Survey and Mining Authority (EGSMA) (1981) Geological map of Egypt, Ministry of Industrial and Mineral Resources

El-Naggar ZR (1970) On a proposed lithostratigraphic subdivision of the late cretaceous–early paleogene succession in the Nile Valley, Egypt, U.A. R 7th. Arab Petrol Congr Kuwait 64(B-3):1–50

El Shazly EM (1977) The geology of the Egyptian region. In: Nairn EM, Kanes WH, Stehli FG (eds) The ocean basins and margins 4A. Plenum, New York, pp 379–444

El-Qady G, Hafez M, Abdella MA, Ushijima K (2005a) Imaging subsurface cavities using geoelectric tomography and ground-penetrating radar. J Cave Karst Stud 67(3):174–181

El-Qady G, Nishimura Y, Kinjyo K, Ushijima K (2005b) Integrated archaeological exploration using geophysical methods at Futenma area, Okinawa prefecture Japan. BUTSURI-TANSA (Geophysical Exploration) 58(5):491–502

Fat-Helbary RE-S, El-Faragawy KO, Hamed A (2019) Application of HVSR technique in the site effects estimation at the South of Marsa Alam City, Egypt. J Afr Earth Sc 154:89–100. https://doi.org/10.1016/j.jafrearsci.2019.03.015

Fat-Helbary RES, El-Faragawy KO, Hamed A (2019) Soil geotechnical characteristics for seismic risk mitigation at the southern extension of Marsa Alam city, Egypt. NRIAG J Astron Geophys 8(1):1–14

Forte E, Dossi M, Pipan M, Colucci RR (2014) Velocity analysis from common offset GPR data inversion: theory and application to synthetic and real data. Geophys J Int 197(3):1471–1483. https://doi.org/10.1093/gji/ggu103

Foti S, Hollender F, Garofalo F, Albarello D, Asten M, Bard YP, Comina C, Cornou C, Cox B, Giulio G, Forbriger T, Hayashi K, Lunedei E, Martin A, Mercerat D, Ohrnberger M, Poggi V, Renalier F, Sicilia D, Socco S (2018) Guidelines for the good practice of surface wave analysis: a product of the InterPACIFIC project. Bull Earthquake Eng 16:2367–2420. https://doi.org/10.1007/s10518-017-0206-7

Golden Software, LLC. 809 14th Street, Golden, Colorado 80401. www.goldensoftware.com

Goodman D, Nishimura Y, Rogers JD (1995) GPR time-slices in archaeological prospection. Archaeol Prospect 2:85–89

Hamed A (2019) Seismic microzonation for earthquake risk mitigation at the Southern extension of Marsa Alam City. Ph.D.'s Thesis, Aswan University, Aswan, Egypt

Hemeda S (2018) Engineering failure analysis and design of support system for ancient Egyptian monuments in Valley of the Kings, Luxor, Egypt. Geoenviron Disasters 5:12. https://doi.org/10.1186/s40677-018-0100-x

Imai T, Yoshimura Y (1975) The relation of mechanical properties of soils to P and S-wave velocities for ground in Japan. OYO Corporation, Technical Note

Issawi B (1969): The geology of Kurkur-Dungul area. Geol Surv Egypt Paper 46:102

Issawi B, Osman RA (1993) Tectonic–sedimentary synthesis of the paleozoic-cretaceous clastics, southwest Aswan. Egypt. Sediment Egypt 1:11–21

Issawi B, Osman RA (1996) The sandstone enigma of South Egypt. Cairo University, Third international conference on geology of the Arab World, pp 359–380

Issawi B (2005a) Archean-phanerozoic birth and development of the Egyptian land. In: Geology of the Tethys. Proceedings of the first international conference on the geology of the Tethys, Cairo University. pp 339–380
Isswai B (2005) Glacial and interglacial phases during the late Ordovician-Early Silurian in the Gabgaba Formation, South Eastern Dessert. Egypt J Sedim 13:407–410
Ivanov J, Park CB, Miller RD, Xia J (2000) Mapping poisson's ratio of unconsolidated materials from a joint analysis of surface-wave and refraction events. In: Proceedings of the symposium on the application of geophysics to engineering and environmental problems (SAGEEP 2000). pp 11–19
Jaeger JC, Cook NGW (1969) Fundamentals of rock mechanics. Elsevier Publ. Comp, New York
Kamai T (2015) Applications of the multi-channel analysis of surface waves exploration for archeological investigations–cases of three castle ruins. In: 16th century Japan–near-Surface Asia Pacific Conference. Waikoloa, Hawaii. https://doi.org/10.1190/nsapc2015-072
Kansas Geological Survey (KGS), 2006 Kansas Geological Survey (KGS) (2006) SurfSeis [computer software], Ver 2 05
Keary P, Brook M (2002) An introduction to geeophysical exploration, 4th edn. Blackwell Science
Kinnaer D (2003) The ancient Egyptian site. http://www.ancient-egypt.org/index.html
Kraus JD (1950) Antennas. McGraw-Hill, New York
Khedr ES, Youssef AAE, Abou Elmagd K, Khozyem HM (2010) Tectono-stratigraphic subdivision of the clastic sequence in Aswan area, southern Egypt. In: Proceedings of the fifth international conference on the geology of the Tethys Realm. South Valley University, pp 197–216
Maijala P (1992) Application of some seismic data processing methods to ground penetrating radar data. In: Hanninen P, Autio S (eds) Fourth international on conference on ground penetrating radar. Geological survey in finland special paper 16. Rovaniemi, Finland, pp 103–110
Majed M (2019) Graeco-Roman era tomb discovered in Upper Egypt [Blog post]. https://egyptindependent.com/egyptian-italian-archeological-mission-unearths-greco-roman-tomb-in-aswan/
Malagodi S, Orlando L, Piro S (1996) Approaches to increase resolution of radar signal. In: Proceedings of the sixth international conference on ground penetrating radar. Department of Geoscience and Technology, Tohoku University, Sendai, Japan, pp 283–288
McDowell PW (1981) Recent developments in geophysical techniques for the rapid location of near-surface anomalous ground conditions by geophysical methods. Ground Eng 14:20–23
Miller RD, Xia J (1999a) Using MASW to map bedrock in Olathe, Kansa Geological Survey Open-file Report 99–9
Milligan R, Atkin M (1993) The use of ground-probing radar within a digital environment on archaeological sites. In: Andresen J, Madsen T, Scollar I (eds) Computing the past: computer applications and quantitative methods in archaeology. Aarhus University Press, pp 21–33
Mohamed AME, El-Hussain I, Deif A, Araffa SAS, Mansour K, Al-Rawas G (2019) Integrated ground penetrating radar, electrical resistivity tomography and multichannel analysis of surface waves for detecting near-surface caverns at Duqm area, Sultanate of Oman. Near Surface Geophys. https://doi.org/10.1002/nsg.12054
Nolan JJ, Sloan SD, Broadfoot SW, McKenna J, Metheny OM (2011) Near-surface void identification using MASW and refraction tomography techniques. In: SEG technical program expanded abstracts. pp 1401 1405
Osman R, Ahmed AS, Khater TM (2002) The stratigraphy and facies associations of Wadi Gabgaba and its surroundings with an emphasis on the lower Paleozoic glaciations. In: 6th International Conference on Geology. Arab World Cairo University. vol. 2, pp 469 482
Osman R, Ahmed AS, Khater TM (2005) The geological development of Wadi Gabgaba, Eastern Desert, Egypt. In: 1st Int. Conf. on the Geol. Of Tethys, Cairo Unive. pp 465–476
Othman AA (2005) Construed geotechnical characteristics of foundation beds by seismic measurements. J Geophys Eng 2(2):2005126–2005138
Park CB, Xia J, Miller RD (1998a) Ground roll as a tool to image near surface anomaly [Exp. Abs.]. Soc Explor Geophys 874–877

Park CB, Xia J, Miller RD (1998b) Imaging dispersion curves of surface waves on multi-channel record. In: 6th annual international SEG technical program expanded abstracts. pp 1377–1380
Park CB, Miller RD, Xia J (1999a) Multichannel analysis of surface waves. Geophysics 64:800–808
Park CB, Miller RD, Xia J, Hunter JA, Harris JB (1999b) Higher mode observation by the MASW Method. In: SEG technical program expanded abstracts. Society of Exploration Geophysicists
Park CB, Miller RD, Xia J, Ivanov J (2007) Multichannel analysis of surface waves (MASW)—active and passive methods. Lead Edge 26:60–64
Papadopoulos N, Sarris A, Yi M, Kim J (2009) Urban archaeological investigations using surface 3D ground penetrating radar and electrical resistivity tomography methods. Explor Geophys 40:56–68
Rojansky V (1979) Electromagnetic fields and waves. Dover, Mineola, N.Y.
Shaaban FA, Shaaban FF, Abbas AM, El-Essaway AH (2003) Mapping of buried archaeological relics using GPR survey at the Isis Temple, bahbeit el hegara area, Egypt. In: Bulletin of National Research Institute of Astronomy and Geophysics (NRIAG). pp 111–133
Sandmeier KJ (2001) Program for processing and interpretation of reflection and transmition data, D-76227. Karlsruha, Germany
Shawarby A, Fathy E, Sadek M, Amin N, Yousri R, Kayser, S (2009) National inventory and database of ancient stone-quarry landscapes in Egypt. In: Abu-Jaber N, Bloxam EG, Degryse P, Heldal T (eds) QuarryScapes: ancient stone quarry landscapes in the eastern mediterranean, geological survey of norway special publication. vol. 12, pp 155–163
Shih SF, Doolittle JA (1984) Using radar to investigate organic soil thickness in the Florida evergaldes. Soil Sci Soc Am J 48:651–656
Sitek D (2003) Ancient Egypt: history and Chronology. http://www.narmer.Pl/index.html
Sternberg BK, McGill JW (1995) Archaeology studies in Southern Arizona using ground penetrating radar. J Appl Geophys 33:209–225
Sloan SD, Nolan JJ, Broadfoot SW, McKenna JR, Metheny OM (2013) Using near-surface seismic refraction tomography and multichannel analysis of surface waves to detect shallow tunnels: a feasibility study. J Appl Geophys https://doi.org/10.1016/j.jappgeo.2013.10.004
Storemyre P (2010) http://www.per-storemyr.net/gharb-aswan-archaeology/
Stuempel H (1984) The use of seismic shear waves and compress signal waves for lithological problems of shallow sediments. Geophys Prospect 32:662–675
Tsokas GN, Papazachos CB, Vafidis A, Loucoyannakis MZ, Vargemezis G, Tzimeas K (1995) The detection monumental tombs buried in tumuli by seismic refraction. Geophysics 60:1735–1742
Yilmaz O (1987) Seismic data processing. In: Doherty SM (ed) Investigations in geophysics, no. 2.
Yoshimura S, Kondo J, Hasegawa S, Sakata T, Etaya M, Nakagawa T, Nishimoto S (1996) Preliminary report of the general survey at Dahshur north, Egypt. http://www.waseda.jp/prj-egypt/sites/Dhshr-E.html
Xia J, Miller RD, Park CB (1998) Construction of vertical seismic section of near-surface shear-wave velocity from ground roll [Exp.Abs.]. Soc Explor Geophys AEGE/CPS, BEIJING 29–33

Ground Penetrating Radar for Investigating Painted Walls and Floors of Ancient Buildings

Massimiliano Pieraccini and Lapo Miccinesi

Abstract In this paper the Authors report performances, application, and case studies of two different ground penetrating radars specifically designed for Cultural heritage investigations. The first radar has been designed for investigating painted walls. It operates without contact (up to 3 m of distance), it transmits 4 GHz bandwidth (that means a theoretical resolution of about 2 cm) and its mechanical positioner is able to scan a surface 1.4 m wide and 1.9 m high. The aim of this radar is to gather information about the shallow layers (the plaster structure and possible plaster voids affecting the stability of paintings) up to 0.5 m depth. The second radar operates in contact with the floor. It has 900 MHz bandwidth, and it is able to penetrate up to several meters in a structure that consists of masonry walls/floors and void spaces. Its aim is to detect girders, vaults cavities, and other buried structures like tombs or crypts. Both the sensors have been used for investigating the world-famous Great Hall (named also "Hall of five hundred") in "Palazzo Vecchio", the city hall of Florence, Italy. These radar sensors have provided valuable clues and factual elements for reconstructing the structure of the hall in the centuries, that is a question often disputed by the scholars.

Keywords Bartolomeo Ammannati · Cultural heritage · Fresco · Ground penetrating radar · Leonardo da Vinci · Mural painting · Masonry · Non-destructive testing · Renaissance

1 Introduction

Ground penetrating radar has already been demonstrated a powerful tool for non-digging, non-destructive investigation in archaeology (Catapano et al. 2012; Conyers 2013; Daniels 2000; Sambuelli et al. 2019) and in structural analysis of ancient buildings (Masini et al. 2010; Catapano et al. 2018).

M. Pieraccini · L. Miccinesi (✉)
Department of Information Engineering, University of Florence, via Santa Marta 3, 50139 Firenze, Italy
e-mail: lapo.miccinesi@unifi.it

G. M. El-Qady and C. Margottini (eds.), *Sustainable Conservation of UNESCO and Other Heritage Sites Through Proactive Geosciences*, Springer Geology,
https://doi.org/10.1007/978-3-031-13810-2_12

One of the Authors of this paper has worked in designing and in-field application of ground penetrating radar since at least fifteen years. This paper reports performances, application, and case studies of two different ground penetrating radars specifically designed for cultural heritage investigations. Both are based on Continuous Wave Step Frequency (CWSF) technology, that allows the highest possible Signal-to-Noise Ratio with other conditions being equal (Pieraccini 2017).

The first radar has been designed for investigating painted walls. It operates without contact. The aim of this radar is to gather information about the shallow layers (the plaster structure and possible plaster voids affecting the stability of paintings). The second radar has been designed to operate in contact with the floor. Both have been used for investigating the world-famous Great Hall "Sala Grande", named also "Salone dei Cinquecento" (Hall of five hundred) as it can hold up to 500 people. It is the largest and most known room of "Palazzo Vecchio" (Old Palace), the City Hall of Florence. This impressive hall is 54 m long, 23 m wide, and 18 m high (see Fig. 1).

In effect, the no-contact radar mentioned was specifically designed just for investigating the walls of the Great Hall. In 2003 one of the Authors operated this no-contact radar in several measurement sessions (lasted about three years) in the Great Hall providing impressive images of the wall interior (Pieraccini et al. 2005).

Fig. 1 Great Hall "Sala Grande", named also "Salone dei Cinquecento" (Hall of five hundred) as it can hold up to 500 people. It is the largest and most known room of "Palazzo Vecchio" (Old Palace), the City Hall of Florence

The aim of this extensive measurement campaign was to gather information about the possible location of remains of the legendary lost mural painting "Battle of Anghiari" that Leonardo da Vinci was commissioned to paint in one of the walls of the Great Hall.

Recently (in 2019) the authors of this paper have been once again required to investigate with a ground penetrating radar the Great Hall. The aim of this measurement campaign was to find traces of another famous masterpiece: the fountain of Juno by Bartolomeo Ammannati.

In 1555 the Florentine sculptor and architect Bartolomeo Ammannati was commissioned by Cosimo I de'Medici to create a monumental fountain in the Great Hall known as Fountain of Juno (Ferretti 2011). We would like to point out that the Great Hall is not at ground floor, but at the second floor. The access to the hall is via two long flights of stairs. The construction of a monumental fountain in the Great Hall would have been a remarkable engineering feat.

Unfortunately, this artwork seems that it never was installed on its own collocation in the Great Hall. The curved statues were collocated in the Pratolino Villa (on the hills near Florence) and later brought to the Boboli's Garden. Only recently (in 2011) the statues have been united in the courtyard of the Bargello museum (Fig. 2) Catapano et al. (2018).

Currently there is not any trace of the fountain in the Great Hall, therefore many open questions are disputed by the scholars: (a) where exactly the fountain should

Fig. 2 Reconstruction of Ammannati fountain at Bargello Museum, Florence, Italy

have been collocated in the Great Hall? (see Fig. 3) (b) was the fountain temporarily installed in the Great Hall? Were preliminary works carried out in the Great Hall for installing the fountain?

A ground penetrating radar investigation on the floor of the hall could give some clues about possible foundation works still under the paving or remains of pipes.

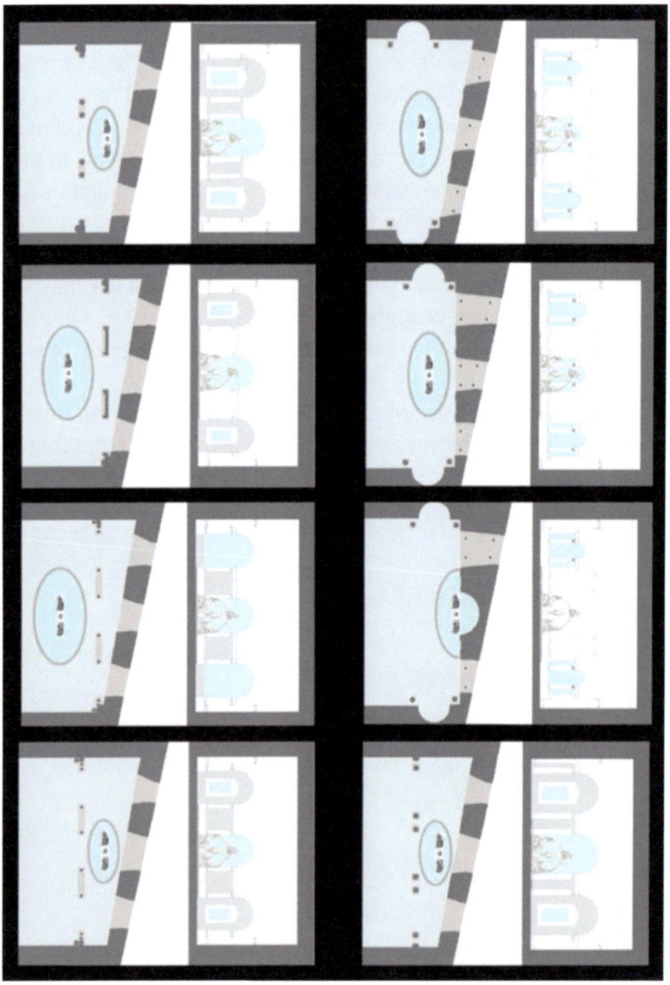

Fig. 3 Hypothetical collocations of Juno fountain (Ferretti 2011)

2 The Radars

A sketch of the most recent version of the high-frequency no-contact radar mentioned above is shown in Fig. 4. The mechanical frame is positioned in front of the wall under investigation and it scans 1.4 m in the horizontal axis and 1.9 m in vertical axis. Two step-by-step motors controlled by the acquisition software move the radar-head. The radar operates a Step Frequency Continuous Wave (SFCW) modulation. The central frequency is 10 GHz. The bandwidth is 4 GHz. A Vector Network Analyzer (HP 8720A) operates as transceiver. The radar-head provides an internal calibration path and amplification up to 19 dBm. The acquired raw data are focused by using a back-projection algorithm that takes into account the electromagnetic path between the antennas and any image point inside the wall (Pieraccini and Miccinesi 2018). Figure 5 shows a radar image obtained with this equipment in a controlled environment for evaluating its performances. The masonry test was a wall 40 cm thick. The image shows how the radar is able to detect the electrical conduits embedded in the wall masonry on the opposite side of the wall.

The other radar that the Authors operated in the Great Hall is a contact GPR. It is a prototype named ORFEUS (Parrini 2007; Pieraccini et al. 2019). The operative frequency range is 100 MHz–1 GHz (so the bandwidth is 900 MHz), the dynamic range is about 100 dB, the unambiguous range is 60 m, the transmitted power is 1 mW, the scan speed is up to 200 scan/s. Its antennas are two co-polarized loaded bowties. With respect to other commercial equipment this radar has a penetration capability

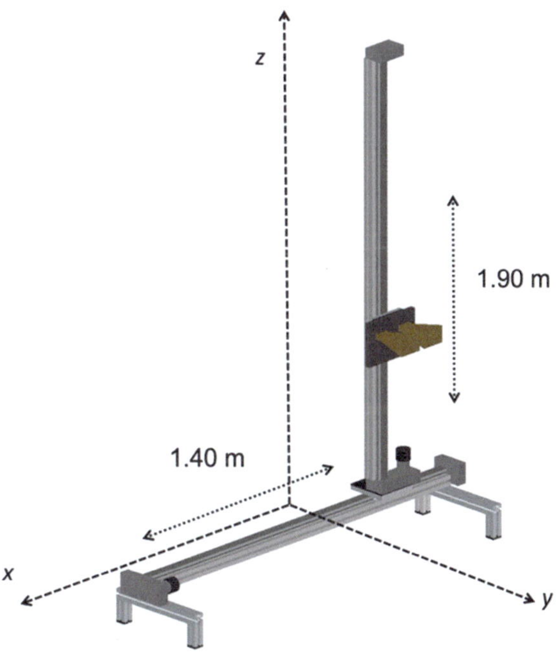

Fig. 4 Sketch of the no-contact high-frequency radar equipment

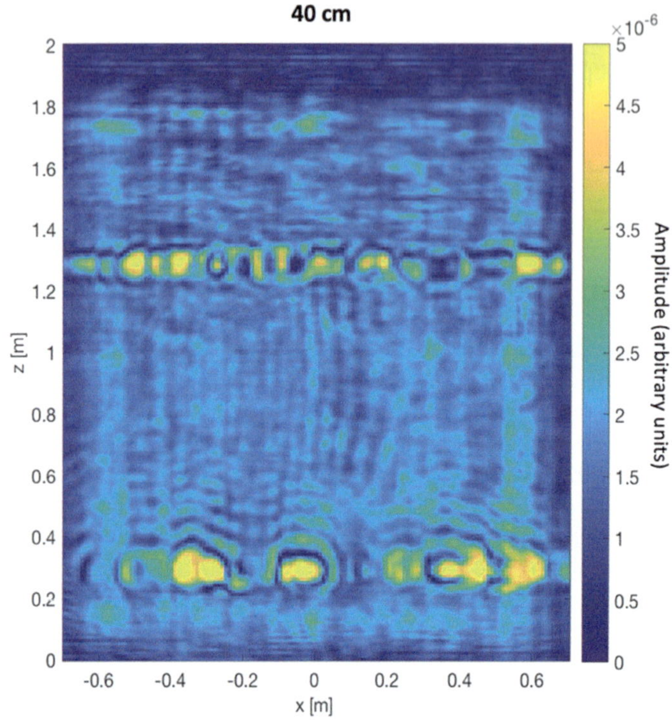

Fig. 5 Radar image obtained with the no-contact high frequency GPR scanner on a test wall 40 cm thick

30% larger and efficiency 20 dB better than stroboscopic pulse GPR (Parrini 2007). Figure 6 shows a picture of the equipment operating in the Great Hall.

3 No-Contact High Frequency GPR: The Case Study of the Walls of the Great Hall in "Palazzo Vecchio", the City Hall of Florence, Italy

A version of the no-contact radar mentioned above with only vertical scan and lower power (1 mW) was used for investigating the walls of the Great Hall. As the walls were painted by Giorgio Vasari in the sixteenth century, the no-contact operation was a mandatory requirement (Fig. 7).

Figures 8, 9, 10 and 11 show the obtained radar images. They are vertical scans. In all the images the interface air-wall is well-evident at about 3 m distance from the radar. Even the back-interface wall-air is clearly detectable at about 3.7 m. It means that the radar is able to penetrate the whole thickness of the wall.

Fig. 6 ORFEUS Ground Penetrating Radar scanning the floor of the Great Hall

About the interior of the walls, a cavity should give two signals (the interface masonry-void and the interface void-masonry). In the East wall we see a clear signal 15 cm inside. But, as we do not see the second signal, we argue that it is a discontinuity between a shallow layer (bricks) and the underlying stone wall. In the West wall this signal is less evident. Probably because the stone wall is an internal wall built with more regular stones. No evidence of a cavity able to preserve the painting of Leonardi da Vinci was found.

4 ORFEUS GPR: The Case Study of the Floor of the Great Hall in "Palazzo Vecchio", the City Hall of Florence, Italy

The scholars (Ferretti 2011) agree that the Juno fountain should have been collocated on the South side of the hall. Therefore, we investigated an area of the floor 21 m wide and 16 m large located as shown in Fig. 12.

The depth of the radar investigation has been evaluated assuming the speed of electromagnetic wave constant in the masonry and equal to 150 cm/ns.

Figure 13 shows the radar image (C-scan) at 0.11 m depth superimposed to the map of the second floor and the map of the ground floor. The frames of "Serena" rock between the floor sections of "Cotto" bricks are clearly detected.

Fig. 7 No-contact high frequency GPR operating on the east wall of the Great Hall

Figure 14 shows the radar image (C-scan) at 0.23 m depth superimposed to the maps of the second floor and the map of the ground floor. At this depth we are able to recognize a long girder (we suppose it, as it is not visible from ground floor) that crosses the Great Hall and that rests on the walls of a room at ground floor (see Fig. 15).

On the basis of these experimental results, we do not notice any feature that can be associated to water pipes, although a fountain should necessarily have them.

Nevertheless, we have clearly detected a long transversal girder that was not previously known. It stands exactly where the fountain was supposed to be collocated. It is possible that the aim of the girder was just to sustain the weight of the fountain.

5 Conclusions

The GPR is today a fundamental equipment in modern diagnostics of Cultural Heritage. It is able to perform non-destructive investigations of walls and floors. The standard GPR equipment operates in contact with the surface under test, but painted walls should be investigated without any contact. For this reason a special

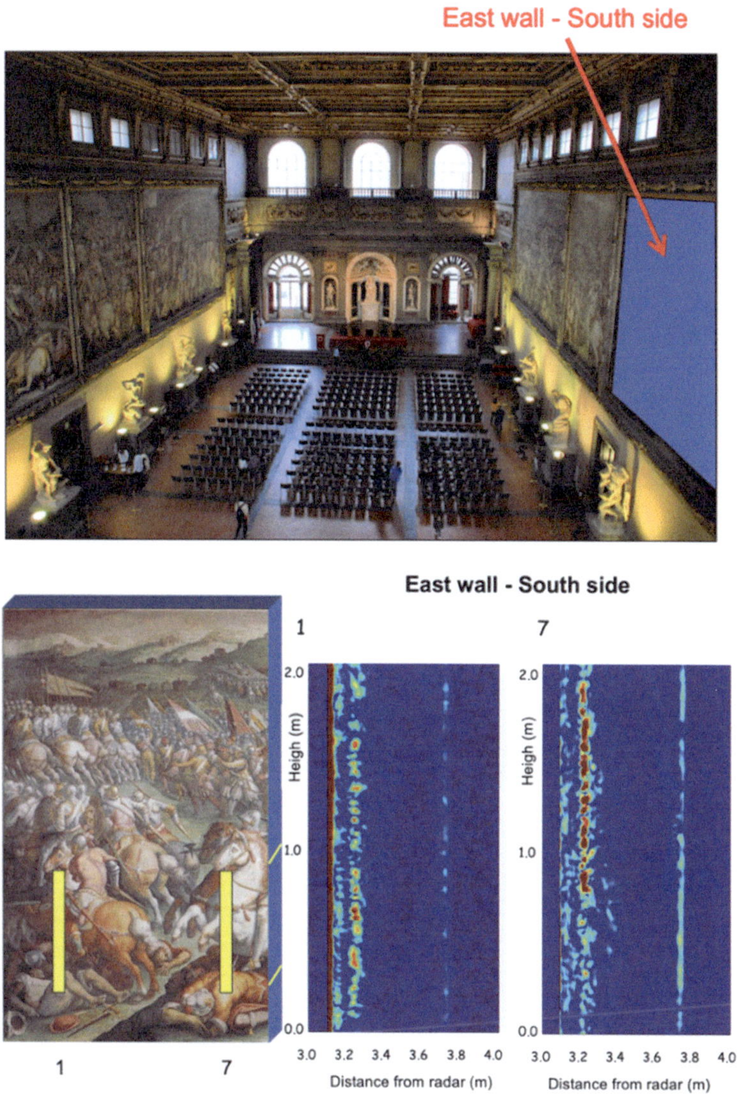

Fig. 8 Radar image of the East wall—South side

radar has been designed: its radar-head, moved by a mechanical scanner, produces images without direct contact with the surface under test.

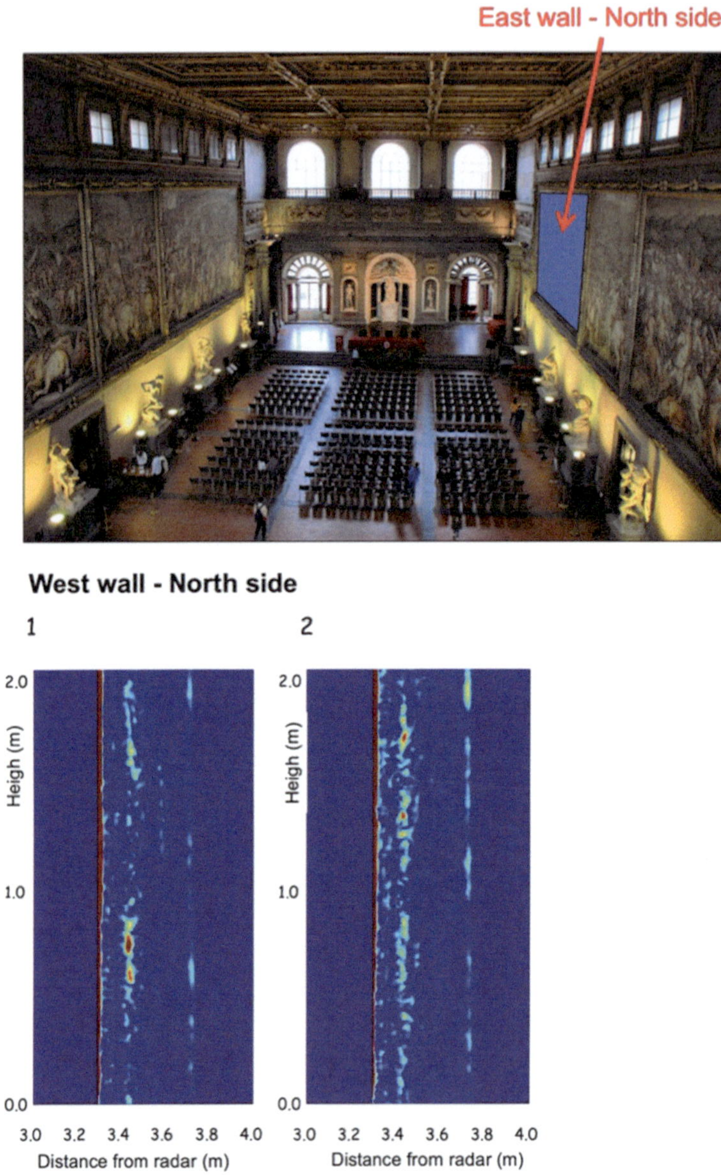

Fig. 9 Radar image of the East wall—North side

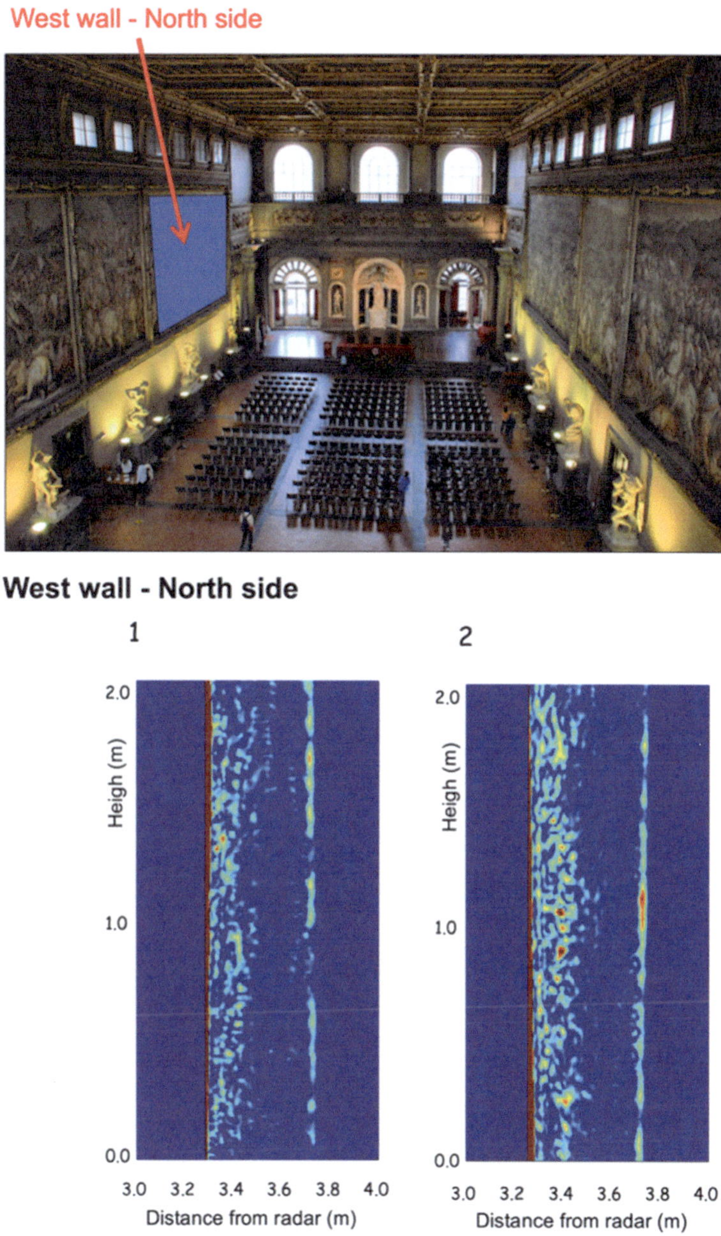

Fig. 10 Radar image of the West wall—North side

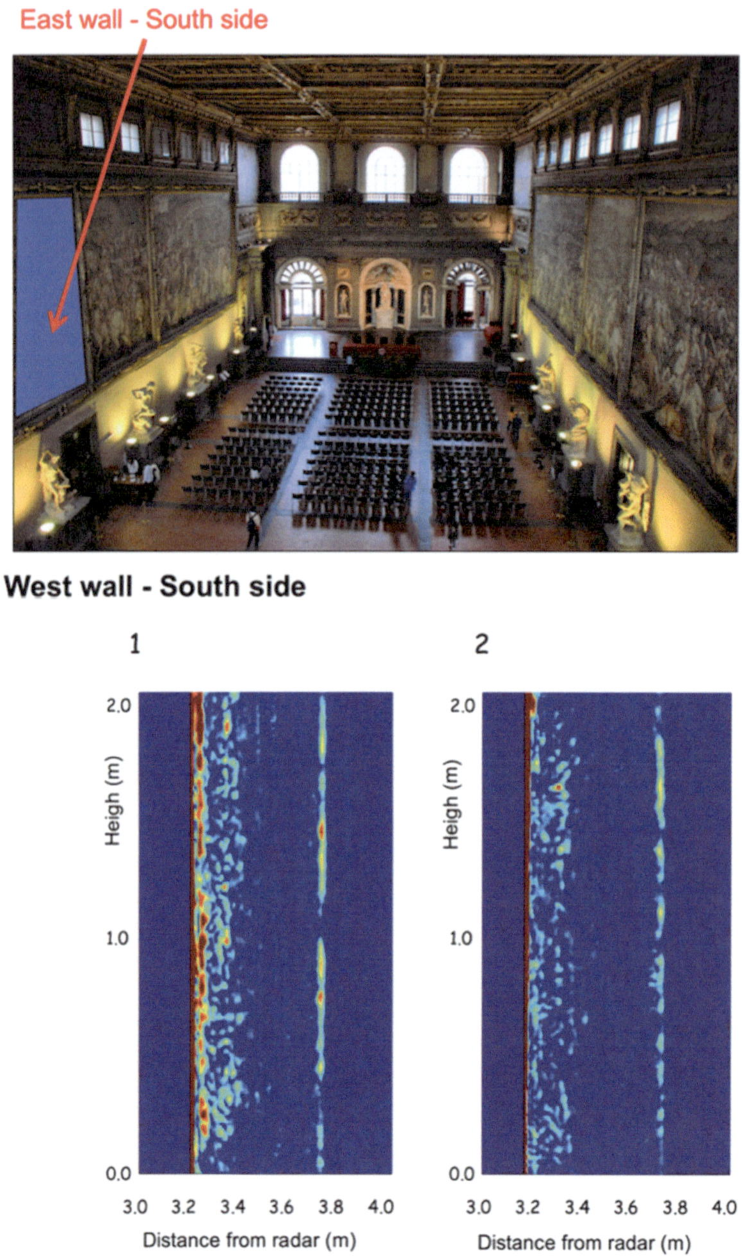

Fig. 11 Radar image of the West wall—South side

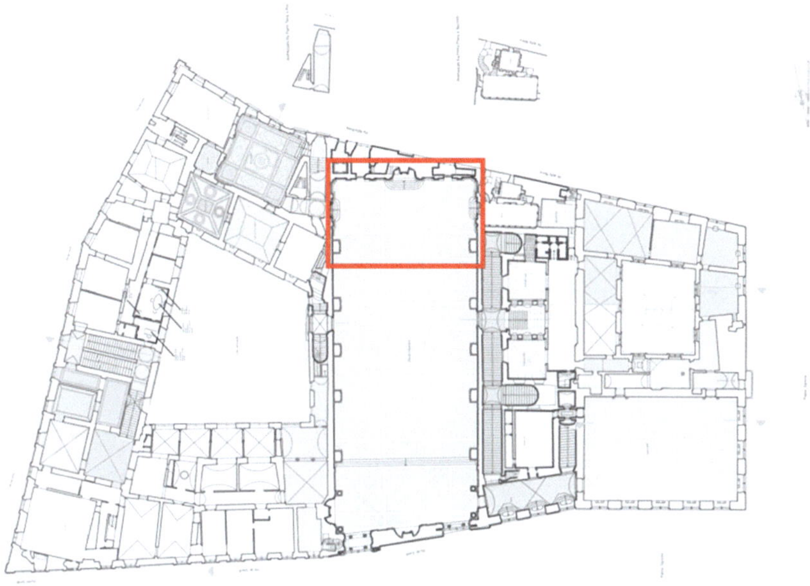

Fig. 12 Investigated area of the floor of the Great Hall

Fig. 13 Radar image of the floor of the Great Hall at 0.11 depth

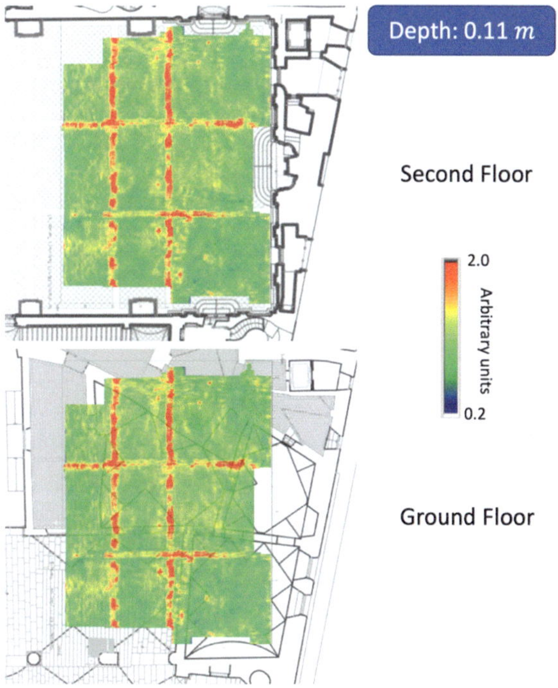

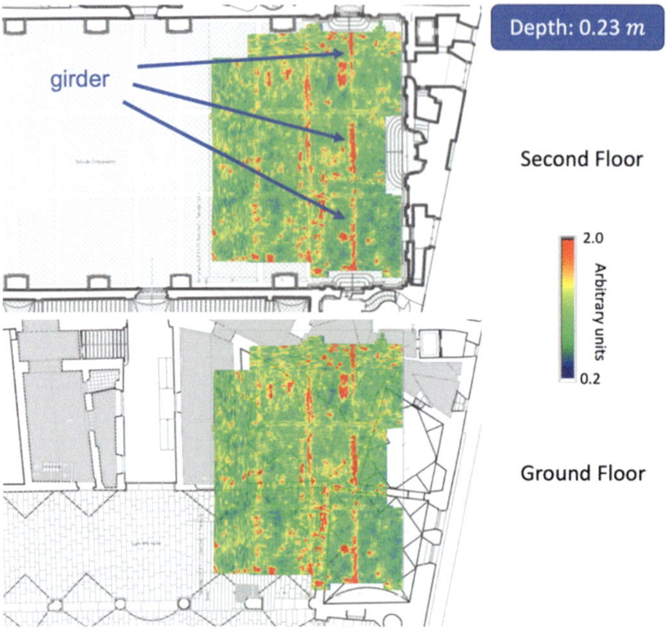

Fig. 14 Radar image of the floor of the Great Hall at 0.23

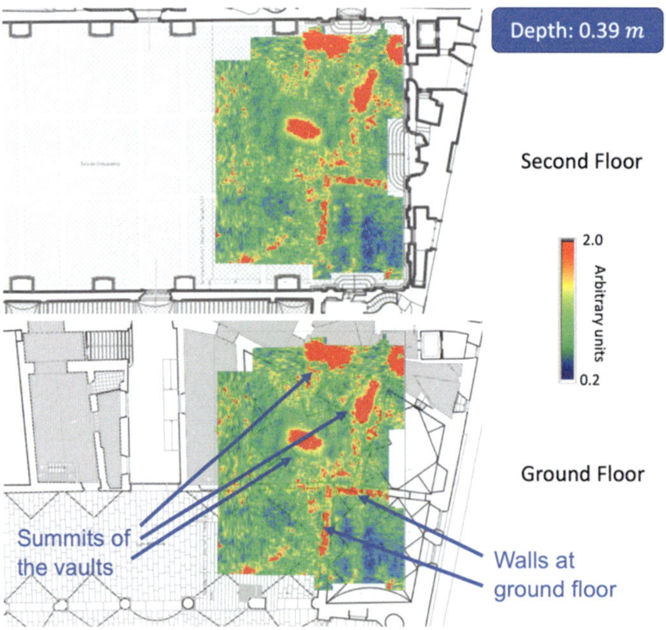

Fig. 15 Radar image of the floor of the Great Hall at 0.39

References

Catapano I, Crocco L, Di Napoli R, Soldovieri F, Brancaccio A, Pesando F, Aiello A (2012) Microwave tomography enhanced GPR surveys in Centaur's Domus, Regio VI of Pompeii, Italy. J Geophys Eng 9(4):S92–S99

Catapano I, Ludeno G, Soldovieri F, Tosti F, Padeletti G (2018) Structural assessment via ground penetrating radar at the Consoli Palace of Gubbio (Italy). Remote Sens 10(1):45

Conyers LB (2013) Ground-penetrating radar for archaeology, 3rd edn. AltaMira Press

Daniels JJ (2000) Ground penetrating radar for imaging archeological objects. In: Proceedings of the New Millennium international forum on conservation of cultural property. pp 247–265

Ferretti E (2011) Bartolomeo Ammannati, la Fontana di Sala Grande e le trasformazioni del Salone dei Cinquecento da Cosimo I a Ferdinando I, In: Paolozzi Strozzi B, Zikos D (eds) L'acqua, la Pietra, il fuoco. Bartolomeo Ammannati Scultore, ed. Giunti, Firenze

Masini N, Persico R, Rizzo E (2010) Some examples of GPR prospecting for monitoring of the monumental heritage. J Geophys Eng 7(2):190–199

Pieraccini M (2017) Noise performance comparison between continuous wave and stroboscopic pulse ground penetrating radar. IEEE Geosci Remote Sens Lett 15(2):222–226

Pieraccini M, Mecatti D, Luzi G, Seracini M, Pinelli G, Atzeni C (2005) Non-contact intrawall penetrating radar for heritage survey: the search of the 'Battle of Anghiari'by Leonardo da Vinci. NDT E Int 38(2):151–157

Pieraccini M, Miccinesi L (2018) No-contact GPR for investigating painted walls. In: 2018 17th International Conference on Ground Penetrating Radar (GPR). IEEE, pp 1–6

Pieraccini M, Miccinesi L, Garcia Canizares H (2019) Critical verification of the underground cartography of the municipality using a high performance ground penetrating radar. In: 10th international workshop on advanced ground penetrating radar. The Hague, The Netherlands

Sambuelli L, Comina C, Catanzariti G, Barsuglia F, Morelli G, Porcelli F (2019) The third KV62 radar scan: searching for hidden chambers adjacent to Tutankhamun's tomb. J Cult Herit

Laser Scanning and Virtual Reality

Developments in Digital Documentation of Cultural Heritage

Heinz Rüther, Roshan Bhurtha, Ralph Schroeder, and Bruce McDonald

1 Introduction

Man's relationship to the world and the environment, the understanding of self and religious views find expression in art and architecture. The design of buildings and the layout of settlements are guided by social structure and hierarchies as well as the need for protection. Architectural styles and the location and spatial relationships between residential, religious buildings, palaces and fortresses and battlements reflect historical development and difference between cultures. Spatially documenting monuments and sites contributes to the understanding of these phenomena and their relevance for the future.

Early travelers and explorers did not only report on their discoveries, they also drew sketches of monuments and diagrams of site layouts. These representations were often, if not mostly, without dimensions and not always realistic. Technological developments have been largely replaced these, often, artful drawings as documentation tools with highly detailed, realistic and authentic representations of historical monuments and human settlements.

Modern Spatial documentation of heritage sites in the form of plans, elevations, 3D models and Geographic Information Systems contribute to the narrative of human history in an educational environment, to research and to the planning of physical conservation and restoration of sites.

Spatial documentation of heritage monuments and sites has developed over the past two centuries following mainstream trends in surveying, mapping and 3D modelling. Beginning with conventional measurements using tape, theodolite and close-range photogrammetry for the creation of maps and elevations, the technology advanced to multi-station as well as aerial photogrammetry based on analogue,

---

H. Rüther (✉) · R. Bhurtha · R. Schroeder · B. McDonald
University of Cape Town, UCT, Cape Town, South Africa
e-mail: heinz.ruther@uct.ac.za

G. M. El-Qady and C. Margottini (eds.), *Sustainable Conservation of UNESCO and Other Heritage Sites Through Proactive Geosciences*, Springer Geology,
https://doi.org/10.1007/978-3-031-13810-2_13

analytical and digital methods. A significant paradigm shift in heritage documentation was the advance from 2 to 3D documentation. The following discusses some of the recent developments in spatial heritage documentation and comments on these based-on experiences gained by the Zamani team of the University of Cape Town (Rüther et al. 2012) during numerous documentation missions.

2 Technologies–Past and Present

Laser scanning and so-called Structure-from-Motion (SfM) photogrammetry (Barazzetti et al. 2010), employed individually or together, have become the principal tools of state-of-the-art spatial heritage documentation. Rapidly evolving hardware and software have resulted in significantly faster fieldwork, enabled data fusion and improved 3D modelling. Today's terrestrial scanners (Fig. 1) acquire data at speeds ten times faster than early models at higher spatial resolution. Instruments are smaller and lighter and, through onboard scanning-control-panels, large storage spaces, and powerful built-in batteries, more efficient on site. On board Wi-Fi allows remote control via mobile phones and a single tablet can simultaneously register scans from multiple scanners on the fly.

Real-time scan registration reduces time in the field and subsequent processing time considerably, while on-board high resolution High Dynamic Range cameras improves vertex coloring and help in the alignment of photos to laser scans.

Fig. 1 Z+F 5010X Laser scanner at Kua Ruins, Juani Island, Tanzania

3 Unmanned Aerial Vehicles

Unmanned Aerial Vehicles (UAVs) or drones have revolutionised the way in which cultural heritage sites are documented. UAVs are affordable, reliable, non-invasive and can capture data relatively quickly compared to conventional survey techniques or traditional manned aerial photography methods (Chiabrando et al. 2017; Themistocleous 2020). Sensor sizes are decreasing and therefore ideal for use on UAVs with various payload options (Themistocleous 2020). Visual spectrum, multispectral and LiDAR sensors are being deployed on UAVs to acquire data in various formats which can provide for a more complete study and better understanding of a subject area.

UAVs can capture nadir and oblique images, which significantly improves the coverage of the data captured on site (Themistocleous 2020).

There are various applications available for smart phones and tablets that allow a UAV pilot to plan and execute autonomous flights. The user has many planning options for such flights including survey extent, image overlap, required ground sampling distance (GSD) and, significantly, altitude. Autonomous flight would seem to be a high-risk operation, but this risk has been reduced by features such as multidirectional collision detection, GNSS positioning and emergency-return function, which allows the UAV to return to its take-off location in case the pilot loses connection to the UAV.

All the above factors have made UAVs a valuable contributor to the multi-sensor data acquisition tool set for cultural heritage documentation. There are, however, many challenges in the application of UAVs for heritage documentation, most prominent among these are stringent regulations and restrictions governing the use of UAVs in most areas of the world.

4 Challenges and Improvements in the Processing of 3D Data

4.1 Challenges

Large scale cultural heritage documentation projects require multi-sensor data capture on a large scale (Figs. 2 and 3). Laser scans, photographs and RTK-GNSS measurements all need to be combined to produce the required results. Production of 3D computer models is complex and takes many hours, days or weeks. Thousands of images and hundreds if not thousands of laser scans need to be captured and combined in order to produce a 3D model. There are challenges in every step of the documentation process, whether it be choosing the appropriate sensors, hardware or software or assuring that the final result is in accordance with the required technical specifications and can be displayed and interacted with fluently (Remondino and Rizzi 2010).

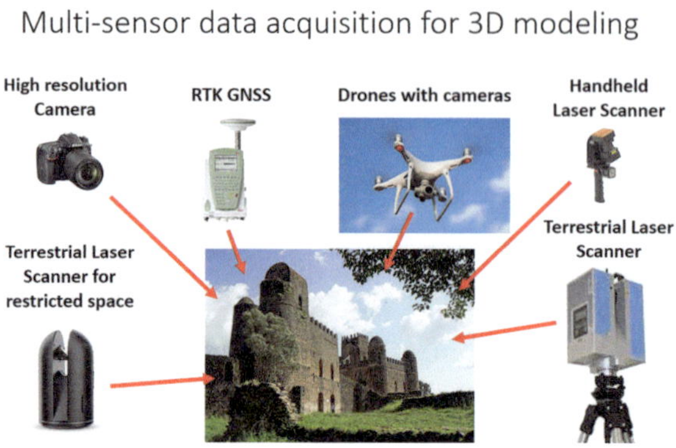

Fig. 2 Multi-sensor data acquisition

Fig. 3 Screenshot from Reality Capture processing software showing the positions of photos and laser scans needed to capture the site's geometry and texture the monuments. Medirigiriya, Sri Lanka

Time consuming and often complex final steps in the processing pipeline of 3D data are hole filling, decimation and cleaning of 3D models.

3D models are increasingly viewed using mobile devices instead of desktop PC's, which makes it necessary to create 3D content that is viewable on these platforms. This means that additional time must be spent optimising 3D data for efficient viewing on a variety of devices. Generated 3D models may initially consist of billions of polygons. In the experience of the authors, this can result in absurd scenarios where the full model only exists during the processing stage and where substantial portions of the original geometric detail, sometimes more than 95%, have to be discarded to make the model viewable and available for analysis or planning of conservation

Fig. 4 House of Wonders, Stone Town Zanzibar. Decimated to 10 million surfaces or 0.2% of the originally acquired data for practical use

interventions (Fig. 4). In such cases the original data set is retained and can be recovered once technology has advanced. Currently the capability of hard and software to create increasingly larger models develops faster than the ability to display and manipulate such models (Remondino and Rizzi 2010).

It must be noted that, although extremely large models can be generated, their creation poses hardware challenges. Large datasets require expensive high capacity solid state disks to provide improved processing performance. These disks are not immune to the intensive demands of processing 3D data. Running high end PC's permanently at full capacity takes its toll on the hardware.

4.2 *Improvements*

Among the most significant recent developments towards improved efficiency while scanning in the field is the on-the-fly registration of laser scans. This real time registration is realised through the use of a tablet (Fig. 5) which is wirelessly tethered to the laser scanner. The alignment is achieved automatically by Visual Inertial Navigation (VINS) (Huang 2019) aided by built-in GNSS and SLAM technologies continuously keeping track of the pose of the scanner. Should tracking be interrupted while moving the instrument between stations, then registration can be achieved by manually aligning the scans displayed on the touch screen of the tablet. The Zamani team has successfully employed real time registration capabilities of Z+F scanners on several sites where this technology proved especially useful where several scanners were used simultaneously. With the completed scans of all instruments being displayed together on the tablet it is possible to dynamically manage the scanning operation and avoid duplications and omissions. The principal advantage of registration in the field is, however, the dramatic reduction in processing time in the office as most of the registration work is already completed in the field.

Similarly, improvements in the speed of image alignments with Structure-from-Motion software make it possible to approximately align images for photogrammetric

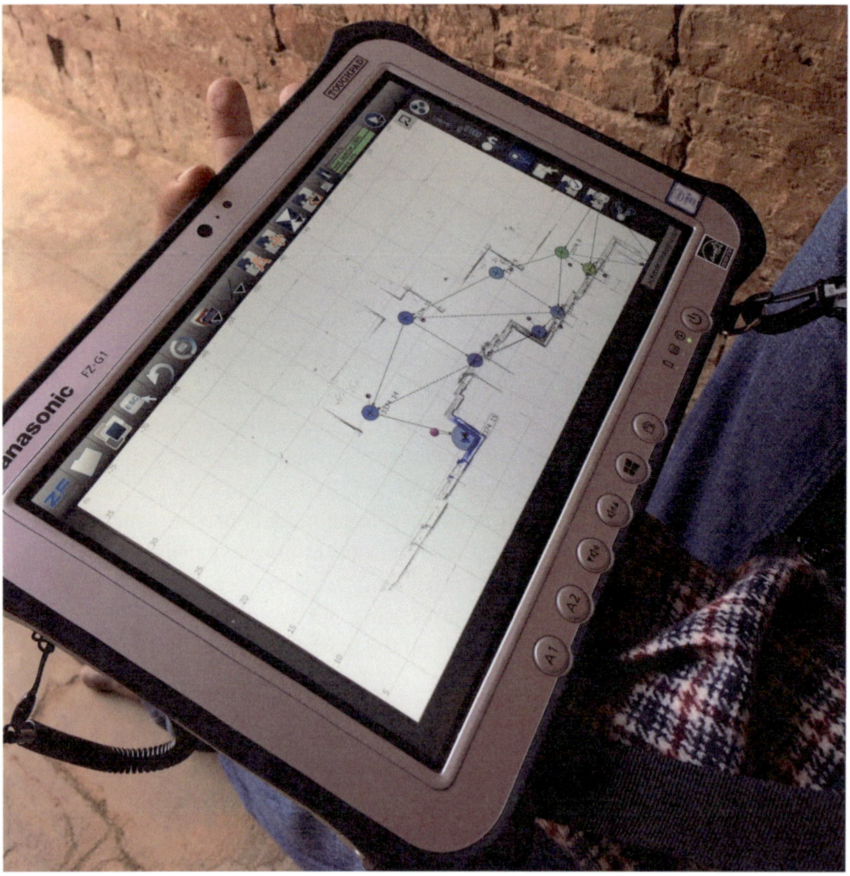

Fig. 5 Real-time registration in the field on tablet using Scout Z+F software (ZF 2019)

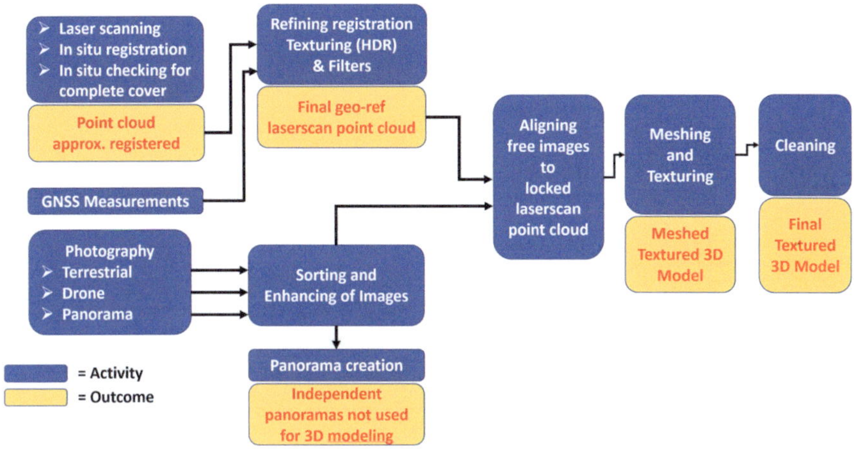

Fig. 6 High level summary of the Zamani Project's current 3D processing pipeline

generation of 3D models or for the texturing of meshed models in situ. This ensures that enough photographs have been captured while still on site.

Despite these significant developments, data acquisition and, more so, processing of multi-sensor data remain complex and time consuming as shown by the diagram of the current 3D processing pipeline of the Zamani project (Fig. 6).

5 Developments in Virtual Reality (VR)

The availability and improvements of software (Bekele et al. 2018) for the development and display of, and interaction with, Virtual Reality (VR) scenarios has not only made 3D models more accessible to scholars and heritage practitioners, it also, and possibly more importantly so, provides new opportunities to develop applications for education and virtual tourism. This has the potential to allow virtual visits of sites and opens new worlds to those who would otherwise have no access to heritage monuments because of limited means, physical disabilities or remoteness of sites.

In the first decades after its inception in the 1960s VR was commercially unsuccessful. It did, however, already then, show potential as an educational tool causing students to engage for longer time periods with a topic and experiencing enjoyment while learning. This in turn motivated and results in deeper learning and long-term retention of information (Kavanagh et al. 2017). In spite of this potential, VR has not been adopted yet by the heritage documentation community as widely as it deserves due to its costs and the logistics of its implementation (Kavanagh et al. 2017). This is now changing, as the latest VR Head Mounted Devices (HMDs)(Suárez-Warden et al. 2015) are becoming more affordable and more portable with some operating now without a physical connection to a PC. These advances make fully immersive cultural heritage VR applications more viable for museums, visitor centers and for Augmented Reality for site visits.

6 Issues in the Use of 3D Data

An important group of potential users of 3D data are heritage experts. While the use of 3D data by heritage practitioners and conservators has increased significantly in recent years, the full potential of this data type in the heritage domain remains to be explored. The efficient use of 3D data is still impeded by a lack of communication between disciplines involved in their creation and their use respectively.

The long-term adoption and use of 3D data, especially in the heritage domain, can only be truly successful if there is harmony and active information flow between the creator of the data and the end user. Creators of digital models, and in fact of all spatial data, need a clear understanding of the needs of the heritage experts, and the end-user needs to be cognizant of the potential of spatial data as well as in the limitations in the generation of such data. This mutual understanding seems not yet fully developed, potentially resulting in inefficient data generation, unrealistic expectations and wasted potential of laboriously created data.

An important step in establishing synergy between creator and user is the discussion of user requirements prior to field campaigns and data processing. The following questions need to be answered in the planning process to assist with planning and to minimize time and cost:

- Is there a need for accurate colour information?
- What geometric resolution is required of 3D models?
- What texture resolution is required of 3D models?
- Should a mesh or a point cloud be generated?
- Should the models of large monuments be delivered as a monolithic unit or in blocks?
- Do 3D data need to be cleaned of unwanted elements?
- Do 3D models need to be geo-referenced?

These parameters determine choice of equipment, method of data acquisition and processing, the size of the field team, duration of field work and processing, and ultimately cost. A relevant aspect of the decision regarding the documentation method is the choice between laser scanning, photogrammetry (SfM) or their combination. The advantages and shortcomings of these approaches and their respective accuracies are beyond the intent of this paper and require a separate discussion.

The above-mentioned developments in hardware and software have made the creation of high-resolution photorealistic 3D models possible. These models are appealing in their appearance, but their file size, sometimes in gigabytes, make them unwieldy and difficult to transfer and view. The transmission and online viewing of 3D objects is a challenge for those without high bandwidth internet connections. This can make the 3D models impractical and raises the bar very high for those needing access to the data. A few institutions, commercial and non-commercial, have created internet streaming software to showcase large models online (Umbra-Umbra Scene Stream 2020; 3DHOP-Home2020; Unreal Engine 5-Unreal Engine 2020; Sketchfab-Publish & find 3D models online 2020). Currently the commercial solutions have steep price tags. However, the display and on-screen manipulation of large data sets is an active area of research and more affordable solutions can be expected in the not too distant future, thus making 3D models more useful and relevant.

7 Training on Using 3D Data

A once off training on 3D data usage is most certainly going to be a loss of resources, as have been seen from experience. Technical questions will always arise in such work and multiple partners are needed to discuss ideas and solve problems. One possible way of solving this problem is to train multiple local groups over a period of time, so that a critical expertise pool size is reached and starts to expand on its own.

8 Metadata

Metadata, which literally means "data about data", is a vital part in Spatial Heritage Documentation. Without metadata, objects, whether physical or digital, would be difficult or impossible to locate in a database.

Any metadata object has three basic features–content, context and structure (Baca 2016). The *content* refers to information that is intrinsic to the object. The *context* relates to the who, what, where and why aspects of the object's creation, which are extrinsic data. The *structure* relates to a relational information among a set of objects, which can be both intrinsic and extrinsic. Many metadata standards exist, such as the Dublin Core (DCMI: Home 2020, 2020) and Carare 2.0 (About CARARE [CARARE Pro] 2020). Carare 2.0 has been developed specially to address metadata creation for 3D cultural objects.

Creation of good metadata and its dissemination can be a tedious task and often needs collaboration between multiple parties. This can pave the way towards increasing the visibility of the data, as is the case of the Zamani Project (University of Cape Town research repository 2020).

9 Conclusion

The spatial documentation of heritage sites has benefited greatly from recent developments in hardware and software. Documentation has become easier and is no longer the sole domain of experts. This overall positive development can, however, also lead to the indiscriminate and uncritical production of site data without quality control. In these cases, 3D models are generated simply because it is possible, without consideration of authenticity, accuracy or intended use. In this respect better communication between the producers of spatial data and the user needs to be refined.

Technical developments in the creation of 3D representation and Virtual Reality environments have contributed to a wider distribution of heritage information and made heritage sites and monuments accessible to many who previously were, due to financial and social constraints, not exposed to other cultures. Heritage documentation can thus, besides its more obvious impact and worth, also contribute to the creation of cultural identity and mutual understanding and intercultural harmony.

References

A first look at Unreal Engine 5-Unreal Engine (2020). https://www.unrealengine.com/en-US/blog/a-first-look-at-unreal-engine-5. Accessed 16 Jun 2020

About CARARE [CARARE Pro] (2020) https://pro.carare.eu/doku.php?id=start. Accessed 13 Jun 2020

Baca M (2016) Introduction to metadata: intro to metadata. https://www.getty.edu/publications/intrometadata/. Accessed 13 Jun 2020

Barazzetti L et al (2010) Orientation and 3d modelling from markerless terrestrial images: combining accuracy with automation. Photogram Rec 25(December):356–381

Bekele MK et al (2018) A survey of augmented, virtual, and mixed reality. ACM J Comput Cult Herit 11(2):36. https://doi.org/10.1145/3145534

Chiabrando F et al (2017) UAV oblique photogrammetry and LiDAR data acquisition for 3D documentation of the Hercules fountain. Virtual Archaeol Rev 8(16):83–96. https://doi.org/10.4995/var.2017.5961

DCMI: Home (2020). https://www.dublincore.org/. Accessed 13 Jun 2020

3DHOP-Home (2020). http://3dhop.net/. Accessed 16 Jun 2020

Huang G (2019) Visual-inertial navigation: a concise review

Kavanagh S et al (2017) A systematic review of virtual reality in education. Themes Sci Technol Educ 10(2):85–119

Remondino F, Rizzi A (2010) Reality-based 3D documentation of natural and cultural heritage sites-techniques, problems, and examples. Appl Geomat 2(3):85–100. https://doi.org/10.1007/s12518-010-0025-x

Rüther H et al (2012) Laser scanning in heritage documentation: the scanning pipeline and its challenges. Photogramm Eng Remote Sens 78(4)

Sketchfab-Publish & find 3D models online (2020). https://sketchfab.com. Accessed 16 Jun 2020

Suárez-Warden F et al (2015) Small sample size for test of training time by augmented reality: an aeronautical case. In: Procedia computer science, vol 75(Vare). Elsevier Masson SAS, pp 17–27. https://doi.org/10.1016/j.procs.2015.12.190

Themistocleous K (2020) 'The use of UAVs for cultural heritage and archaeology, (June), pp 241–269. https://doi.org/10.1007/978-3-030-10979-0_14

Umbra-Umbra SceneStream (2020). https://umbra3d.com. Accessed 16 Jun 2020

University of Cape Town research repository (2020) https://zivahub.uct.ac.za/. Accessed 16 Jun 2020

ZF (2019) ZF-Laser-Z+F LaserControl® Scout. https://www.zf-laser.com/Z-F-LaserControl-R.laserscanner_software_2.0.html?&L=1. Accessed 13 Feb 2019

Usage of Laser Scanning Systems to Document the Cultural and Historical Heritage

Abdel-Monem Sayed Mohamed, Gad Mohamed El-Qady, Abass Mohamed Abass, Abdel-Hamid Mohammed Elbshbeshy, and Ahmed Elhadi Sherif

1 Introduction

Today several methods are being used for the documentation, quantification and surveying of cultural and architecture heritage. These methods may be listed as, conventional instruments surveys, topographic and photogrammetric surveys and laser scanning surveys. Laser scanning technology, both terrestrial and airborne, has encountered great technological advance during past few years. It gives us the opportunity to acquire great amounts of precisely measured data in very short time. The great challenge is to deal with these huge datasets thoroughly and produce a competent and rather automated workflow, from data capture up to the desired product. This product comprises most of the times a Digital Surface Model (DSM) of the investigated object. Once the DSM is created many follow-up products can be derived without difficulties. 3D Laser Scanning is a non-contact, non-destructive technology that digitally captures the shape of physical objects using a line of laser light. 3D laser scanners create "point clouds" of data from the surface of an object. In other words, 3D laser scanning is a way to capture a physical object's exact size and shape into the computer world as a digital 3-dimensional representation. 3DLaser scanning technology is one of the latest and most accurate cadastral techniques used in: 1- Make 3D anthropomorphic dimension for targets observed. 2- Getting massive laser point cloud coordinates information. 3-The ability to locate cracks and deformations in buildings and tunnels. 4-Used for Recording and Documentation in Archaeology. 5- The ability to define the edges of roads and lighting poles. 6-3D modeling of heritage buildings to document and preserve the historical value. 3D laser scanner technology is used in the development of tourism industry in some tourist places in Egypt, where this technology plays major roles in various fields that contribute to

A.-M. S. Mohamed (✉) · G. M. El-Qady · A. M. Abass · A.-H. M. Elbshbeshy · A. E. Sherif
National Research Institute of Astronomy and Geophysics, Cairo, Egypt
e-mail: abdelmonem@nriag.sci.eg

G. M. El-Qady and C. Margottini (eds.), *Sustainable Conservation of UNESCO and Other Heritage Sites Through Proactive Geosciences*, Springer Geology,
https://doi.org/10.1007/978-3-031-13810-2_14

the industry of tourism, archaeology and preservation of ancient heritage. National Research Institute of Astronomy and Geophysics (NRIAG) has a great experience in the field of study and documenting monuments in Egypt. This technology is used, but not limited to (Saqqara, Bahariya oasis in the Golden Mummies Valley area, NRIAG buildings, Dahshour, Ahmed Hamdy Tunnel, Qa'at Mohib el-Din Abu el-Tayeb, Maqa'adMamai El-Saifiand the Vault of Bayt el-Qady and Hawara pyramid… etc.). Our institute has taken over the responsibility of protecting antiquities in Egypt.

2 Methodology

Laser scanning surveys are used for documentation. The use of digital imaging or 3D laser scanning devices is a technology that has been applied in recent years. The beam sent from the laser scanning device to the object or surface across, returns as millions of point data in the computer environment and can be seen as an object formed by the point cloud in digital media. This object or surface is now coordinated according to its ground level, in its actual size and in high accuracy. In this work we are using Mobile Laser Scanner "Trimble MX2", Terrestrial Laser Scanner "Trimble TX6", Total Station "Trimble M3DR5" and Global Navigation Satellite System (GNSS) Devices "Trimble receivers".

3 Laser Scanning Range Principles

A laser scanner is defined as any device that collects 3D coordinates of an object's surface automatically at a high rate and delivers results in nearly real time as a point cloud (Fig. 1). A set of points converted from range and angle measurements to a common Cartesian coordinate system (X, Y, Z).

4 Mobile Laser Scanner

Mobile Laser Scanner is a vehicle-mounted spatial imaging system which combines high resolution laser scanning and precise positioning to collect geo-referenced point clouds for a wide range of requirements. A vehicle mounted mobile laser scanner 'MX2' that contains a state of art two head laser scanner, six 12mb cameras that covers 360°, two GNSS antennas coupled with highly sensible IMU unit (Pfennigbauer and Ullrich 2010). The system can be rapidly deployed onto on- and off-road vehicles of all sizes. It also significantly reduces project field time and operator skill levels compared to traditional techniques. The two laser scanners scan both side collecting 36.000 points/second and rotate 20 cycle/second. The panorama camera composed by 6 individual cameras (5 horizontals and 1 to the top) creates a 30 Mpx panorama

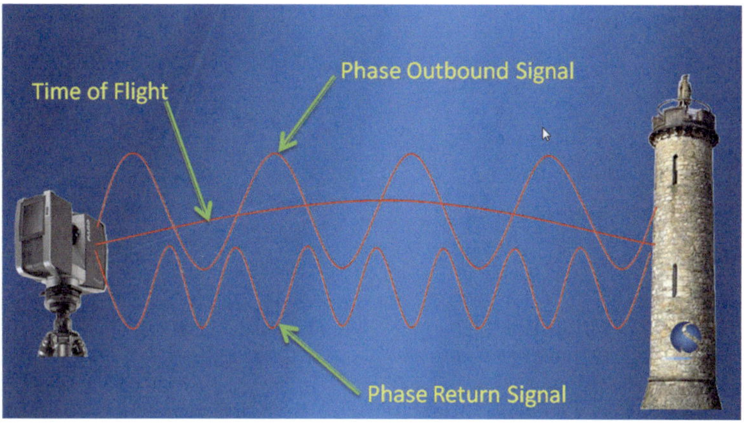

Fig. 1 Principle of laser scanner

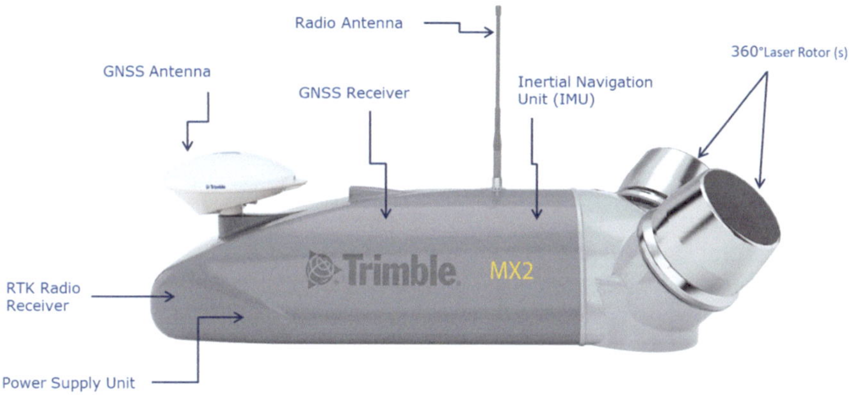

Fig. 2 Mobile laser scanner device components "Trimble MX2"

picture each 5 m (Fig. 2). This cutting-edge instrument is the only mobile laser scanner in Egypt that delivers high quality of 3D point cloud (Fig. 3).

5 Terrestrial Laser Scanner

The laser scanner is a cost-effective 3D scanning solution (Buckley et al. 2008). Trimble TX6 is high scan speed and accuracy (Fig. 4). Trimble TX6 is a terrestrial mapping system that includes: 500,000 points per second with no compromise on scan quality or range. Integrated HDR camera to colorize scans. Easy to use onboard interface with no complex parameters. Integrated WLAN for remote operation from

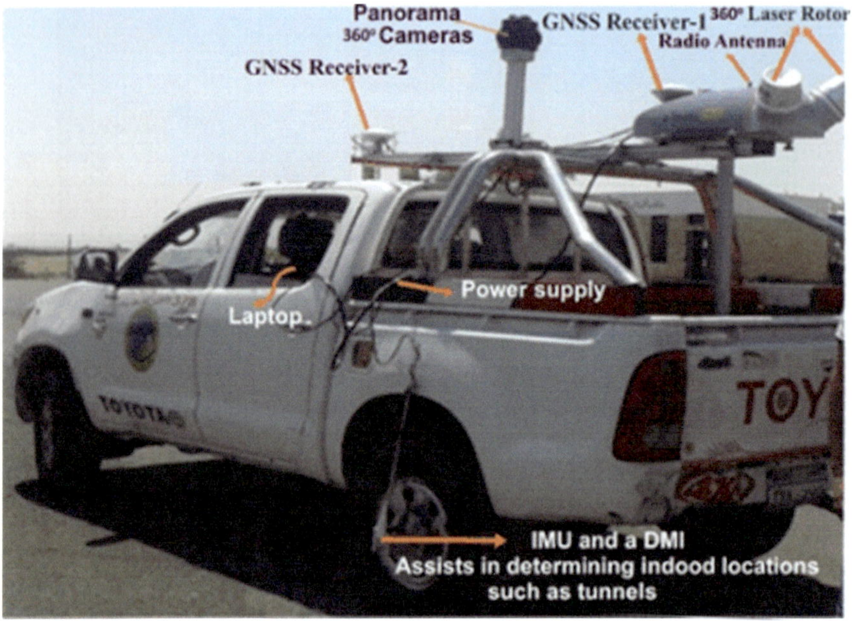

Fig. 3 Initialization Trimble MX2 system component on the car before starting work

any mobile device. Upgradeable from 80 m standard to 120 m extended range. Data from the Trimble TX6 loads directly into Trimble Real Works and Trimble Scan Explorer software for automatic scan colorization and registration. Produce powerful deliverables or export data to CAD software.

6 Total Station

Total station is an electronic/optical instrument used in modern surveying and building construction that uses electronic transit theodolite in the conjunction with electronic distance meter (EDM) (Fig. 5) (http://www.przybilla.biz/bv/Praktikum/TrimbleM3-UserGuide.pdf). It is also integrated with microprocessor, electronic data collector and storage system. The instrument is used to measure sloping distance of object to the instrument, horizontal angles and vertical angles. This microprocessor unit enables for computation of data collected to further calculate the horizontal distance, coordinates of a point and reduced level of point. Angles and distances are measured from the total station to points under survey, and the coordinates (X, Y, and Z or northing, easting and elevation) of surveyed points relative to the total station position are calculated using trigonometry and triangulation. Data can be downloaded from the total station to a computer and application software used to compute results and generate a map of the surveyed area.

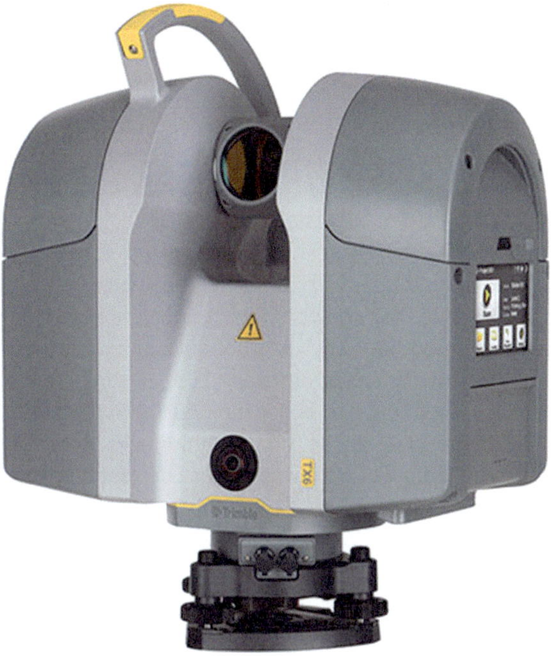

Fig. 4 Terrestrial laser scanner "Trimble TX6"

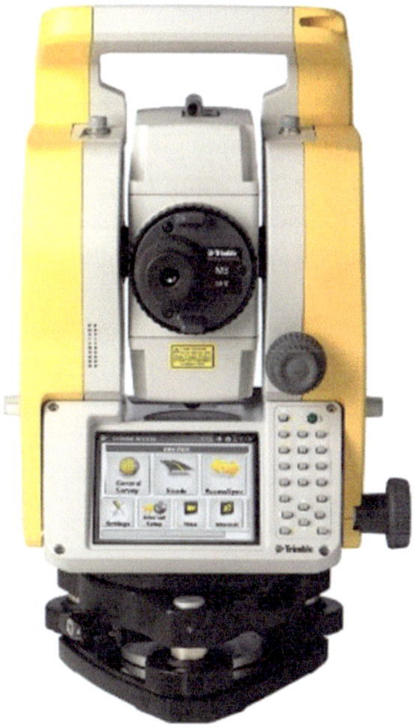

Fig. 5 Total Station Instrument "Trimble M3DR5"

7 Global Navigation Satellite System (GNSS)

GNSS surveys require different planning, execution, and processing techniques. The optimum planning of GNSS surveying must consider several parameters such as site or satellite configurations, satellites status, and the number and type of receivers to be used. In 2006, the National Research Institute of Astronomy and Geophysics (NRIAG) started the establishment of the Egyptian Permanent GNNSS Network (EPGN) (Fig. 6). Basically, the site selection not only aimed to cover geographically all the Egyptian territory but also considers the tectonic setting of Egypt. Also, the chosen places for constructing these stations fulfilled the required criteria, such as clear view without any obstructions, away from any electromagnetic sources and accessibility. This network started only with 4 GPS stations and gradually expanded to reach 30 stations to cover most of the Egyptian territory. Due to the importance of the type and quality of the monument, which directly affects the stability of these stations. The majority of EPGN stations were installed on concrete pillars (Fig. 7a); but in a few cases some stations were installed on the roof of buildings (Fig. 7b). Some of these stations have been installed near the seismological stations where the data has been transmitted to the main center in Helwan automatically via the satellite and the others were installed separately where it has been sent the data through Sim Card and received directly to the Geodesy data center in Helwan. Geodetic receivers which perform precise baseline measurements were Dual-frequency Trimble receivers 5700, NETR5, Trimble R7, Trimble R8 and NETR9 are shown in Fig. 8a–c with sampling interval 0 and 30 s. The antenna at all sessions of measurements was centered above the marker and directed to the North. It was fixed vertically on the benchmark as shown in Fig. 7a. The measurements will be performed by using GNSS receiver's type Trimble. Processing of baselines analyses were performed using Trimble Business Centre software package and other programs for adjustment and deformation parameters calculations. The sampling interval and elevation were fixed throughout the survey at 1^0 and 5^0 respectively. The IGS precise ephemeris was applied in the calculation of the baselines. The reference stations are tied to (IGS) stations to compute the precise coordinates in International Terrestrial Reference Frame.

8 Data Processing

The data processing and elaboration of the scanned laser scan data, can be implemented with the following steps:

1-The processing data gathered from Mobile Laser Scanner "Trimble MX2". After downloaded GNSS data from the control points, the data were analyses using a range of programs to get the exact coordinates of those points to be used during the analysis of laser scanner data. Then, the laser scanner data analysis stage began using a set of programs and analysis steps were as follows:

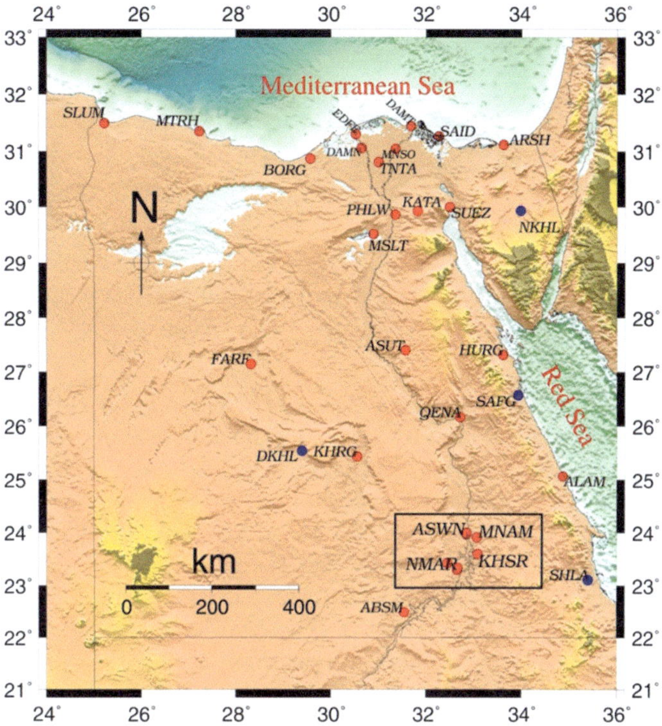

Fig. 6 Geographic Distribution of Permanent GPS Network (EPGN). around Egypt

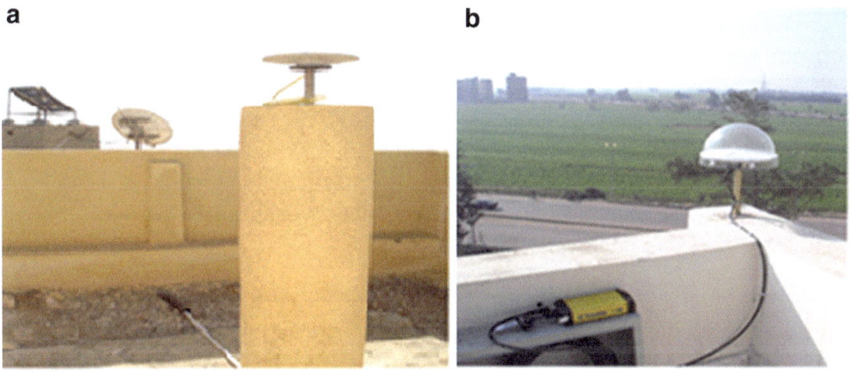

Fig. 7 **a** and **b** Construction of GPS geodetic network

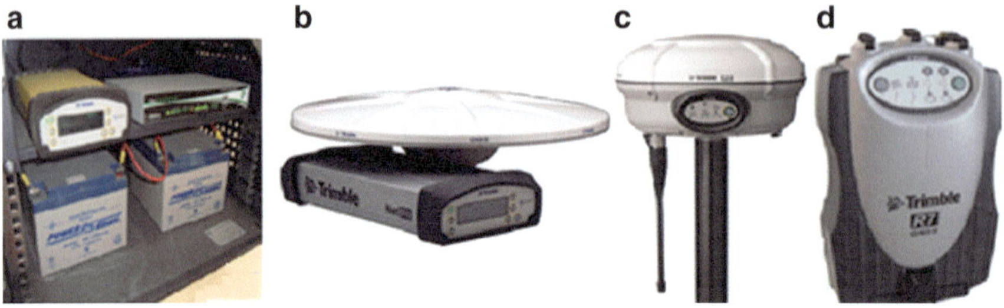

Fig. 8 Trimble receiver NETR5 (**a**), Trimble receiver NETR9 (**b**), Trimble receiver R8 (**c**) and Trimble receiver R7 (**d**)

- Laser scanner trajectory correction observed by GNSS is installed by using the Pospac and use control points for getting the highest accuracy.
- The monitored points cloud patch work using laser head by Trident program (pre-registration). The Trident interface is designed for robust object positioning, measurement, point cloud classification, and data layer creation–ideal for the analysis of geo-referenced imagery and laser scanner data. Industry-leading functions accelerate projects and increase productivity, including key automated processes such as surface modeling, roadway sign and pole detection, lane marking detection, edge and break line detection, road geometry, and clearance measurements.
- In the end the Registration data was made using images taken from cameras in the machine and the control points that have been monitored after this step we managed to get the diorama of the audited laser dot coordinates.

2-The processing data gathered from Terrestrial Laser Scanner "Trimble TX6". Software tool uses is called Trimble Real Works. This software tool provides you with a set of tools for processing 3D point clouds and 2D images to obtain the necessary information for the projects. Generally, this processing can be divided into three modes: field Survey, registration and modeling. Trimble RealWorks software enables you to register, visualize, explore and manipulate as built or scene point cloud data collected with virtually any laser scanner. Advanced, but very easy to use, Trimble RealWorks allows to:

- Perform smart measurement–semiauto clearance, projected vertical and horizontal Communicate your results via video generation and Google Earth export (kml, kmz format).
- Perform automatic registration with or without targets.
- Efficiently integrate data from Trimble GNSS, Optical, and 3D Scanning instruments
- Easily export to the CAD design packages.
- Publish self-cntained project packages for standalone viewing, exploration, measuring and annotating While Trimble RealWork is powerful enough to handle large datasets with ease, it is also straightforward to use. The software

is self-guiding through sophisticated data management and manipulation tasks step-by-step to ensure reaching the objectives and needs.

- Also, by using Trimble RealWork software we are making mesh model for our study then this step the contour map and elevation will be created. Also, it is possible to produce cross-sections for the pyramids. This software allows calculating volume for models.

9 Saqqara Site

Saqqara is one of the most important and richest necropolises in Egypt. It is located on the western bank of the Nile about 20 kmsouth of Cairo and on the same limestone plateau that extends north to Giza. The history of the Egyptian pyramids began from 2650 B.C. in Saqqara. A detailed map of the Saqqara area is drawn up showing all the archaeological sites and exploratory mission sites operating in Saqqara area (Fig. 9). In this map we provide a simplified explanation of each archaeological landmark and each expedition in Saqqara area. This map can be placed on the enterface Saqqara to be a key information to any visitor. Saqqara area is divided into control and registration points (Fig. 10). The data acquisition was carried out with the phase-shift based scanner. The data capturing procedure was executed in mobile and terrestrial laser scanning. Trajectory of 27 tracks in Saqqara area is covering the whole area as shown in Fig. 11. Figure 12 shows an example of the trajectory for eight separated tracks (1, 2, 3, 4, 5, 6, 7 and 8) in Saqqara site. Where they start from Saqqara entrance. The control points are used for getting the highest accuracy within (2 cm) in the tracks. The monitored points cloud patch work using laser head by Trident program. Where in the end the registration data was made using images taken from cameras in the machine and the control points that we have been monitored. Tracks results gave a highly topographic map with accuracy within centimeter. Figure 13 shows a topographic contour map using y-axis and X-axis. The topographic contour map is resulting from GNSS data analysis, is plotted in Fig. 14. Figure 15 shows the topographic of the area under study in three dimensions. We managed to get the audited laser dot coordinates and we will present the example of the features in the tracks. See the following Fig. 16.

10 Saqqara Pyramids

In this chapter, we provide a simplified explanation of each archaeological landmark and each expedition in the Saqqara area. We present a simplified presentation of the pyramids of Saqqara (Fig. 17). Unas, Djoser, Userkaf and Teti pyramid pyramids were observed by the Terrestrial laser scanner. In this way, the physical body of the pyramids is transformed into digital forms that are preserved for future generations. The 3D modeling and digitization of the pyramids at a very high resolution (2 cm

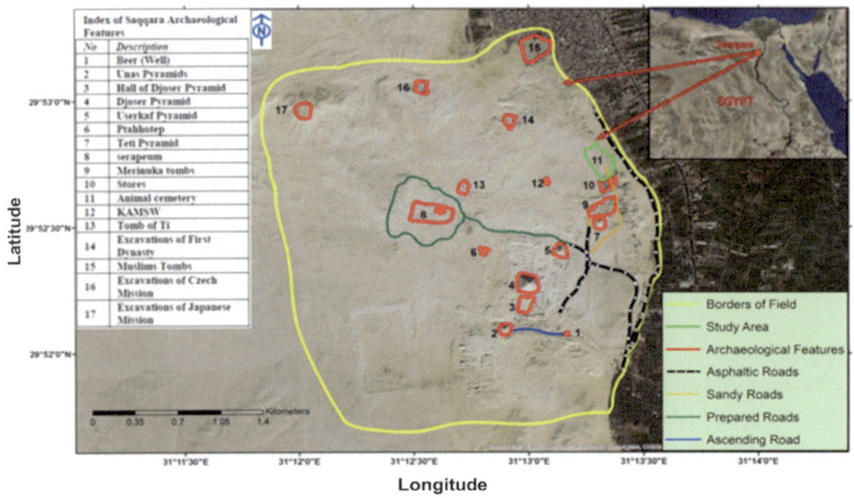

Fig. 9 A detailed map is showing all the archaeological sites and exploratory mission sites operating in Saqqara site

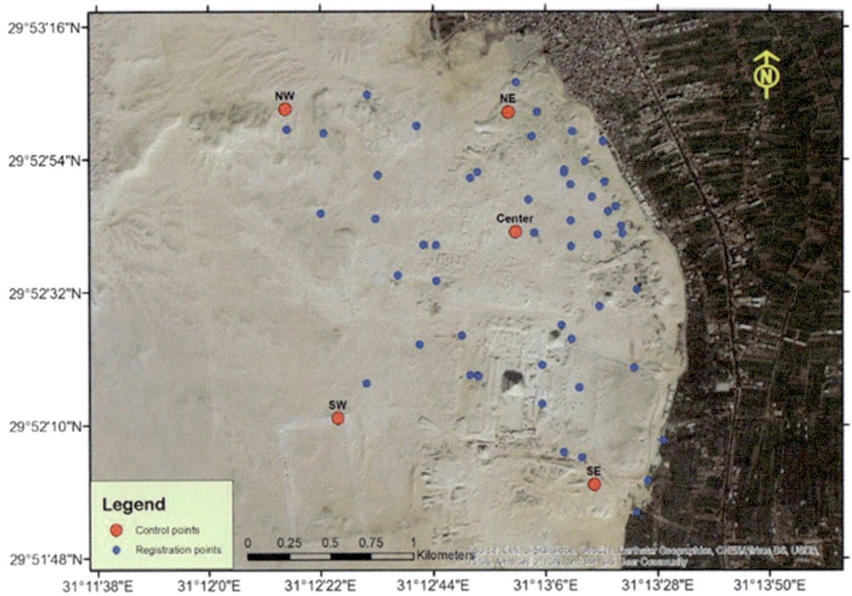

Fig. 10 Registration and control points

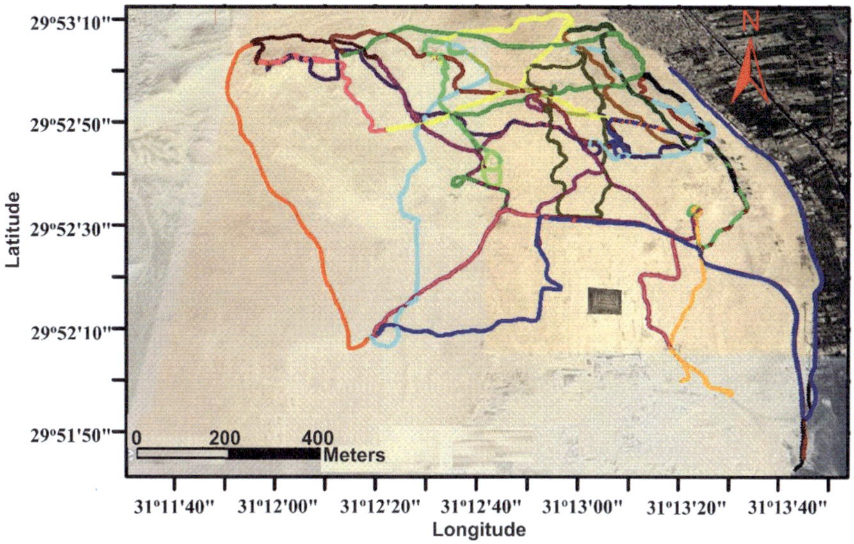

Fig. 11 Trajectory for 27 tracks in Saqqara site

accuracy) helps document them for future generations and preserve and restore their historical value. Figures 18 and 19 show the modelling for Saqqara pyramids in point cloud and point cloud with intensity, the contour map, the Mesh model.

10.1 Unas Pyramid

The Pyramid Complex of Unas is located in the pyramid field at Saqqara. It was built for pharaoh Unas, the ninth and final king of the fifth dynasty in the mid of twenty-fourth century BC. Its ancient name, Nefer Isut Unas, means "Beautiful are the places of Unas" (Fernandez et al. 2013). Unas Pyramid in the point cloud, Contour map, Mesh modelling and Point cloud with intensity is shown in Fig. 20.

10.2 Djoser Pyramid

Djoser pyramid or step pyramid is an archeological remain in the Saqqara necropolis, Egypt, northwest of the city of Memphis. It was built during the twenty-seventh century BC for the burial of Pharaoh Djoser by his vizier Imhotep. Djoser was the first or second king of the 3rd Dynasty (2667 to 2648 BC) of the Egyptian Old Kingdom (2686 to 2125 BC) (Harry 2007) (Fernandez et al. 2013). Djoser Pyramid in the point cloud, Contour map, Mesh modelling and Point cloud with intensity is shown in Fig. 21.

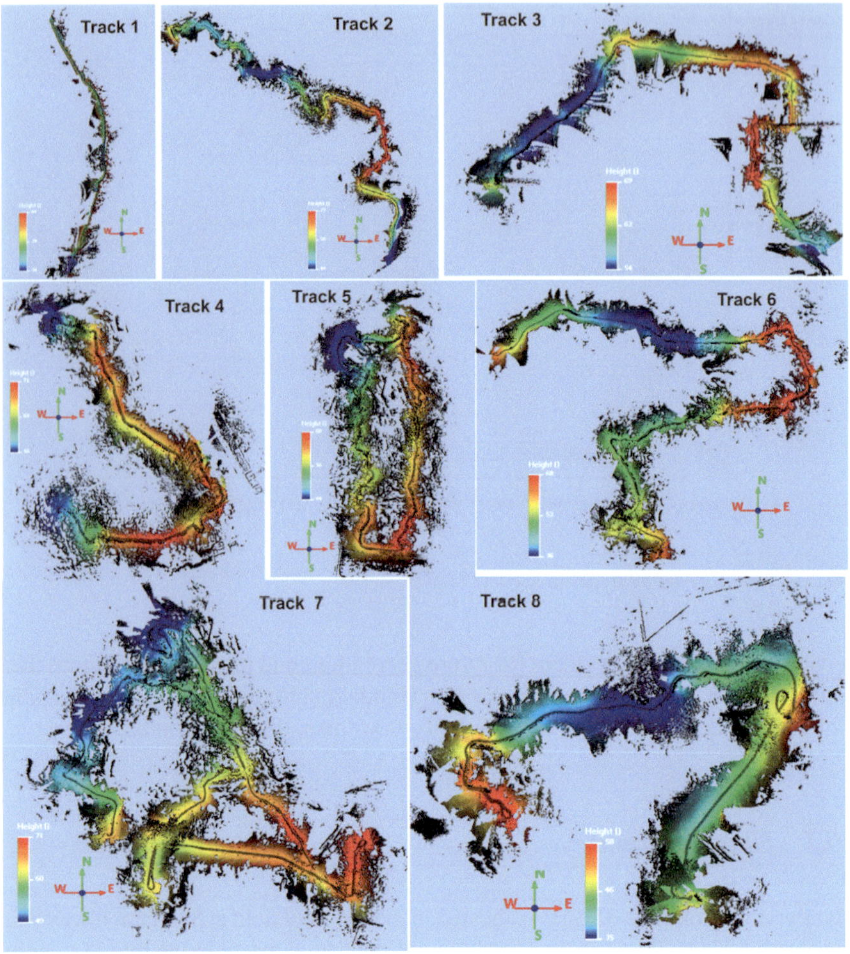

Fig. 12 Trajectory for eight separated tracks (1, 2, 3, 4, 5, 6, 7 and 8) in Saqqara site

10.3 Userkaf Pyramid

The pyramid was built 2490 BC for the pharaoh Userkaf (2494–2487 BC), founder of the 5th dynasty of Egypt (2494–2345 BC). It is located in the pyramid field at Saqqara, on the north-east of the step pyramid of Djoser (Fernandez et al. 2013). Userkaf Pyramid in the point cloud, Contour map, Mesh modelling and Point cloud with intensity is shown in Fig. 22.

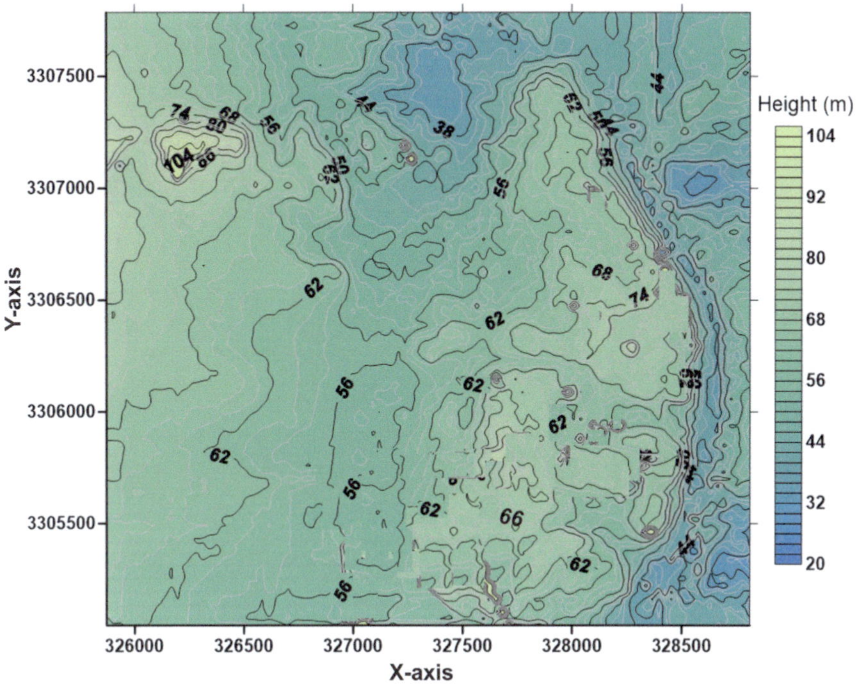

Fig. 13 Topographic map using y-axis and X-axis, is resulting from Mobile Laser Scanner

10.4 Teti Pyramid

The Pyramid was built by Teti, the first king of the sixth Dynasty who ruled Egypt 32 years. This pyramid was built as the ancient Egyptians believe in resurrection. The pyramid of Teti was constructed at North Saqqara on the only remaining spot, south of the 1st and 2nd Dynasty mastabas and to the northeast of the pyramid. The Pyramid is a smooth-sided pyramid situated in the pyramid field at Saqqara. It is historically the second known pyramid containing pyramid texts. It was originally called Teti's Places Are Enduring (Fernandez et al. 2013). The preservation above ground is very poor, and it now resembles a small hill. Below ground the chambers and corridors are very well preserved. Teti Pyramid in the point cloud, Contour map and Mesh modelling is shown in Fig. 23.

11 Bit Al Qadi Site

The Al-Qadi House is an urban fabric includes a small square overlooking three main buildings, the Mahabb Al-Din Hall, Prince Mamay Al-Saifi's House and the entrance of Al-Qadi's House. These buildings are integrated as components of the Al-Qadi House (Fig. 24).

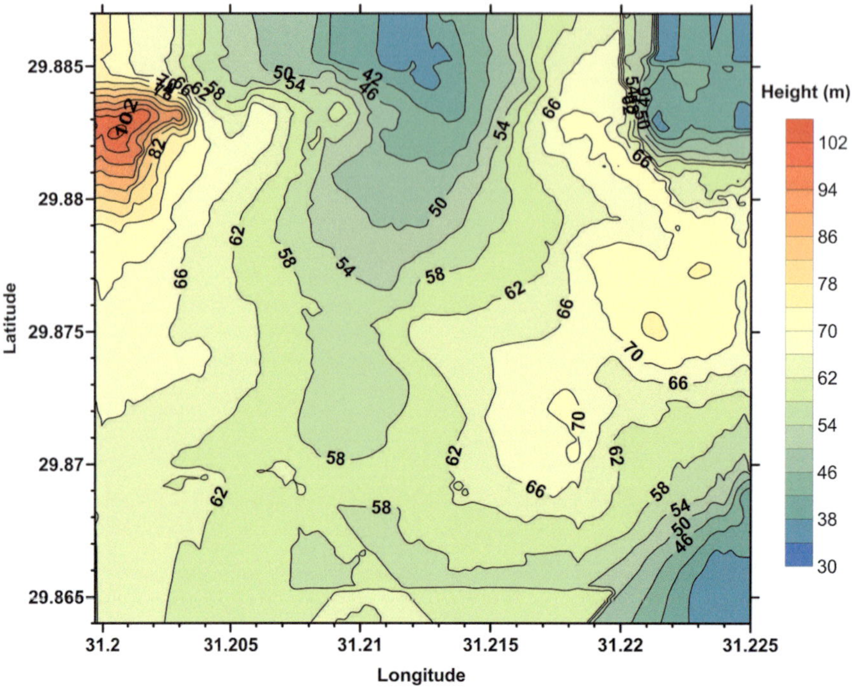

Fig. 14 Topographic map using longitude and latitude, is resulting from GPS

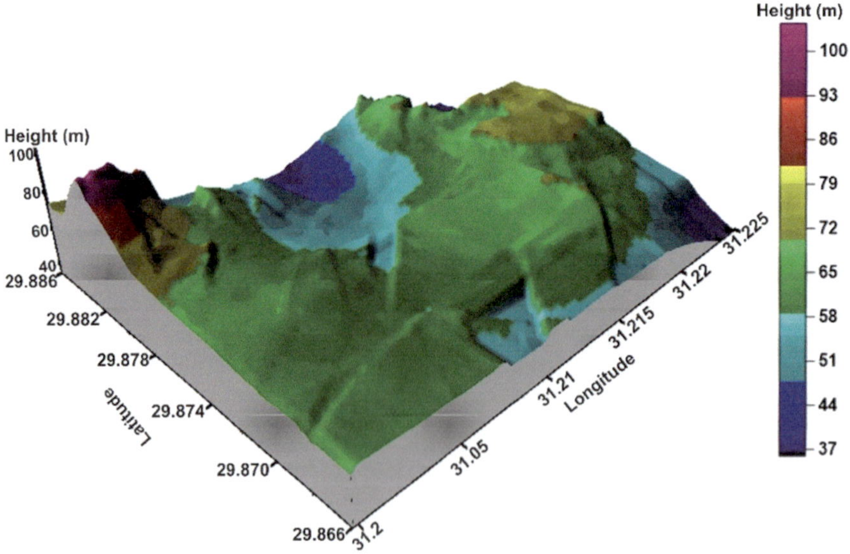

Fig. 15 Topographic map in three dimensions

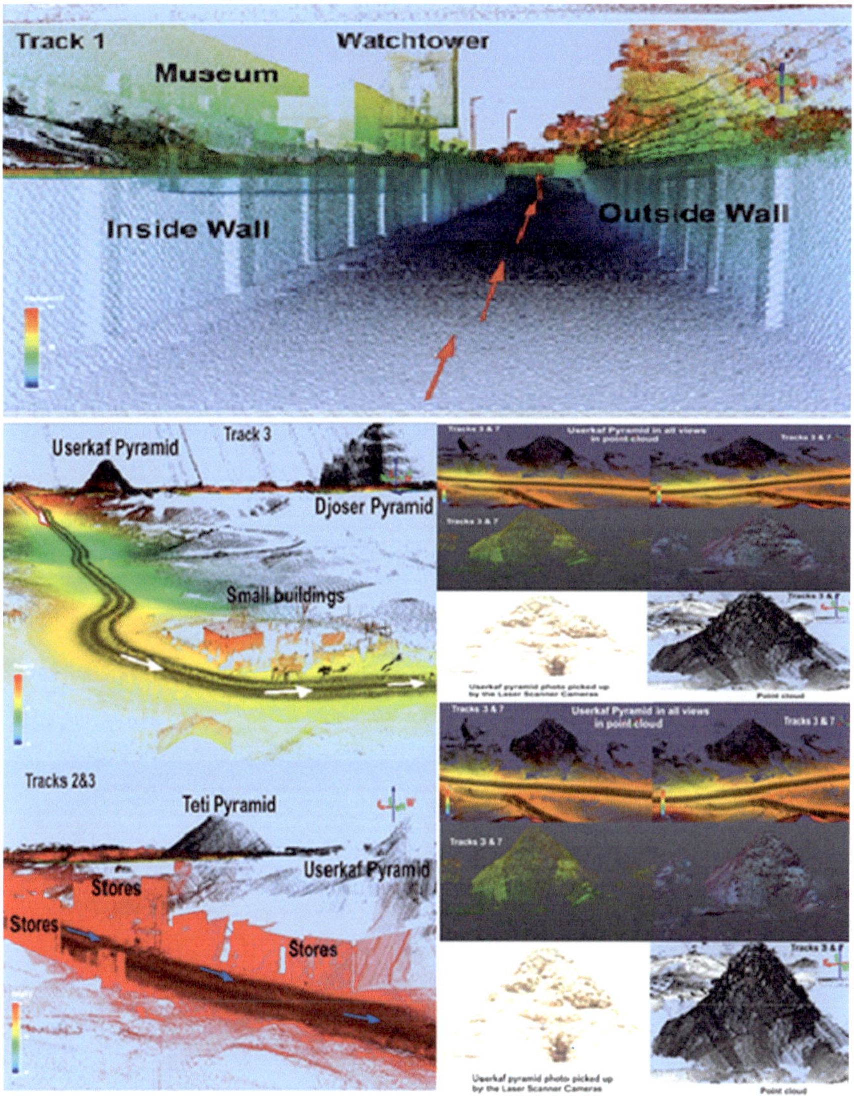

Fig. 16 An example results of the trajectory of some tracks

11.1 Mahabb Al-Din Hall

Mahabb Al-Din Hall located in the middle west of Bet Al-Qadi Street. It was established in 1350 A.D. It has been restored for the first time in 1730 A.D. It has been restored for the second time in 1911 A.D (Fig. 25). Point clouds samples for Mahabb Al-Din Hall (Fig. 26). Figure 27 shows the 2D Cross Sections for Mahabb Al-Din Hall.

Fig. 17 Saqqara pyramids

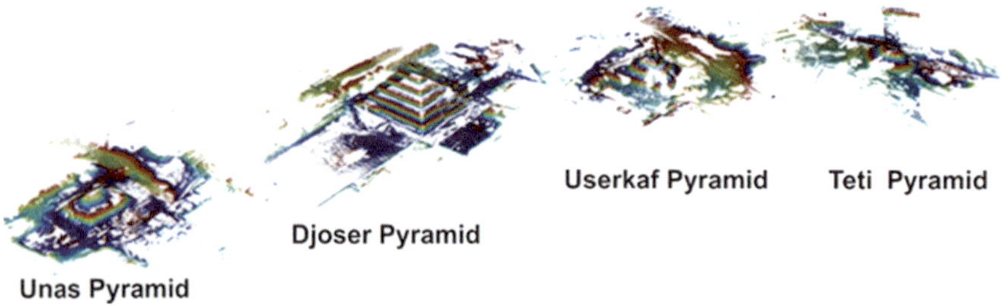

Fig. 18 The modeling for Saqqara pyramids in point cloud with intensity

11.2 Prince Mamay Al-Saifi's House

Prince Mamay Al-Saifi's House was established in 1496 A.D. (Fig. 28). Point clouds samples for Prince Mamay Al-Saifi's House (Figs. 29 and 30). Figure 31 shows the 2D Cross Sections for Prince Mamay Al-Saifi's House.

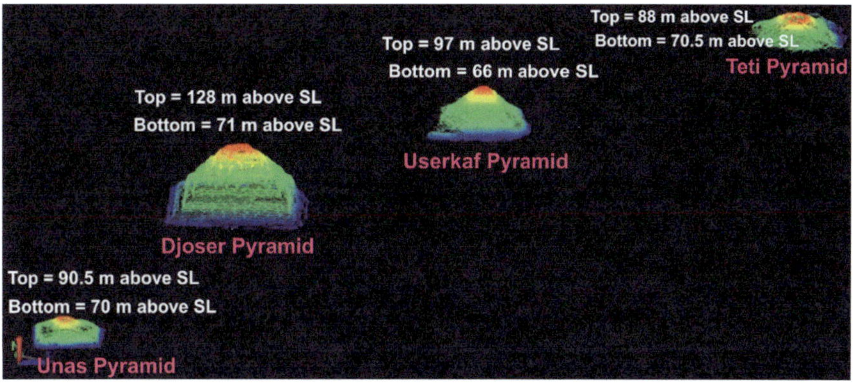

Fig. 19 Saqqara pyramids in the contour map

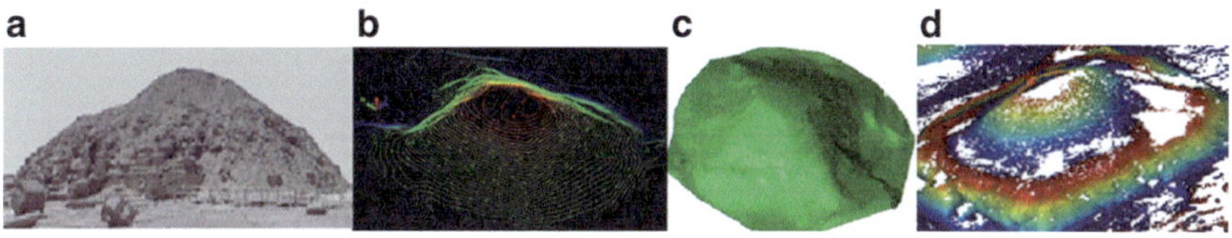

Fig. 20 Unas pyramid in the **a** Point cloud, **b** Contour map, **c** Mesh modelling and **d** Point cloud with intensity

11.3 The Entrance of Al-Qadi's House

The entrance of Al-Qadi's House was established in 1873 A.D. as part of planning for new historical city (Fig. 32). Point clouds samples for the entrance of Al-Qadi's House (Fig. 33). Figure 34 shows the 2D Cross Sections the entrance for Al-Qadi's House.

12 Hawara Site

Hawara is an archaeological site of Ancient Egypt, south of the site of Crocodilopolis (Arsinoe) at the entrance to the depression of the Faiyum oasis (Fig. 35). The first excavations at the site were made by Lepsius (1843). William Flinders Petrie excavated later at Hawara (Petrie 1888). He had found papyri of the first and second centuries and north of the pyramid, a vast necropolis where he found 146 portraits on coffins dating to the Roman period, famous as being among the very few surviving examples of painted portraits from Classical Antiquity, the "Faiyum portraits" illustrated in Roman history textbooks. The entrance to the pyramid is flooded to a depth of 6 m as a result of the water from the Bahr Yusuf (Joseph's Canal) canal, which

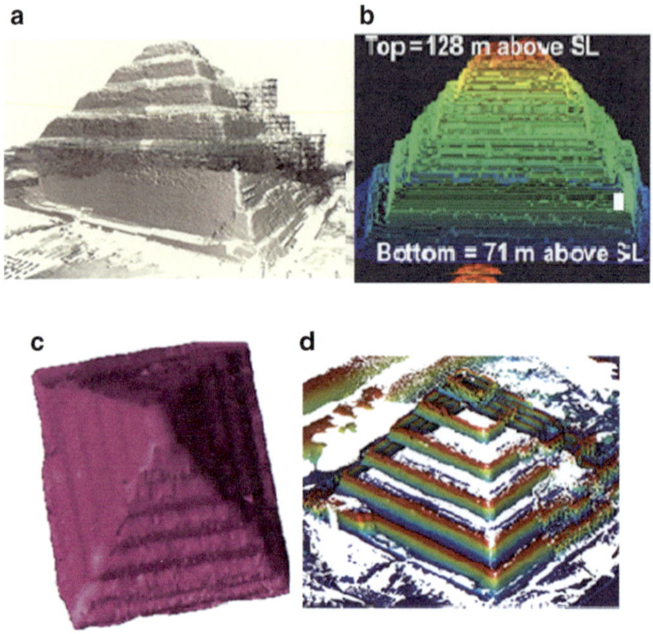

Fig. 21 Djoser pyramid in the **a** Point cloud, **b** Contour map, **c** Mesh modelling and **d** Point cloud with intensity

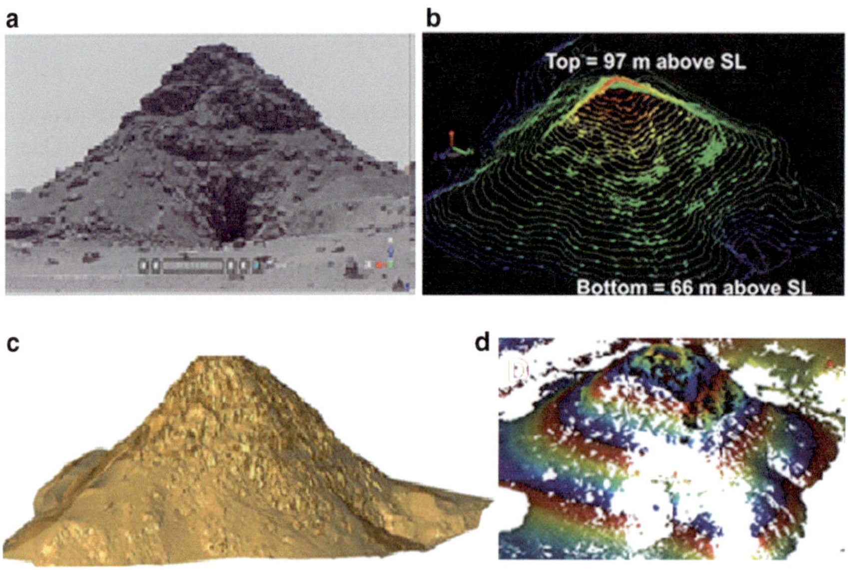

Fig. 22 Userkaf pyramid in the **a** Point cloud, **b** Contour map, **c** Mesh modelling and **d** Point cloud with intensity

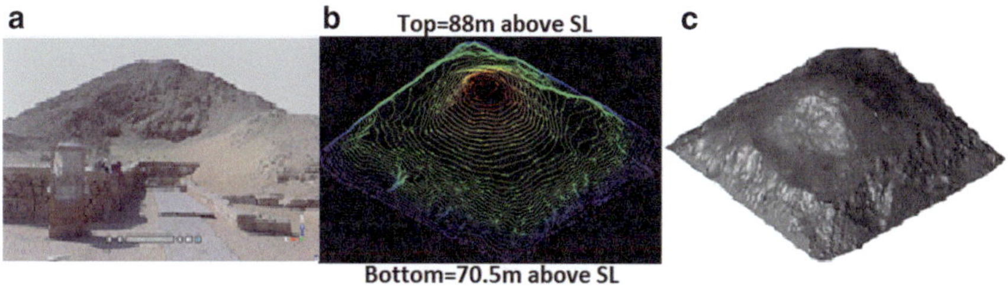

Fig. 23 Teti pyramid in the **a** Point cloud, **b** Contour map and **c** Mesh modelling

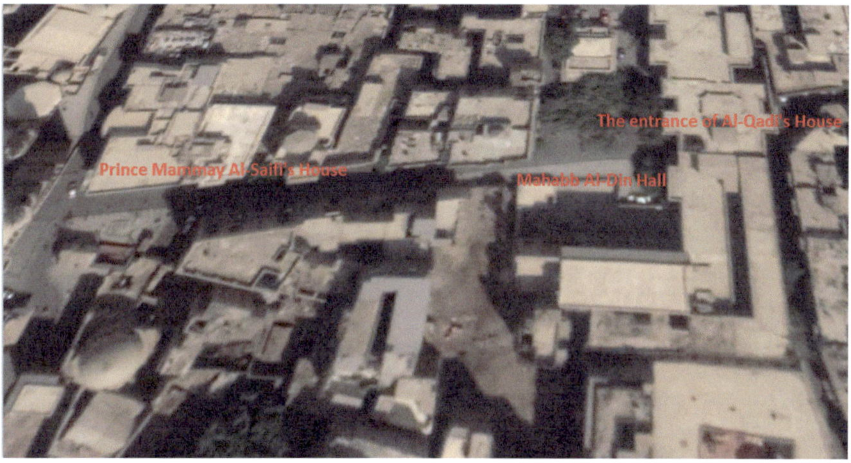

Fig. 24 Mahabb Al-Din Hall, Al-Qadi's House and Al-Saifi's House

flows around two sides of the site and passes within 30 m of the pyramid and the under groundwater also one of the main problems effect on the pyramid. The agricultural lands around the pyramid of Hawara rise the underground water that causes negative effects on it. Nowadays, Hawara area is entirely abandoned and the underground water has a bad effect on the foundation of Hawara Pyramid. The area requires significant efforts to conserve this historical site from desertion. In addition, the cracks appearance in the entrance of the object due to the water inside it.

To cover the Hawara pyramid from all sides, 31 scans were required. The data acquired from 3D laser scanning devices is an image which consists of millions of points. In this image, empty and full surfaces, all the curves, indents and juts can be observed as sharp lines formed by the points. As each point in the image has its own coordinate value, the lines formed by these points can be gripped and combined to produce drawings in CAD environment. The object or the surface can be defined in millions of 3-dimensional coordinates in several minutes. All the points have 3-dimensional coordinates (x, y, z) values. Point clouds combined with the scan images enhancement a detailed 3D model and collect a great data amount of

Fig. 25 Mahabb Al-Din Hall

Fig. 26 Point Clouds Samples for Mahabb Al-Din Hall

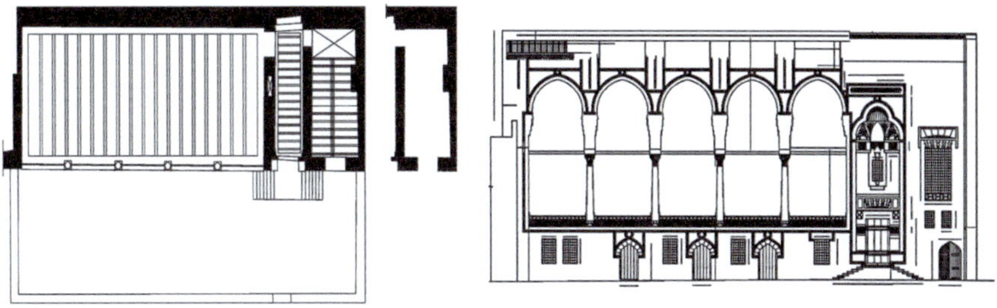

Fig. 27 2D Cross Sections for Mahabb Al-Din Hall

Fig. 28 Prince Mamay Al-Saifi's house

Fig. 29 Prince Mamay Al-Saifi's house

Fig. 30 Point clouds samples for Prince Mamay Al-Saifi's House

Fig. 31 2D Cross Sections for Prince Mamay Al-Saifi's house

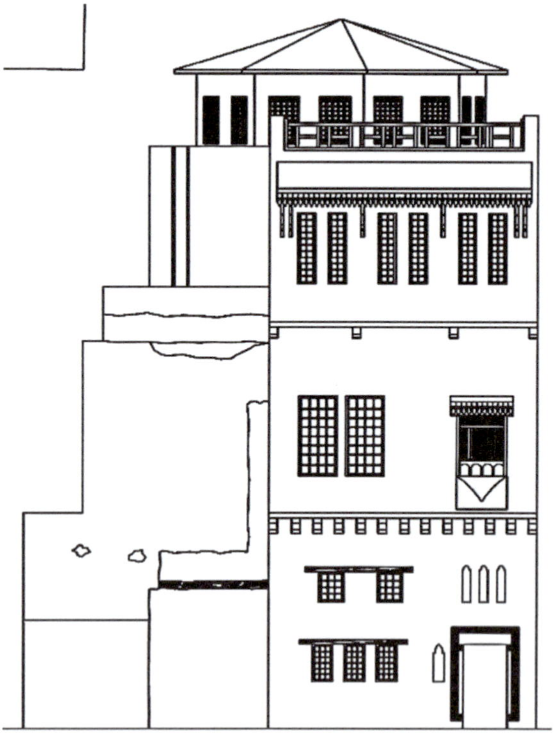

Fig. 32 The entrance of Al-Qadi's house

Fig. 33 Point clouds samples for the entrance of Al-Qadi's house

the Hawara pyramid (Fig. 36). The top of the Hawara pyramid hasn't recorded data that why the beam sent from the laser scanning device doesn't reach to the top of the object. Elevation range from 41.61 to more than 73.61 m above sea level. The interval contouring is 2 m (Fig. 37). Figure 38 shows positive Volume [Cut]: 91,741.78 m3, negative Volume fill: 121.95 m3 and cut plus fill: 91,863.73 m3.

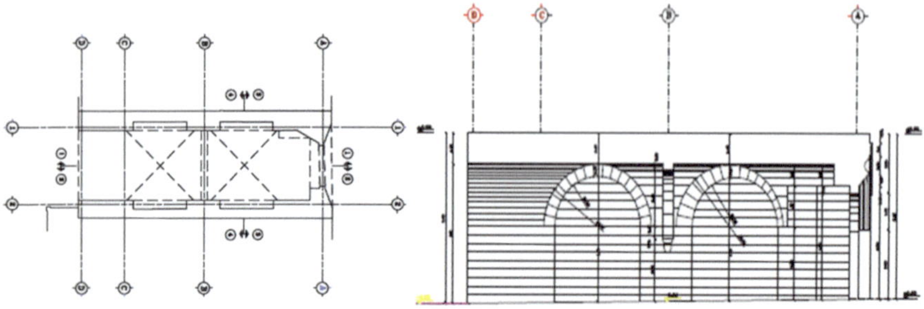

Fig. 34 The 2D Cross Sections for the entrance of Al-Qadi's house

Fig. 35 Hawara pyramid

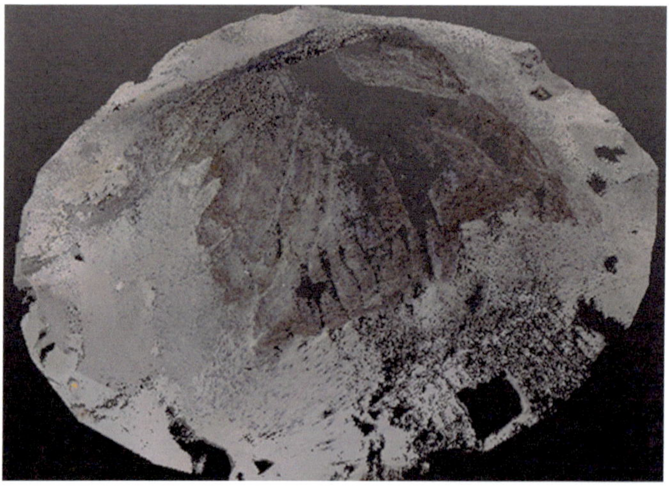

Fig. 36 Hawara pyramid in point cloud

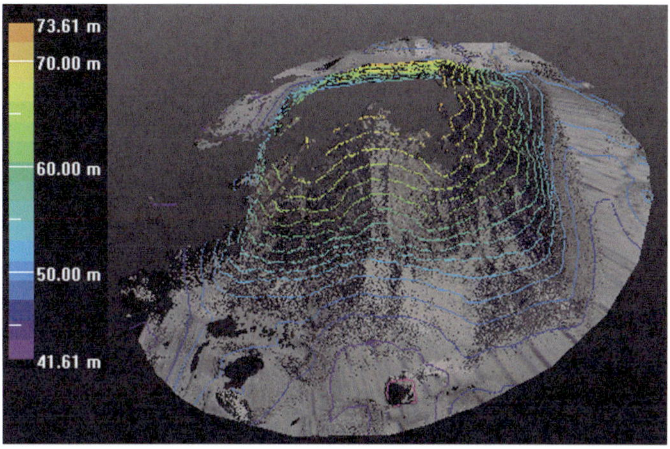

Fig. 37 Contour map for Hawara pyramid

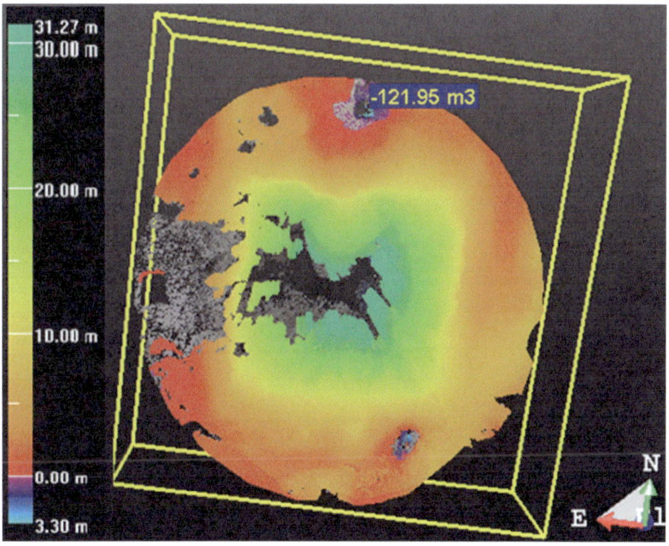

Fig. 38 CUT and fill volumes

References

Buckley SJ, Howell JA, Enge HD, Kurz T (2008) Terrestrial laser scanning in geology: data acquisition, processing and accuracy considerations. J Geol Soc 165(3):625–638

Fernandez I, Becker J, Gillies S (2013) Pleiades. Accessed 22 Mar 2013. ("Saqqara")

http://www.przybilla.biz/bv/Praktikum/TrimbleM3-UserGuide.pdf

Harry Adès A (2007) Traveller's History of Egypt. Chastleton Travel/Interlink, 2007, p 48. ISBN 1-905214-01-4

Lepsius R (1843) Denkmeler aus Egyptien und Ethiopien 11, 1–30. Leipzig,J.C. Hinrichs
Petrie WMF (1888) Hawara, Biahmu, and Arsinoe. Field and Tuer, London
Pfennigbauer M, Ullrich A (2010) Improving the quality of laser scanning data acquisition through calibrated amplitude and pulse deviation measurement. In: Proceedings of SPIE 7684

HBIM Framework for Rehabilitation of Heritage Buildings

Mohamed Marzouk

Abstract Building Information Modelling (BIM) has radically changed the design and documentation processes in architecture, engineering and construction (AEC) industry. BIM coupled with Light Detection and Ranging Technology (LiDAR) technologies revolutionized the built environment documentation methods. Many efforts were directed toward utilizing these technologies in the documentation and restoration of heritage buildings, adopting Heritage Building Information Modelling (HBIM). Heritage documentation is the basic step of any restoration project. LiDAR technology depends on high-speed 3D laser scanners, which send laser beams to the scanned objects in great intensity. HBIM is the heritage layer add-in library that is attached on top of traditional BIM applications. It poses all the power and benefits of BIM along with historical/heritage information. Generating heritage buildings/structures in a 3D HBIM environment are of substantial benefits, such as: remote visualization of the interior and exterior of structures; better understanding of the geometry of the structure; a repository of geometric and historical information; better evaluation of renovation strategies. This chapter presents a framework that integrates LiDAR technology with HBIM in the field of heritage documentation to achieve the following objectives: (1) identifying optimum laser scanner positions, (2) performing structural analysis, and (3) optimizing thermal and visual comfort. The main features of the proposed framework are demonstrated through an actual study.

Keywords Heritage building information modelling (HBIM) · Laser scanning · Structural analysis · Thermal and visual comfort

---

M. Marzouk (✉)
Professor of Construction Engineering and Management, Faculty of Engineering, Structural Engineering Department, Cairo University, Giza, Egypt
e-mail: mmarzouk@cu.edu.eg

G. M. El-Qady and C. Margottini (eds.), *Sustainable Conservation of UNESCO and Other Heritage Sites Through Proactive Geosciences*, Springer Geology,
https://doi.org/10.1007/978-3-031-13810-2_15

1 Introduction

The requirements for the HBIM framework which is essentially needed to support the full range from cultural heritage landscape to specific structure, including (Counsell and Taylor 2017).

- accurately surveyed data with: recording of value, and significance; analysis of materials, structure and pathology; as well as how the use of the structure has responded to and reflects environmental, social, cultural and economic change through time;
- development and construction projects as well as ongoing facilities management' (the wider definition of BIM level 2 in the UK);
- the ephemeral and ambiguous, lacking set boundaries, consistently moving and changing, including infrastructure, trees, planting and water courses;
- the 'Internet of Sensors', responsive systems, and related actuators;
- location-specific 'intangible cultural heritage' and virtual heritage;
- secondary interpretation, community engagement and community-based bottom-up readings, and alternative interpretations, based on potentially ambiguous 'fuzzy' evidence;
- a structure and standards that support continual retrieval and reuse for decades if not centuries.

Barazzetti et al. (2015) described the procedures of developing the HBIM model starting from point cloud data. The developed models were then reformatted to be displayed in mobile apps using augmented and virtual reality visualization technologies. Their purpose was to deliver HBIM models through widely available platforms for both expert and non-expert operators, which is very valuable in the tourism and heritage dissemination sectors. Dore and Murphy (2012) outlined the use of the Dublin institute of technology Historic BIM (HBIM) in modelling historic structures. This approach depends on acquiring special data using 3d laser scanning and photogrammetry to develop an HBIM model using modelling objects from libraries of historical buildings elements. Moreover, the developed model is integrated with GIS data through the City Geography Markup Language (CityGML) framework, thus creating in accurate 3D GIS models. Such approach bridges the gap between 3D GIS, and HBIM in documenting heritage sites. Murphy et al. (2009) outlined in details the procedures of historical BIM (HBIM) system which starts with point cloud and digital imagery and results in a textured 3D parametric model. They extended the ability of the geometric descriptive language (GDL) to generate 3D parametric objects, which they used to develop a library for historical 3D parametric objects of historical buildings' elements. Bruno et al. (2017) investigated an eighteenth century nobel residence with various diagnostic investigation that included photographic survey, cracks mapping and using thermography and radar technologies to demonstrate the constructive material properties for efficient building diagnosis. In terms of retrofitting, BIM tools was found useful to provide the elaborative model, read tables and charts for assessing the building current properties. As well as being useful to validate the criteria digitization tool for integrated building diagnosis. Marzouk et al. (2020a) presented an expert system to identify the most effective method of repair for each specific building material and to propose the appropriate conservation methods for resolving different types of damages. The researchers implemented Artificial Intelligence (AI) to provide a systematic problem-solving technique that saves time and provides the most efficient conservation and preservation method for heritage building elements.

The 3D laser scanning advances have been presented in the field of surveying and can obtain 3D data about physical objects of different shapes and sizes in a cost and time effective way. Laser scanners enable a large number of points to be recorded in a few minutes. As a result of their common sense and flexibility, these sorts of instruments can possibly be broadly utilized as a part of the field of architectural, archeological and environmental surveying (Valanis and Tsakiri 2004). Nex and Rinaudo (2010) worked on an algorithm that automatically combine information gathered from digital imaging and LiDAR to automatically segment and detect heritage building features, such as building break lines. The result of their automation algorithm has the potential to save time and human intervention in the process of documenting heritage buildings using high density point clouds. Hesse (2010) developed a data processing method for the extraction of Local Relief Models from airborne LiDAR high density point clouds, which results in color-coded maps of local Relief Models. The proposed method was applied to LiDAR data (acquired by airborne 3D laser scanners) of Baden-Württemberg and proved valuable and accurate in mapping large archeological sites scanned with airborne 3D laser scanners. Haala et al. (2008) utilized a kinematic terrestrial laser scanning approach to capture the dense point cloud of a historical town. The approach "Street Mapper" depends on four vehicle-mounted 2D laser scanners and a high performance GNXX/inertial navigation system which accurately provides the required dereferenced information which enables the automatic combination of the four 2D dense point clouds. The proposed approach achieved a 30 mm level of accuracy when good GPS conditions were available, thus providing a more feasible 3D laser scanning approach for large heritage sites with an accepted level of accuracy for many urban mapping applications of historical towns.

Coren et al. (2005) integrated LiDAR data with hyper-spectral data to evaluate irregular behavior of major ground indices which in turn improve the discovery of new archeological sites. LiDAR provided the accurate surface geometrical data while the hyper-spectral survey provided the specific humidity, vegetation, and thermal conditions of the target area. The integration of both technologies generated an accurate Digital Terrain Model (DTM). Sánchez-Aparicio et al. (2018) diagnosed the historical structures through multilayered point cloud analysis using the technology of laser scanning to observe the built pathologies. In which, radiometric and geometric data provide a 3D methodology to extract and quantify the different deformations and biological conquest on the masonry. The Fortress of Almeida, in Portugal was examined for restoration activity, where the complex diagnostic results of the radiometer were validated and compared with realistic photographs. Salts and moisture were concluded from the deformation layers within the analysis results that revealed the direct relation between the visual diagnosis and that of the point cloud.

2 Proposed Framework

The proposed innovative framework integrates LiDAR technology with HBIM proved valuable in the field of heritage documentation, achieving the following objectives (see Fig. 1): (1) identifying optimum laser scanner positions, (2) performing structural analysis, and (3) optimizing thermal and visual comfort. Omar Tosson Palace in Egypt is considered as a case study to demonstrate the use of the framework in achieving these objectives.

A methodology is introduced to optimize the terrestrial Laser scanner positions as well as the scanner field of view in order to improve the point cloud quality while reducing scanning and fulfill quality constraints with respect to point cloud such as allowable incidence angle, minimum/maximum scan range, and the field of view. The framework is capable of utilizing processed point clouds using 3D laser scanning to create different purpose BIM models at the different levels of development to simulate the structural performances under different types of actions (see Fig. 2). The research proposes energy and daylight enhancement through different skylight configurations with the usage of relevant technologies in which those technologies are explored, tested, and validated prior to performing optimization for sustainable building reuse strategy. Parameters that affect the thermal and visual behavior of the heritage buildings are explored as well as their relative effect on the skylight energy and daylight optimization as shown in Fig. 3.

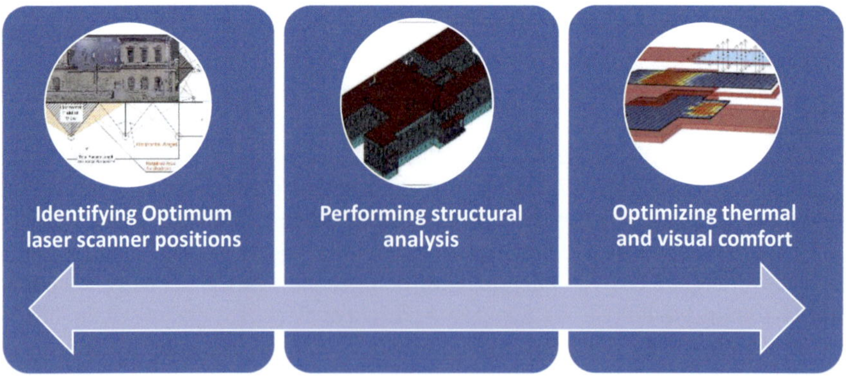

Fig. 1 Proposed HBIM framework

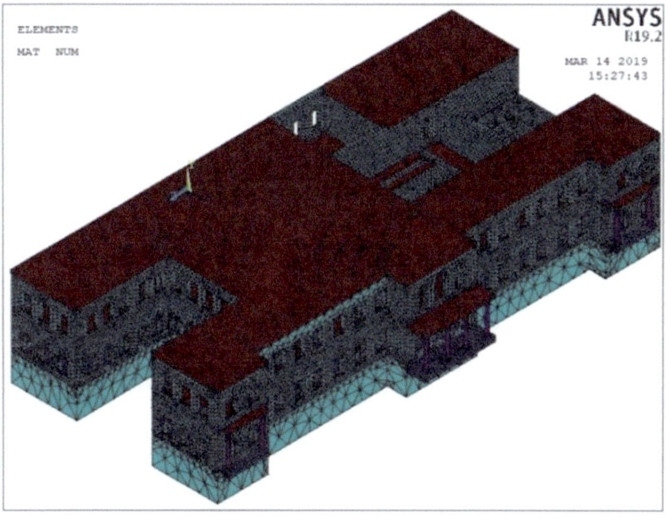

Fig. 2 Geometry of the finite element model

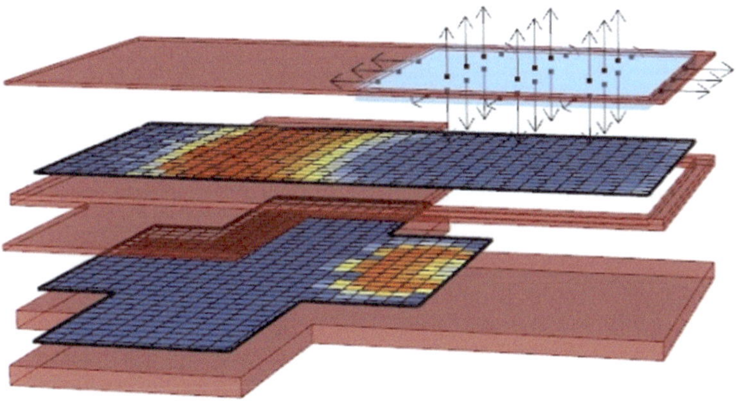

Fig. 3 Examining skylight energy and daylight optimization via simulation

3 Results and Discussion

3.1 Identifying Optimum Laser Scanner Positions

The first objective was applied on Omar Tosson Palace to optimize the scanner locations and the scanner field of view to increase the point cloud quality and shorten the scanning time while guaranteeing a set of quality constraints for the point cloud

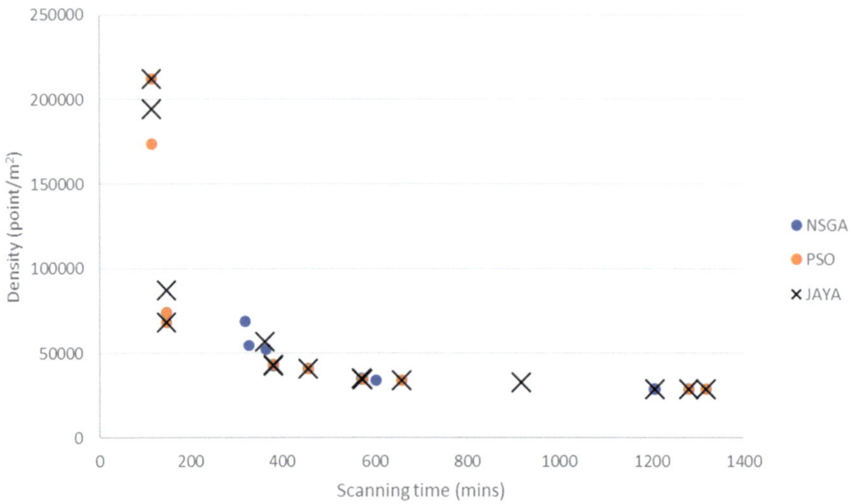

Fig. 4 Sample of the Pareto frontier solutions for the three optimization algorithms

(Metawie and Marzouk 2020). Figure 4 shows a sample of Pareto frontier solutions for three optimization algorithms. Thirty-eight Pareto frontier solutions were obtained as follows; 13 solutions from the genetic algorithm, 15 solutions from JAYA algorithm, and 10 solutions from the particle swarm optimization. Weighed sum model (WSM) technique was implemented to choose the optimum solution from the generated Pareto frontier solutions. All Pareto frontier solutions, generated from the three algorithms, were considered as alternatives. Then, the scanning time and the density attributes were normalized according to their maximum values. Then, the associated weights were considered 60%, and 40% for the scanning time, and the point density attributes respectively. The optimum solution in WSM had scanning time of 146.28 min, a points density of 68,648 point/m^2, with max spacing 5.77 mm, and a max incidence angle of 44.23°.

3.2 *Performing Structural Analysis*

The results of the structural analysis of Omar Tosson Palace focused on estimating the stresses in various structural elements (Marzouk 2020). The assessment was performed considering vertical and lateral loading. Evaluation of tensile stresses in the walls is performed using the finite element method. The geometry of the wall, as well as the mechanical properties were used to produce the numerical models representing the various structural elements of the Palace. The stresses were computed from the various load combinations and were compared with the acceptable levels of stress for such construction. Figure 5 illustrates the equivalent stress (Von-Mises) and

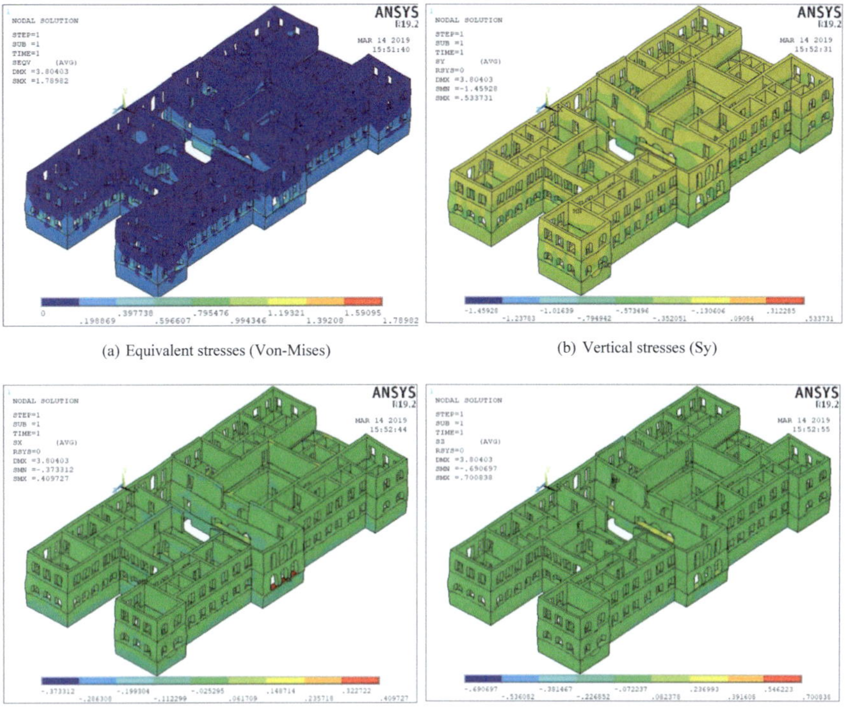

(a) Equivalent stresses (Von-Mises)

(b) Vertical stresses (Sy)

(c) Horizontal stresses (Sx)

(d) Horizontal stresses (Sz)

Fig. 5 Stresses under static loads effect

the stresses in the three principal directions (X, Y, and Z) under vertical loading. Similarly, the equivalent stress (Von-Mises), the stresses in the three principal directions (X, Y, and Z) under seismic loading in X-direction and Z-direction were obtained.

3.3 *Optimizing Thermal and Visual Comfort*

The influence of the skylight shape and opening ratio (SFR) were investigated prior to the optimization process. Exploratory primary analysis of different skylight shapes was performed to draw a conclusion on the most suitable one in terms of daylighting. Suggested Skylight shapes include flat, vaulted, and prismatic. In which the three shapes were compared in respect to their effect on the building space light energy usage in kWh and the space daylight adequacy in Spatial Daylight Autonomy (sDA) and Annual Sun Exposure (ASE) percentages (Marzouk et al. 2020b). Figure 6 illustrates a sample of four inclination cases out of the 33 cases in total. The inclination angle start with an obsolete vertical single side of the prism and the top prism vertex rotates with a magnitude of 0.5 m until the other side becomes absolute vertical in

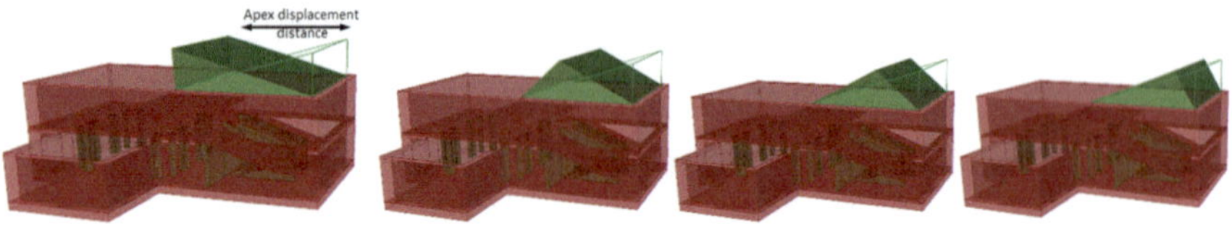

Fig. 6 Skylight to Floor ratio (SFR) change according to skylight apex displacement

shape while the starting vertical side becomes in maximum oblique configuration. Analyzing the points obtained from the optimization algorithm, it is clear the points are clustered automatically into two clusters. Reinstating example points, it has been clearer that the two distinct clusters represent different glazing orientation of the examined skylight. Each cluster points' value varies within a different range of both Useful Daylight Illuminance (UDI) and cooling load values. It is clear form Fig. 7 that the objectives are better achieved with preferable higher value ranges for the southern oriented skylights with enhanced UDI and cooling load values.

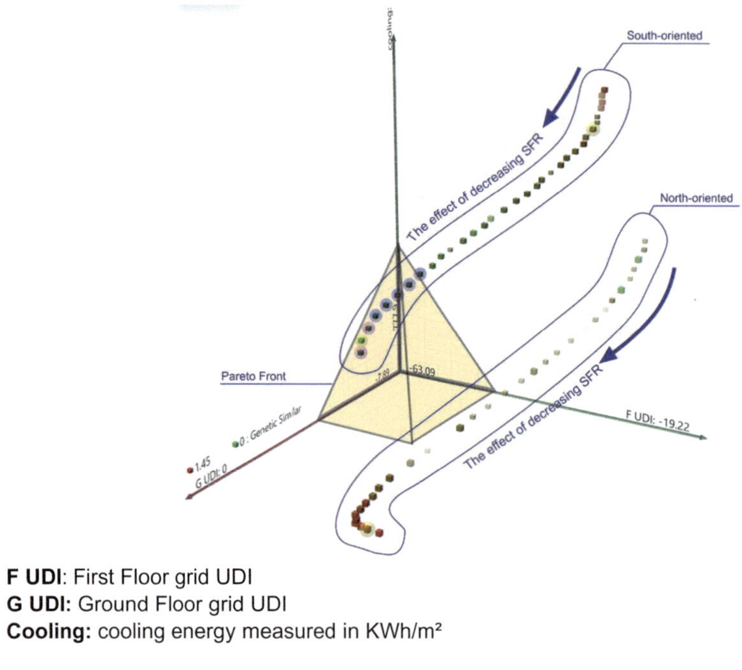

Fig. 7 3D preview for the optimization results

4 Conclusion

This paper presented a methodology to integrate BIM with 3D laser scanning (LiDAR) technologies by proposing a framework that adopts Heritage Building Information Modelling (HBIM). The proposed framework takes into consideration: (1) identifying optimum the laser scanner positions, (2) performing structural analysis, and (3) optimizing thermal and visual comfort. An optimization and decision-making model was implemented to minimize scanning time and density while achieving certain quality constraints. The terrestrial laser scanner positions are determined based on the distance between the laser scanner and the scanned surface, the required horizontal and vertical angles, and the required scanner resolution. Subsequently, structural analysis was performed using the finite element model. The model was subjected to loads and the stresses are computed for various load combination scenarios. Finally, optimizing thermal and visual comfort was implemented by studying a parametric configuration approach to redesign the skylight. Daylight and energy simulation together with genetic optimization were used to evaluate and optimize the skylight performance for heritage retrofitting. The research can be extended in the future utilizing robust artificial intelligence techniques to detect the problems that exist in scanned elements of heritage buildings in an automated manner.

References

Barazzetti L, Banfi F, Brumana R, Oreni D, Previtali M, Roncoroni F (2015) Hbim and augmented information: towards a wider user community of image and range-based reconstructions. Int Arch Photogramm Remote Sens Spat Inf Sci 40(5):35

Bruno S, De Fino M, Fatiguso F (2017) Historic building information modeling towards building diagnostic data management. A case study. Tema: Technol Eng Mater Archit 3(2):99–110

Coren F, Visintini D, Prearo G, Sterzai P (2005) Integrating LiDAR intensity measures and hyperspectral data for extracting of cultural heritage. In: Proceedings of Italy–Canada 2005 workshop on 3D digital imaging and modeling: applications of heritage, industry, medicine and land, Padova, Italy. (May 2005)

Counsell J, Taylor T (2017) 3W V What are the goals of HBIM?. Heritage building information modelling

Dore C, Murphy M (2012) Integration of Historic Building Information Modeling (HBIM) and 3D GIS for recording and managing cultural heritage sites. In: 2012 18th international conference on virtual systems and multimedia (VSMM). IEEE, pp 369–376. (September 2012)

Haala N, Peter M, Cefalu A, Kremer J (2008) Mobile lidar mapping for urban data capture. In: Proceedings of the 14th international conference on virtual systems and multimedia, Limassol, Cyprus, vol 2025, p 95100. (Oct 2008)

Hesse R (2010) Extraction of archaeological features from high-resolution LIDAR data. In: Börner W, Uhlirz S, pp 636–642

Marzouk M (2020) Using 3D laser scanning to analyze heritage structures: the case study of egyptian palace. J Civ Eng Manag 26(1):53–65

Marzouk M, ElSharkawy M, Elsayed P, Eissa A (2020a) Resolving deterioration of heritage building elements using an expert system. Int J Build Pathol Adapt 38(5):721–735

Marzouk M, ElSharkawy M, Eissa A (2020b) Optimizing thermal and visual efficiency using parametric configuration of skylights in heritage buildings. J Build Eng 31:101385
Metawie M, Marzouk M (2020) Optimizing laser scanning positions in buildings exteriors: heritage building application. J Civ Eng Manag 26(3):304–314
Murphy M, McGovern E, Pavia S (2009) Historic building information modelling (HBIM). Struct Surv 27(4):311–327
Nex F, Rinaudo F (2010) Photogrammetric and LiDAR integration for the cultural heritage metric surveys. Int Arch Photogramm Remote Sens Spat Inf Sci 38(Part 5):490–495
Sánchez-Aparicio LJ, Del Pozo S, Ramos LF, Arce A, Fernandes FM (2018) Heritage site preservation with combined radiometric and geometric analysis of TLS data. Autom Constr 85:24–39
Valanis A, Tsakiri M (2004) Automatic target identification for laser scanners. In: Proceedings of XXth ISPRS congress, Istanbul, Turkey. (July 2004)

Hydrogeology

Geophysics’ Role in Investigating and Mitigating Groundwater Hazards on Archaeological Sites: Case Studies from “Sphinx-Giza, Kom Ombo Temple-Aswan and Hawara Pyramid-Fayoum”

Abbas Mohamed Abbas, Usama Massoud, Hany S. Mesbah, Ayman I. Taha, Mohamed Gamal Abdelmonem, and Gad Mohamed El-Qady

Abstract Ancient Egypt left behind numerous landmarks such as pyramids, temples, statues, tombs spread along the shores of the River Nile, most of which are invaluable man-made archaeological treasures. Placed at uniquely chosen archaeological sites and having survived several millennia, they are increasingly devastated by the rising water table, soil composition, and subsurface water flows that have a destructive impact on their foundations and structural integrity. Much of that change took place over the past few centuries due to intensive farming, irrigation plans, climate change, growing urbanization, malfunctioning sewer systems, and changing environmental and soil conditions in the surrounding areas. In particular, intense irrigation, increasing plantation and deficiency of the drainage networks in the vicinity of the archaeological sites are the most significant risks and a major cause for recorded deterioration of subsurface structures across the past fifty years. The problem has been exasperated by rising groundwater containing high dissolved salts that chemically interact with and affect subsurface bases. In dry seasons water evaporates, condensed salt crystallizes and creates multiple pore holes that, following prolonged exposure, cause stone blocks to disintegrate. Geophysical studies emerged, in recent decades, as critical for the diagnostic investigation of subsoil and hydrogeological conditions of archaeological sites and mapping subsurface structures and water flows. Several geophysical techniques are used as non-intrusive methods to generate multiple datasets for underground water management and determining mitigation and dewatering plans. This paper provides a detailed empirical study using essential geophysics methods and applications for three major archaeological sites in Egypt

A. M. Abbas (✉) · U. Massoud · H. S. Mesbah · A. I. Taha · G. M. El-Qady
National Research Institute of Astronomy and Geophysics “NRIAG”, Helwan, Cairo 11421, Egypt
e-mail: abbas@nriag.sci.eg

M. G. Abdelmonem
School of Architecture, Design and the Built Environment, Nottingham Trent University, Nottingham NG1 4FQ, UK

G. M. El-Qady and C. Margottini (eds.), *Sustainable Conservation of UNESCO and Other Heritage Sites Through Proactive Geosciences*, Springer Geology,
https://doi.org/10.1007/978-3-031-13810-2_16

with distinctive groundwater threats and environmental conditions and prevention plans. These include Sphinx-Giza, Kom Ombo Temple-Aswan and Hawara Pyramid-Fayoum. Three geophysical methods have been used to highlight the characteristics of underground water flows and impact and provided a roadmap for solutions and dewatering plans in each case.

Keywords Geophysics · Archaeology · Groundwater · Preservation · Egypt

1 Introduction

Groundwater impact has become a fundamental threat to many Egyptian archaeological sites and remains. It is attributed to the rise of groundwater table due to desert plantation, water channels and cultivation zones coming into close proximity to important ancient Egyptian and archaeological sites throughout the country and along the Nile valley (Ahmed and Fogg 2014). Geophysical methods emerged as rapid, non-destructive methods that provide detailed diagnostic and forensic analysis that directs and determines planned intervention to preserve and save those sites. So, they have been used with increasing frequency in archaeo-prospection, as well as preservation of the archaeological sites from environmental impacts. Geophysics is a science that functions adequately with several disciplines to fulfil such complex assignments (Abbas et al. 2005, 2012a, b, 2015; Pellicer and Gibson 2011; Massoud et al. 2010; Khalil et al. 2010; Deiana et al. 2018; Barone et al. 2019).

Geophysical assessment could be considered as the first stage of restoration in order to limit the hazardous effect of groundwater. In this article, we have selected three archaeological sites from Giza, Aswan and Fayoum Governorates to be representative examples, where each site has its own geological and environmental conditions and requires specific diagnostic tools and treatment plans. In the cases discussed hereafter, the geophysical studies had provided a fundamental vision in the treatment plan to save the endangered antiquities.

2 Geophysical Techniques in Archaeology

Geophysical methods have been used in archaeology since 1946, while aerial photography has been used since 1919 (Toushmalani 2010). At present, the most commonly used geophysical methods are electrical resistivity, magnetic, ground-probing radar and electromagnetic soil-conductivity systems, which are being increasingly used because of their very high rate of data acquisition. Less commonly used methods include self-potential, microgravity, radiometric, thermal infrared imagery, and sonic or seismic techniques. Each geophysical technique has its own advantages and limitations, making it necessary to use more than one geophysical tool to study the conditions of an archaeological site (Toushmalani 2010; Abbas et al. 2012a, b). Integrated application of multiple methods will lead to accurate and coherent results. In the present case studies, three geophysical methods, including electrical resistivity, ground probing radar, and transient electromagnetic techniques, were applied in various deployments with different data acquisition parameters. A short idea about each method will be given in the following sections.

3 Electrical Resistivity (ER)

Electrical Resistivity (ER) of the soil/rock is used as a proxy for the spatial and temporal variability of many other soil physical properties (i.e. structure, clay content, water content, or fluid composition). Because the method is non-destructive and very sensitive, it offers adequate datasets that describe the subsurface properties and differentiate between objects through contrasting readings of their electrical response (e.g. rock structures stone blocks will give a relatively high resistivity response, while ditches and pits retaining high moisture content will give low resistivity values). The application of the method requires an electric current to be injected into the ground by a pair of surface electrodes (current electrodes). The resulting potential field (voltage) is measured at the surface between the second pair of electrodes (potential electrodes). The subsurface resistivity can be calculated by knowing the electrode geometry, the applied current, and the measured voltage. The 1-D resistivity measurements can be conducted in two modes; the vertical electrical sounding and the horizontal profiling. One of the new developments in recent years is using 2-D resistivity tomography (ERT) surveys to map areas with moderately complex geology (Griffiths and Barker 1993). Such surveys are usually carried out using a large number of electrodes, 25 or more, connected to a multi-core cable. A laptop microcomputer together with an electronic switching unit is used to automatically select the relevant four electrodes for each measurement, where the data are measured automatically and stored for further processing and analysis. The apparent pseudo-sections of 2D profiles are inverted to actual resistivity images and depth values (Loke and Barker 1996; Loke 2015; Loke and Dahlin 2002).

4 Ground Penetrating Radar (GPR)

GPR is a high-resolution geophysical method based on the propagation of high-frequency electromagnetic waves. It is a powerful non-invasive tool used to image the shallow subsurface structures based on changes in their dielectric properties. Standard GPR systems consist of a transmitting and a receiving antenna, where high-frequency electromagnetic pulses are emitted into the ground by the transmitter. The radar wavelet propagates through the soil, where the velocity of the wavelet depends on the dielectric properties of the ground. At interfaces, e.g., boundaries of different soil layers or distinct objects, where the dielectric properties of the different medium change erratically, the electromagnetic wave is partially reflected. The travel time and amplitude of the wavelet are recorded by the receiving antenna. The amplitude of the reflected wave can provide information about the properties of the boundary zone. Then, the reflected signals are used to produce a continuous cross-sectional picture of the shallow subsurface medium. Attenuation of the radar wave is related to the conductivity of the ground, where the wave has completely vanished in the highly conductive media or when it touches subsurface water. Therefore, GPR is a powerful tool used in environmental investigations for outlining the conductive soils and mapping the shallow groundwater table (Annan 2005; Saintenoy and Hopmans 2011; Boghdady 2012; Hengari et al. 2013; Peterson and Doliber 2019).

5 Transient Electromagnetic Method (TEM)

The Transient Electromagnetic (TEM) technique is an inductive method that utilizes a strong direct current passing through a square loop commonly laid on the ground surface. The flow of this current in the surface loop will create a primary magnetic field that spreads out into the ground. When the applied current is abruptly terminated, the primary magnetic field starts to decay with time. According to Faraday's law, decay of the primary magnetic field with time will induce eddy currents that flow outward and downward with decreasing velocity and diminishing amplitude with time, rather like smoke rings. The rate of change of these currents and of their respective secondary magnetic field depends on the size, shape and conductivity of subsurface conductors.

As the other electromagnetic methods, the TEM exploration method is controlled by the laws of EM induction, which means that the TEM method is sensitive to conductive media and relatively insensitive to resistive ones as the induction process is strong enough in the low resistivity media (Christensen and Sorensen, 1996). It can resolve conductive structures quite well, and the depth to the good conductors can be well-determined. So, the method has been used extensively for hydrogeophysical and environmental applications, including environmental issues and geotechnical site assessment, water resource management and groundwater investigations, determination of the depth of water table and the overburden thickness, determination of the extent of saltwater intrusions into the coastal aquifers, and location and monitoring of groundwater pollution (e.g. Massoud et al. 2009, 2014; Porsani et al. 2012; Khalil et al. 2013; Metwaly et al. 2014; Abdul Razak et al. 2019).

6 Case Studies

6.1 Sphinx-Giza

This study was running since 2000 and continued periodically. Sphinx is of primary interest because it is endangered by groundwater, which has a significant effect on the disintegration of its rocks. Sphinx is situated in a basin (16 m asl), surrounded by the Pyramids' plateau to the West (about 80 m asl) and Nazlet El-Samman to the East (19–20 m asl). Figure 1 shows the location of the Sphinx and the boundary of the Pyramids cliff. The Sphinx is located at the south-eastern flank of the Giza plateau and was carved in-situ from the Middle Eocene limestone of the Mokattam Formation. The general stratigraphic section at the Pyramids area has been concluded in Fig. 2. As indicated in Fig. 2, the Sphinx rocks consist of three members; Rosetau member, Setepet member and Akhet member from the base to the top, as depicted in Fig. 3.

The Sphinx rocks are undergoing the danger of deterioration due to groundwater invasion. Restoration operations have been applied to it through different periods of time (Pharaonic, Greco-Roman and twentieth century). Unfortunately, they were

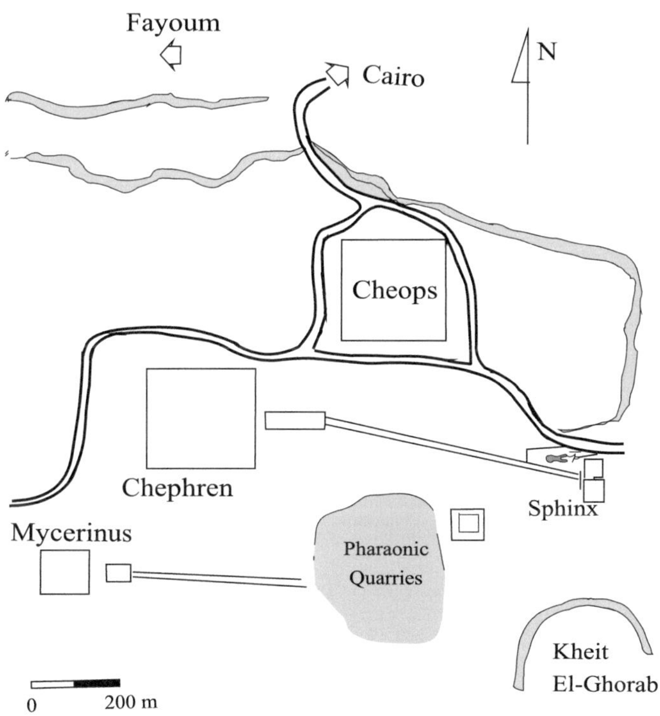

Fig. 1 Location of the Sphinx and the boundary of the Pyramids cliff

Stratigraphy	Formation	Remarkes	Lithology	
Quaternary	Aeolian Sand Nile Terraces Abbassia Fr.		Sand Clay, Gravel and Sand Gravel	
Pliocene			Mainly sand and sandy limestone	
Miocene			Sandstone and Marl	
Oligocene	Mina House Fr.			
Upper Eocene	Maadi Group: Asqobia Fr. (Ain Musa Mb.; Carolia Bed) Wadi Garawi Fr. El-Qura Fr.		Dolomitized limestone Carolia Placunoides biohorizon Shally marl layers and thin bedded marly limestone Argillaceous limestone and gypseous marl	Sphinx
Middle Eocene	Observatory Formation Mokattam Formation: Upper Middle Lower	Akhet Mb. Setepet Mb. Rosetau Mb.	White gray marly limestone Limestone marl complex sequence/Nummulite Bank dark gray dolomitic limestone Nummulitic white limestone	

Fig. 2 Stratigraphic column of the study area

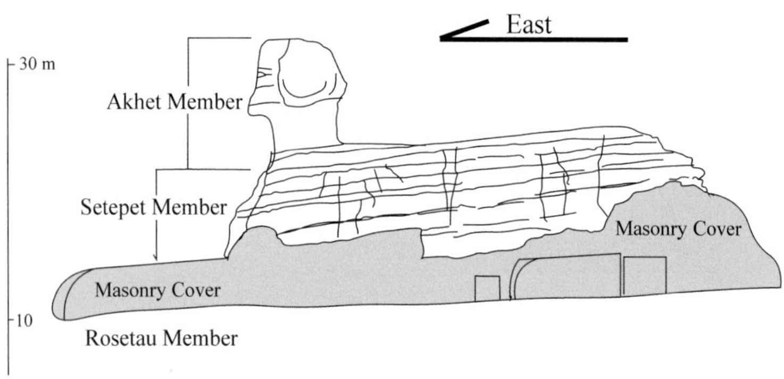

Fig. 3 Sphinx statue: members of Mokattam Formation

unsuccessful due to the presence of salts and the fluctuation of groundwater levels. These salts, originally found in the blocks of different generations of restoration and in the parent rock from which the core of the Sphinx has been carved, have been accumulated by the water (surface or subsurface). Their repeated solution during the condensation of moisture at night and crystallization due to the evaporation of water during the day mechanically disintegrate the stone. Also, these salts weaken the stone, whereby it becomes more venerable to wind erosion (Abbas 1999).

Several geophysical techniques were applied at the Sphinx and its surroundings to assess the groundwater behaviour. Vertical Electrical Soundings (VES), Electromagnetic (EM) and 2D Electrical Resistivity Tomography (ERT) were conducted to determine and to map the changes in groundwater level and the sub-soil characteristics. As an example, Fig. 4 illustrates the distribution of the VES points throughout the site of the Sphinx. The results of the geophysical works were combined with the available borehole data to build a hydrological model for the study area.

Two shallow aquifers were delineated from the geophysical outcomes. The first aquifer layer is unconfined with an average thickness of about 5.5 m. Then, a clay layer is working as a confining layer (Aquitard) with an average thickness of about 7.5 m and the second sandstone layer is a confined aquifer with an average thickness of about 20 m. Modflow© and Modinv© programs were used to construct a three-dimensional hydrological model following the flow chart shown in Fig. 5.

In order to study a possible strategy to reduce the level of water in the surrounding of the Sphinx, two cut-off diaphragms were investigated, cutting the source layers around the Sphinx region as shown in Fig. 6. In the simulation process, two pumping wells were considered active. The first one is the golf club pumping station, and the second is in Zaghloul Street (Nazlet El-Samman), with pumping rates of 982 m3/day and 1968 m3/day, respectively. The MODFLOW© program has been used to simulate the effect of the assumed wall on the two aquifers at two different distances (50 and 100 m) away from the Sphinx.

The first trial was done to simulate the wall at a distance of about 50 m away from the Sphinx statue. The wall effect on the groundwater level of the upper and lower aquifers can be noticed in Fig. 7a and b.

Another simulation has been done as the cut-off wall was at a distance of 100 m from the Sphinx statue (Fig. 8a and b).

From the previous simulations, the following aspects could be concluded:

- There are few differences between the simulated pumping from the first and second aquifers. This could be due to the lithological window between the two layers.
- It is also noticed that after a specific pumping rate, the further improvement or decrease in water level is not comparable with the work done to achieve such high pumping rates.
- The pumping from the first layer with a rate of 1000 m^3/day is economically enough and sufficient to reduce the water head by about 2 m around the Sphinx.
- The application of a cut-off wall would lead to a significant lowering of the groundwater level under the Sphinx, especially if it is 50 m away from the Sphinx.

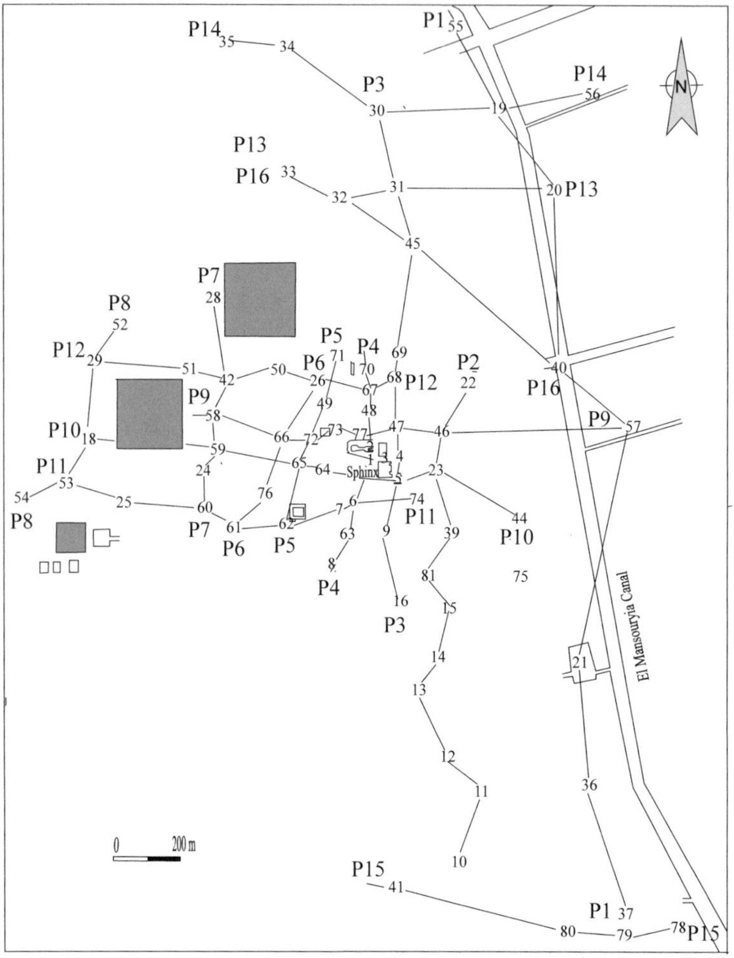

Fig. 4 Location of the VES measuring points in the study area

7 Kom Ombo Temple—Aswan

The archaeological site of Kom Ombo covers an area of approximately 575 m (NW–SE) by 300 m (NE-SW), as shown in Fig. 9. The site includes the standing monuments of the Ptolemaic and Roman period temple, a substantial ancient mud-brick enclosure wall bounding the temple to the northeast and southeast and a modern mud brick enclosure wall to the northwest.

In 2010, the National Research Institute of Astronomy and Geophysics (NRIAG) had collaborated with the Supreme Council of Antiquities to conduct a comprehensive geophysical survey at Kom Ombo temple and its immediate surroundings (NRIAG 2010).The surveys aimed at the study and follow up the destructive

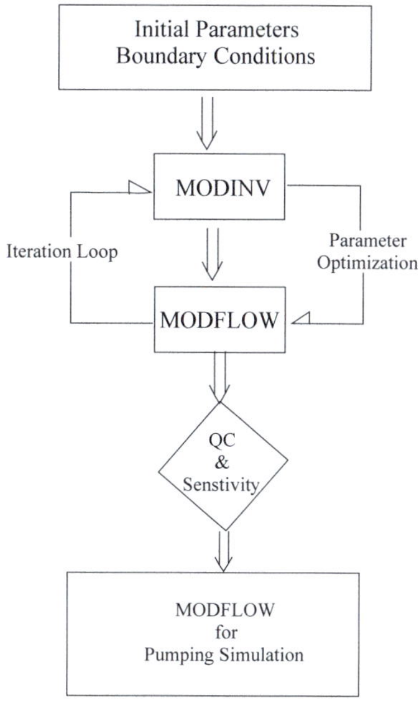

Fig. 5 Flow chart of modelling steps

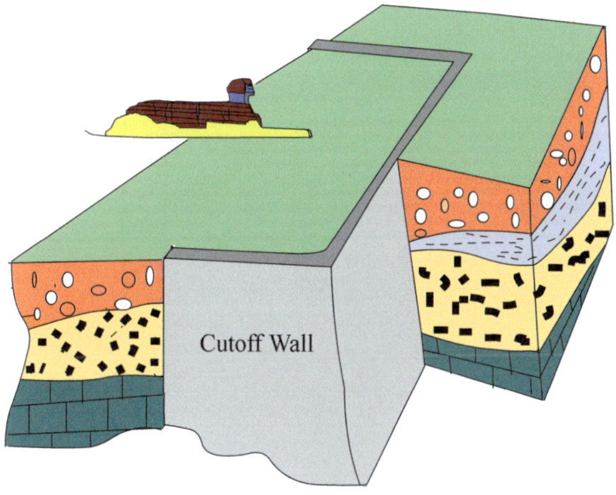

Fig. 6 Schematic sketch for the cut-off wall around the Sphinx

human and environmental impacts on the temple site from the polluted subterraneous/groundwater level rise and invasion from the vicinity area due to the lateral urban sprawl and increasing cultivation activities.

In this survey, two geophysical methods were applied; the Ground Penetrating Radar (GPR) and the Electrical Resistivity Tomography (ERT). Interpretation of the collected data and the results of these studies are described below.

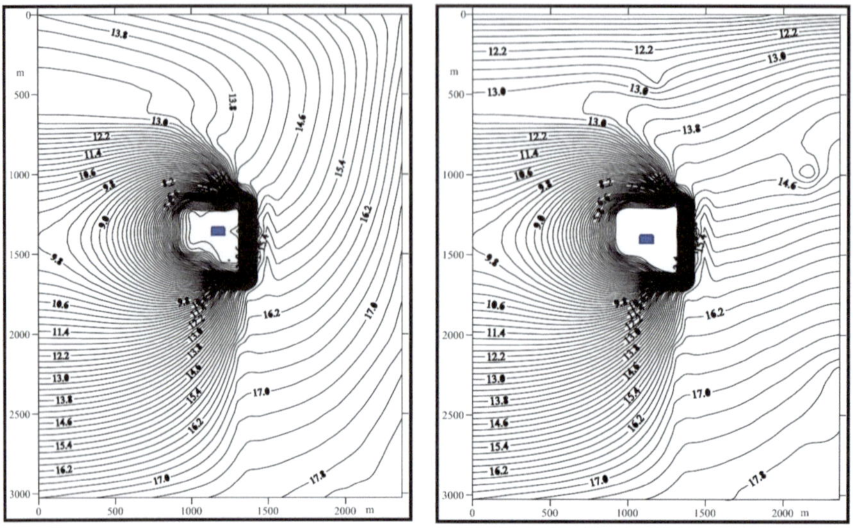

(a) (b)

Fig. 7 Groundwater level of the **a** first (upper) aquifer, **b** second (lower) aquifer affected by cut-off wall 50 m away from the Sphinx Statue

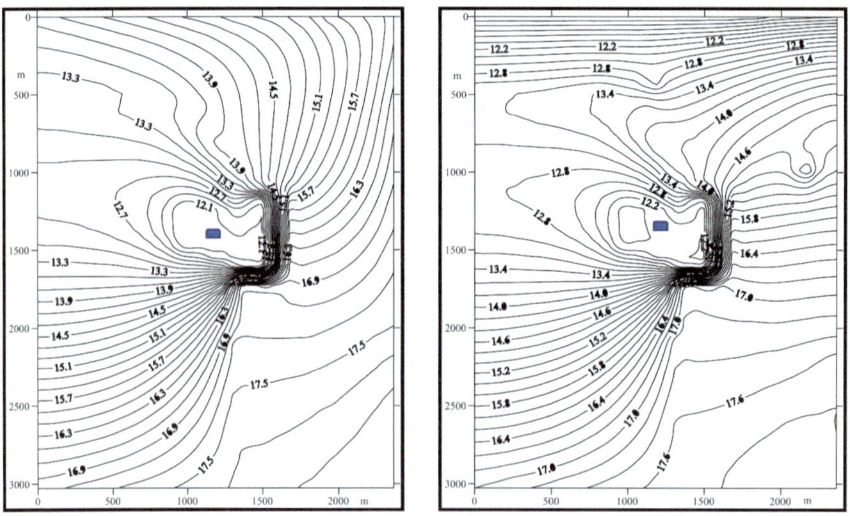

(a) (b)

Fig. 8 Groundwater level of the **a** first aquifer, **b** the second aquifer affected by cut-off wall 100 m away from the Sphinx Statue

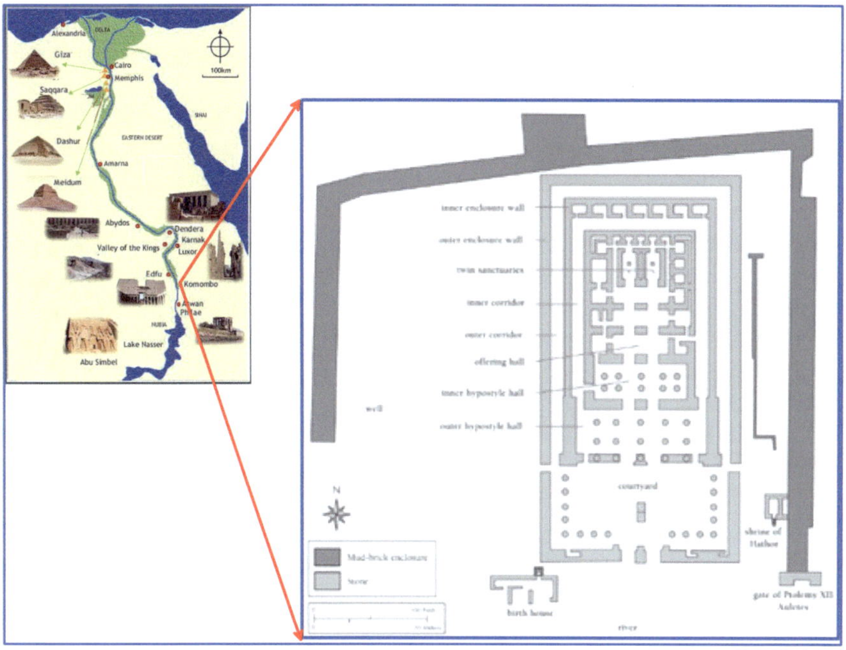

Fig. 9 Location and layout of Kom Ombo temple

In the GPR survey, we used shielded antennae of central frequency 270 and 200 MHz. Also, low-frequency unshielded antennae of frequency 80, 48 and 32 MHz were used to cover the whole temple area in longitudinal traverses. As an example, Fig. 10 shows the used multi-frequency antenna (16–80 MHz) and the distribution of GPR profiles of antenna 200 MHz along the temple area.

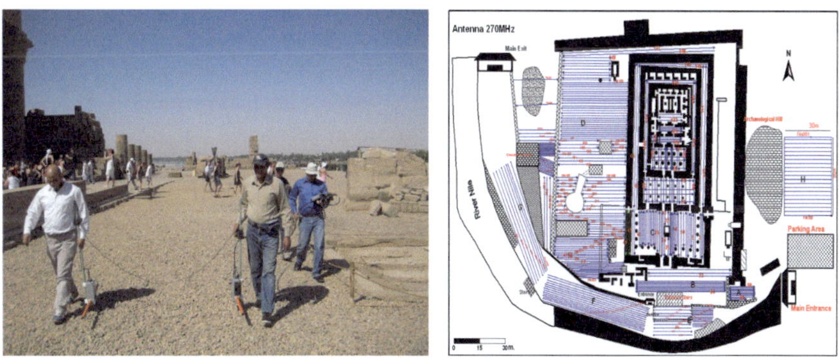

Fig. 10 The used multi-frequency antenna (left) and the distribution of 200 MHz profiles (right) throughout the temple and its surroundings

This survey was suitable for imaging near-surface anomalies associated with buried archaeological features and to image subsurface anomalies associated with soil/rock characteristics and groundwater conditions.

The collected data were processed using the REFLEXW© program V. 7.0 (Sandmeier 2010). The GPR data measured with 200 and 270 MHz antennae revealed the subsurface utilities (e.g. electrical cables) that occupies the shallow sub-surface soil (Fig. 11). Also, shallow restoration work was traced out at depths between 0.80 and 1.60 m. The survey at the Mud-Tell to the north-eastern corner of the temple has indicated some archaeological remains. Meanwhile, the 80, 48 and 32 MHz measuring antennae identified homogenous linear anomalies interpreted as deeper past-human activities (archaeological structures and restoration work indications). The survey at the southwestern (the entrance of the temple) has revealed filling materials and relatively saturated bedrock.

As for the ERT survey, twenty-four 2D imaging profiles were measured, as shown in Fig. 12. The 2D inverted geo-electric sections showed some isolated surficial high/very high resistive anomalies at shallow depths within both the inner and outer temple yards (Fig. 13). The separate zones of high resistivity within the whole foundation layer, from 2.30 to 6.30 m deep, are repeatedly encountered along with some profiles. These anomalies indicate the possibility of archaeological remains at these locations (Fig. 14). Deeper through the imaging profiles, the resistivity values getting lower, reflecting the rising of groundwater levels (Fig. 15).

.

The integration of the GPR and ERT results has indicated some critical findings that could be concluded below:

- There are near-surface archaeological structures/restoration works. It would be appropriate to inspect the delineated spots that might be of potential archaeological interest (Fig. 14, left).
- There is a subsurface tunnel/passage connecting the Nilo-meter with the Nile (Fig. 14, right).
- The admittance area of the temple and the eastern area are highly saturated with groundwater, and these areas represent a weak and unstable subsurface zone (Fig. 15).
- It is recommended to perform a dewatering plan to keep the groundwater level at −7 m. Also, sub-soil stabilization works are to be carried out along the temple entrance zone.

8 Hawara Pyramid-Fayoum

Hawara is an archaeological site of Ancient Egypt (Fig. 16), located at the southern entrance to Fayoum oasis. Hawara Pyramid was built by Amnemhat III, the last powerful ruler of the 12th Dynasty. It was made up of mud bricks surround a core of limestone passages, burial chambers and faced by limestone cover. In 1888, William Flinders Petrie conducted excavations in the Hawara area and found papyri of the

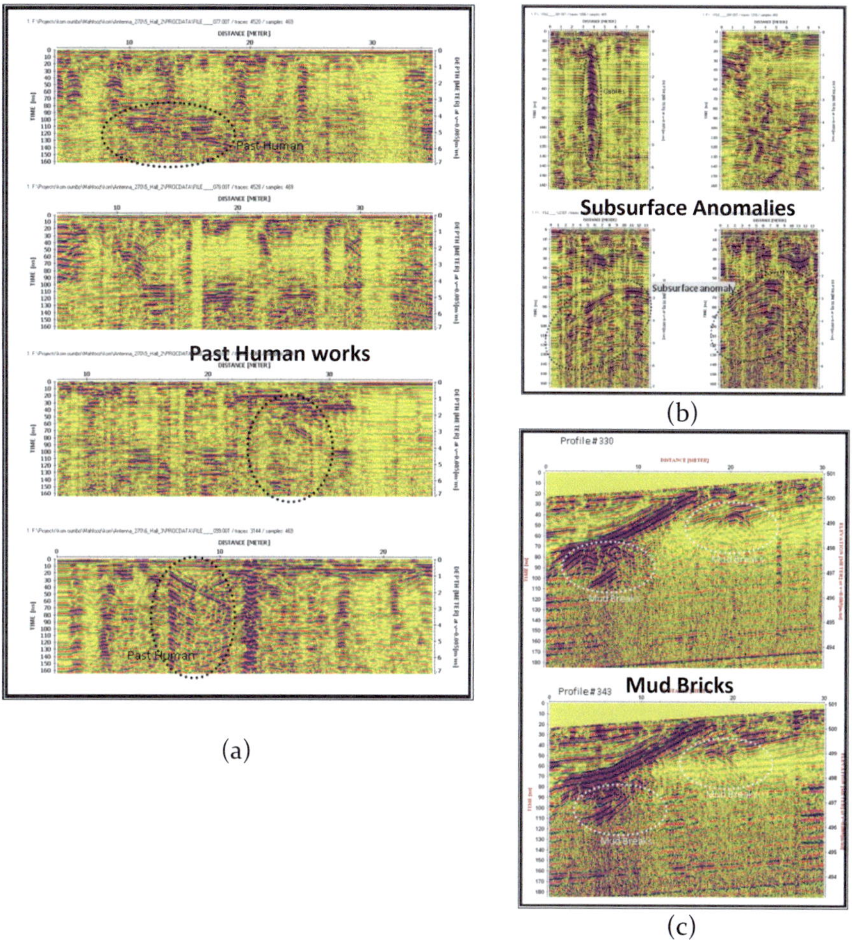

Fig. 11 Some GPR profiles indicating the outlined features at Kom Ombo temple (infra-structure, old-human activities, old restoration,...etc.)

first and second centuries. To the north of the pyramid, he found a vast necropolis with 146 portraits on coffins dating to the Roman period, famous as being among the very few surviving examples of painted portraits from Classical Antiquity, the "Fayoum portraits" illustrated in Roman history (Edwards 1947, 1986; Siliotti and Hawass 1997). The huge mortuary temple that originally stood adjacent to the south of this pyramid is believed to have formed the bases of the complex of buildings with galleries and courtyards called a "Labyrinth" by Herodotus. The demolition of the Labyrinth may date in part to the reign of Ptolemy II, under whom the Pharaonic city of Shedet.

The area of Hawara, including Hawara Pyramid and the Labyrinth ruins, was invaded by groundwater through Bahr Wahba (Wahba Canal), which cuts through

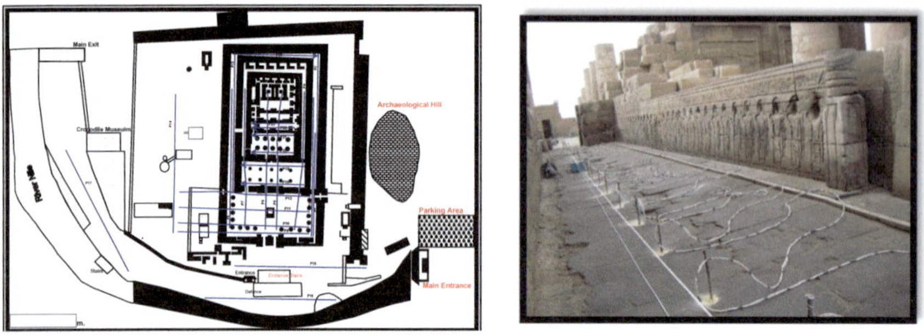

Fig. 12 Distribution of ERT profiles on and around Kom Ombo Temple(left) and ERT profile spreads along one corridor of the temple (right)

the ancient Labyrinth complex. Another significant water contribution is the creeping of agricultural land around the archaeological region (Fig. 16).

In 1889, groundwater was found only in the bottom chambers of the tomb below the main pyramid foundation structure (Petrie et al. 1912). By time and particularly in the twenty-first century, water had flooded into all chambers and the tunnel complex (Fig. 17, up). Currently, the pyramid's internal structures are fully submerged, and at the time of the study, water was existing within 3.0 m from the entrance level (Temraz and Khallaf 2016), as shown in Fig. 17, down.

The most significant risk from groundwater is that all maintenance and conservation works are stopped due to the groundwater invasion. This water can lead to the following negative impacts:

- High salinity in the water accelerates the process of fragmentation of foundation rocks.
- Fluctuation in water levels (high and low) is accelerating rock liquefying processes.
- Pyramid base instability may be realized due to the aforementioned risks.

Various geophysical techniques for investigation and protection were applied to the site. Just the approaches related to groundwater impact assessment will be considered herein. These approaches include Vertical Electrical Sounding (VES), Electrical Resistivity Tomography (ERT) and Transient Electromagnetic (TEM) sounding. Twenty-one Vertical Electrical Soundings (VES) have been conducted around Hawara Pyramid by applying the standard Schlumberger array with half-current electrode spacing varying from 200 and 500 m to identify subsurface layers down to about 100 m below ground surface (Fig. 18). Locations of the VES data measuring points are shown in Fig. 19.

The collected data have been processed using adequate 1D inversion programs to get the final geoelectrical model that matches the geologic model of the Hawara site and its surroundings. Figure 20 shows some examples of the 1D models obtained from VES data inversion. The individual geoelectrical models have been used to construct two sets of geoelectrical cross-sections in E-W and N-S directions, as shown in

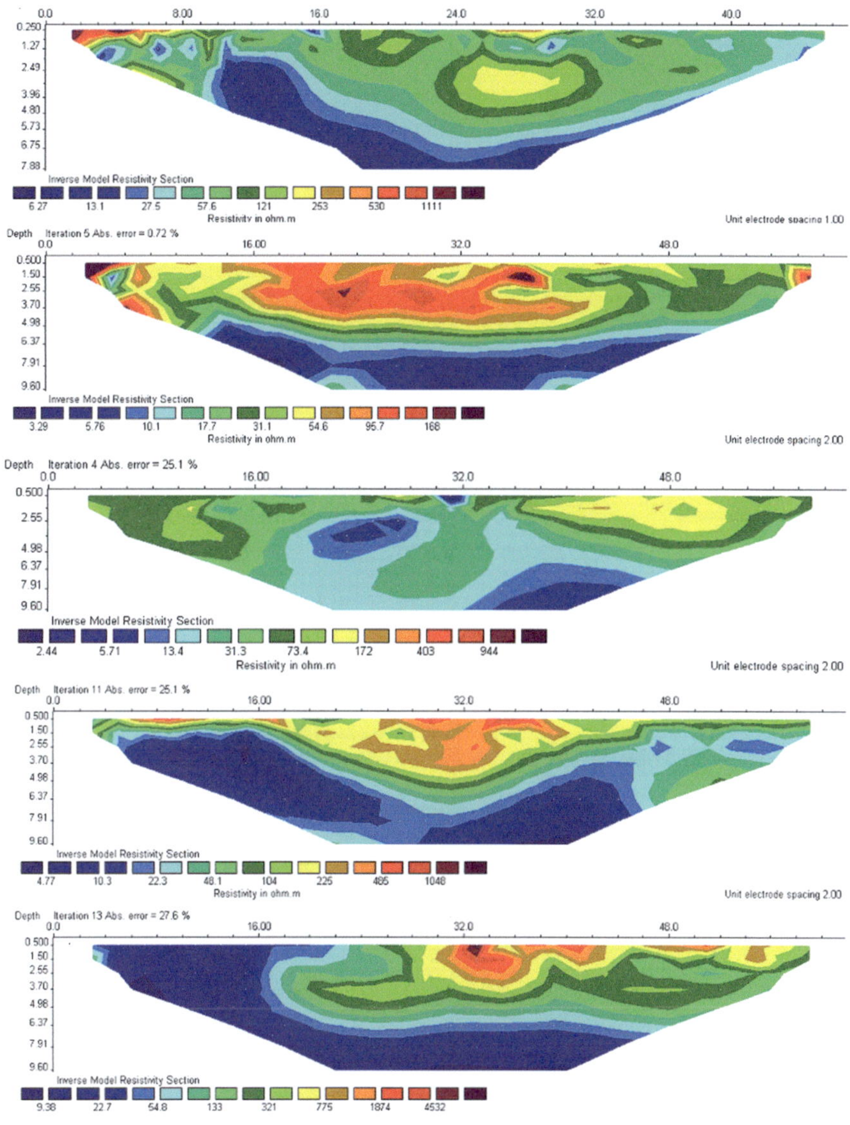

Fig. 13 Inversion results of some ERT profiles

Fig. 19. The program SURFER© (Golden Software Inc. 2014) was used to plot the cross-sections. These cross-sections revealed the vertical and the lateral variations in the resistivity values, which could be attributed to lithologic changes, the existence of some archaeological remains, and due to groundwater invasion. Examples of the constructed cross-sections are given in Figs. 21 and 22.

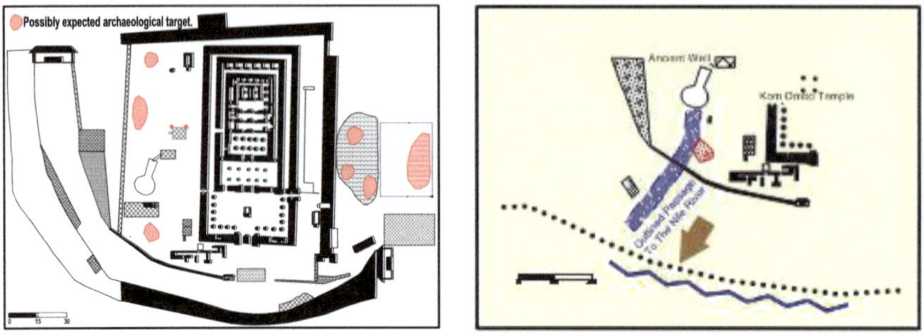

Fig. 14 Archaeological remains (left) and the connection between the Nile and the old-well inside the temple (right), as detected from the integrated GPR and ERT results

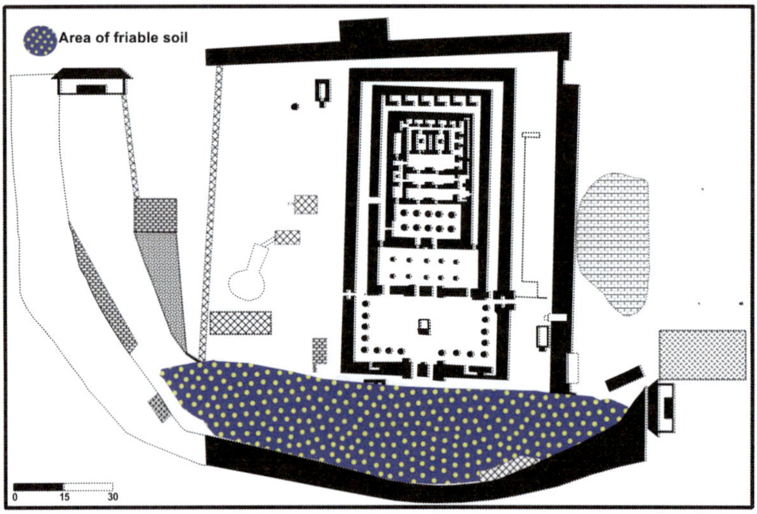

Fig. 15 The filling materials which cause sub-soil instability due to their constitution and high-water saturation

The VES models and the built cross-sections revealed three geoelectrical layers in the investigated section. The uppermost (surface) layer exists in all geoelectrical sections with thickness values ranging from 0.25 to 3.4 m. The resistivity values of this layer are strongly affected by surface conditions. So, it exhibits high resistivity values in the desert area and low values towards the cultivated land. It corresponds to intercalations of sand, weathered clay and gravel. The second layer is represented in all cross-sections with a variable thickness ranging from 0.45 to 6.6 m. This layer is attributed to intercalations of clays and partially saturated sand. The third geoelectrical layer is intensely affected by the infiltrated irrigation water from the nearby cultivated land. This layer could be interpreted as fully saturated sandy clay. Spots of significant-high resistivity were outlined specifically on VES-2 and VES-10

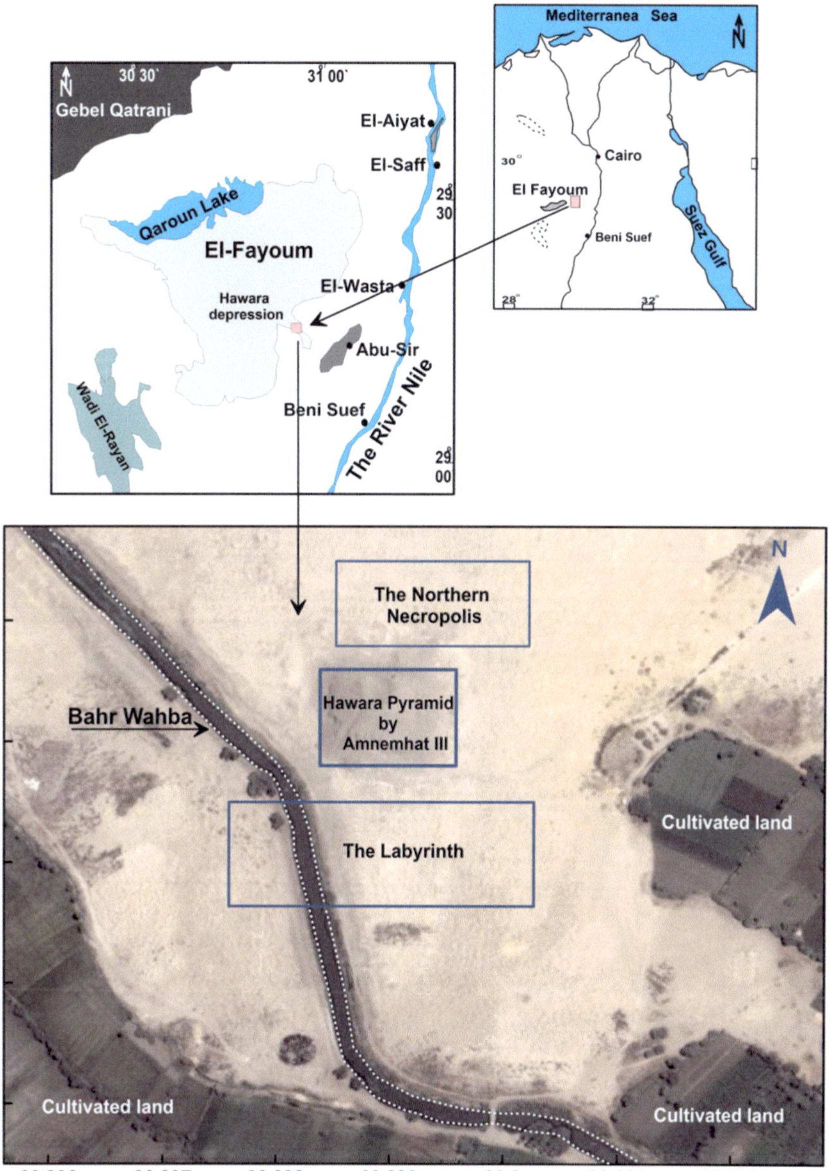

Fig. 16 Location and layout of Hawara archaeological site

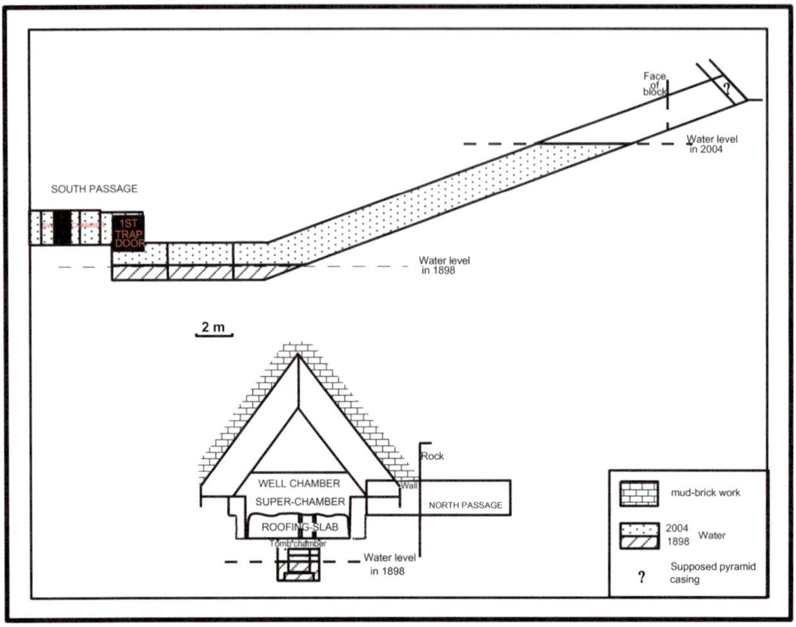

Fig. 17 Rising of groundwater inside the pyramid since 1898 (Up), and water filling the pyramid's entrance in 2008 (Down)

Fig. 18 VES data acquisition at Hawara Pyramid

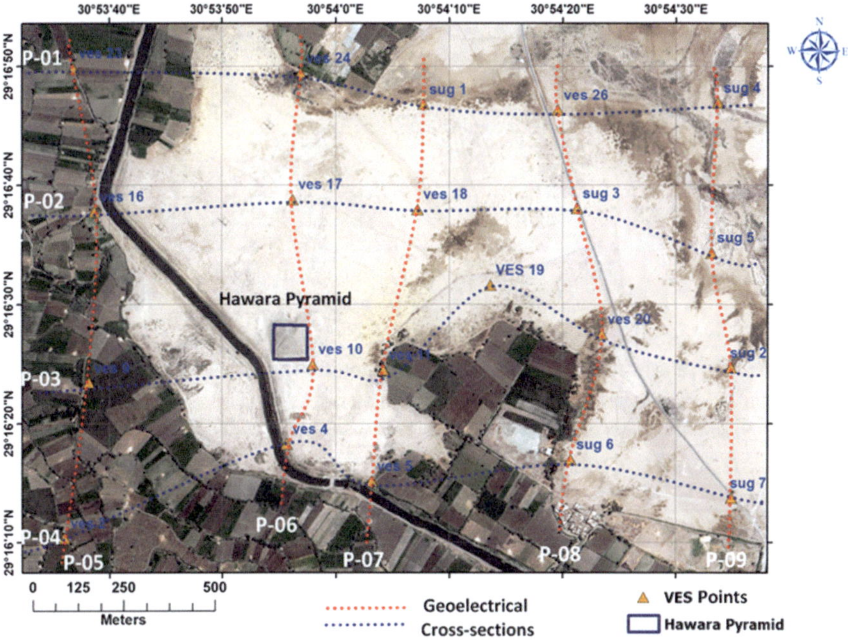

Fig. 19 Location of the VES points and geoelectrical cross-sections around Hawara Pyramid

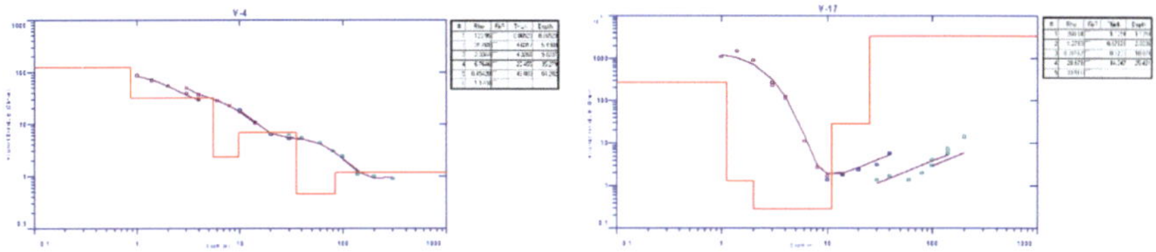

Fig. 20 The VES-4 (left) and VES-17 (right) and their corresponding geoelectrical model

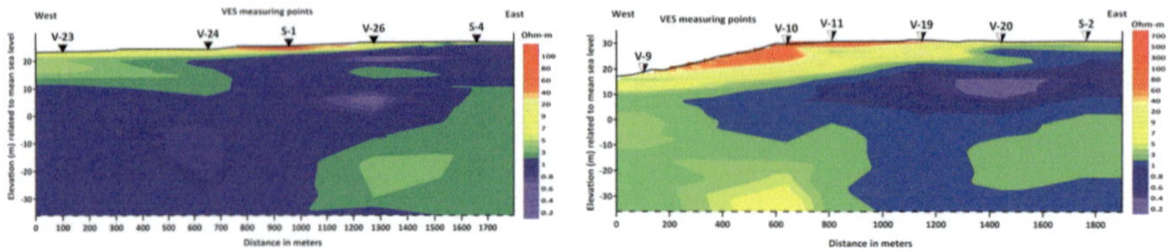

Fig. 21 Geoelectrical cross-sections P-01(left) and P-03 (right)

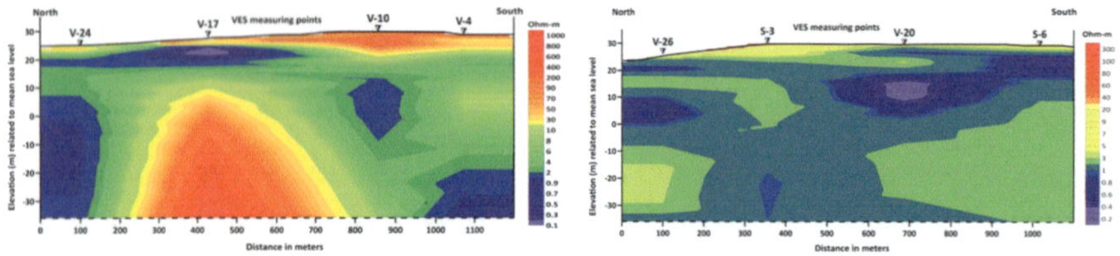

Fig. 22 Geoelectrical cross-sectionsP-06 (left) and P-08 (right)

that could have archaeological measures or sudden change in the sub-surface soil characteristics.

Twenty-nine ERT profiles have been conducted at Hawara Pyramid and its surroundings (Fig. 23) with different spacing between electrodes. The adequate sequence is prepared by the operator according to the required coverage depths and the desired resolution that matches the expected objects.

Using an advanced processing software (RES2DINV©) (Loke 2015; GEOTOMO Software 2014), the collected data have been visualized and modified to produce the optimum images that provide essential and accurate information about the subsurface conditions. The subsequent Figs. 24, 25 and 26 are samples of the resultant 2-D ERT profiles that have been measured at Hawara Pyramid and the Labyrinth area.

The Transient Electromagnetic (TEM) data were measured at 74 stations distributed throughout the studied area (Fig. 27), applying the single loop configuration with a loop side length of 25 and 50 m. In this study, the data were collected with high accuracy and high signal-to-noise ratio due to the highly conductive subsurface medium and absence of cultural noise sources in this area.

TEM data were inverted in 1-D scheme using TEMIX XL 4 (2000) program, assuming layered-earth models. To select the most appropriate starting model for the TEM data inversion, several criteria have been tested, including the number and thickness of the layers. The robust model was the one that provides minimum error, maximum fit and best matching with the available geological information. The final output of the data inversion was a set of models; each of them describes the geoelectrical parameters of the subsurface section at its respective site. Figure 28 shows examples of the obtained models.

The obtained models have been used for the construction of geoelectrical cross-sections. Examples of such cross-sections are shown in Figs. 29 and 30.

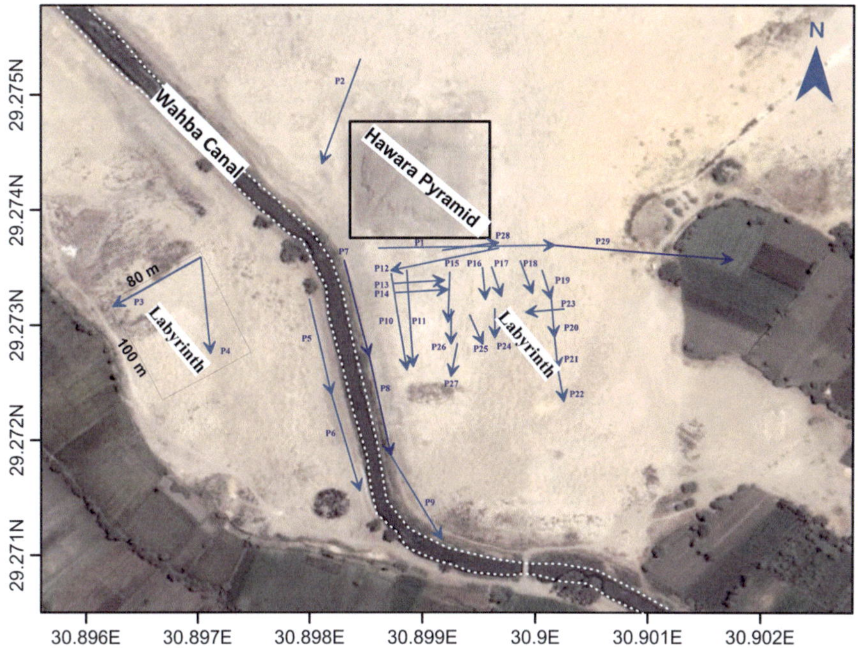

Fig. 23 Location map showing the conducted ERT profiles at the Hawara site

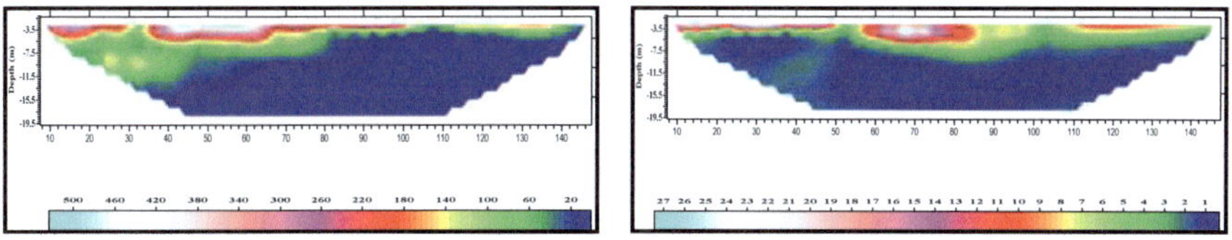

Fig. 24 ERT profile No. 28 parallel to the southern edge of the pyramid and oriented W-E (left), and ERT profile No. 29 is an eastward extension of P28 and trending W-E from the pyramid to the eastern cultivated land (right)

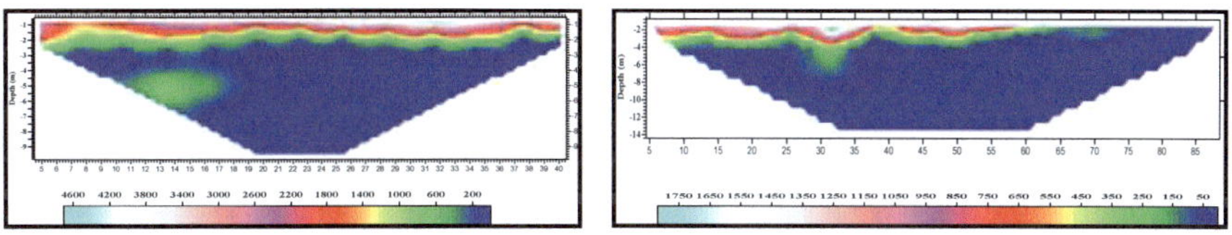

Fig. 25 ERT profile No. 12 inside the Labyrinth and parallel to the southern edge of the pyramid and directed E-W (left), and ERT profile No. 8 parallel to the bank of the canal and directed NW-SW (right)

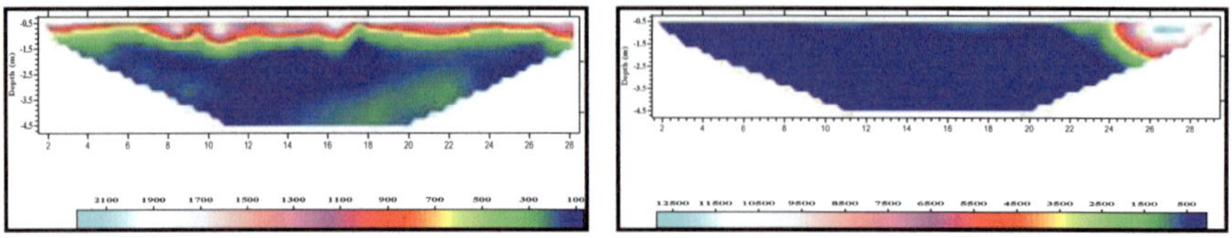

Fig. 26 ERT profile No. 16 inside the Labyrinth and directed N-S (left), and ERT profile No. 22 inside the Labyrinth and directed N-S (right)

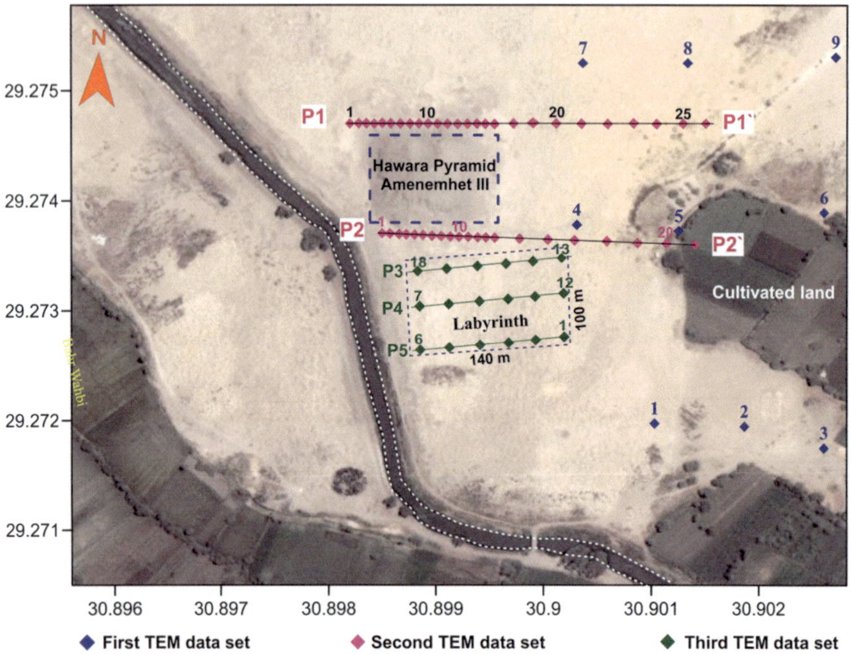

Fig. 27 Location map of the TEM data stations

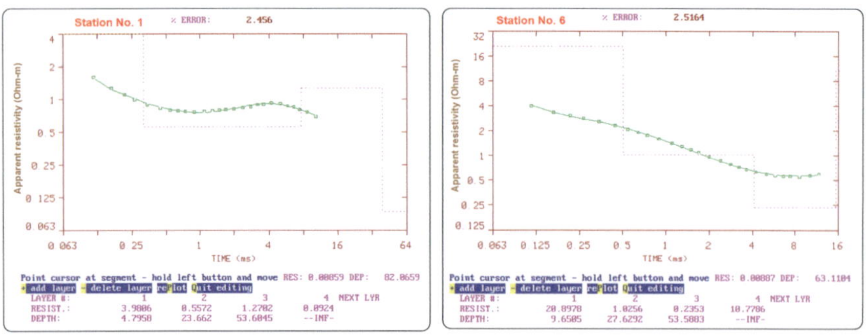

Fig. 28 Results of 1-D inversion of TEM data at station 1, profile P1 (left) and station 6, profile P2 (right)

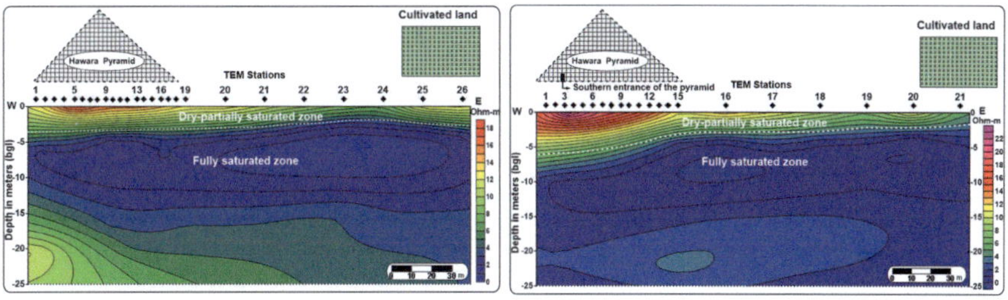

Fig. 29 Geoelectrical cross-sections along profile P1 (left) north of the pyramid and profile P2 (right) south of the pyramid

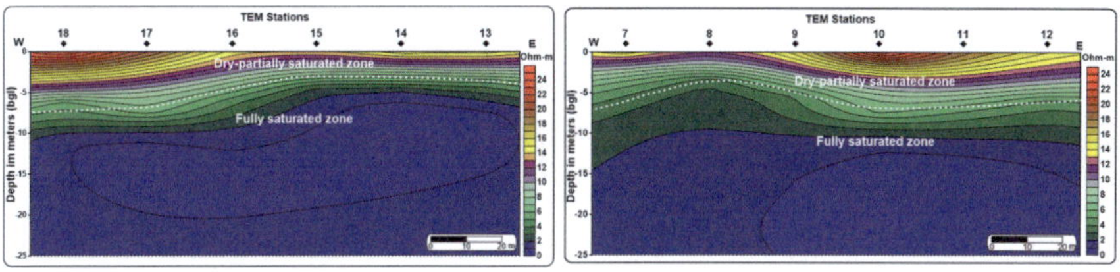

Fig. 30 Geoelectrical cross-sections along with profile P3 (left) and profile P4 (right) measured in the Labyrinth

Investigation of these cross-sections revealed the following features.

The cross-section constructed along with the profile (P1), which is located relatively away from the cultivated land at the northern side of the pyramid, differs from the remaining profiles (P2–P5), which are located close to the cultivated land south of the pyramid.

In profile (P1), the upper surface of the water-saturated zone was detected almost at the same depth along the entire profile length. The profile P2 (southward of the pyramid) and profiles P3–P5 (in the Labyrinth) are similar to each other and show the same behaviour. Inspection of profile (P2) which is started at the southern entrance of the pyramid and extended eastward to the cultivated land, indicates that the upper surface of the water-saturated zone could be detected at a shallow depth (about 2 m bgs at the cultivated land and at about 8 m bgs at the entrance of the pyramid. This behaviour could be explained as excessive irrigation water is infiltrated downward at the cultivated land. Then, the groundwater table was detected at a shallow depth. By moving westward, the water level gradually deepens, where the water continues to move laterally and vertically. This behaviour indicates that the flooded water at the pyramid's entrance comes from irrigation water applied to the nearby cultivations.

Integrating the outcomes of the three techniques could outline the flow direction of the underground water through the area of study. It is proved that the groundwater source inside the Hawara pyramid originates from the cultivated area around the archaeological site. A dewatering plan to protect the area from groundwater leakage could be established in 4 steps:

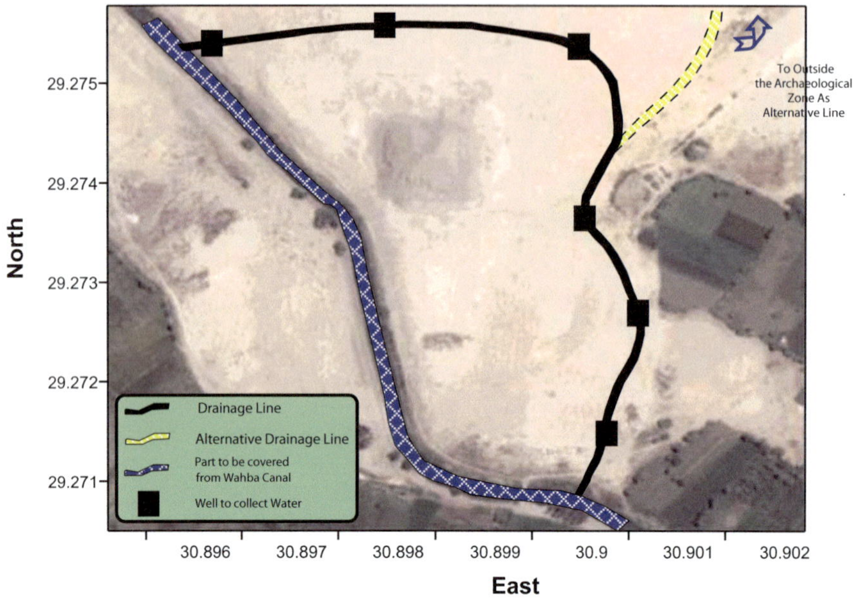

Fig. 31 The suggested drainage system for the dewatering plan

- Managing the water source to reduce the irrigation water quantity by replacing the flood system with spray or dropping systems.
- Pumping the water from the surface.
- Installing an underground drainage system surrounding the archaeological area as shown by the red line in Fig. 31.
- Covering the part of the Bahr Wahba canal that passes through the archaeological area as illustrated by the blue line in Fig. 31.

Acknowledgements The authors direct their sincere gratitude to the staff members of Geo-electric and geomagnetic laboratory, the national Research Institute of Astronomy and Geophysics (NRIAG), for their excellent assistance to acquire the geophysical data used in this work. The study on Kom Ombo was part of a collaborated project between NRIAG and the Supreme Council of Antiquities, Egypt. The work conducted on the Hawara Pyramid was part of the Labyrinth Project, undertaken in collaboration with the Centre for Architecture, Urbanism and Global Heritage (CAUGH) at Nottingham Trent University, as part of a Joint Grant from the Science and Technology and Development Fund (STDF) in Egypt and the Arts and Humanities Research Association in the United Kingdom.

References

Abbas AM (1999) Assessment of groundwater hazard on the Sphinx and Pyramids area, Giza, Egypt. Ph.D., Ain Sham University

Abbas AM, Kamei H, Helal A, Atya MA, Shaaban FA (2005) Contribution of geophysics to outlining the foundation structure of the Islamic Museum, Cairo, Egypt. Archaeol Prospect (12):167–176

Abbas AM, Khalil MA, Massoud U, Santos FM, Mesbah HS, Lethy A, Soliman M, Ragab EA (2012a) The implementation of multi-task geophysical survey to locate Cleopatra Tomb at Tap-Osiris Magna, Borg El-Arab, Alexandria, Egypt "Phase II." NRIAG J Astron Geophys 1:1–11

Abbas AM, Shaaban F, El-Sayed EA, Abdel Hafez T (2012b) Uncovering the Pyramids-Giza Plateau in a search for archaeological relics by utilizing ground penetrating radar. J Am Sci 8(2):168–174

Abbas AM, Salah H, Massoud U, Fouad M, Abdel-Hafez M (2015) GPR scan assessment at Mekaad Radwan Ottoman—Cairo, Egypt. NRIAG J Astron Geophys 4:106–116

Abdul Razak MF, Said MAM, Abustan I, Mohamad MH (2019) Application of TEM sounding method on riverbank filtration site subsurface suitability. Water Supply 19(8):2206–2213. https://doi.org/10.2166/ws.2019.100

Ahmed A, Fogg G (2014) The impact of groundwater and agricultural expansion on the archaeological sites at Luxor, Egypt. J Afric Earth Sci. https://doi.org/10.1016/j.jafrearsci.2014.02.007

Annan P (2005) GPR methods for hydrogeological studies, Ch. 7. In: Rubin Y, Hubbard SS (eds) Hydrogeophysics. Springer, Dordrecht, The Netherlands, 532 p

Barone PM, Ruffell A, Tsokas GN, Rizzo E (2019) Geophysical surveys for archaeology and cultural heritage preservation. Heritage 2:2814–2817. https://doi.org/10.3390/heritage2040174

Boghdady SEA (2012) Ground penetrating radar and electrical resistivity evaluation on Kom Ombo Temple Hazard, Aswan, Egypt. M.Sc., Geology Department, Faculty of Science, Alexandria University, 165 p

Christensen NB, Sorensen K (1996) Pulled array continuous vertical electrical soundings (PA-CVES) with an additional inductive source. In: Proceedings of the symposium on the application of geophysics to engineering and environmental problems, Keystone, Colorado, USA. EEGS, pp 1–10

Deiana R, Leucci G, Martorana R (2018) New perspectives on geophysics for archaeology: a special Issue. Surv Geophys 39:1035–1038. https://doi.org/10.1007/s10712-018-9500-4

Edwards IES (1947, 1986) The pyramids of Egypt (Penguin Archaeology). Published February 4th 1986 by Penguin Books (first published 1947), Revised edn., 352 p

Geotomo Software (2014) RES2DINV ver. 6.1 for windows, Rapid 2-D Resistivity & IP inversion using the least- squares method (Wenner, dipole-dipole, inline pole-pole, pole-dipole, equatorial dipole-dipole, Wenner-Schlumberger and non-conventional arrays) on land, underwater and cross-borehole surveys. www.geoelectrical.com

Golden Software Inc. (2014) Surfer for windows, version 11, powerful contouring, gridding and 3D-surfer mapping

Griffiths DH, Barker RD (1993) Two-dimensional resistivity imaging and modelling in areas of complex geology. J Appl Geophys 29:211–226

Hengari GM, Hall CR, Kozusko TJ, Bostater CR (2013) Use of ground penetrating radar for determination of water table depth and subsurface soil characteristics at Kennedy Space Center. In: Michel U, Civco DL, Schulz K, Ehlers M, Nikolakopoulos KG (eds) Proceedings of SPIE, earth resources and environmental remote sensing/GIS applications IV, vol 8893, 889318, © 2013 SPIE, CCC code: 02770-786X/13/$18. https://doi.org/10.1117/12.2030023

Khalil MA, Abbas AM, Santos F, Massoud U, Salah H (2013) Application of VES and TDEM techniques to investigate seawater intrusion in Sidi Abdel Rahman area, northwestern coast of Egypt. Arab J Geosci 6:3093–3101

Khalil M, Abbas AM, Santos F, Mesbah HS, Massoud U (2010) VLF-EM study for archaeological investigation of the Labyrinth mortuary temple complex at Hawara area, Egypt. Near Surf Geophys 8(3):203–212. https://doi.org/10.3997/1873-0604.2010004

Loke MH (2015) Tutorial: 2D and 3D electrical imaging surveys, pp 4–31
Loke MH, Barker RD (1996) Practical techniques for 3D resistivity surveys and data inversion. Geophys. Prospect 44:499–523
Loke MH, Dahlin T (2002) A comparison of the Gauss-Newton and quasi-Newton methods in resistivity imaging inversion. J Appl Geophys 49:149–162
Massoud U, Abbas AM, Mesbah HS, Ragab EA, Taha A (2010) Mapping of subsoil water level and its impacts on Hawara archaeological site by a transient and multi-frequency electromagnetic survey. Chin J Geopys 53(3):638–645. https://doi.org/10.3969/j.issn.0001-5733.2010.03.018
Massoud U, El Qady G, Metwaly M, Santos F (2009) Delineation of shallow subsurface structure by azimuthal resistivity sounding and joint inversion of VES–TEM data: case study near Lake Qaroun, El Fayoum, Egypt. Pure Appl Geophys 166:701–719
Massoud U, Kenawy AA, Ragab EA, Abbas AM, El-Kosery HM (2014) Characterization of the groundwater aquifers at El Sadat City by joint inversion of VES and TEM data. NRIAG J Astron Geophys 3:137–149
Metwaly M, Elawadi E, Moustafa SSR, Al-Arifi N (2014) Combined inversion of electrical resistivity and transient electromagnetic soundings for mapping groundwater contamination plumes in Al Quwy'yia Area, Saudi Arabia. JEEG 19(1):45–52
NRIAG (2010) Evaluation of groundwater impact and stability of Kom Ombo temple. NRIAG Internal report in cooperation with Supreme Council of Antiquities
Pellicer XM, Gibson P (2011) Electrical resistivity and Ground Penetrating Radar for the characterization of the internal architecture of quaternary sediments in the Midlands of Ireland. J Appl Geophys (75):638–647
Peterson CD, Doliber SR (2019) Groundwater Surface (GWS) Mapping by Ground Penetrating Radar (GPR) for use in protecting freshwater habitats, water quality, and active dune landscapes, in the Florence Coastal Dune Sheet, Oregon, USA. J Geogr Geol 11(1). ISSN 1916-9779, E-ISSN 1916-9787. Published by Canadian Center of Science and Education
Petrie WMF, Wainwright GA, MacKay E (1912) The Labyrinth, Gerzeh, and Mazghuneh. Bernard Quartich, London
Porsani JL, Almeida ER, Bortolozo CA, Santos FAM (2012) TDEM survey in an area of seismicity induced by water wells in Paraná sedimentary basin, Northern São Paulo State, Brazil. J Appl Geophys 82:75–83
Saintenoy A, Hopmans JW (2011) Ground Penetrating Radar: water table detection sensitivity to soil water retention properties. IEEE J Sel Top Appl Earth Observat Remote Sens 4(4):748–753. https://doi.org/10.1109/JSTARS.2011.2171920
Sandmeier KJ (2010) ReflexW 7.0 program for the processing of seismic, acoustic or electromagnetic reflection, refraction and transmission data. Zipser Straße1, D-76227 Karlsruhe, Germany
Siliotti A, Hawass Z (1997) Guide to the Pyramids of Egypt. Publisher, Barnes & Noble; 1997th edn., January 1, 1997, 168 p
TEMIX XL 4 (2000) TEMIX V. 4 user's manual, Interpex. Colorado, USA, 468 p
Temraz MG, Khallaf MK (2016) Weathering behaviour investigations and treatment of Kom Ombo temple sandstone, Egypt- Based on their sedimentological and petrographical information, J Afric Earth Sci (113):194–204
Toushmalani R (2010) Application and limitation of geophysical techniques in archaeology. Austr J Basic Appl Sci 4(12):6440–6449. ISSN 1991-8178 © 2010, INSInet Publication

Strengthening the Conservation and Management of Lumbini, the Birthplace of Lord Buddha, World Heritage Property (Lumbini, Nepal)

Claudio Margottini, Daniele Spizzichino, Paolo Pagnin, and Luca Maria Puzzilli

Abstract Maya Devi Temple is the major archaeological heritage in the Lumbini Sacred Garden (Nepal). The Temple is currently suffering from the rising dampness in the foundation and general weathering of the masonry bricks. In the period June 2015–November 2015 a manual monitoring of the water table (WT) was conducted on a daily basis on existing 5 shafts and 3 pipes. The data, integrated with previous WT survey, showed the rapid uprising of the WT, right after the rainfall, but also its declining and slow disappearing in the dry season. Geotechnical and geophysical investigations were carried out in spring 2016, defining the geological stratigraphy of the area and the possible aquifer. Based on the 2017 last field mission, it was possible to establish (clearly and unequivocally) the direct relationship between rainfall and WT level in the surrounding of the Maya Devi Temple. The water table showed an ephemeral character, appearing only in the monsoon period and located in a so-called "Cultural Deposit" (historical and modern fill containing large volumes of brick, boulders, and a clay matrix). This is a material mobilized during the construction period to raise the topography in order to protect against riverine floods. Some alternative solutions to mitigate the impact of the WT fluctuations have been proposed in the present paper. Each of them has both positive and negative features, to be discussed with the local Institution, in order to find the most suitable solution.

Keywords UNESCO · Cultural Heritage · Lumbini · Temple foundation · Water table

C. Margottini (✉)
Former Scientific Attaché, Embassy of Italy in Egypt, 15, Abd El-Rahman Fahmy Str., Garden City, Il Cairo, Egypt
e-mail: claudio.margottini@gmail.com

UNESCO Chair on Prevention and Sustainable Management of Geo-Hydrological Hazards, University of Florence, Florence, Italy

D. Spizzichino · L. M. Puzzilli
ISPRA, Geological Survey of Italy, Via V. Brancati 48, 00144 Rome, Italy

P. Pagnin
Lithos, Venice, Italy

G. M. El-Qady and C. Margottini (eds.), *Sustainable Conservation of UNESCO and Other Heritage Sites Through Proactive Geosciences*, Springer Geology,
https://doi.org/10.1007/978-3-031-13810-2_17

1 Introduction

Lumbini, the Birthplace of the Lord Buddha, lies within the Nepal Terai region, a subtropical chain of forests, marshes, and grasslands, now intensively cultivated. This is located between the Indian border and the Siwalik Range of the Himalayas (Fig. 1). This gently sloping plateau is cut by the Ganga's tributaries, creating alluvial fans and meanders. The site of Lumbini includes numerous artefacts, archaeological structures, and a series of gardens (Coningham et al., 2011).

Most relevant is the Maya Devi Temple, protected since 2002 by a new cover structure from the action of weathering effects (Bidari, 2007). This is a large, enclosed structure, composed mainly of iron beams and concrete, and equipped with two entrance doors and some windows for ventilation (Fig. 2).

Lumbini has been on the UNESCO World Heritage List since 1997 (Bhaddamanika et al., 2019). The site was subject to periodical flooding, since historical time, due to the heavy rainfalls which occur during the monsoon period. To solve the problem, before inhabiting the site, the place where the urban settlements were to be built was filled with earth and debris, in order to raise the topographic level above the potential flood level. In April 1967 the UN Secretary-General U Thant visited Lumbini, and proposed the development of Lumbini into a major center of pilgrimage.

In 1972, Kenzo Tange was in charge of preparing a development Master Plan for Lumbini. With this master plan, a new circular embankment (levee) was constructed

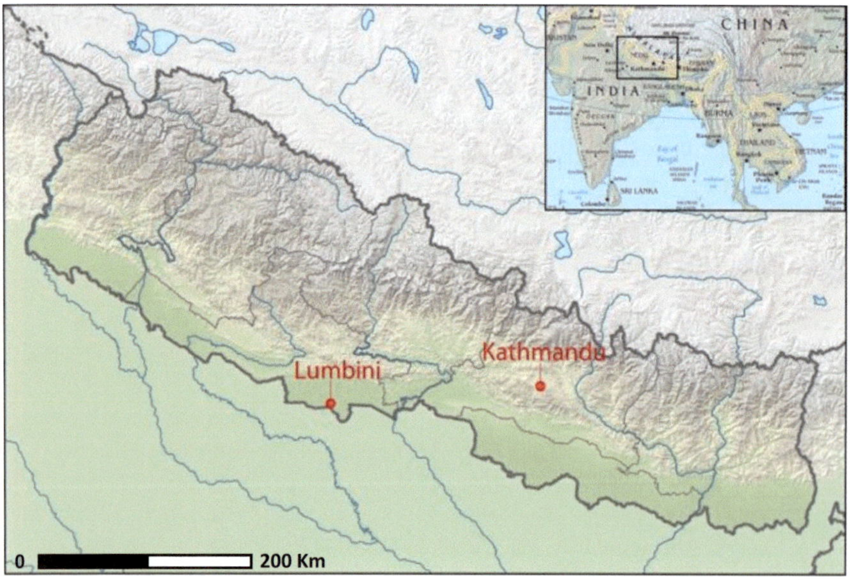

Fig. 1 Location of the investigated site

Fig. 2 Maya Devi Temple, the sacred garden, and a detail geological section marked A-B (in red)

to protect Lumbini from flooding (Kenzo Tange e Urtek, 1968, 1981). This solution certainly protected from external floods, but also did not allow the rainwater which fell between the temple and the levee, to flow properly out of the site. The raising of the water table was also evident in the Maya Devi temple, especially in rainy seasons, affecting the conservation of the bricks, and consequently the archaeological structures. Inside the temple, the archaeological walls, made of fired brick and mud mortar, are conserved. These archaeological structures are directly affected by phenomena of fluctuation of the groundwater level and are therefore subject to continuous variations in the quantity of moisture, as well as the effect of condensation, due to the microclimate inside the temple. Under these conditions of continuous hygrometric variability, significant deterioration phenomena occur, among which the most serious are:

1. formation of saline crystallization, which could cause the deterioration of the bricks (salts dissolved in the groundwater migrate, by capillarity, into the walls and can crystallize on the surface by evaporation);
2. deformation of the wall structures made with mud mortars, due to the volumetric variation of the clay materials when in the presence of strong hygrometric variations;
3. growth of microorganisms onto the structures.

The management of the water table below the Maya Devi Temple is the major topic of the present paper. Geological, geotechnical, and geophysical investigations were performed to identify the geometry of the aquifer; as well the investigation of seasonal fluctuation was implemented by daily monitoring the piezometric level and rainfall in the period June–October 2015.

2 Geological Setting

Lumbini is located in the Terai region, which is a morphologically flat area, along the foothills of Himalayas. The Terai plan is composed of alluvial deposits of Holocene age, which also include channel sand and gravel deposits and outwash deposits. More in detail, the alluvial sediments in the plain are mainly clay and silts intercalated with layers and lenses of gravel and sand.

On a regional scale, sediments tend to get finer southward Very complex is also the distribution and bedding of different strata, since distinctive layers are generated by local floods with confined spatial distribution, at times excavated and then filled by palaeo-channels. Available data in Lumbini are not enough to allow the 3D reconstruction of different geological strata or the respective hydro-geological properties.

Some general data are drawn from Kenzo Tange and Urtec (1968) and (1981). Later on, some more detailed information was obtained through the archaeological investigation developed by Coningham et al. (2011) and Coningham and Acharya (2011, 2012, 2013).

Nonetheless, even with the above information, it is not clear which geological formation comprises the aquifer (a geologic formation from which significant amounts of groundwater can be extracted), *aquiclude* (a geological formation that acts as such a water barrier) as well as *aquitard* (a geological formation that acts as such a minor water barrier and still permits the flow of water).

A preliminary hypothesis based on the auger drilling developed by Coningham and Acharya (2011), is reported in the following Fig. 3.

From the above figure, it is possible to synthesize the five major geological formations, even if they are more differentiated from the archaeological point of view. Green stands for compact loamy clay topsoil; orange for modern fill containing large volumes of brick, boulders, and a clay matrix (so-called *"Cultural Deposit"*); yellow for alluvial deposits, consisting of dense clays; light brown for dense clay with trace elements of kankar; grey for compact clay with kankar inclusions.

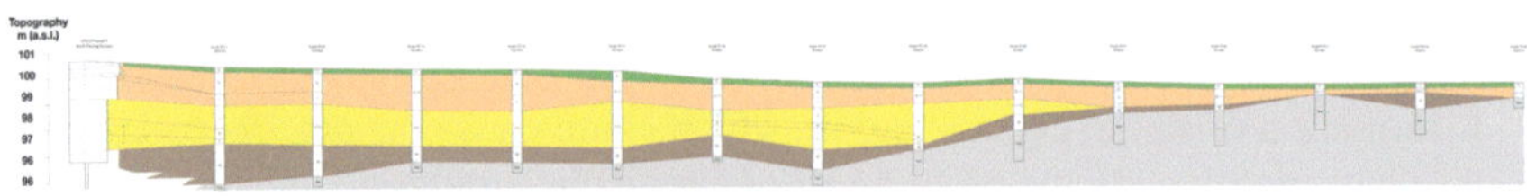

Fig. 3 Geological section (A-B) of Maya Devi temple (redrawn after Coningham et al. 2019). Green is compact loamy clay topsoil; orange is modern fill containing large volumes of brick, boulders and a clay matrix; yellow is alluvial deposits, consisting of dense clays; light brown dense clay with trace elements of kankar; grey is compact clay with kankar inclusions

2.1 Geotechnical Boreholes

In order to increase and detail the available lithological and hydrogeological information, three boreholes have been planned, jointly with a geophysical prospection based on a passive seismic analysis. The position of the boreholes and the geophysical prospection is reported in the following Fig. 4. The boreholes were executed far away from the Maya Devi Temple, according to the risk map for cultural heritages prepared by Coningham and Acharya (2013). In this way, we are probably missing the real stratigraphy below the Temple, but we did not damage any valuable heritage. All the boreholes were executed with the fundamental support of archaeologists of the *Lumbini Development Trust*.

In one case, the boreholes found some brick remains, and it was agreed to abandon this position (n. 4 in Fig. 4). The three boreholes resulted in similar alluvial materials with limited vertical variability. The first borehole drilled an alternation of alluvial deposits, mainly silt, reflecting different flooding periods. Water appeared at 4,4 m. below the ground level. The second borehole penetrated the modern fill (cultural deposit, Cunnhighan 2011 and 2012), containing some ceramics, boulders, and a clayey matrix, when alternation of alluvial deposit started appearing. Two piezometric surfaces were recovered, the first at 6,1 and the second at 8,2 from the surface. The third borehole, in the Northern part, exhibited the same alternation of alluvial deposit, but without water until the depth of 7,5 m from the surface.

The stratigraphy of the three boreholes, the water table, the samples for laboratory tests (disturbed and undisturbed) and the location of permeability tests are reported in the Fig. 5.

The following Fig. 6 is an example of material recovered in borehole n. 1. It is possible to notice, in a predominant silty material, the different colours of layers and the presence of small gravel. Similarly, looking at physical properties obtained from

Fig. 4 Geotechnical and geophysical prospection implemented in Lumbini in April 2017

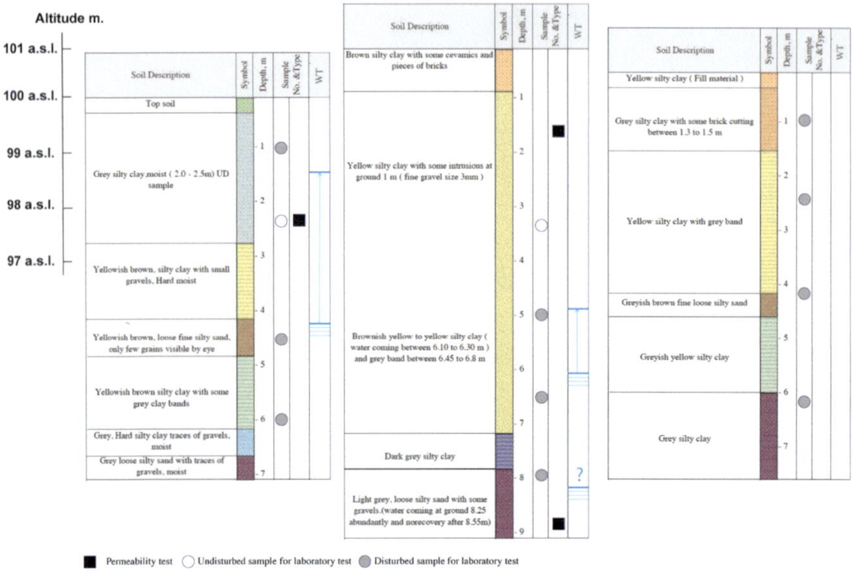

Fig. 5 Stratigraphy of the three boreholes with indication of water table, samples for laboratory tests (disturbed and undisturbed) and location of permeability tests

laboratory tests in all three boreholes, there is no large variability among the different samples.

In Fig. 7, the Plasticity chart for 9 of the investigated 11 samples, coming from the three boreholes, is reported. The material is relatively homogenous, falling in the category of inorganic silts with clay and very fine sands, with light to no plasticity, slow to rapid dilatancy.

Fig. 6 The stratigraphy in borehole n. 1 (box n.1)

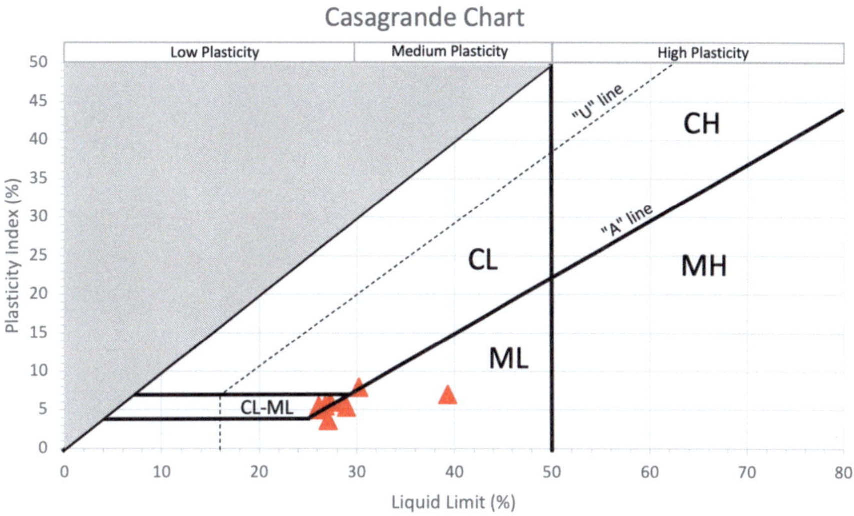

Fig. 7 Plasticity chart for the investigated samples

The same homogeneity can be found in the grain size distribution of all samples (Fig. 8). There are no major differences between the investigated samples. Only in one case, a layer of well classed sand was clearly identified in Borehole n 3 (−4,3 m from surface). In conclusion, the area features cross-bedded layers, with palaeo-channels, but the geotechnical properties are almost constant among the different samples. Some small layers of sand may appear but with very limited thickness. Such sand layers are responsible for the limited water tables we found in borehole 1 and 2. Most likely, the water table of borehole 1 and the upper water table of borehole 2 can belong to the same aquifer.

The homogeneity of stratigraphy, composed by silty alluvial deposit, with a limited difference among the detected lithotypes, at least in the first 10 m depth, is also confirmed by the Pocket Penetrometer and Vane Test investigations which, despite of the low accuracy, may provide an information about the vertical variability of geological materials inside the boreholes (Fig. 9).

3 Active and Passive Seismic Analysis

3.1 Active Seismic Analysis

In order to implement and integrate information coming from boreholes, a geophysical (seismic) survey was implemented. An active seismic survey by means of the Surface Waves method was performed using a single 3-component geophone (Fig. 10 left) along with a profile with 8 shot points, simulating a multichannel dataset using

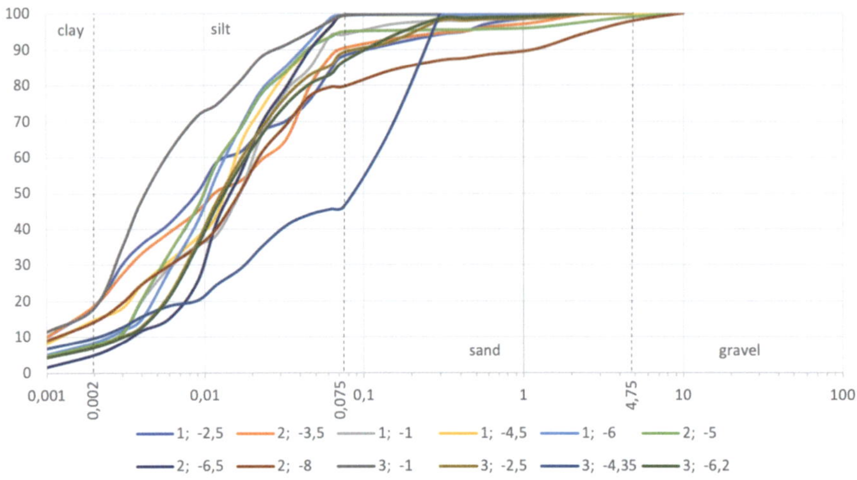

Fig. 8 Grain size distribution of the collected samples. The numbers in legend refers to the borehole and depth of sample

a single receiver. The relative phase velocity spectrum is reported in the following figure.

It is worth noting that the thickness of each geological stratum has been not used as a constrain in the versus modeling. This implies that a very good match in thickness (differences less than 10%) between seismic layers and geological ones could be representative of a different degree of stiffness of each geological unit.

In Fig. 11 the final versus model is reported and compared with S2 borehole stratigraphy.

The first 3 layers seem to not differ very much in terms of stiffness as the increase in Vs value is less than 50 m/s in about 4 m, from 175 to 220 m/s. This is also a confirmation of the homogeneity revealed by the boreholes and laboratory tests. A little more significant increase in Vs value is located just below 7 m in depth where the shear wave velocity reaches 350 m/s (blue arrow).

The deeper seismic layers are poorly constrained but the low-frequency peak around 1.5 Hz suggests that the seismic bedrock should be located deeper, probably more than 100 m in depth. It also possible that the silt and sand formation, detected in all boreholes, will continue until 35–40 m in depth (Fig. 12).

3.2 *Passive Seismic Analysis*

32 noise-ambient vibration recordings have been acquired for a fast and preliminary inspection of seismic site response (April 2017 filed survey). The location is reported in Fig. 4. The horizontal-to-vertical spectral ratio curves have been elaborated for all the recordings (Fig. 9).

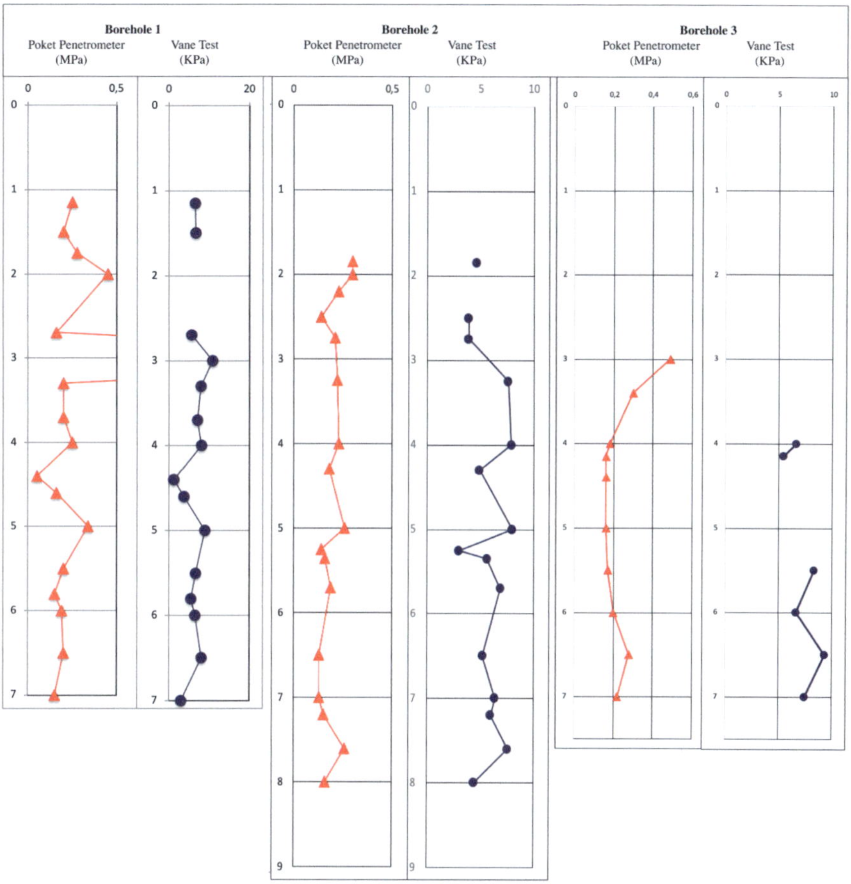

Fig. 9 Pocket Penetrometer values and Vane Tests in the cores obtained by the three boreholes

On the basis of this survey, the seismic response of the whole area can be sketched as a simplified 1D geotechnical model (single layer over bedrock), supporting the main outcomes derived from the 3 boreholes, the laboratory tests, and geotechnical characterization. Also, the spectral ratio curves are quite similar along the whole area; no significant variation in shape has been observed even when considering every single horizontal component with the respect to the vertical one (no directionality).

The bedrock is responsible for the main peak frequency, located at about 1.3 Hz mean HVSR curves in Fig. 13. This is suggesting the presence of a seismic impedance contrast located very deep, most likely >100 m, in agreement with MASW elaboration. Thus, the mean amplitude of the peak does not exceed the value of 2 (i.e. it does not meet the SESAME 2004 criteria).

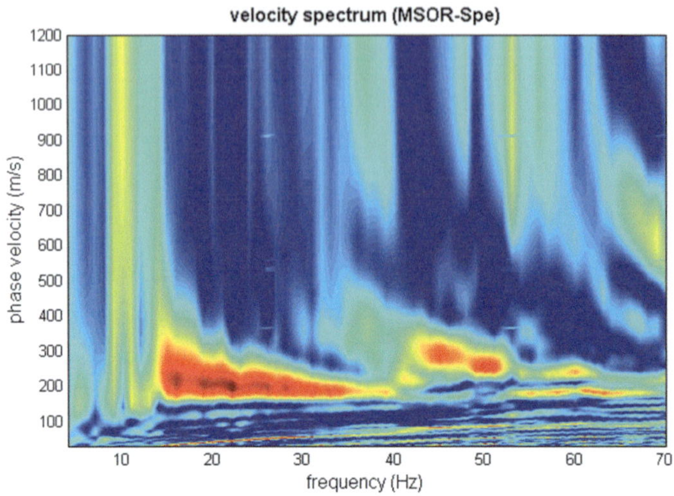

Fig. 10 Phase velocity spectrum of the MSOR (Multichannel Simulation with One Receiver) considering the Rayleigh-waves vertical component

Since the survey is based only on ambient vibration recordings, the outcomes should be confirmed by means of further investigations (e.g. HVSR using earthquakes recordings calibration) in order to: (1) estimate the fundamental frequency of the site and (2) assess the possibility of stratigraphic amplification of ground motion.

4 Hydrological and Hydrogeological Condition of the Maya Devi Temple

The Terai plain, Lumbini zone, is underlain by alluvial deposits of Holocene age which also include channel sand and gravel deposits and outwash deposits. The streams in the Terai are characterised by frequent shifting of their courses sometimes by a few kilometers. As a consequence, the sediments are cross-bedded, eroded reworked and redeposited. Aquifer materials show good sorting (Krishna Rao et al., 1996).

In the depth range of less than 100 to about 200 m drilled for tube well (Krishna Rao et al., 1996), aquifer zones constitute a thickness of 5–70 m or slightly more. The normal thickness is 50–60 m. The lithological sections reveal that the aquifers are highly localised in the form of lenses and layers. This is due to frequent shifting of river courses (Krishna Rao et al., 1996).

The alternance of level of alluvial deposits, cross-bedded, eroded and reworked is also a characteristics of near sub-surface deposit in the Lumbini area. To this, it must be added that the Maya Devi Temple is founded on a so called “Cultural

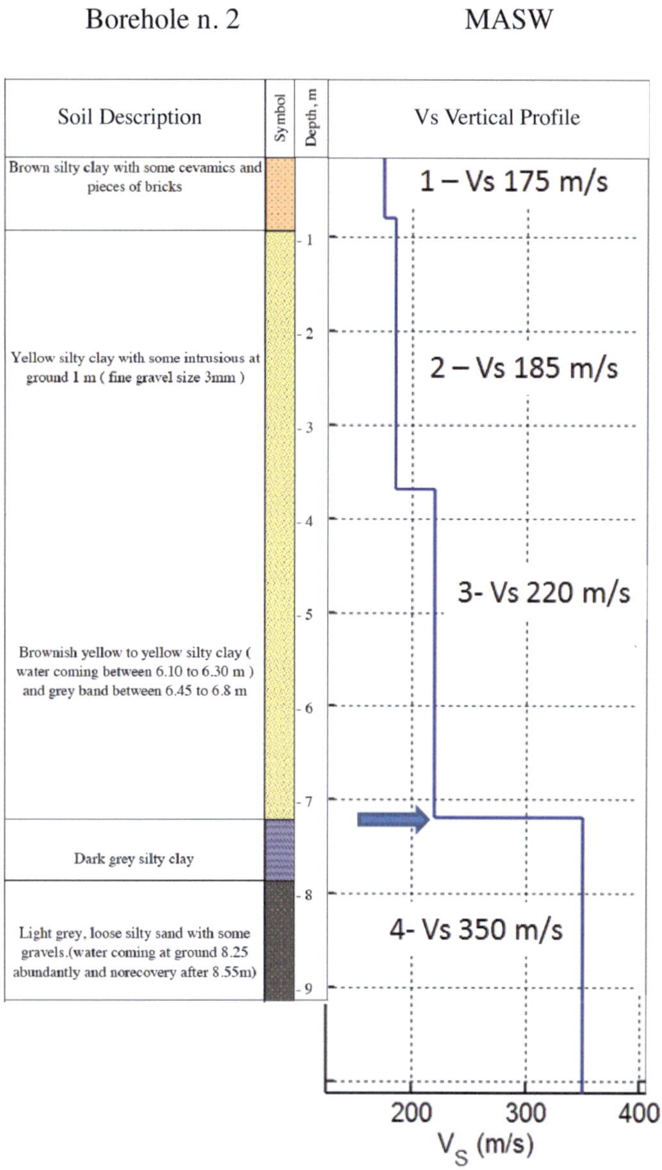

Fig. 11 S2 borehole stratigraphy compared to the Vs final model

Deposit" (Coningham and Acharya, 2011, 2012), most likely constructed to rise the topography of the area, to protect the Temple from the continuous floodings.

As a consequence, the presence of the two (or three) water tables detected in boreholes 1 and 2 in a dry season, are likely to be associated to small lens of silty sand alluvial strata (multi-aquifer system), also with a confined aquifer. This is making the

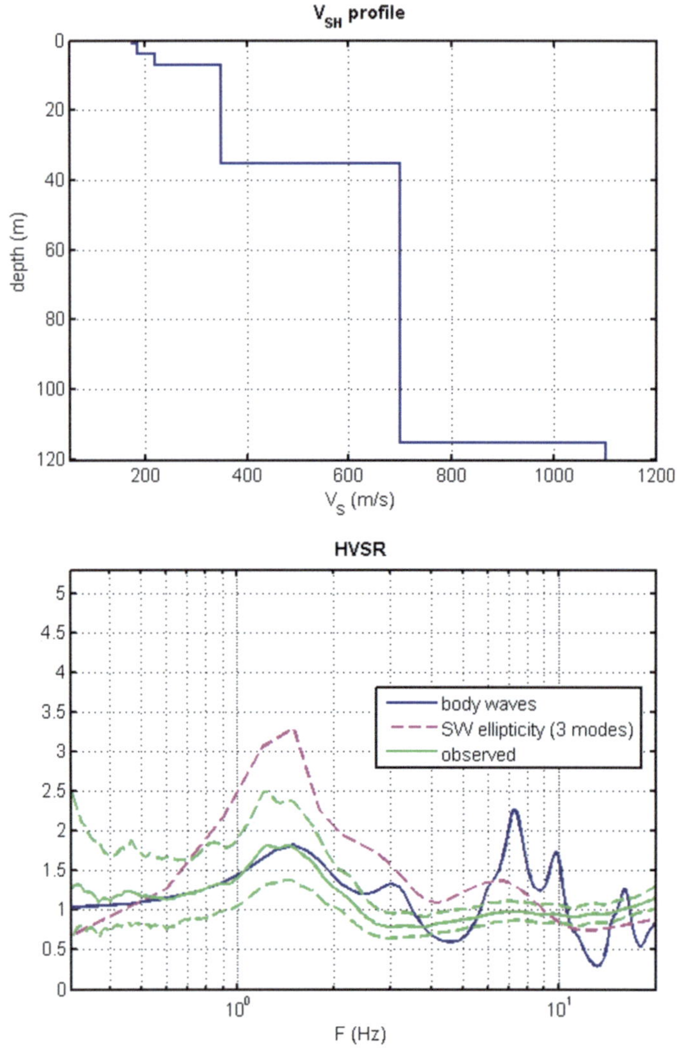

Fig. 12 Seismic layers until the bedrock detected with Surface Waves method (upper) and used model (lower)

hydro-geological situation non-uniform, and strongly dependent by local geology. As general note, not water table can be assumed in the area in the first 7–10 m. The main aquifer is deeper. As previously reported, below this material investigated by boreholes, a unit of alluvial sediments is detected by MASW until 35 m depth (Vs 350 m/s) and, lower, a second unit until about 110–120 m (Vs 700 m/s). The bedrock is likely identified from HVSR at >100 m, in agreement with MASW.

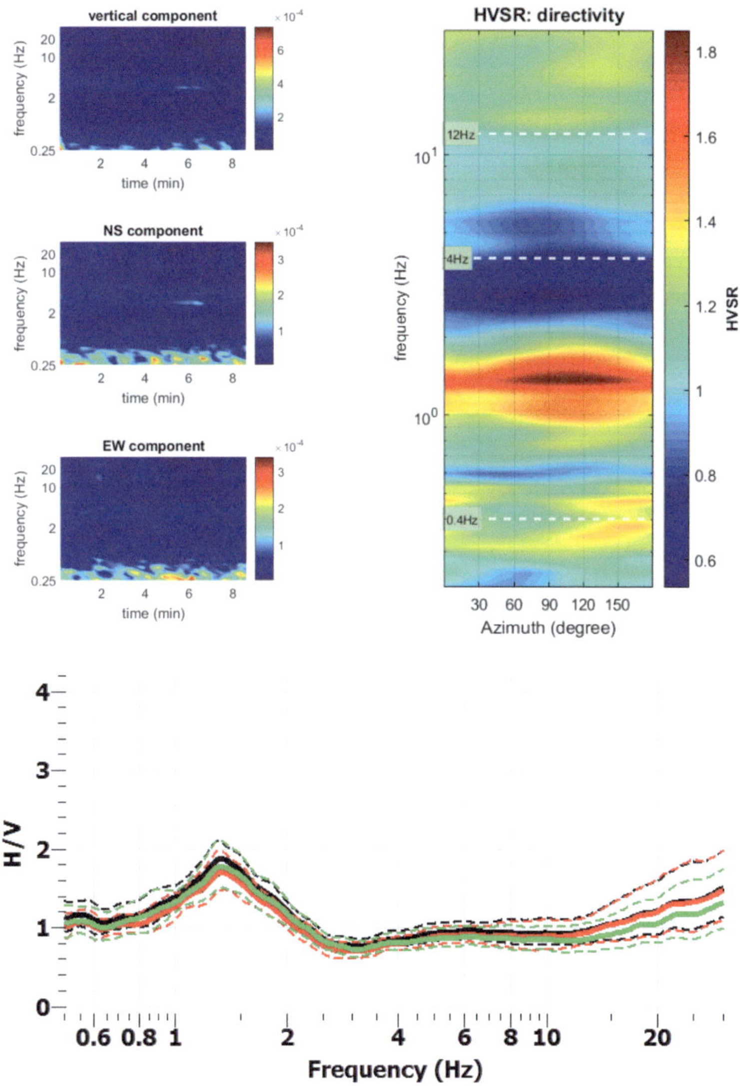

Fig. 13 Example of HVSR processing: directivity analysis (up) and mean HVSR curves from the 32 ambient recordings while considering each horizontal component (NVSR, EVSR) also separately (down)

Three falling heads permeability tests were executed in the boreholes. In this method, the water level in the test hole is allowed to fall and the equivalent permeability is computed from the data of the rate of fall of the water level. The simplest interpretion of piezometer recovery data is that of Hvorslev (1951), later on reported in the standards of Associazione Geotecnica Italiana (1977). The test may be conducted both above and below water table but is considered more accurate below

Table 1 Falling head permeability tests

Test n	Borehole	Depth from surface (m)	Permeability (m/sec)
1	1	– 2,5	$4{,}92 \times 10^{-8}$
2	2	– 1,5	$8{,}85 \times 10^{-8}$
3	2	– 9,0	$2{,}92 \times 10^{-8}$

water table. It is applicable for strata in which the hole below the casing pipe can stand and has low permeability; otherwise the rate of fall of the head may be so high that it may be difficult to measure.

Determining permeability using the Falling Head method involves the following formula:

$$K = \frac{A}{Cl(t2 - t1)} ln \frac{h1}{h2}$$

where:

K = permeability (m/sec);

A = boreholes area (m^2);

h_1 and h_2 = the water level at time t_2 and t_1, above the water table or from the bottom of borehole (m);

t_2 and t_1 = time for measurement of h_1 and h_2.

Cl = shape factor, depending from the filtering section.

The three permeability tests were executed in borehole 1 ($-2, 5$ from surface) and borehole 2 ($-1, 5$ and $-9, 0$ from surface). The obtained results are in the following Table 1 and the drawdown curves in Fig. 14.

Considering the grain size distribution, a preliminary permeability, using the Hazen formula (Hazen, 1911) was also determined. From the 11 sieve analysis an average value of $7{,}8 \times 10^{-8}$ m/sec was obtained.

Clearly, Hazen's rule gives approximate hydraulic conductivity estimates only. The formula's applicability is generally limited to the range of D = 0.01–0,3 cm and in this case was used only for comparison with falling head permeabilities in boreholes.

The new Maya Devi Temple was built in 2002 to protect the archaeological structures from the action of weathering; it is a large, enclosed structure, composed mainly of iron beams and concrete, and equipped with two entrance doors and some windows for ventilation. The state of conservation of the building is very poor. In particular, oxidation of some iron elements and traces of rainwater infiltration from the roof are highly evident.

Rising dampness was also detected, with a water table located below the Maya Devi Temple and with a negative impact on archaeological remains. The relationship of such water table with local aquifers was not clear. During the year 2015 a proper

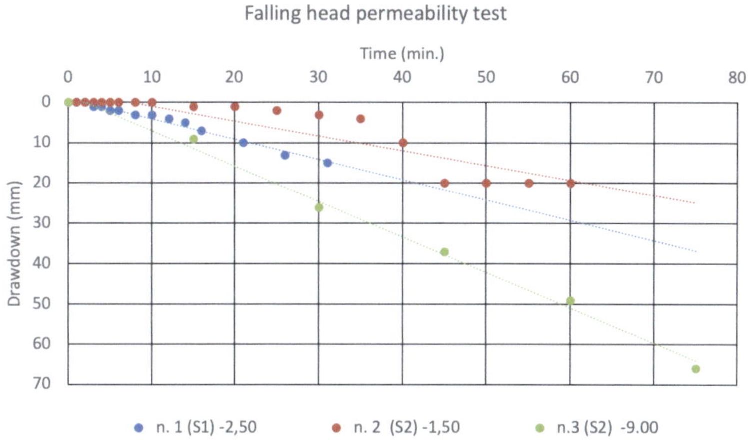

Fig. 14 The drawdown curves in the falling-head tests

monitoring of the water table in 5 small shafts (outside but close to the temple) and three plastic pipes (inside the temple), was successfully implemented by the Lumbini Development Trust (LDT) (Fig. 15). Measurements were collected on a daily basis. Similarly, the rainfall gauge (pluviometer) located in the sacred garden has been reactivated by the Lumbini Development Trust, in charge of conducting daily measurement.

This database is of basic relevance to understand, for the first time, the relationship between rainfall, water table rise, and impact on the temple (Fig. 16).

The major achievements of the monitoring system in 2015 are:

1. the water table shows an ephemeral character and directly correlated to monsoon period. During the dry season, there is not piezometric level in the underground;
2. one day is generally enough to transfer precipitation into a piezometric level (water table level rise);
3. the water level is always the same in all the 8 inspection points (5 shafts and 3 plastic pipes);

The major conclusion from the analysis of monitoring data is that there is an aquifer (a layer of sediment that can store and transmit significant quantities of water under an ordinary hydraulic gradient) just below the Maya Devi Temple.

Geotechnical investigation has identified a complex alluvial deposit, composed by silt with clay and sand, an area mainly covered by the s.c. "cultural deposit" (Coningham and Acharya, 2011, 2012). The distribution of such a "cultural deposit" is reported in the following Fig. 17.

From an hydrogeological point of view, it is possible to consider the "cultural deposit", a silty clay unit with ceramics and bricks, or part of it, as an aquifer, overlapping an aquiclude (an impermeable stratum of sediment that acts as a barrier to the flow of water) composed by the alluvial deposits. Unfortunately, due to the high archaeological relevance of the site, it was not possible to execute boreholes very

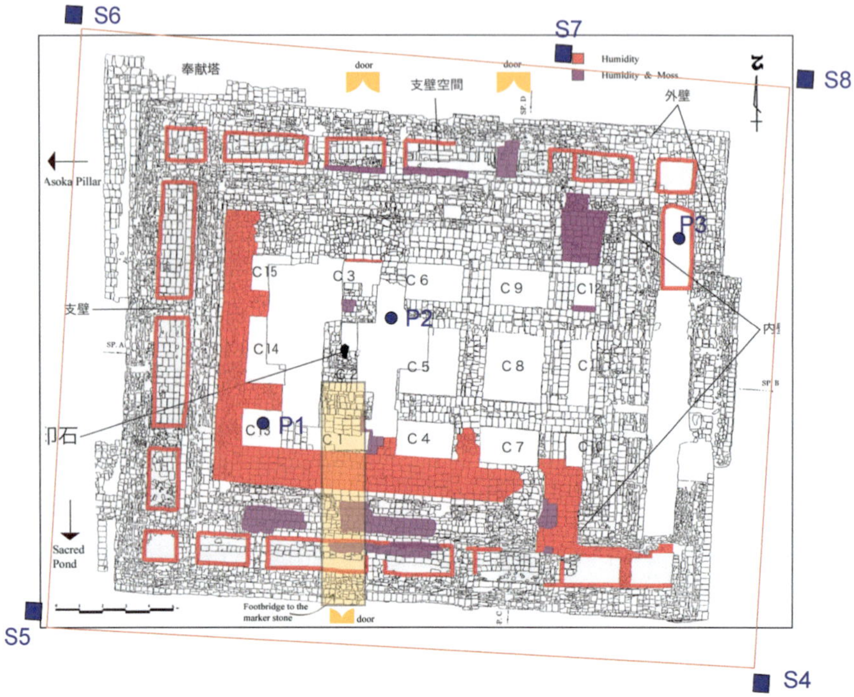

Fig. 15 The monitoring network around the Maya Devi Temple location, composed of 5 shafts (blue squares) and the 3 small pipes (blue circle). (Foreground image from Atzori et al. 2006)

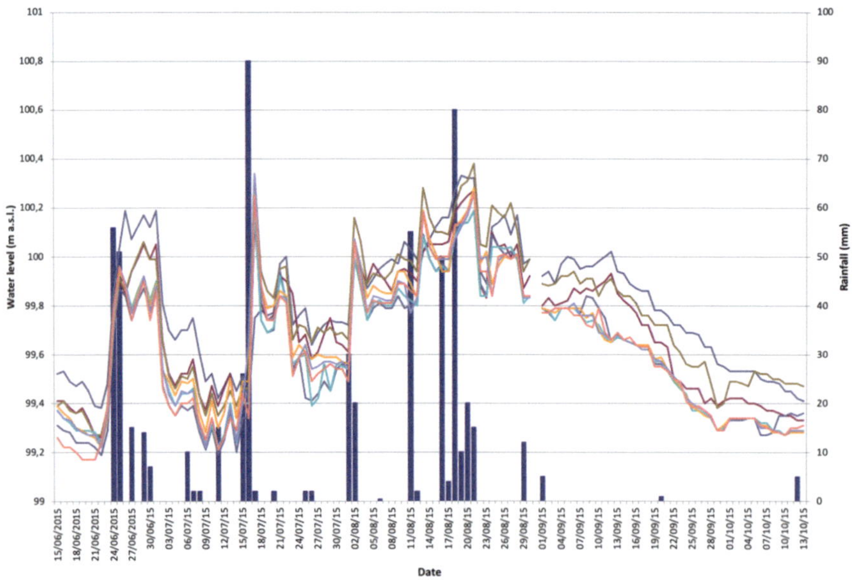

Fig. 16 Daily water table oscillation and daily rainfall measurement during 2015

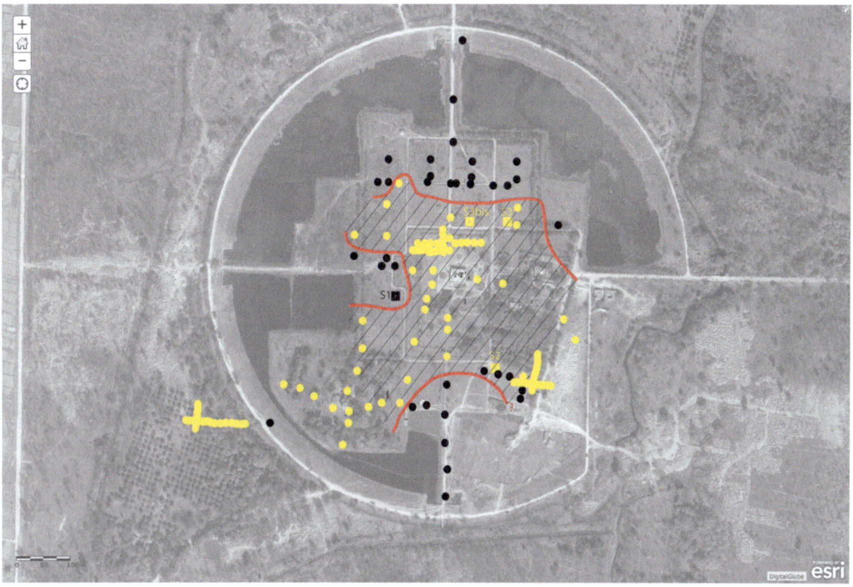

Fig. 17 Distribution of the modern fill (dashed area) containing large volumes of brick, boulders, and a clay matrix (Cultural Deposit). Yellow dots are archaeological pits where the upper "*Cultural Deposit*" is identified, while the black dots stand for areas where it does not appear. (data from Coningham and Acharya (2011, 2012)

close to the Temple. This is necessary not to confirm the presence of the "cultural deposit", which is verified by Coningham and Acharya (2011, 2012) as well as by geophysical investigation, but to measure the permeability in site, providing a value suitable to transfer the rainfall in an aquifer in one day. That is because the permeability tests executed in the boreholes are too low and only marginally confirming such hypothesis.

A possible explanation is that close to Temple the soil is more permeable than in the surroundings, due to more relevant anthropic works to rise topographically the site, before the execution of the first archaeological site.

In addition, no drainage system was properly designed and implemented for the rainwater collected from the top terrace and roof in modern Temple. This is one of the major critical issues, especially during the monsoon season when the intensity of rainfall is very high.

As a consequence, a large amount of water collected on the roof is now discharging close to the border of the Maya Devi construction, an area that is likely very easy to infiltrate. This mechanism, which is also most likely connected with the infiltration from the surrounding of the temple, can explains the emerging of an ephemeral water table inside the cultural deposit, as well as the almost simultaneous to rainfall.

5 Conclusion and Possible Mitigation Actions

The present paper summarizes the main activities carried out to investigate the origin of the water table affecting the archaeological remains of the original Maya Devi site, currently protected with a new Temple built in 2002.

Understanding the flooding mechanism for Maya Devi Temple is requiring a proper knowledge about hydrogeological properties of underground geological materials, which is depending from an accurate 3d reconstruction of the subsoil.

The archaeological site is located in a gentle artificial hill and likewise apparently an amalgamation of a natural clay and accumulated sediments comprised of building remains, debris, and occupation surfaces. According to Coningham and Acharya (2011), the sacred garden site and the Village Mound were chosen for occupation because they represented a relative haven from the seasonal flooding of the plains. A modern shelter was constructed in 2002 to protect the remains.

The site is currently affected by rising damping which heavily impact on the conservation of the heritage.

Three boreholes and geophysical survey were executed, demonstrating a complex alluvial stratigraphy, mainly dominated by silty-clay component. In the upper part a silty clay unit with ceramics and bricks was identified.

The upper part of the area is characterized by a multi-aquifer system, with strata and layers very often confined. The top of the sacred garden is covered by a silty clay unit (cultural deposit) with archaeological remains and debris, also with low permeability. Thus, the direct relationship between rainfall and water table below the Temple is suggesting the hypothesis that in this area, the deposit may be more permeable, and constituted by accumulated sediments comprised of building remains, debris, and occupation surfaces (Coningham and Acharya, 2011). This stratum is most likely acting as aquifer, within the "cultural deposit".

According to the above described data, the water table affecting the Maya Devi Temple is likely very shallow and ephemeral.

Possible mitigation measures should consider the collection of rainy or underground water and dispersion of them into the nearby river, in connection with a continuous monitoring of water table and rainfall. Promising solutions may include:

1. pumping the water out of the site, by using the existing shafts and a proper pipe network. In case of further needs, additional pumps can be installed;
2. the realization of a drainage system starting from a new rainfall evacuation system from the top of the roof; re-connecting all existing pipes to avoid dispersion into the soil; draining outside of the sacred Garden (Cultural Deposit) as much as possible;
3. draining water, by gravity, into the deep aquifer (to be studied in detail).

Clearly the proposed strategy must be defined with local community, realized and maintained by local institutions, in order to make the solution sustainable and long lasting.

Acknowledgements The present work is an advancement of data gathered and elaborated for the UNESCO Project Strengthening the Conservation and Management of Lumbini, the Birthplace of the Lord Buddha, World Heritage Property (Phase II). Maya Devi Temple and Archaeological Remains. The authors are very grateful to UNESCO Kathmandu Office in Nepal, namely the Director Mr. Christian Manhart, the project officers Ms. Shrestha, Nipuna, and Ms. Nabha Basnyat-Thapa for their efforts on project implementation and continuous support in the field activities. We would like to express our special thanks to the Lumbini Development Trust and the Department of Archaeology for their support and assistance for our project activities. A special thanks to Dr. Maurizio Guerra for his help in investigating the hydrogeological conditions of the site. Last but not least, our friend Prof. Robin Coningham, for the continuous discussions and suggestions, and for sharing all the available data.

References

Associazione Geotecnica Italiana (1977) Raccomandazioni sulla programmazione ed esecuzione di indagini geotecniche

Atzori A, Cüppers C, Ghimire HL, Rai R, Suwal R, Ukesh B, Weise K (2006) Lumbini. Present status and future challenges. Kathmandu, UNESCO. http://unesdoc.unesco.org/images/0014/001471/147105e.pdf. Accessed 2 Oct 2013

Bhaddamanika S, Bidari B, Choegyal L, Coningham RAE, Cuppers C, Rai G (2019) Balancing competing requirements of faith and preservation. In: The sacred garden of Lumbini. UNESCO, pp 210–223

Bidari B (2007) LUMBINI A haven of Sacred Refuge, Hill Side Press (P) Ltd. Kathmandu, Nepal

Coningham RAE, Acharya KP (2011) Identifying, evaluating and interpreting the physical signature of Lumbini for presentation, management and long-term protection: report of season one and two of activity 2. UNESCO Internal Report, Kathmandu

Coningham RAE, Schmidt A, Strickland KMS (2011) A cultural and environmental monitoring of the UNESCO world heritage site of Lumbini, Nepal. In Ancient Nepal, Number 176, March 2011

Coningham R, Acharya KP (2012) Identifying, evaluating and interpreting the physical signature of lumbini for presentation, management and long-term protection. Interim report of the second season of field activities. UNESCO internal report

Coningham R, Acharya KP (2013) Identifying, evaluating and interpreting the physical signature of lumbini for presentation, management and long-term protection. Report of the third season of field activities. UNESCO internal report

Coningham RAE, Acharya KP, Tremblay J (2019) Archaeology and site interpretation. In: The sacred garden of Lumbini. UNESCO, pp 50–103

Hazen A (1911) Discussion of dams on sand foundations. In: Koenig AC (ed) Transactions of the American society of civil engineers, vol 73, no 3, pp 199–203

Hvorslev MJ (1951) Time lag and soil permeability in ground water levels and pressures. US Army Engineer Waterways Experiment Station. MS, Bulletin, pp 36

Kenzo Tange & Urtec (1968) Final outline design for Lumbini

Kenzo Tange & Urtec (1981) Master design for the Lumbini garden

Margottini C, Spizzichino D, Puzzilli LM, Pagnin P, Fiorin A (2019) Water table fluctuation and impact on the remains of Maya Devi Temple, the birthplace of the Lord Buddha, UNESCO World Heritage Site (Lumbini, Nepal). In: Proceedings of the XVII ECSMGE-2019. Geotechnical Engineering foundation of the future. ISBN 978-9935-9436-1-3. © The authors and IGS: All rights reserved. https://doi.org/10.32075/17ECSMGE-2019-0303.

Krishna Rao G, Rama S, Vaidya YL (1996) Hydrogeological conditions in the terai plain of Rupanoehi district, Lumbini zone, Nepal with special emphasis on groundwater recharge. In: Singh V, Kumar (eds) Subsurface-water hydrogeology. Kluver Academic Publishers, pp 131–149

Water Infiltration and Waterproofing of Susan-Ri Tomb (North Korea)

Claudio Margottini, Ugo Castellotti, Debi Ghoshal, and Rodolfo Lujan

Abstract The Susan-ri Tomb is located about 35 km Southwest of Pyongyang (North Korea) and it was strongly affected by water infiltration, generally during the rainfall season. A detail geological survey and the monitoring of rainfall, water table and water infiltration into the tomb were performed. Rainfall monitoring were collected from the nearest station and 6 piezometers were installed. According to the collected data it is possible to verify that the inundation into the Tomb is correlated to two different factors: the saturation of the mound and the rising of an ephemeral (temporary) water table. To ensure the long-term safeguarding of the Tomb and to prevent flooding it was necessary (1) to waterproof the earthen mound and (2) to maintain the ephemeral water table below the level of the Tomb floor. These are not separate points but have to be considered as elements of the same project. In order to satisfy the above requirements, the possibility of lowering the ephemeral water table through the use of a drainage trench and of waterproofing the earth mound with a filter allowing the water to freely flow outside the Tomb area was investigated and realized. The project was completed in Spring 2011 and since then no more infiltration was registered inside the tomb.

Keywords Earth mound · Infiltration · Water table · Waterproofing · Drainage · Susan-ri Tomb (North Korea)

C. Margottini (✉)
Former Scientific Attaché, Embassy of Italy in Egypt, 15, Abd El-Rahman Fahmy Str., Garden City, Il Cairo, Egypt
e-mail: claudio.margottini@gmail.com

UNESCO Chair On Prevention and Sustainable Management of Geo-Hydrological Hazards, University of Florence, Florence, Italy

U. Castellotti · D. Ghoshal
Officine Maccaferri Asia, Petaling Jaya, Selangor, Malaysia

C. Margottini · R. Lujan
UNESCO Consultant, Rome, Italy

G. M. El-Qady and C. Margottini (eds.), *Sustainable Conservation of UNESCO and Other Heritage Sites Through Proactive Geosciences*, Springer Geology,
https://doi.org/10.1007/978-3-031-13810-2_18

1 Introduction

The Susan-ri Tomb is located about 35 km Southwest of Pyongyang. It is placed on top of a small hill and surrounded by pine trees. The archaeological remains can be classified as a stone-built (with stone blocks) chamber tomb covered with an earthen mound (Fig. 1).

Despite water infiltration, the stone part is well preserved. The earth mound, according to photographs taken in the 1970s, it appeared to be highly eroded. The present earthen mound was reconstructed during the earlier phases of restoration in the last century (1970–1980 approx) (Fig. 2).

The Tomb was strongly affected by water infiltration, generally during the rainfall season. The temporal relationship with rainfall was clear, but the cause of the infiltration inside the Tomb needed to be understood. The Tomb also has a concrete horizontal slab, apparently covering the whole perimeter, adding to the problem of water infiltration.

Due to the above uncertainties, the project focused on the evaluation of the geological condition of the site. Monitoring of the water-table fluctuations by setting up a piezometer system and of the water infiltration by collecting reliable rainfall data was carried out and the gathered information analysed in order to understand the reasons for the water infiltration and provide a waterproofing solution for the long-term safeguarding of the Tomb.

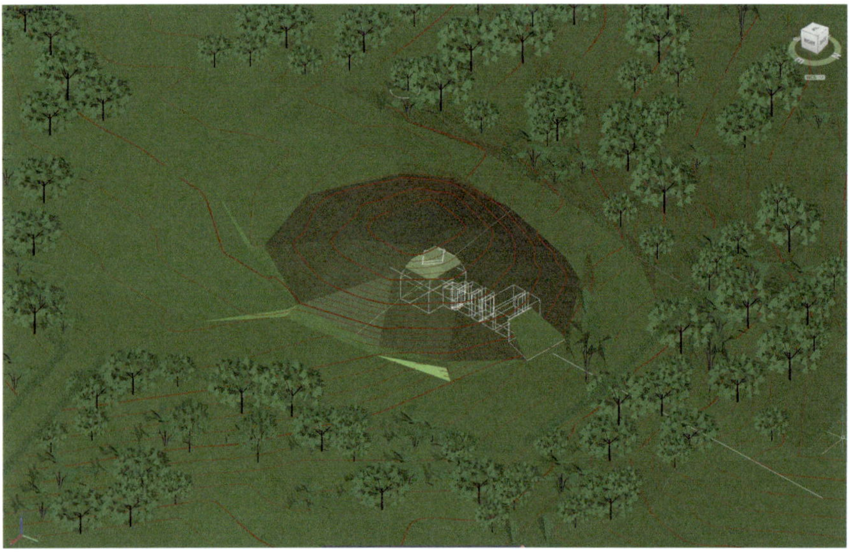

Fig. 1 3D detailed view of the Susan-ri site

Fig. 2 The mound above the Susan-ri Tomb in the 1970s

2 The Problem

The Tomb was mainly affected in the past by rising damp, the origin of which was not very clear. The first available information dates back to 2005. The last flooding was in 2011. Many years were necessary in order to investigate the local water table, the saturation of the mound, the condensation of humid air on the Tomb's walls and the combination of these factors.

Table 1 lists the flooding events affecting the Tomb of since 2005.

3 Geological Condition of the Site

According to the geographical coordinates derived from the UNESCO Website, the Susan-ri Tomb is located in the Tanchon Complex, a geological formation composed of biotite granite, granodiorite and fine-grained granite (Paek et al. 1996). Its origin is related to the magmatism–the development and movement of magma and its solidification into igneous rocks–that took place in the Jurassic Period in the middle of the region during the magmatic activity that affected the north of the Korean Peninsula in the Mesozoic Era.

In this part of the country, the geological arrangement is composed of three intrusive phases, and most of it consists of a granitic rock of the second phase, together with a certain amount of fine-grained alaskites (a plutonic rock consisting of oligoclase,

Table 1 Flooding events in the Susan-ri Tomb

Date	Event	Source
July–August 2005	400 buckets of water (1 bucket = 10 lt. equal to perhaps 10 kg) were removed from the Tomb's floor	Tomb keeper
14/07/2006	The beginning of heavy rain	Tomb keeper in Lujan (2006)
16/07/2006	Water began to infiltrate from the floor	Tomb keeper in Lujan (2006)
18/07/2006	300 buckets of water were removed from the Tomb's floor (1 bucket = 10 lt.). The maximum height reached by the water inside the Tomb was 27 cm	Tomb keeper in Lujan (2006)
19/07/2006	Rain stopped but 200 buckets of water were removed	Tomb keeper in Lujan (2006)
20/07/2006	150 buckets of water were removed	Tomb keeper in Lujan (2006)
21/07/2006	70 buckets of water were removed	Tomb keeper in Lujan (2006)
	According to the Tomb keeper, in the period 14–21 July 2006, 720 buckets of water were removed, equal to 7,200 L. No more rainwater has infiltrated since then	Tomb keeper
11/08/2007	300 buckets of water were removed = 3,000 lt	Tomb keeper
12/08/2007	600 buckets = 6,000 lt	Tomb keeper
13/08/2007	500 buckets = 5,000 lt	Tomb keeper
14/08/2007	500 buckets = 5,000 lt	Tomb keeper
15/08/2007	350 buckets = 3,500 lt	Tomb keeper
16/08/2007	350 buckets = 3,500 lt	Tomb keeper
19/08/2007	200 buckets = 2,000 lt	Tomb keeper
20/08/2007	100 buckets = 1,000 lt	Tomb keeper
	Water inside the Tomb up to 35 cm from the floor and persisting until October even if pumped away. The water was flowing from a small hole in the west wall of two cm in diameter 20 cm from the floor	Tomb keeper
August to April 2008	No more rainwater infiltrated from August 2008 to April 2009	Tomb keeper in Lujan (2008/a)
2008	No water infiltration. According to KCPC experts the infiltration occurs only after three days of continuous rainfall	KCPC experts, personal communication

(continued)

Table 1 (continued)

Date	Event	Source
2009 July 18	Water started to infiltrate from the junction between the west wall and the floor. 34 cm of water	Tomb keeper in Lujan (2009) and KCPC
19	Some water remaining inside	Tomb keeper in Lujan (2009) and KCPC
20	Flooding level was 32 cm	Tomb keeper in Lujan (2009) and KCPC
21	3.3 lt of water removed from the Tomb floor	Tomb keeper in Lujan (2009) and KCPC
2010 August 27	20 cm of water. The Tomb keeper took away 160 buckets of water	Tomb keeper in KCPC
28	20 cm of water. The Tomb keeper took away 200 buckets of water. The rain stops on 28 August	Tomb keeper in KCPC
29	20 cm of water. The Tomb keeper did not take away any water	Tomb keeper in KCPC
30		Tomb keeper in KCPC
31	The water level lowers to 15 cm	Tomb keeper in KCPC

microcline and quartz, with subordinate muscovite and few or no mafic constituents) of the third phase and a very few (rare) mafic and intermediate rocks generally having a silica content of 54–65%, e.g. syenite and diorite, from the first phase. (Intermediate is one subdivision of a widely used system for classifying igneous rocks on the basis of their silica content; the other subdivisions are acidic, basic, and ultrabasic).

The following Figure shows the geological map of the region (Fig. 3).

In order to understand the reason for the rising damp, six boreholes and one pit were executed in the area in 2009. The position is shown in Fig. 4. The reason for the drilling was the need to identify the stratigraphy of the site and to install an open standpipe piezometer inside each of the boreholes. An additional measurement/piezometer point was considered for the well of the Tomb keeper. The boreholes were executed using a mixed rotary-percussion machine and using rotary techniques in soil and/or soft rock and percussion in hard rock (Figs. 5 and 6). A manual pit was also executed inside the Tomb, exactly where the water was detected in the past. In this case a piezometer was temporarily installed inside the ditch.

The collected information allowed the following observations to be made:

1. The stratigraphy of the terrain is composed, from bottom to the top, as follows:

    a. the upper part of the bedrock is of a granite/schistose formation (Tanchon Complex), a geological formation composed of biotite granite, granodiorite and fine grained granite (Paek et al. 1996). This material can be considered of low permeability in terms of porosity and of medium permeability when fractured;

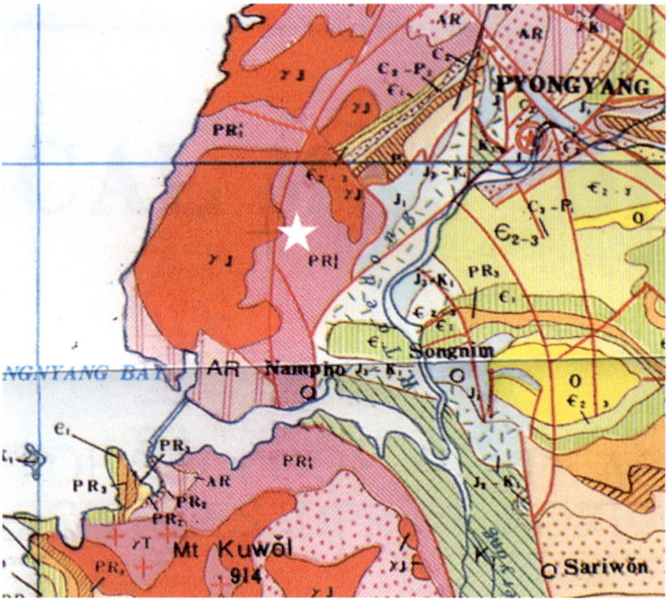

Fig. 3 Geological map of the Kangso area and location of the Susan-ri Tomb. The PR formation where the tomb is located (white star) belongs to the Jungsan group: Cristalline schist, amphibolite (from Paek et al. 1996)

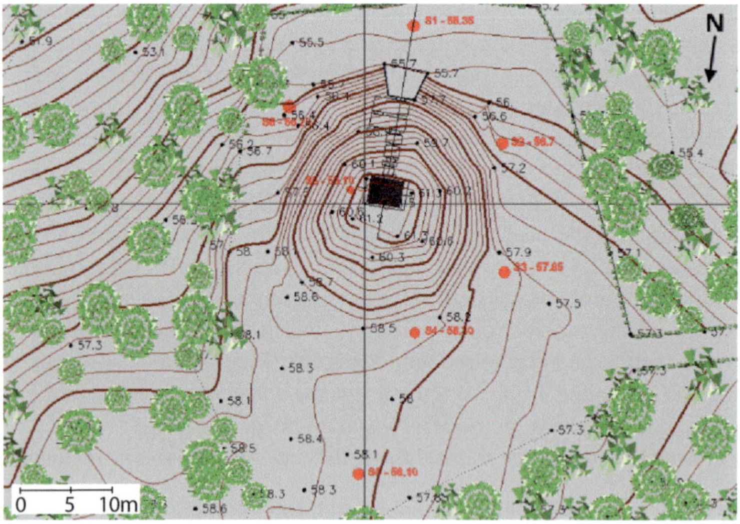

Fig. 4 Boreholes (red dots) and the pit inside the Susan-ri Tomb (S8). The piezometer/measurement point in the Tomb keeper's well is outside the map, in North-West direction, about 60 m from the tomb

Fig. 5 The rotary/percussion machine in borehole S1

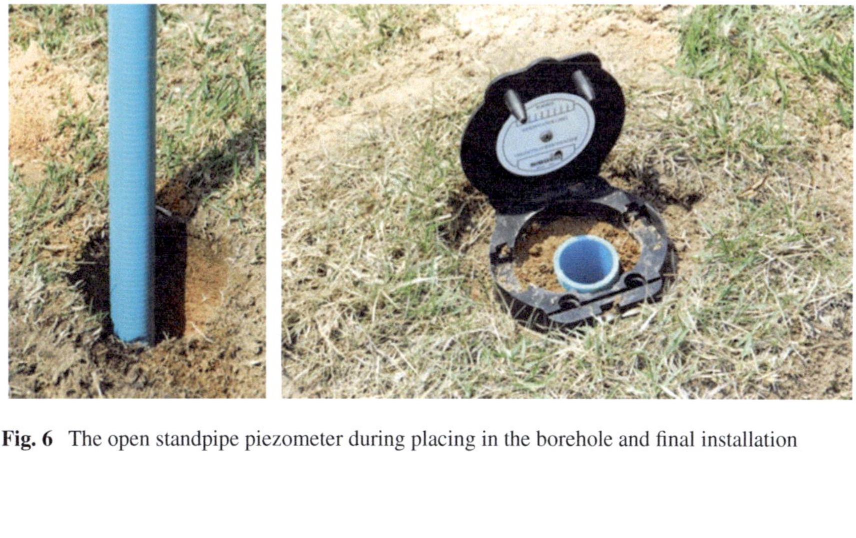

Fig. 6 The open standpipe piezometer during placing in the borehole and final installation

b. weathered granite with relevant clay component (low permeability due to medium porosity);
c. a regolith deposit (the fragmented and unconsolidated rock material, whether residual or transported, that nearly everywhere forms the surface of the land and overlies the bedrock) produced by the alteration of the granite/schistose formation. This material can be characterized by medium to high permeability due to medium porosity and the presence of non-cohesive soil; it is composed of the alternating of slightly different materials, with various degrees of weathering, and sometimes, as in borehole S1, also with an important occurrence of clay;
d. Earth fill mainly composed of yellow granitic sand; this material can be classified as of high permeability due to the presence of non-cohesive soil and high porosity.

2. The water table was not detected during the dry season

   The following Figs. 7 and 8 show the geological map, a section of the mound and the Tomb. From the latter, it is possible to notice the sequence of the different materials composing the terrain, as well as a rough idea of permeability. The granite formation outcropping in the northern part of the area and gently dipping towards the south. This material, in case of no joint as in the present condition, can be considered of low permeability. On the other side, the upper regolith formation has a higher value of permeability for porosity. Such this contrast of permeabilities can generate a water table during heavy rainfall. This ephemeral (temporary) water table was supposed to be the main cause of water infiltration into the Tomb from the floor and into the lower part of it in 2009. In fact, the limit between the permeable regolith deposit and the impermeable granite/schistose formation is only 70 cm, as was revealed in borehole S8, and even less in the northern part of the Tomb (see geological section).

4 The Mechanism of Water Infiltration into the Tomb

As a result of the geological setting, during heavy rainfall infiltrating water is blocked by the impermeable bedrock and stagnates within the regolith permeable deposit. This hypothesis is confirmed from the Tomb keeper's well, which has been excavated from the regolith deposit to the granite/schistose formation. In this well the water is quite abundant during the rainy season, and it is scarce during the dry period. The refilling of the well is quite fast during the rainy period, thus confirming the idea that water is generously diffused into the terrain and consequently into the aquifer. In the meantime, the saturation of the mound due to rainfall generates water infiltration into the Tomb. The way water inundated the Tomb in 2010 verified the above-mentioned hypothesis. The full model of water infiltration into the Tomb is given in the following Figure (Fig. 9).

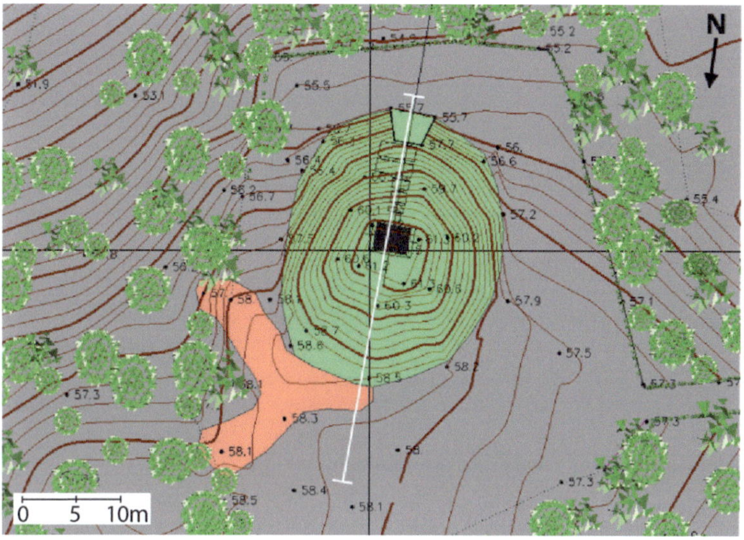

Fig. 7 Geological map of the area. Orange is granite/schistose formation; dark grey is regolith deposit produced by the alteration of the granite/schistose formation; green is earth fill mainly composed of yellow granitic sand. The white line is the tracing of the geological section portrayed in Fig. 8

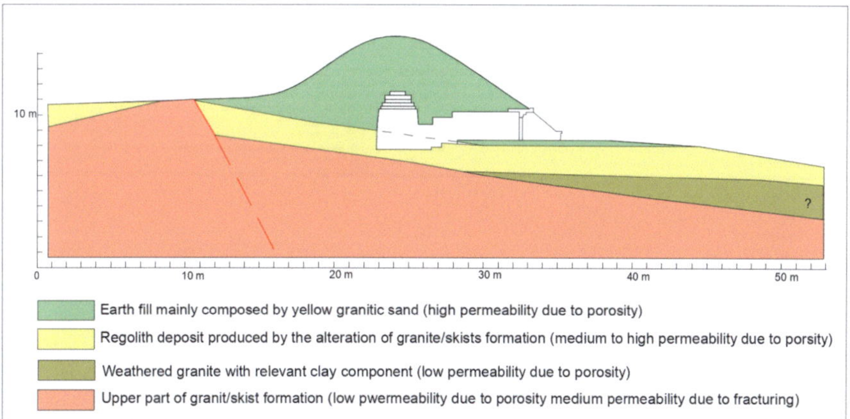

Fig. 8 Geological section of the area and the earthen mound

The two following figures illustrate the thickness of the permeable material (regolith deposit) and the topography of the granite formation on which an ephemeral water table is formed if completely saturated. From these it is possible to theoretically describe the flux of water underground (Figs. 10 and 11).

A definitive confirmation for this model comes from the analyses of the piezometer data obtained in 2009–11 and the comparison with the rainfall amounts measured

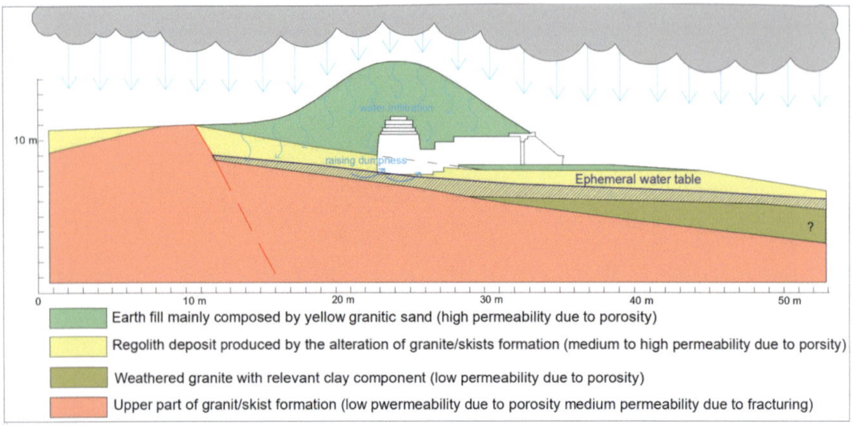

Fig. 9 Mechanism of water infiltration into the Tomb

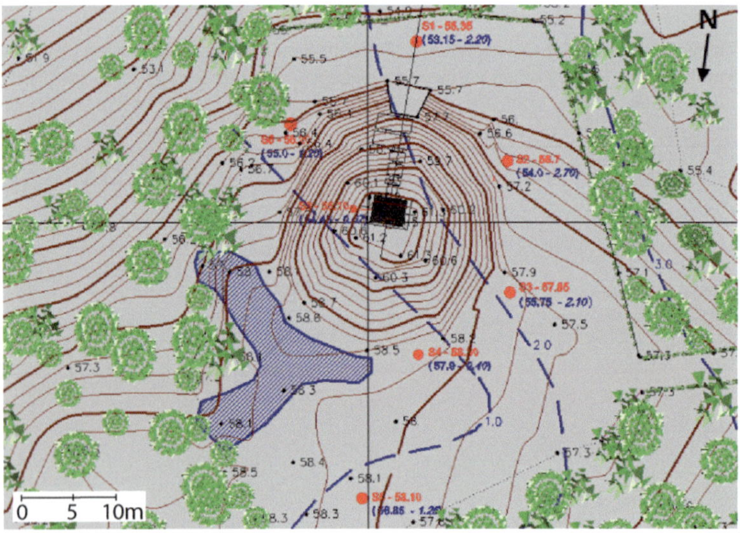

Fig. 10 Thickness of the permeable material. For any borehole is reported in blue, the lower quote of permeable material (a.s.l.) and the thickness

during the same period. Meteorological data were collected by Kangso Station, about 10–12 km from the site.

By comparing the piezometer and rainfall data of July 2009 when important flooding occurred in the Tomb, it is possible to notice the following:

1. The rainfall of 45.6 mm of 12 July did not affect the water level in the piezometers or in the Tomb keeper's well;

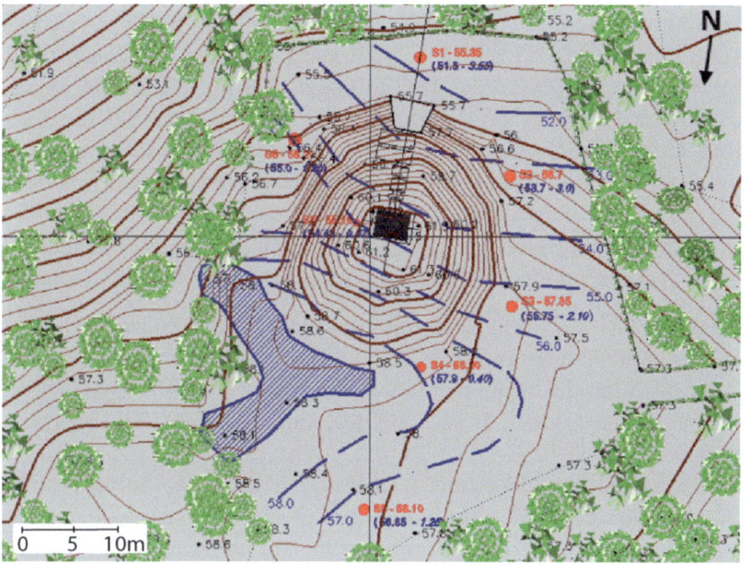

Fig. 11 Topography of the granite formation. For any borehole is reported in blue, the lower quote of granite (a.s.l.) and the thickness

2. A rise of the water level in the piezometers on 16 July was recorded, two days after intense precipitation of about 111.4 mm occurred on 14 July. There is no piezometer data for 15 July, but the Tomb keeper's well was filled by this heavy rainfall;
3. 158.5 mm of rainfall occurred on 18 July, and the measurements on the following day showed the maximum level of the water table;
4. Water arrived at the same time in all the piezometers but not in the one inside the Tomb. The delay was about three days (ex. 16–19 July 2009);
5. The maximum level of water recorded in the piezometers was in the most topographically elevated points of the granite formation, e.g. the northern side of the mound;
6. The Tomb was flooded one day before it reached the bottom of piezometer S8 located in the burial chamber;
7. The maximum peak of water contained inside the piezometers was reached on 19 July and then the amount of water began to decrease.

The above points explain the rising of the ephemeral water table in the regolith formation as a consequence of water infiltration from the surface due to heavy rainfall. Water infiltrates into the soil and is blocked by the granite impermeable bedrock. The distribution of water that reaches the higher levels at the most elevated piezometers reflects the topography of the granite bedrock. Only when the water fills the area below the Tomb, due to lateral infiltration, it rises and inundate the burial chamber.

As first intervention, in spring 2010 a water-collecting trench was constructed around the Tomb (Figs. 17 and 18). No waterproofing of the earth mound was improved in this year, due to the need to understand the efficacy of the drainage trench, with respect to the infiltration from the mound. Thus, the Tomb was also flooded in August–September due to very heavy rainfall, and the data collected in 2010 was compared with that of 2009, meaning that the water-collecting trench was useful.

By comparing the piezometer data and the rainfall of August–September 2010, when some flooding occurred in the Tomb, it is possible to notice the following:

1. The rainfall of 106.70 mm of 26 August soaked the mound above the Tomb;
2. An increase of the water level was found in four of the piezometers on 27 August, just one day after the heavy rainfall of 26 August;
3. The tomb flooded on 27 August, and the water remained for 16 days until 11 September;
4. The increase of the water level in piezometer S8 in the burial chamber was recorded seven days after the heavy rainfall of 26 August and six days after the inundation of the Tomb. Comparing the similar rainfall and flooding data collected in 2009 (Fig. 12), the infiltration behaviour is different because there is a delay in reaching the floor of the Tomb, clearly demonstrating the efficacy of the drainage channel as well as the need for surface protection/drainage of the earth mound;
5. The water remained in the piezometers out of the Tomb for seven days, until 2 September, while in the Tomb it remained for 16 days (Fig. 13); the piezometer in the burial chamber contained water until 9 September.

The different behaviour of the 2010 inundation with respect to the 2009 one, despite of the realization of the drainage trench, can be explained by the different the total amount of precipitation and distribution during many days. The following Fig. 13 shows the rainfall, piezometer and inundation data of 2010 described above.

A similar trend to 2010 was registered also in 2011, before the waterproofing of the earth mound implemented in 2012. The rainy season of July–August was not exhibiting any high intensity peak. After three days of precipitation the water appeared in the tomb but not in the piezometer inside the same tomb (n. 8). The water table was measured only 3 days after the flooding of the tomb floor and 6 days after the beginning of the rainy days (Fig. 14).

With respect to the triggering mechanism of inundation in the Tomb, the relationship with rainfall is quite evident. What is more complex is the identification of a possible threshold value above which flooding occurs. From the gathered information it is possible to identify the mechanism that generates water infiltration and an ephemeral water table, limited to the specific case studies of 2009 and 2010. These effects (infiltration and the water table) are connected to significant precipitation (>100 mm) in a scenario of cumulative rainfall more than to a single high-intensity event. This amount of water is necessary (1) to saturate the earth mound and (2) to saturate the regolith formation and generate the water table.

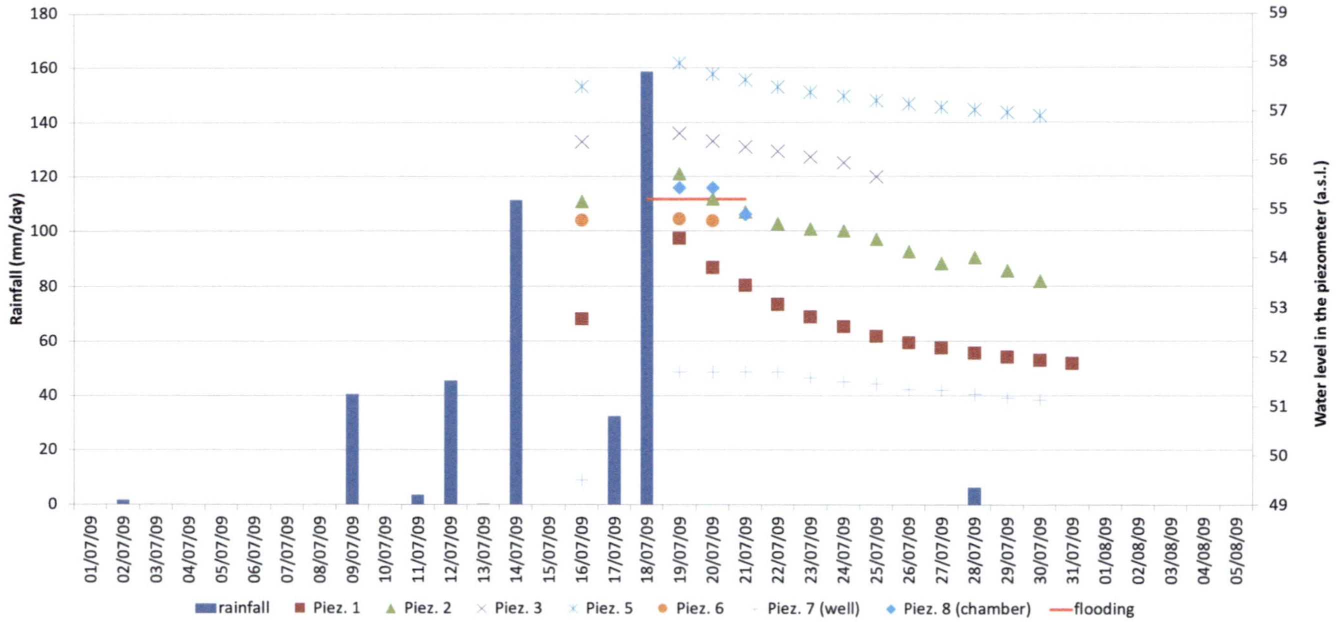

Fig. 12 Comparison of the 2009 data for rainfall (mm/day), piezometers and flooding inside the Tomb (red line). On the left of the chart the rainfall mm/day is marked and on the right side is the altitude (a.s.l.) of the water level in the piezometers

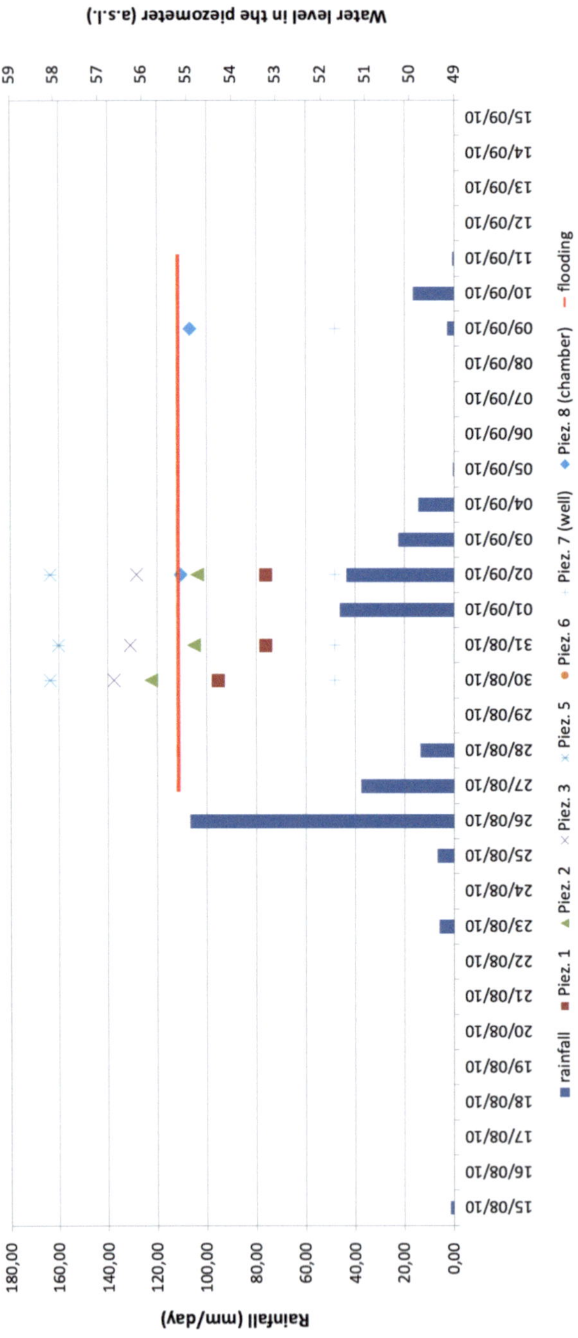

Fig. 13 Comparison of the 2010 data for rainfall, piezometers and flooding inside the Tomb (red line)

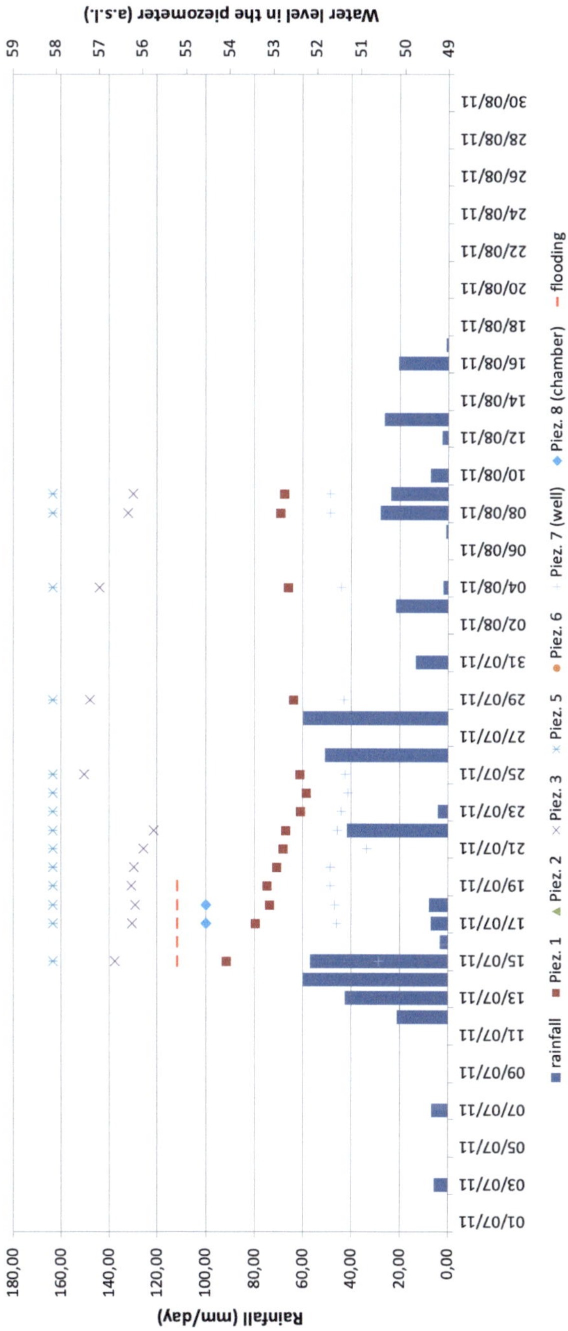

Fig. 14 Comparison of the 2011 data for rainfall, piezometers and flooding inside the Tomb (red line)

Only one day of extreme precipitation, as on 19 July 2010, was apparently not enough to satisfy the conditions for inundating the Tomb. The following Fig. 15 illustrates the rainfall, piezometer levels and flooding of the Tomb for the period 2009–2011, showing the importance of two/three days of cumulative precipitation rather than one day of high intensity (mm/hour) rainfall.

Upon saturation of the regolith layer, the excess water refills the water table with a transfer of time falling within the order of hours and not days. The demonstration of this is that the heavy storm of 18 July 2009 (158.5 mm) produced a maximum peak of groundwater in the piezometers but did not influence the water table during the following days. The water level inside the piezometers slowly decreased day by day, due to the lack of subsequent water recharge, or no rainfall. This means that after the saturation of the regolith layer and within some hours of rainfall or even very extreme precipitation, water infiltrates almost immediately into the soil and does not just form surface run-off. This behaviour is also connected to the very shallow aquifer (water table) we are dealing with.

After the construction of the drainage channel in the spring of 2010, the above phenomena was reduced in favour of the saturation of the earth mound. This was clearly demonstrated by the short delay between the rainfall and the flooding into the Tomb in 2010 and, in the meantime, the delay in the rise of water in the piezometers located in the Tomb. This can easily be explained by the major contribution made by the water that is now coming from the saturation of the mound as a consequence of the efficacy of the drainage system that is significantly lowering the level of the ephemeral water table.

Finally, when considering the mechanism of water inundation into the Tomb it is possible to say that the concrete slab above, inside the earth mound, it does not influence water infiltration directly. On the other hand, constant (spring and winter) water vapour condensation on the lanternendecke ceiling of the Tomb due to the cooler surface has been observed (information kindly provided by R. Luján), supporting the idea of the limited influence of the slab in reducing the infiltration of water.

5 Waterproofing the Tomb

According to the geological investigation and the meteorological and hydrological monitoring of 2009 and 2010, it is possible to say that rising damp in the Tomb is activated by the saturation of the earth mound and the generation of an ephemeral water table below the Tomb. The first infiltrates into the walls of the Tomb, while the latter produces the inundation from the ground.

In order to ensure the long-term safeguarding of the Tomb and to prevent flooding, it is necessary (1) to waterproof the earth mound and (2) to maintain the ephemeral water table below the level of the Tomb floor.

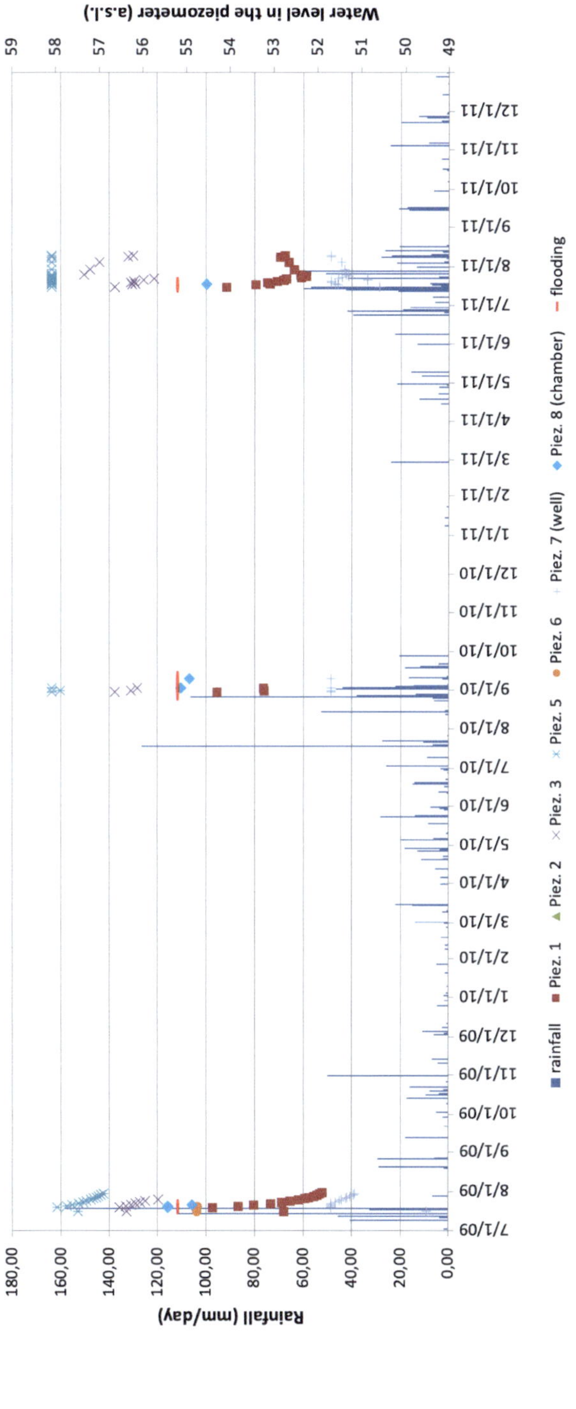

Fig. 15 Comparison of rainfall data (Kangso Meteorological Station), water levels in the piezometers and flooding in the Tomb (red line) for the period 2009–2011

These actions were not considered as separate points, but were considered as closely related elements of the same project. In order to satisfy the above requirements, the possibility of lowering the level of the ephemeral water table with a drainage trench and waterproofing the earth mound with a filter, allowing the water to freely flow outside the Tomb area, was investigated.

6 The Drainage Trench

The drainage trench was constructed in 2010 around the mound, according to the scheme given in Fig. 16. The initial points "A" and "D" were selected in the place where the regolith deposit showed a thickness of no more than 0.7–0.8 m. After the construction of the first part of the trench, its extension from point A towards the east was carried out (yellow dotted line) since the area of this granite/schistose formation shows a sharp cut, likely a fault in the terrains' morphology or paleomorphology.

The drainage trench has an average gradient of 9%, with a minimum value of 1.3% outside the area to be fully drained. Generally, an average value of two to three

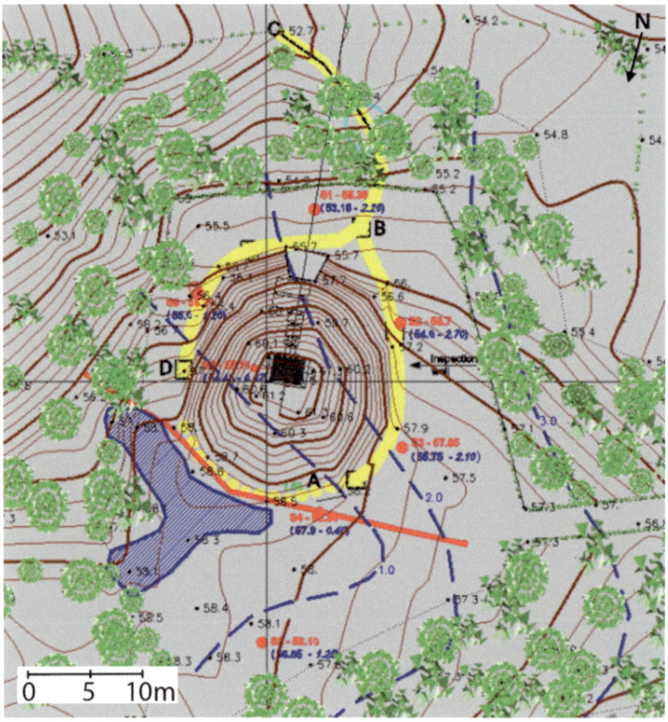

Fig. 16 Ground plan of the drainage trench built in spring 2010 (Yellow). Blue lines are the thickness of permeable material and red line is a fault, most likely affecting the granite bedrock

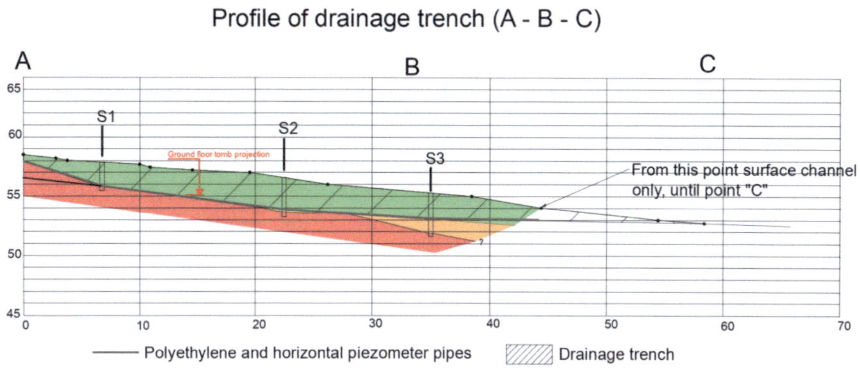

Fig. 17 Topographic profile of drainage trench (west side). Values are in metres; A,B,C have been specified in Fig. 15

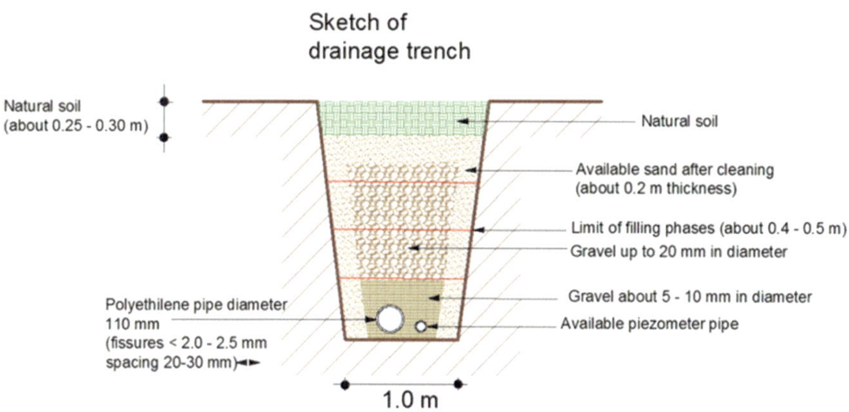

Fig. 18 Section of the drainage trench

per cent is recommended. The trench was constructed on the contact point between permeable and non-permeable materials, as shown in the following Fig. 17. The last 10–12 m were supposed to be built in the open air, as a channel because the maximum depth was only one metre. A waterway was executed by hand at the end of the trench, such that water flowing downwards would not be freely dispersed in the field.

The maximum depth was about 2.8 m from the surface, suitable for the NBCPC workers, according to national experts. In any case, it was recommended that the excavation be protected with wooden planks, as in the proposed scheme. Unfortunately, the available equipment did not allow the digging of a 2.8 m excavation. The trench was executed according to the scheme reported in Fig. 18.

The standardized 0.2 mm diameter grain size of sand was obtained with metal mesh sieves. Once the trench was excavated around the mound of the Tomb, polyethylene pipes were placed in the bottom. It was proposed by the expert that along with

Fig. 19 Construction of the drainage trench

these polyethylene pipes, a second line of pipes be installed utilizing the already existing piezometer pipes, available since the previous mission in 2009. This is to increase the efficacy of the drainage system. Finally, different sizes of gravel and sand were then put into the trench, according to the scheme of Fig. 18.

Details of the construction techniques and specifications are given in Margottini (2010). Two photographs of the construction period are given below (Fig. 19).

7 Waterproofing the Earthen Mound

The identification of the best solution was not simple since there is almost no experience in Asia of the conservation of earthen mounds, and the technique of construction and waterproofing of burial earthen mounds is not very precise in Korea (Onitsuka et al. 2011). Nevertheless, it has to be emphasized that earthen mounds are an integral part of the cultural heritage and not an additional component of the Tomb. They were designed and built to ensure the conservation of the Tombs for as long as possible in the climatic and meteorological condition of each site. Seen in this light, they are elements of the historic tradition of those who built the magnificent Koguryo Tombs. In this specific case, the mound was reconstructed after the excavation of the Tomb in '60, and so the archaeological concern is of minor importance (see Fig. 1).

Preventing the saturation of the earth mound with water is a complex task that involves many aspects. Discussions among experts mainly concentrated on two different ones: the complete waterproofing of the earthen mound or the realization of a filter allowing water to freely flow outside the Tomb area. There are advantages and disadvantages to both systems. Full waterproofing (e.g. a stratum of clay or polyethylene) of the earthen mound, perhaps 25/50 cm below the surface, does not require any maintenance but may affect the 'transpiration' (water vapor exchange) of the Tomb, affecting the mural painting conservation. The construction of a drainage stratum located 50/100 cm below the surface level avoids the problem of transpiration but needs periodical maintenance. Sometimes the filter could be filled by soil particles, limiting effectiveness. In both cases, the proposed materials require that no sliding of natural soil above the 25/50 cm layer occurs.

After investigating all possible alternatives, the best solution was identified as the drainage surface filter 25/50 cm below the present surface (about 800 sq. m in area). This would allow the water on the surface to flow easily away in a short time in the direction of the drainage trench. The latter can be constructed in different ways: geotextile coupled with mini-drains regularly spaced; geotextile pack with highly permeable material in the middle; geotextile pack with sand strata in the middle; etc. In any case, the drainage surface and related collector of water needs to be connected to the existing drainage trench around the mound, like in the sheltering of tunnels.

The final solution, also taking into account the possibility of easily available material, was considered to be the use of a geotextile pack with highly permeable material in the middle, working as a filter and allowing the water to flow into the drainage trench. The filter also needs to be coupled with a special metal retention net, in order to avoid the sliding of the material. The drainage and retention sheet need to be connected with the drainage trench, in order to allow all the collected water to flow away from the Tomb area. The following Fig. 20 demonstrates the proposed scheme, implemented in 2012.

The filter is placed 25–50 cm below the natural soil, as in the following Fig. 21. Above the filter, on the most sloping sections of the earthen mound, and in order to avoid the sliding of the earth over the filter when this is completely saturated, it was proposed to install a metal net with special horizontal ridges to support the soil's weight (Fig. 22).

After the water proofing, the year 2012, in occasion of rainy season, no inundation occurred inside the tomb and even the relative humidity was under control, with the average value of 97% (s.d. 1,7%), outside the glass barrier separating the paintings

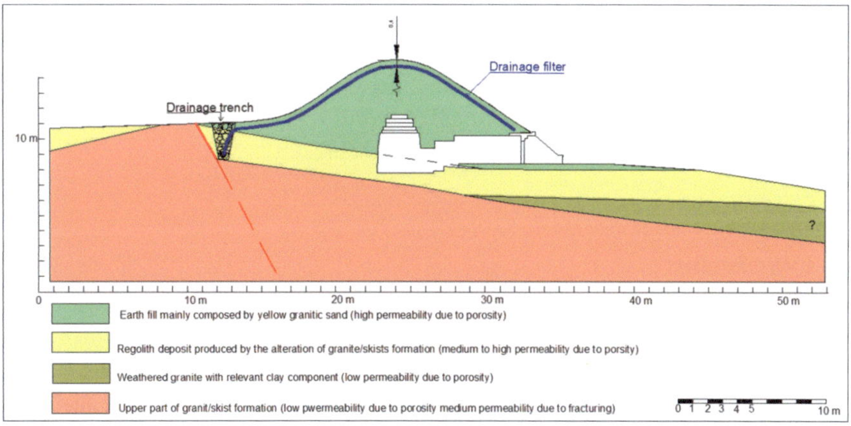

Fig. 20 Waterproofing the Tomb

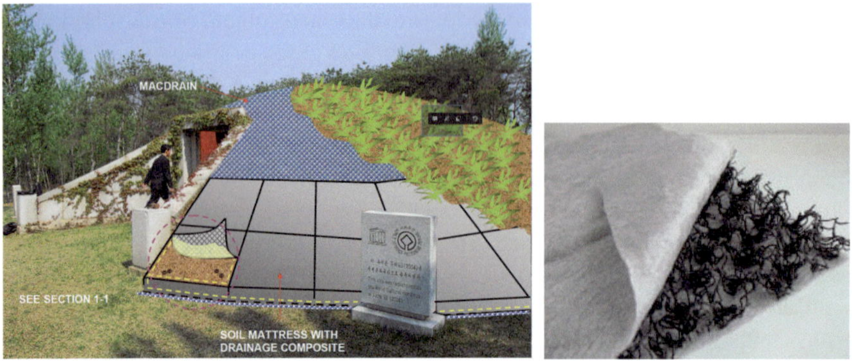

Fig. 21 Example of the position of the filter under the soil (courtesy of Maccaferri, Asia) and detail of the filter

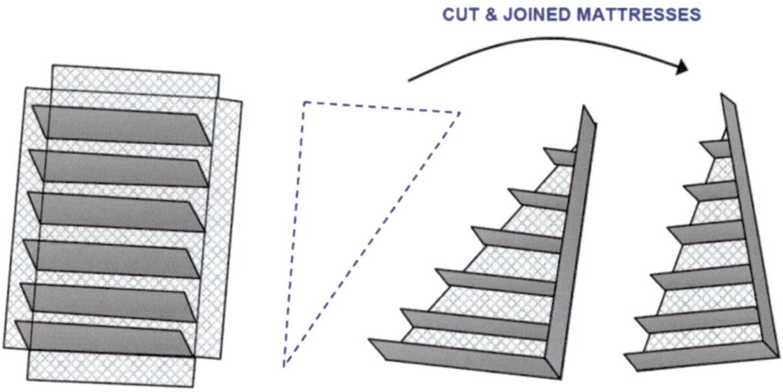

Fig. 22 The metal retention sheet to hold the soil on the inclined sides of the mound (courtesy of Maccaferri, Asia). The metal retention net is placed over the drainage filter

from the main chamber. The relative humidity is an important factor since its low value may generate salt formation. For this reason, not all the surface was covered with the high permeable filter but a small part of it was left to allow a little amount of water to infiltrate freely into the underground (Fig. 23).

8 Conclusion

In recent years much information has been collected to allow the long-term waterproofing of the Tomb. The available information includes a new topographical map, geological map and sections, data from permeability tests on the boreholes, meteorological data from the Kangso Meteorological Station, piezometer information from

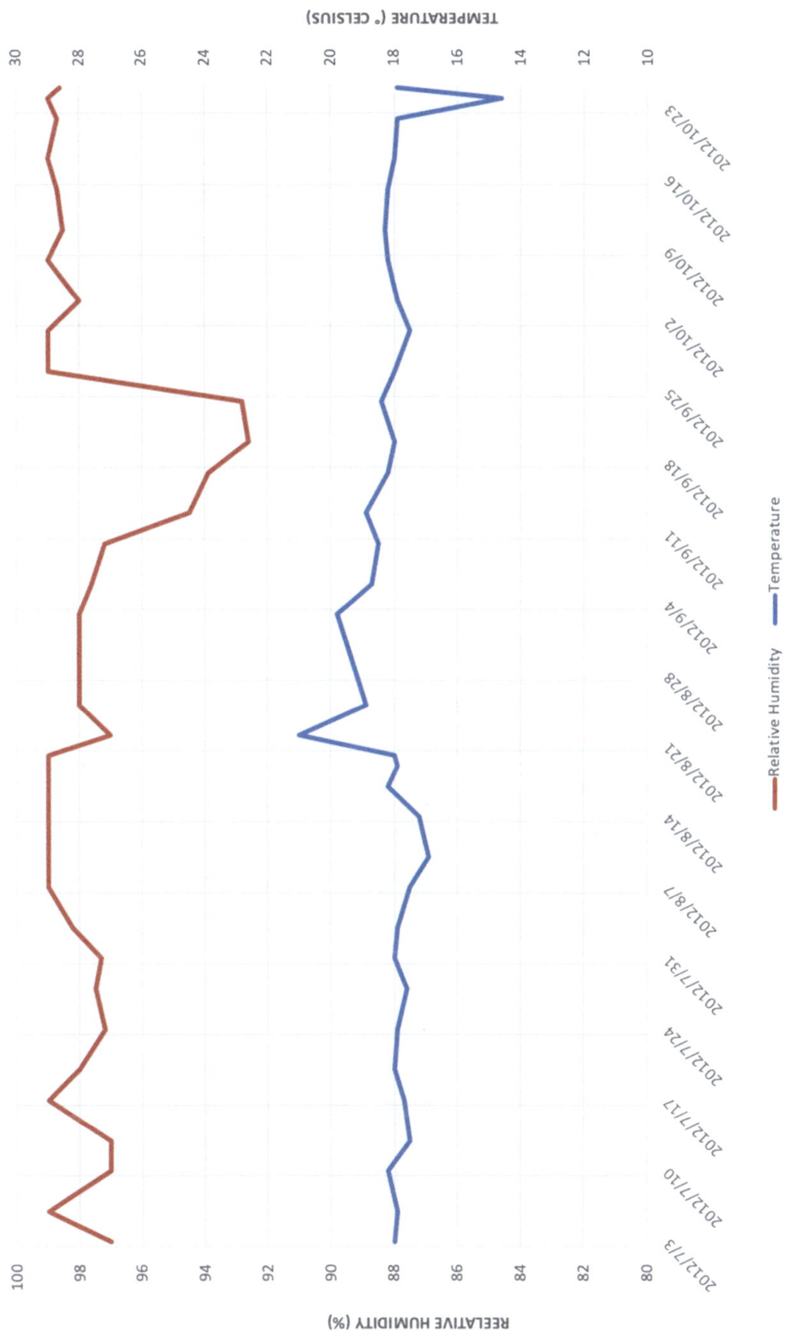

Fig. 23 Relative Humidity and Temperature during the rainy season of 2012 *Source* DPRK, Ministry of Culture

the instruments placed in 2009 and direct observation of inundation inside the Tomb. In the period 2009–2010, two inundation events were studied.

According to the above information it is possible to say that the inundation in the Tomb is correlated to two different factors: the saturation of the mound and the rising of an ephemeral water table. In order to ensure the long-term safeguarding of the Tomb and to prevent flooding it was necessary (1) to waterproof the earthen mound and (2) to maintain the ephemeral water table below the level of the Tomb floor. These are not separate points but have to be considered as elements of the same project.

In order to satisfy the above requirements, the possibility of lowering the ephemeral water table through the use of a drainage trench and of waterproofing the earth mound with a filter allowing the water to freely flow outside the Tomb area was investigated. The project was completed in Spring 2012.

Acknowledgements The authors are deeply acknowledging the UNESCO Office in Beijing and UNESCO headquarter in Paris (France) for supporting the present investigations. A special thanks is to the many friends of DPRK Ministry of Culture, who deeply contributed to the field works and mitigation projects.

References

Luján RL (2006) Report on mission. UNESCO internal document

Luján RL (2008) Report on mission. UNESCO internal document

Luján RL (2009) Report on mission. UNESCO internal document

Margottini C (2010) Report on mission. UNESCO internal document

Onitsuka K, Hara Y, Lu J, Tang X, Chen P (2011) Roots in China of construction methods of ancient Yoshinogari burial mound. Special session ATC19, GeoEngineering for conservation of cultural heritage and historical sites. In: ATC19/Asian technical committee 19 14th Asian regional conference, ISSMGE, Hong Kong.

Paek RJ, Kang HG, Jon GP (1996) Geology of Korea. Institute of Geology, State Academy of Sciences DPR of Korea. Pyon Yang

Structural and Environmental Monitoring

Monitoring Cultural Heritage Sites Affected by Geohazards in Cyprus Using Earth Observation

Kyriacos Themistocleous

Abstract Monitoring geo-hazards of cultural heritage sites and structures as well as the surrounding archeo- landscapes facilitates the early recognition of potential risks and enables effective conservation, monitoring and planning. Landslides, earthquakes and erosion are the predominant geo-hazards in Cyprus. In order to identify and monitor geohazards and environmental displacements, Earth Observation techniques can be used in combination with long-term low-impact monitoring systems, such as UAVs and geodetic techniques to monitor and assess the risk from geohazards on cultural heritage sites and structures to evaluate potential changes and risks.

Keywords Cultural heritage · Geo-hazards · Earth observation monitoring

1 Introduction

Tangible cultural heritage sites are continuously impacted and weathered by several environmental factors, including climate change, precipitation, natural hazards, etc. (Themistocleous et al. 2016a; Agapiou et al. 2015, 2016; Margottini et al. 2016). Cultural heritage sites are highly vulnerable to geological disasters induced by earthquakes, volcanic activity, floods and catastrophic landslides, as well as non-catastrophic geo-hazards that can progressively affect its structural integrity and accessibility including slow-motion landslides, subsidence, sinkholes, ground settlement and active tectonics (Themistocleous 2018; Agapiou et al. 2015, 2016; Margottini et al. 2016; Themistocleous et al. 2016a). Several studies have focused on local-scale assessments of various geohazards on UNESCO World Heritage Sites (Howard 2013, Philips 2015, 2014; Howard et al. 2016; Wang 2015; Schmidt and Rudolff 2013; Lollino et al. 2015; Lanza 2003; Hapciuc et al. 2016; Vojinovic et al. 2016). Pavlova

K. Themistocleous (✉)
Department of Civil Engineering and Geomatics, Cyprus Univeristy of Technology, Limassol, Cyprus
e-mail: k.themistocleous@cut.ac.cy

Eratosthenes Centre of Excellence, Limassol, Cyprus

G. M. El-Qady and C. Margottini (eds.), *Sustainable Conservation of UNESCO and Other Heritage Sites Through Proactive Geosciences*, Springer Geology,
https://doi.org/10.1007/978-3-031-13810-2_19

et al. (2017) identified 60% of the sites on the World Heritage list as being exposed to at least one geological disaster.

In order to examine the major risks to cultural heritage, it is vital to conduct a risk identification and assessment, which includes the history, topography, geology, land use, urban development and any other major risks that may affect the site. Although additional information regarding the geomorphological and geological processes of the area can improve understanding of the risk of natural hazards on cultural heritage sites (Howard 2013), most studies on long- term vulnerability studies on cultural heritage sites do not focus these variables (Tang et al. 2016). A recurring theme of studies on geo-hazards affecting cultural heritage sites is the need to strengthen monitoring and maintenance of vulnerable cultural heritage sites (Sesana et al. 2018). According to Tang et al. (2016), the adoption of combined structural and ground deformation monitoring methodologies facilitates the early recognition of potential risks that enables effective conservation planning. However, although deformation monitoring is essential for preserving significant cultural heritage sites, the published results are sparse (Gutiérrez and Cooper 2002; Rohn et al. 2005; Canuti et al. 2009).

Ecological risk assessment (ERA) can be used to assess the potential geo-hazards that are occurring at a specific cultural heritage site (UNESCO et al. 2010; Yanes et al. 2018). ERA focuses on the potential for producing a hazard and the vulnerability of exposure to a hazard (UNESCO et al. 2010). Therefore, the hazard is the external source of a disaster, and the vulnerability is the inherent weakness (UNESCO et al. 2010). To minimize these consequences and achieve sustainability, ERA can analyse potential and extreme adverse impacts that can be caused by geohazards, making ERA a powerful technical tool (Xu et al. 2015; Liu et al. 2018; Popescu et al. 2015).

In Cyprus, current infrastructure for monitoring and studying these natural hazards is limited to conventional equipment, such as seismographs, geophones, and inclinometers (Geological Survey Department). Thus, no systematic research has been conducted to monitor ground deformation with high accuracy and dense spatial resolution in a timely manner. Several geotechnical field investigations have been conducted by the Geological Survey Department (GSD) of Cyprus in collaboration with the British Geological Survey between 1985 and 1986 in Paphos District, and especially in the villages of Agios Photios, Kannaviou, Pendalia, and Simou, which are areas prone to landslides (Northmore et al. 1986; Northmore et al. 1988). Extensive research has been conducted to study several slope failures in Cyprus through extensive geological and geotechnical investigations as well as interpretation of aerial photographs and high- resolution satellite imagery for observation and verification of the results (Tzouvaras et al. 2020; Charalambous et al. 1997; Hart et al. 2010, 2013).

2 Types of Geohazards

CypruS is in an area of high susceptibility to seismic activities and landslides. The main geotechnical problems that take place are landslides, rock falls, ground subsidence, earthquakes and erosion, which pose serious disadvantages to the development of heritage sites and safety for tourism (Tzouvaras et al. 2020; Liu et al. 2019; Hagenlocher et al. 2018).

2.1 Landslides

Landslides are a major hazard in Cyprus, with large landslides taking place in the south-west part of the island, primarily in the Paphos district (Hart et al. 2013) as seen in Fig. 1. Extreme weather conditions such as drought and heavy rainfall can lead to amplification of the soil erosion processes (Tzouvaras et al. 2020).

In some parts of the Paphos District landslides cover as much as 70 per cent of the landscape. The landslides have been mapped by the Geological Survey Department of Cyprus in collaboration with BGS (Northmore et al. 1987; Northmore et al. 1988) and more recently as part of Scott Wilson's mapping programme. Figure 2 features the Digitized Locations of Landslides in Cyprus as developed by the Geological Survey Department of the Government of Cyprus.

Fig. 1 Landslide, Paphos-Limassol Highway in Cyprus

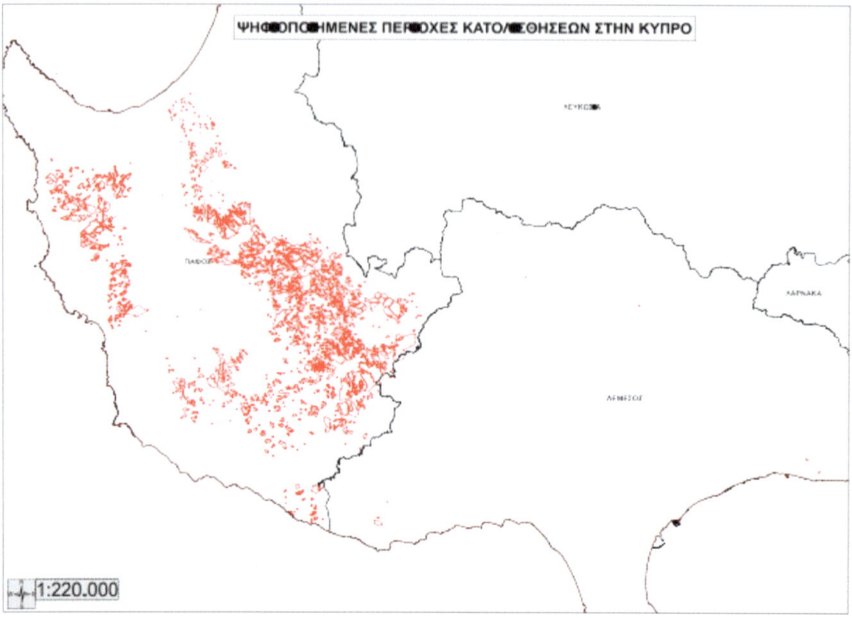

Fig. 2 Digitized Locations of Landslides in Cyprus (*Source* Geological Survey Department)

Landslides have resulted in significant damage to the Pissouri area in the south-west area in Cyprus (Fig. 3). The south western side of Cyprus has a long history of slope instability problems due to its morphological and hydrological conditions. Various factors have contributed to the current pattern of slope instability, such as climate change with intensive rainfall, hydrological conditions, land cover and land use changes (Tzouvaras et al. 2020). In this area, the remains of former sea-floor deposits and massive submarine slides, are exposed in the Paphos District situated between the Troodos Mountains and the sea. These deposits tend to be heavily deformed and are rich in the types of clay minerals that are prone to landslides (Tzouvaras et al. 2020). This tendency is exacerbated by the steep terrain and the long history of powerful earthquakes in the region (British Geological Survey). The terrain is characterised by steep-sided plateaux made up of 'Melange' and 'Kannaviou Clay' capped by thick chalk sequences.

Rockfalls are observed mostly in the mountainous areas of Cyprus and in natural and man-made slopes. They are mainly related to one or more causes of geological, topographical and rainfall conditions. Earthquakes often contribute to the fracturing of the rock mass and eventually to rock falls. Areas covered with loose deposits, including most of the coastal areas are more vulnerable to destruction in contrast to areas where the rocks are massive, like the Troodos range which is mostly covered with igneous rocks. Due to the nature of the main soil types in Cyprus ground which is composed of marls, clays, gypsum and sulfates, ground subsidence is a frequent phenomenon, as it is highly related to the geological conditions and the mechanical properties of the ground itself.

Fig. 3 Landslide in Pissouri area in Cyprus

2.2 *Seismic Hazards*

Cyprus is frequently affected by catastrophic earthquakes, as it lies within the Alpine-Himalayan seismic zone, in which about 15% of the world earthquakes occur (Cagnan and Tanircan 2009). The entire island is considered an earthquake vulnerable area. The seismicity of Cyprus is thought to be due to the "Cyprus Arc", which constitutes the tectonic boundary between the African and Eurasian lithospheric plates in the region, which is situated in the sea to the west and south of Cyprus (Mart and Ryan 2002). Between 1896–2004, more than 400 earthquakes occurred which had their epicentres on Cyprus. As seen in Fig. 4, most earthquakes have their epicentres in the west and south, which affects the coastal areas in the western and southern areas of the island. A large number of medium to strong earthquakes have their epicentres in the sea, several tens of kilometres away from inhabited areas and the damage that may cause is small to negligible, as the intensity with which an area is struck depends on the epicentral distance.

2.3 *Erosion*

In Cyprus, extreme weather conditions such as drought and heavy rainfall can lead to amplification of erosion processes. Rocks eroded through physical erosion often form clastic sediments. Clastic sediments are composed of fragments of older rocks that have been transported from their place of origin (National Geographic 2020). Softer stones such as limestone and soapstone are eroded much more quickly (Sensana et al.

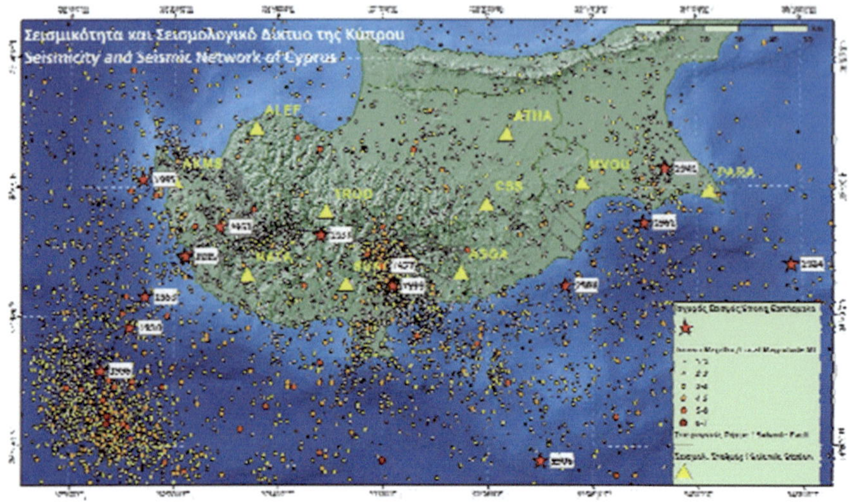

Fig. 4 The Seismicity of Cyprus up to 2018 (*Source* Department of Geological Survey 2018)

2018). Changes in atmospheric acidity, landform development, and vegetation cover rapidly affect the intensity and rate of the erosion process (Fig. 5).

Stone weathering has been investigated on several stone monuments, including studies on the on-going weathering rate are especially effective to evaluate the present condition of stone monuments in order to identify the extent of erosion on the cultural heritage site (Doehne and Price 2010; Winkler 1986). To understand the present weathering condition of target monuments, the first step is to measure the precise shape of the stone surfaces. Erosion rates on rocks can be obtained easily since most stone-built cultural heritages sites tend to have a geometrical shape. As well, such sites are more commonly located in humid temperate areas. North-facing surfaces tend to have lower rates of erosion than surfaces facing in other directions because each surface has different temperature and moisture conditions due to insolation.

Coastal sites have long been a significant focus of archaeological research, particularly in the Mediterranean (Knapp and Demesticha 2017). Several ancient civilizations have existed in the Mediterranean area (Canzenave 2014; Benoit and Comeau 2005; EEA 2014), resulting in a high concentration of cultural heritage sites in the Mediterranean coastal areas. Despite recent theoretical and methodological advances in coastal archaeology, on-going erosion remains a serious predicament for both the preservation and interpretation of coastal sites.

The coastal zone of Cyprus is a valuable and vulnerable area. This zone, in which most urban development and economic activity takes place, covers 23% of the total country's area, 50% of total population and 90% of the tourism industry (EU 2018). According the European Commission's report on climate change in Cyprus, 30% of the coastline under control of the Republic of Cyprus is subject to erosion (EU 2018; Papadaskalopoulou 2014). 38% of the coastline is already subject to erosion, mostly

Fig. 5 Erosion in the Tomb of the Kings region, Paphos (*Source* A photograph taken by Kyriacos Themistocleous in 2017)

the result of human activities such as beach mining, dam and illegal breakwater construction and urbanization (EU 2018).

South-central Cyprus is experiencing the highest rates of coastal erosion on the island (EU 2018; Papadaskalopoulou 2014). Although archaeological sites in this region have received increased attention as a result of systematic surveys (Manning et al. 1994; Manning et al. 2000; Georgiou 2013; Andreou and Sewell 2015; Demesticha 2015), the impact of coastal erosion has only recently been investigated from a cultural heritage perspective (Andreou et al. 2017). Studies such as the Cyprus Ancient Shoreline Project (CASP) are focused on monitoring the impact of coastal erosion on archaeological sites (Andreou 2018). Since 2014, the CASP has compiled archaeological and geographical information through aerial photographs to reconstruct historical coastlines in this region digitally, and to map sites that are under threat. More recently, this dataset was examined to classify the severity of coastal erosion, and to highlight the areas that require more frequent recording (Thieler et al. 2009; Radosavljevic et al. 2016; O'Rourke 2017).

3 The Case Study of Choirokoitia, Cyprus

The documentation site is the UNESCO World Heritage Site of Choirokoitia in Cyprus (Fig. 6). The Neolithic settlement of Choirokoitia is considered one of the most important prehistoric sites in the Eastern Mediterranean (UNESCO n.d.) as it is one of the best-preserved settlements of this period in Cyprus and the Eastern Mediterranean. The Neolithic settlement of Choirokoitia is located in the District of Larnaka, 6 km from the southern coast of Cyprus and lies on the slopes of a hill.

The site, which was occupied from the 7th to the 5th millennium B.C., is a village that covers an area of approximately 3 ha at its maximum extent. The site is an example of the Aceramic Neolithic period in of Cyprus at its pinnacle, where the first organized human community was developed. The site was initially.

populated by farmers originating from the Near East mainland around the beginning of the 9th millennium.

Since only part of the site has been excavated, it forms an exceptional archaeological reserve for future study. To date, 20 houses have been excavated which were constructed with limestone, clay and brick (Fig. 7). According to UNESCO, the site was officially abandoned in the 4th millennium BC. The reason for this abandonment remains unknown (UNESCO n.d.).

Fig. 6 Aerial photo of the Choirokoitia, Cyprus

Fig. 7 Excavated houses in Choirokoitia

3.1 Methods

In order to identify and monitor geo-hazards and their severity, local scale monitoring is used. Local scale monitoring provides the opportunity to identify deformation phenomena resulting from geo-hazards for monitoring, assessing and predicting geo-hazards using field survey techniques to measure and document the extent of damage on the cultural heritage site. This method integrates field monitoring techniques such as UAVs, laser scanning, total stations and GPS as well as satellite techniques such as InSar data (Margottini et al. 2015, 2018; Novellino et al. 2018; Themistocleous 2017; Themistocleous et al. 2017a).

Monitoring cultural heritage sites and monuments in Europe has traditionally used on-site observation. However, this procedure can be time consuming, difficult and expensive, especially in large or remote areas, since it includes periodical observations, field surveying and ground-based data collection (Themistocleous et al. 2016a). Deformation monitoring in cultural heritage sites is often carried out by installing electrical sensors in selected structures with automatic systems for data acquisition and recording or by using portable instruments with manual reading of data taken at fixed time intervals (Zhou et al. 2015; Garziera et al. 2007; Glisic and Inaudi 2008). However, such methods can only acquire data of the monitored structure within the cultural heritage sites and does not include the entire area of the site and its surrounding landscape (Zhou et al. 2015). Moreover, the installation of monitoring devices, such as optical targets, permanent GNSS stations or inclinometers, on the heritage sites and monuments can affect the integrity the heritage site.

Space-based Earth Observation (EO) monitoring techniques can be used to identify natural hazards (Casagli et al. 2016). Research indicates that the integration of satellite data and in-situ data offers the best solution for monitoring geo-hazards in cultural heritage sites (Margottini et al. 2015, 2018; Novellino et al. 2018). Local scale monitoring can be used to assess the severity of these geo-hazards by using integrated field monitoring techniques while satellite radar interferometry is capable of monitoring surface deformation with high accuracy using precise ground measurements. Once cultural heritage sites vulnerable to geo-hazards are identified using InSAR satellite imagery, local-scale monitoring can be used to assess the severity and monitor the cultural heritage sites over time. Ground-based geotechnical monitoring and surveying models are used to measure deformation parameters (e.g. strain and inclination) over a relatively short measurement base. In-situ measurements using UAV and GPS are then used to measure possible movements that may result from geo-hazards (Themistocleous 2017a, b, Gikas 2012; Themistocleous 2017) Field monitoring can be combined with UAVs for documentation purposes and 3D modeling comparison. The aerial imagery obtained from UAVs can be processed using photogrammetry methods, such as Structure in Motion to generate highly accurate point cloud models in order to document and monitor the extent of progressing geo-hazards. In this way, areas exposed to geo-hazards and their evolution over time can be monitored and crucial information can be provided to decision makers in order to protect cultural and heritage sites from natural hazards.

3.2 Methodology

A methodology for local-scale monitoring of cultural heritage sites was developed during the PROTHEGO study that would assess the risk from natural hazards on the archaeological sites and monuments from a geospatial perspective (Fig. 8) (Margottini et al. 2016, 2018; Themistocleous et al. 2017a). A multi-criteria analysis of the UNESCO sites was conducted to estimate the severity of each geo- hazard (Silvestrou and Themistocleous 2018). The main objective of the methodology was the development of long-term low-impact monitoring systems as well as indirect analysis of environmental contexts to investigate changes and decay of structure, material and landscape (Themistocleous et al. 2016a; Themistocleous 2018a).

In order to test the PROTHEGO study, the methodology developed was applied to the Choirokoitia World Heritage site in Cyprus to assess the risk from geo- hazards on the archaeological sites and monuments from a geospatial perspective. The methodology focused on long-term, low-impact monitoring systems as well as indirect analysis of environmental context to investigate changes and decay of a structure or landscape (Themistocleous et al. 2016b). The methodology for the local scale monitoring begins with the use of InSAR images to identify natural hazards in the UNESCO World Heritage demonstration sites. When InSAR ground motion data indicates that

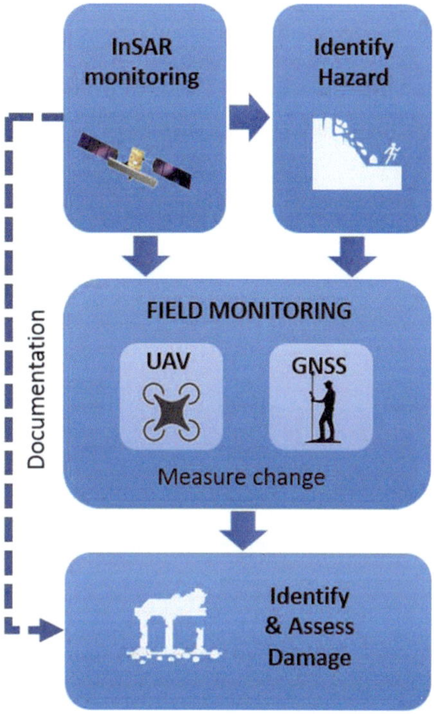

Fig. 8 PROTHEGO methodology

a geo-hazard is evident at or near the demonstration site, field monitoring and verification is used to document and measure the extent of change caused by the natural and/or geo-hazards, if any. Documentation of the damage can be performed, using UAVs and photogrammetry.

A GNSS network was established at the Choirokoitia site in order to monitor the site from ground displacement (Fig. 9). The geodetic network consisted of four points in total. The main control point was established on solid ground outside the area of interest, whilst the remaining three points were set up in carefully selected locations (i.e. on top of rocks or ridge lines) after consulting geologists and archaeologists to address high-risk areas within the site.

Two types of GNSS receivers were used for data acquisition; (a) 3 Trimble R9s equipped with Zephyr 2 Geodetic GNSS antennas, and a Leica GS15 Smart GNSS receiver were used to monitor displacement using satellite (GNSS) and combined with conventional techniques (via high precision total stations and levels) to identify potential ground displacements with respect to the network reference points, during the life-span of the monitoring activity. Horizontal displacements were measured using an industrial-grade total station (Topcon® MS05AXII), which bears nominal accuracy in directions and ranges of 0.5″ and 0.5 mm respectively when combined with specifically designed topographic prisms and reflective targets. Vertical motion was measured using a high-precision digital level (Leica Geosystems® DNA03). The

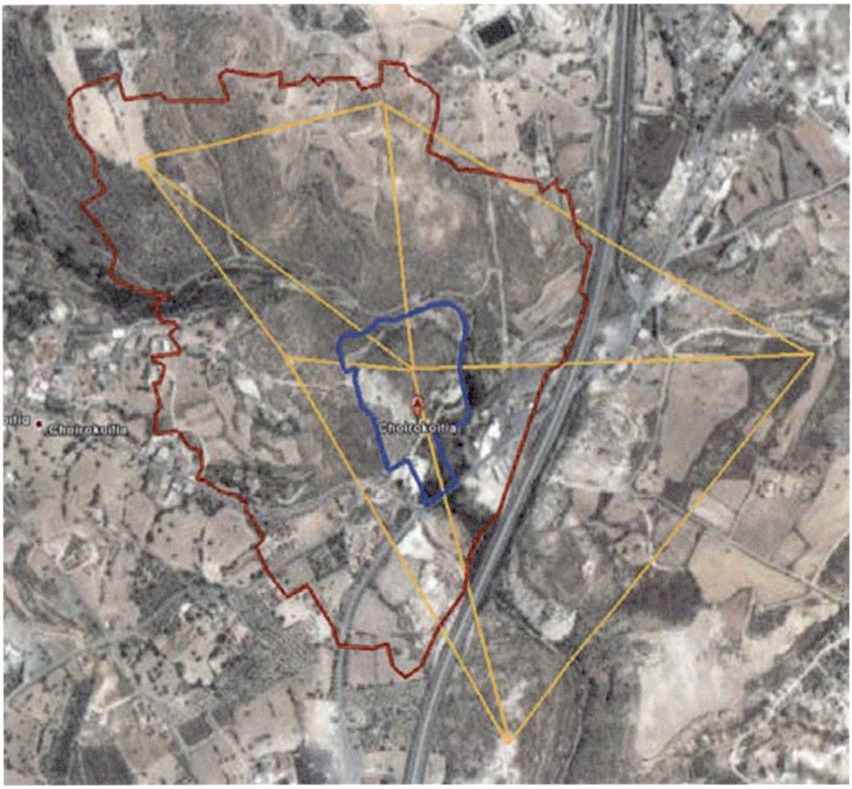

Fig. 9 Geodetic Network at the Choirokoitia area

leveling campaign were carried out using Invar Barcode Staffs, achieving a vertical accuracy at the order of 0.3 mm/km. Measurements for calibration of these products are taken using GNSS and total station. After the change is identified using field verification, InSAR images are again used to verify and assess the extent of the damage to the cultural heritage site (Themistocleous et al. 2018b).

Images generated from UAVs are used to document cultural heritage sites under threat from geo-hazards (Fig. 10). Hundred of UAV images are used to create orthophotos, dense point clouds, 3D model and Digital Elevation Models (Themistocleous 2017). It is recommended that UAVs be equipped with a high- resolution camera to acquire images over a site with fixed ground control points for geo-referencing in order to produce a photogrammetric ortho-image and point.

Fig. 10 UAV flight over Choirokoitia area

3.3 Results

To support field monitoring, geometric documentation of the area was performed using RTK GPS measurements (ground control points) with images taken by UAV systems and processed using photogrammetry techniques. This data was supported and geo-referenced using a geodetic network measured via contemporary GPS/GNSS receivers by reconstructing the cross-sections over the identified areas of the site in order to investigate possible changes in the vertical and horizontal profiles of the cultural heritage site. As part of the local-scale monitoring process of the Choirokoitia demonstration site, a UAV fitted with a 20 MP camera was used to acquire images of the Choirokoitia site over the course of 13 months that were processed using photogrammetry to produce Digital Elevation Models (DEM) and ortho-images for comparison over temporal intervals (Themistocleous 2019).

The digital images acquired by the UAV flight were interpolated in order to create high resolution, scaled, geo-referenced 3D model based on photogrammetric techniques. Using images from the UAV flights ortho- photos were created. Following the generation of point clouds from the UAV images using a distribution of ground control points (GCP) throughout the site, the model had sub-centimeter accuracy, which was critical in order to compare the 3D model with the GNSS measurements.

Following, Digital Elevation Models (DEMs) (Fig. 11) and Elevation Contour Maps were generated to examine changes and deformation in the Choirokoitia cultural heritage site over time. The DEMs were generated from the images acquired from the UAV were processed based on the position of the GCPs and compare the results with the GNSS measurements and the InSAR images. During the 13-month monitoring period, all the measurements were recorded and showed a slight shift at the top peak of the hill, at the PILR.

During the two GPS/GNSS campaigns of the geodetic network in the area of interest within a 5-month period, GNSS observations of 1 Hz were collected for a timespan of six hours. The four GNSS network points were measured to compute displacement in the East (DE), North (DN) and Up (DU) directions. The displacements were derived from topocentric coordinates with respect to the main network control point. The results of the GPS/GNSS campaigns indicate a change of 2 cm in

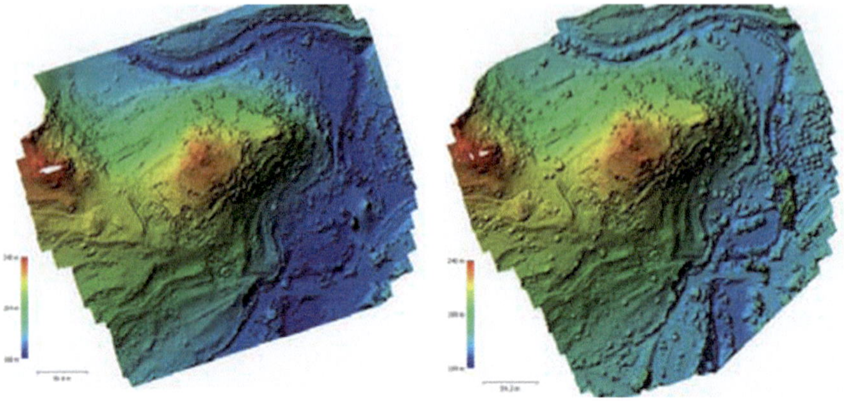

Fig. 11 DEM of Choirokoitia site 11/11/2017 and 7/3/2018

the Up component (vertical) within the 5- month period. Furthermore, a PSI (Persistent Scatterer Interferometry) analysis was conducted on the Choirokoitia broader area to determine potential displacements using 26 Cosmos Skymed SAR images from the years 2011–2017. During this time span, the points exhibit an average velocity of 3.3 cm per year. The results of the PSI analysis reveal a similar displacement pattern at the broader area occupied by the GNSS control network. As well, a displacement of 2 cm was identified using a UAV model cross section comparison for the UAV images generated from 2.

February 2017 until 11 November 2017 feature. Longer- term monitoring of the cultural heritage site is required in order to identify and correlate the two techniques and measure and monitor the severity of the displacements (Fig. 12).

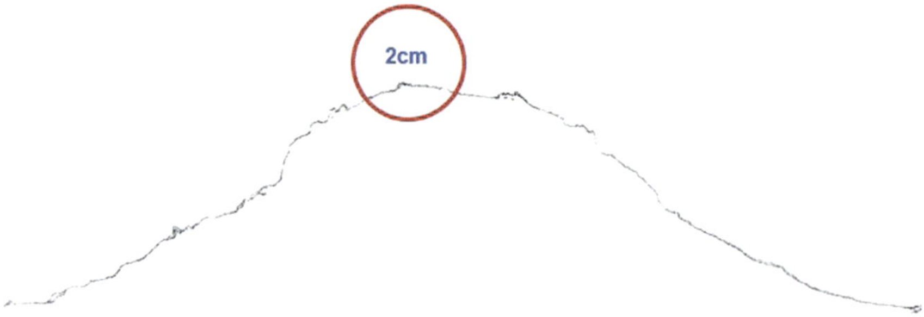

Fig. 12 UAV model comparison of displacement (*Source* An image created by Kyriacos Themistocleous in 2019)

4 Conclusions

The case study at Choirokoitia, Cyprus provides an example of detecting and analyzing geo-hazard induced ground deformation based on InSAR ground motion data and field survey techniques for cultural heritage applications. InSAR data, satellite positioning and conventional surveying techniques were employed to measure the micro-movements, while the UAV and photogrammetry were used for documentation purposes and 3D modeling comparisons. A correlation is evident between the geodetic techniques and SAR images, as the PSI analysis and GNSS Control Network of the Choirokoitia site exhibited similar levels of displacement suggesting that longer-term monitoring of the site is required to diagnose the severity of the problem. Furthermore, local-scale monitoring data forms the basis for the development of geological and geotechnical models of the kinematics of the investigated sites. Such information can provide detailed models highlighting the evolution of deformation processes affecting the heritage sites and thereby identifying the best mitigation strategy and to evaluate the effectiveness of these actions for the test site. Future research can further develop the methodology for identifying geo-hazards by using multi- sensor co-located configurations, including permanent GNSS reference stations, weather stations, tiltmeters as well as specifically designed Corner Reflectors established throughout Cyprus to estimate deformation with dense spatial resolution at the cm- to mm-level using corner reflectors. This will enable effective and accurate surveillance of geo-hazards, which can be used to provide early warning services, risk management, and mitigation of the impact of natural hazards on cultural natural heritage sites. In addition, Ecological risk assessment can be used to provide a model of the vulnerability of the cultural heritage site as well as the severity of the impact of geo-hazards on the site.

Acknowledgements This paper was developed within the framework of the EXCELSIOR project, that has received funding from the European Union's Horizon 2020 Research and Innovation Programme, under Grant Agreement no 723583.

The "PROTection of European Cultural HEritage from GeO-hazards (PROTHEGO)" project HERITAGE PLUS/0314/36 is funded in the framework of the Joint Programming Initiative on Cultural Heritage and Global Change (JPICH)—HERITAGE PLUS under ERA-NET Plus and the Seventh Framework Programme (FP7) of the European Commission and the Cyprus Research Promotion Foundation, contract KOINA/ΠΚΠ-HERITAGE PLUS/0314/36.

References

Agapiou A, Lysandrou V, Themistocleous K, Hadjimitsis DG (2016) Risk assessment of cultural heritage sites clusters using satellite imagery and gis: the case study of Paphos District, Cyprus. J Int Soc Preven Mitigation Nat Haz 2:1–15. https://doi.org/10.1007/s11069-016-2211-6

Agapiou A, Lysandrou V, Alexakis DD, Themistocleous K, Cuca B, Sarris A, Argyrou N, Hadjimitsis DG (2015) Cultural heritage management and monitoring using remote sensing data and gis: the case study of Paphos area, Cyprus. Comput Environ Urban Syst 54:230–239. https://doi.org/10.1016/j.compenvurbsys.2015.09.003

Andreou GM, Sewell DA (2015) Tochni-Lakkia revealed: reconsidering settlement patterns in the Vasilikos and Maroni Valleys, Cyprus. In: Matthäus H, Morstadt B, Vonhoff C (ed) PoCA (Postgraduate Cypriot Archaeology) 2012, pp 198–219. Newcastle: Cambridge Scholars

Andreou GM, Opitz R, Manning SW, Fisher KD, Sewell DA, Georgiou A, Urban T (2017) Integrated methods for understanding and monitoring the loss of coastal archaeological sites: the case of Tochni-Lakkia, Cyprus. J Archaeol Sci Rep 12:197–208. https://doi.org/10.1016/j.jasrep.2017.01.025

Andreou G (2018) Monitoring the impact of coastal erosion on archaeological sites: the Cyprus ancient shoreline project. Antiquity 92(361):E4. https://doi.org/10.15184/aqy.2018.1

Benoit G, Comeau A (2005) A sustainable future for the Mediterranean. The Blue Plan's environment and development outlook (Earthscan, London, 2005)

British Geological Survey, Landslides in Cyprus. https://www.bgs.ac.uk/landslides/cyprus.html

Cagnan Z, Tanircan G (2009) Seismic hazard assessment for Cyprus. J Seismolog 14:225–246. https://doi.org/10.1007/s10950-009-9163-1

Canuti P, Margottini C, Fanti R, Bromhead EN (2009) Cultural heritage and landslides: research for risk prevention and conservation. Landslides–Disaster Risk Reduction. Springer: Berlin Heidelberg, pp 401–433

Cazenave A (2014) Anthropogenic global warming threatens world cultural heritage. Environ Res Lett 9:51001

Casagli N, Cigna F, Bianchini S, Hölbling D, Füreder P, Righini G, Del Conte S, Friedl B, Schneiderbauer S, Iasio C, Vlcko J, Greif V, Proske H, Granica K, Falco S, Lozzi S, Mora O, Arnaud A, Novali F, Bianchi M (2016) Landslide mapping and monitoring by using radar and optical remote sensing: examples from the EC-FP7 project SAFER. Remote Sens Appl Soc Environ 4:92–108. https://doi.org/10.1016/j.rsase.2016.07.001

Charalambous M, Petrides G (1997) Contribution of engineering geology in the planning and development of landslide prone rural areas in Cyprus. In: Engineering geology and the environment; Balkema: London, UK, pp 1205–1210

Demesticha S (2015) Gone with the waves: scattered Roman amphorae in shallow waters around Cape Kiti, Cyprus. In: Demesticha S (ed) Per terram, per mare: seaborne trade and the distribution of Roman amphorae in the Eastern and Central Mediterranean. Uppsala: Åström, pp 55–76

Doehne E, Price CA (2010) Stone conservation: an overview of current (second edition). The Getty Conservation Institute, Los Angels

European Commission Report (2018) Cyprus Climate Change

European Environment Agency (EEA). Horizon 2020 Mediterranean report. Toward shared environmental information systems—EEA-UNEP/MAP joint report (Publications Office of the European Union, Lanza, S. G. Flood hazard threat on cultural heritage in Luxembourg, 2014). the town of Genoa (Italy). J Cult Herit 4:159–167 (2003)

Garziera R, Amabili M, Collini L (2007) Structural health monitoring techniques for historical buildings. In: Proceedings of the 4th Pan-American conference for non-destructive testing, Buenos Aires, Argentina, 22–26 October 2007

Georgiou G (2013) An Early Christian baptistery in the south coast of Cyprus. Cahier Du Centre D'études Chypriotes 43:117–126. https://doi.org/10.3406/cchyp2013.1058

Gikas V (2012) 3D terrestrial laser scanning for geometry documentation and construction management of highway tunnels during excavation. Sensors 12:11249–11270

Glisic B, Inaudi D (2008) Fiber optic methods for structural health monitoring. Wiley, Hoboken, NJ, USA

Gutiérrez F, Cooper AH (2002) Evaporite dissolution subsidence in the historical city of Calatayud, Spain: damage appraisal and prevention. Nat Haz 25(3):259–288

Hagenlocher M, Renaud FG, Haas S, Sebesvari Z (2018) Vulnerability and risk of deltaic social-ecological systems exposed to multiple hazards. Sci Total Environ 631–632:71–80

Hapciuc O-E et al (2016) Flood susceptibility analysis of the cultural heritage in the Sucevita catchment (Romania). Int J Conserv Sci 7:501–510

Hart AB, Ruse ME, Hobbs PRN, Efthymiou M, Hadjicharalambous K (2010) Assessment of landslide susceptibility in Paphos District, Cyprus. In: GRSG-AGM geoenvironmental remote sensing conference; the geological society of London: London, UK
Howard AJ et al (2016) Assessing riverine threats to heritage assets posed by future climate change through a geomorphological approach and predictive modelling in the Derwent Valley Mills WHS. UK. J Cult Herit 19:387–394
Howard AJ (2013) Managing global heritage in the face of future climate change. The importance of understanding geological and geomorphological processes and hazards. Int J Herit Stud 19:632–658
Knapp AB, Demesticha S (2017) Mediterranean connections: maritime transport containers and seaborne trade in the Bronze and Early Iron Ages. Routledge, London
Liu Q, Yang Z, Shi H, Wang Z (2019) Ecological risk assessment of geohazards in Natural World Heritage Sites: an empirical analysis of Bogda. Tianshan, Open Geosci 11(1):327–340. https://doi.org/10.1515/geo-2019-0026
Liu D, Liang XY, Chen H, Zhang H, Mao NZ (2018) A quantitative assessment of comprehensive ecological risk for a loess erosion gully: a case study of Dujiashi Gully, Northern Shaanxi Province, China. Sustainability 10(9)
Lollino G et al (eds) (2015) Engineering geology for society and territory—Volume 8. Springer International Publishing, Cham
Manning SW, Monks SJ, Sewell DA, Demesticha S (2000) Late Roman type 1A amphora production at the Late Roman site of Zygi-Petrini, Cyprus. Report of the Department of Antiquities, Cyprus, 233–57
Manning SW, Collon D, Conwell DH, Jansen H-G, Steel L, Swinton A (1994) Tsaroukkas, Mycenaeans and Trade' project: preliminary report on the 1993 season. Report of the Department of Antiquities, Cyprus, 83–106
Margottini C, Antidze N, Corominas J, Crosta GB, Frattini P, Gigli G, Giordan D, Iwasaky I, Lollino G, Manconi A, Marinos P, Scavia C, Sonnessa A, Spizzichino D, Vacheishvili N (2015) Landslide hazard, monitoring and conservation strategy for the safeguard of Vardzia Byzantine monastery complex. Landslides 12(1):193–204. https://doi.org/10.1007/s10346-014-0548-z
Margottini C, Spizzichino D, Cigna F, Crosta GB, Frattini P, Themistocleous K, Fernandez Merodo JA (2016) European UNESCO Cultural Heritage and Geo–Hazards: The PROTHEGO Project. In: Proceedings of the fourth international conference on remote sensing and geo-information of environment, 4–8 April, 2016, Paphos, Cyprus
Margottini C, Spizzichino D, Leoni G, Bee EJ, Crosta GB, Frattini P, Themistocleous K, Fernandez Merodo JA (2018) Satellite monitoring applied to natural hazards and cultural heritage: the PROTHEGO project. Sixth international conference on remote sensing and geo-information of environment, 26–29 March, 2018, Paphos, Cyprus
Mart Y, Ryan W (2002) The complex tectonic regime of the Cyprus Arc: a short review. Israel J Earth Sci 51:117–134. https://doi.org/10.1560/DCF4-08Q2-UF1U-6QK5, https://www.nationalgeographic.org/encyclopedia/erosion/. Accessed 25 June 2020
Northmore KJ, Charalambous M, Hobbs PRN, Petrides G (1986) Engineering geology of the Kannaviou, 'Melange' and Mamonia complex formations—Phiti/Statos Area, S W Cyprus: engineering geology of cohesive soils associated with ophiolites, with particular reference to Cyprus; report of the EGARP research group British geological survey, No. EGARP-KW 86/4; Keyworth, Nottinghamshire, UK; Report of the Geological Survey Department of Cyprus, No. G/EG/15; Nicosia, Cyprus
Northmore KJ, Charalambous M, Hobbs PRN, Petrides G (1988) Complex Landslides in the Kannaviou, Melange, and Mamonia Formations of South-West Cyprus. In: Bonnard C (ed) 5th International Symposium on Landslides. Balkema Publishers, Brookfield, WI, pp 263–268
Novellino A, Harrison A, Bee E, Wang L, Hobbs P, Margottini C, Themistocleous K, Crosta GB, Fernandez Merodo JA (2018) Characterising and modelling geohazard susceptibility in the UNESCO World Heritage Site of the Derwent Valley Mills (UK). Geophysical Research Abstracts, vol 20, EGU2018- 13316-1, 2018, EGU General Assem-bly 2018, Vienna, 8–13 April 2018

O'Rourke MJE (2017) Archaeological site vulnerability modelling: the influence of high impact storm events on models of shoreline erosion in the western Canadian Arctic

Papadaskalopoulou C (2014) Assessment of Cyprus vulnerability to climate change, ADAPTto-CLIMATE Conference, NIcosia, March 27 2014

Pavlova I, Makarigakis A, Depret T, Jomelli V (2017) Global overview of the geological hazard exposure and disaster risk awareness at world heritage sites. J Cult Herit 28:151–157

Phillips H (2015) The capacity to adapt to climate change at heritage sites—the development of a conceptual framework. Environ Sci Policy 47:118–125

Phillips H (2014) Adaptation to climate change at UK World Heritage Sites. Progress and challenges. Hist Environ Policy Pract 5:288–299

Popescu ME, Trandafir AC, Federico A (2015) Risk assessment of slope instability related Geohazards. In: Numerical methods for reliability and safety assessment: multiscale and multiphysics systems. Kadry S, El Hami A (Eds). Springer International Publishing: Cham, pp 243–269

Radosavljevic B, Lantuit H, Pollard W, Overduin P, Coutur N, Sachs T, Helm V, Fritz M (2016) Erosion and flooding—threats to coastal infrastructure in the Arctic: a case study from Herschel Island, Yukon Territory, Canada. Estuaries Coasts 39:900–915. https://doi.org/10.1007/s12237-015-0046-0

Rohn J, Ehret D, Moser M, Czurda K (2005) Prehistoric and recent mass movements of the World Cultural Heritage Site Hallstatt, Austria. Env Geol 47(5):702–714

Schmidt M, Rudolff B (2013) Climate change and cultural heritage: findings of a rapid vulnerability assessment of cultural heritage in the Republic of Macedonia and recommendations towards National Strategies for Cultural Heritage Protection in the context of Climate Change. Project report. http://docshare04.docshare.tips/files/24286/242864272.pdf

Sesana E, Gagnon A, Bertolin C, Hughes J (2018) Adapting cultural heritage to climate change risks: perspectives of cultural heritage experts in Europe. Geosciences 8:305. https://doi.org/10.3390/geosciences8080305

Silvestrou A, Themistocleous K (2018) Multi-critera analysis of UNESCO sites in Cyprus: The case study of Choirokoitia. Sixth international conference on remote sensing and geo-information of environment, 26–29 March 2018, Paphos, Cyprus

Tang P, Chen F, Zhu X, Zhou W (2016) Monitoring cultural heritage sites with advanced multi-temporal InSAR technique: the case study of the summer palace. Remote Sens 8:432. https://doi.org/10.3390/rs8050432

Themistocleous K, Agapiou A, Hadjimitsis DG (2016a) Experiencing cultural heritage sites using 3D modeling for the visually impaired. In:: Proceedings of Euromed 2016: digital heritage, progress in cultural heritage: documentation, preservation, and protection, Lecture Notes in Computer Science, vol 10059, 171–177, 31 October–4 November 2016 Nicosia Cyprus

Themistocleous K, Cuca B, Agapiou A, Lysandrou V, Tzouvaras M, Michaelides S, Hadjimitsis DG, Margottini C, Cigna F, Crosta G, Fernandez Merodo JA (2016b) The protection of cultural heritage sites from geo-hazards. In: Proceedings of the European geosciences union general assembly 2016, 17–22 April 2016, Vienna, Austria

Themistocleous K (2017) The use Of UAVs to monitor archeological sites: the case study of Choirokoitia within the PROTHEGO Project. Fifth international conference on remote sensing and geo-information of environment, 20–23 March 2017, Paphos, Cyprus

Themistocleous K, Danezis C, Mendonidis E, Lymperopoulou E (2017) Monitoring ground deformation of cultural heritage sites using UAVs and geodetic techniques: the case study of Choirokoitia, JPI PROTHEGO project. Proceedings of the SPIE remote sensing conference, vol 10428, 104280Q, Earth resources and environmental remote sensing/GIS applications VIII, 11-–14 September 2017. Warsaw, Poland. https://doi.org/10.1117/12.2279478

Themistocleous K, Hadjimitsis DG, Michaelides S, Spizzichino DJ, Crosta GB, Fernandez Merodo JA, Bee E (2018a) Best practices for monitoring, mitigation and preservation of cultural heritage sites affected by geo-hazards. Geophys Res Abstracts, vol 20, EGU2018a-18220-1, 2018a EGU general assembly 2018, Vienna, 8–13 April 2018

Themistocleous K, Michaelides S, Hadjimitsis DG (2018b) The integrated use of space technologies, UAVs and field measurements intended for cultural heritage in Cyprus. Geophys Res Abstracts, vol 20, EGU2018b-5282-1, 2018b EGU General Assembly 2018, Vienna

Themistocleous K (2019) DEM modeling using RGB- Based Vegetation Indices from UAV Images. Seventh international conference on remote sensing and geo- information of environment, 18–21 March, 2019, Paphos, Cyprus

Thieler ER, Himmelstoss EA, Zichichi JL, Ayhan E (2009) Digital shoreline analysis system (DSAS) version 4.0—an ArcGIS extension for calculating shoreline change: US geological survey open-file report 2008–1278. Reston (VA): US Geological Survey

Tzouvaras M, Danezis C, Hadjimitsis D (2020) Small scale landslide detection using Sentinel-1 interferometric SAR coherence. Remote Sensing 12:1560. https://doi.org/10.3390/rs12101560

UNESCO, ICCROM, I.U.C.N. ICOMOS, Managing Disaster Risks for World Heritage, UNESCO, Paris, France (2010)

Vojinovic Z et al (2016) Holistic approach to flood risk assessment in areas with cultural heritage. A practical application in Ayutthaya, Thailand. Nat Hazards 81:589–616

Wang J-J (2015) Flood risk maps to cultural heritage. Meas. Process J Cult Herit 16:210–220

Yanes A, Botero CM, Arrizabalaga M, Vásquez JG (2018) Methodological proposal for ecological risk assessment of the coastal zone of Antioquia, Colombia. Ecol Eng

Xu XG, Xu LF, Yan L, Ma LY, Lu YL (2015) Integrated regional ecological risk assessment of multi-ecosystems under multi-disasters: a case study of China. Environ Earth Sci 74(1):747–758

Zhou W, Chen F, Guo H (2015) Differential radar interferometry for structural and ground deformation monitoring: a new tool for the conservation and sustainability of cultural heritage sites. Sustainability (Basel), ISSN 2071-1050, 1712–1729 https://doi.org/10.3390/su7021712

Innovative Structural Monitoring as Tool of Preservation and Valorisation of Monumental Architectures: The Case of Neptune Temple in Paestum (Salerno, Southern Italy)

Petti Luigi, Barone Fabrizio, Domenico Greco, and Gabriel Zuchtriegel

Abstract The knowledge of the real conditions and their evolution over time is the basis for maintenance processes aimed at the protection and conservation of monumental architectures. In complex conditions, including zones prone, in example, to seismic or hydrogeological hazard, the knowledge of the static and dynamic behaviour of the structures can be achieved using vibration monitoring systems by means of innovative devices and real-time analysis procedures. Currently, the real limit of this methodology is the need to have distributed and modular monitoring systems, adaptable and optimized for the specific structure, with non-invasive sensors (materials, dimensions and weights) characterized by high bandwidths and sensitivities. The paper describes the innovative monitoring project of the Neptune Temple carried out by the Archaeological Park of Paestum and Velia (PAEVE) and by the University of Salerno (UNISA), based on high sensitivity large band sensors developed at UNISA for a large variety of applications of monitoring and control. The monitoring system is a distributed modular continuous data acquisition and storage system, designed to provide real-time information on the dynamic response of monumental architectures due both to natural and human-induced phenomena. It will lead to carry out a research framework aimed at a synergic centralization of the contribution of researchers on the analysis of complex aspects that describe monumental structures as support tool of management plans.

Keywords Structural monitoring · Monumental architecture · Maintenance management

P. Luigi (✉) · B. Fabrizio · D. Greco
University of Salerno, Via Giovanni Paolo II, 132 – 84084, Fisciano, Italy
e-mail: petti@unisa.it

G. Zuchtriegel
Archaeological Park of Paestum and Velia (PAEVE), Capaccio - Paestum Via Magna Grecia 919, Capaccio, Italy

G. M. El-Qady and C. Margottini (eds.), *Sustainable Conservation of UNESCO and Other Heritage Sites Through Proactive Geosciences*, Springer Geology,
https://doi.org/10.1007/978-3-031-13810-2_20

1 Introduction

The recent negative natural and human-induced events, i.e., include Typhoon in West Japan in 2018, Earthquakes in Central Mexico in 2017, Kumamoto in Japan, Central Italy and Myanmar in 2016, Nepal earthquake in 2015, UK floods in 2015, Balkan floods in 2014 (UNESCO 2018), highlight the relevance of the maintenance process management, which should be approached with a dynamic vision, especially by considering the climate change impacts as well (UNESCO 2017a, b). For this purpose, the knowledge of the goodness (artifacts, monuments, sites, etc.) behaviour evolution plays the main role in the assessment and decision process (Petti et al. 2018c; Baratta et al. 2019). Therefore, real time monitoring techniques are of fundamental importance to support the assessment process of the behaviour of the Cultural Heritage continuously over time, from structural point of view as well.

Furthermore, the recent technological progress in this discipline lead to develop excellent broadband high sensitivity mechanical sensors.

Recently, a wide experimental campaign was carried out on the Trajan arch in Benevento (Italy) (Petti et al. 2017a, b), whose results were applied to the design of an innovative monitoring network of the temple of Neptune in the Archaeological Park of Paestum and Velia (Italy) (Petti et al. 2018a, 2019a).

The monitoring system implemented for the Trajan arch is an application of a general adaptive assessment strategy, with the aim to get an in-deep knowledge of the monument construction details by means of non-destructive investigations. This knowledge was obtained with an optimized design and implementation of a continuous and distributed monitoring of all its important structural elements, in connection with a careful analysis based on an adaptive structural FEM (Finite Element Method) dynamic model of the Arch (Fig. 1). With this process, it was possible to assess both the dynamic and static behaviours of the arch by using anthropic (traffic) and natural (wind) noise, such as to provide reliable description and clarification of the existing damage pattern on its sculptures and structural elements.

This above-mentioned strategy required the implementation and optimization of a numerical finite element model (FEM) describing the dynamical behaviour of the Trajan Arch based on measurements obtained with an adaptive monitoring system (Barone et al. 2015/1) as described in the following Fig. 2.

In the case of the Neptune temple in Paestum, several on site and laboratory tests were carried out to design the best strategy to monitor the Temple by means of an advanced sensors network. The tests, part of a joint research among the University of Salerno (IT), the University of Kassel (D) (DFG Project, "Tendon System for Protecting Ancient Column Structures", *TeSSPACS*), the University of Roma "La Sapienza" (IT) and the Archaeological Park of Paestum and Velia (PAEVE), consider on site accelerometric and seismic measures, laboratory medium and small-scale tests on columns and blocks sub-structures (Obón Santacana et al. 2017; Petti et al. 2018b, 2019b).

With regard to the on-site monitoring investigations, the structure of the Temple and the soil conditions were mainly investigated respectively in the 2016 and 2019

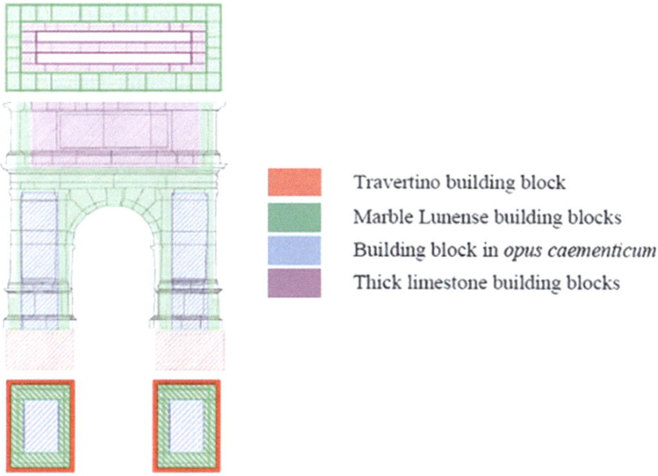

Fig. 1 Assessment of materials and construction details (Petti et al. 2017a, b)

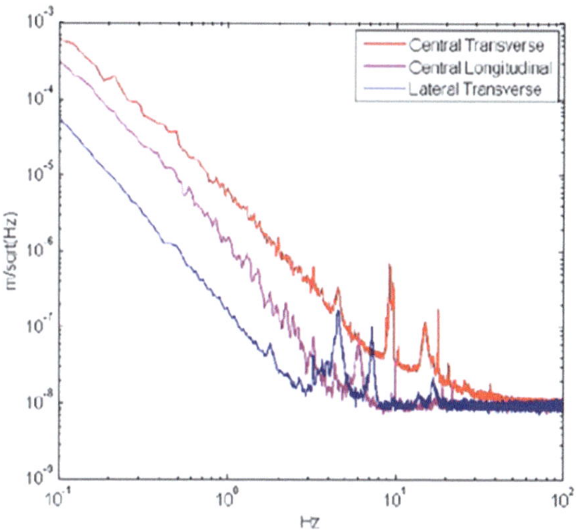

Fig. 2 Example of displacement spectral density of selected measurement points on the top of Trajan Arch (I test)

(Fig. 3), using triaxial accelerometric sensors, characterised by 0-200 Hz frequency range and 0.25 g maximum acceleration scale range.

Two test layouts were considered, and, in both cases, the accelerometers were installed on the north side of the Temple. In details, they were positioned between the fourth and fifth columns of the aforementioned side, numbering the columns from East to West. Figure 4 describes a test example.

The recorded signals analysis highlighted several peaks showing a complex dynamic behaviour due to the construction complexity.

Fig. 3 Monitoring campaign of Neptune temple, 2016

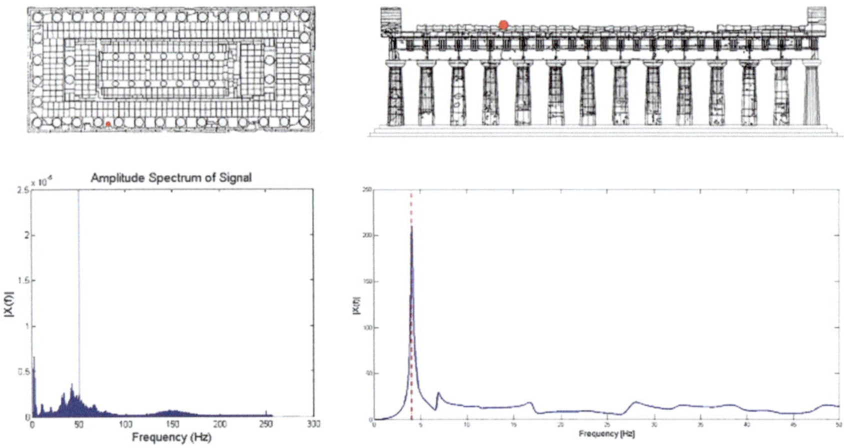

Fig. 4 Example of transversal accelerometric spectrum and Fourier transform in the frequency domain

Moreover, several tests (Petti et al., 2019a; Grelle et al. 2014) were carried out in the 2019 to assess the local seismic characteristics of the soil. In this context, indeed, it is fundamental to underline that the knowledge of a monument has to be based not only on the architectural and structural aspects but also on aspects concerning the structure-soil interactions.

For the Archaeological Park of Paestum and Velia, the particular conformation of the soil, geologically characterized by a complex process of evolution of the coast and the river branches, coming from the carbonate Apennine chains, strongly affect the local seismic demand.

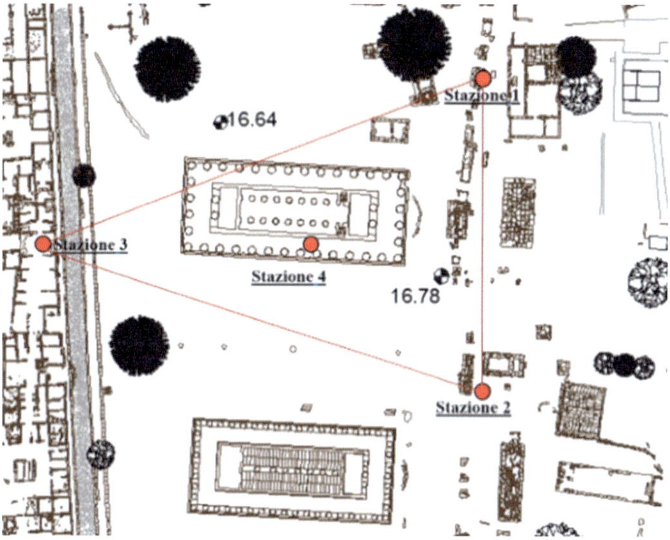

Fig. 5 On-site experimentation set-up, 2019

This seismic soil characterization highlighted a significant speed inversion of shear waves between the rigid superficial travertine bench, with an overall thickness exceeding 10 m, and the underlying loose deposited, presumably characterized by important pyroclastic fall and/or sedimentation contributions. This behaviour of the stratigraphy realizes a natural physical filter on wide frequency ranges, which presents most of the main natural modes of vibration of the structural elements and in particular in the frequency potentially damaging the temple.

Figures 5 and 6 describe the implemented seismic network and the main obtained results, respectively.

Moreover, the research activities carried out investigated the possibilities of using remote sensing technologies, as the satellite interferometric one (Bianchini et al. 2015; Casagli et al. 2017) that can be advantageous considered to assess long-time (weeks-years) global movements and deformations. These techniques allow to measure surface displacements of the ground or structures with millimetre precision thanks to the use of processing techniques called "interferometric", which consider the phase difference between SAR images (Synthetic Aperture Radar).

2 Neptune Temple Monitoring Network Framework

As part of the joint conservation activities between the Department of Civil Engineering (DICIV) of the University of Salerno and the Archaeological Park of Paestum

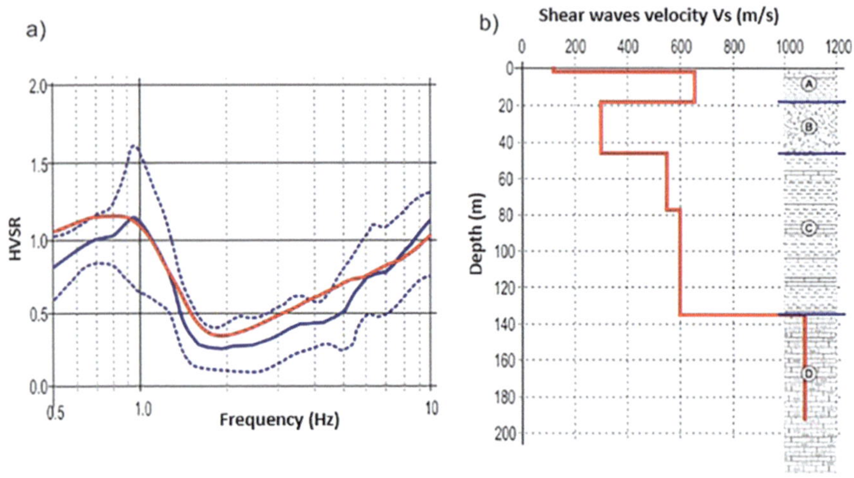

Fig. 6 On-site campaign: **a** result of inversion analysis between the experimental HVSR curve (blue) and the theoretical one (red); **b** speed profile of shear waves and relative stratigraphic column of reference: A = Travertine; B = River-Dunali deposits mixed with pyroclastic deposits; C = Compact or very thickened deposits of probable fluvial and marine origin; D = Basic Seismic Bedrock

and Velia (PAEVE), an agreement was stipulated in 2018, aimed at the direct monitoring and the analysis of the safety of the archaeological structures present in the Park, which envisages the realization of an executive project of a dynamic and seismic monitoring network of the Neptune temple. This network has as its main purpose the monitoring of displacements and accelerations induced by natural and human-induced events on the structure, in order to obtain a more in-depth knowledge of its behaviour, and also for the purpose of evaluating the safety of the monument and visitors. In details, modular biaxial and triaxial sensors, based on high sensitivity and large band and monolithic mechanical transducers (Acernese et al. 2010; Barone et al. 2015), are implemented.

As shown in Fig. 7, the 12 points indicated in the map of the Temple will be monitored. The data acquisition is fully locally distributed. An information structure (Totem), that will be installed near the Temple to show details to inform visitors about monitoring activities, to act also as a rack for the management of the equipment, data collection and storage, for an effective real-time pre-analysis and data transfer.

The passage of the monitoring network cables from the top of the Temple to its base is carried out by means of a cable-duct/sheath with a diameter of 7 cm along one of the columns of the West side of the Temple (Fig. 8).

More in details, the monitoring system of the Neptune temple in Paestum was conceived starting from the consideration that an effective evaluation and correlation of the static and dynamic structural behaviour of structures and sites, with natural and anthropic actions, requires the optimization and validation of adaptive structural finite elements models (FEM). Those models need a high quality and accurate experimental

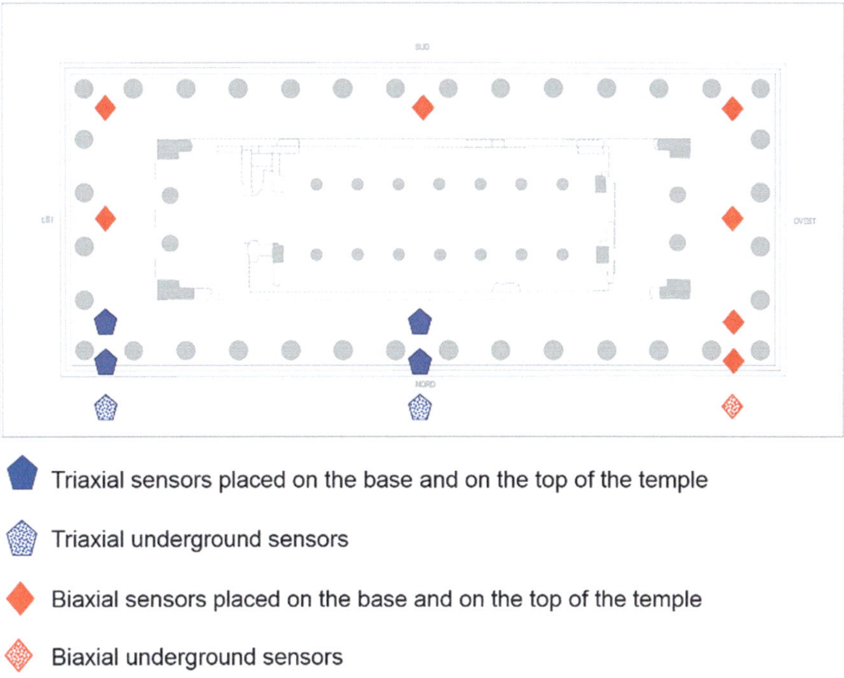

Fig. 7 Allocation of the mechanical sensors on the top, on the base and on the foundation layer of the Neptune temple

data, the latter acquired with modular distributed monitoring systems, adaptive both in terms of architecture and in terms of sensitivity and bandwidth of sensors. This innovative approach required a dual design action, aimed at both the implementation of the monitoring system and at the selection and specialization of sensors.

The monitoring system was designed to ensure not only compliance with the band and sensitivity requirements of the initial system for the different degrees of freedom (horizontal and vertical), but also with expandability and versatility (e.g., increase of the sensors number and/or repositioning), if the structural behaviour analysis requires actions aimed at the optimization of its performance. In order to optimize the signal-to-noise ratio, the acquisition will be obtained near the data production points (sensors), making, in this way, the system not sensitive to couplings with environmental noises. Figure 9 shows the block diagram of the data acquisition, storage, distribution and analysis system designed for the Neptune temple in Paestum.

The key element of the monitoring system is the local, compact and robust DAQ (Data Acquisition), based on the FieldDAQ™ technology of National Instruments, state of-the-art in the field both in terms of data quality and system robustness. The system consists of 34 degrees of freedom (sensors) acquired through 5 FieldDAQ™ voltage modules FD-11603 (8 voltage channels at ± 10 V, resolution up to 24 bits,

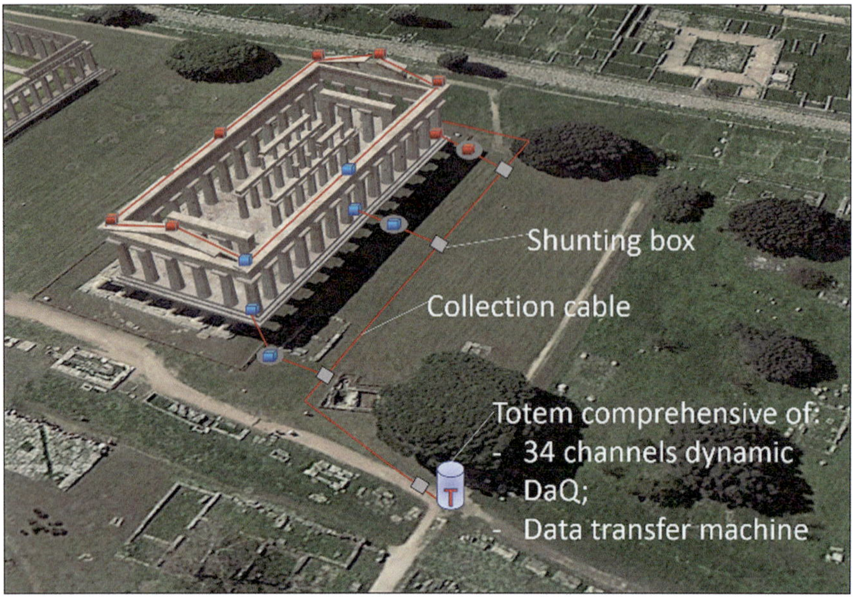

Fig. 8 3D view of the monitoring system set-up

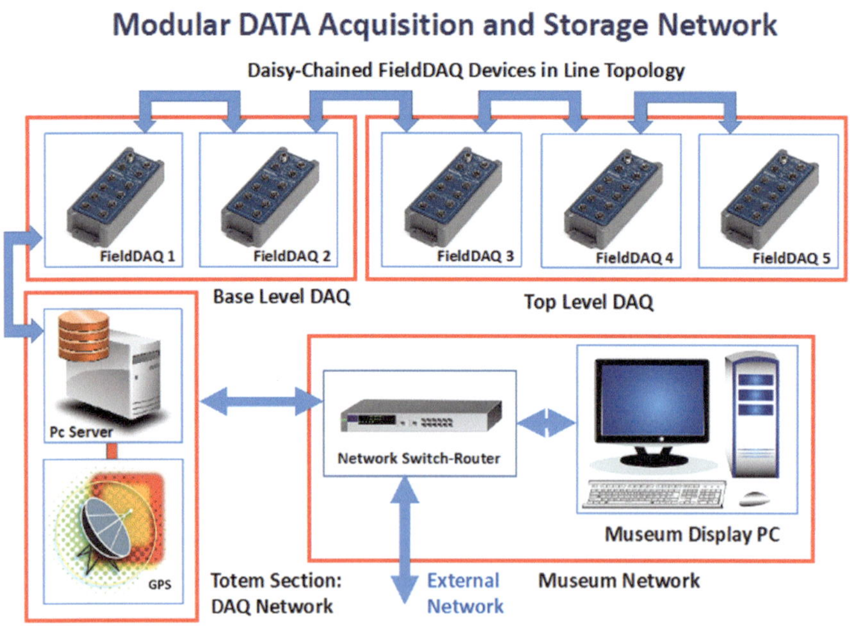

Fig. 9 Block scheme of the monitoring system of the Temple of Neptune in Paestum

simultaneous sampling frequency up to 100 kS/s, absolute accuracy of 0.019 V from −40 to 85 °C, less than 0.2% of the total measuring range).

In order to satisfy possible new requirements, the monitoring system can be easily expanded to the measurement of environmental conditions (temperature, pressure, humidity, wind speed and direction), relevant for the study of the dynamic slow-motion behaviour of the Neptune temple structure by cross-correlation analysis.

The data acquisition and storage processes are managed and synchronized by a Pc server (4 Tbyte) acting as frame builder and data storage system, downloading the data on local disks for backup, data distribution and presentation. The synchronization of the acquisition, better than 1 μs on the whole network of DAQ nodes, is carried out via Ethernet by the Pc server using the NI-DAQmx software. The time is provided by GPS card allowing, in this way, the definition of a timestamp on each frame, necessary for a correct reconstruction of signals.

The network topology, based on the IEEE 802.1AS profile (Ethernet TSN), is a daisy chain, chosen in consideration of the relatively low number of initial degrees of freedom (34) and of the relatively small distances among sensors. This network topology guarantees direct expansion of the number of sensors by increasing the number of DAQs, up to a maximum of 15 for a total of 120 channels. The simple addition of a TSN switch (compliant with the IEEE 802.1Q standard to support the Rapid Spanning Tree protocol (RSTP) and the IEEE 802.1AS standard to support synchronization of DAQs on individual branches) allows the extension of data acquisition and synchronization to other temples and structures of the historical area of Paestum, through the creation of a star configuration of daisy chains.

A direct connection to the museum Ethernet network is implemented mounting remotely the PC server storage disks through NFS, allowing both real time data backup and data presentation on monitors in museum rooms, using open software presentation standards (e.g., GEOPSY, typically used in seismology).

This technological approach leads possible also an easy development of dedicated presentation software, offering the visitor the opportunity to visualize via mobile apps in an effective and direct (even interactive) way the global and/or local dynamical behaviour of the temple of Neptune, also in connection with the daily evolution of the environmental conditions.

For what concerns sensors, the key element of the monitoring system, as it will be clear in the following, they have also been adapted to this specific application, which requires high sensitivity ($10^{-8} m/\sqrt{Hz}$) in the band ($1 mHz - 100 Hz$), together with small dimensions, weights and costs.

The design and implementation philosophy characterizing the data acquisition, storage, analysis and presentation system was extended to the sensors as well.

The mechanical architecture of the sensors is based on an innovative technological platform, developed by the Applied Physics Research Group of the University of Salerno (UNISA) Folded Pendulum, international state-of-the-art platform for the design and implementation of monolithic mechanical oscillators (Acernese et al. 2010; Barone et al. 2015).

This innovative technological platform, already protected by four families of international patents, has allowed the development of linear and angular (tiltmeter) position, speed and acceleration state-of-the-art sensors. These sensors, very innovative in the field of seismology, can be calibrated in frequency, are fully scalable in terms of sizes and weights with band in the low frequency region as large as $\left(10^{-7}-10^{2}\text{Hz}\right)$, can be designed to guarantee very high sensitivities $\left(< 10^{-12} m/\sqrt{Hz}\right)$ in the band, (10 mHz – 100 Hz), even lower than the minimum seismic noise that can be measured on the earth's surface defined by the classical models (Peterson 1993; McNamara and Buland 2004). Moreover, they can operate in ultra-high vacuum (UHV) and cryogenic environments (<3 K), still remaining simple, compact, light and, with suitable designs, applicable to very different fields, including the cultural heritage one (Barone et al 2015; Barone and Giordano 2017).

The advantage of using a particularly sensitive and broadband instrumentation is that it allows a continuous evaluation of the dynamic behaviour of the structures, highlighting also effects of environmental forcing, such as those due to wind, to ground heating, to thermal expansion of the structures due to the different illumination from the sun during the day.

The sensors used in the initial monitoring network of the Neptune temple in Paestum, models SE-10H (horizontal) and SE-10 V (vertical) with a LVDT readout module, were slow motion seismometers (no feedback) in uniaxial configuration. These new sensors, produced by a privately owned company that develops sensors and systems for scientific and engineering applications for monitoring and control called ADV3S Company (Advanced Scientific Sensors and Systems, http://www.adv3s.com/en/homepage/), are based on the model of an aluminium folded pendulum monolithic oscillator internationally known as GE15, a choice that guarantees the implementation of biaxial or triaxial configurations by means of suitable combinations of uniaxial sensors Fig. 10.

Being the GE15 mechanical oscillator limited by its thermal noise only, its sensitivity is practically determined by the readout system, that in this specific case is a commercial LVDT. Nevertheless, a LVDT readout upgrade with an optical readout allows to get a sensitivity improvement of three orders of magnitude $(< 10^{-11} m/\sqrt{Hz})$ and a band extension in the low frequency region.

Figure 10 shows the transfer function of the GE15 mechanical oscillator calibrated at the resonance frequency of 3.75 Hz, the different sensitivities in acceleration achievable with sensors based on GE15 monolithic oscillator according to the type of readout used and a picture of the basic architecture of the seismometers SE-10H and SE-10 V, based on the GE15 oscillator and on a commercial LVDT (Barone et al, 2015; Barone and Giordano 2017).

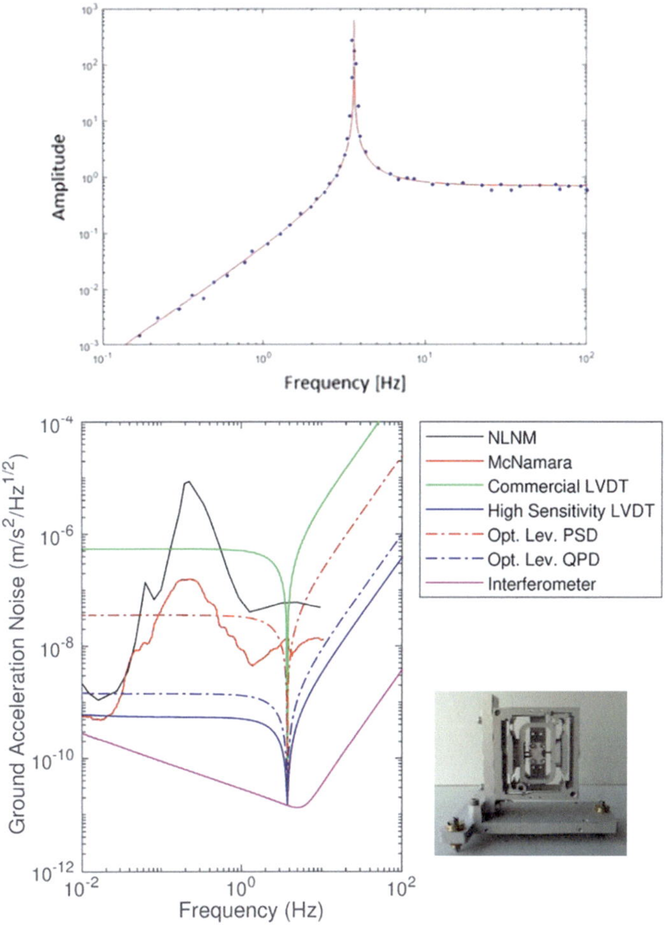

Fig. 10 GE15 mechanical oscillator transfer function calibrated at the frequency of 3.75 Hz (left); sensitivity (acceleration) of the GE15 monolithic oscillator equipped with different types of readouts, compared with the minimum levels of earth's seismic noise (Peterson (NLNM) and McNamara models) (right); GE15 sensor equipped commercial LVDT readout system (bottom right)

3 Conclusions

The paper presents an innovative network for monitoring the dynamic behaviour of the local seismic response of the Neptune temple in Paestum. The system uses an excellent broadband high sensitivity mechanical sensors network that allows to carry out an effective assessment of the static and dynamic behaviour of the temple structure, also in connection with the soil-structure interaction, also monitoring in real-time the effects due to human-induced and natural events, including, in example, those related to climate change phenomena. The network will allow a real-time data collection and storage, guaranteeing an effective dissemination of the information through open source platforms, key element for carrying out a research framework aimed at a synergic centralization of he contribution of researchers on the analysis of the complex aspects that describe monumental structures and for a share of further research activities.

Acknowledgements The research has been carried out with the support of Archeological Site of Paestum and Velia in Italy (Gabriel Zuchtriegel, Director; Antonella Manzo, Architect) and PRIN2015 National Research Project "Protecting the cultural heritage from water-soil interaction related threats" n. prot. 2015EAM9S5.

References

Acernese F, Giordano G, Romano R, De Rosa R, Barone F (2010) Tunable mechanical monolithic sensor with interferometric readout for low frequency seismic noise measurement. Nucl Instrum Meth A 617:457–458. https://doi.org/10.1016/j.nima.2009.10.112

Baratta I, Corbi O, Corbi L, De Andreis L, Marchetti L, Greco D, Petti L (2019) Heritage preservation: Challenges and perspective, STC 2019 International Conference. Salerno

Barone F, Giordano G, Acernese F, Romano R (2015) Watt's linkage based large band low frequency sensors for scientific applications. Nucl Instrum Meth A. https://doi.org/10.1016/j.nima.2015.11.015

Barone F, Giordano G (2017) The UNISA folded pendulum: a very versatile class of low frequency high sensitive sensors. Measurementhttps://doi.org/10.1016/j.measurement.2017.09.001

Bianchini S, Pratesi F, Nolesini T, Casagli N (2015). Building deformation assessment by means of persistent scatterer interferometry analysis on a landslide-affected area: the Volterra (Italy) case study. Remote Sens 7(4):4678–4701

Casagli N, Frodella W, Morelli S, Tofani V, Ciampalini A, Intrieri E, Raspini F, Rossi G, Tanteri L, Lu P (2017) Spaceborne, UAV and ground-based remote sensing techniques for landslide mapping, monitoring and early warning. Geoenvironmental Disasters 4(1):

Grelle G, Bonito L, Revellino P, Guerriero L, Guadagno FM (2014) A hybrid model for mapping simplified seismic response via a GIS-metamodel approach. Nat Hazards Earth Syst Sci 14(7):1703–1718

McNamara DE, Buland RP (2004) Ambient noise levels in the continental United States. Bull Seism Soc Am 94:1517–1527. https://doi.org/10.1785/012003001

Obón Santacana F, Dorka UE, Nguyen CKL, Petti L (2017) Behaviour of greek columns during earthquakes with and without tendon system, Lisbon

Peterson J (1993) Observations and modelling of background seismic noise, Report 93–322, U.S. Geological Survey, Albuquerque, New Mexico. https://doi.org/10.3133/ofr93322

Petti L, Sicignano F, Greco D (2018a) Valutazione della risposta dinamica e sismica del tempio di Nettuno: risultati preliminari. L'emblema dell'eternità. Edizioni ETS, Paestum. ISBN-13: 978–8846754639, pp 177–189

Petti L, Grelle G, Greco D (2019a) Procedure innovative pe l'analisi della risposta sismica delle costruzioni monumentali realizzate a blocchi rigidi. Il caso del tempio di Nettuno. Zuchtriegel, Carter, Oddo. Paestum. POSEIDONIA E L'ACQUA. Archeologia e cambiamenti climatici. ISBN: 978–88–87744–86–6, pp 123–133

Petti L, Sicignano F, Greco D (2017a) Dynamic behaviour assessment of ancient columns through experimental analysis. XVII Convegno ANIDIS, Pistoia

Petti L, Sicignano F, Greco D (2018b) Seismic and dynamic assessment of monumental structures made up of rigid clocks: the Neptune temple. In: 16th European Conference on Earthquake Engineering (ECEE), 18th -21st June 2018b. Thessaloniki

Petti L, Greco D, Mammone A, Di Muro C (2018c) New multi–scale risk governance and management approach of natural, cultural and artistical preserved areas. The case studies of the Amalfi coast and the Cilento National Park. 8th ICBR Lisbon

Petti L, Greco D, Nuccio F, (2019b). Seismic behaviour assessment of columns made up of rigid blocks. XVIII Convegno ANIDIS, 15–19 September 2019b. Ascoli Piceno

Petti L, Trillo C, Di Mauro M (2017b) Heritage and recostruction: Different perspectives. living under the threat of earthquakes, pp 101–116. ISBN: 978–3–319–68043–9. https://doi.org/10.1007/978-3-319-68044-6

UNESCO (2017a) Strategy for action on climate change, 39 C (2017a). UNESCO General Assembly, 39th session, Paris

UNESCO (2018) Proceedings of UNESCO chair program on cultural heritage and risk management. International Training Course on Disaster Risk Management of Cultural Heritage. Kyoto and Kobe, Japan, 29th August–19th September

UNESCO (2018) Proceedings of UNESCO chair program on cultural heritage and risk management. International Training Course on Disaster Risk Management of Cultural Heritage. Kyoto and Kobe, Japan, 29th August–19th September

Structure Stability Analyses of Chapels Dedicated to Wives of Amun Using Non-Invasive Techniques, Case Study "Madinet Habu Temple—Luxor—Egypt"

Ayman Hamed, Nevin Aly, and Mathias Lang

Abstract During the last few decades, multidisciplinary research and new technologies have been introduced in the monument preservation activities. In this context, non-destructive and non-contact techniques become highly demanded. This study first aims to develope such methodologies to investigate the Chapels Dedicated to Wives of Amun at Medinet Habu temple (the huge Mortuary Temple of Ramses III) and specifically the Saite chapels of the Divine Adoratrices of Amun and to diagnose the damage and its possible related degradation mechanism. The Saite chapels are constructed mainly from the Nubian sandstone, which is siliceous sandstone, quarried from Gebel El Silsila (50 km north of Aswan- Egypt). The Nubian sandstone has been used for constructing most of the cultural heritage architectures in Upper Egypt and still employed in recent restoration works. In this study, 3DTerrestrial Laser Scanning (TLS), was used to scan the superstructure, Ground Penetrating Radar (GPR) was used to investigate the substructure. Output results are used, as a valuable tool, to analyse and accurately measure façade displacement, explained by the rediscovered tombs underneath the chapels.

Keywords Medinet Habu temple · 3D Laser scan · Ground Penetrating Radar

1 Description of the Case Study and Historical Background

This article is part of a research study financed by the Egyptian Ministry of Antiquity to safeguard the saite chapels of the Intermediate Period, 25th and the 26th Dynasties (Asteris 2008; Twenty-third et al. n.d.; Ayad 2012). South-East chapel of Amenirdis

A. Hamed (✉) · N. Aly
Faculty of Petroleum and Mining Engineering, Suez University, Suez 43512, Egypt
e-mail: ayman.hamed@suezuni.edu.eg

N. Aly
e-mail: nevin.aly@suezuniv.edu.eg

M. Lang
eScience Center, Tubingen University, Tubingen, Germany

G. M. El-Qady and C. Margottini (eds.), *Sustainable Conservation of UNESCO and Other Heritage Sites Through Proactive Geosciences*, Springer Geology,
https://doi.org/10.1007/978-3-031-13810-2_21

II is the best preserved with an inner arched mortuary chapel preserving the original painting colors. To North-West Nitocris II followed by Shepenwepet (II), both are arched chapels with wall engravings and traces of antique colors. At most North-West the roofless Mehetnusekhet chapel with standalone inscribed walls. Since the beginning of this research work, authors have no clue about the underneath adoratrices crypts, it was an absent information, Fig. 1.

The Saite chapels are located at the western shore of the Nile at Thebes, inside an archaeological site once called Djemet, serving as a necropolis and monument depository, Fig. 2. Hathpsut (r. 1473–1458 B.C.E.) and Tuthmosis III (1479–1425 B.C.E.) erected a temple honoring the god Amun at Medinet Habu. The dominating monument, however, is a fortified temple complex erected by Ramsesses III (1194–1163 B.C.E.) (Bryan 2003).

The original temple foundation dating to the Eighteenth Dynasty was started by Tuthmosis I (r. 1504–1492 B.C.E.) and was called "Splendor of the West" or "Amun is Splendid in Thrones." Hatshepsut directed much of the construction of the temple,

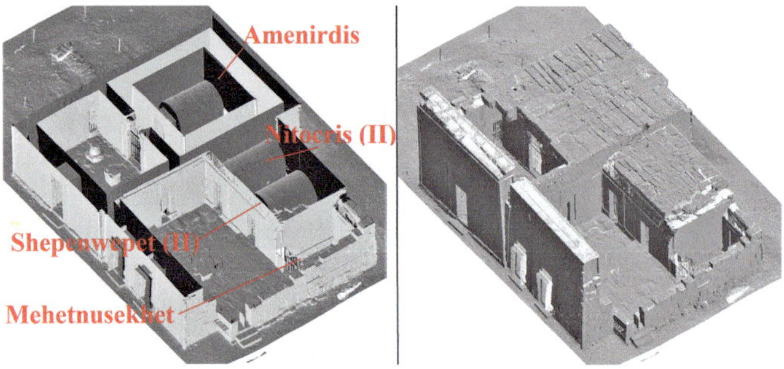

Fig. 1 Right, general view for the Saite chapels. Left horizontal section shows the arched, TLS

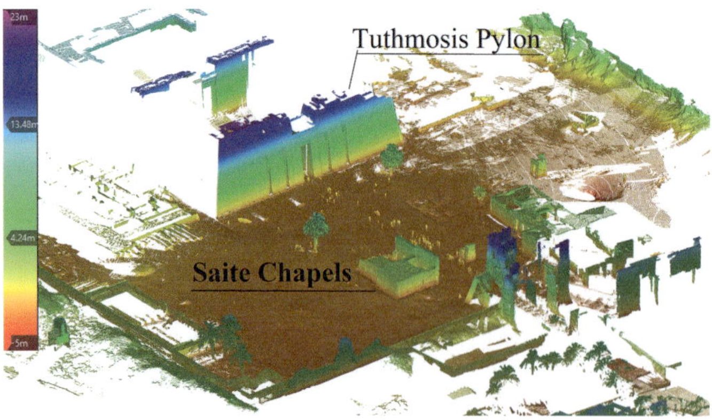

Fig. 2 Landscape of Medinet Habu including Saite chapel, using TLS indicated levels

Fig. 3 One of the crack systems initiated on walls of chapels

but the dedication and opening of the site dates to the reign of Tuthmosis III (Nelson n.d.).

2 Problem Definition and Objective

In 2017 inspectors of the site proclaimed a critical crack in chapels walls. Present work is objected toward a practical methodology using modern techniques to assess structure stability of archaeological monuments. Accurate data measurements, collection and analyses of the edifice may give a true evidence of certain facts of inclinations and displacements. While ground penetrating radar (GPR), data analyses present a factual response toward the probable causes of structure instability, Fig. 3.

3 Scope of Work

TLS documentation of the studied chapels presented a fit detailed architecture deformation as well as other types of material deformation. The scope of this work is only concentrated on architecture anomalies which are caused by severe stress redistribution. Analyses of the crack system directing to collect as much relevant information as possible both on the damage itself and on its context.

4 Material and Methods

In general, there are two types of laser scanners that are based on the principle of time measurement: phase-based laser scanners and pulse-based scanners. Pulse-based laser scanners are based on emitting pulses of a laser beam, while phase-based laser scanners are based on a continuous laser beam. with a higher accuracy at short distances (Nuttens et al. 2010).

Using Leica Scan Station P30/P40 with the WFD technology is a kind of time-of-flight measurement system but combines advantages of the phase shift and the time-of-flight method into one single system (Walsh 2015), with the following specifications (Table 1) (Leica Geosystems 2017)

The approach adopted in this work to register all the individual scans acquired in the field, relies on aligns scans based on overlapping surfaces using well established variants of the Iterated Closest Point (ICP) algorithm.

iSTAR 360 Degree Rapid Imaging Panoramic Camera is also used, which has 4 × pre-calibrated sensors delivering a 50MP full spherical accurate image with excellent high dynamic range. iSTAR is fast at capturing images with typical capture times from around 5 s. Typically, 50 to 60 panoramas can be captured on one battery charge. This can be increased to over 600 when using the optional external battery

Table 1 The specifications of the used scan station

Range accuracy Angular accuracy 3D position accuracy	1.2 mm + 10 ppm over full range 8 "horizontal; 8" vertical 3 mm at 50 m; 6 mm at 100 m			
Target acquisition	2 mm standard deviation at 50 m			
Dual-axis compensator	Liquid sensor with real-time onboard compensation, selectable on/off, resolution $1''$, dynamic range $\pm$ 5', accuracy $1.5''$			
Wavelength	1550 nm (invisible) / 658 nm (visible)			
Beam divergence	<0.23 mrad (FWHM, full angle)			
Beam diameter at front window	$\leq$3.5 mm (FWHM)			
Range and reflectivity Minimum range 0.4 m	Minimum range 0.4 m			
	Maximum range at reflectivity			
		120 m	180 m	270 m
	P30	18%	–	–
	P40	8%	18%	34%
Scan rate	Up to 1,000,000 points per second			
Range noise	0.4 mm rms at 10 m 0.5 mm rms at 50 m			
	0.5 mm rms at 50 m			

Fig. 4 3D archaeometric point cloud model for the chapels

allowing you to capture throughout the full day ("ISTAR 360 Degree Measurement Module Integrated by Imaging Companies 2021).

Subsurface investigation was conducted using, MALA GPR X3M Control Unit, and MALA GPR Shielded Antenna System 250 MHz, were used during this study (Mala 2017). Survey parameters selected were 0.03 m trace-interval with a 112ηs total time window with number of samples 500 and 4449 MHz sample frequency.

4.1 Archaeometric Model

When the dimensions from definitive drawings are completely redacted in a computer model with no loss of accuracy, this could be a fundamental part of archaeometry reconstruction, Fig. 4, contrasted with "sketch" or "pedagogical" models, which only incorporate a subset of the authoritative dimensions, or which simplify the geometric forms in published drawings for reasons of expediency (Johnson 2008).

4.2 Archaeometric Measurements

Rüther and Held (2011) concluded that terrestrial laser scanner (TLS) is a powerful tool which makes it possible to collect valuable and accurate spatial information in relatively short field time. It played a prominent role in the acquisition of data for the African Heritage Database and it is unlikely that other techniques would have been able to acquire similar data volumes, they also stated that, it is important to explore new ways of using laser scanned models, besides for visual presentation, to justify the effort and cost spent on their creation. Gordon and Lichti (2007), stated that, it has been shown that with simple processing strategies, normally coarse accuracy (TLS) instruments can be used to measure small deformations (less than 3 mm in some circumstances, with high accuracy. Comparing with the traditional monitoring methods, TLS monitoring has better flexibility and especially in real projects. Because some embedded sensors failed after running several years, other means of structural deformation monitoring are required (Yang et al. 2018). However, the real possibility of acquiring dense geometric information by TLS allows us to analyse range dataset through different statistical and modelling approaches not considered until now (Gonzalez-Aguilera et al. 2008). While (Monserrat and Crosetto 2008), proposed a procedure for measuring land deformation using repeated TLS acquisitions, which proved an ideal scenario for achieving highly precise deformation measurements.

According to (ICOMOS 2003), structural damage occurs when the stresses produced by one or more actions exceed the strength of a material, either because the actions themselves have increased or because strength has been reduced. Kieker (2008) is a European research project "ESECMaSE" discussed the correlation between numerically modelled structures using a finite-element- programme and masonry houses, where the project shows the complex behaviour of masonry walls under in plane loading. Nohutcu et al. (2015) defined the dynamic characteristics of the historical structure, in a realistic way by performing destructive and NDT tests on the structure, the study suggests that seismic, may incorporate calibrated numerical models, should further be conducted for sustainable structures. In other words numerical models are not reliable by itself, however it should be calibrated by field investigations. Borri et al. (2015) proposed a method for the classification and analysis of historical masonry typologies allows important parameters to be quantitatively, produce a value for final index representing the quality of the masonry, MQI, ranges from 0 to a maximum value of 10, a correlation logarithmic curve has been introduced to calculate the main mechanical properties of masonry from the value of MQI. Pattern recognition is an approach proposed by (Balen 1996), collecting a different types of information using the questionnaire and other related summary sheets of an interesting collection of degradations of different types. Those examples are useful for the elaboration of the damage atlas. The most sight informative and easily applied from the first visit of inspection are studied and proposed by (VENT, Ilse Anne Elisabeth DE 2011).

The method of analyses is essentially depending on the way masonry walls are cracked and fail. In a way it could be described as reverse engineering analyses Fig. 5.

Point cloud archaeometric model of the chapels, shows different instability features of inclination. NE façade 'entrance pylons' measures bidirectional dispositioning. Angle between vertical axe and silhouette of pylons is 0.65° in one side, NW, the opposite direction measures an angle of 1.64° toward SE. The two pylons are tilting far from each other, in the same way as a rotation around aground axe perpendicular façade plan, Fig. 6.

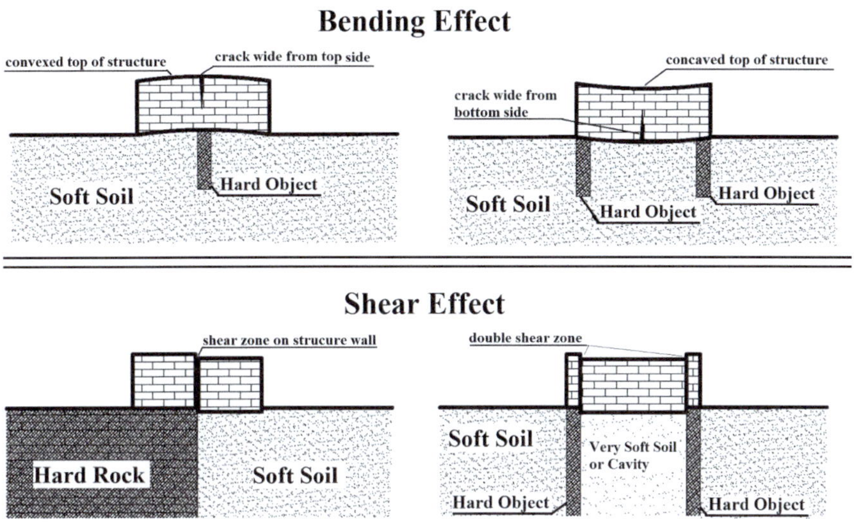

Fig. 5 Inference of trough location and length from settlement damage

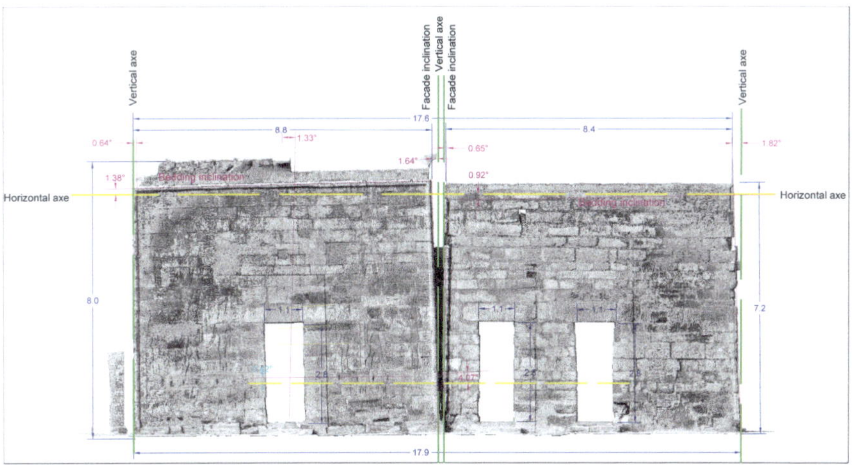

Fig. 6 NE façade shows 'entrance pylons', dispositioning angles both in NW and SE directions

While, SW façade shows a set of vertical cracks and bedding inclination, both are a typical demonstration of complex structure deformation, Fig. 7. Bedding inclination measures 1.34 to 1.36° to SW, while NE side measures −0.17 to−0.36°, which mean a bedding inclination in opposite directions. The pylon shows toppling displacement toward NE direction measuring 1.5°, in reversed direction. Considering Facade inclination of Tuthmosis pylon (9.6°), Figs. 8 and 9.

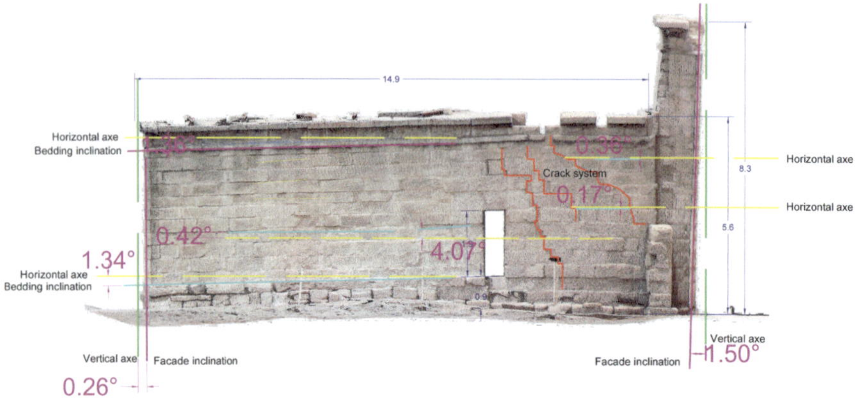

Fig. 7 SW façade shows a set of vertical cracks and bedding inclination

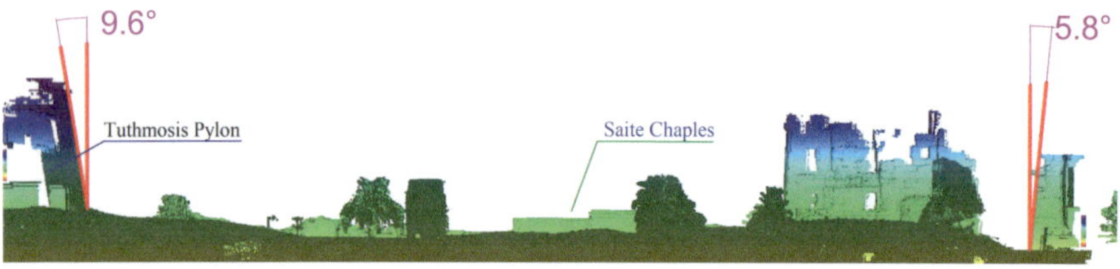

Fig. 8 Measuring inclination of Tuthmosis pylon using TLS

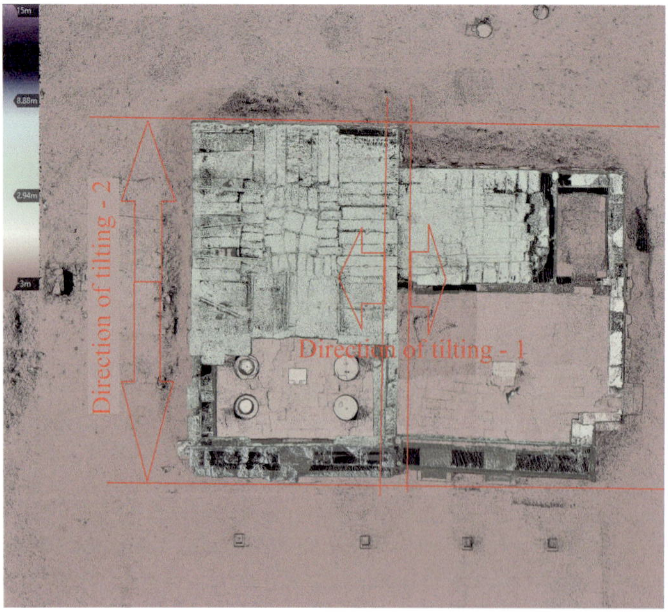

Fig. 9 Directions of movement based on crack and tilting pattern of chapels walls

5 Subsurface Investigations

Since earth ground is always in a dynamic condition, earthquakes, traffic, or groundwater may cause soil disturbance in soil material easier than rocky foundations. Taking into consideration the fact that masonry buildings, are elements connected with its subsurface ground and cannot move freely, then we must assume by evidence that there are no static structures elements as they react to changes in their environment (Hamed 2009). This reaction may involve dimensional changes. In other words, materials can shrink or swell. Depending on subsurface earth composition either homogeneous or differential settlements well act.

Archaeometric measurements, suggest a multidirectional displacement of chapels walls. Walls tilting and cracking suggest a source problem beneath ground. Soil investigation using open pit, showed a shallow footing over a dark brown mud, like the one used in the surrounding agricultural region, Fig. 10, which may explain vertical displacement of walls, but not rotational.

Further investigations using GPR survey on chapels ground floor using the 250 MHz antenna, with profile lines-oriented SE-NE (Fig. 11), showed depth response exceeds five meters deep in the ground. Produced GPR data, are filtered using bandpass frequency filtering to remove unwanted frequency components from the data. Bandpass filtering is usually applied to remove global background and system noise, which can result in the improvement of signal-to-noise ratio. Filtering method selects between finite impulse response (FIR) and (infinite impulse response (IIR) filters which is also known as recursive filters. Upper cut-off frequency defines the highest frequency that is passed through the filter without attenuation, and lower

Fig. 10 Open pit showing shallow foundation of chapels walls

Fig. 11 Location a long GPR profiles

cut off frequency defines the lowest frequency. FIR band-pass filtering can help interpreters customize their frequency range to filter GPR data referring to the dominant frequency or a signal of interest (Lijn et al. 2003). GPR data analyses require an understanding of the following:

(1) The physics and chemistry of material properties as they describe behavior in response to electromagnetic fields
(2) The physics of electromagnetic wave propagation
(3) The physics, geometry, and processes of scattering (Olhoeft 2014)

A minimal processing was realized before analyzing the radar sections. GPR profile sections present different zones of wave reflections, where the signal was highly attenuated, with the presence of weak reflection surfaces. GPR profiles through chapels, where Figs. 12, 13 and 14 present zones with similar behavior, shows two partially hyperbolic shaped anomaly, caused by breaking of waves on a buried wall edge. In between, a disturbed wave signals returned from depth, indicate an underground structure. In further investigations, archaeological excavations re-discovered the tombs of deities' underneath chapels, zone (C). Anomalies lines produced by wave transition from high attenuation to high reflection, are indicating subsurface disturbance, Zone (D) (Fig. 15).

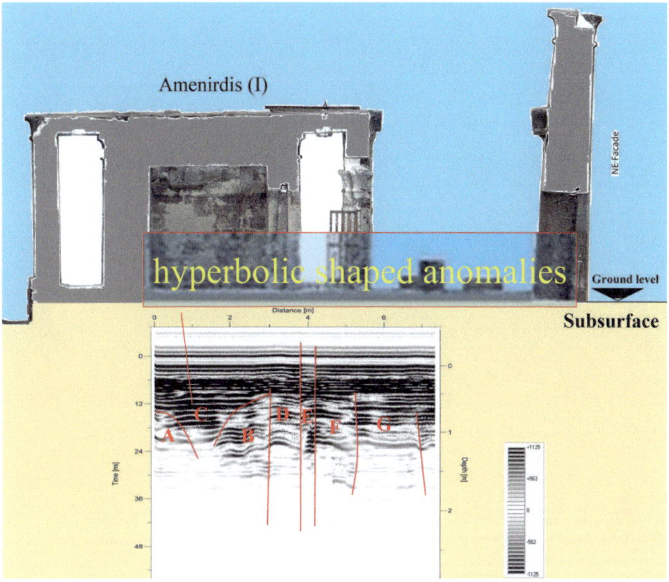

Fig. 12 Radargram showing possible cavity under Amenirdis (I) chapel

6 Conclusions

The use of non-invasive techniques is a powerful tool in understanding the variation in the ground conditions and the facades integrity of the archaeological buildings and to reduce the risk and the cost of the site investigation. Both GPR and TLS proved being an effective method in determine the structure deformation of the investigated chapels (Fig. 16).

The GPR inspection of the chapels ground showed, the direct relation between the subsurface findings and the structural problems of the investigated chapels. The multiple structures that have been re-discovered underneath chapels walls are located exactly at the centre of each chapel (Fig. 14). As a result, the possible stresses have been redistributed toward external walls which have shallow footing on disturbed soil. The detailed GPR survey revealed clearly the potential inner damage in the base of the chapels and as well as the stress concentration points that could affect the safety of the structure.

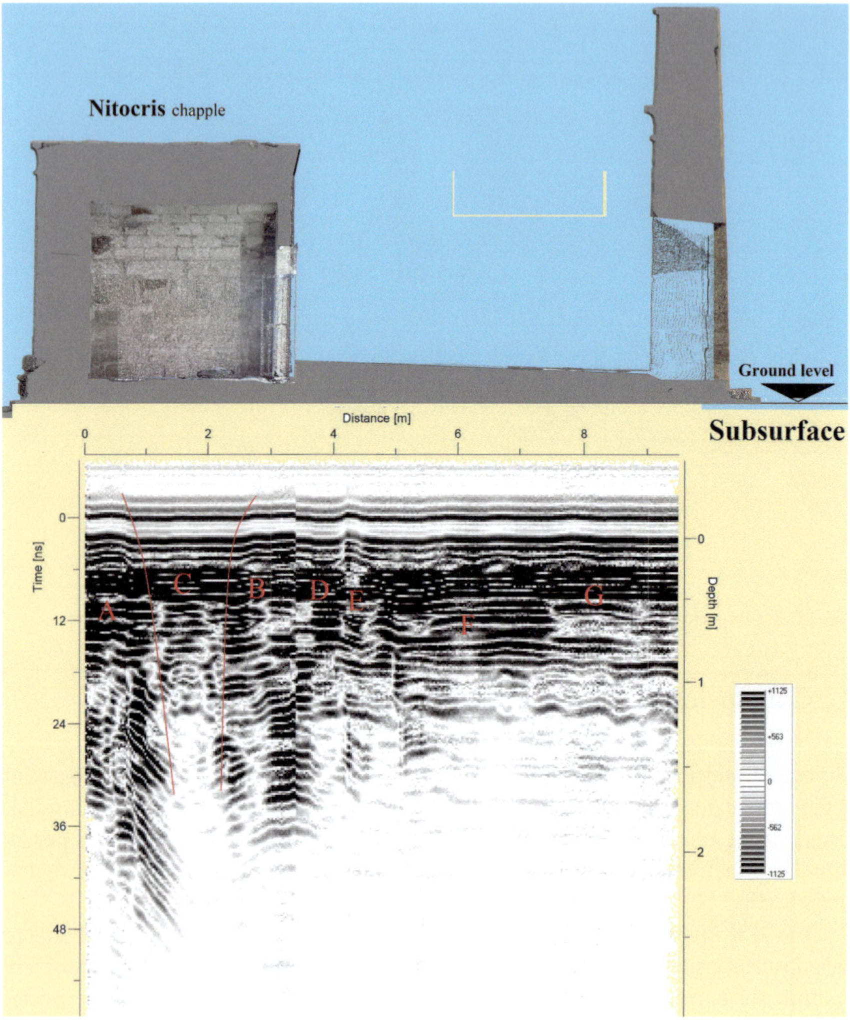

Fig. 13 Radargram showing possible cavity under Shepenwepet (II) chapel

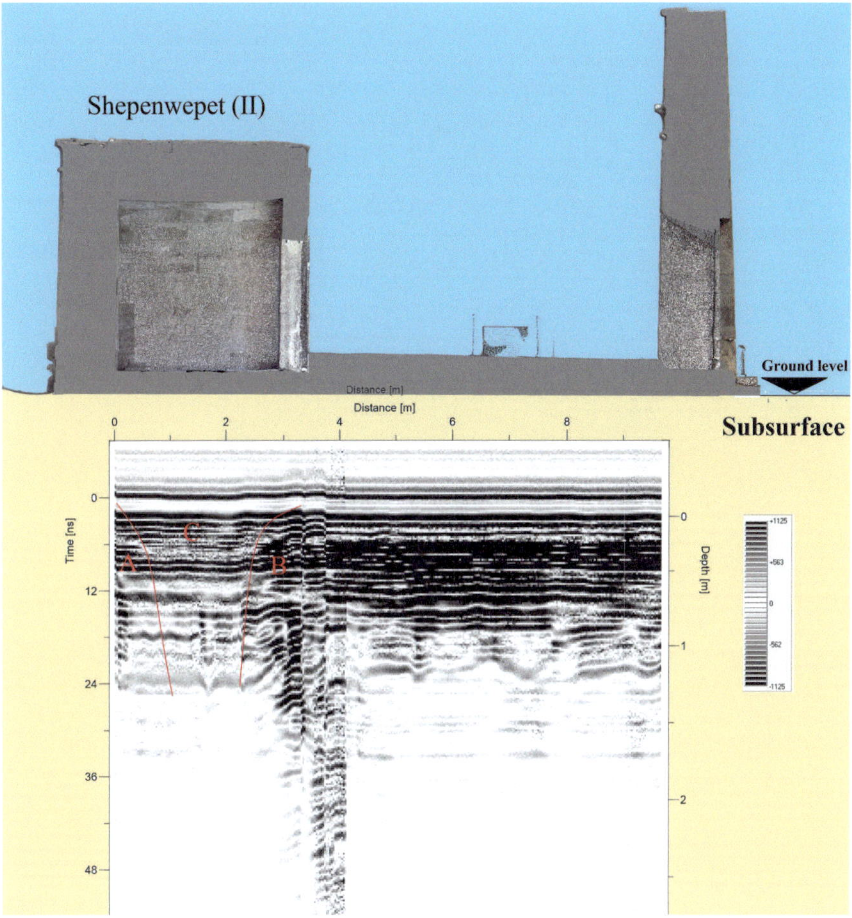

Fig. 14 Radargram showing possible cavity under Nitocris (II) chapel

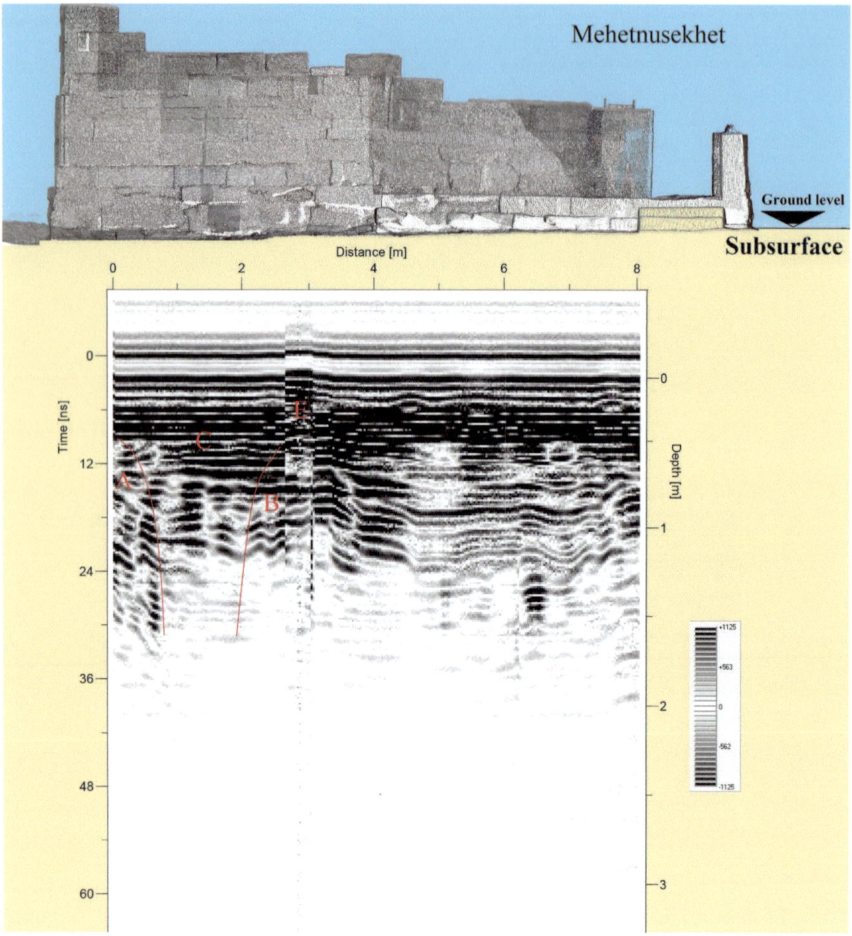

Fig. 15 Radargram showing possible cavity under Mehetnusekhet chapel

Fig. 16 Cross section through chapels, showing relation between underground tombs and superstructure walls

Acknowledgements Authors wish to thank colleagues in Egyptian Ministry of Antiquities, as well as Dr. Ramadan Hussein from Germany's Eberhard Karls University of Tübingen

References

Asteris PG (2008) On the structural analysis and seismic protection of historical masonry structures. Open Constr Build Technol J 2(1):124–133. https://doi.org/10.2174/1874836800802010124

Ayad M (2012) God's wife of amun. The encyclopedia of ancient history, pp 1–2. https://doi.org/10.1002/9781444338386.wbeah15176

Balen Van K (1996) Expert system for evaluation of deterioration of ancient brick masonry structures. Sci Total Environ 189–190:247–254. https://doi.org/10.1016/0048-9697(96)05215-1

Borri A, Corradi M, Castori G, De Maria A (2015) A method for the analysis and classification of historic masonry. Bull Earthq Eng 13(9):2647–2665. https://doi.org/10.1007/s10518-015-9731-4

Bryan B (2003) Property and the {Gods} {Wives} of {Amun}. Women and Property. Cambridge, Center of Hellenic Studies, University, 1–15

Gonzalez-Aguilera D, Gomez-Lahoz J, Munoz-Nieto A, Herrero-Pascual J (2008) Monitoring the health of an emblematic monument from terrestrial laser scanner. Nondestruct Test Eval 23(4):301–315. https://doi.org/10.1080/10589750802259000

Gordon SJ, Lichti DD (2007) Modeling terrestrial laser scanner data for precise structural deformation measurement. J Surv Eng 133(2):72–80. https://doi.org/10.1061/(ASCE)0733-9453(2007)133:2(72)

Hamed A (2009) Hamed: Monitoring of masonry tower (Case study of - Google Scholar). In: 1st International congress, The building techniques. https://scholar.google.com.eg/scholar?cites=7376788562418175855&as_sdt=2005&sciodt=0,5&hl=en

ICOMOS (2003) ICOMOS (2003) - ICOMOS charter. Principles for the analysis, Conservation and structural restoration of architectural heritage, pp 3–6

ISTAR 360 Degree Measurement Module Integrated by Imaging Companies. n.d. Accessed July 25, 2021. https://www.nctechimaging.com/istar-360-degree-measurement-module-integrated-by-imaging-companies/

Johnson D (2008) Architectural drafting standards in archaeological computer modeling: reconstructions from drawings and surveys of the metropolitan museum of art Egyptian Expedition. Perception, pp 1–5

Kieker J (2008) Enhanced safety and efficient construction of masonry structures in Europe collective research final activity report period

Leica Geosystems AG (2017) Leica ScanStation P30/P40 Because Every Detail Matters, 2

van der Lijn F, Roth F, Verhaegen M (2003) Estimating the impulse response of buried objects from ground-penetrating radar signals. Detect Remediat Technol Mines Minelike Targets VIII 5089:387. https://doi.org/10.1117/12.486979

MALA (2017) Mala X3M GPR system - MALA Ground Penetrating Radar–GPR. https://www.malagpr.com.au/x3m-system.html

Monserrat O, Crosetto M (2008) Deformation measurement using terrestrial laser scanning data and least squares 3D surface matching. ISPRS J Photogramm Remote Sens 63(1):142–154. https://doi.org/10.1016/j.isprsjprs.2007.07.008

Nelson HH (n.d) BETWEEN Episode 26, Where the Im- 30

Nohutcu H, Demir A, Ercan E, Hokelekli E, Altintas G (2015) Investigation of a historic masonry structure by numerical and operational modal analyses. Struct Design Tall Spec Build 24(13):821–834. https://doi.org/10.1002/tal.1213

Nuttens T, Wulf ADE, Bral L, De Wit B, Carlier L, De Ryck M, Stal C et al. (2010) High resolution terrestrial laser scanning for tunnel deformation measurements. In: FIG working week, vol 2010, 11–16

Olhoeft GR (2014) Electrical, magnetic, and geometric properties that determine ground penetrating radar performance
Rüther H, Held C (2011) Challenges in heritage documentation with terrestrial laser scanning. In: Proceedings of AfricaGeo. http://africageodownloads.info/122_ruther.pdf
Twenty-third, Egyptian, Middle Kingdom, Mariam F Ayad, Egyptian Art. n.d. G O d s Wi F e , g O d ' s Se RVant
VENT, Ilse Anne Elisabeth DE (2011) Structural damage in masonry
Walsh G (2015) Leica Scanstation White Paper
Yang H, Xu X, Neumann I (2018) Deformation behavior analysis of composite structures under monotonic loads based on terrestrial laser scanning technology. Compos Struct 183 (1): 594–99. https://doi.org/10.1016/j.compstruct.2017.07.011

Use of Integrated Regional Data and Local Low Impact Investigations for the Geo-Mechanical Characterisation of Rupestrian Cultural Heritage in Apulia (Southern Italy)

Ilenia Argentiero, Maria Dolores Fidelibus, Roberta Pellicani, and Giuseppe Spilotro

Abstract Over time, human activities have modified natural systems generating hybrid systems. The natural/anthropic interaction makes hybrid systems more vulnerable, and less resilient than natural ones. As a consequence, the equilibrium of hybrid systems can evolve through various critical paths culminating in disastrous events, such as local or total collapses, which may lead to the loss of cultural heritages but also of human lives. In such contexts, the assessment of signals, which emerge due to the resilience decrease, could be useful to manage connected hazards. Puglia (Southern Italy) includes widespread rupestrian settlements located on the flanks of typical ephemeral fluvial valleys where calcarenite rock masses outcrop. These settlements of historical and artistic importance well represent hybrid systems, which are characterized by the integration of natural and man-made caves, and urban fabrics. From a geo-mechanical standpoint, calcarenite rocks are characterized by fragility, i.e., under critical stress fields, by sudden failure without significant strains or displacements. This makes it difficult to appreciate signals useful for an early warning and then, for safe management of rupestrian cultural heritages. In the technical testing of cultural heritages, to gain an accurate knowledge of the site stress and deformation the use of non-Destructive Tests (NDTs) is recommended as an alternative to Destructive Tests (DTs). This research analyses emerging signals in three hybrid cave systems (Villa Giustiniani and Santa Candida in Bari and one private property in Ginosa) with in-situ NDTs (temperature, humidity index, ultrasound velocity, rebound number). The NDT results are consistent with the literature and archive DT data concerning

I. Argentiero · G. Spilotro (✉)
IREA - Institute for Electromagnetic Sensing of the Environment, CNR - Research National Council of Italy, Via Amendola 122/D, 70126 Bari, Italy
e-mail: g.spilotro@tin.it

M. D. Fidelibus
DICATECh - Department of Civil, Environmental Land, Building Engineering and Chemistry, Polytechnic University of Bari, Via Edoardo Orabona 4, 70126 Bari, Italy

R. Pellicani
DiCEM - Department of European and Mediterranean Cultures, University of Basilicata, Via Lanera 20, 75100 Matera, Italy

G. M. El-Qady and C. Margottini (eds.), *Sustainable Conservation of UNESCO and Other Heritage Sites Through Proactive Geosciences*, Springer Geology,
https://doi.org/10.1007/978-3-031-13810-2_22

calcarenite rocks of Puglia and the neighbouring Basilicata region. The application of a combined NDT method, as an alternative to DTs, has proved expeditious, easy to perform, and cheap; results can be useful in supporting the assessment and management of deterioration of cultural heritage.

Keywords Cultural heritage · Hybrid system · NDT · Deterioration · Management

1 Introduction

In the Mediterranean region, rupestrian civilizations took the dimension of a cultural phenomenon resulting from a conscious choice. Firstly, men choose these places to live in natural caves. Subsequently, according to the technical advancement, they were able to model caves according to their needs, creating complex settlement systems (Giordano 1992; Caprara and Dell'Aquila 2008; Musotto 2017).

Several favourable conditions allowed a wide diffusion of this precious cultural heritage in the Puglia region (southern Italy, Fig. 1); there, the rupestrian civilizations perfectly fitted in the local tradition and the natural environment.

The development of the civilizations was favoured by the presence of soft and easily workable calcarenite rocks, and the location of their outcrops at the flank top of "*gravine*" and "*lame*" (local name of deep and shallower incisions, respectively). They are crossed by ephemeral streams; the former is characterized by steep flanks and the latter by gradual sloping flanks.

These cultural heritages perfectly represent "hybrid systems", in which the interaction between natural and human factors causes an increase in the vulnerability of systems and therefore a decrease of their resilience. Since hybrid systems behave as complex systems, such conditions allow approaching tipping points, so triggering an abrupt shift from a metastable equilibrium state to an alternative equilibrium state. An estimate of the resilience variation over time of a generic complex system can be performed by identifying the related emerging signals (Pellicani et al. 2017). Thus, for a hybrid (complex) system as a cultural heritage, we consider that the assessment of emerging signals and their evolution over time is fundamental. If in some rupestrian settlements instability phenomena have already occurred, in other settlements only emerging signals, mainly expressed by the variations of the physical parameters of materials (as in the geo-mechanical behaviour of the natural or man-made caves), can be observed. Sudden variations of measured parameters and their critical combinations can lead to local or progressive collapse (Pellicani et al. 2017). The excellent characteristics of calcarenites (strength behaviour and workability) are influenced by a series of factors (such as weathering, syngenetic and post-depositional history, Spilotro et al. 1993) that may cause strength degradation and instability phenomena. So, variations of technical parameters should be regularly monitored over time.

The physical and mechanical behaviour of the rock mass can be determined by laboratory tests, but the sample extraction from natural/man-made caves of great cultural value, i.e. in the presence of architectural structures and frescoes, is one

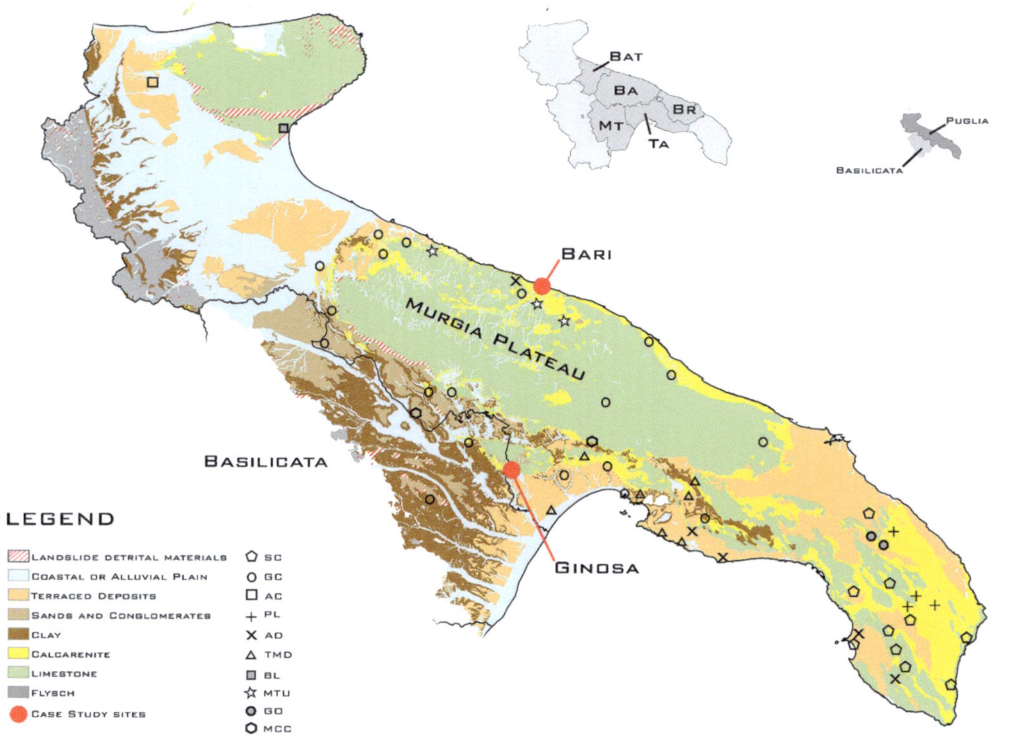

Fig. 1 Schematic geological map of Puglia showing the location of study sites. Geographical distribution of DT samples per calcarenite category: SC: Salento Calcarenites; GC: Gravina Calcarenites; AC: Apricena Calcarenites; PL: Pietra Leccese; AD: Ancient Duna; TMD: Terraced Marine Deposits; BL: Briozoi Limestone; MTU: Murge Tuffs; GO: Calcarenite of Gallipoli Outcrops; MCC: Monte Castiglione Calcarenites

of the major issues (Moropoulou et al. 2013). Therefore, to evaluate the properties of materials and their potential role on the statics of rupestrian systems it becomes essential to use NDTs. This way, it is feasible to detect emerging signals, such as local variation of physical–mechanical parameters, otherwise impossible to assess by destructive testing (DT) only.

Our research firstly included the collection of data about DTs already completed in Puglia and the neighbouring Basilicata region (Fig. 1) on calcarenite samples. Then, NDT techniques as the thermo-hygrometric, ultrasound velocity, and rebound hammer were selected. These ND techniques are widely used in the laboratory, but literature shows only a few examples of their application to natural or man-made caves. For performing the ND tests, three study sites were identified as representative of the hybrid systems in Puglia: Villa Giustiniani and Santa Candida in Bari and one private property in Ginosa (Fig. 1). Finally, collected NDT data were compared with those of DT database.

2 Regional Data and Local Case Studies

Rupestrian settlements were established within sites characterised by both the presence of calcarenite rocks, and a favourable condition for human life. Calcarenite is widespread both in Puglia and at its western side, bordering the Basilicata region (Fig. 1). The settlements are more extensive in *gravine* and *lame* within the outcrops of calcarenite. In the *gravine*, the vertical flanks have favoured either the modelling of natural pre-existent caves or the excavation of new chambers, with horizontal development perpendicular to the slope of *gravine*, often on several levels. On the contrary, since the lame are shallower than *gravine* and show a gradual slope, only ancillary structures as necropolis and churches developed.

These rupestrian civilizations appear as a dynamic phenomenon: the natural and man-made caves are adapted from time to time to new needs, sometimes even disrupting the original plans of life (Bertelli 2007). This gave rise to complex rupestrian systems characterized by the fusion of natural and artificial elements, where natural and man-made caves coexist with buildings made with the excavated material (Grassi et al. 2006). In more recent times, often urban expansion did not consider the complexity of these systems due to the lack of specific technical regulations and adequate control of competent authorities. Indeed, reductions in the thickness of the calcarenite walls of rupestrian structures and the consequent loss of lateral confining pressure (unloading) or the lack of adequate rainwater drainage systems (with consequent problems of water absorption and infiltration), have often led to local and progressive collapses.

2.1 *Geological Setting*

Apart from factors due to human interventions, the behaviour of calcarenites is influenced by their syngenetic and post-depositional history.

Calcarenite rocks are located on the flanks of the Murgia Plateau, an elongated calcareous structure bounded to the East by the Adriatic Sea and to the West by a deep tectonic trough filled by marine clastic sediments (Fig. 1). On both flanks of the Murgia Plateau, the marine degradation of limestones, bioclastic sediments, and other minor mineralogical components determined large sedimentation of calcarenite rocks. They are composed of major granular elements, matrix, and cement, which proportion controls their mechanical behaviour. When the granular elements dominate the rock behaviour, it exhibits a skeleton structure with a high void ratio and brittleness. When matrix content is high, the void ratio is reduced and the matrix works as a stabilizer. However, the most important component of calcarenite rocks is carbonate cement, which is structured according to different morphologies and cementation phases.

The physical properties of calcarenite rocks, in terms of fabrics, petrophysical properties, and anisotropy, are influenced by weathering processes (i.e., water infiltration, environmental, climatic, and microclimatic conditions, constant load, or tension release), acting on the long-term by altering mechanical properties and inducing a variation of the stress–strain behaviour.

Therefore, rupestrian settlements can be affected by failure even centuries after their construction (Dell Anna et al. 1968; Cotecchia et al. 1985; Spilotro et al. 1993, 2015, 2016; Ciantia and Castellanza 2016).

2.2 *Local Case Studies*

The NDTs were performed on three sites in Puglia. The three sites, which differ as to their history and boundary conditions, are representative of hybrid systems.

Villa Giustiniani (VG) is a rupestrian settlement, formerly a Benedictine farm from the 9th to the twelfth century, located in the urban context of the municipality of Bari; it develops below a building from the late nineteenth century, used until recently as a cowshed and deposit (Fig. 2).

The rupestrian church of Santa Candida (SC) is located in the municipality of Bari and dates back to the X–XI century. It is excavated on the upper part of the right side of the Picone *lama*. The discovery at an ancient riverbed of artifacts and architectural remains such as walls, steps and wells confirm the presence of a settlement in the area. SC is the largest rupestrian basilica of Puglia (about 120 square meters) and has a complex fan-shaped plan (Fig. 3).

The third investigated site is a private property (CP) in the historic centre of Ginosa, located on the slope of the *gravina* close to the border of a collapsed area (Fig. 4). The town of Ginosa is a good example of natural/man-made complexes, where rupestrian

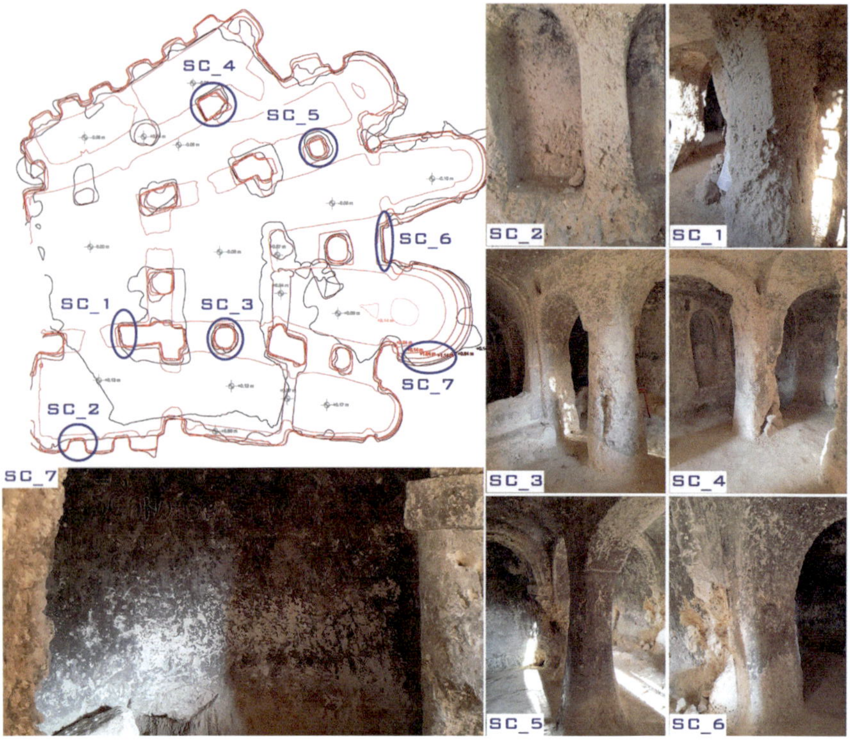

Fig. 2 Layout of Villa Giustiniani and analysed areas

structures have been completely integrated into more recent buildings. All the human modifications have reduced in the time the resilience of this hybrid system, to the extent that, in January 2014, the site suffered a progressive collapse with damage to structures and infrastructures fortunately without victims.

3 Collection of Data

The research consisted of three phases. In the first phase, DT data were collected to create a calcarenite DT database from several sources. These data were analysed and organised by physical and mechanical parameters. In the second phase, several activities were executed in the study sites to obtain physical–mechanical and thermal-hygrometric data. These data were analysed and collected in a NDT database. The third and final phase concerned the comparison between DT and NDT databases.

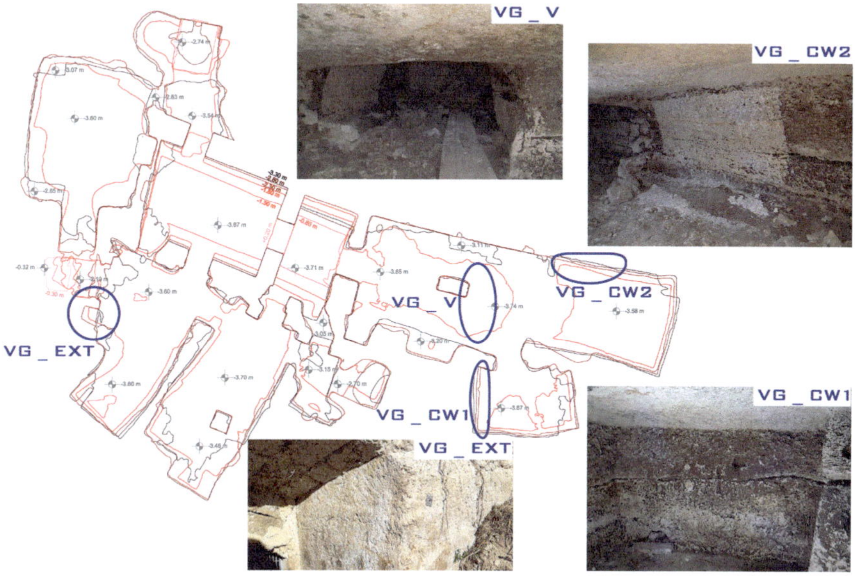

Fig. 3 Layout of Santa Candida and analysed areas

Fig. 4 Location of the private cavity in Ginosa and collapsed area

3.1 Regional Database of Destructive Tests

The physical–mechanical data of calcarenite obtained with DT were acquired from archives and reports (Puglia 1984). The final database consists of physical and mechanical information concerning 384 samples collected in Puglia and its western side bordering the Basilicata region. They were obtained both as cores from drillings and from in situ samplings (from quarries, natural and man-made caves, and natural outcrops).

First, data were subject to standardization concerning used techniques, physical–mechanical parameters, and the measurement units. Subsequently, samples were geographically identified and classified. The merge of DT databases allowed the identification of 10 calcarenite categories (Fig. 1), which are mainly distinguished because of their age and fabric. In particular, calcarenite rocks of SC and VG study sites (located at the Adriatic border of Puglia and of Plio-Pleistocene age) belong to the Murge Tuff category (MTU, where tuff is a local term for calcarenite). They are characterized by frequent fossiliferous levels and sometimes by a predominant pelitic fraction (Azzaroli and Valduga 1967; Cotecchia et al. 1985). CP calcarenite rocks, of Plio-Pleistocene age and located on the Murgia Plateau and at the western side of Puglia bordering the Basilicata region, belong to Gravina Calcarenite (GC) category. GC calcarenite shows a variable grain size and carbonate content between 90 and 99% (Andriani and Walsh 2002; Azzaroli and Valduga 1967). The above category classification has been preserved and the samples were not reclassified according to the more recent classifications (Moretti et al. 2011; Pieri et al. 2011; Ricchetti et al. 2013).

The following parameters have been considered: Bulk Unit Weight, γ_b (kN/m^3), the weight per unit volume of the rock sample in the natural state (AGI 1994); Imbibition Coefficient, IC (%), percentage ratio between the mass of water retained by the sample brought to saturation and the mass of the dried sample (CNR BU 137/92); Uniaxial Compressive Strength, UCS (MPa), obtained by unconfined compression tests and corresponding to the maximum axial compressive stress that the sample can withstand under unconfined conditions (UNI EN ISO 17892–7: 2018); Young's modulus, E (MPa), obtained by unconfined compression tests that measure elastic behaviour of the sample (UNI EN ISO 17892–7: 2018).

3.2 Local Non-destructive Tests

The NDTs were planned considering the aim of the research: economic and expeditious tests that could characterize the rupestrian structures, whose results could be also comparable with the DT data. Thus, the following tests were chosen: point measurements of temperature and humidity, ultrasonic and rebound hammer tests.

A standard procedure was adopted to perform the measurements. First, the locations on a given surface to be analysed were identified: in case of walls or vaults, equidistant (generally 15 cm) points for measures were identified on the related diagonals (UNI EN 12,504–4:2005); in case of columns, the points for measures were generally chosen on the same circumference based on presence and characteristics of fractures or deteriorated areas (direct and semi-direct methods).

Ultrasound Pulse Velocity (UPV) was measured with the ultrasonic equipment Solgeo CMS. Data processing was performed with the software Datasonic System. The rebound hammer test was carried out by Geohammer (DRC); correlation curves were developed between the Rebound Number (RN) and the compression strength, calculated from data of different types of rocks, including Pietra Leccese category (PL).

Thermo-hygrometric measurements have been carried out with Trotec-BP25 and Trotec-BM20. Compressive strength was estimated using the SONREB (SONic + REBound) combined method, created for concrete to reduce the uncertainties, which occur separately using the UPV local values and RNs with the relation (1) (Di Leo and Pascale 1994)

$$\text{Compressive strength} = 1.2 \cdot 10^{-9} \cdot \text{RN}^{1.058} \cdot \text{UPV}^{2.446} \tag{1}$$

where compressive strength is expressed in [N/mm^2] and UPV in [m/s]

The dynamic Young's modulus E_d was calculated by the following equations

$$E_d = \rho_{bulk} V_P^2 \frac{(1-2\nu)(1+\nu)}{(1-\nu)} \tag{2}$$

$$\nu = \frac{V_P^2 - V_S^2}{2\left(V_P^2 - V_S^2\right)} \tag{3}$$

where, ν is the Poisson coefficient, V_p and V_s are the compressional and shear ultrasonic wave velocities, respectively, and ρ_{bulk} is the bulk density of the material.

Some physical parameters as ν and ρ_{bulk}, required for the application of the Eq. (2), cannot be obtained from the NDT tests. In this research, we used the values from the DT database associated with samples referred to the study sites.

4 Results and Discussion

The 384 records of the DT database allowed classifying the samples (Fig. 5) according to the calcarenite classification of Andriani and Walsh (2010). Most of the samples are distributed in the classes from A3 to A1, i.e. from very to moderately soft rocks; only a few samples are extremely soft. Regrettably, this classification does not include all the database records due to the lack of E values.

DT records show wide variability in the values of the selected physical and mechanical parameters. This variability is justified by the variation of calcarenite fabrics: as an example, the more massive the carbonate structure with fewer voids than

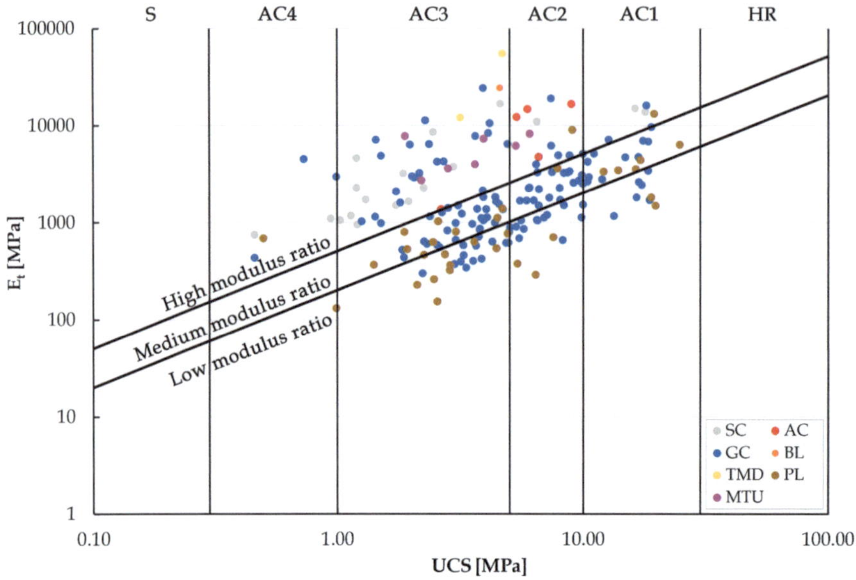

Fig. 5 Classification of calcarenite (Fig. 1) according to Deere and Miller (1966), ISRM (1978) and Andriani and Walsh (2010) per UCS range: Soil (S), Extremely soft (AC4), Very soft (AC3), Soft (AC2), Moderately soft (AC1), Hard Rock (HR)

the finer fragments and more delicate shells, the greater the rock strength (Andriani and Walsh 2007). The DT database analysis also highlights variations within each calcarenite category, which can be traced back to the single features as syngenetic properties and post-depositional diagenesis.

The GC category shows a marked difference between the samples collected in Puglia (in the provinces of Bari—BA, Brindisi—BR, Barletta-Andria-Trani—BAT, Taranto—TA) and those collected in its western border (province of Matera—MT, Basilicata) (Fig. 6). Despite being syngenetic formations, it can be assumed that their different behaviour is due to the different post-depositional history. The GC-MT calcarenite rocks, which outcrop on the western Puglia border and in the surrounding territories of Basilicata, were subjected to a higher lithostatic weight (marine clastic covers) than the other GC outcrops: the weak fabrics were destroyed, while the strong ones were reinforced by cement, so showing a harder strength because of an increase in density and a decrease in porosity of the rock (Festa et al. 2018).

The PL category shows a random variability in the range 13–20 kN/m^3 of γ_b (Fig. 7). The variation of the geotechnical characteristics within the same category can be attributed to anisotropy and petrographic features of the samples: the presence of lithoclasts, and the amount and the type of calcite cement are factors that influence the UCS of carbonate soft rocks as stated by Andriani and Walsh (2010), Ciantia and Castellanza (2016), and Festa et al. (2018).

The mechanical behaviour of calcarenite rocks is strongly dependent on the water content: the presence of interstitial water significantly reduces the strength, especially

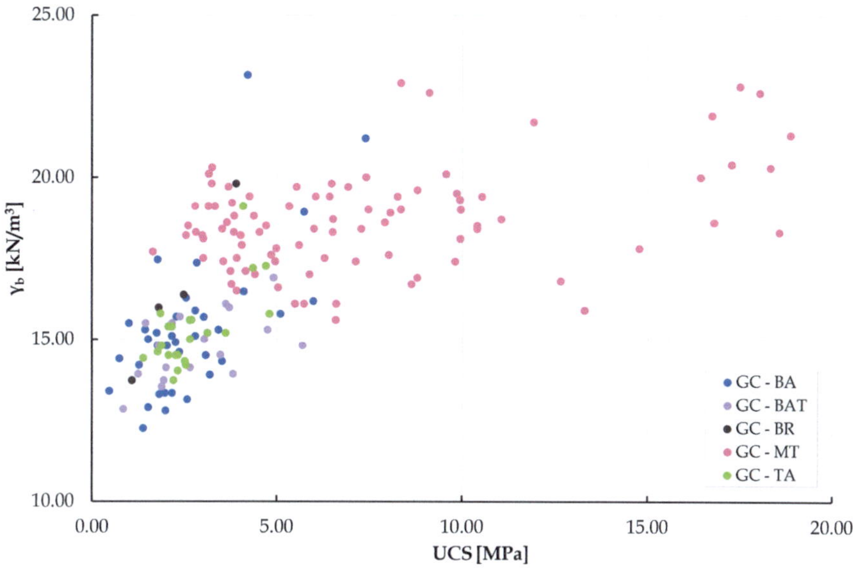

Fig. 6 GC category, divided by provinces (Bari—BA, Brindisi—BR, Barletta-Andria-Trani—BAT, Taranto—TA; Matera—MT): plot of γ_b and UCS

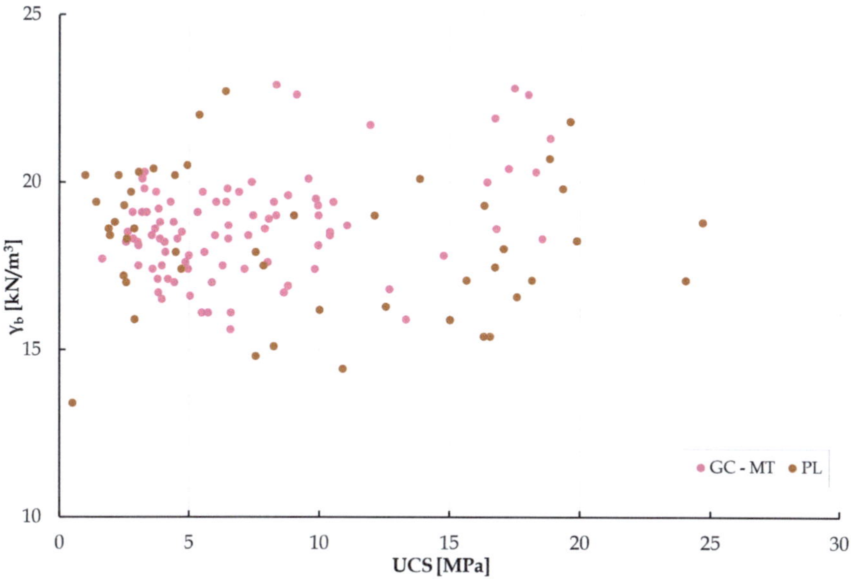

Fig. 7 Comparison between PL and GC-MT categories: plot of γ_b and UCS

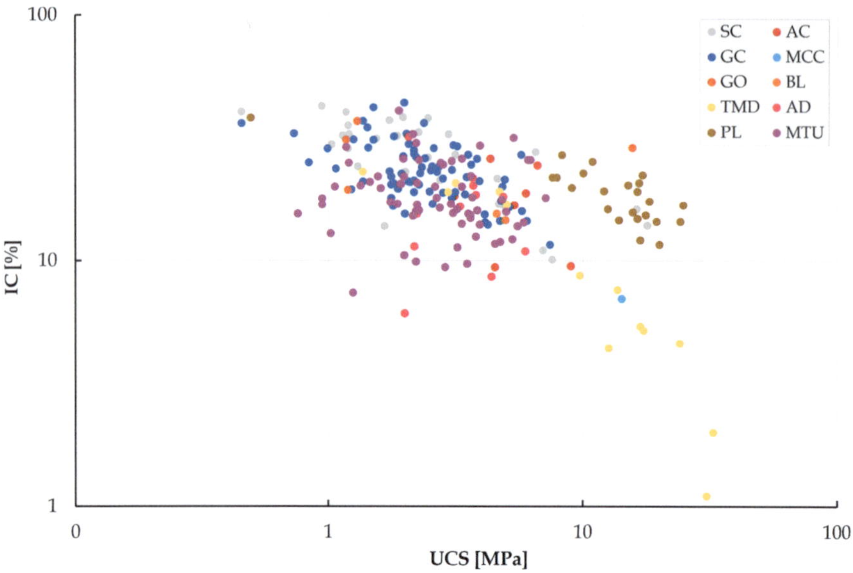

Fig. 8 Lithological units: logarithmic plot of IC and UCS

in weak cemented and fine-grained rocks (Andriani and Walsh 2007). Figure 8 shows a decrease of UCS with increasing IC for the calcarenite DT samples. No clear relation between IC and UCS can be deduced from the study of the samples belonging to the MTU category. The MTU deposits are generally cemented; however, they show clayey marl and fossiliferous levels, where often the pelitic part may be predominant, thus conditioning compression strength and water absorption.

The analysis of the DT database made clear that it is difficult to identify unique physical and mechanical properties representative of calcarenite rocks (Spilotro et al. 1993; Cherubini et al. 2007; Festa et al. 2018). Table 1 shows the range of variation of γ_b, UCS, IC, and E for the 10 calcarenite categories. Unluckily, the collected DT records do not include a description of weathering state.

As to the NDT data, Fig. 9 shows the values of temperature, T, and humidity index, HI, referred to the three study sites. They result different because of their boundary conditions. In the CP site, the back wall (a1, blue box) has a very low temperature and high humidity, in contrast with the other walls (b1, blue box), which are closer to the entry; the SC site shows a variation between the measurements made closer to the openings (a2, green box) and those made at the far end of the chambers (b2, green box). The VG site, on the other hand, does not show any T and HI variability within it (a3, red box), probably because it is completely developed underground; however, there is a remarkable difference between the internal and external T and HI (b3, red box). Other aspects distinguish the sites. The VG site is below street level and therefore is influenced not only by capillary rise but also by weather conditions and humidity that comes from the street level; humidity on the vault and the walls is also favoured by the action of root proliferation. On the contrary, the SC site is completely

Table 1 Variation range of the physical and mechanical properties of the calcarenite categories

Cat.	Samples	Range			
		γ_b	UCS	E	IC
	No	kN/m^3	Mpa	Mpa	%
SC	37	12.0–20.5	0.5–17.8	666–17,388	10.1–42.5
AC	8	14.8–19.7	2.6–8.9	970–17,665	9.4–26
GC	181	12.3–23.1	0.5–18.8	487–21,063	11.6–43.9
MCC	2	16.5–18.5	5.5–14.2	–	7–13.8
GO	4	12.0–19.1	1.2–15.6	–	19.4–36.9
BL	2	15.7–19.1	4.6–4.9	–	14.6–15.5
TMD	20	12.9–24.5	1.4–30.6	15,449–55,915	1.1–23.0
AD	11	14.5–21.6	2.0 –5.9	–	6.1–31.9
PL	54	13.4–20.1	0.5–24.7	640–13,243	8.5–38.1
MTU	65	12.4–19.0	0.7–7.1	2050–10,549	7.4–40.7

exposed to the atmosphere and well ventilated, thus showing higher temperature and lower humidity compared to the other sites.

Figure 10 shows the relation between UPV and CS values obtained from Eq. (1): sample distribution shows a clear gap between the CS values obtained from tests carried out on intact walls or columns (values higher than 1.2 MPa) from those that show fractures or sedimentary discontinuities (values lower than 1.2 MPa). Therefore,

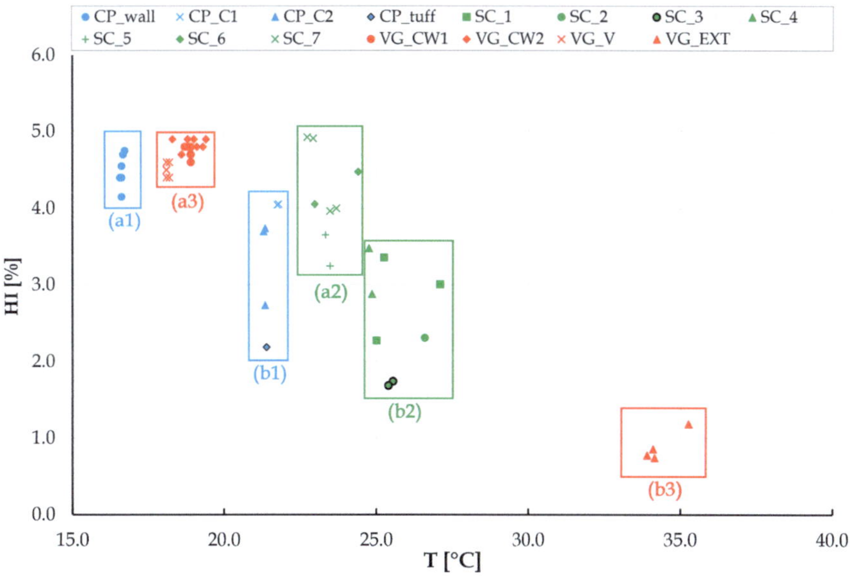

Fig. 9 NDT measures: plot of HI versus T

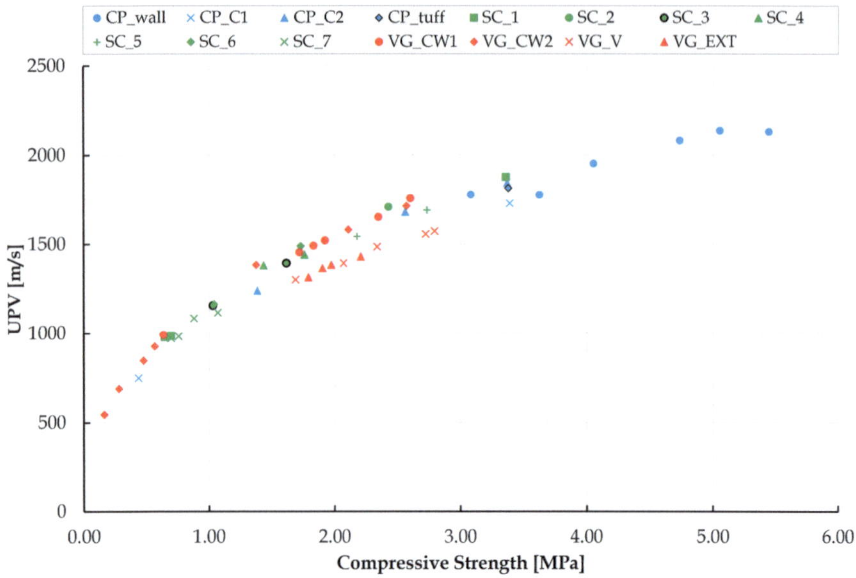

Fig. 10 NDT data analysis: plot of UPV versus Compressive strength

the presence of fractures or sedimentary discontinuities greatly influences the UPV (Varma et al. 2017; Wang et al. 2018; Vasanelli et al. 2014, 2015), that in some cases is halved.

4.1 DT and NDT Results Comparison

The results of the analyses carried out on the single DT and NDT databases have been compared to verify the effectiveness of the NDT approach.

All available NDT and DT data were considered for group them according to the classification proposed in Fig. 5. Figure 11 shows the plot of elastic modulus (calculated with Eqs. 2 and 3) versus the compressive strength (calculated with Eq. 1), both referred to NDT data; the same Fig. 11 shows the relation between E_s and UCS from the DT database. Figure 11 indicates that only the values related to the ND test VG_CW2 (belonging to the VC site, location of the sample in Fig. 2) fall in class S, therefore classified as soil. The ND data standing in the AC4 class (extremely soft calcarenite) refer to measures carried out on surfaces with fractures and sedimentary discontinuities; the rest of the samples is the AC3 class (very soft calcarenite), with only a few samples in AC2 (soft calcarenite). NDT data well describe the mechanical features of MTU and GC categories that characterize the site themselves. The ranges of both DT and NDT data confirm the high variability of calcarenite rock properties.

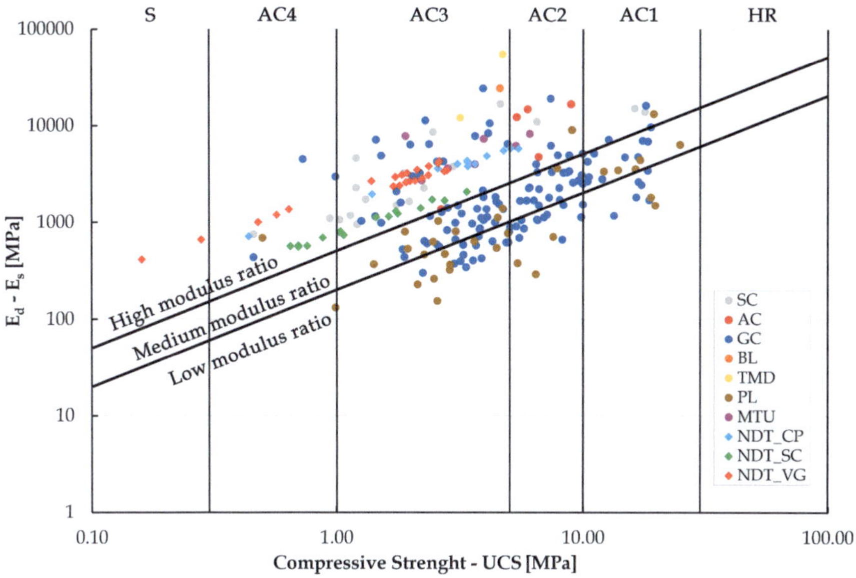

Fig. 11 Classification of calcarenite rocks according to Deere and Miller (1966), ISRM (1978) and Andriani and Walsh (2010). NDT values (E_d and compressive strength) derive from application of Eqs. (1), (2) and (3). (E_s and UCS data) refer to all DT samples

For further detail, the elastic modulus and the compressive strength as defined for Fig. 11 were used to compare the only DT data concerning the samples referred to the three study sites and the NDT data collected in the same sites (Fig. 12). Considered the assumptions previously done for the calculated values, we can assert that for the study sites the NDT measures give good estimations of the concerned mechanical parameters comparable to those related to the DT tests.

The comparability between static and dynamic elastic modulus that emerges from Figs. 11 and 12 is however contradicted by the results obtained from ultrasonic tests performed on samples in the laboratory. In this case, the difference between the two modules is due to the presence of fractures, joints, voids, planes of weakness, and foliation in the specimens, which affect both ultrasonic wave propagation and uniaxial compression tests (Martínez-Martínez et al. 2012; Vasanelli et al. 2015). Many authors have proposed relations between the UPV and the dynamic module E_d, but they are not widely used because they are not practical for engineering purposes (Shen et al. 2017). However, no results were found in the literature regarding the comparison between the dynamic elastic modulus calculated from in situ ultrasound velocity tests and the static elastic modulus obtained from uniaxial compressive tests.

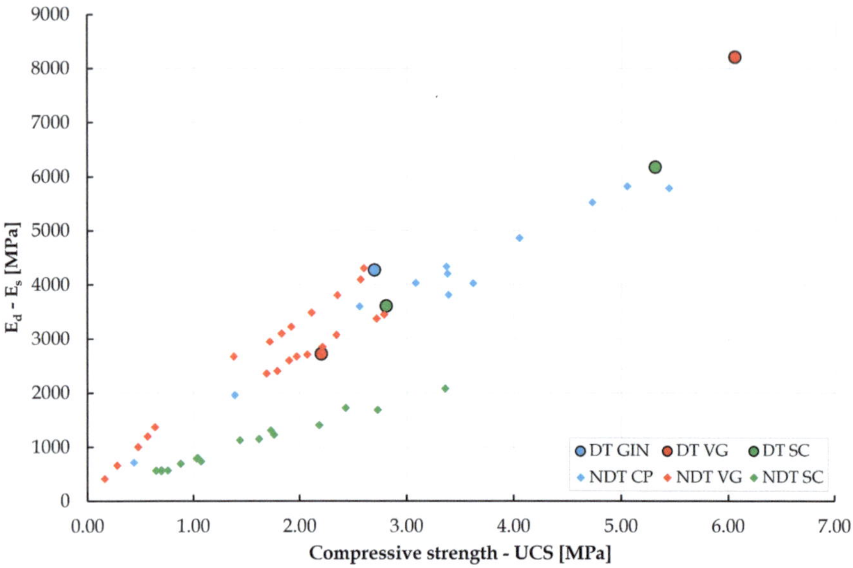

Fig. 12 NDT and DT data comparison: plot of elastic modulus (E) versus Compressive strength. NDT values (E_d and compressive strength) derive from application of Eqs. (1), (2) and (3). DT values are E_s and UCS data of samples taken in study sites

5 Conclusions

This research aimed at verifying the effectiveness of the application of ND techniques as an alternative to DTs in the deterioration assessment of rupestrian hybrid structures consisting of natural and man-made caves developed within calcarenite rocks of Puglia.

The study focused on the analyses of two databases, the first obtained from the collection of DT data concerning the physical and mechanical properties of calcarenite rocks present in Puglia and the second relating to the in situ ND measures carried out in three rupestrian sites developed in calcarenite rocks belonging to the same region; then, DT and NDT data were compared to verify the equivalence between the respective information.

In general, the study proves that NDT data give information congruent with that of DT data. The study also suggests that the application of the UPV test and the SonReb method is very useful for characterizing natural and man-made caves in calcarenite rocks since the reduction in wave velocity is related to the presence of fractures and sedimentary discontinuities and to the material degradation degree. Temperature and humidity evaluation support in understanding the whole deterioration status. The NDT combined approach is expeditious, easy to perform, and economical: an entire site can be investigated avoiding extraction of samples, therefore guaranteeing the conservation of the site. Physical properties not obtainable from ND tests can be obtained from literature and archives.

Certainly, the combined NDT approach needs improvement both in terms of timing and integration with other methods. Current NDT data refer, indeed, to one only period (spring–summer 2019) but a more efficient approach should involve a periodical monitoring to follow the possible deterioration evolution of the sites. For sure the seasonal variation of meteorological conditions in the study sites is likely to cause variations in the internal thermo-hygrometric environment; the variations of such properties, in turn, may modify the results of UPV and rebound hammer tests. Further progress in the assessment of deterioration state of the rupestrian structures could be obtained by performing NDT measures, whenever feasible, on the whole of the exposed surfaces of the sites; with the support of Laser scanner techniques, properties related to NDT measures could be rendered in a three-dimensional model, including purposed geo-mechanical information as well.

A 3D model of rupestrian structures, continuously updated through careful monitoring, could be not only a tool for evaluating the progress over time of deterioration state to set the potential risk and recovery interventions but also a means for supporting the needed action of management and promotion.

Acknowledgements The authors wish to thank the Puglia Regional Government for having shared the calcarenite DT data (PO-FERS Puglia 2007-2013, Fondo Europeo di Sviluppo Regionale—Asse II _ Linea di intervento 2.3—Azione 2.3.6 "Miglioramento del sistema dell'informazione, del monitoraggio e del controllo nel settore della Difesa del Suolo"), and the Bari Municipality and Mr. M. Pastore for having provided access to Villa Giustiniani and Santa Candida, and CP study sites, respectively.

References

AGI - Associazione Geotecnica Italiana (1994) Raccomandazioni AGI 94

Andriani GF, Walsh N (2002) Physical properties and textural parameters of calcarenitic rocks: Qualitative and quantitative evaluations. Eng Geol 67:5–15. https://doi.org/10.1016/S0013-7952(02)00106-0

Andriani GF, Walsh N (2007) The effects of wetting and drying, and marine salt crystallization on calcarenite rocks used as building material in historic monuments. Geol Soc London, Spec Publ 271:179–188. https://doi.org/10.1144/GSL.SP.2007.271.01.18

Andriani GF, Walsh N (2010) Petrophysical and mechanical properties of soft and porous building rocks used in Apulian monuments (south Italy). Geol Soc Spec Publ 333:129–141. https://doi.org/10.1144/SP333.13

Azzaroli A, Valduga A (1967) Note illustrative della Carta Geologica d'Italia alla scala 1:100000 - Foglio 177 e 178 Bari e Mola di Bari. Italy, Rome

Bertelli G (2007) Strutture e morfologie degli insediamenti rupestri. Alcune riflessioni su Lama d'Antico, S. Lorenzo, S. Giovanni, Lamalunga e la Lama di Seppannibale in agro di Fasano. In: Puglia tra grotte e borghi - Atti del III convegno internazionale sulla civiltà rupestre. 24 - 26 Novembre 2005, Menestò, E. Centro Italiano di Studi sull'Alto Medioevo, Savelletri di Fasano (Br), pp 93–117

Caprara R, Dell'Aquila F (2008) Note sull'organizzazione urbanistica degli insediamenti rupestri. Tra Puglia e Mediterraneo. In: de Minicis E (ed) Insediamenti rupestri di età medievale: abitazioni e strutture produttive. Italia centrale e meridionale. Atti del Convegno di studio Grottaferrata, 27–29 ottobre 2005. Centro Italiano di Studi sull'Alto Medioevo, Spoleto

Cherubini C, Reina A, Bruno D (2007) Le rocce tenere del salento: proposta di classificazione con l'uso delle caratteristiche tecniche e meccaniche. Geol e Territ 2:37–47
Ciantia MO, Castellanza R (2016) Modelling weathering effects on the mechanical behaviour of rocks. Eur J Environ Civ Eng 20:1054–1082. https://doi.org/10.1080/19648189.2015.1030086
CNR (1992) Norme sugli aggregati—Determinazione del coefficiente di imbibizione—B.U. 137/92
Cotecchia V, Calò G, Spilotro G (1985) Caratterizzazione geolitologica e tecnica delle calcareniti Pugliesi. Quarr Constr
Deere DU, Miller RP (1966) Engineering classification and index properties of rock. Albuquerque, New Mexico
Dell'Anna L, Garavelli CL, Nuovo G (1968) Sui cosiddetti "Tufi Calcarei" della Regione Pugliese-Lucana (Ricerche preliminari). Period Di Mineral 3:657–715
Di Leo A, Pascale G (1994) Prove non distruttive sulle costruzioni in c.a. Il G delle prove non distruttive, vol 4
DRC - Diagnostic Research Company - Non Destructive Testing (2014) Geohammer - Instruction Manual. pp 1–60
Festa V, Fiore A, Luisi M et al (2018) Petrographic features influencing basic geotechnical parameters of carbonate soft rocks from Apulia (southern Italy). Eng Geol. https://doi.org/10.1016/j.enggeo.2017.12.009
Giordano D (1992) Il comprensorio rupestre appulo-lucano: casali e chiese da Gravina al Bradano. Levante Editori, Bari
Grassi D, Grimaldi S, Simeone V (2006) Localizzazione e problemi di stabilità dei siti rupestri dell'area pugliese. G Di Geol Appl 4:65–72. https://doi.org/10.1474/GGA.2006-04.0-08.0136
ISRM (1978) Suggested methods for the quantitative description of discontinuities in rock masses. Int J Rock Mech Min Sci 15:319–368
Martínez-Martínez J, Benavente D, García-del-Cura MA (2012) Comparison of the static and dynamic elastic modulus in carbonate rocks. Bull Eng Geol Env 71:263–268. https://doi.org/10.1007/s10064-011-0399-y
Moretti M, Pieri P, Ricchetti G, Spalluto L (2011) Note illustrative della Carta Geologica d'Italia alla scala 1:50000 - Foglio 396 - San Severo. Italy, Rome
Moropoulou A, Labropoulos KC, Delegou ET et al (2013) Non-destructive techniques as a tool for the protection of built cultural heritage. Constr Build Mater 48:1222–1239. https://doi.org/10.1016/j.conbuildmat.2013.03.044
Musotto L (2017) Caratteristiche, storia ed evoluzione delle soluzioni insediative ipogee e rupestri in area mediterranea. Geologia dell'Ambiente 4:2–15
Pellicani R, Argentiero I, Parisi A, et al (2017) Resilience modification and dynamic risk assessment in hybrid systems: Study cases in underground settlements of murgia edge (Apulia, Southern Italy). In: Lecture Notes in Computer Science (including subseries lecture notes in artificial intelligence and lecture notes in bioinformatics). pp 230–245
Pieri P, Sabato L, Spalluto L, Tropeano M (2011) Note illustrative della Carta Geologica d'Italia alla scala 1:5000 - Foglio 438 - Bari. Bari
Regione Puglia - BARI (1984) Indagini sui materiali da cava (Calcari, Calcareniti e Argille) della Puglia ai fini di favorirne l'utilizzazione, la valorizzazione e la possibilità di nuovi procedimenti di lavorazione ed estrazione. In: Regione Puglia - Assessorato all'Industria al C e all'Artigianato (ed). Bari, pp 109–214
Ricchetti G, Ciaranfi N, Bortone U, et al (2013) Note illustrative della Carta Geologica d'Italia alla scala 1:50.000. - Foglio 537 Capo S. Maria di Leuca. Rome, Italy
Shen X, Chen M, Lu W, Li L (2017) Using P wave modulus to estimate the mechanical parameters of rock mass. Bull Eng Geol Env 76:1461–1470. https://doi.org/10.1007/s10064-016-0932-0
Spilotro G, Fidelibus MD, Fidelibus C, Zinco MR (1993) Lithological and geotechnical features of the calcarenites in the West of the murgian platform
Spilotro G, Fidelibus MD, Pellicani R, et al (2016) Il patrimonio architettonico di Matera e i materiali naturali da costruzione: nel tufo e col tufo. Caratterizzazione tecnica delle calcareniti e variazioni

per condizioni ambientali. Geol Territ Ambient:10–24. https://doi.org/10.13140/RG.2.1.3994.1209
Spilotro G, Qeraxhiu L, Pellicani R, Argentiero I (2015) Caratteristiche tecniche delle rocce calcarenitiche e loro variabilità in relazione all'ambiente di esposizione. In: Colonna A, Conte A, Di Ginosa FP (eds) Laboratorio di pratiche della conoscenza nei Sassi di Matera. ARCHIVIA, Matera, pp 81–84
Trotec Trotec BP25—Operating manual
Trotec Trotec BM20—Operating manual
UNI—Ente Nazionale Italiano di Unificazione (2018) UNI EN ISO 17892–7:2018—Indagini e prove geotecniche—Prove di laboratorio sui terreni—Parte 7: Prova di compressione non confinata
UNI (ENTE ITALIANO DI NORMAZIONE) (2005) UNI EN 12504–4
Varma M, Maji VB, Boominathan A (2017) A study on ultrasonic wave propagation across fractures in jointed rocks. In: 51st US Rock Mech/Geomechanics Symposium 2017, vol 3, pp 1792–1796
Vasanelli E, Colangiuli D, Calia A et al (2015) Ultrasonic pulse velocity for the evaluation of physical and mechanical properties of a highly porous building limestone. Ultrasonics 60:33–40. https://doi.org/10.1016/j.ultras.2015.02.010
Vasanelli E, Sileo M, Leucci G et al (2014) Mechanical characterization of building stones through DT and NDT tests: Research of correlations for the In Situ analysis of ancient masonry. Key Eng Mater 628:85–89. https://doi.org/10.4028/www.scientific.net/kem.628.85
Wang Z, Wang J, Yang S (2018) An ultrasonic-based method for longwall top-coal cavability assessment. Int J Rock Mech Min Sci 112:209–225. https://doi.org/10.1016/j.ijrmms.2018.10.019

Geological Factors Controlling Evolution of Theban Tomb Stability, Luxor

Andrea Wolter, Martin Ziegler, Rachael Collldeweih, Andrea Loprieno-Gnirs, Rodrigo Alcaino-Olivares, and Matthew Perras

Abstract Many factors can influence the stability of natural and cultural heritage sites, including geohazards such as flooding, seismic activity and mass movement, as well as rock weathering, swelling and thermal fatigue. Understanding these phenomena enables more efficient and effective preservation of important locations. The so-called Theban nobles' tombs at Sheikh 'Abd el-Qurna (SAQ) in Luxor, Egypt, showcase the geological complexity that affects tomb stability. While some of these rock-cut tombs were excavated during the early Middle Kingdom (twentieth century BC), tomb construction at SAQ peaked during the 18th Dynasty (15th–14th centuries BC) and included large tombs with pillared halls. The cemetery occupies a paleo-landslide deposit comprising mainly the Thebes Limestone Formation, originating from the high cliffs of the Theban Plateau above the Nile valley and overlying the weaker Esna Shale Formation. This paper explores the geological and geotechnical factors that contribute to the evolution of the stability of the tombs. We address how resistant-over-recessive lithology, mass movement, and rock mass properties possibly influenced tomb construction procedures and strategies. We provide a geological map of the investigated archaeological area, describe discontinuity properties, and classify the rock mass using the Rock Mass Rating (RMR) and Geological Strength Index (GSI) systems, based on field traverses and the analysis of landscape and local photogrammetric and laser scanning datasets. We also characterise the stability of rock pillars within tombs, using a rating system. Results show that significant modifications to design of large tombs were rare when tomb builders encountered weaker

A. Wolter (✉)
Engineering Geology, GNS Science, 1 Fairway Ave, Lower Hutt, New Zealand
e-mail: a.wolter@gns.cri.nz

M. Ziegler · R. Collldeweih
Department of Earth Sciences, ETH Zurich, Sonneggstrasse 5, Zurich, Switzerland

A. Loprieno-Gnirs
Department of Ancient Civilizations, University of Basel, Petersgraben 51, Basel, Switzerland

R. Alcaino-Olivares · M. Perras
Department of Civil Engineering, Lassonde School of Engineering, York University, 11 Aboretum Lane, Toronto, Canada

G. M. El-Qady and C. Margottini (eds.), *Sustainable Conservation of UNESCO and Other Heritage Sites Through Proactive Geosciences*, Springer Geology,
https://doi.org/10.1007/978-3-031-13810-2_23

rock masses. Rock failures during and after tomb construction occurred because of unfavourably oriented tension cracks and faults or disintegration of weathered nodular limestone. We are currently monitoring a pillared tomb at SAQ (TT95A), using non-invasive seismometer, extensometer, temperature and relative humidity sensors, and recommend the monitoring of metastable pillars on a broader scale to identify long-term failure processes and prevent further damage and collapse.

Keywords Rock mechanics · Pillar stability · Geological control · Environmental monitoring · Theban tombs · Sheikh 'Abd el-Qurna

1 Introduction

The Sheikh 'Abd el-Qurna (SAQ) cemetery, part of the UNESCO World Heritage Site of Ancient Thebes, exemplifies how complex geological settings influence preservation of cultural heritage sites. The tombs were excavated in shallow rock masses for high dignitaries and their families, and commonly included open courtyards and decorated subsurface halls, with long shafts leading to remote burial chambers. Larger complexes were often constructed with spacious pillared halls. Most tombs at SAQ were constructed during the 15th–14th century BC (18th Dynasty), but several tombs date back to the early Middle Kingdom (twentieth century BC).

This paper summarises the key findings of Ziegler et al. (2019), part of the "Life Histories of Theban Tombs" interdisciplinary project led by the University of Basel, Switzerland, and presents geological and geotechnical investigations of the western and central SAQ area. It focusses on the evolution of the tombs after construction, highlighting modern tomb instabilities and conservation efforts. The geomorphological and geological context of the SAQ cemetery are highlighted in a geological map of the study area, followed by an assessment of pillar stability and ongoing monitoring activities at one specific tomb (TT95A). Detailed methods and supplementary materials can be found in Ziegler et al. (2019).

2 Geology and Rock Mass Quality

2.1 Geomorphological and Geological Context

The SAQ hill is located west of Luxor, Egypt, below the high cliffs at the edge of the Theban Plateau (Fig. 1a). The "hill" is in fact an ancient landslide deposit, presumably slumping from the Theban Plateau during the wetter Pleistocene (Dupuis et al. 2011). This paleo-slumping phenomenon is relatively common along the edge of the steep Theban Plateau cliffs, as evidenced by other slump blocks adjacent to SAQ and in the foothills along the Nile River valley (Karakhanyan et al. 2010) (Fig. 1b).

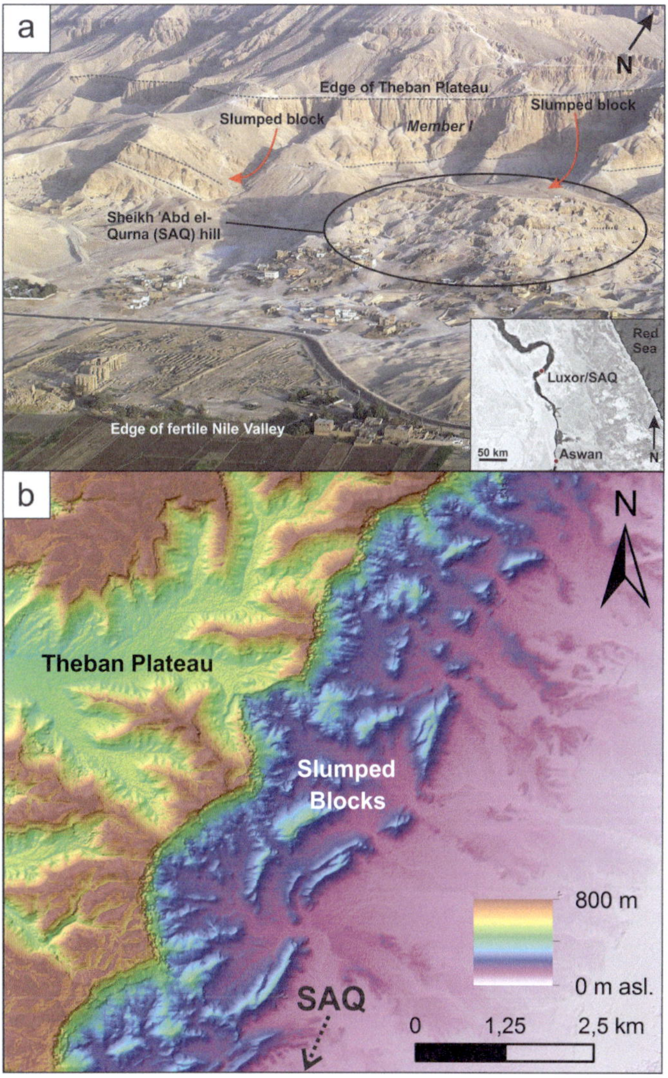

Fig. 1 **a** Setting of the Sheikh 'Abd el-Qurna (SAQ) hillside cemetery at the edge of the Theban Plateau near Luxor, Egypt (image: Raimond Spekking, 25 July 2008). **b** Elevation model based on TerraSAR-X data showing displaced, slumped blocks at the edge of the Theban Plateau

The local stratigraphy comprises the massive Thebes Limestone Formation overlying the weak Esna Shale Formation (cf. Curtis 1979; Lazar 1995; Wuest 1995; Dupuis et al. 2011; King et al. 2017 for detailed stratigraphy in the area). This resistant-over-recessive sequence is one of the main preconditions for instability in the area.

After deposition in a shallow basin environment from the Late Paleocene to early Eocene, the local rock formations were uplifted and affected by the opening of the Red Sea in the Oligocene and erosion of the Paleo-Nile basin in the Miocene (Youssef 1968; Lazar 1995; Wuest 1995; Shaaban 2004). A transtensional regime dominates the recent tectonic history (Wuest 1995). Lazar (1995) identified four dominant tectonic discontinuity sets in the Thebes Limestone Formation: two subvertical joint sets striking N-S and WNW-ESE, and two fault sets dipping 60–65° to the SW and NE. Local seismicity is moderate and earthquakes likely affected ancient tombs (Maamoun et al. 1984; Karakhanyan et al. 2010; Sawires et al. 2015).

Due to their high clay content (29–50%; Rutherford 1990; Wuest and McLane 2000; Tawfik et al. 2011), the marlstones in the Esna Shale and base of the Thebes Limestone formations are prone to swelling, which can lead to an increase in local stress and rock damage through fracture formation. For example, increased fracture frequency can be seen in the marlstones of the Thebes Limestone Formation, overlying the Esna Shale.

Considering the age of the tombs, an understanding of the effects of long-term degradation processes in a changing environment and of the geological setting on their stability is fundamental. Episodic heavy precipitation leading to water infiltration, flooding, seismic activity, and thermo-hydro-mechanical damage related to temperature and humidity cycles may contribute to rock mass strength degradation and instability.

2.2 Geological Mapping and Rock Mass Characterisation

We mapped the geology of western SAQ based on high-resolution LiDAR and photogrammetry data and two surface and near-surface field transects. We focussed on geotechnical parameters of rock units, such as rock mass strength, fracture density and condition, and properties of persistent faults and tension cracks. To evaluate the rock mass, we applied discontinuity characterisation along scanlines, Schmidt hammer testing, Geological Strength Index (GSI; Marinos and Hoek 2000; Cai etal. 2004), and Rock Mass Rating (RMR; Bieniawski 1973, 1989) at several tombs. Note that large areas of SAQ are covered by debris, resulting in uncertainty of unit thicknesses and locations of geological contacts.

The detailed mapping of west and central SAQ (Fig. 2) indicates that the rock mass can be separated into six rock mass domains: (1) massive limestone, (2) three thin marlstone to argillaceous limestone beds, (3) lower, intermediate, and upper nodular limestones, (4) two coquinite (hard) beds, (5) bedded limestone, and (6) bedded limestone with tabular chert interbeds.

Three dominant discontinuity sets were identified at SAQ. Bedding dips 20–60° with a dip direction of 324–043°; mean bedding orientation is 36°/349°. Tension cracks and faults mapped at SAQ have two main strike orientations: (i) WSW-ENE and (ii) NNW-SSE. The latter set agrees with regional tectonic orientations and is likely related to extension during the opening of the Red Sea; the former set is

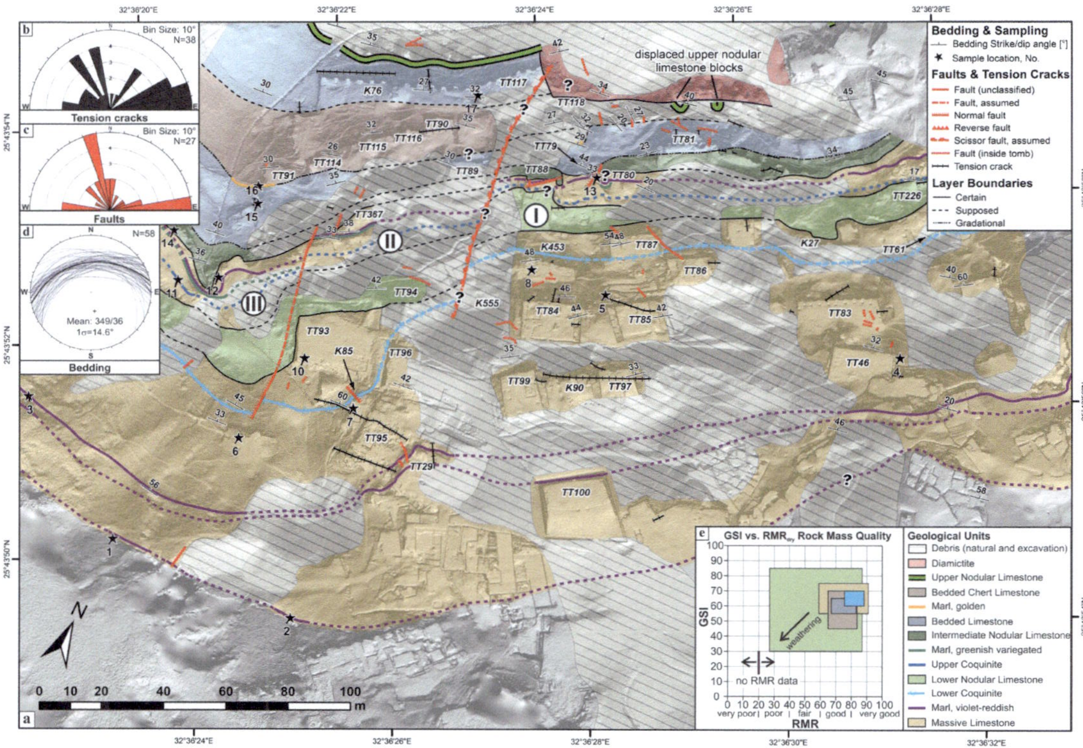

Fig. 2 **a** Geological map of western SAQ showing possible slump compartments (I–III) separated by possible gravitational faults (red lines). **b**, **c** Rose plots of tension crack and fault orientations, with dominant sets oriented ENE-WSW and NNW-SSE. **d** Stereonet of bedding measurements; mean bedding oriented 36°/349° (dip/dip direction). **e** Plot of Geological Strength Index (GSI) vs Rock Mass Rating in dry conditions (RMR_{dry}) for the six rock mass domains determined during mapping

Table 1 Summary of rock mass quality by rock mass domain using the Rock Mass Rating (RMR) and Geological Strength Index (GSI) systems. RMR_{wet} approximates rock mass condition during periods of flash flooding. For more details, please refer to Ziegler et al. (2019)

Domain	RMR_{dry}	RMR_{wet}	GSI
1	54–91 (good-very good)	39–76 (fair-good)	55–75
2	NA	NA	15–30
3	27–87 (poor-very good)	12–72 (very poor-very good)	30–85
4	74–88 (good-very good)	59–73 (good)	60–70
5	65–83 (good-very good)	50–68 (fair-good)	55–65
6	65–84 (good-very good)	50–69 (fair-good)	45–70

most likely related to slumping. Tension cracks have up to 50 m persistence and 0.3 m apertures, and can contain no, unconsolidated or cemented infill. Faults show a few decimetres to maximum 10 m apparent displacement. Where movement type could be determined, most faults were classified as normal faults with a strike-slip component. Both inherited tectonic faults and gravitational faults are present at SAQ, highlighting a complex evolutionary history. Faulting and slumping have altered the stratigraphy at SAQ. For example, there seem to be two slumping-related discontinuities striking SSW, with apparent offsets of about 10 m, possibly delineating landslide compartments (I–III on Fig. 2).

Rock mass quality for each of the domains is summarised in Table 1. Most of the domains include good to very good quality rock masses in dry conditions. Only the nodular limestones and marlstones have poor rock mass quality in places. The nodular limestones are high quality when unweathered, but dramatically decrease in strength when weathered. The marlstones are weak and disintegrated at surface; hence, RMR could not be determined for this domain. In wet conditions, such as during flash flooding, the RMR values decrease by 15.

3 Pillar Stability

After completing geological mapping and rock mass characterisation of the lithologies found at SAQ, we assessed the stability of 82 individual pillars in the halls of eleven rock-cut tombs. All pillars are near-surface and their stability is dominated by geological structures; hence, we did not consider rock stresses in our evaluation.

After mapping pillar dimensions and fracture density, we applied Esterhuizen et al.'s (2005) pillar structure rating system. In this visual system, pillar shape is assessed based on the influence of geological structures—Class I pillars show little damage and Class V pillars are bisected by fully persistent structures. We added an additional class for missing or failed pillars (Class VI) (Fig. 3).

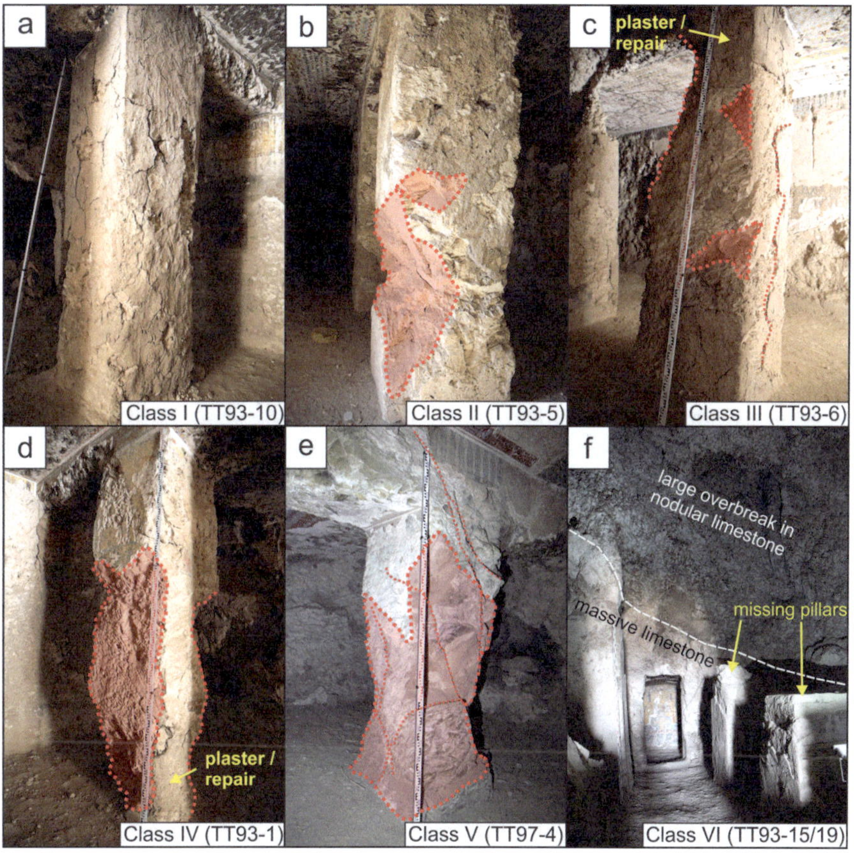

Fig. 3 Examples of the pillar classes applied to the SAQ tomb pillars, after Esterhuizen et al. (2005). Tomb and pillar numbers are indicated (cf. Ziegler et al. 2019 for locations). Thick red lines and red areas show fall-out, and thin red dashed lines show fracture traces

Most of the pillars assessed have a low width/height ratio (53% have a W/H ratio < 0.35), indicating slim pillars, and show no to minor damage (19% were designated Class I, and 55% Class II of Esterhuizen et al. 2005) (Fig. 4a, b). Two pillars in TT97 were designated as Class V pillars, as they were bisected by faults dipping > 35°. Eight pillars were missing or completely collapsed, one from each of TT95, TT94 and TT367 and five from TT93.

Discontinuities intersecting the pillars are typically smooth to slightly rough, tight (<1 mm) to open (30 mm), and moderately to highly weathered. Where persistent faults and tension cracks intersect pillars, pillar stability is low (e.g. Figure 5a).

Other factors affecting pillar stability are lithology and pillar dimensions. Pillars constructed in nodular limestone (e.g. those in TT93) are more prone to collapse. Weak marl layers intersecting pillars also decrease pillar stability. There is a weak positive correlation between pillar W/H and stability, suggesting that stout pillars are more stable (Fig. 4c).

4 Current Environmental Monitoring

To assess environmental factors affecting the long-term stability of tombs, researchers from the University of Basel and ETH Zurich, Switzerland and York University, Canada are conducting an ongoing monitoring program. The focus to date is on seasonal and diurnal temperature and humidity fluctuations outside and above tomb KV42 in the Valley of the Kings (about 1 km NNW of SAQ; cf. Alcaino-Olivares et al. 2019, 2020), and inside tomb TT95A in the SAQ study area (pilot project). Installations at the shallow chapel of TT95A, which is dissected by faults and tension cracks, include temperature and relative humidity sensors, extensometers, and a seismometer to monitor the thermo- and hydro-mechanical behaviour of a fractured pillar in the undecorated section of the first pillared hall (Fig. 5a). The goal of the installed monitoring equipment is to investigate possible reversible and irreversible displacements that may be triggered by climatic variations inside the tomb and/or natural creep processes, affecting the preservation of the tomb.

Preliminary results (Fig. 5b) show that average daily air temperature within the pillared hall increases during hot summer months from 30 °C to 40 °C, with large diurnal fluctuations up to 10 °C (not displayed). Relative humidity does not show such dramatic fluctuations, with values within 10% of the average daily records for summer 2019. Diurnal fluctuations in relative humidity are within 5% of the average daily value.

The next stages of this research include further analysis of heat energy dissipation and accumulation within the rock mass and thermo-mechanical effects on pillar strain and potential fracturing. This will be done through back-analysis numerical modelling once a full year of strain measurements has been recorded in TT95A.

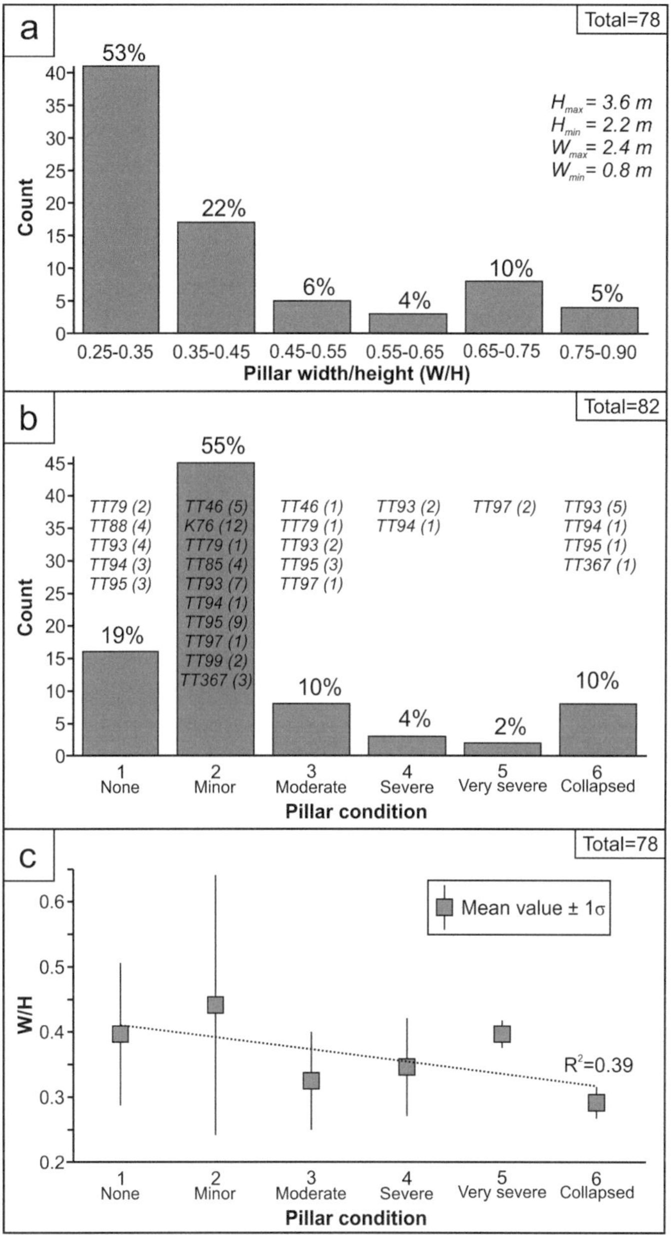

Fig. 4 **a** Histogram of pillar width-to-height (W/H) ratio for pillars of Class I–V. **b** Histogram of pillar condition classes with tomb number (pillar number) listed (Class VI pillars have already collapsed). **c** W/H vs pillar condition, showing a slight positive correlation between pillar dimensions and condition

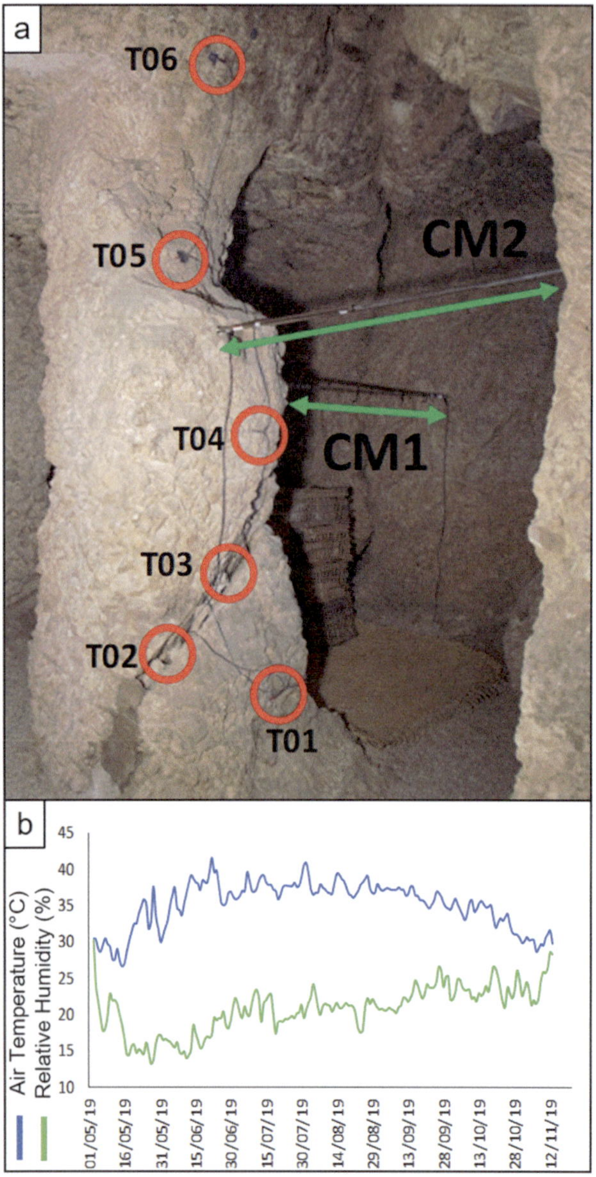

Fig. 5 **a** On-site monitoring system installed by Basel-ETH-York team at tomb TT95A. Two extensometers (CM), six temperature string sensors (T) on pillar 8, and two wireless thermometer/hygrometers, one on the roof of the pillared hall and one on the façade of the tomb, were installed in April 2019, in addition to an existing seismometer inside the tomb, installed in April 2018. Note discontinuity bisecting the pillar. **b** Preliminary averaged results of the monitoring

5 Discussion

The complex geological evolution of SAQ has influenced tomb construction and conservation. The landslide deposits of SAQ are more disintegrated and deformed than neighbouring slump blocks, as indicated by three possible compartments separated by scissor faults (I, II and III on Fig. 2). The rock masses are dissected by tectonic and gravity-driven fractures, faults, and tension cracks. Tension crack infill has been useful to distinguish possible relative age relationships. At least two phases of tension crack formation are proposed: (i) tension cracks oriented NNW-SSE with cemented infill indicate tectonic extension due to the Red Sea opening or early slumping, and (ii) tension cracks oriented E-W to ENE-WSW with unconsolidated or no infill possibly formed during a later stage of slumping. Most tension cracks existed before tomb building, as workers filled them during construction.

Our evaluation of the rock masses at SAQ indicates six rock mass domains, with rock mass quality ranging from poor (marlstones and weathered nodular limestones) to very good (bedded and massive limestone, unweathered nodular limestone, coquinite beds). It seems tomb builders preferred the higher quality massive limestones for smaller tomb entrances (cf. Karlshausen and Dupuis 2014). They avoided weathered nodular limestone at the hill's surface, encountering much less weathered, more competent nodular limestone at depth.

Generally, however, ancient stone masons and miners did not avoid complex or lower quality rock masses. Rather, they found solutions to geotechnical challenges. Restored pillars, tomb design changes such as artificial ceilings, filled fractures, and wooden support beams across unfilled tension cracks are some of the examples we observed. In certain cases, tomb builders even exploited weak rock masses and geological structures to expedite tomb excavation. For example, the courtyard and descending corridor of tomb TT95C and the façade of tomb K85 were constructed parallel to tension cracks.

Pillars are critical to ancient and modern tomb stability (cf. Aydan et al. 2008). Our evaluation shows that most pillars (74%) are in good to very good condition, even thousands of years after excavation. Most collapsed pillars (in TT93) are in nodular limestone and failed during tomb construction. Failure of individual pillars can lead to larger tomb damage, such as ceiling collapse at tombs TT81, TT95, and TT226. Conservation of pillars at SAQ has so far concentrated only on severe cases (Fig. 6). We recently began monitoring the strain in one pillar in TT95A with extensometers to understand current environmental conditions and environmentally controlled fatigue processes that could affect pillar and tomb stability. Our research contributes to a more systematic and process-based approach to conserving this important cultural heritage site.

Fig. 6 Example of pillar conservation efforts in TT331 at SAQ

Acknowledgements Funding for this project was provided through an SNF grant supporting the larger "Life Histories of Theban Tombs" project. The German Aerospace Centre (DLR) kindly provided TerraSAR-X digital elevation data. The ongoing monitoring project is sponsored by the Nature Science and Engineering Research Council of Canada through the Discovery Grant Program (RGPIN-2018–05,918), and sensors were subsidized by the manufacturer Yield Point Inc.

We greatly appreciate the approval and support of the Permanent Committee of the Supreme Council of Antiquities in Cairo including Secretary General Dr Moustafa Waziri and Director of the Department of Foreign Missions Dr Nashwa Gaber, the General Directors of Upper Egypt and the Luxor Inspectorate Mr Mohammad Yahya and Mr El-Kazzafy El- 'Azab, the General Director of Western Thebes Mr Fathy Yaseen and Second Director of Western Thebes Mr Bahaa Abd el-Gabir, the Director of Foreign Missions of Western Thebes Mr Ramadan Ahmed Ali and the Director of the Middle Area of Western Thebes Mr Ezz El-Din, as well as their predecessors and staff. We thank Mr Mahmoud Ibrahim and *Raïs* 'Abd el-Hameed Othman for their administrative, logistical and technical support during our field campaigns.

We are also indebted to Dr K. Powroznik for kindly capturing aerial images using the project's helikite, and to Prof. Andreas Wieser and his staff and students (P. Theiler, E. Friedli, D. Steinmann, M. Martinoni, L. Kaiser, K. Henggeler, A. Baumann, C. Zhou, Z. Gojcic) at ETH Zurich for providing TLS-based 3D models. Discussions with Andrew Hyett and Pierre Ballester on the monitoring system are greatly appreciated.

References

Alcaíno-Olivares R, Perras MA, Ziegler M, Maissen J (2019) Cliff stability at tomb KV42 in the Valley of the Kings, Egypt: a first approach to numerical modelling and site investigation. In: ARMA 201,9 pp 1–10

Alcaíno-Olivares R, Perras M A, Ziegler M, Leith K (2020) Thermo-mechanical cliff stability at tomb KV42 in the Valley of the Kings, Egypt. In: EUROCK 2020, pp 1–8. (in prep)

Aydan O, Tano H, Genis M, Sakamoto I, Hamada M, Yoshimura S (2008) Environmental and rock mechanics investigations for the restoration of the tomb of Amenophis III. Japan-Egypt joint symposium new horizons in geotechnical and geoenvironmental engineering, Tanta University, Egypt, pp 151–162

Bieniawski ZT (1973) Engineering classification of jointed rock masses. Civil Engineer in South Africa 15(12):335–343

Bieniawski ZT (1989) Engineering rock mass classifications: a complete manual for engineers and geologists in mining, civil, and petroleum engineering. Wiley

Cai M, Kaiser PK, Uno H, Tasaka Y, Minami M (2004) Estimation of rock mass deformation modulus and strength of jointed hard rock masses using the GSI system. Int J Rock Mech Min Sci 41(1):3–19

Curtis GH (1979) The geology of the Valley of the Kings, Thebes, Egypt. Theban Royal Tomb Project, The Brooklyn Museum Theban Expedition, vol 28. Unpublished report, Brooklyn Museum

Dupuis C, Aubry MP, King C, Knox RW, Berggren WA, Youssef M, Galal WF, Roche M (2011) Genesis and geometry of tilted blocks in the Theban Hills, near Luxor (Upper Egypt). J Afr Earth Sci 61(3):245–267. https://doi.org/10.1016/j.jafrearsci.2011.06.001

Esterhuizen GS, Innancchione AT, Ellenberger JL, Dolinar DR (2005) Pillar stability issues based on a survey of pillar performance in underground limestone mines. In: Peng S, Mark C, Finfinger G L, Tadolini S, Khair A W, Heasley K A, Luo Y (eds) Proceedings of the 25th international conference on ground control in mining, West Virginia University, Morgantown, WV, pp 354–361

Karakhanyan A, Avagyan A, Sourouzian H (2010) Archaeoseismological studies at the temple of Amenhotep III, Luxor, Egypt. In: Sintubin M, Stewart IS, Niemi TM, Altunel E (eds) Ancient earthquakes (Geological Society of America Special Paper 471). Geological Society of America, Boulder, pp 199–222

Karlshausen C, Dupuis C (2014) Architectes et tailleurs de pierre à l'épreuve du terrain. Réflexions géoarchéologiques sur la colline de Cheikh Abd el-Gourna. Bulletin de l'Institut Français d'Archéologie Orientale 114(1):261–289

King C, Dupuis C, Aubry MP, Berggren WA, Knox RB, Galal WF, Baele JM (2017) Anatomy of a mountain: the Thebes Limestone Formation (Lower Eocene) at Gebel Gurnah, Luxor, Nile Valley, Upper Egypt. J Afr Earth Sci 136:1–48

Lazar J (1995) Geologisch-geotechnische Untersuchungen im Thebanischen Gebirge, Teil Nord, Luxor, Ägypten. Diploma thesis, University of Bern, Geological Institute

Maamoun M, Megahed A, Allam A (1984) Seismicity of Egypt. Bull HIAG 4:109–160

Marinos P, Hoek E (2000) GSI: a geologically friendly tool for rock mass strength estimation. International Society for Rock Mechanics and Rock Engineering, In ISRM international symposium

Rutherford JB (1990) Geotechnical causes of ancient tomb damage, valley of the kings, Egypt. In: Balasubramaniam AS et al (eds) Proceedings of the symposium on geotechnical aspects of restoration works on infrastructures and monuments, Bangkok, December 1988. A.A. Balkema, Rotterdam, pp 3–15

Sawires R, Peláez JA, Fat-Helbary RE, Ibrahim HA, García Hernández MT (2015) An updated seismic source model for Egypt. In: Earthquake engineering, IntechOpen. https://doi.org/10.5772/58971

Shaaban MN (2004) Diagenesis of the lower Eocene Thebes formation, Gebel Rewagen area, Eastern Desert. Egypt. Sediment Geol 165:53–65

Tawfik HA, Zahran EK, Abdel-Hameed AT, Maejima W (2011) Mineralogy, petrography, and biostratigraphy of the Lower Eocene succession at Gebel El-Qurn, West Luxor. Southern Egypt. Arab J Geosci 4(3–4):517–534. https://doi.org/10.1007/s12517-010-0158-6

Wuest RAJ (1995) Geologisch-geotechnische Untersuchungen im thebanischen Gebirge, Teil Süd, Luxor, Ägypten. Diploma thesis, University Bern, Geological Institute

Wuest RAJ, McLane J (2000) Rock deterioration in the Royal Tomb of Seti I, Valley of the Kings, Luxor. Egypt. Eng Geol 58(2):163–190. https://doi.org/10.1016/S0013-7952(00)00057-0

Youssef MI (1968) Structural pattern of Egypt and its interpretation. AAPG Bull 52(4):601–614

Ziegler M, Collldeweih R, Wolter A, Loprieno-Gnirs A (2019) Rock mass quality and preliminary analysis of the stability of ancient rock-cut Theban tombs at Sheikh ‘Abd el-Qurna. Egypt. Bull Eng Geol Env 78:6179–6205

Seismic Hazard and Archeoseismology

Archeoseismology and the Lost Villages in Northern Syria, the Impact of Large Earthquakes on Cultural Heritage

Mustapha Meghraoui and Reda Sbeinati

Abstract Archeoseismology and historical seismology are powerful tools to study earthquake damage and evaluate its impact to cultural sites. We present results of archeoseismic studies of the Lost Cities located nearby the northern segments of the Dead Sea Fault (DSF) in Syria, in the frame of the EC-Funded APAME Project (Archeo-PAleoseismology for the protection of cultural sites in the Middle East). Field investigations and description of the Lost Cities as archeological sites affected by large historical earthquakes are documented and provide consistent typology of earthquake damage. Further archeoseismic investigations were conducted in the lost village of Deir Dahess, and the city of Qalaât Samaan show displaced arches with shifted keystones, waved and collapsed walls, buried artefacts, turned columns and horizontal shifting of building stones. Detailed investigations are supplemented with precise C^{14} dating in all sites along the DSF and provide the age of past earthquakes that affected the two localities. The fallen buildings of Deir Dahess reveal two main dates of earthquake destruction in AD 427–602 and AD 1426–1486, the earlier is correlated with the AD 526 destructive earthquake (M 6.5) and the latter coincide with the 1408 earthquake (M 7.4), both reported in the seismicity catalogue. The detailed archeoseismic studies, coupled with historical data along the DSF document the effect of past earthquakes on buildings and show the long-term faulting behaviour associated with clustering of large seismic events in the antiquity and Medieval periods. Earthquake sequences in northern Syria and their return period associated to the collapse of monuments and historical buildings constitute the basis for a realistic seismic hazard assessment. Archeoseismology contributes to the protection of cultural heritage and social development through the careful study of past earthquake damage.

Keywords Archeoseismology · Paleoseismology · Dead Sea Fault · Seismicity · Old cities · Cultural heritage

M. Meghraoui (✉)
ITES, University of Strasbourg, Strasbourg, France
e-mail: m.meghraoui@unistra.fr

R. Sbeinati
SAEC, Damas, Syria

G. M. El-Qady and C. Margottini (eds.), *Sustainable Conservation of UNESCO and Other Heritage Sites Through Proactive Geosciences*, Springer Geology,
https://doi.org/10.1007/978-3-031-13810-2_24

1 Introduction

Archeoseismology is the study of earthquake damage in archeological sites, and paleoseismology addresses the investigation of earthquake records in geological units. The European-Funded APAME project (Working Group on APAME Project 2007) was dedicated to the Archeo-PAleoseismological study of the impact of past earthquakes in the Middle East (Jordan, Lebanon, Syria and Turkey). Among regions with a rich pre-instrumental seismicity catalogue is the Middle East, and Syria in particular (Ambraseys 2009; Sbeinati et al. 2005). Several written reports on northern Syria describe with details the seismic activity including the number of felt aftershocks, and severe damage to buildings associated to large historical earthquakes (Figs. 1a, b, 2). Among the striking effects is the damage to antique and medieval villages also called "The Dead Cities" and among them, 40 localities are classified as UNESCO World Heritage. They consist in more than 90 small localities, villages and cities located for the most part on a limestone massif east of the Ghab depression west and SW of Aleppo (Figs. 1a and 3). Although severely damaged, the lost cities and villages dated to the late Antiquity until the Byzantine period show earthquake damage (Fig. 2; Tchalenko 1953; Sbeinati and Darawcheh 1998). The lost villages were abandoned around the tenth century, after a sequence of large and disastrous earthquakes in northern Syria.

Previous studies of written documents and field investigations of archeological sites revealed the occurrence of earthquake sequences along the DSF (Sbeinati et al. 2005; Meghraoui 2015). Indirect earthquake features are, however, very often

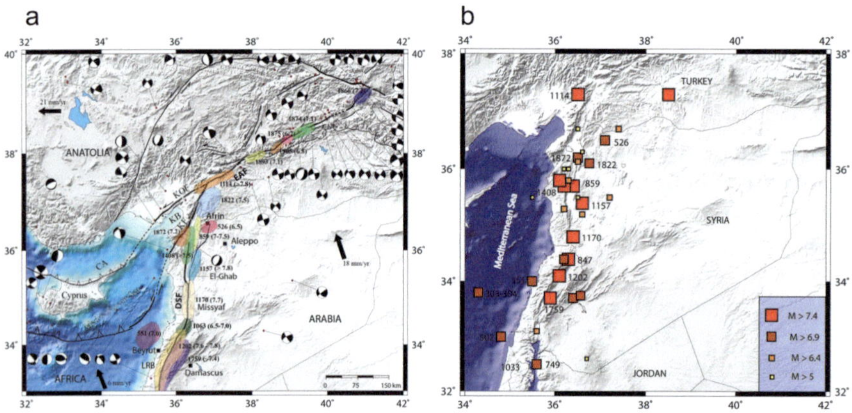

Fig. 1 **a** Seismotectonic context of the Northern Dead Sea fault (DSF, black lines) and its connection with the East Anatolian fault (EAF); focal mechanisms are Harvard CMT (Meghraoui 2015). Area of maximum damage for each historical earthquake are indicated in color (Guidoboni et al. 1994; Ambraseys 2009; Sbeinati et al. 2005) KF: Karasu Fault; KOF: Karasu-Osmaniye Fault; AfF: Aafrin Fault. **b** Epicenter location of large historical earthquakes along the DSF and SW EAF (Sbeinati et al. 2005). Topography data is from SRTM (3-arc-second) and bathymetry from GEBCO

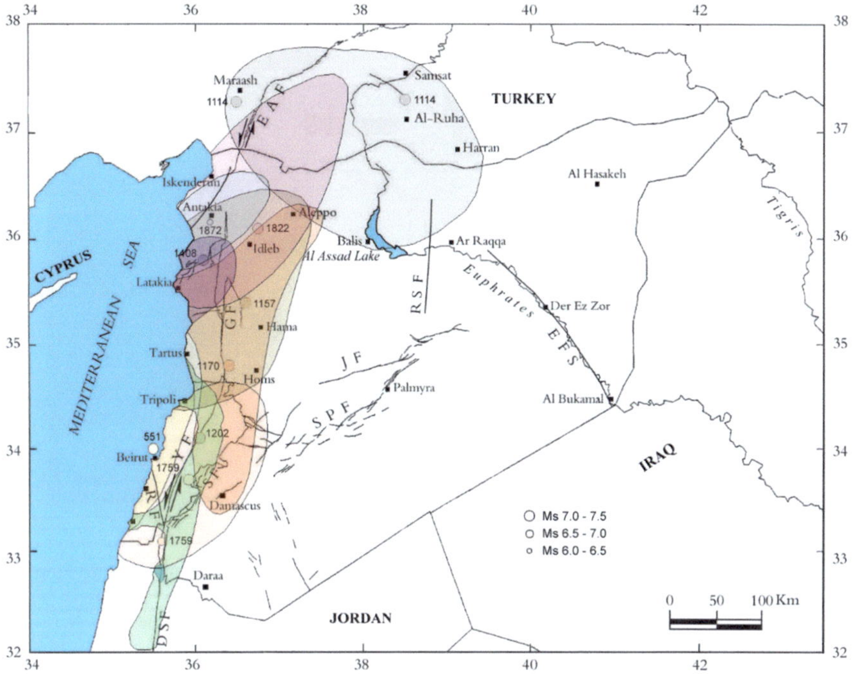

Fig. 2 Major historical earthquakes and areas of maximum damage as recorded along the northern DSF in the Medieval and more recent times (Sbeinati et al. 2005). The area of maximum damage for the Ad 1114, AD 1157, AD 1170 and AD 1408 large earthquakes have affected the lost villages (see also Fig. 3)

problematic and unless dedicated to the specific study of known historical earthquake damage (Stiros and Jones 1996), most of archeological reports can hardly provide usable earthquake parameters (Ambraseys and Melville 1995). Recent studies that combine archeoseismic excavations and paleoseismic trenching provide some constraints of the left-lateral strike-slip movements and related past earthquake events (Meghraoui et al. 2003; Sbeinati et al. 2010). North of our study area, archeological sites are largely spread in the Amik Basin where the fault crosses the ~5000 BC Tell Sicantarla and reveals 42.4 ± 1.5 m cumulative left-lateral movement thus yielding 6.0 ± 0.2 mm/yr slip rate (Altunel et al. 2009).

In addition of the indirect identification of archeological periods, different isotopic dating techniques were used in the APAME project in order to determine the age and timing of past earthquakes. However, the most recent radiocarbon measurement using accelerator mass spectrometry (AMS) applied to the collection of organic matter, tufa, charcoal, bones, fossil test and wood, is the most widely used isotopic method due to the time frame of radioactive counting part and accuracy (often with a maximum uncertainty of ± 100 for ages between 300 and 5000 years Before Present). The laboratory dating and corresponding calibration uncertainties (sometimes as low as _15 years) are generally expressed in AD or BP (AD for *Ad Dominum*, and BP for Before

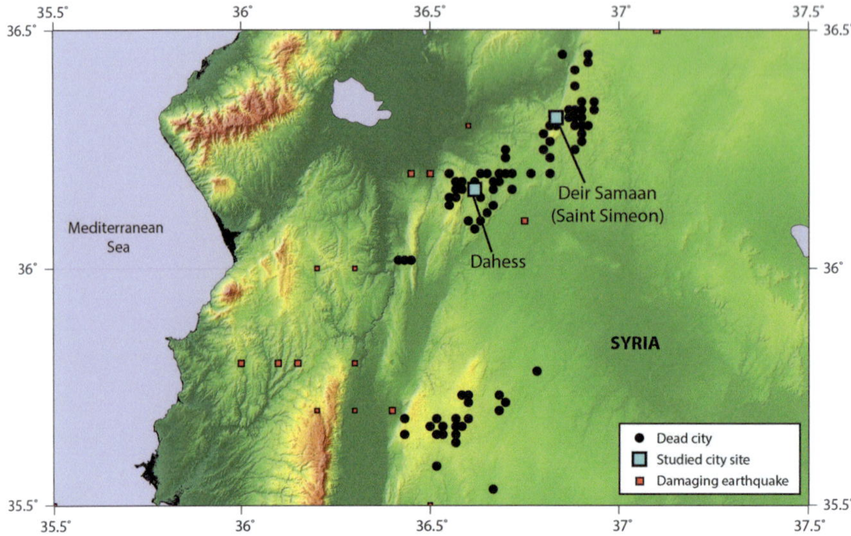

Fig. 3 Location of lost (Dead) cities in northern Syria along the DSF (UNESCO World Heritage and Tchalenko 1953). See also Figs. 1a and 2 for the tectonic and seismological context

Present; Bronk Ramsey 2009) and can be correlated with the historical or archeological data which may provide more accurate dates. All radiocarbon dating were calibrated (2σ range, 95.4% probability density) using Oxcal v4.0 (Bronk Ramsey 2001) and INTCAL04 calibration curve of Reimer et al. (2004). Often reaching a laboratory accuracy of ±25 years, radiocarbon dates are given with a statistical uncertainty of one or two standard deviations (1sigma or 2sigma) if a layer has several dates (Beers 1962). More than 200 samples of organic matter, charcoal fragments, and tufa core pieces were collected for radiocarbon analysis in order to characterize the timing of past earthquakes along the DSF in northeast Syria.

The richness of the historical archives allows us to accurately locate the earthquake activity and extent of damage distribution. The correlation between the historical accounts and field investigations in active tectonics, paleoseismology and archeoseismology in Syria and Lebanon points out at seismic sources along the DSF which corresponds to a plate boundary between Arabia and Africa (Fig. 1; Meghraoui 2015). The DSF is a N-S trending and ~1000-km-long plate boundary that accommodates ~5 mm/yr left-lateral slip. Field investigations in earthquake geology and paleo-archeoseismology conducted in the frame of the EC-Funded APAME Project (2003–2007) targeted the identification of past earthquakes and their impact on cultural sites in the Middle-East (Working Group on APAME Project—Final Report 2007). The temporal seismic gap is a characteristic of the large earthquake activity in these regions and while the historical period reports the occurrence of large seismic events with M > 6.5, no large earthquake has occurred in the last centuries. In particular, seismic gaps have been identified along with long-term temporal quiescence reaching 987 and 850 years (as per the year 2020) along fault segments of the DSF in Syria

(Meghraoui 2015). The repetition of seismic events and related earthquake faulting parameters suggest a high level of seismic hazard and risk along the DSF.

In this article, we first present the location of the lost villages and their proximity to the epicentral area of the large historical earthquakes in northern Syria. The seismic and archeological setting is followed by a typology of earthquake damage in archeological buildings of lost villages described with attention to the age and periods of destructions. Specific archeoseismic studies with detailed damage description, dated artefacts and correlation to historical earthquakes are devoted to the dead cities of Deir Dahess and Qalaat Samaân (Saint Simeon Monastery). The impact of earthquakes on buildings is presented along with reasons of abandonment of the lost cities. A discussion addresses possible economic and environmental factors for the lost cities.

2 Archeological and Seismological Setting of "Dead Cities"

2.1 Archeoseismology and Typology of Buildings Exposed to Earthquakes

Previous works on the impact of large earthquakes in Northern Syria have collected a large number of documents from epigraphic, archaeological and historical literature covering more than 3000 years of information concerning earthquakes (Guidoboni et al. 1994; Ambraseys and Jackson 1998; Sbeinati et al. 2005). The documents contain names of localities that are relatively easy to locate in the field and often with reported number of victims and description of damaged buildings. The remaining dwellings, churches and temples, monuments, cisterns, bathhouses of lost cities, ruined buildings are remarkably well preserved in northern Syria. The region includes military compounds, protective walls of agricultural parcels for mainly olive trees for which the region is well known since the Phoenician times (BC 1500–300).

Typology of earthquake damage in archeological buildings have been characterized from detailed accounts of ancient buildings supported by archeological studies (Stiros ad Jones 1996; Marco 2008). From field investigations in lost cities we recognize horizontal shifting of large building blocks, downward sliding of one or several blocks from masonry arches, collapse of heavy and stably-built walls, chipping of corners of building blocks, aligned falling of walls and columns. We also identify specific Earthquake-induced damage features such as (Fig. 4):

- Cracked, rotated and waived walls and columns.
- Human skeletons buried under ruins.
- Diagonal cracks in wall stones and displaced keystone
- Triangular holes in the corners of brick-walls.
- Collapse of several columns lying parallel.

(a) (b) (c) (d) (e) (f)

Fig. 4 Earthquake-induced damage features (arrows) as identified in the remains of lost cities. **a** Qalaât Samaan (Saint Simeon), **b** Qalaât Samaan, **c** Barisha, **d** Kherbet Maez, **e** El Bara, **f** Deir Dahess

In addition, inscriptions referring to a past catastrophe with reconstruction and renovation of buildings that often require funding can be retrieved in archeological sites. Other damage features may include fractures that cut across several structures, leaning walls and columns, warps and bulges in walls.

2.2 *Rate of Earthquakes and Active Deformation*

The continuous monitoring of instrumental seismicity and several years of investigations from historical documents allow for the inventory of major earthquakes (moment magnitude Mw > 7) along the DSF (Sbeinati et al. 2005; Ambraseys 2009). The seismicity distribution and known patterns of active faulting indicate that the generation of large and damaging earthquake occur along a relatively narrow zone of ~10-km-wide and may spread along fault branches of northern Syria (Fig. 2). The apparent quiescence and the lack of major seismic events with Mw > 6.0 on most fault segments in the last centuries are in contradiction with the historical catalogue and related report of faulting events over the last 3,000 years or so along the continental DSF (Guidoboni et al. 1994; Meghraoui 2015).

The rate of seismicity depends on tectonic movements measured by off fault geodetic GPS networks that show 3 to 5 mm/yr (Reilinger et al. 2006; Mahmoud et al. 2013) which can be compared to active tectonic movement rate of 6 $\pm$ 0.5 mm/yr obtained over the last 3000 years (Meghraoui et al. 2003; Altunel et al. 2009; Sbeinati et al 2010). The Arabia-Africa Euler pole calculated from GPS measurements predicts plate motions from 4 to 5 mm/yr along the southern DSF to about 6 to 7 mm/yr along the northern DSF (Mahmoud et al. 2013). This is in agreement with tectonic results. However, GPS velocities depend on the off-fault measurements and modelling of crustal deformation while the tectonic slip rate derived from tectonic geomorphology and paleoseismology with measurements of offset markers in trenches and landscape that result from coseismic slip. Field studies of coseismic slip based on faulted late Pleistocene and Holocene units and archeological sites along fault branches of northern Syria give a slip-rate between 0.2 and 0.6 mm/yr (Sbeinati et al. 2005; Meghraoui et al. 2003; Karakhanian et al. 2008).

3 Major Earthquakes in the Area of Lost Cities

The northern section of the DSF (i.e., in Lebanon and Syria) is among the main seismogenic zones in the region since it has a long (since 1365 BC), rich and well-documented history of large destructive earthquakes that severely damaged many ancient cities (Fig. 2; Ambraseys and Melville 1995; Guidoboni 2004a, b; Sbeinati et al. 2005). In contrast, the instrumental seismicity during the last century along the plate boundary is of low level and does not reflect the hazardous nature of the fault (Ambraseys 2009). The long-term faulting behaviour needs to be investigated and a better constrain on the rate of active faulting is required for the seismic hazard assessment.

Sbeinati et al. (2005) updated a historical earthquake catalogue of Syria and neighboring regions where large shocks are represented in Fig. 1b and Table 1 with parametric estimations of the textual descriptions of historical earthquake damage. The catalogue is presented as a list of earthquakes with its parameters ordered by date, time, epicenter coordinates, estimated maximum intensity, calculated magnitude, intensity at affected localities, and the natural coseismic features. In Table 1 the historical catalogue of the region of lost cities in northern Syria covers the period from AD 53 to 1872 and contains important information on the occurrence of large earthquakes and related damage to buildings.

Earthquakes along the northern DSF have been the source of numerous large earthquakes with surface faulting in the historical time (Ambraseys and Jackson 1998). The present-day seismic quiescence along the northern DSF does not reflect its seismogenic character in particular when we consider the large historical earthquakes of Table 1. Although no recent surface ruptures have been observed in recent times, the combined analyses in historical seismology, paleoseismology and archeoseismology resulted in a better understanding of the relationship between large historical earthquakes ($Mw > 7$) and fault segments (Meghraoui 2015). Major historical earthquake-faulting include from north to south (Fig. 1b), the 1114, 526, 1408, 847, 1157, 1170 earthquake (Ambraseys and Melville 1995; Meghraoui et al. 2003; Sbeinati et al. 2005). Most of studies referring to historical earthquakes confirm that nearly all of them are associated with left-lateral surface faulting ruptures. Figures 1b and 2 shows the historical earthquakes of magnitude ≥ 6.5 as in Table 1 associated with the DSF in Syria. Even though the low level of instrumental seismicity and the absence of strong motion records, seismic parameters and fault activity along the northern DSF related to the historical earthquake data has an important role in estimating the seismic hazard for the region. A repetition of seismic events matching to those of the past, nowadays, will cause enormous damage to the cultural heritage, modern buildings and to the population.

4 Archeoseismology of in Two Lost Cities: Case Studies

Field investigations in active tectonics and archeoseismology and the identification (dating) of past earthquakes and correlation with the seismicity catalogue were conducted as a close collaboration between the Directorate General of Antiquities and Museums and SAEC (Syria), Tubingen university (Germany) and the IPG of Strasbourg (France), in the frame of the APAME project. In particular, Deir Dahess and Qalaât Samaan (Saint Simeon) lost cities and related archeological sites were

Table 1 Main destructive earthquakes with Ms ≥ 6.5 that may have affected the lost cities along the DSF (Ambraseys and Melville 1995; Sbeinati et al. 2005)

Date (dd.mm.yyyy)	Long. (E°)	Lat. (N°)	Major affected localities	I0 (EMS-92)	H (km)	Ms
53	36.50	36.20	Antioch, Afamia, Manbej, Lattakia	VIII	30	6.6
12.12.115	36.30	35.80	Antioch, Aleppo	IX	20	7.1
494	36.30	35.80	Antioch, Tripoli, Lattakia	VII-VIII	25	6.5
25.09.526	37.00	36.50	Aleppo, Afrin	VIII-IX	30	6.5
13.08.847	36.80	36.80	In and around Damascus, Antioch, Al-Mosel	IX	35	7.5
30.12.859 29.01.860	36.40	35.70	Antioch, Lattakia, Jableh, Homs, Palmyra, Tarsus, Balis, Damascus, Adana, Ar-Raqqa	IX	33	7.4
29.11.1114	37.20	37.50	Maskaneh, Marash, Antioch	IX	40	7.4
01.10.1138	36.7	36.20	Aleppo, Atharib, Azrab	VIII-IX		6.5
12.08.1157	36.40	35.30	Shaizar, Kafar Tab, Afamia, Hama, Arqa, Aleppo, Homs, Lattakia, Tripoli, Antioch, Qalaat Al-Hosn, Maarret Annooman	IX-X	15	7.2
29.06.1170	36.40	34.80	Missyaf, Homs, Hama, Lattakia, Baalbak, Shaizar, Barin, Aleppo	IX	35	7.3
29. 12.1408	36.30	35.90	Shugr, Bkas, Blatnes, Lattakia, Jableh, Antioch, Syrian coast	IX-X	25	7.4
21.01.1626	37.10	36.50	Aleppo, Gaziantab, Hama	IX	20	7.3
26.04.1796	36.20	35.30	Qalaat Al-Marqeb, Al-Qadmous, Nahr Al-Kabir, Jableh, Bkas, Lattakia	VIII-IX	20	6.8

(continued)

Table 1 (continued)

Date (dd.mm.yyyy)	Long. (E°)	Lat. (N°)	Major affected localities	I0 (EMS-92)	H (km)	Ms
13.08.1822	36.75	36.10	Jisr Ash'Shoughour, Quseir, Aleppo, Darkoush, Antioch, Iskenderun, Idleb, Kelless, Armanaz, Sarmada, Lattakia, Homs, Hama, Maraash, Ram Hamadan, Bennesh, Maarret Missrin Safad	IX	18	7.0
03.04.1872	36.50	36.20	Harem, Armanaz, Lake of Al-Amq, Antioch, Aleppo, Suaidiya, Izaz, Idleb, Iskenderun	VIII-IX	10	7.2

the focus for further investigations with excavations and dating of charcoal samples and other artefacts (poteries, bones, coins, …).

4.1 Deir Dahess

Excavation activities started In September 2005 with a first investigation in the Djebel Barisha at Deir Dahess. The site was selected during an earlier field reconnaissance shortly before been selected on a round trip with Reda Sbeinati, who showed several possible objects in the Ghab and in the Djebel Barisha. Archeological excavations allowed the collection of charcoal and bone samples in the domestic house area of the monasterial complex (Figs. 5a, b). In addition, coins from early Byzantine period (sixth century) were found among the buried artefacts below collapsed walls (Figs. 5c, d). The results of AMS dating give two groups: one for the period AD 410–539, and the second AD 528–645 (2σ – 95% probability density function).

A second site shows the destruction of the annexed room and provide dating ranging betwen 1426 to 1486 AD. These two groups correspond with the two investigated areas with destructions that can be correlated to the sixth century A.D. large earthquake and likely the 29 May 526 (M 6.5) and the fifteenth century large earthquake of the 29 December 1408 (M 7.4). Archeological excavations show that the Deir Dahess suffered twice from earthquake damage leaving the archeological site as ruins which correspond to a lost city.

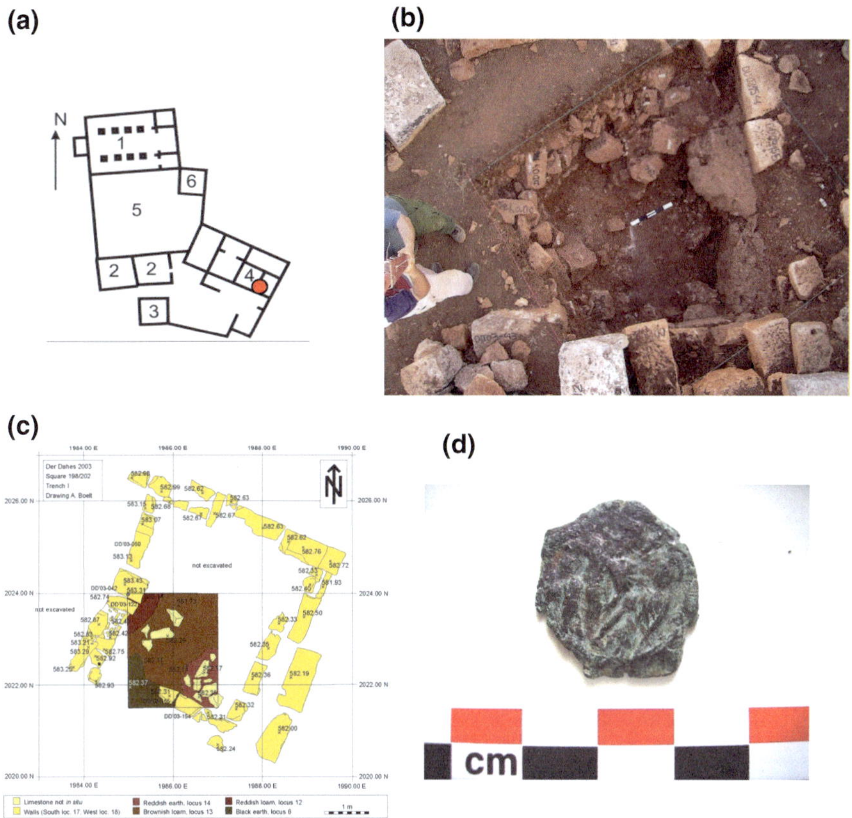

Fig. 5 Archeoseismological investigations at the lost city of Deir Dahess. **a** Detailed plan of the archeological sections; **b** Excavations of Sect. 4 (in a) with collapsed walls and buried artefacts; **c** Illustration of walls and excavated area (in brown) with collapsed walls and buried artefacts; **d** Coins from early Byzantine period (sixth century) found among the buried artefacts below collapsed walls. These results are obtained in the frame of the APAME Project from a collaboration between the University of Tuebingen and GDAM-SAEC Damascus (Working Group on APAME Project 2007)

4.2 Earthquake Damage at Qalaât Samaan (Saint Simeon Church)

This Church—Basilica is located on a limestone hill, about 60 km northwest of the city of Aleppo. It was erected at the end of fifth century in the honour of St. Simeon (390–459 AD.), a shepherd and priest from northern Syria. After his death, his body was moved to Antioch and the emperor Zenon ordered that a church of cross shape be built at the same place named after him. The building was one of the most beautiful cathedrals in the Middle East region. Later-on a monastery and many houses for pilgrims were built around to form a village. Following that, walls and towers were built around the church which was turned into a fortress known as Qalaat Samaan. It became the site of conflict between the Byzantines and the Hamadani kingdom until the year 986 AD when the son of the Prince Saif Al-Dawlah Al-Hamadani finally conquered it.

The Basilica is made of medium to large size blocks of limestone, but it was affected by earthquakes visible in displaced and distorted arches and fractured walls (Fig. 6). Earthquake damage consists in displaced arches with shifted keystones, waved walls, turned columns and horizontal shifting of building stones. The cathedral was partially restored recently. Due to its location at the junction of the two major

Fig. 6 Earthquake damage on arches of the Qalaât Samaan (Saint Simeon) Basilica. Earthquake damage (arrows) consists in displaced arches with shifted keystones, waved walls, turned columns and horizontal shifting of building stones. Written accounts and archeological investigations correlate the main earlier damage to the 29 May 526 earthquake (M 6.5, Table 1)

seismogenic faults (East Anatolian Fault and DSF, see Fig. 1a), this site was affected by the historical earthquakes of AD 526, 1138, 1170, 1408, 1822 and 1870 (Table 1).

4.3 The Impact of Earthquakes on Buildings

The identification of earthquake damage in archeological sites needs an analysis of structural characteristics of buildings and a comparison with present-day destruction of seismic origin. The difficulty is even greater if one would study damage distribution at local and regional scale and extract macroseismic information. Although historical documents contain valuable descriptions of earthquake damage, macroseismic information in the study area is rather poor and may be subjected to misinterpretation (Ambraseys and Melville 1995). Although not that frequent, the coseismic faulting of archeological sites provide the unambiguous information on the occurrence of past earthquakes, related archeoseismic damage and earthquake severity (Meghraoui et al. 2003). The large number of damaged lost cities often appears as ruins and it is sometimes difficult to extract macroseismic information that allows assessment of intensity. However, due to the style of buildings (mostly from the pre-Roman, Roman and Byzantine periods) that uses heavy stones, it is likely that ground deformations are mostly of tectonic origin. Due to the local geology and geomorphology of the lost cities region built on a limestone nearly flat plateau, ground effects and accelerations that may generate landslides and lateral spreading liquefaction are unlikely.

Man-made structures and related designed buildings in the area of lost cities have no paraseismic frames, may vary greatly in vulnerability and are not standard buildings that comply with the characteristics of old or modern intensity scales. Whether located on hills and mountainous regions or on alluvial plains, ancient buildings and houses with for instance two to three stories and subject to strong shaking and ground acceleration are easy to collapse. For most lost cities, field investigations on buildings and houses show severe damage and collapse (Figs. 4 and 6). Open vertical cracks in walls and differential settlements in pavements of archaeological sites can be due to non-seismic causes. For instance, small incipient tilting and compaction of buildings may result from weathering or leaching of its foundation materials, which can be brought about by climatic changes and fluctuations of the underground water table over the ages. This is a particularly true when ancient walls and other man-made structures are built on thick layers of cultural debris. In archeoseismology, assigning a degree of macroseismic intensity needs a clear distinction between damage caused by vibrational, dynamic or inertia loading, and damage caused by indirect, secondary effects, such as foundation spreading, liquefaction, slides, rock falls and aftershocks (Ambraseys 2009).

5 Discussion and Conclusion

The study of ancient earthquakes and their damage to houses, buildings, bridges, dams, and lifelines (road, aqueduct, etc., …) is of importance for the safety of a present-day society. The EC-Funded APAME project was dedicated to the protection of cultural sites throughout the study of paleoseismology and archeoseismology. A major challenge was the identification and characterization of past earthquakes in buildings and landscapes. The outcome of the project revealed that important historical and pre-historical earthquakes affected the Middle East regions and northern Syria in particular. The study of large distant earthquakes confirm that ground motions have much longer duration and period and hence responsible of damage to large buildings with columns. The studied extensive damage incurred to the Lost Cities (also called Dead Cities) constitutes evidence to the impact of large earthquakes in northern Syria and along the DSF. Nevertheless, the abandonment of an urban site may have other origins such as floods, acts of warfare, military operations and deliberate damage that may be misinterpreted as earthquake-induced. In contrast with wars, epidemics and other long lasting calamities that have serious and prolonged effects, earthquakes, no matter how large, seem to have had little long-term impact on Man. Therefore, the study of written accounts needs to be cross-correlated with archeoseismic and paleoseismic investigations in order to discriminate between earthquake-induced damage and anthropogenic effects. Although the impact of large earthquakes may be severe, we observe through our experience and existing literature that extensive repairs of city walls, public buildings, aqueduct were performed after earthquake damage (Sbeinati et al. 2010). The occurrence of a large earthquake with damage implies changes in the construction with more resistant structures. Some early monuments, mostly religious, damaged or destroyed by an earthquake were quickly rebuilt with updated engineering knowledge.

In parallel with the impact of large earthquakes on cultural sites, the presence of human skeletons under collapsed buildings indicates the nature of victims. Archaeological studies often report the loss of human lives buried under the debris of fallen buildings. Earthquake victims imply that the collapse came suddenly, without warning, and caught them in their sleep or lying down.

One of the striking conclusions that can be drawn from field observations of ancient earthquakes (in antiquity and medieval times) is the identification and inventory of seismic damage from field investigations compared to the rich parametric earthquake catalogue. The correlation between archeoseismology and paleoseismology with the history of ancient cities and related earthquake catalogue is documented throughout specific examples of field studies in Deir Dahess and Qalaât Samaan. Evidence of earthquake damage in both sites indicates the occurrence of severe damage during the 29 May 526 and 29 December 1408 along the DSF in northern Syria. The occurrence of large earthquakes with magnitude Mw > 6.5 (Table 1) is evidenced not only from the seismicity catalogue but also from field investigations (Fig. 7). Sequences of large seismic events with southward seismic migrations were generated along the DSF with severe damage to monuments, buildings, palaces and important economic edifices (aqueduct).

The assessment of long-term seismic hazards in northern Syria requires a completed seismicity catalogue of major (destructive) earthquakes. Assigning a reliable intensity data from damage distribution with estimated magnitude remain an important source of data for estimating the size, location and tectonic implications of past and future earthquakes. The detailed study of the DSF, related mechanism, segmentation, slip rate and estimate of maximum magnitude with a return period of maximum ground displacement is the basis for a new model for the seismic hazard evaluation. The recurrence time of large earthquakes along the Jordan Valley-Dead Sea and Missyaf segments of the DSF shows relatively short periods (few tens of years) of seismic sequences alternating with relatively long-period of quiescence reaching ~1000 years (Fig. 7). A comparison with a G-R relation (LogN = a-bM)

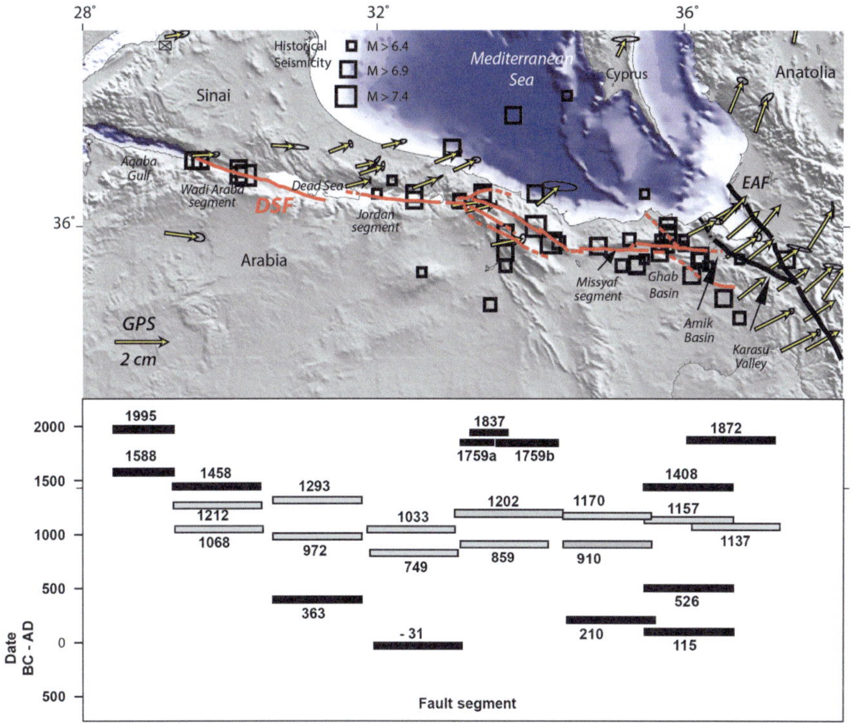

Fig. 7 Upper: The DSF trace (in red) drawn on a topography from GEBCO 1 km resolution. The fault extends from the Gulf of Aqaba in South, crosses the Dead Sea and Lebanese Mountains, and reaches to the East Anatolian Fault (EAF) in the North. Lower: Historical large earthquakes (M > 6.9) correlated with each fault segment according to the detailed study of earthquake damage area along the DSF (Guidoboni et al. 1994; Ambraseys and Jackson, 1998; Sbeinati et al., 2005; APAME, 2007; Ambraseys, 2009; Meghraoui, 2015). Note the southward migration of the remarkable large earthquake sequences from 1138 to 1202 AD along the DSF

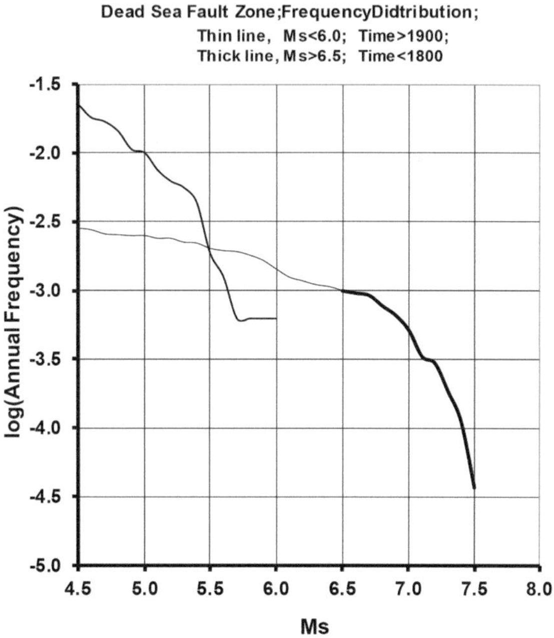

Fig. 8 Annual frequency distribution of earthquakes per square degree in the Dead Sea region. Thin line for the twentieth century and thick line is for the period 1 to 2000 years (Ambraseys 2009)

provide a b-value of 0.9 and a return period of ~1000 years for Mw 7.3 for the same segment (Fig. 8). These estimates of seismicity rate requires an extended catalogue that includes paleoseismic and archeoseismic data that takes into account the occurrence of earthquake sequences with two or three earthquakes separated by a long period of quiescence (Meghraoui 2015).

Acknowledgements This research was funded by the European Commission–funded APAME Project (contract ICA3-CT-2002-10024) and by the UMR 7516 of Centre National de la Recherche Scientifique in Strasbourg (France). Field investigations were supported by the Syrian Atomic Energy Commission (SAEC) and the Directorate General of Antiquities and Museums (DGAM) in Damascus. We are grateful to Ibrahim Osman (director general of the Atomic Energy Commission of Syria) for his continuous support. We are grateful to Gad El-Qady and Claudio Margottini for the invitation to the 2019 conference on "*Sustainable Conservation of Heritage Sites Through Proactive Geoscience*" (Luxor–Nile, 10-12 December 2019). We also thank two anonymous reviewers who significantly helped to improve the presentation of our manuscript.

References

Altunel E, Meghraoui M, Karabacak V, Aky€uz S, Ferry M, Yalçiner Ç, Munschy M (2009) Archaeological sites (Tell and Road) offset by the Dead Sea fault in the Amik Basin, Southern Turkey. Geophys J Int 179:1313–1329

Ambraseys NN, Melville CP (1995) Historical evidence of faulting in Eastern Anatolia and Northern Syria. Ann Geophys XXXVIII 3–4:337–343

Ambraseys NN, Jackson JA (1998) Faulting associated with historical and recent earthquakes in the Eastern Mediterranean region. Geophys J Int 133(2):390–406

Ambraseys NN (2009) Earthquakes in the Mediterranean and Middle East: a multidisciplinary study of seismicity up to 1900. Cambridge University Press, Cambridge, MA, p 947

Bronk Ramsey C (2009) Bayesian analysis of radiocarbon dates. Radiocarbon 51(1):337–360

Guidoboni E, Comastri A, Traina G (1994) Catalogue of ancient earthquakes in the Mediterranean area up to the 10th century. ING Roma-SGA, Bologna, p 504

Guidoboni E, Bernardini F, Comastri A (2004a) The 1138–1139 and 1156–1159 destructive seismic crisis in Syria, south-eastern Turkey and northern Lebanon. J Seismol 8:105–127. https://doi.org/10.1023/B:JOSE.0000009502.58351.06

Guidoboni E, Bernardini F, Comastri A, Boschi E (2004b) The large earthquake on 29 June 1170 (Syria, Lebanon, and central southern Turkey). J Geophys Res 109:B07304. https://doi.org/10.1029/2003JB002523

Karakhanian AS, Trifonov VG, Ivanova TP, Avagyan A, Rukieh M, Minini H, Dodonov AE, Bachmanov DM (2008) Seismic deformation in the St. Simeon Monasteries (Qal'at Sim'an). Northw Syria, Tectonophysics 453:122–147

Mahmoud Y, Masson F, Meghraoui M, Cakir Z, Alchalbi A, Yavasoglu H, Yönlü O, Daoud M, Ergintav S, Inan S (2013) Kinematic study at the junction of the East Anatolian fault and the Dead Sea fault from GPS measurements. J Geodyn 67:30–39

Marco S (2008) Recognition of earthquake-related damage in archaeological sites: examples from the Dead Sea fault zone: Teconophysics, v. 453, pp 148–156

Meghraoui M, Gomez F, Sbeinati R, Van der Woerd J, Mouty M, Darkal A, Radwan Y, Layyous I, Najjar H, Darawcheh R, Hijazi F, Al-Ghazzi R, Barazangi M (2003) Evidence for 830 years of seismic quiescence from paleoseismology, archeoseismology and historical seismicity along the Dead Sea fault in Syria. Earth Planet Sci Lett 210:35–52. https://doi.org/10.1016/S0012-821X(03)00144-4

Meghraoui M (2015) Paleoseismic history of the Dead Sea Fault zone. In: Encyclopedia of earthquake engineering, p 20, Berlin Heidelberg: Springer. https://doi.org/10.1007/978-3-642-36197-5_40-1

Reilinger R et al (2006) GPS constraints on continental deformation in the Africa Arabia-Eurasia continental collision zone and implications for the dynamics of plate interactions. J Geophys Res 111:B05411. https://doi.org/10.1029/2005JB004051

Sbeinati MR, Darawcheh R (1998) Archaeological Evidence of Earthquake Damage in Syria, Report No. 2 in "Seismic Data for Siting and Site-Revalidation of Nuclear Facility" (contract No. 6247/R3/RB), Atomic Energy Commission, Damascus, 33 pp

Sbeinati MR, Darawcheh R, Mouty M (2005) The historical earthquakes of Syria: an analysis of large seismic events from 1365 B.C. to1900 A.D. Ann Geophys 48:347–435

Sbeinati MR, Meghraoui M, Suleyman G, Gomez F, Grootes P, Nadeau M, Al Najjar H, Al-Ghazzi R (2010) Timing of earthquake ruptures at the Al Harif Roman Aqueduct (Dead Sea fault, Syria) from archeoseismology and paleoseismology, Special volume "Archaeoseismology and paleoseismology. In: Sintubin M, Stewart IS, Niemi TM, Altunel E (eds) Ancient earthquakes: geological society of America special paper, 471. https://doi.org/10.1130/2010.2471(20)

Stiros S, Jones RE (eds) (1996) Archaeoseismology, British School at Athens, Fitch Laboratory Occasional Paper 7, pp 268

Tchalenko G (1953) Villages antiques de la Syrie du Nord, Le Massif du Belus à l'époque Romaine, Edition Institut Français d'Archéologie de Beyrouth, Librairie Orientaliste Paul Geuthner, Paris

Working Group on APAME Project—Final Report (2007) Archeoseismology and paleoseismology for the protection of cultural heritage and archeological sites in the Middle East: the impact of large earthquakes on the archaeological sites and cultural heritage in the Middle East (Jordan, Lebanon, Syria and Turkey, APAME EC project—ICA-CT-2002-10024—Coordinator Mustapha Meghraoui, DG XII Brussels

Integrated Geoscience Investigations in Hittite Imperial Sites Affected by Earthquakes

Mahmut Göktuğ Drahor, Ökmen Sümer, Meriç Aziz Berge, Caner Öztürk, Atilla Ongar, Aygül Süel, and Andreas Schachner

Abstract Earthquakes caused great changes in ancient societies, affecting human civilizations. More recently, important historical earthquake traces have been identified by archaeo-seismological investigations conducted during archaeological excavations. It is possible to document contemporary and future earthquake risk by investigating past earthquake traces through integrated geoscience studies in archaeological sites located in important tectonic zones. Two Hittite cities, which are located near a major tectonic zone in the northern Central Anatolia region of Turkey, are the main subjects of this study. Integrated geology studies in these Hittite cities were undertaken to determine the effects and consequences of ancient earthquakes, which could have impacted the Hittite Period. Thus, we tried to determine the traces of an old earthquake that may have occurred in Ḫattuša, the capital of the Hittite Empire, and another capital city, Šapinuwa. Therefore, in light of geological risk, this project tries to determine the presence of ancient earthquake traces, which may have originated from the Sungurlu Fault (SF) and the North Anatolian Fault Zone (NAFZ) in these two Hittite cities. In this context, extensive archaeo-seismological studies have

Director of Šapinuwa archaeological excavation.
Director of Ḫattuša archaeological excavation.

M. G. Drahor (✉) · M. A. Berge · A. Ongar
Engineering Faculty, Department of Geophysical Engineering, Dokuz Eylül University, Tınaztepe Campus, 35390 Buca-İzmir, Turkey
e-mail: goktug.drahor@deu.edu.tr

Ö. Sümer
Engineering Faculty, Department of Geological Engineering, Dokuz Eylül University, Tınaztepe Campus, 35390 Buca-İzmir, Turkey

C. Öztürk
Geoim Engineering, Consultant, Software and Construction Ltd, Karacaoğlan Mahallesi, 6253 Sokak Deligöz İş Merkezi No: 8/19, 35070 Bornova-İzmir, Turkey

A. Süel
Faculty of Arts and Sciences, Department of Archaeology, Hitit University, Çorum, Turkey

A. Schachner
Deutsches Archäologisches Institut, Gümüşsuyu, İstanbul, Turkey

G. M. El-Qady and C. Margottini (eds.), *Sustainable Conservation of UNESCO and Other Heritage Sites Through Proactive Geosciences*, Springer Geology,
https://doi.org/10.1007/978-3-031-13810-2_25

been conducted at both archaeological sites. In addition, the soil-earthquake-structure relationships were investigated using geophysical methods in certain regions of the Šapinuwa archaeological site. As a result, we aimed to reveal the future earthquake risks of historical cities in this region by determining the earthquakes that occurred during the Hittite Period with the help of integrated geology studies. Therefore, by predicting future earthquakes that may affect the city of Ḫattuša, which has been on the UNESCO World Heritage List since 1986, along with the city of Šapinuwa, we aimed to start sustainable studies of these cities with the goal of having those studies continue in the long term.

Keywords Ancient Earthquakes · Archaeo-seismology · Central Anatolia · Fault Zones · Hittite Period

1 Introduction

The Hittites settled in the Hatti culture region, which was located around the Halys River in northern Central Anatolia, in approximately 1700 BC. The Hittite community, which also mixed with the Hattian and Hurrian peoples, developed the first state in the region in Anatolian history. Although there are different views of the chronology, the Hittite state is defined according to three historical periods: Old Hittite (1650–1500 BC), Middle Hittite (1500–1400 BC) and Empire (1400–1180 BC) (Schachner 2019). The Empire period reached its peak between c. 1344–1295 BC. By spreading its territory to the west, south and southeast, it got close to Egypt. Especially in the written documents (Amarna letters), which are known from the Amarna Period (c. mid-fourteenth century BC) and remain from the period when diplBomatic relations were intensified between Near Eastern societies, it is revealed that there was correspondence between the Hittite kings and Egyptian pharaohs. One of the most important events between the Hittites and Egyptians was the Battle of Kadesh in 1274 BC. However, neither side achieved a decisive victory. Then, the first peace treaty between the states in which they accepted each other as equal, was signed between the Hittites and the Egyptians (Bryce 1998; Klengel 1999). At the end of this generally successful period, the Hittite Empire suddenly disappeared from the historical scene around 1180 BC, during or after the period of Suppiluliuma II. Although many historians associate this phenomenon with the arrival of sea peoples, experts on the subject have recently come to think that the process resulted from the overlap of many factors, including throne struggles within the empire, drought resulting from a shortage of crops because of climate change, and, possibly, the arrival of sea peoples and major earthquakes (Gurney 1990; Drews 1993; Cline 2014; Schachner 2019, 2020a).

Today, many researchers argue that the Bronze Age cultures in the Eastern Mediterranean collapsed due to catastrophes triggered by several successive events, such as emerging drought due to climatic changes, collapse of the trade network, immigration of sea peoples into the coastal areas of Anatolia, civil wars, system collapse and big earthquakes. According to some of these researchers, data regarding the existence of large earthquakes during the Late Bronze Age in the region up to the Levant, Anatolia, Crete and Mycenai have been presented (e.g. Schaeffer 1948; Blegen 1953; Schaeffer et al. 1968; Soren 1982; Åström 1985; Nur and Cline 2000; French 2009; Stockhammer 2009; Cline 2014). In addition, the existence of important climate records showing that there was gradual climate change in the Late Bronze Age suggests that natural processes were active in bringing about this change (deMenocal 2001; Riehl et al. 2008; Drake 2012; Weiss 2016; Psomiadis et al. 2018). Excavations carried out at many archaeological sites in the Mediterranean region showed that there was significant stagnation and decline in the Late Bronze Age civilizations between the thirteenth and twelfth centuries BC, which was followed by sudden social change in the civilizations, except partly in Egypt. Although there is no consensus among scientists about the timing of this event, in approximately 1250–1177 BC, the major cities in the Levant and Anatolia were destroyed, and one of these was Ḫattuša. After this date, the Bronze Age civilization disappeared, and diplomatic and commercial relations were broken between the Mediterranean societies. In addition, the disappearance of traditional cuneiform writing systems and the emergence of a new alphabetic writing system (Phoenician) along the Mediterranean coast in the following periods is another important cultural datum showing the end of this age (Cline 2014).

Natural events (climate change, earthquakes, and other similar processes) are also thought to have had an important impact on the collapse of the Bronze Age cultures. Thus, the pressures on important Bronze Age cities may have also triggered a period of destruction, especially with the emergence of events such as harvest shortages and diseases due to climatic changes. This situation must have caused regional migration, unrest, and social and political imbalances in the Bronze Age territory. As a result, the chaos caused by regional migration may have led to social collapse and the deterioration of diplomatic and commercial relations between societies due to the impact of population imbalances in the Bronze Age territory. Thus, a resulting domino effect, which could have affected all Bronze Age civilizations in the region, may have caused social collapse and civilizational change. Especially when it comes to the claim (Nur and Ron 1997; Nur and Burgess 2008; Nur and Cline 2000, 2001) that cities along the Mediterranean coast were destroyed by a F earthquake storm in the thirteenth and twelfth centuries BC, this view reveals extreme tectonic activity in the region over a short period of time. Only very limited philological information is given by Ünal (1977) about the earthquake(s) in Anatolian lands during the Bronze Age. However, there is still no strong evidence for the existence of such an effect. This may be due to the lack of historical earthquake data from the limited archaeological excavations related to this period in Western and Central Anatolia. Considering the intense earthquake activities in the instrumental period, especially in the same regions, it is thought that the earthquake phenomena in these regions during the historical period were also caused by local tectonic events rather than a catastrophic event. In the present day, these regions are currently active, and many major earthquakes have occurred during the last century (Armijo et al. 1986). However, these major earthquakes are not at a level that affects all civilizations around the

Mediterranean. In order for earthquakes to affect a civilization, they must be in the form of a regional earthquake storm and therefore cause a significant part of the population to be displaced in response to great destruction that has occurred. Given the other impacts to be added to these events, major changes may occur in societies at a regional level. After all, these are just assumptions. For this reason, it is extremely important to examine the great earthquakes that occurred during the Bronze Age in the Mediterranean region and to consider their effects on the cities in the region in light of geological studies. Thus, these views will either be supported or become unnecessary as new data emerge. This study aims to contribute to the understanding of earthquake phenomena through earth science studies conducted in two cities in the Hittite region.

Figure 1 shows the epicenters of earthquakes of magnitude 6.5 and above that occurred between 1900 and 1980, overlapping with the location of ancient cities in Levant, Anatolia, Greece, Crete and Cyprus at the end of the Late Bronze Age (1250–1177). As can be seen from the figure, most of the cities destroyed in the Late Bronze Age were located on or near active fault systems. Comparing the earthquake events of the last century with the destruction at the end of the Bronze Age makes clear that there is a gradual overlap between the two, except for the Halys River and the region to the south. This region is today called structurally the Amasya Shear Zone (ASZ) (Fig. 2) and can produce large and consecutive earthquakes at regular intervals along with the North Anatolian Fault Zone. This situation indicates that active tectonic zones in the region during the Hittite period may have produced earthquakes.

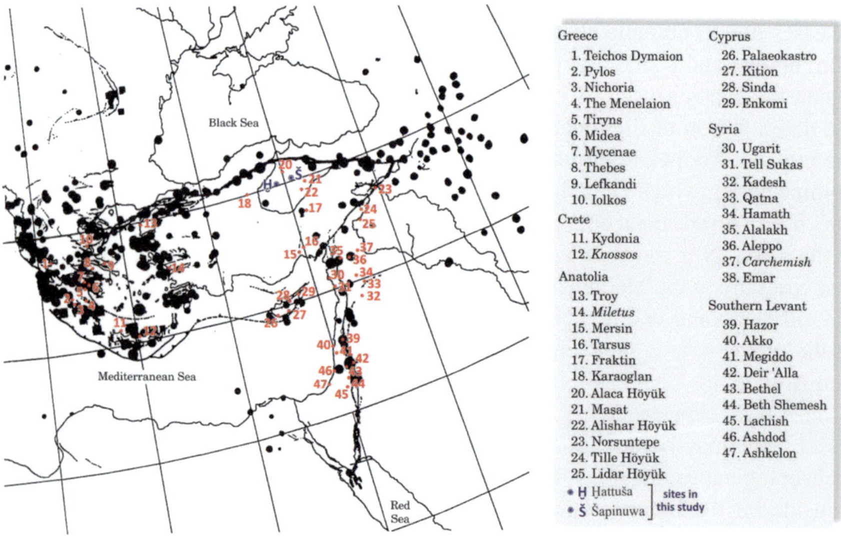

Fig. 1 Ancient settlements destroyed from Aegean to Eastern Mediterranean region during the end of the Bronze Age between 1250 and 1177 BC together with the epicenter locations of earthquakes (magnitudes of 6.5 or higher) occurred between 1900 and 1980 (Armijo et al. 1986; Drews 1993; Nur and Cline 2000)

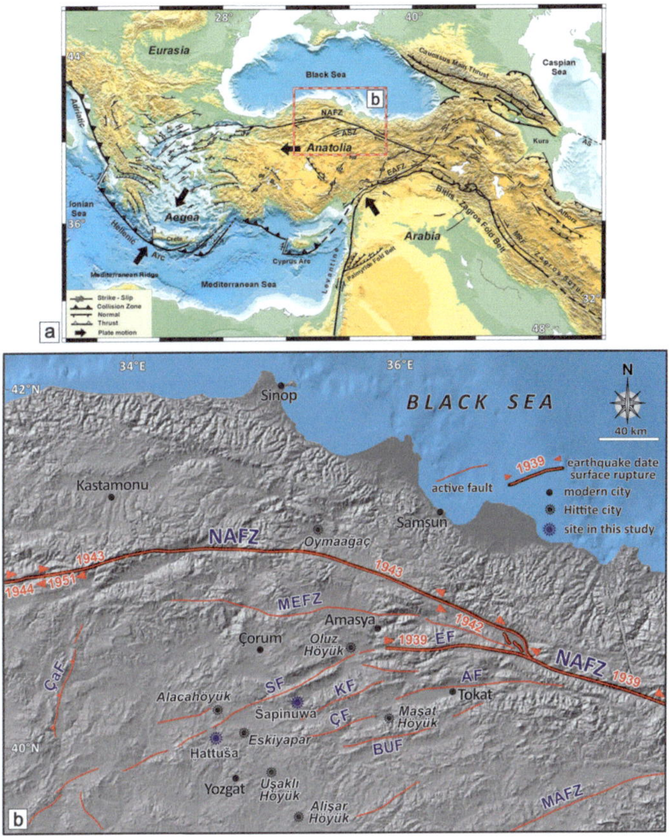

Fig. 2 **a** Relief tectonic map of the faulting and bathymetry in the Eastern Mediterranean region (simplified from Taymaz et al. (2007) and reference therein). **b** Morphotectonic map of the Northern part of Central Turkey and distribution of Ḫattuša, Šapinuwa and other Hittite cities (Faults and their names are taken from Koçyiğit (2003) and Emre et al. (2013). Surface rupture of instrumental earthquakes are taken from Barka and Kadinsky-Cade, 1988 and Emre et al. (2013). Abbreviations of the faults are North Anatolian Fault Zone (NAFZ); East Anatolian Fault Zone (EAFZ); Amasya Shear Zone (ASZ); Middle Anatolian Fault Zone (MAFZ); Merzifon—Esençay Fault Zone (MEFZ); Sungurlu Fault (SF); Çankırı Fault (ÇaF); Buğdaylı Fault (BUF); Almus Fault (AF); Çekerek Fault (ÇF); Kazankaya Fault (KF) and Ezinepazarı Fault (EF)

In light of this interpretation, the authors came up with the idea of building an integrated project considering the geoarchaeological, archaeo-seismological and geophysical contexts in two important Hittite cities. Ultimately, the geoscience studies were conducted in Šapinuwa, another capital city of the Hittite Empire, between 2015 and 2019, and Ḫattuša, the capital of the Hittite Empire, between 2018 and 2019. This study presents the preliminary results of the geoscience studies and reveals what should be focused on or avoided in future studies of earthquakes in these archaeological sites.

2 Geology and Active Tectonic Zones of Hittite Territory

Most Hittite cities spread across the central part of Anatolia located on the border between the Pontides and Anatolides geological belts described by Ketin (1966). All of the Hittite cities explored here, except Oymaağaç (Nerik), are located south of the North Anatolian Fault Zone (NAFZ), one of the best known strike-slip structural discontinuities in the world. Most of the Hittite cities were founded on different rock assemblages, and in terms of both lithological and geological age, they are distributed across an area called the Amasya Shear Zone (ASZ) (Erturaç and Tüysüz 2012), which is structurally separated from the NAFZ as an extensional horsetail splays form (Fig. 2). The ASZ, which is approximately 250 km long and 100 km wide, mostly consists of dextral strike-slip fault segments formed in a transtensional stress regime and varying in length from 230 to 25 km. These structural discontinuities follow an approximately NE-SW trend in a convex geometry between the Middle Anatolian Fault Zone (MAFZ) in the south and the NAFZ in the north. The faults here respectively consist of, from the south to north, the Buğdaylı Fault (BUF), Almus Fault (AF), Çekerek Fault (ÇF), Kazankaya Fault (KF), Sungurlu Fault (SF), Ezinepazarı Fault (EF), and Merzifon-Esençay Fault Zone (MEFZ). The Hittite cities generally show a NE-oriented sequence parallel to the extension of this shear zone. Two important Hittite cities (Ḫattuša and Šapinuwa) were located on the southern block of the Sungurlu Fault, which is the most important structural discontinuity in the ASZ (Fig. 2).

Ḫattuša was founded on the Ankara Mélange, which is one of the basic geological formations, exposed in the İzmir-Ankara-Erzincan Suture Zone (Sümer et al. 2019). The Ankara Mélange of Ḫattuša and its environs is dominantly represented by recrystallized limestone blocks in a clastic-carbonate deformed matrix as an olistostromal facies that includes sandstone, siltstone, marl, and intervening mesoscopic-scaled metaserpantinite and metabasite olistoliths. On the other hand, an N-S-trending serpentinite unit is exposed as a structural slice geometry within the mélange matrix at the western termination of the city. Sümer et al. (2019) state that if it is considered in terms of geology and the rock formation type and features of the structural elements, the city has an important position in terms of habitation and construction.

The Šapinuwa was founded on probable Plio-Quaternary aged terrestrial clastic deposits, which rest on an angular unconformity with clastic sediments deposited in flysch facies in the Eocene age. This Plio-Quaternary unit is generally composed of coarse clastic sediments forming the ancient river channels deposited along the NNW axis. Although sedimentary facies are predominantly fluvial type, the alluvial fan facies is also observed at the northern border of the city. The contact between Eocene sequences and the Plio-Quaternary unit is locally represented by a thick hardened natural cement of calcium carbonate rich caliche level. Most of the old building structures of the city were built on the cutting process of caliche profile. In addition, geological units used in the main architectural materials of the city are medium to coarse-grained, massive, calcareous cemented, well-compacted conglomerates and thick to medium-bedded micritic limestones belonging to Eocene formations.

3 Archaeo-Seismological Studies

Important Hittite cities are located in the Amasya Shear Zone associated with the North Anatolian Fault Zone. These are Ḫattuša, Alacahöyük, Šapinuwa, Eskiyapar, Uşaklı Höyük, Alişar Höyük, Maşat Höyük, and Oluz Höyük. Only a Hittite city called Oymaağaç (Nerik) is located to the north of the NAFZ and outside of ASZ (Fig. 2b). Hittite cities located in the south of four important segments of NAFZ, which is one of the most active fault systems in the world, have been under the influence of these highly active segments throughout history. In the last century, 1939 Erzincan (M: 7.9), 1942 Niksar-Erbaa (M: 7.1), 1943 Tosya-Ladik (M: 7.9) and 1951 Kurşunlu-Ilgaz (M: 6.9) earthquakes occurred along these segments (Fig. 2b) (Barka and Kadinsky-Cade 1988). These fault segments, which were ruptured within the last century, caused four major earthquakes with magnitudes of 7 or higher over the course of 12 years. In addition, there are many active faults in the Amasya Shear Zone. The most important of these is the Sungurlu Fault, which connects to the Ezinepazar segment of the NAFZ and is thought to have continued until Kırıkkale. In addition, the Merzifon-Esençay Fault Zone is another important zone in terms of its length. In the region, there are other active faults with small segment lengths given in Fig. 2b. The Hittite cities established in this region, which is highly effective in terms of seismicity, should have been under constant threat of the NAFZ during this period. The rupture intervals of second-order faults inside the ASZ are not known precisely, but these faults should be seen as other risks in the region. In particular, the rupture intervals of these faults should have much longer durations than the NAFZ and could produce significant earthquake(s). As a matter of fact, when the Hittite structures in the region are examined, it can be thought that they may have some information about the earthquake intensity of the period in which the Hittites lived. Thus, it can be assumed that the Hittites tried to build their important structures with seismically isolated building systems.

Seismically isolated (anti-seismic) buildings are a form of construction that has been applied at different stages of history for a long time by communities living in important earthquake zones of the world. These building systems, which are made with large megalithic stones and designed to be intertwined, had shown resistance to very large earthquakes. The best examples of such structures are those built by the Inca community living in and around the Andes, where earthquakes of magnitude 7 and higher were experienced (Fig. 3a, b). Also, a similar building technique in Japan, which has been an earthquake country throughout history, was applied during the medieval ages (Fig. 3c). Actually, some Hittite monumental buildings also have similar architectural patterns like these worldwide examples (Fig. 3d). Contrary to the prevalent view, Neve, Schirmer and Naumann (former excavation director and investigators of Ḫattuša) claimed this building technique was preferred for representative purposes (Naumann 1971; Schachner 2020b). However, it is of great importance to study this special subject in the future in order to attain a more comprehensive understanding of the link between ancient engineering on earthquake-resistant building design and soil mechanic parameters on Hittite monumental structures.

Fig. 3 Seismically isolated wall examples. Inca walls from **a** Sacsayhuamán **b** Cusco (https://paleoseismicity.org/active-faults-around-cusco-peru-field-work-on-paleoseismology-and-archaeoseismology-for-project-cusco-pata/, **c** Tokyo castle wall from Medieval Age (https://www.worktravelshoot.com/2018/11/Walkthough-guide-tokyo-imperial-palace-east-gardens.html), and **d** Hittite wall from Alacahöyük (https://mapio.net/pic/p-12656965/)

An archaeo-seismology project was conducted in Šapinuwa (2015–2019) and Ḫattuša (2018–2019) cities, which were excavated in the region. The aim of these projects was to reveal what kind of major earthquakes the Hittites, who gave great importance to construction activities, were exposed to. The first archaeo-seismological studies began in the city of Šapinuwa, 80 km northeast of Ḫattuša, the capital of the Hittite Empire. The Šapinuwa archaeological site is located in the Çorum region of Central Anatolia, Turkey. The capital city was an important religious, military, and governmental centre during the Empire Period of Hittite. In 1990, systematic scientific excavations were initiated by Mustafa Süel (archaeologist) and Aygül Süel (Hittitologist) at this site, which is found 3 km southwest of the Ortaköy district of Çorum. In these excavations, which have been going on for 29 years, more than 4000 Hittite, Hurrian, Hattian and Akkadian cuneiform tablets were unearthed. The archive of Šapinuwa is the second-largest cuneiform archive discovered in Anatolia during the Hittite period and found after Ḫattuša (Boğazköy). On the tablets uncovered, it was seen that the city was called Šapinuwa during the Hittite period. Numerous religious, administrative and industrial structures were unearthed during excavations. In addition, many figurative molds and important unique artifacts associated with the Hittite civilization were found. Šapinuwa was ruled in the early fourteenth century BC by the Great King Tuthaliya II and his queen Tavananna Taduhepa (Süel 1998; Süel and Süel 2008). The city was also used by the Great King Mursili II to deploy his army around Šapinuwa in the thirteenth century BC, according to information in the cuneiform tablets (Süel and Süel 2017).

Today, the archaeological excavations continue in two areas in Šapinuwa. These are the sacred area named as Ağılönü and the urban settlement located in the Tepelerarası region (Süel 2002; Drahor et al. 2015). In the Tepelerarası area, there are structures consisting of large stone blocks made of limestone, sandstone and conglomerate. The largest of these is called building "A" and located on a prominent hill. One of the richest tablet archives of the Hittite period was unearthed in this structure (around 3500 cuneiform). There are fortifications to the north and south of the building. In the southeast of building "A", the building named "B" is located, and during excavations, many large pithos were found in this building, which are thought to have contained food and drink. This structure is followed by the "C" and "D" structures, respectively, and the "D" structure is thought to have had a religious function. In addition, buildings called "E", "F", "G", and "H" with different functional properties have emerged during archaeological excavations (Fig. 4). The "G" structure is a workshop area and various bronze materials, molds used in metalwork and other workshop materials were found inside this building during the excavations between 2015 and 2019. Ağılönü is a ritual area of Šapinuwa. There is a dry riverbed separating the area from the Tepelerarası area. In this region, a building of approximately 2000 m^2, called "Taş döşem-stone pavement", was uncovered. The structure is one of the largest structures built in Anatolia during the Hittite period. In front of the building, the pits that include the different sacrificial materials and the bones of animals presented to the gods were revealed. Therefore, the archaeological excavations performed in this area showed that the Ağılönü site was an important sanctuary area of the city (Fig. 4). In addition, Drahor et al. (2015) conducted integrated geophysical studies in this area.

Archaeo-seismological studies in the Šapinuwa archaeological site have been conducted in the "A, B, C, D, and G" buildings in the Tepelerarası region and in the buildings in the Ağılönü site (Fig. 4). Building "A" is the largest building in the Tepelerarası area, and its walls were built by placing stones with seismically isolated walls. This structure, which was built in an earthquake-resistant manner, was unfortunately significantly affected by the earthquakes. Therefore, important deformations on the structural elements with the soil subsidence have occurred in the different parts of the structure. In Fig. 5, the archaeo-seismological data added on the drawing produced during the 1990s excavations shows that there was significant damage in the whole area; unfortunately, it was found that the larger damages were concentrated in the western and southern parts. There was also an intense fire in this section. In the archaeo-seismological studies conducted on this building, important seismo-gravitational deformation structures such as dilatation, tilting, rotation, dislocation and shifting were determined (Fig. 6). In addition, the earthquake waves based antiform- and synform-shaped surface irregularities have been identified in the northwest and western parts of the area (Fig. 7). In the southern section of the area, there is a part where the intense destruction traces and significant inclination on the surface due to the earthquake(s). All of these important archaeo-seismologic deformational structures are also presented in Fig. 5. In addition, because of a large and intense fire, the sandstones and conglomerates on the walls of the building were transformed into red, and there are also explosions and cracks on the walls, which are made

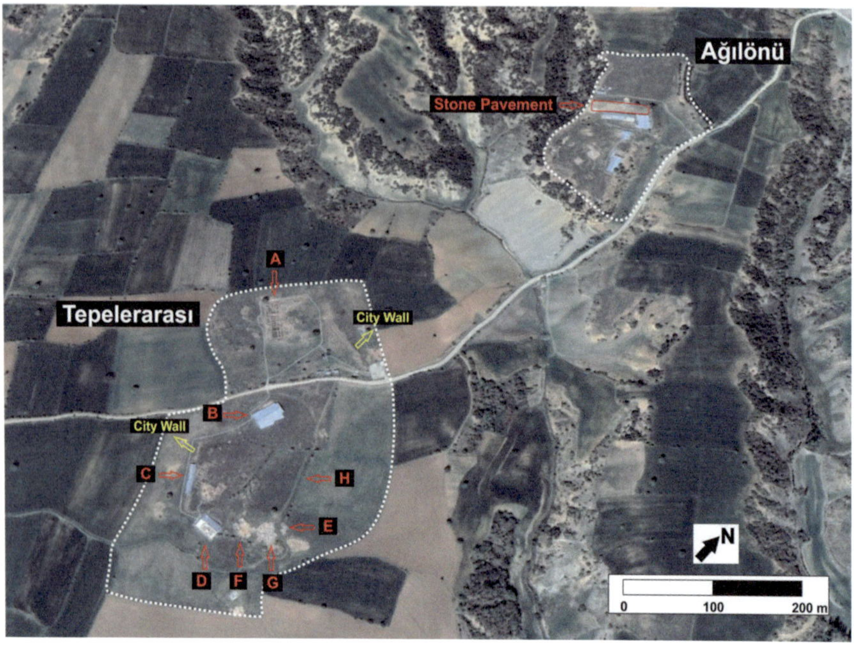

Fig. 4 Satellite image of the Šapinuwa archaeological site. Investigation sites are depicted in satellite images with the capitals

up of mainly limestone blocks. In conclusion, the high deformation characteristics observed in this building indicate the presence of the important earthquake(s).

There are also widespread deformations in building "B", which is thought to have been an official storage area and have contained many pythos. In addition to phenomena such as displacement, rotation, shifting, dislocation, tilting and dilatation on the walls, there are also displacements reaching 8 cm horizontally and vertically in this area. In addition, a snake-shaped undulation is found on the wall in the southeast part of the area. Besides, it was determined that something of important subsidence had been present on the soil of this section. A collapse toward the northeast was also observed on the walls in the northern section. As a result, significant deformation traces that could be associated with the earthquake were identified in this area (Fig. 8).

The building "C" contains a large group of structures located in the northwest of the site and with residences within the city walls. During the field studies, in this building and its surrounding area, seismo-tectonic and seismo-gravitational deformation structures were observed (Fig. 9). A morphological scarp with 70 cm approximately was found during the archaeo-seismological studies conducted in this area. This scarp with the direction of N62E intersects the building, and intense deformations are seen on the walls of the structures around this area. In addition, there is a vertical displacement of a 21 cm length cutting the fortification wall, which is parallel to the previous scarp. The direction of this displacement is N50E. Also, it is available

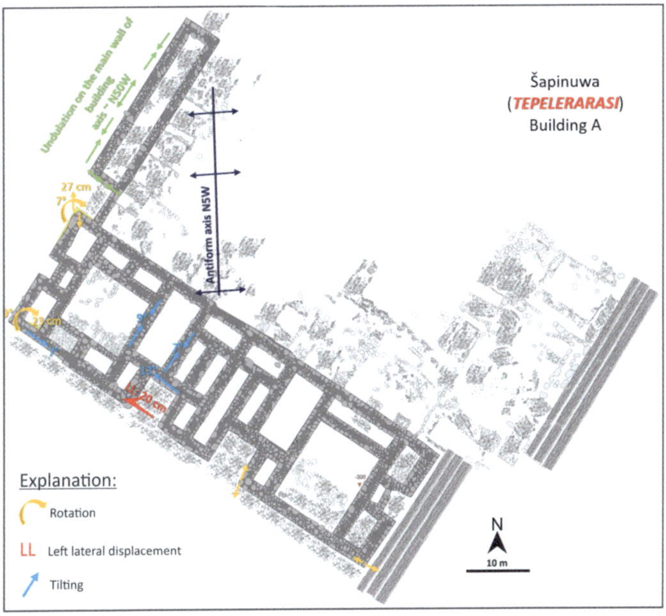

Fig. 5 An overview of the archaeo-seismological features observed in the building "A"

in a left-lateral horizontal displacement up to 12 cm. Ground subsidence can also be observed in different parts of the city wall and buildings.

Area "D" is characterized by high deformation, and there was also a significant fire there. The rotated and tilted burned mudbrick walls of varying degrees above the stone foundations characterize this area. Archaeo-seismological studies on this area revealed the high rates of undulations and tilting on the stone and mudbrick walls (Fig. 10). In addition, a significant collapse was observed on the ground subsidence of the southern part of the walls, where distortion was observed. There is a vertical displacement of 28 cm in this section. In addition, the N15W oriented axes were observed in the undulated wall. The tilting of the mudbrick walls in distinct directions indicates an earthquake. The damage rate throughout the building is very high, and it especially increases toward the south.

An important workshop building was unearthed during the archaeological excavations carried out between 2015 and 2019 in the area determined by the magnetic gradiometry and Ground Penetrating Radar (GPR) studies conducted in 2014 and 2015 in Şapinuwa. This is called building "G". Since no data related to the earthquake were collected in the previous archaeological excavations made in the sections described above, although archaeo-seismological studies were conducted in the area, no direct archaeological context information about the earthquake could be gained. However, since the archaeological excavations in the building "G" were conducted together with the archaeo-seismological studies in the region, the data obtained from this area revealed critical information about the ancient earthquake that happened

Fig. 6 Seismo-gravitational deformation structures on the building "A", **a** some rotation, dislocation, tilting, dilatation and antiform type and **b** synform, shifting type deformations

there. All the traces associated with the earthquake in this area, which yielded extremely important results related to the ancient earthquake, were protected in order not to deteriorate due to natural causes. Archaeological excavations in the area were completed within three years and important earthquake data were collected through coordinated archaeo-seismological studies. These studies yielded results showing the presence of a major earthquake in the workshop complex. In particular, the formation of a wave shape on the ground and buildings overlapping of waveform deformations in the structural elements gives important results in terms of determining the magnitude of the earthquake and the ground deformation feature. Important undulations, rotations, vertical and horizontal displacements and dilatations were observed on

Fig. 7 Earthquake waves based antiforms, synforms and parallel undulations

the walls of the building complex (Fig. 11). The vertical displacement on the walls reaches 67 cm in the north and the horizontal displacement reaches 12 cm in the center. Three undulation axes were determined in the building and their directions were measured as N54E, N56E, and N62E, respectively. These directions are consistent with those of other buildings and show the presence of the same event. As a result, the studies carried out in this building provided important information about the earthquake in the region. In addition, the presence of an important fire in the buildings in this area has also been revealed. Although it is argued that some traces found under this structure may be related to another cultural layer, it is also thought that context interference may occur due to the high vertical displacement of the same layer due to the earthquake effect.

In the studies conducted in the area of the Ağılönü portion of the Šapinuwa archaeological site, important results related to the same earthquake were obtained. The

Fig. 8 Archaeo-seismologic deformational structures on the building "B", tilting, shifting, snake-type lateral undulations, dislocations and rotations

Fig. 9 Deformational structures on the building "C" and its surrounding, tilting, vertical displacement on the building wall and morphological scarp

structure, which has burnt material in the middle and contains significant deformations on its walls, characterizes the important earthquake in the city, especially with the aspects of the rotated mudbricks in the middle part of the building (Fig. 12). In addition, vertical and horizontal displacement in the walls in the area, orientations in a certain direction of the walls and displacement of the floor were seen as important earthquake data. Studies on the structural elements have also revealed that the axes in the places where deformation occurs are compatible with the Tepelerarası area.

The results of archaeo-seismological studies obtained in Šapinuwa revealed that such studies should also be conducted in other Hittite cities. For this purpose, the second study was carried out between 2018 and 2019 in Ḫattuša, the Hittite capital. First, a detailed geological map was prepared to show the geological units and tectonic

Fig. 10 Deformational structures and significant fire tracks on the building "D"

lines of the city (Fig. 13). The geological map of Ḫattuša shows the two main geological groups, from older to younger, respectively, Ankara Mélange and Miocene—Pliocene continental rocks. The Ankara Mélange, on which the main city was established, consists of three main rock units. These are a matrix of the mélange, blocks, and olistoliths of limestone and tectonic slice of serpentines. The olistostromes of the Ankara Mélange had various benefits for Ḫattuša. According to geological observations, separate limestone blocks are located in the city, and this situation provides, besides panoramic views, effective protection and secure construction grounds for important buildings such as Yenicekale, Sarıkale and Büyükkale. In addition, the

Fig. 11 Archaeo-seismologic deformational structures on the building "G", rotation, vertical and lateral displacements and earthquake waves based undulations

lithological components of the Ankara Mélange have a property of water retention, and the structural discontinuities of this geological unit also facilitate the water supply. The use of these olistoliths as quarries for construction of the city minimizes the effort necessary to move building blocks from the outside. The combination of these geological features explains why Ḫattuša's location is in this place. The geological map prepared for the first time in Ḫattuša shows several fault structures, which cut across the city within the orientation of a general NW–SE direction. Various types of faults were described: thrust, dextral/sinistral strike-slip and high-angle dip/oblique-slip normal faults. On the other hand, especially the deformation phases, which

Fig. 12 Rotation, tilting, vertical displacement, shifting, undulation, uplift and complex destruction indications observed in the buildings located in the Ağılönü area

are defined by high-angle faults, are characterized by conjugated structures with N10-20E and N50-60 W directions, which appear throughout the city (Sümer et al. 2019).

The city was geographically divided into two parts; the archaeo-seismological observations were conducted on archaeological structures in the Upper and Lower parts of the city. In the Upper City, some deformed structures were present, including the Lion Gate, King's Gate, Yerkapı and the Temple district (Fig. 14). The presence of a deformation toward N65E was found in the fortification wall made of cyclopean-shaped stones at Lion Gate (Fig. 15). The stones have dilatations and separations in this direction and some stones have cracks. There is also a low level of displacement

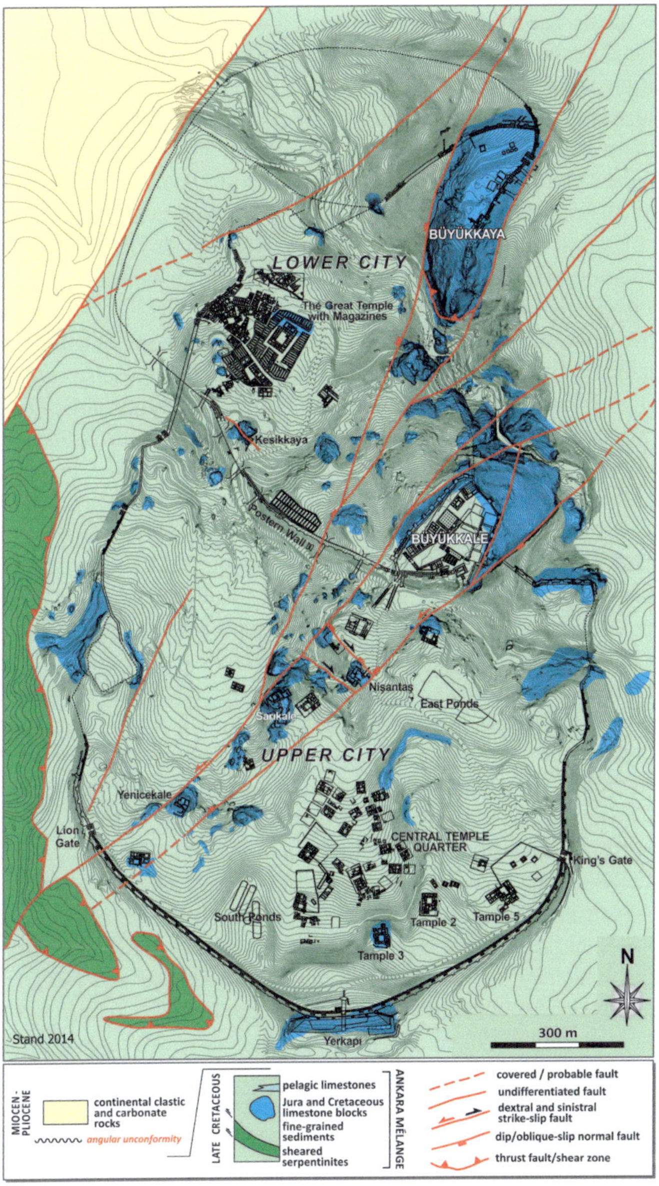

Fig. 13 Geological map of Ḫattuša archaeological site from Sümer et al. (2019) (Topographical contour resolution is 1 m; scale 1/10.000)

at the gate. In the King's Gate, the deformation with the direction of N73W was found. It was observed that there were dilatation, slopes and shifts on the city walls in the direction of deformation (Fig. 16). There is an important potern structure at the Yerkapı and a lateral shift with 17 cm that was measured inside the potern located in the southern part of the Upper City. In addition, significant deformations were observed in the same locations at the East and West stairs of the Yerkapı (Fig. 17). In the Temples' district (Fig. 14), incidents such as undulations, lateral displacements, tilting and dilatation were found on the temples' walls. In particular, as shown clearly in the last photograph in Fig. 18, there was also 15 cm dilatation in the western direction seismically isolated wall of the Temple 3.

The most remarkable archaeo-seismologic deformation structures were also observed in the Lower City of Ḫattuša (Fig. 14). In the archaeo-seismological studies conducted in this part of the city, which includes the buildings in the Great Temple and the Lower City built in the late 17th or sixteenth century BC (Schachner 2020b), various deformation elements such as systematic cracks, offsets, extensions, rotations/flexions, camber and antiformal structures were observed (Fig. 19). Especially the presence of antiform and synform-shaped undulations and snake-type sprains in and around the magazine part of the Great Temple area shows that there could have important movement on the ground during the earthquake (Fig. 20). Furthermore, in the same area, lateral shifts/dilatations were also observed (Fig. 21). Also, Sümer et al. (2019) specified numerous deformational structures in the Great Temple (Fig. 22). The systematic cracks, which have dimensions varying between 1 and 2 m in length, were defined on the stable large blocks of the temple in 18 locations. Particularly, the systematic cracks in a similar orientation observed in the sill blocks of the temple entrance are remarkable. In addition, a smoothly-antiformal deformation pattern was observed on the giant sill blocks and a temple wall block. The main axes of these antiforms are also aligned with the general trend of the systematic crack direction (Fig. 23). One of the important results is the presence of dilatation of 32 cm in length, which is similar to a NW–SE extension observed in the area, between two wall blocks at the western side of the temple entrance (Fig. 24). The direction of the cracks, which cut across the Great Temple, range between N20E and N60E. All the data observed on the Great Temple presented by Sümer et al. (2019) reveal that an approximately N45E-trending deformation zone affected this monumental building (Fig. 22). The deformation model shows that these traces were revealed by an earthquake that occurred after the Great Temple was built. This deformation zone generally trends N45E, and it is quite compatible with the orientation of Sungurlu Fault (SF), which is located only 13 km north of the Ḫattuša archaeological site.

Findings from the excavations of living-quarters in the northern lower town of Ḫattuša probably indicate destruction by an earthquake due to the lintel position of the walls and warping of the foundations (Schachner 2020a). Radiocarbon dating suggests that the buildings of the destroyed layer were constructed in the late 16th or early fifteenth century BC. The destruction can only be dated in terms of the archaeological finds, which suggest that the building was destroyed at the latest at the turn of the 15th to the fourteenth century BC. In any case, there are no finds of the Great Kingdom period. This preliminary result points to a severe earthquake at

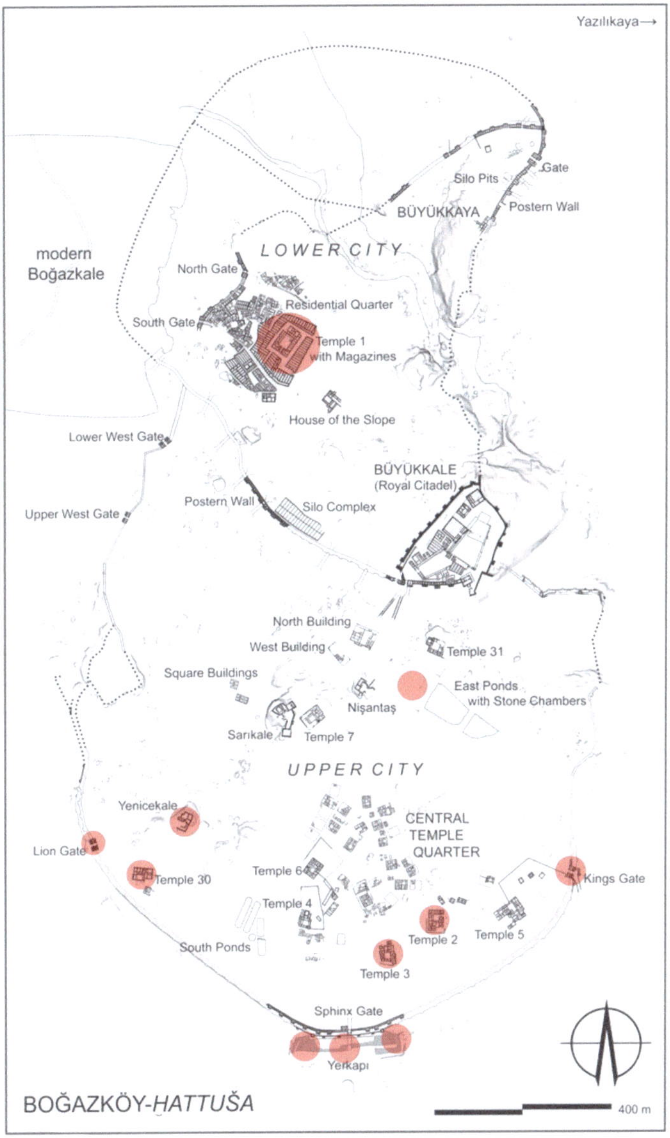

Fig. 14 Distribution of archaeo-seismologic deformational structures within the Ḫattuša (city plan is taken from Deutsches Archäologisches Institut Archive)

the end of the middle Hittite period, but of course, it is unclear whether this was responsible for the destruction observed at these monumental buildings.

Archaeo-seismological studies in two Hittite cities (Šapinuwa and Ḫattuša) in Central Anatolia revealed data showing at least one major earthquake during the Hittite period. For this reason, it is necessary to expand these studies to other Hittite

Fig. 15 Archaeo-seismologic deformational structures observed in the Lion Gate

cities in the region, which would be useful in determining the important damages to the Hittite civilization caused by earthquakes. Thus, toward the end of the Hittite civilization, important data will be obtained regarding the presence or absence of a major earthquake. In addition, it is thought that with the help of the orientation of the deformation zones to be determined from these areas, the possible fault zone that led to the major earthquake(s) will be determined. Thus, through the geological, geophysical and paleoseismological trenching works that will be conducted in the possible fault zone, the major earthquakes of the Hittite period in that region would be defined.

4 Geophysical Studies

Integrated geophysical studies have been carried out around the excavated sites and buildings that have been damaged from old earthquake(s) in parallel to the geoarchaeological and archaeo-seismological studies in the Šapinuwa archaeological site. The aim of this study was to reveal the ground effect of important damages based on the earthquake(s), and even in the structures seismically isolated, in the study area. In fact, it may have resulted in an increase in earthquake damage in the area, given the geological problems in the ground along with possible construction errors in the structures. In this context, the aforementioned areas were examined using two-dimensional electrical resistivity tomography (ERT), seismic refraction P wave tomography (SRT) and multi-channel analysis of surface waves tomography (MASWT) methods. In this

Fig. 16 Undulation, shifting and dislocation found on the King's Gate

study, the results of the integrated geophysical studies conducted around building "A", the largest structure in the region, are given.

Figure 25 shows the position on the satellite image of the two-dimensional integrated geophysical investigation lines positioned around building "A". The names on the lines were made according to geographical directions and they were named as NE, NW, SW and SE lines. The length of the lines around the NW and SE walls is 69 m, while the lengths of the lines around the NE and SW walls are 81 m. All ERT studies were performed using the lines of 60 m length. The ERT study on these lines was conducted using the Wenner-Schlumberger configuration. Two-dimensional model sections were obtained using the robust algorithm (L1 norm) during the inversion process.

In the ERT model cross-section of SE line, there is a high resistive environment (80–150 Ωm) to a depth of approximately 2 to 2.5 m. Below this depth, there is a zone with low resistivity below 30 Ωm. This zone is a geological environment and should show the Eocene clays where the archaeological structures were built (Fig. 26a). These clays have high consolidation. In the ERT model section of the NW line, the

Fig. 17 Horizontal displacements both side stairs and 17 cm lateral displacement potern structure at Yerkapı

Fig. 18 Tilting, shifting, undulation, dilatation and dislocation type deformations in The Temple's district

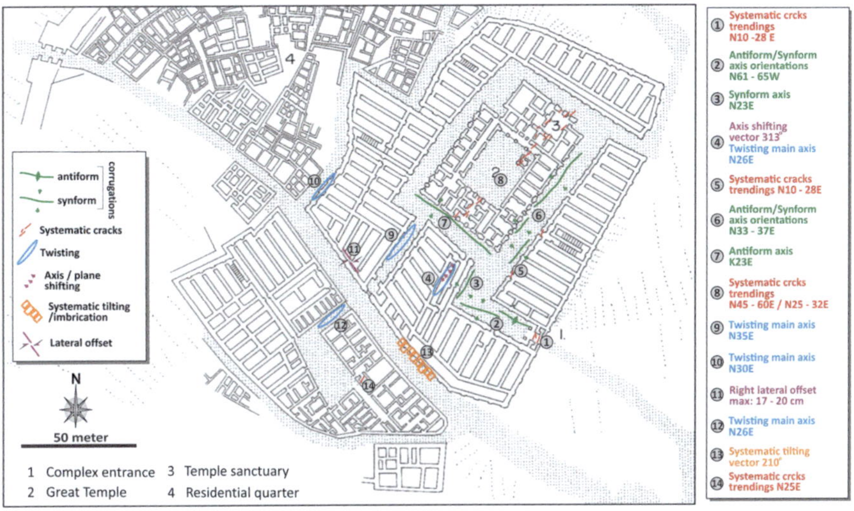

Fig. 19 Archaeo-seismologic deformational structures observed in the Lower City of Ḫattuša

Fig. 20 Antiform and synform deformational structures and snake-type undulations observed in the Lower City of Ḫattuša

Fig. 21 Lateral displacement and dilatation type deformations in the Lower City of Ḫattuša

high resistive subsurface changes close to the surface extend up to a depth of 2.5 m. Below the high-resistive environment in the southwestern part of the line, there is a medium-resistive zone (25–50 m) that extends to a depth of about 7 m, and this zone is thought to be related to deep archaeological structures. This environment suddenly disappears at 35 m of the line toward the northeast. In the northeast part and below the layer with medium resistivity in the model, resistivity values decrease. Thus, it is thought that the clayey bedrock was entered in this section (Fig. 26a).

The NE section model is different from the others. To the southeast of the section, as in the previous section, there is a high-resistive environment extending to a depth of about 7 m. The uneven distribution of the high resistivities suggests that there may have been buried archaeological layers or a different geological layer. The high resistivity in the central part of the line to a depth of about 2 m may indicate a ground improvement associated with the structure. While high resistivities close to the surface are dominant in the northwest of the line, a medium resistivity layer is observed between the 15th and 25th meters of the line toward the southeast. In the southeast of this layer, low resistivity bedrock emerges. The sudden vertical change between the two environments is interesting here, and such a sudden change must be either related to the geological environment (perhaps a fault) or to an archaeological layer. If the high-to-medium resistivity environment in the northwest reflects an archaeological layer, it can be speculated that a ground improvement related to the structure may have been made in this section in the Hittite period (Fig. 26a).

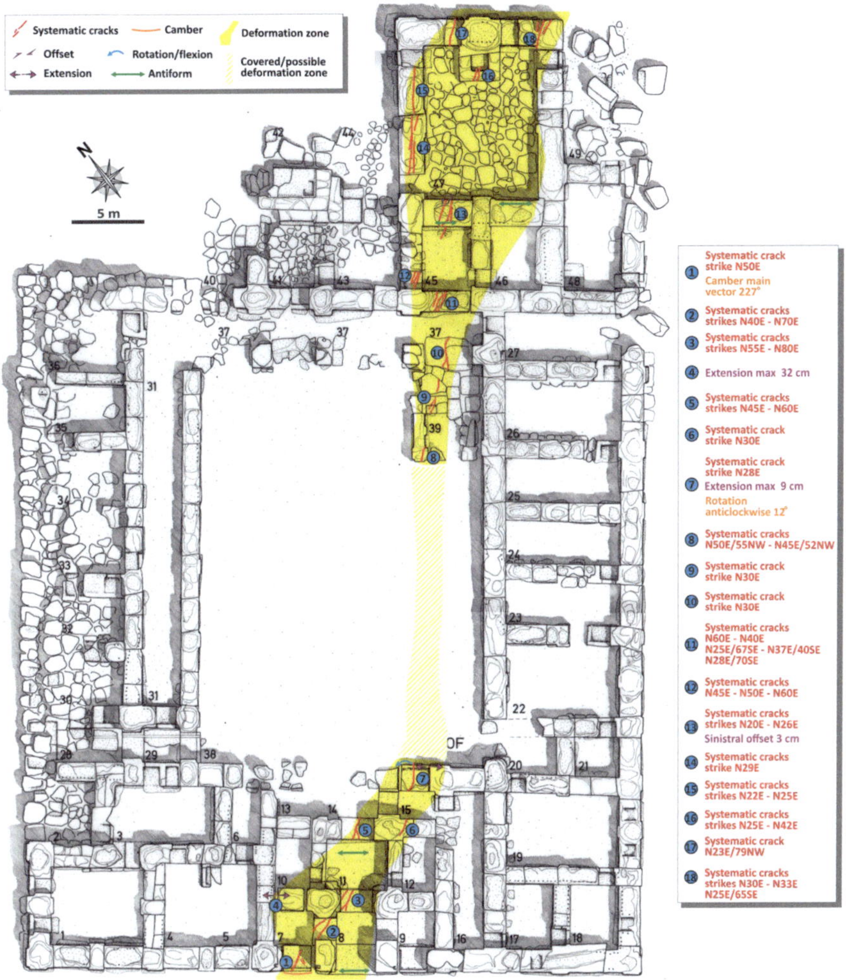

Fig. 22 Deformational structures observed by Sümer et al. (2019) on the Great Temple in the Lower City of Ḫattuša (Stone plan is of the Great Temple is taken from Deutsches Archäologisches Institut Archive; scale 1/400)

The final section of the ERT model shows the result of the SW line in the southern part of the building. The highest deformation in the area is in the southwestern part of the building, and this data is particularly important in relation to the parts with highly deformed structures. In the southeastern part of the SW line, there is an environment with high resistivity up to a depth of 7 m. This medium must be associated with the embedded structural elements in this section. Although some other building elements are observed in the middle part reaching a depth of 2.5 m, high resistive environments toward NW become very shallow here. However, at the end of the line, a high-resistive environment, which deepens again, is formed. Below

Fig. 23 NE-trending systematic cracks observed on the Great Temple and the magazine section of the Lower City

the resistive environment, clayey bedrock characterized by 12 and 30 Ωm can be seen. However, between the 25th and 30th meters of the line, there is also a slope varying between 1 and 1.5 m from northwest to southeast on the surface (Fig. 26a).

Fig. 24 Systematic cracks and dilatations on the Great Temple in the Lower City of Ḫattuša

Fig. 25 Plan showing the measuring lines of the ERT, SRT and MASWT studies around the building "A". The arrows indicate the direction of measurement. NW, NE, SW and SE identify the measuring lines

The tomographic model sections obtained from the results of SRT studies on the same lines are given in Fig. 26b. Seismic P velocity values start from 300 m/sec at the surface and reach 1500 m/sec at the bottom. In the sections, there is generally an environment with low P wave velocities (350–550 m/sec), which extends to a depth of approximately 5 m. This environment indicates that the ground beneath the archaeological strata has weak soil character. However, the low velocity values

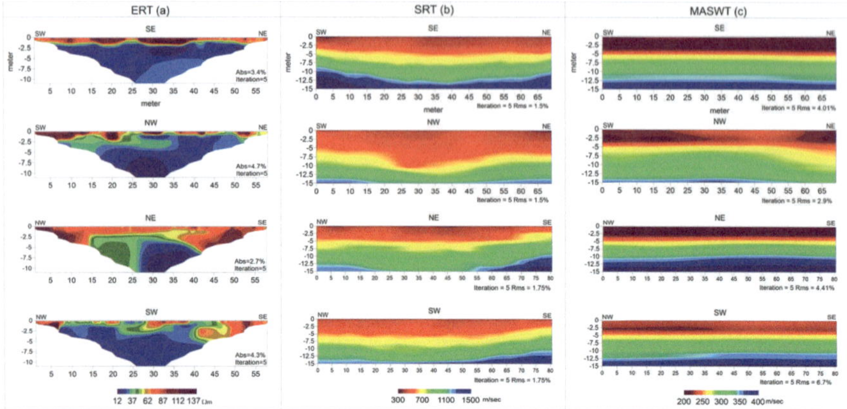

Fig. 26 The tomographic model sections of integrated geophysical investigations conducted around the building "A". **a** ERT, **b** SRT and **c** MASWT

in the NW line are different and continue up to 10 m depth. In particular, the low velocities have a V-shape between 20 and 30 m of the line. This situation may be related to a deformation that occurred during the earthquake. The fact that this part is located where the walls are inclined and where the antiform-synform structures are located reveals a weak zone under the surface, when the seismic P velocities are considered. This weak zone may have been effective in increasing the damage in this part during the earthquake. The medium velocities are found at 5 to 13 m in NE, SW and SE lines, whereas there is a different distribution only in the NW line. In the NW line, medium-velocity values start at 10 m and show a U-shaped distribution. This situation shows that the subsidence in the middle of the line extends deeper. As a result, the high-velocity environment in the east-southeast part of the area starts from a depth of 10 m, while in the west-northwest parts it starts from 15 m. This situation reveals that the seismic velocity character is extended toward different depths in both parts of the area where the building is settled. This difference may be related to a geological environment or may be characterized by ground subsidence that occurred due to an earthquake (Fig. 26b).

Seismic shear wave velocities of the measuring lines surrounding building "A" vary between 210 and 400 m/sec. These values are generally low, and the low shear wave velocities between 210 and 270 m/sec, especially seen in the first 5 m, indicate that the environment should be an earth fill. Although all cross-sections are similar, there is an undulation on the SW line as in the P wave model. However, the direction of the undulation is the opposite of the P wave. This case indicates a change in the environment between 7 and 10 m. Especially, these changes occur in the part where antiform and synform undulations occur. High shear wave velocities are seen in sections after 12 to 13 m except for the NW line (Fig. 26c).

Integrated geophysical studies indicate that a thick fill layer may be found under the place where building "A" is settled, especially in the area where high deformations occur in the west of the building. It should also be noted that the western and eastern parts of the area show a different character. In particular, data on the presence of a fault in the ERT section indicate that there are important ground problems in the area where the structure is located.

In 2D seismic P and S velocity models, the classification of the environments starting from the surface up to a depth of 15 m in terms of velocity values has been carried out separately for each line. Three seismic zones were located in all sections. These are shown as low, medium and high velocity zones. In Table 1, these zones are given depending on the measurement lines. The dynamic parameters of soil were calculated according to this classification. In Table 1, the poisson ratio, density, V_p/V_s ratio and elasticity, bulk and shear modulus are given together with the P and S values. When these values are examined, comments for each line are as follows.

Seismic P velocities are generally close to each other in all lines. However, the velocities in the NW line are lower in all zones, except for the surface. The same situation applies to S wave velocities. Thus, there is a different subsurface distribution in terms of seismic velocity in the NW part of building "A". Considering the V_p/V_s ratio, while the ratios are generally close to each other, there is a partial increase in the ratios in the SE part. This situation reveals that there is a geological unit or various earth fills in this part of the area, which create a velocity difference in the subsurface. The increase in poisson values in the 3rd zone of the SE line indicates that the clay may be closer to the surface. The NW line also exhibits a different distribution character in density values. Thus, debris and earth fills with lower density are thought to be more common in this part. Elasticity, bulk and shear modules have also shown the lowest values in the NW line (Table 1). These results reveal that there is an earth fill environment until approximately 10 m depth, and this fill is thicker in the NW and SE regions than the other parts of building "A". This may have resulted from a case of soil improvement and earth filling done during the Hittite period. Therefore, these differences in some locations in the area were thought to have caused damage to building "A" during the earthquake. As a result, it should be declared that the environments with low values found in the NW line and the SE line with high V_p/V_s ratio are consistent with the findings in the archaeo-seismological studies. Therefore,

Table 1 Seismic velocities and geotechnical parameters obtained from building "A". For profile directions of seismic lines see in Fig. 25

	Zone	Vp (m/sec)	Vs (m/sec)	Vp/Vs	Poisson	Density (kg/m^3)	Elasticity Modulus (E) (MPa)	Bulk Modulus (K) (MPa)	Shear Modulus (G) (MPa)
NW	z1	347–680	210–260	1.5–2.1	0.10–0.20	1338–1430	17–40	5–80	6–14
	z2	600–1000	260–330	2.1–3.2	0.20–0.36	1430–1700	40–75	80–210	14–20
	z3	1000–1279	330–370	3.2–4.0	0.36–0.47	1700–1853	75–92	210–274	20–24
NE	z1	348–700	205–280	1.7–2.2	0.20–0.28	1345–1490	16–50	6–100	5–14
	z2	700–1200	280–390	2.2–3.1	0.28–0.38	1490–1790	50–115	100–280	14–30
	z3	1200–1525	390–456	3.1–3.6	0.38–0.46	1790–1931	115–151	280–384	30–39
SW	z1	381–750	222–280	1.5–2.3	0.16–0.24	1338–1520	17–50	4–120	6–14
	z2	750–1200	280–370	2.3–3.3	0.24–0.38	1520–1780	50–95	120–280	14–26
	z3	1200–1451	370–412	3.3–3.9	0.38–0.47	1780–1940	95–124	280–394	26–32
SE	z1	348–750	209–270	1.7–2.5	0.22–0.30	1353–1550	17–45	2–140	5–13
	z2	750–1250	270–350	2.5–3.7	0.30–0.41	1550–1850	45–90	140–340	13–24
	z3	1250–1575	350–400	3.7–4.4	0.41–0.47	1850–1962	90–118	340–449	24–30

the most severe damages and deformations occurred in the western and southwestern parts of the building. It can be thought that these changes caused the formation of the undulations of the antiform and synform types, and thus the damage in the building might have been increased.

5 Discussion and Conclusion

Ḫattuša and Šapinuwa archaeological settlements are located in the northern part of the Central Anatolia region, which shows relatively low values of active deformational slip rates compared to other regions of Turkey (less than 2 mm/year according to the GPS data of Reilinger et al. (2006)). This study reveals the archaeo-seismological studies conducted for the first time in two Hittite cities and the existence of damages caused by one or more earthquakes as a result. As a result of the integrated geoscience studies carried out in these areas, it was concluded that high deformations were seen in both areas and that the buildings related to them were destroyed. However, these studies exclude information about the fault or faults, earthquake(s), time and magnitude. According to Stein et al. (1997), who conducted detailed studies of NAFZ, earthquakes exhibit a successive distribution (Fig. 27a). In the last century, there is mostly a movement from east to west in the occurrence of earthquakes. The most damaging earthquakes that occurred in this region in the past century are the 1939 Erzincan and 1943 Ladik-Tosya earthquakes. Especially, the major cumulative right-lateral slip (m) values are generally in the region, where the Hittite cities are established, indicating that the damage could be higher in the settlements in these regions. Also, the rupture of both NAFZ and SF together during the 1939 earthquake shows that both fault systems may have worked together. Thus, the active SF is thought to be a fault zone with the potential to create an important earthquake in the region. However, in the illustrations given in Fig. 27b that reveal the data of the historical earthquakes that occurred in the Ottoman and Byzantine periods in the region, it is seen that there was no significant earthquake in the region between 967 and 1050 AD. However, the earthquakes that occurred in this region during the Ottoman period (August 17, 1668) and the last century (1939–1943) caused significant damages. The earthquake of August 17, 1668 is the biggest known ground motion and this earthquake caused great damage in the region and affected the area between Ankara and Erzincan. The magnitude of this earthquake, which forms an approximately 400 km surface rupture, should be around $M = 8$. Obviously, as Stein et al. (1997) show, a high fracture interval of approximately 450 ± 220 years is important in terms of the increase in earthquake magnitude and consequently the increase in damage in the region. After a gap of approximately 280 years, two major earthquakes occurred in the region in the last century (1939 and 1943). Ultimately, the different rupturing of the fault segments of the NAFZ in the historical process reveals that the system did not experience regular rupture. In the context of these data, a similar event may have occurred in and around the Hittite territory during the Bronze Age. The aforementioned study introduced the existence of an earthquake that caused high damage in both Hittite cities at a date indeterminable with the available data. Archaeo-seismological studies indicate that the deformation aspects of both ancient cities may be related to Sungurlu Fault (SF). Historical records show that there were no major earthquakes on this fault line for at least 1000 years. Only a small part of SF was broken in 1939. Unfortunately, there are no detailed geological data about the western part of this fault. However, the morphology reveals that the fault lies between Ezinepazar and Kırıkkale. In the active fault map produced by the General Directorate of Mineral Research and Exploration of Turkey, this fault is shown to be as active within the Quaternary. This fault line is found approximately 12 km away from Šapinuwa and Ḫattuša and 4 km from Alacahöyük. It is informed that a major earthquake with a magnitude of 7 and over, which may have occurred on this fault, can produce significant damage in the Hittite cities. In this context, it should be stated that the earthquake potential and occurrence intervals of active faults found around Hittite cities and their surroundings should be examined in detail. In addition, the archaeo-seismological studies should be conducted in the Hittite Empire cities, which have been settled in the region for a long time. It is also necessary to examine the suspected faults (especially SF), which can produce major earthquakes in the region, with geophysical and paleoseismological studies, and to reveal their fault characteristics. As a result, the region where the Hittite cities are located is a shear zone, and, considering the existence of faults with long earthquake intervals, there is a possibility of a major earthquake in the NAFZ and SF at the same time. Historical records suggest that there were extremely large earthquakes in the region and that an earthquake similar to those of 1668 (Ms 7.8–8.0), 1939 (Ms 7.9) and 1943 (Ms 7.2) may have forced the Hittite civilization to leave the region. Geological and seismological studies carried out over the last decades in the region and archaeo-seismological studies conducted in this study revealed important new findings related to the past earthquakes and seismicity of this region. Naturally, new findings that emerge will raise many new questions. Therefore, in line with the current problems in the region, studies on detailed geology, active fault detection and determination of their earthquake generation intervals should be conducted. Thus, it is thought that gaining the data related to the Hittite region will make an important contribution to the destruction of this society.

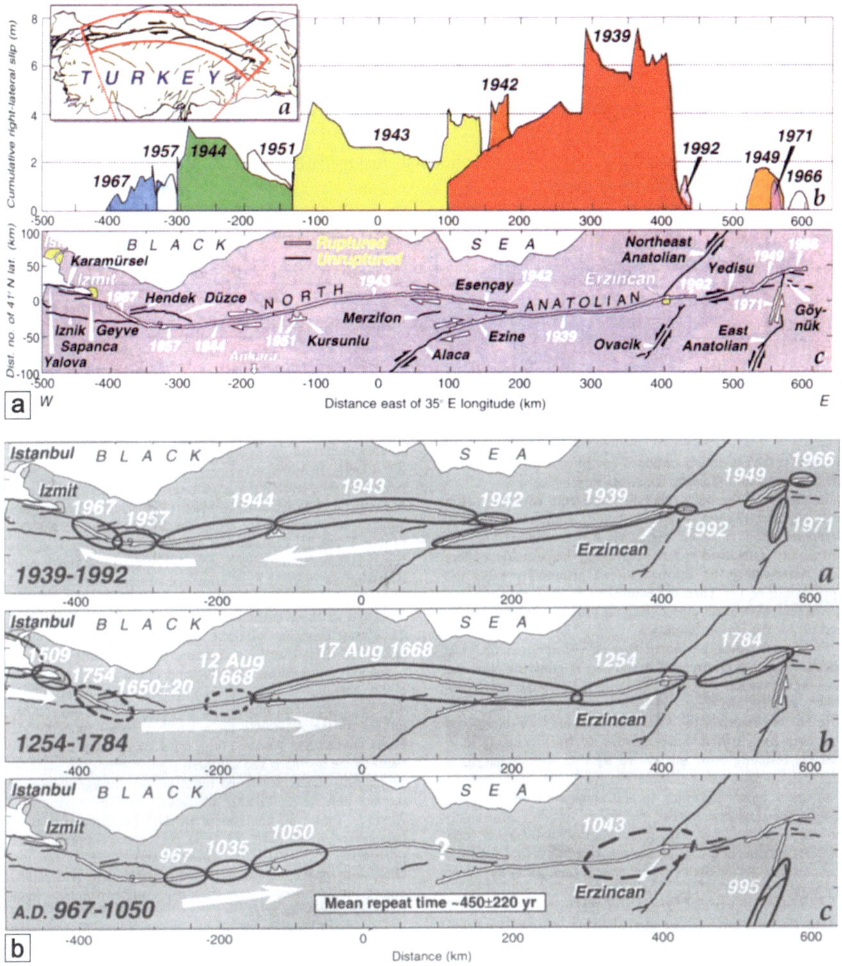

Fig. 27 **a** Showing the portion of the NAFZ in Anatolia and the cumulative right lateral-slip rate in the last century (from Stein et al., 1997). This combined illustration is showing the earthquake storm occurring inside the 38 years in the twentieth century along the North Anatolian Fault Zone and migration of the NAFZ to the west in this century. **b** shows the earthquake activity and surface ruptures in the same region in the last century and Ottoman—Byzantine historical period (from Stein et al. 1997 and reference there in; Ambraseys and Finkel 1988, 1995; Ikeda et al. 1991; Barka 1992)

Acknowledgements These studies were conducted with the contributions of Geoim LLC (Turkey) and we warmly thank them for their support. In addition, we grateful to Geoim LLC engineers Mustafa Sadık Yağlıdere and Buğra Oğuz Kaya who contributed to the figure drawings during the preparation stage of the publication. Finally, our special thanks to the editors of this book Gad El-Qady and Claudio Margottini. We also thank Mohamed ElGabry and other anonymous reviewer for their valuable contributions.

References

Ambraseys NN, Finkel CF (1988) The Anatolian earthquake of 17 August 1668, in Historical Seismograms and Earthquakes of the World, pp 173–180, eds Lee WHK, Meyers H, Shimazaki K, Academic Press, New York, NY

Ambraseys NN, Finkel CF (1995) The seismicity of turkey and adjacent areas: a historical review, 1500–1800, Muhittin Salih Eren, İstanbul

Armijo R, Poirier JP, Deschamps A (1986) Carte sismotectonique de l'Europe et du Bassin méditerranéen. Institut de Physique du Globe de Paris

Åström P (1985) The sea peoples in the light of new excavations. Centre D'etudes Chypriotes, C. III:6–7

Barka AA (1992) The North Anatolian fault zone, Annales Tectonicae, Special Issue—Supplement to Volume VI, 164–195

Barka AA, Kadinsky-Cade K (1988) Strike-slip fault geometry in Turkey and its influence on earthquake activity. Tectonics 7(3):663–684

Blegen CW (1953) Troy: excavations conducted by the University of Cincinnati, Princeton, 330–332.

Bryce T (1998) The Kingdom of the Hittites, Oxford

Cline EH (2014) 1177 B.C. The year civilization collapsed, Princeton, Princeton University Press

deMenocal PB (2001) Cultural responses to climate change during the late Holocene. Science 292(5517):667–673

Drahor MG, Berge MA, Öztürk C, Ortan B (2015) Integrated geophysical investigations at a sacred area in Central Anatolia. Turkey. Near Surface Geophys 13(6):523–543. https://doi.org/10.3997/1873-0604.2015037

Drake BL (2012) The influence of climatic change on the late bronze age collapse and the greek dark ages. J Archaeol Sci 39(6):1862–1870

Drews R (1993) The end of the Bronze Age: changes in warfare and the catastrophe ca. 1200 BC. Princeton University Press, New Jersey

Emre Ö, Duman TY, Özalp S, Elmacı H, Olgun Ş, Şaroğlu F (2013) Açıklamalı Türkiye Diri Fay Haritası, MTA Genel Müdürlüğü, Özel Yayın Serisi—30, Ankara—Türkiye

Erturaç MK, Tüysüz O (2012) Kinematics and basin formation along the Ezinepazar-Sungurlu fault zone, NE Anatolia. Turkey. Turkish J Earth Sci 21(4):497–520

French EB (2009) The significance of changes in spatial usage at Mycenae. In: Bachhuber C, Roberts RG (Eds) Forces of transformation: the end of the bronze age in the mediterranean: proceedings of an international symposium held at St. John's College. Oxbow Books, British Association for Near Eastern Archaeology

Gurney OR (1990) The Hittites, Second edition, revised. Viking-Penguin Books, London

Ikeda Y, Suzuki Y, Herece E, Saroglu F, Isikara AM, Honkura Y (1991) Geological evidence for the last two faulting events on the North Anatolian fault zone in the Mudurnu Valley, western Turkey. Tectonophysics 193:335–345

Ketin İ (1966) Anadolu'nun tektonik birlikleri. Maden Tetkik Ve Arama Dergisi 66:20–34

Klengel H (1999) Geschichte des Hethitischen Reiches (HdO I/34), Leiden—Boston—Köln

Koçyiğit A (2003) General Neotectonic characteristics and seismicity of Central Anatolia. TPJD, Special Publication 5:1–26

Naumann R (1971) Architektur Kleinasiens von ihren Anfängen bis zum Ende der hethitischen. Zeit: Tübingen

Nur A, Cline EH (2001) What triggered the Collapse? Earthquake storms. Archaeology Odyssey 4/5, 31–36, 62–63

Nur A, Cline EH (2000) Poseidon's horses: Plate tectonics and earthquake storms in the Late Bronze Age Aegean and Eastern Mediterranean. J Archaeol Sci 27(1):43–63

Nur A, Burgess D (2008) Apocalypse: earthquakes, archaeology, and the wrath of God. Princeton University Press, Oxford

Nur A, Ron H (1997) Armageddon's earthquakes. Int Geol Rev 39:532–541

Psomiadis D, Dotsika E, Albanakis K, Ghaleb B, Hillaire-Marcel C (2018) Speleothem record of climatic changes in the northern Aegean region (Greece) from the Bronze Age to the collapse of the Roman Empire. Palaeogeogr Palaeoclimatol Palaeoecol 489:272–283

Reilinger R, McClusky S, Vernant P, Lawrence, et al (2006) GPS constraints on continental deformation in the Africa-Arabia-Eurasia continental collision zone and implications for the dynamics of plate interactions. J Geophys Res 111:B05411

Riehl S, Bryson R, Pustovoytov K (2008) Changing growing conditions for crops during the Near Eastern Bronze Age (3000–1200 BC): the stable carbon isotope evidence. J Archaeol Sci 35:1011–1022

Schachner A (2019) Ḫattuša. Efsanevi Hitit İmparatorluğu'nun İzinde. Homer kitapevi. İstanbul

Schachner A (2020a) Die Ausgrabungen in Boğazköy-Ḫattuša. Archäologischer Anzeiger 1:10–66

Schachner A (2020b) The great temple at Ḫattuša: some preliminary interpretations. In: Görke S, Steitler ChW (eds), Cult, temple, sacred spaces. Cult practices and cult spaces in hittite anatolia and Neighbouring Cultures, (StBoT 66) Wiesbaden, 105–158

Schaeffer CFA (1948) Stratigraphie comparée et chronologie d l'Asie occidentale--.: Syrie, Palestine, Asie Mineure, Chypre, Perse et Caucase. IIIe et IIe millénaires--. Oxford University Press for Griffith Institute, Ashmolean Museum

Schaeffer CF, Nougayrol J, Laroche E, Virolleaud C (1968) Commentaires sur les lettres et documents trouvés dans les bibliothèques privées d'Ugarit. Ugaritica V. Nouveaux Textes Accadiens, Hourrites et Ugaritiques des Archives et Bibliothèques Privées d'Ugarit. Commentaires des Textes Historiques (Première Partie), 607–768

Soren D (1982) An earthquake on Cyprus. New Discoveries from Kourion. Archaeology 38(2):52–59

Süel A (1998) Ortaköy- Šapinuwa Tabletlerinin Tarihlendirilmesi, III. Uluslararası Hititoloji Kongresi Bildirileri, Ankara, 551–558

Süel A (2002) Ortaköy- Šapinuwa, Recent developments in Hittite archaeology and history, Winona Lake, Indiana, 157–165

Süel A, Süel M (2008) Shapinuwa: a capital of the hittite state beyond babylon: art, trade, and diplomacy in the Second Millenium B.C. (Eds. Aruz J, Benzel K, Evans JM), Metropolitan Museum of Art, Newyork, 184–185

Süel A, Süel M (2017) The Discovery of a Hittite city developments in Hittite geography based on the identification of Ortaköy- Šapinuwa. Weeden M, Lee U (eds), Hittite Landscape and Geography. Leiden: BRILL, pp 28–36

Sümer Ö, Drahor MG, Berge MA, Ongar A, Schachner A (2019) Geoarchaeological and Archaeoseismological Observations in Ḫattuša: first evidence of earthquake traces from the Hittite capital. Archäologischer Anzeiger 1:90–96

Stein R, Barka A, Dietrich JH (1997) Progressive failure on the North Anatolian fault since 1939 by earthquake stress triggering. Geophys J Int 128(3):594–604

Stockhammer PW (2009) The change of pottery's social meaning at the end of the Bronze Age: new evidence from Tiryns. In: Bachhuber C, Roberts RG (eds) Forces of transformation: the end of the bronze age in the mediterranean: proceedings of an international symposium held at St. John's College. Oxbow Books, British Association for Near Eastern Archaeology

Taymaz T, Yılmaz Y, Dilek Y (2007) The geodynamics of the Aegean and Anatolia: introduction. Geol Soc London Special Publ 291(1):1–16

Ünal A (1977) M.Ö. II. Binyıl Anadolu'sunda Doğal Afetler, Belleten, 41, 163, Türk Tarih Kurumu Basımevi, Ankara

Weiss H (2016) Global megadrought, societal collapse and resilience at 4.2–3.9 ka BP across the Mediterranean and west Asia. PAGES 24:62–63

Multi-scenario Physics-Based Seismic Hazard Assessment of Cultural Heritage Sites

Marco Fasan, Hany M. Hassan, Andrea Magrin, Franco Vaccari, Fabio Romanelli, and Mohamed ElGabry

Abstract For relevant engineering purposes a viable alternative to standard estimates of seismic hazard is represented by the use of physics-based ground shaking scenarios. The scenarios are characterized in terms of magnitude, distance and faulting style, taking into account the complexity of the kinematic source rupturing process. In fact, ground-shaking scenarios modelled before the occurrence of an earthquake can be of extreme value in any seismic risk study, in particular in sites with priceless cultural heritage. In those places the effect of low occurrence rate—high consequences events can lead to invaluable losses, therefore an accurate evaluation of the expected ground motions is desirable. To this purpose, a web application, with a friendly graphic user interface, has been developed for multi-scenario physics-based seismic zoning and microzoning (considering site effects). Computational examples at different space and detail scales are presented, focussing on historical sites, such as the Dahshur pyramids, the Madrasa of the Princess Tatar al-Higaziya, Saint Catherine's Monastery in Egypt. For all the cases, the acceleration time histories, generated with the knowledge of the physical properties of the earthquake source and of the medium travelled by the seismic waves, can be used by engineers as seismic input for the vulnerability assessment.

Keywords Ground shaking scenarios · Synthetic seismograms · Physics-based hazard assessment · Cultural Heritage Sites

M. Fasan (✉)
Department of Engineering and Architecture, University of Trieste, Trieste, Italy
e-mail: mfasan@units.it

H. M. Hassan · F. Vaccari · M. ElGabry
National Research Institute of Astronomy and Geophysics, Cairo, Egypt

A. Magrin
National Institute of Oceanography and Applied Geophysics, Trieste, Italy

F. Vaccari · F. Romanelli
Department of Mathematics and Geosciences, University of Trieste, Trieste, Italy

G. M. El-Qady and C. Margottini (eds.), *Sustainable Conservation of UNESCO and Other Heritage Sites Through Proactive Geosciences*, Springer Geology,
https://doi.org/10.1007/978-3-031-13810-2_26

1 Introduction

For historical sites and monuments, the economic quantification of damages due to earthquakes can be incalculable. In fact, given their uniqueness it is impossible to establish a level of seismic action for which their loss is acceptable, and an accurate estimate of seismic hazard is needed also for very strong rare earthquakes.

In such conditions standard methods based on the use of Ground Motion Prediction Equations (GMPEs) can be difficult to apply, as there may not be enough observations to constrain their definition. For some countries, there may be no observations at all or available GMPEs.

An alternative to standard methods of seismic hazard evaluation is represented by the use of physics-based scenario earthquakes, characterized at least in terms of magnitude, distance and faulting style, also taking into account the complexity of the source rupturing process. Physics-based ground motion simulations are purely based on the geophysical and seismotectonic features of a region and can supply realistic time series of ground motion readily applicable by engineers in dynamic time-history analyses of structures. Physics-based scenarios can provide information on the expected accelerations also at very short distances, or due to large magnitude, which are usually the conditions in which the observed data are lacking. Therefore, they are useful to plan the preservation of historical monuments and relevant man-made structures, in particular when a Multi-Scenario assessment is applied accounting for all known sources that could affect a site of interest.

This paper shows the application of physics-based ground shaking scenarios to estimate the seismic hazard at cultural heritage sites. As illustrative examples, the methodology is applied to different Egyptian historical sites. The ground shaking scenarios for all the case studies were modelled using a web app (Vaccari 2016; Vaccari and Magrin 2019) that implements the proposed methodology. This methodology has been successfully applied to many urban areas worldwide for the purpose of seismic microzoning, to strategic buildings (Panza et al. 2016), lifelines and cultural heritage sites such as the city of Venice or Rome (e.g. Romanelli and Vaccari 2016; Chieffo et al. 2021) and it has proved reliable even in comparison to real events (Fasan et al. 2016).

2 Method

In this work, to assess the seismic hazard, a "physics-based" methodology was used, in which the ground shaking is represented by broadband accelerograms in the three spatial components, calculated through the tensor product between the tensors of the earthquake source and the Green's function of the medium (including the soil layers) crossed by the seismic waves (Panza et al. 2001, 2012; Fasan 2017). The hazard evaluation is based on the modelling of the propagation of seismic waves starting from the knowledge of the seismic sources and of the structural properties of the earth's lithosphere, allowing to take into account the kinematic complexity of the rupture process of the seismic source as well as site and path effects and, thus, considering the intra and inter-event spatial variability of the ground motion.

In order to assess the ground shaking associated with the hypothesized seismic scenario, the calculation of synthetic accelerograms occurs in two phases:

- 1: simulation of the fault rupture process on the fault plane;
- 2: simulation of wave propagation and calculation of synthetic accelerograms for the sites of interest.

It is therefore necessary to model the properties of the seismic source and of the bedrock-soil structure interposed between the fault and the sites of interest.

The source can be modelled as extended source (ES) or Size and Time Scaled Point Source (STSPS) (Gusev 2011); this modelling allows us to catch the effects related to the kinematic rupture process (i.e. directivity) and, in the near field, to the dislocation (i.e. static displacement—fling step).

When the extended model is used, the source of the earthquake is considered a relative sliding field distributed on the fault surface, on which the rupture process is presumed to occur. This surface is then modelled as a grid of point sub-sources, whose seismic moment is calculated by considering each of them as a component of a realization of a non-stationary random process. Assuming a realistic kinematic description of the rupture process, the extended seismic source model allows us to generate a spectrum (in amplitude and phase) of the temporal function of the source that takes into account both the rupture process and the effects of directivity. The simulation of the space and time evolution of the rupture is performed by the algorithm PULSYN06 (Gusev 2011). For the chosen scenario, different possible realizations of the rupture process can be considered: in this way the stochastic nature of the fault rupture is accounted for. Each realization is characterized by a different slip distribution on the fault plane, nucleation point and time evolution.

When the STSPS approximation is used, the source time functions generated by the distributed (point) subsources are summed up in order to obtain the equivalent single source, representative of the entire space and time structure of the extended source, and the related Green's Function.

The calculation is usually conducted in laterally homogeneous media, i.e. the bedrock-soil structural model is represented by a semi-infinite space in plane and parallel inelastic layers, up to a frequency of 10 Hz and using two different techniques: the MS—modal summation technique (Panza et al. 2001, 2012) and the DWN—discrete wavenumber (Pavlov 2009).

It is also possible to estimate the ground motion in laterally heterogeneous media in order to account for site and topographic effects. In the latter case, the wave-field generated by the MS technique is introduced in the mesh that defines the local heterogeneous area, where it is propagated according to the finite-differences scheme (Fäh and Panza 1994).

3 Web App and Case Studies

The calculation of synthetic accelerograms following this method can be very tedious in the initial phase of input definition. Moreover, the user should handle thousands of information in the postprocess phase. To help users in this process, a web app has been developed in order to make the method available with a user friendly interface, and all the generated data can be easily downloaded. This application is organized in different tabs, each one dedicated to a specific functionality as summarized in Table 1.

In order to show the power of the method and the capability implemented in the web app, the different functions are presented through several case studies having different levels of detail and modelling strategies. Applications aim at showing how physics-based ground motion simulations cab be a helpful tool for the seismic hazard evaluation of cultural heritage sites, where losses due to rare strong earthquakes are invaluable.

The case studies deal with three Egyptian cultural heritage sites. Egypt is characterized by the occurrence of small to moderate intra-plate events, while the large events generally take place farther east, along the northern Red Sea or Gulf of Aqaba, and offshore to the north, the Mediterranean Sea and toward Crete and Cyprus. Egypt is located at the north eastern part of the African continent and is bounded by three main tectonic elements; the African-Eurasian plate margin, the Red Sea plate margin and the Levant transform fault. The present-day tectonic deformation within Egypt is related to interaction and relative motions along these boundaries and their remote effects inside the Egyptian land. The predominant factor in terms of seismic hazard is generally related to the occurrence of moderate size earthquakes near to the population crowded areas (i.e. the Cairo 1992 earthquake, with mb = 5.8 in the ISC catalogue), rather than large earthquakes that occur at large distances along the northern Red Sea and the two Gulfs of Suez and Aqaba (i.e. the Shedwan 1969 earthquake, with mb = 6.1 by USGS, the 1995 Gulf of Aqaba earthquake, with mb = 6.2 in ISC), as well as the Mediterranean offshore earthquakes (i.e. the 1955 Alexandria earthquake, with mb = 6.1 in the USGS catalogue, the Cyprus 1996 earthquake, with mb = 6.3 in the ISC catalogue, the Ras El-Hekma 1998 earthquake, with mb = 5.8 in the ISC catalogue).

Egypt has a very long historical record of earthquakes going back about four millennia. Nevertheless, detailed and reliable information is available only for a few destructive earthquakes, and their parameters have a limited accuracy. Lack of reliable data makes it difficult to apply methods based on the use of GMPEs that cannot be adequately calibrated. This lack can be overcome thanks to the use of physics-based ground motion simulations as proposed in the method section.

Table 1 Description of the main functionalities available in the web app

Tab name	Functionality
Structure	Definition of the properties of the soil layers (thickness, density, V_p and V_s velocities and their attenuations Q_p and Q_s) Generation of Love and Rayleigh modes to be used in the modal summation technique (Fig. 1)
Eigen	Computation of eigenfunctions
Source	Modelling of the source rupturing process. In this panel several realizations of the rupturing process can be generated, each one with different slip distribution, rupture velocity and nucleation point (Fig. 6)
Parametric	Explore the dependency of the ground shaking scenarios on specific parameters of the model
Scenario	Quick generation of a ground shaking scenario around the epicentre of an earthquake. The user can select regional structure, the source rupturing model, the fault mechanism, the minimum and maximum epicentral distances, and the step to be used in the calculation of synthetic seismograms
Fault Scenario	Prediction of the ground motion expected at a specific site using an extended source model (ES). The user can select the structure, the source parameters, the number of realizations to be performed and the site of interest. Synthetic accelerograms and their response spectra are computed at the site for each realization. As output there are statistical analyses, period by period, of the generated response spectra (median values and their variability), the plot of each realization of the fault rupture and the possibility to extract the accelerograms at a given structural period and percentile of interest (Figs. 2, and 3)
2D Scenario	Calculation of seismograms in cases where the effects of lateral heterogeneities cannot be neglected, as implemented by Fäh and Panza (1994)
Sites	This tab allows to run, for a given site along the considered 2D profiles, different realizations of the rupture process. The user can select magnitude, type of rupture, directivity and number of realizations to be performed. As output there are statistical analyses, period by period, of the generated response spectra (median values and their variability) and the possibility to extract the accelerograms
MCSI	Combination, for a given site, of results coming from different scenarios (from 2D and Fault scenarios) (Fig. 7). This tab allows the computation of a Multi-Scenario seismic input, extracting, at each vibrational period, the scenario with the highest median value and its variability. Selection of accelerograms for engineering purposes is performed similarly, selecting the vibrational period, the percentile and the desired number of signals

4 Saint Catherine's Monastery

The first case study deals with the Saint Catherine's Monastery. The monastery lies on the southern part of Sinai Peninsula, at the foot of the highest mountain in Egypt and Sinai, near the town of Saint Catherine. The monastery is named after Catherine of Alexandria, is controlled by the autonomous Church of Sinai, part of the wider Greek Orthodox Church, and is a UNESCO World Heritage Site. Built between 548–565 AC, the monastery is one of the oldest working Christian monasteries in the world.

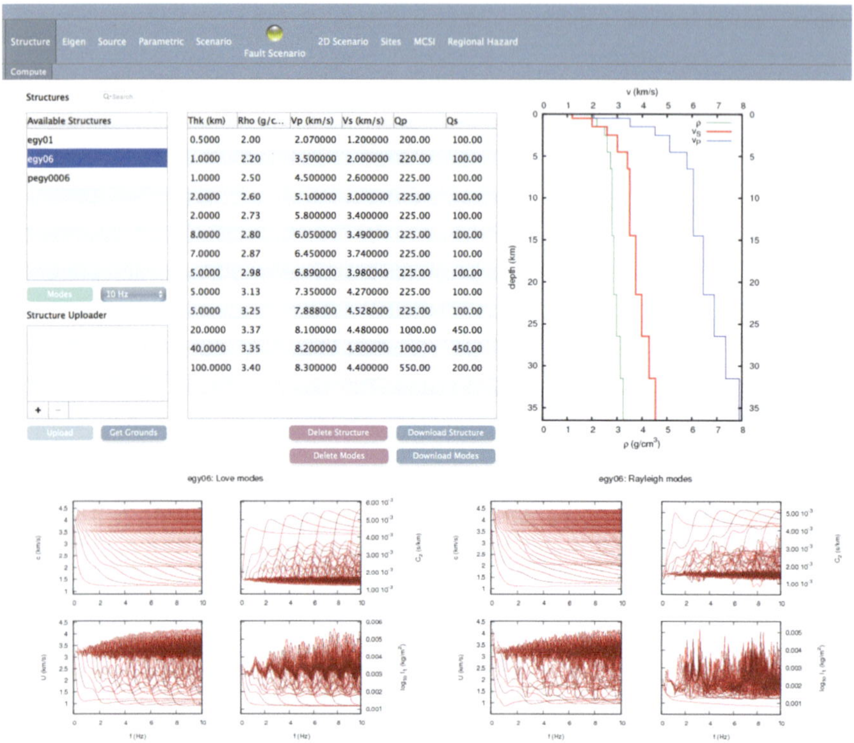

Fig. 1 Panel used for the definition of the layer properties and the computation of the Love and Rayleigh modes. The structure properties are those used for the computations at the sites of Dahshur Pyramids and Madrasa of the Princess Tatar al-Higaziya

The site contains the world's oldest continually operating library. For this site an example of extended fault scenario modelling was performed and the appropriate panel of the web application is shown in Fig. 2. The adopted structural model is shown in Fig. 1 with the associated Love and Rayleigh modes, and is taken from El-Khrepy (2008). The selected scenario is the 22 November, 1995 Gulf of Aqaba event which, with an estimated magnitude Mw = 7.3, is the largest instrumentally observed in Gulf the of Aqaba and Egypt as well. The fault and site positions are shown in the map along with other relevant information about the modelled scenario. Given the scenario, synthetic accelerograms and their response spectra are computed at the site for many realizations of the source rupturing process (up to 100 in this case) in order to account for its stochastic nature. The response spectrum plot shown is then obtained by using a statistical analysis, period by period, of the generated response spectra and represents the variability expected in the spectral acceleration for the selected scenario. The grey area in the plot spans from the median to the 95th percentile. The user can select which realization to be shown in the bottom part of the panel. In the right part of the panel the user can extract and download

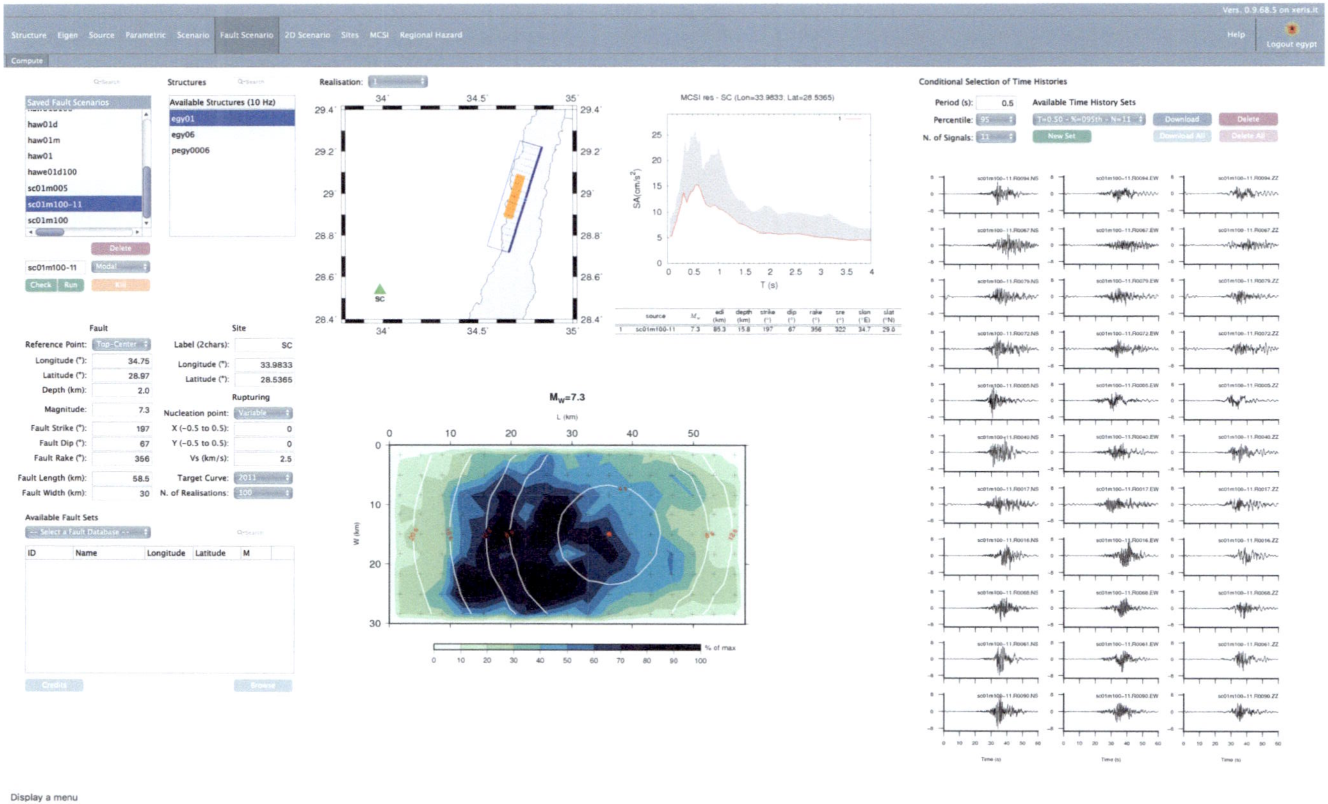

Fig. 2 Computation of multiple ground shaking scenarios at a single site for many realizations of the rupturing process of an extended fault: an example for the case of the Saint Catherine's Monastery

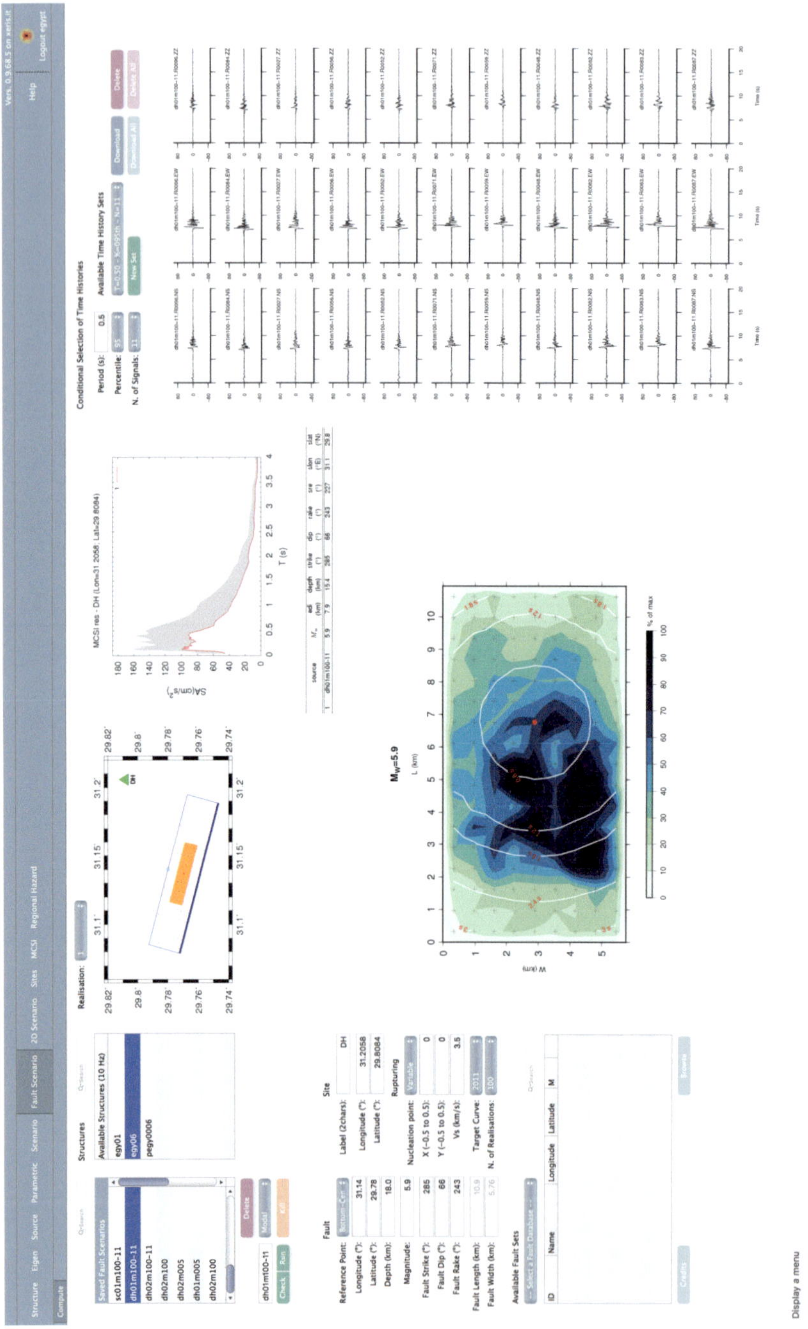

Fig. 3 Fault Scenario example for the site of the Dahshur Pyramids

the generated accelerograms selecting the vibrational period of interest (e.g. the first vibrational period of the building to be analysed), the percentile of interest and the number of desired accelerograms. Both the response spectrum and/or the selected accelerograms can be used by engineers to run a Scenario-Based seismic hazard assessment. All accelerograms are three-components, including the vertical one that can have significant influence on the expected performance of masonry structurers (see Rinaldin et al. 2019).

5 Dahshur Pyramids

The site of the Dahshur pyramids is located at an ancient royal necropolis, located approximately 40 kms south of Cairo. The pyramids were built from 2613–2589 BC and are among the oldest pyramids in the country. For such historical monuments it is essential to evaluate accurately the expected accelerations due to different scenarios. In this case an extended source model was again considered (see Fig. 3), but showing the possibility to combine multiple scenarios in a final unique seismic input definition. Two scenarios were selected: a magnitude Mw = 5.9 at about 8 km, corresponding to the 12 October 1992 Cairo earthquake (see Fig. 4) and a magnitude Mw = 6.7 at about 65 km corresponding to the 22 July 950 lower Egypt earthquake (see Fig. 5), as reported in Ambraseys et al. (1994).

For each of the selected scenarios, 100 realizations of the possible rupture process of the fault were computed. This is performed in order to catch the variability in the expected ground motions due to source effects. Figure 6 shows two of these realizations, one for the Mw = 5.9 scenario and the other for the Mw = 6.7 scenario. A description of the picture is given in the caption.

Results from both scenarios are then combined into a single multi-scenario response spectrum (see Fig. 7) using the MCSI tab available in the web app. For the construction of this response spectrum, at each period the values of the scenario with the highest median value (or a selected percentile) are retained only (see Fasan 2017). The reported variability is the one associated with the winning scenario at each period and is represented by the grey area (values from the median to the 95th

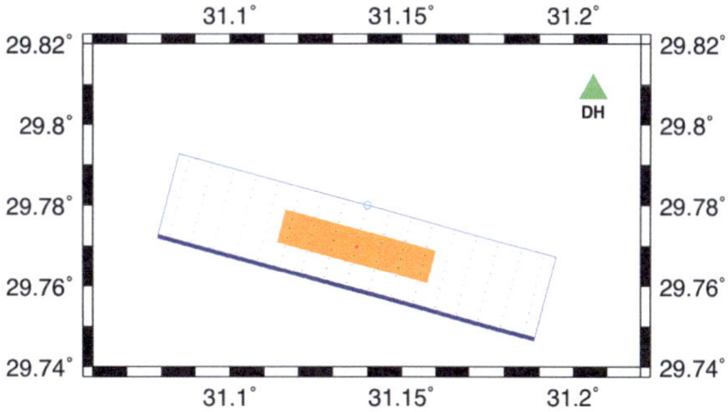

Fig. 4 Dahshur Pyramids case study: fault and site positions for the Mw = 5.9 scenario (12 October 1992 Cairo earthquake)

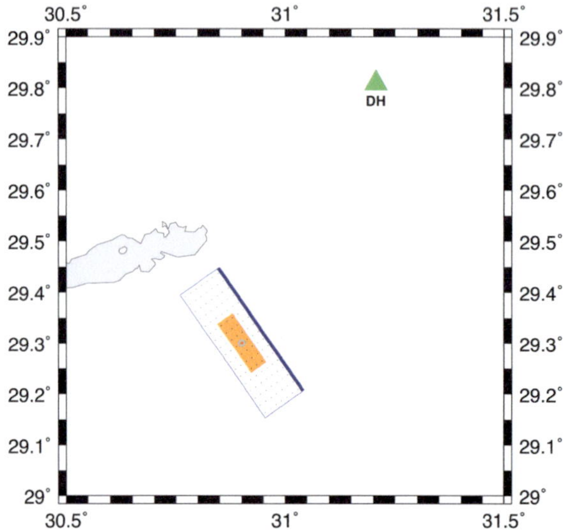

Fig. 5 Dahshur Pyramids case study: fault and site positions for the Mw = 6.7 scenario (22 July 950 lower Egypt earthquake)

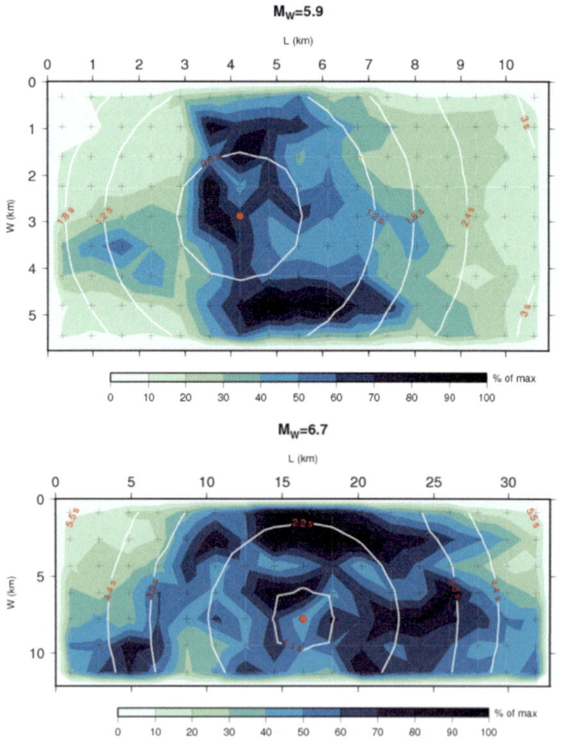

Fig. 6 Two rupture processes for (up) the Mw = 5.9 scenario (12 October 1992 Cairo earthquake) and (bottom) the Mw = 6.7 scenario (22 July 950 lower Egypt earthquake). The dark areas correspond to a large slip on the fault, and the red dot shows the nucleation point of the rupture. The white isochrones describe the time evloution of the rupturing process

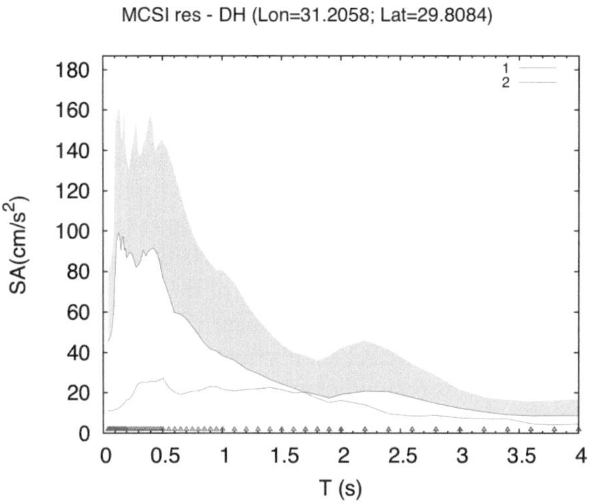

	source	M_w	edi (km)	depth (km)	strike (°)	dip (°)	rake (°)	sre (°)	slon (°E)	slat (°N)
1	dh01m100-11	5.9	7.9	15.4	285	66	243	227	31.1	29.8
2	dh02m100-11	6.7	63.7	10.0	145	32	332	117	30.9	29.3

Fig. 7 Combination of Dahshur scenarios into one single response spectrum. The picture shows the median values for each scenario (red for scenario 1 and blue for scenario 2). The grey area represents the variability of spectral accelerations from the median to the 95th percentile of the winning scenarios. The table reports the main characteristics of the winning scenarios

percentile). In this case study, only two scenarios are combined together. However, the procedure can be adopted with all known possible scenarios coming from faults, seismogenic zones, historical catalogues etc. If all known sources are taken into account, the procedure can be used to estimate the Maximum Credible Seismic Input (MCSI), which represents a reasonable upper bound for the ground motion level at a site of interest (Fasan 2017).

Otherwise, if only a selected number of sources is taken into account, it is proper to call it a Multi-Scenario Seismic Input (MSSI).

6 Madrasa of the Princess Tatar al-Higaziya

The third case study is the Madrasa (meaning school in Arabic) of princess Tatar al-Higaziya in Old Cairo. This building was constructed in 1348 AC as an extension to princess Tatar's house; then, after thirteen years, the palace and the mausoleum was converted into the Madrasa. The Madrasa complex consists of a mausoleum, minaret, and an ablutions court, and it is one of the few schools endowed by a woman in Cairo. The school was built and endowed for educating orphan children and for uplifting the daily prayers, as well as serving as a public library.

Simulations presented in previous subsections were developed assuming a laterally homogeneous layered structure of the crustal and site soil model. When the site of interest lies on a complicated subsurface geology, the effects of lateral heterogeneities cannot be neglected. They can generate, combined with the characteristics of the incoming wave field originated at the source, significant amplifications, and often at very specific frequencies.

For this case study a laterally heterogeneous profile is used (Fig. 8). The profile was developed for the Old Cairo area by Hassan (2018) and a detailed description is reported in Hassan et. al (2020) For this type of modelling, the web application makes use of the computer codes that implement the hybrid approach developed originally by Fäh and Panza (1994). A laterally homogeneous inelastic layered model is defined to represent the average lithospheric properties along the path from the source to the vicinity of the site. In this part of the model wave propagation is modelled by the MS technique. The generated wavefield is then introduced in a mesh that describes the local heterogeneous area characterizing the site of interest (Fig. 8), where it is propagated according to the finite difference scheme. With this hybrid approach, the source, path, and site effects are all taken into account, and the detailed ground shaking scenarios can be efficiently evaluated along a 2D profile even at large distances from the epicentre. In particular, this kind of modelling allows to evaluate amplifications with respect to homogeneous soil conditions and bedrock (Fig. 9) and to account for reflections and refractions due to the interfaces between different layers. These effects are accounted for in the simulated accelerograms. The web application allows to select and download these site-specific accelerograms

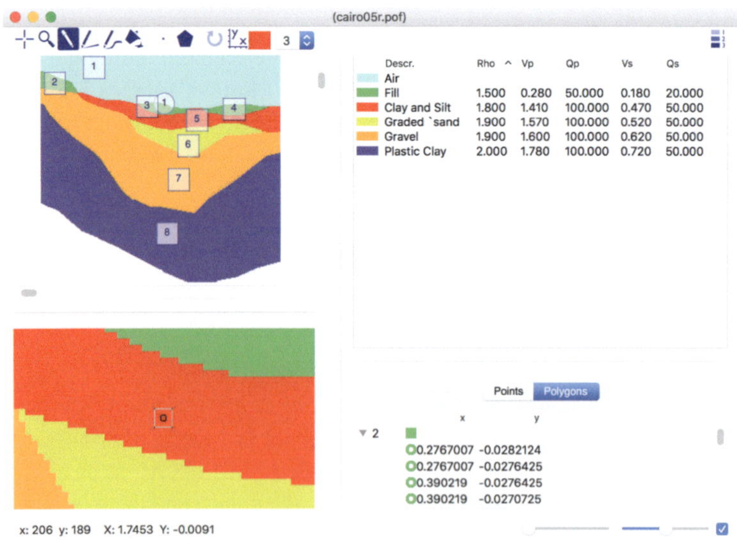

Fig. 8. 2D profile developed for the "Madrasa" case study

Fig. 9 Amplifications (dark red areas) are shown with respect to a laterally homogeneous bedrock model for the "Madrasa" profile. From top to bottom, the amplifications are shown along the profile for the vertical, radial, and transverse components. The abscissa shows the position along the profile, represented at the bottom, while the ordinates in the upper images refer to the frequencies at which the amplifications eventually occur

as shown in Fig. 10. Structural engineers can use these accelerograms in dynamic time history structural analyses to perform site-specific hazard and vulnerability assessments.

7 Conclusions

Physics-based ground motion simulations allow making accurate estimations of expected accelerations even in areas where little or no recorded data are available. Synthetic accelerograms can be generated and used to perform non-linear time history analysis of structures.

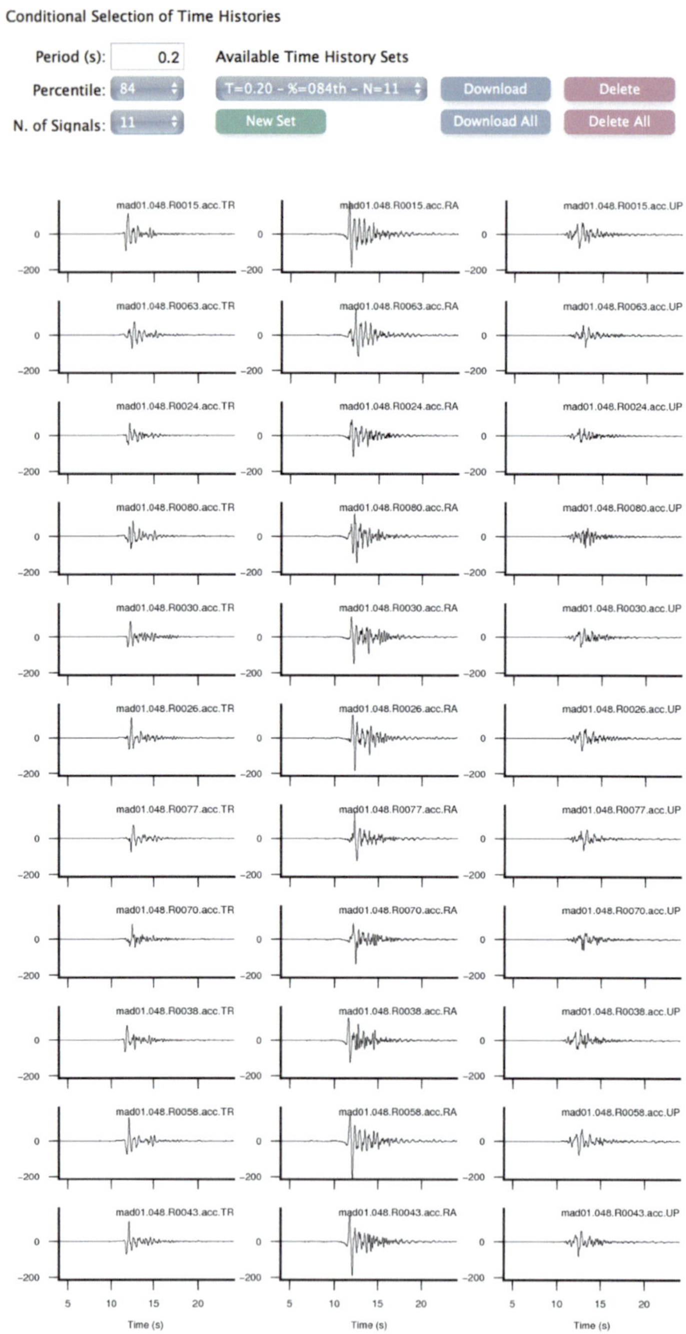

Fig. 10 Accelerograms selection for the site of the Madrasa. 11 three-component accelerograms were selected around the 84th percentile at a vibrational period of 0.2 s

Being based on geophysical and seismotectonic features of a region, physics-based scenarios can be a helpful tool to evaluate the seismic hazard for ancient cultural heritage sites and monuments. In fact, the uniqueness and invaluable loss of this kind of assets makes it necessary to evaluate the hazard even for very strong rare earthquakes.

In this paper, a method for multi-scenario physics-based hazard evaluations is proposed and applied to some Egyptian sites. The generated synthetic seismograms include source, path and site effects. The modelling of ground-shaking scenarios for the considered case studies was done using a web app that implements the proposed method at different scale and level of detail.

References

Ambraseys NN, Melville CP, Adams RD (1994) The seismicity of Egypt. Arabia and the Red Sea. Cambridge University Press, Cambridge, UK, p 182

Chieffo N, Fasan M, Romanelli F, Formisano A, Mochi G (2021) Physics-based ground motion simulations for the prediction of the seismic vulnerability of masonry building compounds in Mirandola (Italy). Buildings 11(12):667. https://doi.org/10.3390/buildings11120667

El-Khrepy S (2008) Detailed study of the seismic waves velocity and attenuation models using local earthquakes in the Northeastern part of Egypt. PhD Thesis, Mansoura Univ., Egypt

Fasan M (2017) Advanced seismological and engineering analysis for structural seismic design. PhD Thesis, University of Trieste, Italy

Fasan M, Magrin A, Amadio C, Romanelli F, Vaccari F, Panza GF (2016) A seismological and engineering perspective on the 2016 Central Italy earthquakes. Int J Earthq Impact Eng 1:395–420

Fäh D, Panza GF (1994) Realistic modelling of observed seismic motion in complex sedimentary basins. Ann Geofis 3:1771–1797

Gusev AA (2011) Broadband kinematic stochastic simulation of an earthquake source: a refined procedure for application in seismic hazard studies. Pure Appl Geophys 168:155–200

Hassan HM (2018) Multi-scale seismic hazard assessment for Egypt-PhD Thesis, University of Trieste, Italy

Hassan HM, Fasan M, Sayed MA, Romanelli F, ElGabry MN, Vaccari F, Hamed A (2020) Site-specific ground motion modeling for a historical Cairo site as a step towards computation of seismic input at cultural heritage sites. Eng Geol, 268. https://doi.org/10.1016/j.enggeo.2020.105524

Panza GF, Romanelli F, Vaccari F (2001) Seismic wave propagation in laterally heterogeneous anelastic media: theory and applications to seismic zonation. Adv Geophys 43:1–95

Panza GF, La Mura C, Peresan A, Romanelli F, Vaccari F (2012) Seismic hazard scenarios as preventive tools for a disaster resilient society. Adv Geophys 53:93–165

Panza GF, Romanelli F, Vaccari F, Altin G, Stolfo P (2016) Vademecum for the seismic verification of existing buildings: application to some relevant buildings of the Trieste Province. Atti dei convegni Lincei. 306:355–363. Bardi Edizioni, Roma

Pavlov VM (2009) Matrix impedance in the problem of the calculation of synthetic seismograms for a layered-homogeneous isotropic elastic medium. Izvestiya, Phys Solid Earth 45:850–860

Romanelli F, Vaccari F (2016) Earthquake scenarios and seismic input for cultural heritage: applications to the cities of Rome and Florence. Atti dei convegni Lincei. 306:157–166. Bardi Edizioni, Roma

Rinaldin G, Fasan M, Amadio C, Noè S (2019) The influence of earthquake vertical component on the seismic response of masonry structures. Eng Struct 185:184–193. https://doi.org/10.1016/j.engstruct.2019.01.138

Vaccari F (2016) A web application prototype for the multiscale modelling of seismic input. Earthquakes and their impact on society. Springer Natural Hazards, 563–583. https://doi.org/10.1007/978-3-319-21753-6_23

Vaccari F, Magrin A (2019) NDSHA—computational Aspects of the neo-deterministic seismic hazard assessment. In: Dobran F (ed) Resilience and sustainability of cities in hazardous environments, pp 202–212. GVES, Napoli—New York, ISBN: 978-88-903183-1-3

Seismic Vulnerability Assessment of Historical Heritage Structures, Analysis and Intervention: Application in Skikda City, Northeast of Algeria

Hamidatou Mouloud, Lebdioui Saad, Hallal Nassim, Hugo Rodrigues, Assia Harbi, Ammouchi Nesrine, Karima Messaoudi, and Beldjoudi Hamoud

Abstract The preservation of historical heritage buildings against natural hazards, especially seismic risk, is extremely challenging. The earthquake is, of all the cataclysms, the one that engages the most deeply the responsibility of the State. Countries presenting medium to moderate seismic risk, like Northeast of Algeria, want to protect this particularly sensitive but also numerous heritages buildings. The objective of this paper is to predict the consequences of a possible earthquake that would strike Skikda city in the future. It will be necessary to elaborate the different phases of the action simulation process of a major earthquake on the urban fabric at the cities scale, and to simulate the extent of the damages caused to any type of structure, and the economic consequences for communities. This will be based on the lessons learned from the experience of earthquakes on August 21 and 22, 1856, which caused significant damage to Skikda city. Our research work, therefore, seeks to meet those two requirements. First, the large-scale of the study deals with a large number of existing structures to be diagnosed across the Skikda city structures at the time of the earthquake of 1856. Then the need to propose a relevant analysis of the seismic behavior of historical heritage masonry, by inexpensive ways. We rely on studying the rich heritage of Skikda city. The second part of this work, a detailed statement of the pathologies observed as well as the characteristics of the historical structures will be established. In the third part, the collected data will allow to implement and identify models for the structures of the site. The pilot site chosen is the agglomeration of Skikda city. We first present the choices we have made and the tools we have developed to meet the identified requirements. Then we develop how we used

---

H. Mouloud (✉) · H. Nassim · A. Harbi · B. Hamoud
Research Centre of Astronomy Astrophysics and Geophysics (CRAAG), Algiers, Algeria
e-mail: mouloudh83@yahoo.com; m.hamidatou@craag.dz

L. Saad · A. Nesrine · K. Messaoudi
Faculty of Technology, University of August 20, 1955, P. O. Box 26, 21000 Skikda, Algeria

H. Rodrigues
Civil Engineering Department, Campus Universitário de Santiago, RISCO, University of Aveiro, 3810-193 Aveiro, Portugal

G. M. El-Qady and C. Margottini (eds.), *Sustainable Conservation of UNESCO and Other Heritage Sites Through Proactive Geosciences*, Springer Geology,
https://doi.org/10.1007/978-3-031-13810-2_27

ambient noise measurements in order to calibrate and validate initial models of unfamiliar complex structures. Finally, we explain how we defined a damage threshold criterion by comparing different approaches, numerical and kinematics.

Keywords Seismic risk · Vulnerability · Heritage structure · Damage · GIS · Skikda city

1 Introduction

Regions with medium to moderate seismic risk, like Algeria, aim to protect their sensitive heritage sites. Therefore, it is necessary to develop methods adapted to areas with moderate seismicity, where the systematic consideration of this risk for the existing structure is recent, with few damage databases (Karbassi and Lestuzzi 2014). In addition, a significant part of the built heritage, which is the case of this study, consists of vernacular architectures in traditional materials, and often with very little maintenance. On the other hand, most of the historic structures built with unreinforced masonry, material whose characterization remains complex even with a significant investment (Valluzzi et al. 2003). It should be also noted that the diagnosis of a complete heritage necessarily involves several buildings and in many cases, it is therefore impossible to treat them all one by one without being arbitrarily ignored. Finally, any intervention on a historical heritage is subject to rules, even for tests, requiring readability, compatibility and reversibility (ICOMOS 2003).

Earthquake activity in North Algeria constitutes a constant threat to human life and property, causing major economic losses and disruption. The losses result from not only great earthquakes, such as the Ms = 7.3 El Asnam event of October 10, 1980, but minor events such as the Ms = 5.9 Constantine earthquake of October 27, 1985 and the May 21, 2003, Zemmouri earthquake (Mw = 6.8). Skikda region is one of the most seismically active regions of Northeast Algeria. The current work consists of regional and local studies, mainly based on seismic risk assessment and seismic hazard (Sbartai and Hamidatou 2013; Hamidatou and Sbartai 2016, 2017; Hamidatou et al. 2019, 2021). Seismic risk assessment has been already well established for major urban centers, but is still of great importance in risk assessment in more sparsely populated areas. For example, it is crucial to examine the potential impact of earthquakes on Skikda city because there are few connecting roads between the city and nearby centers. The accurate assessment of the seismic risk faced by urban areas is needed by public authorities and decision makers, who are responsible for regional planning and urbanization. Geotechnicians, architects, construction supervisors, and public works institutions, among others, must have a good understanding of the nature of the potential for soil liquefaction before making decisions related to development. Here, it is presented the results of a seismic risk scenario for historical heritage established for Skikda city. In order to address these issues, we seek to propose a large-scale seismic vulnerability analysis method. This method of diagnosis, based on the numerical modeling by Finite Elements (EF) of historic structures and their responses to earthquakes, relies on a hierarchy of modeling levels: linear model of all the structures concerned, then more detailed nonlinear model in the case of the structures having shown a level of marked risk during the linear analysis. Given the strong constraints outlined above, the approach must therefore meet.

two main requirements:

- Provide a relevant analysis of earthquake behavior of masonry historic constructions.
- Containing the number of numerical models necessary for the analysis.

The following points therefore constitute the framework of the chosen strategy:

- Define typologies of works, allowing in first approach to work by group.
- Define the choices and modeling hypotheses adapted to the specificities of the studied structure.
- Define the criteria needed to assess the vulnerability of the studied typologies.

In the end, it will be discussed and presented the used assumptions, and the tools advanced to meet these imperatives, with the description on the calibrated and validation models of the structural models used. Finally, we will showed how was define he damage criterion by comparing numerical approaches, linear and nonlinear, and nonlinear kinematic analysis.

2 Seismic Assessment of Historical Buildings: Issues and Key Points

Historical constructions were typically built before the first seismic design rules, and were consequently built in some cases only with the local seismic culture introducing some elements to improve the seismic behavior base on an empiric knowledge. The performance conditions of seismic design codes established for current structures do not include the discriminations of cultural heritage buildings (Laupper 2004). The characteristics to be considered, span the subsequent topics:

- Protection documentation: In circumstance of damage or destruction, the documentation must allow the reconstruction of permanent and transferable cultural property (FOCP 2014).
- Safety of individuals: Past and traditional heritage structures are frequently sacred structures or museums. As a consequence, and unlike for greatest normal structures, the number of persons in or everywhere the cultural heritage structure can diverge mostly among time of day/weekday/season. The number of fatalities in the event of an earthquake has to be limited to satisfactory ranks of individual and shared risks by retrofit processes or by limiting publics' admission to the structure.

- Safeguarding of the original structure: To preserve not only the structure's appearance, but also, its original materials. Interventions must be reserved to a least. If interventions are conducted, they would be preferably reversible.
- Structural safety of the construction: To preserve the structural integrity for a selected return period, it is frequently essential to place retrofit measures in place. Though, these measures frequently meaningfully touch the appearance and the structure's material. Numerous retrofit measures are consequently in straight conflict with the impartial to preserve the original structure.
- Preserving historical constructions in the case of earthquake occurrence consequently a predominantly stimulating task, subsequently, a considerable higher variety of features requirements to be addressed than when standard structures. Besides, several stakeholders are implicated in the preservation of cultural heritage structures. They may be the owner, structural engineer, architect, and curator of monuments, local and national authorities. The work was motivated by the need of a plan for Algerian typical historical heritage buildings. It starts with a brief overview on the classification of Skikda heritage buildings and a review on performance boundary conditions for usual structures. It then appearances at the discriminations of heritage structures and the greatest significant boundary conditions (Calderini et al. 2012). It suggests a performance state matrix for the seismic calculation of heritage structures as a meaning of the importance class of the heritage place considering different preservation approaches and life safety requirements of inhabitants and visitors. With regard to the latter, it focuses on the seismic risk in regions of low to moderate seismicity and extends the innovative method for considering of the least life security requirements (Kölz and Schneider 2005), the retrofit costs and benefits of usual current structures to historical heritage structures.

3 Overview of Skikda City

3.1 Study Area

Skikda city, among most important cities in Algeria, houses important economic, scientific, cultural and historical aspects of Algerian infrastructure. Skikda province occupies a central geographic location within the Algerian Northeast region. The history of Skikda is entirely determined by its position; it is also linked to that of the village of Stora, which in antiquity was its port on an anchorage at the bottom of a sheltered golf course where there is drinking water, at the outlet of the trade route which brings products from the highlands to the sea by the most direct route, it is the meeting point of the sea route and the land route. In addition, a picturesque coast and a pleasant climate have always made it an attractive region (Fig. 1).

The historic center of Skikda covers an area of 40 hectares and an estimated population of over 16,000 inhabitants and which makes it the most populated part of the city with an average density of 400 inhabitants per hectare, while the city

Fig. 1 Location map of Skikda city

average is only 90 people per hectare. Strong pressure is therefore weighing on its Built framework. It is bounded in the North by the port, to the South by the Allees and the Faubourg de l'espérance, to the East the Bouabaz area and to the West the Beni Melek area. The main and unique axis of the city is avenue Didouche Mourad, (the arcades) which crosses the center from end to end and through which all the mechanical flow passes through it in both directions with parallel voices and tracks.

3.2 Seismotectonic Context of Skikda Region

The Tellian Atlas (Northeast Algeria) is an active collision zone between the African and Eurasian plates that is experiencing shortening of 5–6 mm/year (Anderson and Jackson 1985; Argus et al. 1989; DeMets et al. 1990). The present study area location is close to Constantine, where the regional tectonics have been studied previously (Vila 1980; Bounif et al. 1987. Meghraoui 1988; Coiffait et al. 1992; Harbi et al. 2010) (Fig. 2). From a geological point of view, the Constantine region belongs to the external domain of the Tellian Atlas chain, a part of the North African Alps (Maghrebides), built during the main paroxysmal compressional phases of the Eocene and Miocene epochs (Mattauer 1958; Vila 1980) and the Quaternary period (Fig. 3).

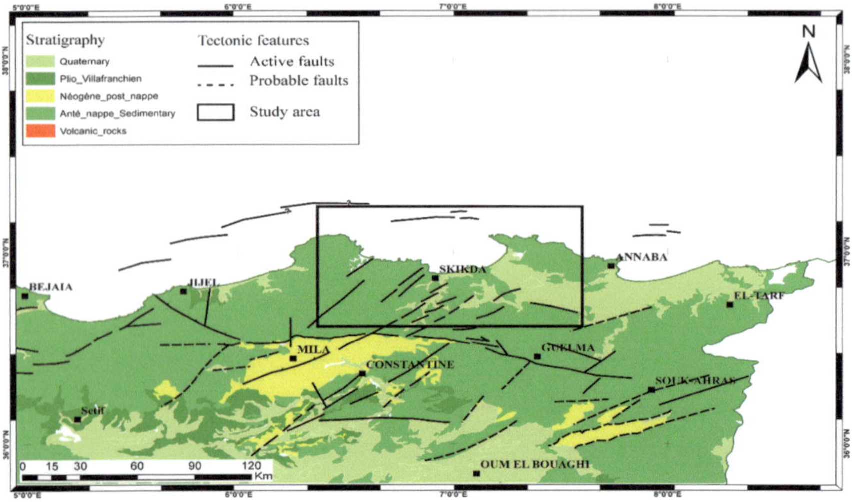

Fig. 2 Tectonic map of the Skikda region (Meghraoui 1988)

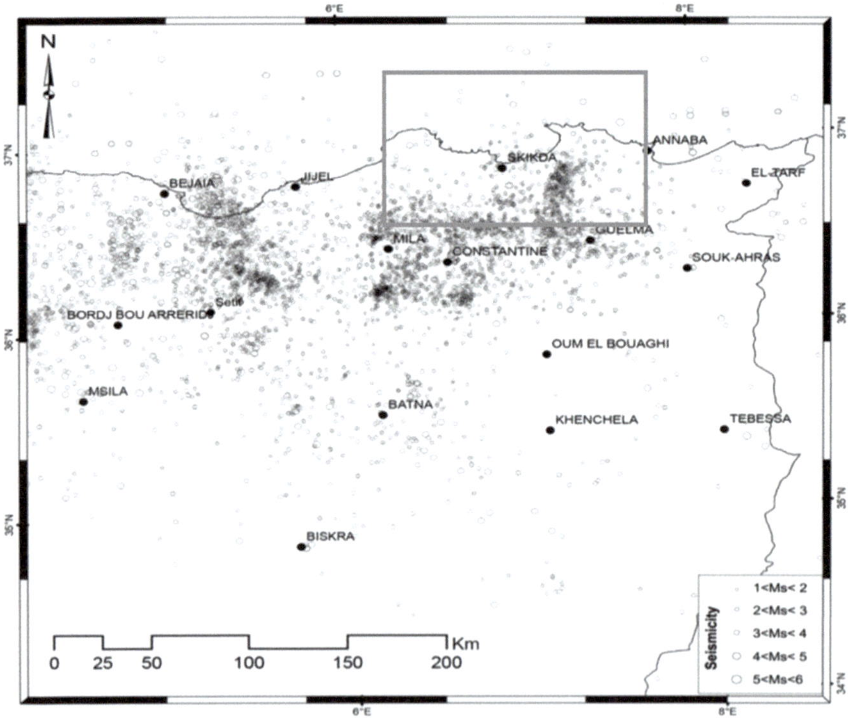

Fig. 3 Spatial distribution of earthquakes in Northeast Algeria from 1300 to 2018 (Hamidatou et al. 2019)

The earthquake data used in this study are largely based on the Hamidatou dataset (2017). The catalogue established by Hamidatou covers portions of eastern Algeria [4°E–8°E, 33°N 38°N] from 1900 to 2017, and in Hamidatou et al. (2019) study updated the dataset for periods from 1357 to 2019. The present study area is located in an active seismic zone within Algeria, which has been shaken by several moderate earthquakes during the last few centuries.

3.3 *Strategic Choices for the Study Scale: Development of a Typological Classification*

Fieldwork and archival research enabled us to identify the historic heritage, built or strongly modified between 1900 and 1930, constituting the case under study (Fig. 4).

To provide a diagnosis of all these historic structures, it is therefore imperative to define relevant tools at this scale, based on geometric indices. However, it cannot

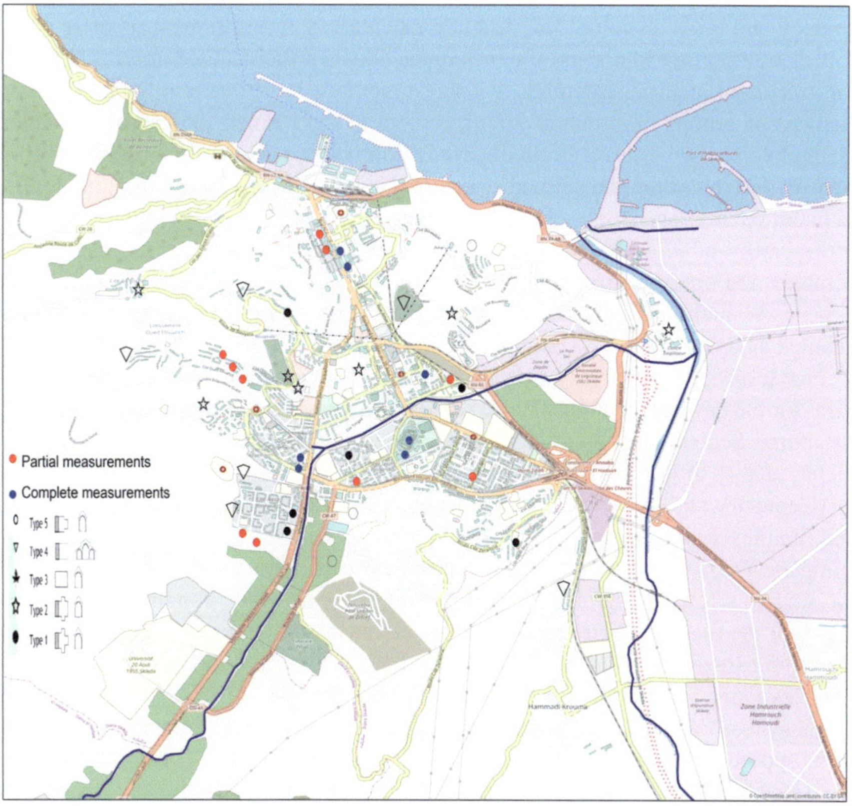

Fig. 4 Map of the structural types of the panel studied and vibration measurements performed

be evaluated neither the level of expected damage nor the behavior of the structure under earthquake (Lourenço and Roque 2006), It were selected vulnerability curves, perfectly adapted to this scale. This probabilistic approach, used on modern structures as on cities, defines a level of damage according to a global criterion (Perrault 2013).

These fragility curves give information on a particular category of structures with the same structural characteristics, which coincides with our goal of limiting the number of models. We conducted a survey campaign on all historical structures. Based on the fieldwork for ancient heritage and Eurocode 8 (2004, 2005), we defined five structural criteria of fragility under earthquake: type of plan regularity, the presence of vertical discontinuities, number of floors, and the presence of a basement. According to their combination, we can classify all the structures of the panel.

4 Vulnerability and Risk Assessment Methodologies

When evaluating the seismic vulnerability of historical heritage structures, it is important to define the purpose, beforehand subsequently choosing the most appropriate strategy and tools, essential for structure calculation and fulfilment of these aims. It is also important to discuss the difference between the detailed approaches used for individual historic structures, and those methods most efficient for larger scale analysis of groups of structures. With respect to the former, the use of a detailed methodology implies a very reliable evaluation with a necessarily in-depth level of information regarding the studied structure. However, when increasing the number of historical heritage structures and enlarging the area to be assessed, the resources and quantity of information required are also increased, and thus the use of less complex and onerous inspection and recording tools is more practical. The use of large-scale methodologies, could point for witch structures my need an additional detailed analysis in order to perform the improvement of the seismic behavior.

First level approaches use a considerable amount of qualitative information and are ideal for the development of seismic vulnerability assessment for historical heritage structures. Second level approaches are based on mechanical models, and rely on a higher quality of information.

(Geometry and mechanical properties), concerning building structure. The third and final level includes the usage of numerical modeling techniques, which need a complete and rigorous survey of individual structures. The description and nature of the studied standards (qualitative and quantitative), naturally condition the formulation of the methodologies, and their respective assessment level, which can vary from the expedite evaluation of structures built on visual observation to the most complex numerical modeling of single structures (see Fig. 5).

Many researchers have developed the systemization of these vulnerability assessment approaches, and therefore differs due to the varying levels of dependence on the following factors: Nature and objective of the assessment, quality and availability of information, characteristics of the inspected building structure, scale of assessment, methodology criteria, the degree of reliability of the expected results, and use

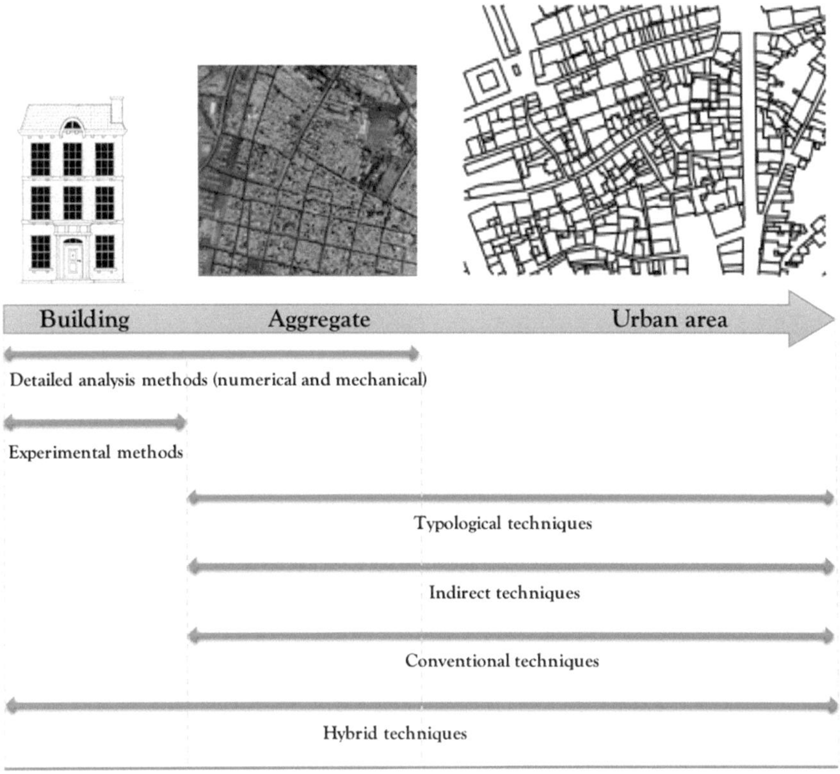

Fig. 5 Global concept of the study method (Romeu et al. 2011)

by the end-user of the information produced. For the reason of these differences, the coherency of consensus regarding this classification is still a contentious issue (Romeu et al. 2011).

The Skikda city and its vicinity, with an area of about 10 km^2, were chosen to illustrate GIS-based risk vulnerability assessment, using data on the structural vulnerabilities of different types of historical heritages. Maps were produced from GIS databases using thematic analysis tools in MapInfo 15.0, spatial analysis tools in ArcGIS 10.3 (Christophe 2008), and other techniques, such as georeferencing and vectorization layers (Bourcherdt et al. 1991; Thierno, 2004).

4.1 3D Mesh of Identified Typologies and Geometric Variability: MicMac and Preprocessor

The Cast3M finite element code was used for numerical studies. Given the large number of structures to be modeled, it was defined an easy was to build the mesh of

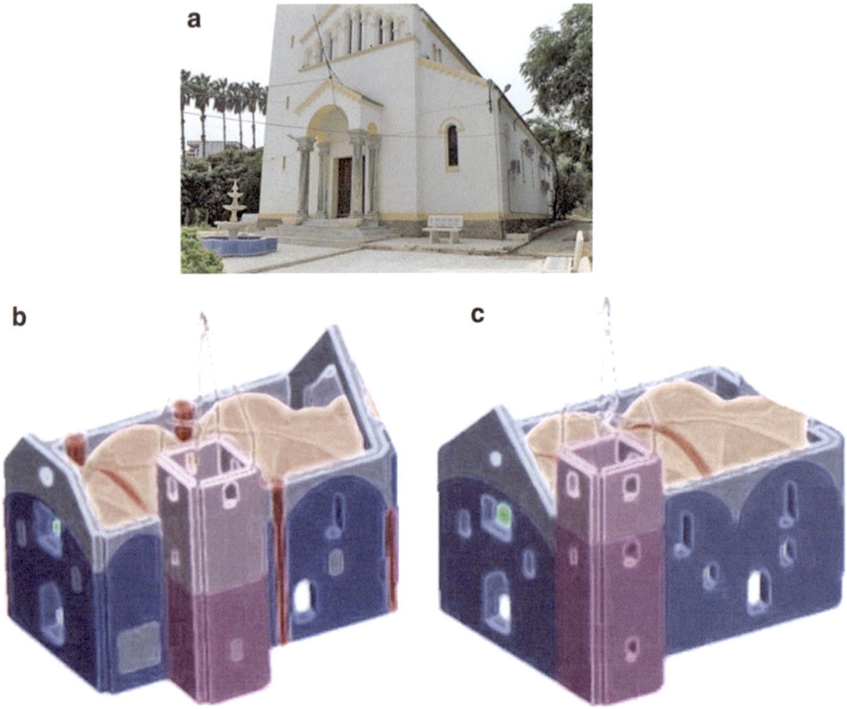

Fig. 6 Former church Ste Theresa: **a** Initial MicMac façade survey, **b** Cast3M mesh and its substructures, **c** variation on the same structural type for another Cast3M mesh

complex structures, but also to vary the characteristics by structural type, consider different geometries, materials and locations. These dissimilarities around the most representative model of each type are crucial for vulnerability studies. It was therefore, developed a preprocessor for Cast3M that responds to these requirements (Fig. 6). In the absence of detailed plans, it is based on photogrammetric surveys carried out in situ using a digital camera.

4.2 Modeling Assumptions: Specific Elements

As it is a very coarse, highly heterogeneous, additional isotropic blocking masonry than a paired masonry with preferred cracking directions, then a macro-modelling approach was used on plate elements. This lightens the model and the construction of parametric meshes. The mechanical characteristics of the initial linear models are taken from the literature (Binda et al. 2013; DPCM 2011).

The geometry of structures require assumptions, sometimes complex, and to improve their domains of validity in order to arrive at a simplified model that is relevant to the nature of the results to be obtained. Three area were identified that focus the most important variation possibilities: foundations, vaults, and frames (Limoge et al. 2014). A parametric study of the different modeling choices allowed quantifying the impact on the results in terms of stiffness, modal shapes and natural frequencies. This allowed us to determine, based on the structure types, the most appropriate modeling assumptions. Moreover, to consider, the material and geometrical specificities of the macro elements, each structure was divided into substructures.

5 Routine Conditions for Structures

Figure 7 illustrates the performance conditions clear defined by FEMA 356 (2000) for the seismic assessment of structures in general. These performance conditions were defined to assure a suitable structure performance under a huge choice of possible ground motions. Contingent on the importance of the structure, dissimilar sets of impartial stages can be designated as signified by the diagonals "Limited Objectives", "Basic Objectives", and "Enhanced Objectives" in Fig. 7.

The "Basic Objectives" serve as a reference level considering (i) the safety of persons for seismic events with a return period of 475 years, (ii) the limit of structural and non-structural damage and consequently economic losses for events with smaller return periods; and (iii) the collapse prevention of the structure for an event with a return period of 2500 years. When designing rehabilitation and strengthening measures, characteristically only the first of the three limit states, i.e., the life safety limit state, are obviously addressed.

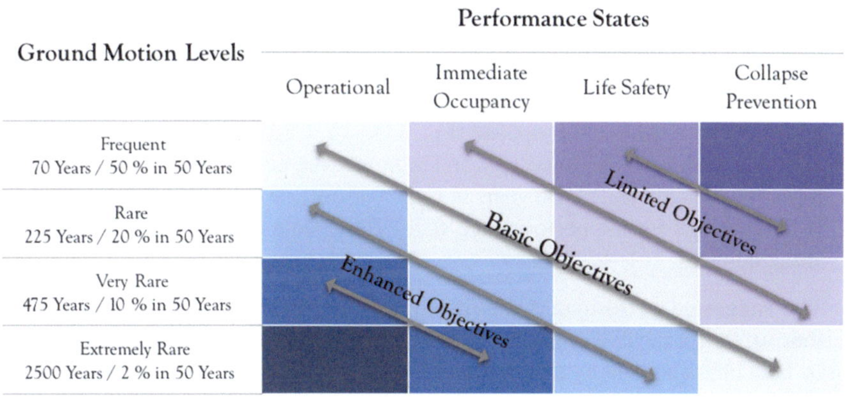

Fig. 7 Restoration objects for structures (adapted from FEMA 356 (2000))

6 Importance Classes of Historical Heritage Structures

To determine the suitable level of seismic safety, historical heritage structures are classified rendering to their importance. The suggested grouping in significance classes shadows the similar well-established norms of guard of cultural heritage for additional risks than seismic one. In Algeria, the ordinance for the safeguard of historical property in the event, provides the following five classes in descending order of importance: (i) Historic property of international importance: Class A, (ii) Historic property of national importance: Class B, (iii) Historic property of regional importance: Class C, (iv) Historic property of local importance: Class D, (v) Historic property of Authorities importance: Class E.

Figure 8 Displays cases of historical heritage structures in Skikda city for diverse importance classes.

7 Performance Conditions for Historical Heritage Structures

The support that in performance-based seismic calculation of historical heritage structures, the preservation and security of persons are evaluated in an essential method, and describe three kinds of performance bounds that interpret for the dissimilar features to be considered: Performance bounds describing the effects on the structure's occupancy/use and life security, (ii) performance bounds related to the structure preservation, and (iii) performance bounds to the preservation of creative assets in the structure (Figs. 9 and 10).

To consider the moderate seismicity in the study area for this type of structures, the subsequent assumprtions are suggested:

- The importance factor ɣk, which adapts the return period for a specific performance level, is directly, associated to the classification of the heritage structure in the five importance classes (A/B/C/D/E), and a matrix of performance conditions according to FEMA 356 proposed (Fig. 11).
- The return period of the lowest ground motion level, i.e., "Frequent", was reduced from 70 years in Fig. 7 to 50 years in Fig. 10 reflecting the minimum condition of individual risk for the performance state "Restorable Damage" of usual current structures with little occupancy.

The suggested performance matrix allows discriminating among the level of seismic safety for the five importance classes (A/B/C/D/E) of cultural heritage structures, both characterized by a diagonal in Fig. 11. For the lowest importance class C of cultural property, the level of seismic safety would at smallest reach the required minimum code level for usual existing structures of importance class I or II according to Algerian earthquake code (2003). For the categories of highest importance, the return periods for the performance levels are scaled to less frequent ground motion

Fig. 8 Five examples of historical heritage buildings in Skikda city: Former church Ste Theresa and the lighthouse of Miramar (Category A, up), Theater of Skikda (Category B, bottom left), Skikda Town Hall (ex-Philippeville) in (Category C, bottom right)

	Use and Human life		Architechtonic assets		Artistic assets	
$T_{RD,PU}/\gamma_k$ (k=U,B,A)	Immediate Occupancy	Life Safety	Significant but restorable damage	Near Collapse	Restorable Damage	Loss Prevention
$72/\gamma_k$	2U				2A	
$475/\gamma_k$		3U	3B			3A
$2475/\gamma_k$				4B		

Fig. 9 PERPETUATE: Return periods for cultural heritage structures. The importance constant ɤk is related to the usage, the archetonic and artistic cost of the structure and its assets (Calderini et al. 2012)

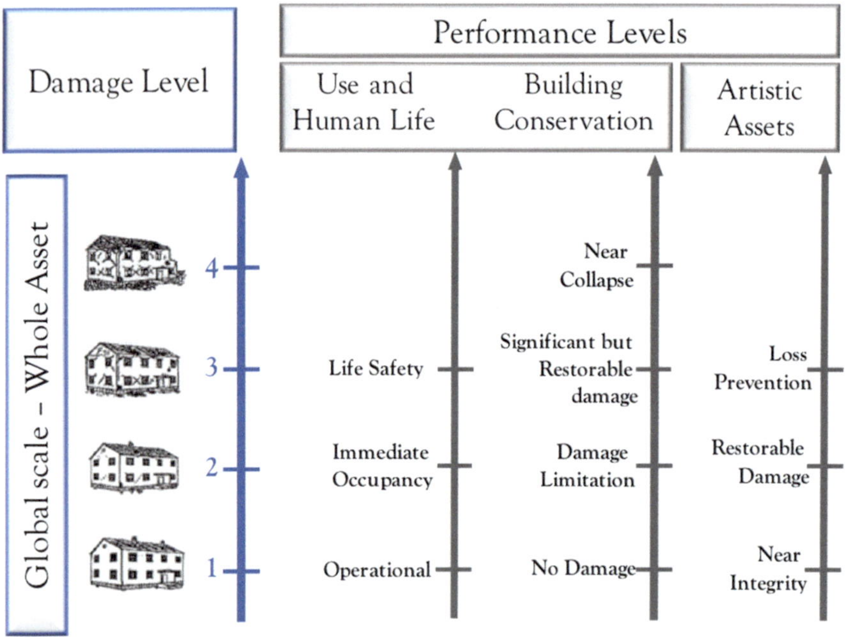

Fig. 10 PERPETUATE: Performance ranks and loss levels for cultural heritage structures (Calderini et al. 2012)

levels. For importance class B of cultural property, the suggested performance state agrees to the required code level for usual new structures, as noticeable in yellow in Fig. 11. For the two uppermost importance classes A and B of cultural property, advanced performance conditions important to "No Damage" for the ground motion levels "Rare" or even for "Very Rare" are suggested.

The maximum diagonal in Fig. 10 characterizes the lowest level of seismic safety for usual existing structures with little occupancy according to RPA (2003). They

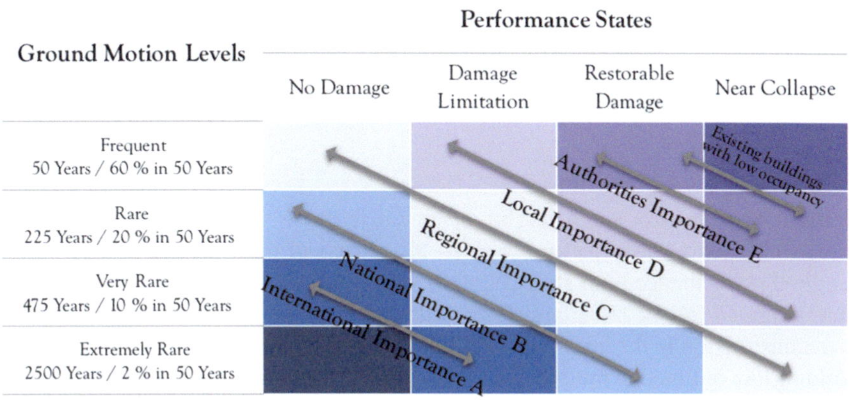

Fig. 11 Suggested performance conditions for cultural heritage structures

have to accomplish the performance state "Restorable Damage" which agrees to the performance state "Life Safety" (Fig. 7) for the ground motion level "Frequent". Then the individual risk is satisfactory as deliberated in the Section "Compliance Factor vs. Return Period". If the seismic calculation of the structure displays that its performance complies with the performance limits defined in Fig. 10, no additional measures are mandatory. Heritage structures of class C are acknowledged as adequately safe smooth however, they do not meet the essential code level for usual new structures reliable with the overall relaxation of code necessities for existing structures (RPA 2003). As they accomplish the performance states for more occasionally ground motion levels than usual existing structures with little occupancy, they can be accepted even with advanced occupancy. For the advanced importance classes (A/B/C) of cultural property, the performance level would then be equivalent or overhead the code level for usual new structures. If the seismic calculation exposes that not even the necessities for usual current structures with low occupancy in Fig. 10 are satisfied, instant measures are essential. Such measures can include retrofit measures or the control of admission to the heritage location. According with the Algerian code RPA 2003, the regular occupancy has to be kept below 0.2 persons and the extreme number of persons, which are waiting in the structure (Hamidatou et al. 2021), has to be kept below 10 persons if the smallest necessities for individual risk are not encountered. If the seismic performance of the structure is overhead the level for usual current structures with little occupancy, nonetheless motionless under the suggested diagonal line of its class in Fig. 10, numerous preservation approaches are thinkable. These approaches are discussed in the subsequent section.

8 Results and Discussions

8.1 Conservation Strategies for Historical Heritage Structures

Seismic conservation strategies for historical heritage structures can be directed towards two opposite objectives, i.e., (i) the retrofit of the structure to achieve the required performance limit state (Fig. 10); or (ii) avoidance of any intervention, but opting instead for the safeguarding documentation of the structure, allowing hence, its reconstruction in the event of an earthquake, that partially or entirely destroys the building. As outlined in the previous section, a limit on the latter is set by life safety considerations, which must comply with ordinary existing structures considering individual and collective risks to persons. In zones of low to moderate seismicity, this limit is rather low and could be satisfied by an important share of historical heritage constructions. If the risk does not satisfy life safety requirements, retrofit measures are necessary to guarantee the same level of life safety protection as for ordinary existing structures. If this is not feasible or too costly, the heritage site must be closed to the public. Hence, the performance of a heritage structure must not be more risky to its occupants or visitors than that of an ordinary existing structure, while the more stringent performance requirements result from its cultural heritage value (categories A/B/C/D/E).

The choice of the conservation strategy depends on considerations on the impact of the required interventions on appearance, fabric, costs and their proportionality. Structural engineers, architects, curators of monuments, and other stakeholders should develop the conservation and retrofitting strategy. During this phase of decision-making, an effective and clear communication between the different parties is essential. Experience has shown that as a tool for the communication between stakeholders, the elaboration of different scenarios works often best. Applying different conservation strategies, the scenarios should illustrate (i) interventions and their costs, and effect on the appearance and the structure's fabric, (ii) the consequence for the use of the building, (iii) the expected damage from seismic events of different return periods, and (iv) the reconstruction costs. Example strategies could for example be:

- No intervention: Conservation of the existing state without any intervention, safeguarding documentation of the structure so that it can be reconstructed. Significant damage expected for relatively short return periods. Restricted use to prevent a larger crowd of people in the structure (e.g. closed to the public) if criteria of personal risks are not met in the existing state.
- Minimum intervention: Minimum interventions which are required to permit the full use (e.g. completely open to the public, use for large assembles), safeguarding documentation of the structure so that it can be reconstructed in the event of rare seismic scenarios. This approach corresponds to that of ordinary structures not protected as cultural heritage.

- Intermediate intervention: Interventions to reach the level of seismic protection provided in the performance matrix for a lower importance category than that of the heritage structure (Fig. 10). Safeguarding documentation of the structure so that it can be reconstructed in the event of rare seismic scenarios.
- Maximum intervention: Interventions to reach the level of seismic protection provided in the performance matrix for its importance category (Fig. 10).

Interventions do not only assure the full use, but also the structure's integrity in the event of very rare seismic scenarios. The conservation strategy should always comprise a certain safeguarding documentation of the structure and its contents. Special consideration must be given to the long-term preservation of the documents in shelters, the formats of the documents, the keeping of several copies, etc. To illustrate the possible choices of conservation strategies, Fig. 12 shows, as an example, the recommended performance states for a category (A) heritage structure in lighter green between the diagonal limits "Ordinary Structures" and "National Importance A". The performance state "Near Collapse" below the minimum requirement for ordinary existing structures, marked in red, is not acceptable due to life safety requirements. The higher performance state of those defined by the diagonal "National Importance A" are, of course, acceptable but usually not reachable with reasonable measures. The range in between the diagonal "Ordinary Structures" and "National Importance A" are feasible. Any significant distance from the diagonal "National Importance A" should be compensated by a safeguarding documentation of the entire structure.

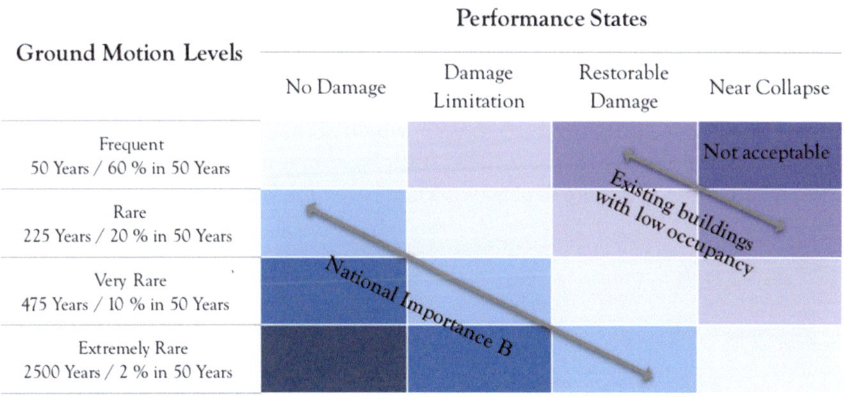

Fig. 12 Proposed performance states for cultural heritage structures

8.2 Complete Measurements: Calibration and Identification of Numerical Models

The case of object of the present study was built from 1949 to 1955, is of type 5 (Figs. 6 and 13). Based on the documentations three important structural changes are registered in its history, adjustment of the bell tower and bays in the nave and the sacristy; 1890 and 1912, heavy rearrange of the frame and the cover, recovery of the decorations. Cracks are again visible (Fig. 13). Coarse masonry with large proportion of mortar and very irregular stones can be observed. Note that the AutoMAC matrix of experimental measurements describes the excellence of the MAC values to be targeted: it displays that certain local modes are not recorded, such as the oscillation of the pediment of the numerical mode 1, well coordinated in frequency with the experimental mode 1, but poorly deformed for lack of a sensor.

The results of the model without optimization are illustrated in Fig. 14a. the value of the MAC criterion of 0.65 on the first four modes illustrates a very bad correlation: lack of data on the ground or foundations. The frequency error exceeds 15%: the masonry of the facades is of less good quality than that of the steeples taken in reference, uncertainties about the mass of the roof. In addition, the MAC criterion is sensitive to extreme values that distort the results of the whole structure. Therefore, we decided to optimize the nave and the bell tower separately.

Bases on the results it was performed decoupled optimizations in three steps. It was considered the missing part by completing the rigidity matrix of each model by supports at the junction points. Their lightness makes it possible to increase the number of parameters: representation of the foundations by supports to calibrate, and take into account the different types of masonry due to the construction method and the successive works. By going from 3 to 35 parameters on the bell tower, the average MAC value goes from 0.5 to 0.92, which allows us to show the importance of the parameter's choice, because their multiplication is not acceptable. It makes the optimization more and more expensive, while being misleading since we will not be

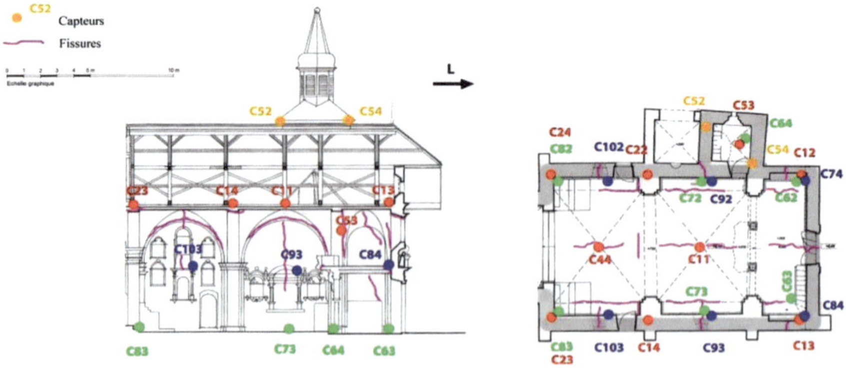

Fig. 13 Plan and section with crack readings (MicMac): position of the sensors

Fig. 14 Former church Ste Theresa: third mode, Y-bending before/after optimization Frequency exp = 5.99 Hz. **a** Frequency num: 4.52 Hz–MAC: 0.64. **b** Frequency num: 5.85. Hz–MAC: 0.97

able to control in-situ the final values. A sensitivity analysis was then carried out on all the possible parameters for the nave (100), and the bell tower (52), in the value ranges of the Italian code and the literature. The Young's modulus and the stiffnesses of supports have the least important density, leading us to preserve the values of the optimization of the steeples and the Italian code. Finally, six different error functions were tested on a nave and bell tower model, combining the value of the MAC criterion and the frequency error. These two indicators per mode vary up to 35%, depending on the choice. It is therefore necessary to adapt the function to what is least known in the structure to be stalled. For the tower model we will give more weight to the MAC because the frequencies are relevant, while bad deformed would be very detrimental to the general MAC. Once the two models have been recalibrated separately, they are assembled into a single model due to rigidities calibrated on those representing the part missing from the decoupled models. In this way, the optimization of the complete model at 36,000° of freedom is carried out with a single parameter, whereas the optimizations with several constraints were made on the partial models. The final MAC criterion on the first four modes varies from 0.99 to 0.90, showing a very good correlation. The frequency error varies from zero to 9%, for an average of 5.4%. The results are therefore very satisfactory (see Fig. 14b), especially for such a structure.

8.3 *Linear and Nonlinear Kinematic Analysis*

Several authors (Casarin 2006) approve that multifaceted models are not the guarantee of clear and dependable consequences. For structures such as churches, the Italian code recommends linear or smooth non-linear kinematic investigation (DPCM 2011). The churches are divided hooked on shares showing distinct earthquake behaviors, important to listed rupturing mechanisms, 23 in total. Macro element by macro element, one determines all the mechanisms of rupture possible in the church studied.

Then, case by case, it were computed the value of the horizontal acceleration that triggers the mechanism, in relation to the state of possible bound loss. This agrees essentially to the horizontal action where the structure can progressively receive throughout the development of the mechanism. By associating the acceleration with that of the location, we can determine the possible mechanisms. For our case study, the fall of the main façade is expected (Fig. 15): these mechanisms are not tested by linear analysis or nonlinear analysis.

If the compare the different results concerning Former church Ste Theresa are compared, it is possible to perceive that the areas where the tensile and shear strength of the masonry is exceeded, according to our definition of the thresholds, correspond well to the local mechanisms of damage of the kinematic approach, since the cracks on the sloped wall (Figs. 13 and 15) correspond to a spill of the main facade (Fig. 15).

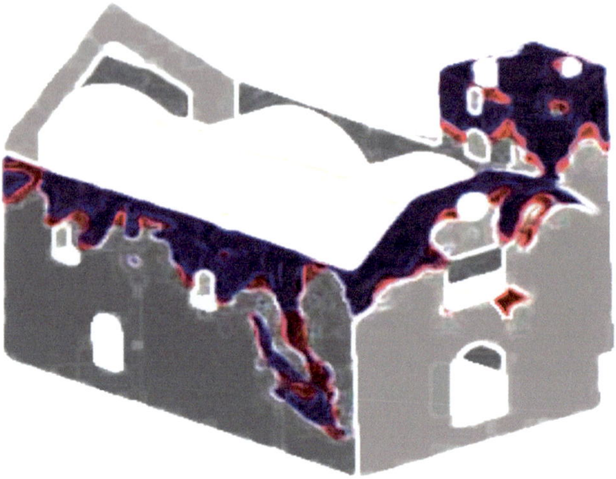

Mécanismes envisagés	Accélération ultime (m/s²)	Accélération exigée (m/s²)	Mécanisme possible
Déversement du fronton	1.57	2.12	Oui
Déversement des parties supérieures plus minces	0.747	0.962	Oui
Rotation de la façade avec arrachement des angles	0.486	1.152	Oui

Fig. 15 Former church Ste Theresa. Areas where SI11 exceed the criterion: nonlinear (red) and linear (blue) models–Kinematic analysis, main facade: ultimate acceleration and site mechanisms

9 Conclusions

The work plans a background for a seismic preservation approach for Algerian historical heritage places. The earlier delivers in specific the performance state matrix for current structures and the main knowledge that it is acceptable that current structures comply with minor performance bounds than innovative structures. The main concepts of this outline relate to: (i) Smallest performance level owing to life security necessities; (ii) Insignificant performance levels as a function of the importance class of the heritage place (authority/local/regional/national/international) importance; (iii) Select of seismic preservation approach for performance levels between smallest and insignificant allowable.

Two different seismic preservation approaches were outlined while numerous middle approaches exist. The first approach purposes at decreasing loss in the event of an earthquake and at attainment the insignificant performance levels that are defined as a function of the importance class of the heritage place. This approach normally consequences in important restoration measures. The second approach requests at keeping the involvements to a complete least possible, i.e., by introducing only those that are essential to reach life security necessities. No additional measures are occupied in instruction to keep the influence on the structure's fabric and appearance to a total smallest. Based on the work developed, it was proposed that the structure performances noteworthy damage for events with return periods as small as 50 years. To account for this risk, the structure must be carefully documented to agree to its renovation.

The inventory and study of a susceptible heritage that was frequently unknown was carried out, establishing a comprehensible and extensive study panel. A humble typical construction way has been established through the description of experimentally authenticated structural categories, the creation of tools for a completely parameterized mesh of multifaceted structures and an influence study of the demonstrating selections in three important zones. A simple process has been suggested to standardize an enormous number of dissimilar structure models on vibratory measurements from insufficient data and authenticate them experimentally. Stress damage criteria founded on the first linear demonstrating step were suggested, and then verified on a non-linear model. It was concluded that associate the consequences of the non-linear models with those of the usually acknowledged kinematic investigation, which allowed us to confirm that their conclusions are in respectable agreement currently.

References

Anderson H, Jackson J (1985) Active tectonics of the adriatic region. Geophys J Int 91(3):937–983. Architectural Heritage, 14th General Assembly of ICOMOS, Victoria Falls. https://doi.org/10.1111/j.1365-246X.1987.tb01675.x

Argus DF, Gordon RG, DeMets C, Stein S (1989) Closure of the Africa-Eurasia-North America plate motion circuit and tectonics of the Gloria fault. J Geophys Res 94 https://doi.org/10.1029/88JB03988. ISSN: 0148-0227. baudenkmalpflege-62

Binda L, Tiraboschi C, Tongini Folli R (2013) On site and laboratory investigation on materials and structure of a bell-tower in Monza Int. Zeitschrift fur Bauinstandsetzen und

Borcherdt R, Wentworth CM, Janssen A, Fumal T, Gibbs J (1991) Methodology for predictive GIS mapping of special study zones for strong ground shaking in the San Francisco bay region, CA. In: Proceedings of the Fourth Int. Conf. on Seismic Zonation, Stanford, California

Bounif MA, Haessler H, Meghraoui M (1987) The Constantine earthquake of October 27, 1985: surface ruptures and aftershock study. Earth Planet Sci Lett 85:451–460

Calderini C, Cattari S, Lagomarsino SM, Brunenghi M (2012) PERPETUATE, performance-based aproach to earthquake protection of cultural heritage in European and mediterranean countries, Final Report D42. In: Seventh Framework Programme, EU Commission, Brussels

Casarin P (2006) Structural assessment and seismic vulnerability analysis of a complex historical building. University of Trento

Christophe L (2008) Highlights of ArcGIS. ESRI Geo informatique SA

Coiffait B, Guellal S, Vila JM (1992) Carte géologique au 1/50.000 d'El Aria. Sonatrach. Division Hydrocarbures, Algeria

Coiffait PHE (1992) Un bassin post-nappe dans son cadre structural: l'exemple du bassin de Constantine (Algérie nord-orientale). PhD Thesis, University of Nancy

DeMets et al., 1990 DeMets C, Gordon RC, Argus DF, Stein S (1990) Current Plate Motion. Geophys J Intern 101:425–478

DPCM (2011) Valutazione e riduzione del rischio sismico del patrimonio culturale con riferimento alle NTC 14 gennaio 2008. DPCM 9 febbraio 2011, G.U.R.I. February 26th 2011

Eurocode 8–1 (2004) Design of structures for earthquake resistance–part 1: general rules, seismic actions and rules for buildings, EN 1998–1. European Committee for Standardization, Brussels

Eurocode 8–3 (2005) Design of structures for earthquake resistance–part 3: assessment and retrofitting of buildings, EN 1998–3. European Committee for Standardization, Brussels

FEMA 356 (2000) Prestandard and commentary for the seismic rehabilitation of buildings. Federal Emergency Management Agency (FEMA), Washington DC

FOCP (2014) Federal office for civil protection. www.bevoelkerungsschutz.admin.ch/internet/bs/en/home/themen/kgs/schutzmassnahmn.html Bern, Switzerland

Hamidatou M, Sbartai B (2016) Deterministic assessment of seismic risk in Constantine city. Northeast Algeria Nat Hazards. https://doi.org/10.1007/s11069-016-2693-2

Hamidatou M, Mohammedi Y, Yelles-Chaouche A, Thallak I, Stromeyer D, Lebdioui S, Fabrice C, Hallal N, Khemici O (2019) Seismic hazard analysis of surface level, using topographic condition in the Northeast of Algeria. Pure Appl Geophys. https://doi.org/10.1007/s00024-019-02399-7

Hamidatou M, Sbartai B (2017) Probabilistic seismic hazard assessment in the Constantine region, Northeast of Algeria. Arabian J Geosci https://doi.org/10.1007/s12517-017-2876-5

Hamidatou M, Mohammedi Y, Hallal N, et al (2021) Reply to the comment on the paper "Seismic hazard analysis of surface level, using topographic condition in Northeast of Algeria" by Mohamed Hamdache and José A. Pelàez. Pure Appl Geophys https://doi.org/10.1007/s00024-020-02644-4

Hamidatou M (2017) Seismic risk assessment in urban areas: Application in Constantine region. Doctorate (Ph.D.) Thesis, University of Skikda, Civil Engineering Department

Harbi A, Peresan A, Panza G (2010) Seismicity of Eastern Algeria: a revised and extended earthquake catalogue. Nat Hazards 54:725–747

ICOMOS (2003) Principles for the analysis, conservation and structural restoration of

Karbassi A, Lestuzzi P (2014) Seismic risk for existing buildings in Switzerland—development of fragility curves for masonry buildings. EPFL, FOEN, Lausanne, Switzerland, p 56

Kölz E, Schneider J (2005) Beurteilung der Erdbebensicherheit bestehender Gebäude—der risikobasierte Ansatz des Schweizer Merkblatts SIA 2018, Bautechnik 82 Heft 8

Laupper H (2004) Expertenbericht: Erdbeben und Kulturgüter, Arbeitsgruppe Erdbeben und Kulturgüter. Bundesamt für Bevölkerungsschutz (BABS), Bern, Switzerland
Limoge C, Giry C, Desprez C, Ragueneau F (2014) Toward a large-scale seismic assessment method for heritage building: vulnerability of masonry baroque churches. EJECE
Lourenço PB, Roque JA (2006) Simplified indexes for the seismic vulnerability of ancient masonry buildings. Constr Build Mater 20(2006):200–208
Mattauer M (1958) Etude géologique de l'Ouarsenis oriental (Algérie). Publ Serv Carte géol Algérie, N.S. Bull 17:534
Meghraoui M (1988) Géologie des zones sismiques du nord de l'Algérie, tectonique active, paléosismologie, et synthèse sismotectonique, PhD Thesis, University of Paris-Orsay, p 356
Perrault M (2013) Evaluation of the seismic vulnerability of buildings from in situ measurements, PhD thesis, University of Grenoble. Professional. ENPC-LVMT, Marne-la-Vallee
Raoult JF (1974) Géologie du centre de la chaine numidique (Nord Constantinois, Algérie). Thesis Paris Mém. Soc. Géol. Fr. 121: 163. Roma, Italy
Vicente R, Parodi S, Lagomarsino S, Varum H, Silva JAR (2011) Seismic vulnerability and risk assessment: Case study of the historic city centre of Coimbra, Portugal. Bull Earthquake Eng 9:1067–1096
RPA (2003) Règles parasismiques algériennes (Rules of Algerian earthquake). CGS, Algeria
Sbartai B, Hamidatou M (2013) Seismic risk assessment of the Constantine city, Algeria. Intelligent systems and decision making for risk analysis and crisisresponse. Risk analysis and crisis response. Taylor and Francis Group, London, ISBN 978–1–138–1–138–00019–3. https://doi.org/10.1201/b16038
Schneider J (2000) Safety—a matter of risk, cost, and consensus. Structural Engineering International, vol.10 No. 4. Special Issue AUGC
Thierno AW (2004) Initiation to mapinfo professional step 1 basic mapping with mapinfo
Valluzzi MR, Porto DAF, Modena C (2003) Structural investigation and strengthening of the Civic Tower in Vicenza. In: Proc of Structural Faults and Repairs, London UK
Vila JM (1980) La chaine alpine d'Algérie orientale et des confins Algérotunisiens, PhD-Thesis. Univ, Paris VI, p 633

Seismic Assessment of a Cultural Heritage Minaret in Cairo

Hany M. Hassan, Mohamed A. Sayed, Marco Fasan, Fabio Romanelli, Claudio Amadio, Ayman Hamed, Mohamed ElGabry, and Islam Hamama

Abstract Dealing with cultural heritage is a sensitive process since each monument has its history, story, conditions, and scharacter. In this work, we assessed and evaluated the seismic vulnerability of a well-preserved cultural heritage structure that is the minaret of the Madrasa of the Princess Tatar al-Higaziya in Cairo. We selected the minaret site's input seismic source based on a physics-based ground motion simulation named multi-scenario seismic input (MCSI). This seismic source was used for the assessment of the dynamic behaviour of the minaret. A detailed numerical model of the minaret was developed in SAP2000. An initial bi-directional response spectrum analysis was performed on the minaret, considering the coefficient of subgrade reaction of soil. Both a record of the 1992 Cairo earthquake and synthetic seismograms were used. The calculations confirm no damage in the case of the 1992 earthquake while, in the worst-case scenario, the minaret could suffer significant tensile stresses that exceed the tensile strength of the limestone material. Results

H. M. Hassan (✉) · M. A. Sayed · M. ElGabry · I. Hamama
National Research Institute of Astronomy and Geophysics, Helwan 11421, Cairo, Egypt
e-mail: hany_hassan@nriag.sci.eg

M. A. Sayed
Department of Civil Engineering, University of Toronto, Toronto, Canada

M. Fasan · C. Amadio
Department of Engineering and Architecture, University of Trieste, Trieste, Italy

F. Romanelli
Department of Mathematics and Geosciences, University of Trieste, Trieste, Italy

H. M. Hassan · M. A. Sayed · M. Fasan · F. Romanelli · C. Amadio · A. Hamed · M. ElGabry · I. Hamama
Faculty of Petroleum and Mining Engineering, Suez University, Suez, Egypt

H. M. Hassan · A. Hamed · M. ElGabry
North Africa Group for Earthquakes and Tsunami Studies (NAGET), Net40/OEA ICTP, Trieste, Italy

I. Hamama
School of Systems Engineering, Kochi University of Technology, 185, Miyanokuchi, Tosayamada, Kami 782-8502, Kochi, Japan

G. M. El-Qady and C. Margottini (eds.), *Sustainable Conservation of UNESCO and Other Heritage Sites Through Proactive Geosciences*, Springer Geology,
https://doi.org/10.1007/978-3-031-13810-2_28

denote enormous cracking and even crushing in the minaret body, particularly at the base and at a geometry transition zone right above the base. Furthermore, the tensile stresses' level predicts collapse or severe minaret damage under the C-MCSI-50% bidirectional response spectrum load. Results were confirmed by time-history analyses performed on the model. The results emphasize the importance of predicting the behaviour of heritage and historical structures against strong earthquakes, especially for those that share similar structural characteristics (e.g., height, construction time and materials) with our case study historical structure.

Keywords Seismic assessment · Cultural heritage minaret · Scenario based approach · Historical Cairo · Seismic input · Dynamical behaviour analysis

1 Introduction

Egypt is located at the conjunction of the Mediterranean basin, Africa, and Asia's continents. It has been known for its history and geography. Egypt has a multicultural heritage (e.g., Ancient Egypt, Coptic, and Islamic monuments) known for its diversity and richness. This heritage, a symbol of identity and an essential element of our memory, holds our shared principles and values and must be transferred to future generations. Cultural heritage is a unique wealth, requiring particular attention such as security, management, restoration, and preservation from all kinds of impacts brought by natural and human activities. Conservation of cultural heritage involves protection and restoration using intervention strategies that effectively maintain a particular property in a condition close to the original one and for as long as possible.

Egypt is considered a country of low to moderate seismicity. The earthquake record infers that inland earthquakes of moderate strength can substantially damage standard and cultural heritage structures and buildings. During the 1992 Cairo earthquake of moment magnitude M_w 5.9, about 212 of 560 monuments in Cairo were reportedly damaged (Sykora et al. 1993). Although the earthquake was of intermediate magnitude, a maximum intensity of VIII on the Modified Mercalli Intensity (MMI) scale was reported in Cairo; while the maximum observed intensity in the historic Cairo area was VII (Sykora et al. 1993). Several seismic hazard assessment studies and reviews have been conducted on Egypt (e.g., Sawires et al. 2016, Hassan et al. 2017b, Gorshkov et al. 2019, Hassan et al. 2017a, ElGabry and Hassan 2021). In this paper, we refer to the results provided by Hassan et al. (2020), where the seismic input in Cairo was computed through physics-based ground motion simulations.

The process of finding and selecting a proper mitigation strategy against seismic action for cultural heritage buildings and structures is based on two factors: the accurate evaluation and estimation of the expected ground motion at the site of interest and understanding the performance of such buildings and structures during the earthquake shaking. The performance status and protection objectives developed for modern conventional (non-heritage) buildings are not directly applicable to heritage buildings since they do not address and share cultural concerns and often different construction techniques. In fact, in the preservation of historical monuments against seismic shaking, it is well known that each building has its characteristics (e.g., structure type, age, construction materials, state of preservation, site conditions, expected ground motion level, and surroundings). Therefore, specific requirements and intervention strategies are required based on a rigorous and detailed evaluation of the factors mentioned above.

Historic Cairo is a UNESCO World CulturalCentre (WHC) site with its famous mosques, madrasas, hammams, and fountains. Although these buildings were built in periods that date back to more than ten centuries, many have survived earthquake events with no or minor damage; others have suffered severe or complete damage. This study presents an assessment of the seismic vulnerability of a well-preserved cultural heritage structure, the minaret of the Madrasa of the Princess Tatar al-Higaziya in the historic Cairo area, hereafter referred to as the minaret. The study's peculiarity is the use of site-specific synthetic accelerograms to represent the seismic demand at the site of interest adequately. These accelerograms are used for an assessment of the minaret's dynamic behaviour, obtained by joining seismological and structural aspects to understand this monument's performance under earthquake loading. This study may contribute to the understanding of the seismic conservation criteria for cultural heritage sites in Egypt.

The results of this work could be necessary for the seismic risk reduction of heritage structures, particularly for those constructed during the same time and having similar structural systems and components. The research findings encourage detecting proper mitigation measures for the minaret and are a step towards a comprehensivemanagementstrategy for this historic structure.

2 Structural Elements of the Madrasa

The Madrasa (meaning a school in Arabic) of princess Tatar al-Higaziyawas constructed in two stages. First, the mausoleum was built in 1348 as an extension to princess Tatar's house, then after thirteen years, the palace and the mausoleum were converted into the Madrasa (Williams 2008). The Madrasa complex consists of the mausoleum, a minaret, and an ablutions court. It is one of few schools endowed by a woman in Cairo. The school was built and endowed to educate orphan children, raise the daily prayers, and serve as a public library. A clear sketch of the school layout, a three-dimensional view, and a recent photo of the minaret are shown in Fig. 1.

The school has been restored several times and more recently by the Egyptian government and the German Cultural Centre in 1980–1982 (Mayer and Speiser 2007). The minaret is one of the most significant school elements, consisting of a 24.16 m vertical shaft above the ground level, and it lacks an upper cap or mabkhara (Mayer and Speiser 2007). The minaret's ground plan is squared, with a width of 3.45 m up to a height of 11 m. There is a transition above the base to an octagonal plan, which is further emphasized using bevelled corners. This minaret has been thoroughly studied after the 1992 Cairo earthquake to inspect the level of damage

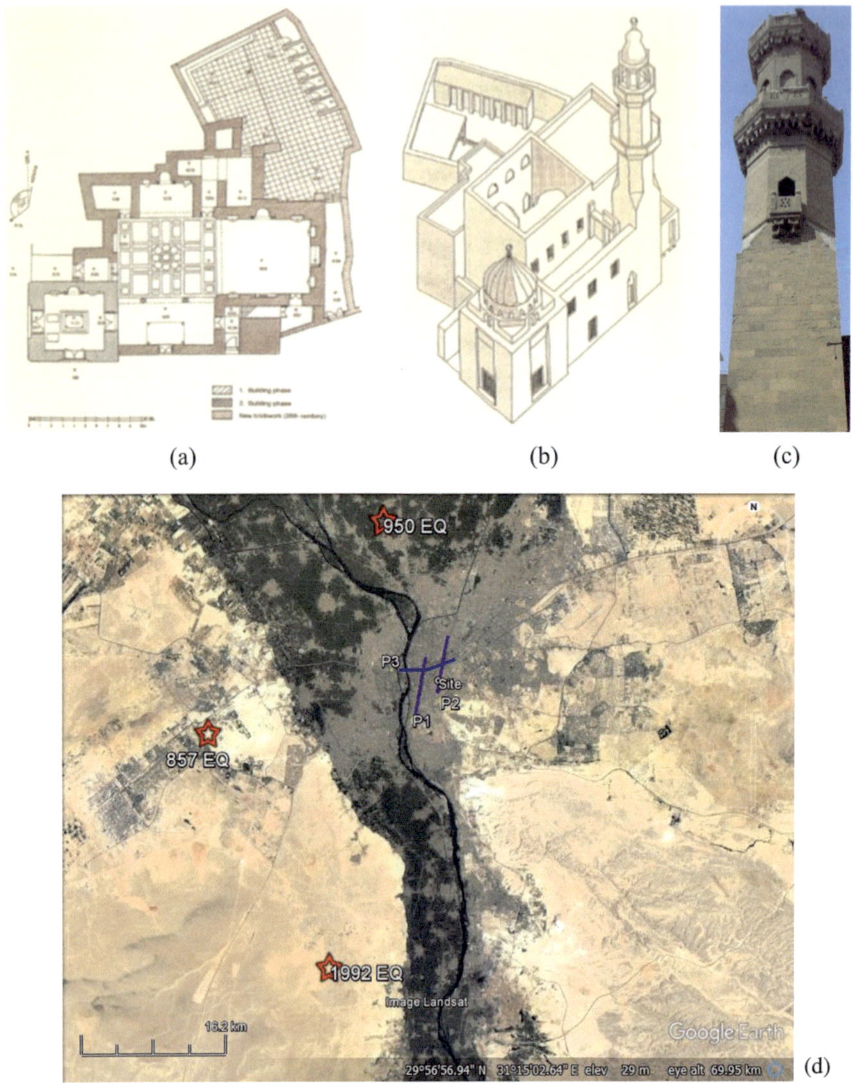

Fig. 1 Madrasa of the princess Tatar al-Higaziya: **a** layout, the filled brown square indicates the position of the minaret; **b** three-dimensional view; **c** recent photo of the minaret by the first author on 28/09/2017 taken from south west direction; **d** Location map of the Madrasa (site), subsurface profiles and earthquake scenarios based on which the MCSI was computed

that occurred to the minaret surviving the earthquake. The minaret's performance under earthquake loading was examined, but important seismological and other input parameters lacked in the studies. In modelling, a simplified shaft model that ignores some critical structural elements without justification was considered.

Additionally, the excitation records used in the study were simply the elastic response spectra of the 475-year return period of the Egyptian Building code, which is not logical since the building age is already beyond that period. The UNESCO project framework conducted the study to restore historic Cairo (Imam 2001). The incomplete assessment of the minaret's seismic performance motivated this study to reevaluatethe minaret's dynamic response using historical and recent seismic input and detailed numerical modelling techniques.

Several previous studies have introduced the numerical modelling and dynamic analysis of historical minarets under earthquake loading. For example, El-Attar et al. (2005) proposed a seismic protection technique for a historic limestone masonry minaret built between 1348 and 1960 AD in historic Cairo. In their study, the minaret was built one year after the minaret was reviewed in this study, with the same construction techniques and limestone material properties. Moreover, they considered a linear elastic finite element model in SAP2000 for the minaret, while the modal analysis results were evaluated and compared with ambient vibration test results. Sezen et al. (2008) have investigated the dynamic analysis and assessment of a reinforced concrete cylinder minaret constructed in Turkey. They developed four finite element models in SAP2000 representing the same minaret while ignoring various structural components in each model, such as openings, balconies, and interior spiral stairs, to arise a simplified numerical modelling approach for similar cylindrical minarets. They concluded that ignoring the spiral stairs' modelling influenced the modal analysis and modal participation factor, reducing the minaret's stresses. In the view of the studies mentioned above, we constructed our model considering all openings, balconies, and geometrical transitions while ignoring the spiral stairs for a conservative estimation.

3 Selection of Response Spectra and Time Histories

The multi-scenario physics-based seismic input (hereafter abbreviated as MCSI) provided in Hassan et al. (2020) represents a useful and conservative response spectrum analysis tool. Hassan et al (2017b) have provided a set of seismic hazard maps computed within the framework of NDSHA procedure at a national scale that may effectively accommodate any reliable new information to adequately compute the ground shaking scenarios maps. At local scale, further investigations can be performed within NDSHA taking into account the source effects and local soil conditions (i.e., surface and subsurface topography, resonance, water content, wave conversions, and geometry and heterogeneity of the sedimentary layers). In the work of Hassan et al. (2020), the map of the seismic sources that contribute to peak ground motion values at the historic Cairo area obtained using the NDSHA approach (Hassan et al. 2017a, b) is used for the definition of the earthquake scenarios that could affect a given site; this is needed to be considered for the detailed SSA studies in order to investigate the modification in the ground motion parameters due to the source, propagation medium, and the possible local site conditions and to obtain the MCSI.

Three earthquake scenarios have been found to be of magnitude 5.9–7.0 Mw earthquakes at a distance between 10 and 25 km, north and west of the site of interest. The selected scenarios comprise two historical events (i.e., 857 and 950 both of them have intensity MMI = IX) and one instrumental (i.e., 1992 of MMI = VIII). 1D laterally non varying structural model was considered to compute the seismic input at the bedrock level. The computed seismic input was feed up into 2D laterally varying model that represent the subsurface condition (Fig. 1d).

MCSI is being recommended for the seismic design of new structures, although it might be less suited to selecting accelerograms that can be used for dynamic analysis of existing structures. Since MCSI is based on the computation of synthetic accelerograms when running time history analyses, a fast and effective accelerogram selection method could be utilized. The number of the selected accelerograms depends on the number of rupture realizations, rupture styles, and directivity angles considered for every simulated scenarios that contributes to the MCSI spectra computed for the site as shown in Fig. 1d (Hassan et al. 2020). However, this can become impractical due to the enormous amount of time histories (on the order of thousands of simulated accelerograms), requiring long computational times, significant analysis efforts and massive machine power.

Figure 2a, b shows the seismic input computed at bedrock (MCSIBD) and considering the site-specific soil stratigraphy ($MCSI_{SS}$) (values of the 50th, 84th and 95th percentiles) compared with the elastic spectra from the Egyptian building code (ECP-201 2011) (Type 1 is devoted for the whole country and Type 2 for the coastal zone along the Mediterranean) for different return periods (recurrence intervals) and different site conditions, i.e., bedrock and soil site of type B (soil with shear wave velocity of 360–800 m/s) according to the soil classification provided by the Egyptian building code (ECP-201 2011). The peak ground acceleration for two return periods (i.e., 475 and 2475) was used to scale the elastic response spectra defined by the Egyptian building code, the PGA value for 475-year spectra (PGA = 0.15 g) is adopted from the Egyptian building code (ECP-201 2011), while the PGA for 2475 return period (PGA = 0.25 g) is taken from a recent study done by Gaber et al. (2018).

Since we are dealing with an already existing structure and our aim is evaluating its dynamic behaviour through time history analyses, there is a need to find an approach for the proper selection of time histories. As suggested by Fasan (2017), when using a multi-scenario physics-based seismic hazard assessment, an approach to select a restricted number of accelerograms could be to limit our selection to the earthquake scenario that controls the seismic hazard at the structural vibrational periods of interest. This seismic input at the fundamental period of the building of interest is defined as a "Conditional" (C-MCSI), as proposed by Fasan (2017). The concept is similar to what is called Conditional Mean Spectrum (CMS), proposed by Baker and Cornell in (2006) as a more realistic alternative to the UHS (Baker 2011; Baker and Cornell 2006).

C-MCSI response spectrum can be defined by considering the most hazardous source's spectral accelerations at the period of interest, contrasting with MCSI that should consider all possible scenarios. In our case, the minaret's fundamental period

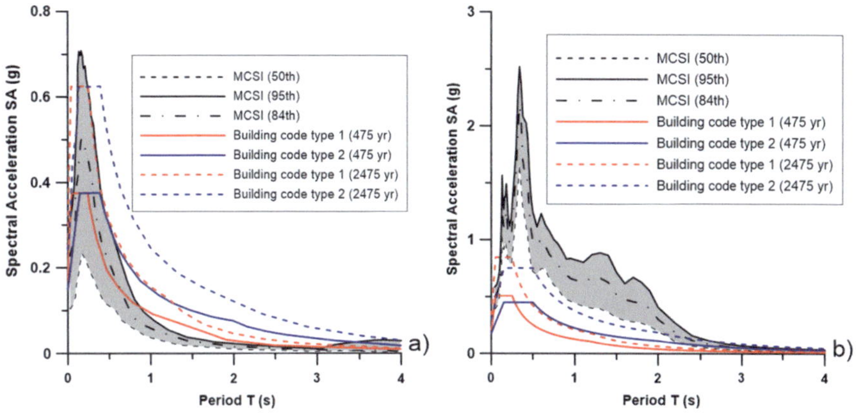

Fig. 2 **a** $MCSI_{BD}$ for the values of the 50, 84 and 95th percentiles, compared to the building code (Type 1 and Type 2) for two different return periods (475 and 2475 years). **b** $MCSI_{ss}$ for the values of the 50, 84 and 95th percentiles, compared to the building code (Type 1 and Type 2) of different return periods (475 and 2475 years) and the adopting recommended site coefficient. Shaded areas represent the range between 50 and 95th percentiles

is 0.5 s, as inferred from the noise measurements described in the next section. Figure 3 shows the C-MCSI for a period of 0.43 s, and the MCSI computed for the minaret site, which is set equal to the value of the 50th percentile and compared to the building code (Type 1 and Type 2) after considering site-effects.Fig. 4 shows the 50th C-MCSI and 1992 Cairo earthquake response spectra computed at 5% damping at the minaret site. They will be adopted to analyze the seismic performance for the minaret structure in the next section. C-MCSI response spectrum is calculated by selecting only the simulations from the scenario with a median spectral acceleration corresponding to the 50th percentile at the period of interest and then choosing the median values of these simulations at each period. The C-MCSI spectra account for the most dangerous scenario's spectral shape at the period of interest (Fasan 2017). Figure 3 indicates that both 50th C-MCSI and $MCSI_{SS}$ spectra vastly exceed the elastic response spectra defined by the building code (Type 1 and 2) for return periods of 475 and 2475 years.

Adopting this approach, the selection of accelerograms for the structural analysis becomes immediate since it merely retrieves the simulations used to define the C-MCSI. A subset of C-MCSI spectrum compatible accelerograms (e.g., seven, as suggested by building codes) are selected to conduct time history analysis. In our case, we have chosen seven simulations with two horizontal components, i.e., 14 accelerograms. The seven selected C-MCSI time histories in EW and NS directions at the minaret site and their corresponding response spectra of the ground motion components are plotted in Fig. 5. Once the accelerograms are selected, they can be considered a seismic input for the minaret's engineering model. Moreover, it is worth to mention that the seismic behaviour of a masonry structure is strongly influenced

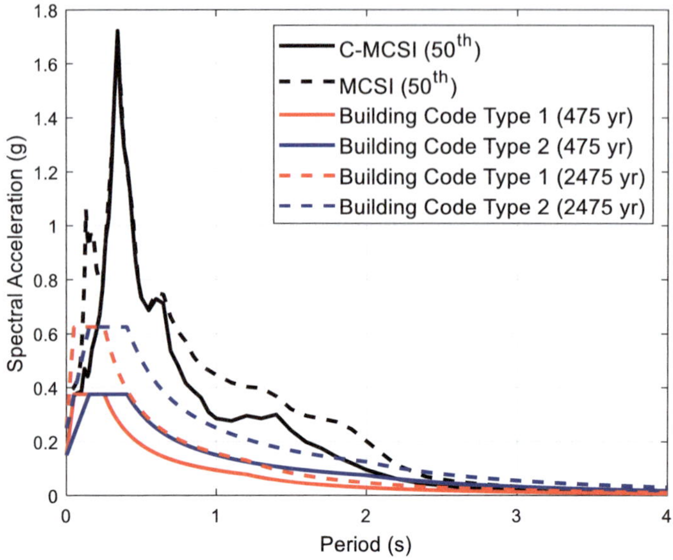

Fig. 3 C-MCSI and MCSI of 50th percentile response spectra compared with Type 1 and Type 2 Building Code with site-effect consideration at 5% damping

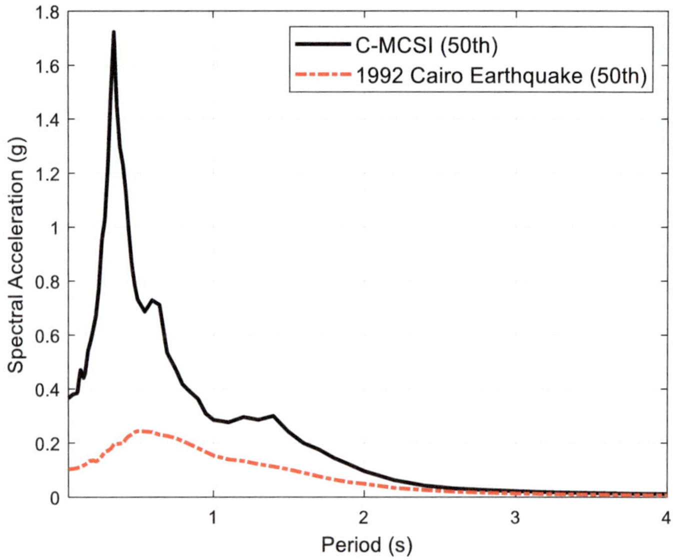

Fig. 4 C-MCSI and 1992 Cairo earthquake response spectra at 5% damping

by axial stresses which can be strongly modified by the vertical component of the seismic load, especially if they are located near the fault (Rinaldin et al. 2019).

4 Minaret Modelling

The minaret consists of a vertical shaft of a total height of 24.16 m. The minaret base is squared, after which there is a transition to an octagonal plan, further emphasized by bevelled corners. There are two tiers of balconies with stone parapets, accessed by an internal stone spiral staircase. A cylindrical shaft containing the inner spiral staircase extends from 7.2 m above the ground to the minaret's crest. The minaret dimensions and geometry is illustrated in Fig. 6.

A previous study performed by Imam (2001) on the structural condition of the Madrasa and its buildings was conducted in 2001 as part of the national historic Cairo project. The study affirmed that the minaret is structurally separated from the Madrasa with a vertical gap or separator. Moreover, five stone cone samples from three different parts of the Madrasa were extracted by (Imam 2001) to assign the different values of the limestone's mechanical properties used in construction. The report found that the un-cracked stone samples have a specific weight of 2.0 Mg/m^3, compressive strength of 27.5 MPa, a tensile strength of 5.4 MPa, and Young's Modulus of 25.5 GPa.

The minaretwas visually and physically inspected to investigate cracking and construction materials condition. Correspondingly, ambient vibration measurements by deploying seismic instruments at different locations inside the minaret body were conducted. The ambient vibration was used to evaluate the minaret's modal frequencies of the generated finite element model. After precise inspection, no visible cracks were detected in the minaret body, around the different openings, or near the joints. Therefore, un-cracked limestone material properties are considered in the minaret modelling with values stated previously.

4.1 Ambient Vibrations Analysis

The ambient vibration measurements were significant for the calibration of the dynamic behaviour of the minaret numerical model. Four tri-axial accelerometers with a range of ± 4 g were installed at different heights on the minaret to record the minaret's ambient vibration response. The sensors' locations are as follows: sensor (U1) was installed at the top balcony, and sensor U2 at ground level, sensors U3 and U4 were placed at the first balcony and the entrance, respectively. The sensors' locations are illustrated in Fig. 6. The McSIES-MT NEO (OYO Corporation) data acquisition instrument for microtremors measurement and vibration monitoring was used for the ambient vibration response analysis.

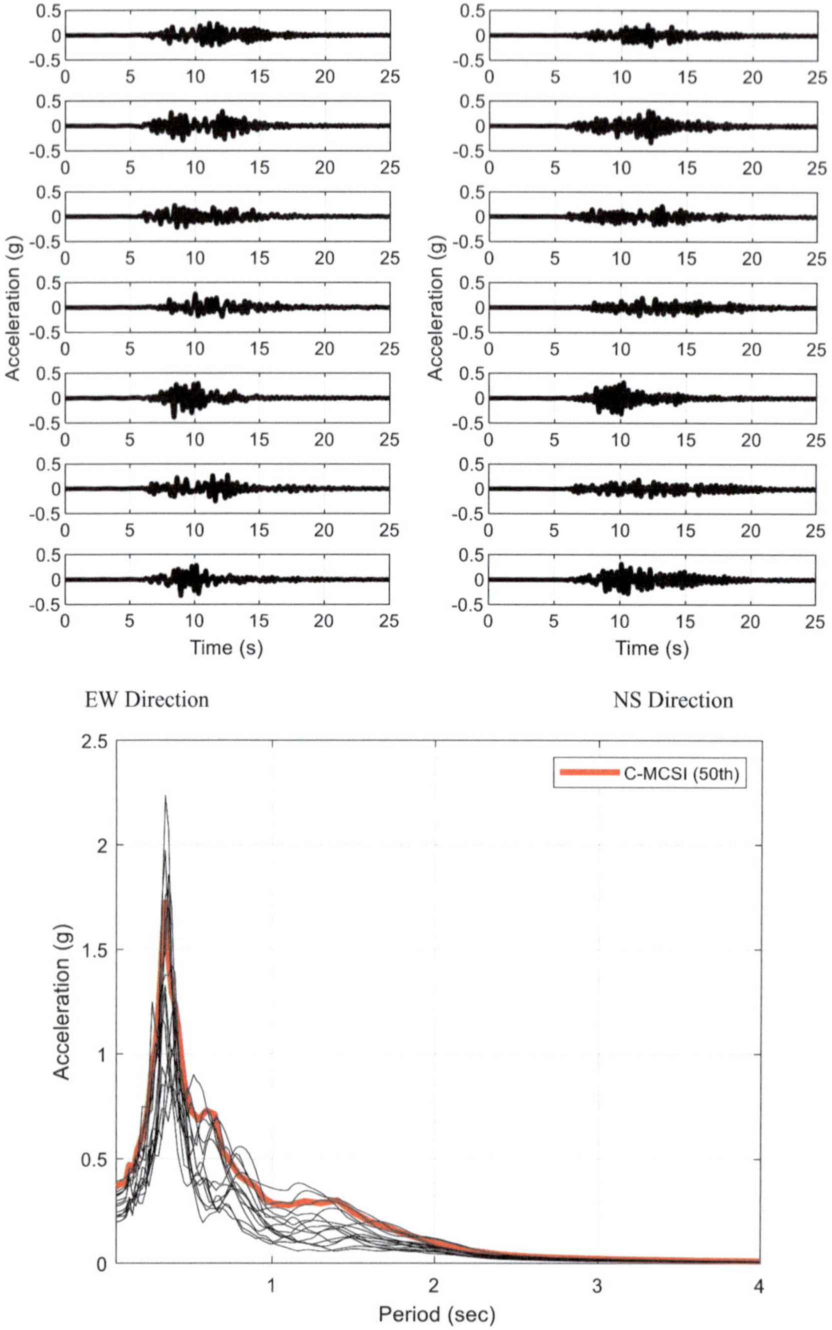

Fig. 5 Seven selected (on the median) C-MCSI time histories in EW and NS directions at the site of the minaret (top panel) and their corresponding response spectra withthe50th C-MCSI at 5% damping (bottom panel)

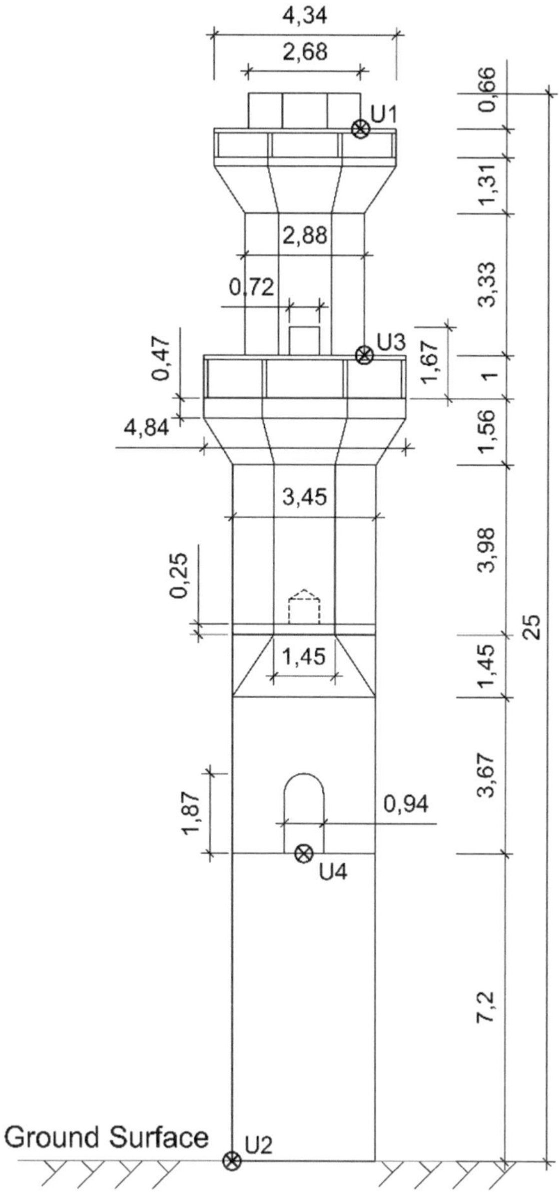

Fig. 6 Minaret geometry and instrumentation locations (all dimensions are in m). Instrument U1 installed at the top balcony; U2 at the ground surface; U3 at the lower balcony; U4 at the main entrance

The ambient vibration recording lasted for 60 min, with a recording sample frequency of 100 Hz. The measurements were recorded on a calm day with no moderate winds to eliminate weather effects on the ambient vibrations. It is also worth noting that the minaret is surrounded by taller buildings from all directions, reducing any wind effect on the minaret's ambient vibration. Post-processing of the acceleration measurements with baseline correction and high-pass filter hasbeen considered. A fourth-order Butterworth high-pass filter was applied, and a corner frequency of 0.2 Hz for the selected filter is calculated as (Sayed et al. 2015):

$$f_c = \frac{1}{T\left[\frac{H_0}{1-H_0^2}\right]^{\frac{1}{2n}}} \tag{1}$$

where, f_c is the corner frequency, n is the high-pass filter order, T is the acceleration recording time, and H_0 is the filter amplitude threshold and selected as 0.02. The waveforms and power spectra at the top balcony (U1) and the lower balcony (U3) are depicted in Figs. 7, 8, 9 and 10.

Figure 7 and 9 illustrate the filtered recorded noise time history of the measurement points at the top and bottom balcony, respectively. Figures 8 and 10 show the power spectrum of the measurement points at the top and bottom balconies. The results show that the observed natural frequency of the minaret is about 2 Hz. Post-processing the different minaret measurement points such as U4confirms that the minaret's fundamental period is approximately 0.5 s. The ambient vibration results are correlated with the finite element model results in the following section.

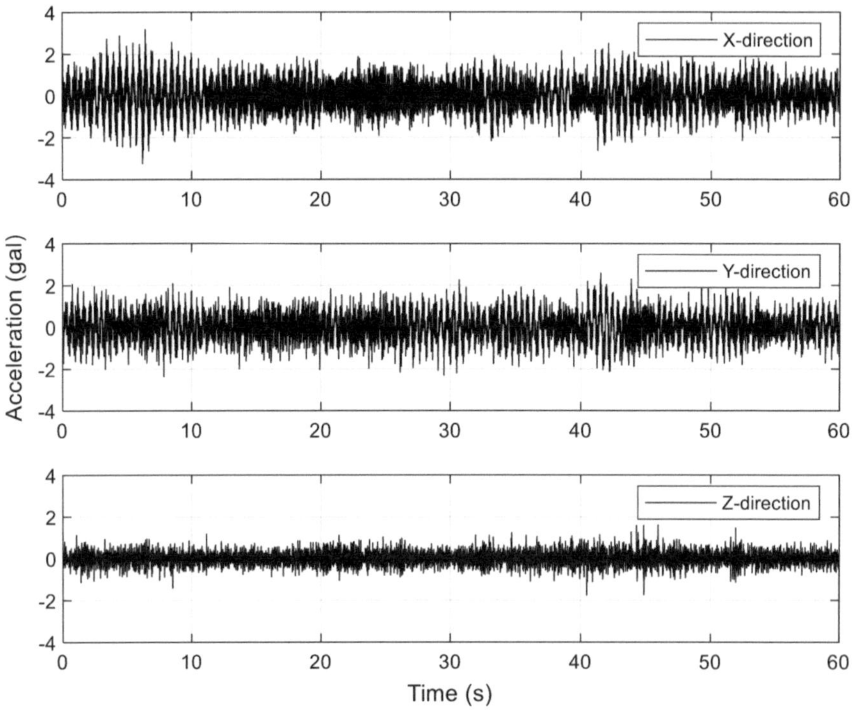

Fig. 7 Ambient noise time series in three directions at the top balcony (U1)

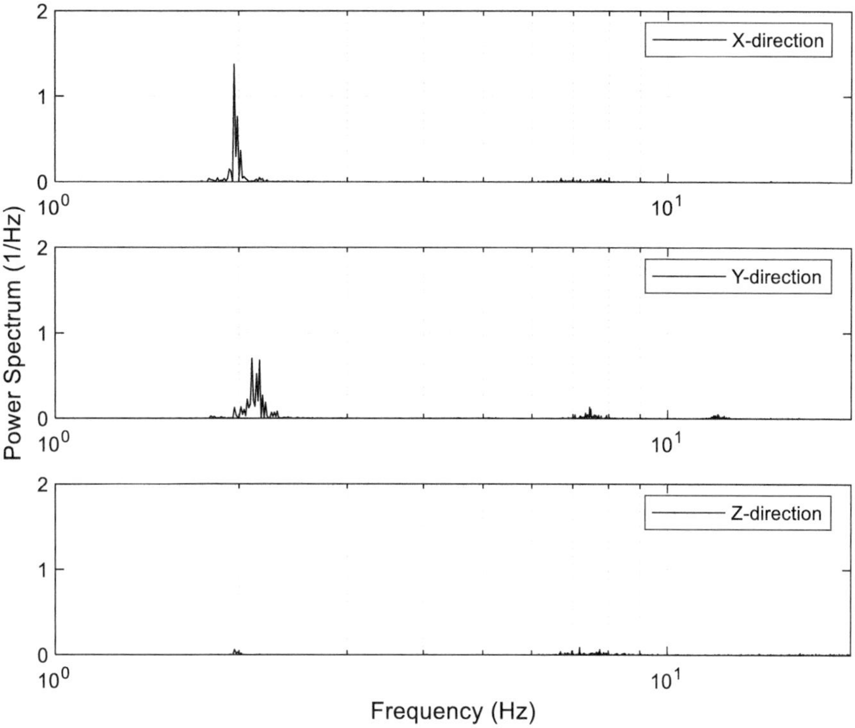

Fig. 8 Power spectra of the time histories in three directions at the top balcony (U1)

4.2 Numerical Model

The finite element (FE) numerical model of the minaret was developed in SAP2000. A linear elastic analysis was conducted for the minaret. The minaret's linear elastic model is a simplification, but it is coherent with preliminary assessments. Limestone minarets show limited dissipation capacity because limestone is typically a brittle material and that minarets are pendulum-like structures. Hence, a brittle failure without a plastic phase is reasonably expected to occur.

Moreover, no cracks were observed during the minaret inspection, suggesting a linear behaviour during the 1992 earthquake. The minaret base and shaft are modelled using eight-node solid elements, while the balcony posts are modelled as frame elements with a squared cross-section of 0.13 × 0.13 m. The balcony walls are modelled as shell elements with a thickness of 0.13 m. All the cross-sectional variations and openings in the minaret were accurately simulated in the numerical model. The spiral staircase was not modelled in this study for conservative modelling since ignoring modelling the spiral stairs insignificantly influenced the modal analysis and reduced the stresses (Sezen et al. 2008). Mesh sensitivity analysis is conducted to the minaret to produce a modal analysis close to the ambient vibration analysis.

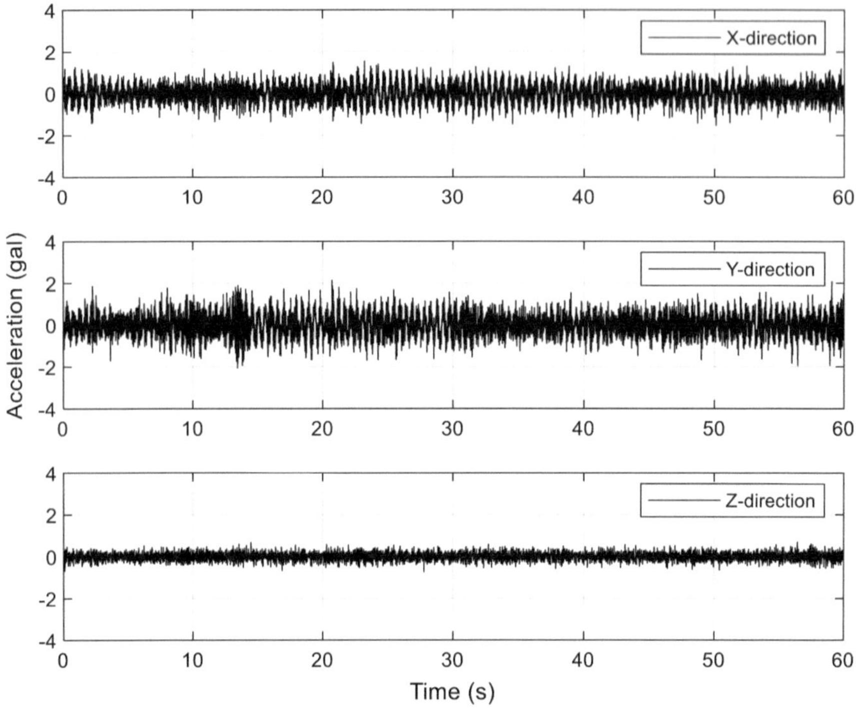

Fig. 9 Ambient noise time series in three directions at the first balcony (U3)

Following the minaret model discretization, the finite element model consists of 80,849 nodes, 64,355 solid elements, 16 frame elements, and 16 area elements. The total weight of the minaret is calculated as 4412.90 kN.

The soil beneath the minaret base is modelled as linear springs with a specified modulus of subgrade reaction. The soil modulus subgrade reaction was calculated using (Vesic 1961) as:

$$K_s = \frac{0.65E_s}{B\left(1-\upsilon_s^2\right)}\sqrt[12]{\frac{E_s B^4}{E_f I_f}} \tag{2}$$

where, K_s is the soil modulus subgrade reaction, E_s is the soil Young's modulus, B is the foundation width, υ_s is soil Poisson ratio, E_f and I_f are Young modulus and moment of inertia of the foundation, respectively. The soil properties beneath the minaret foundation are extracted from (Toni 2012) with a mass density of 1.5 Mg/m^3, shear wave velocity of 180 m/s, Poisson ratio of 0.30, and calculated soil Young modulus of 74.77 MPa. Moreover, the properties of the minaret foundation are considered as B of 3.45 m, E_f of 25.49 GPa and I_f of 11.80 m^4. The calculated soil modulus subgrade reaction is 11.71 N/mm^3. The modal analysis results and the comparison between the ambient vibration results, and the numerical model with

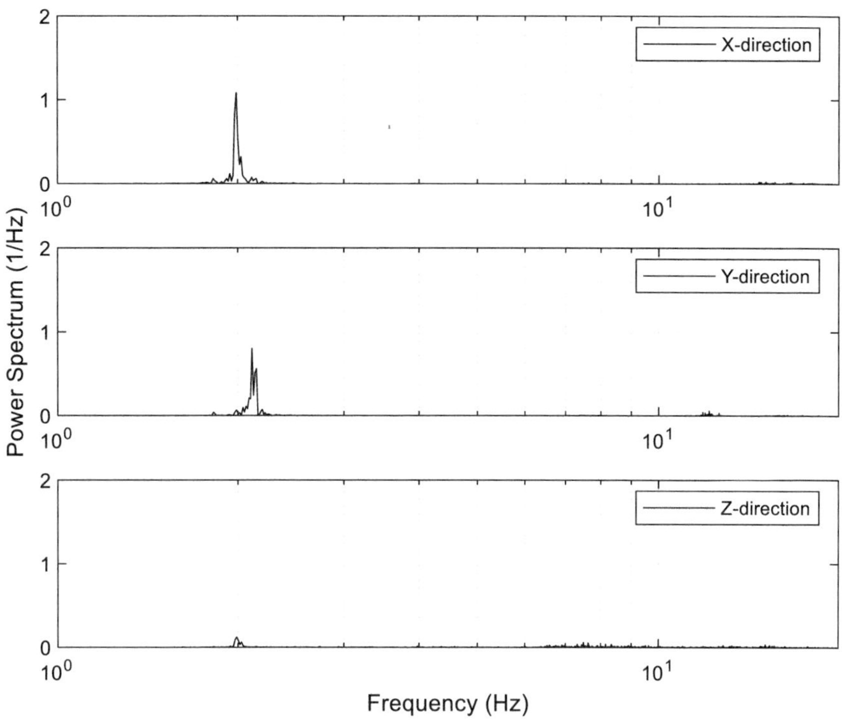

Fig. 10 Power spectra of the time histories in three directions at the first balcony (U3)

a modulus subgrade reaction, are given in Table 1. For comparison, the minaret's modal analysis with considering a fixed base condition is shown in Table 1. The difference between the period obtained from the noise measurements and the FE model is about 12%. The discrepancy could be attributed to the uncertainty in the empirical equation used to estimate the soil subgrade reaction, the absence of accurate measurements for soil properties and the presence of microcracks and local deteriorations that the FE model cannot tackle. Moreover, looking at the power spectrum for the different components of the noise measurements in Figs. 8 and 10, the estimated natural period ranges between 2.0 and 2.2 Hz, close to the FE model of 2.29 Hz. The numerical model of the minaret in SAP2000 is shown in Fig. 11. The results indicate that employing a soil subgrade reaction of $K_s = 11.71$ N/mm^{3}produces a fundamental period close to the ambient vibration. Therefore, all further analyses were investigated considering soil subgrade reaction of $K_s = 11.71$ N/mm^3.

The normalized horizontal displacement extracted from the ambient vibration and the FE model's modal analysis along the minaret height is plotted in Fig. 12. The ambient vibration analysis's normalized displacement represents the minaret's different sensor locations' peak displacements, particularly sensors U1, U3, and U4. Moreover, the modal analysis's normalized displacement stands for the lateral displacement x-direction of Mode 1 and y-direction of Mode 2. Since the governing

Table 1 Modal analysis results of the measured ambient vibration and finite element model with soil subgrade reaction and fixed base conditions

Mode	Frequency (Hz)			Mode description
	Ambient vibration	FFE with fixed base	FE with soil modulus subgrade reaction	
1	2.00	5.59	2.29	(x direction)
2	2.28	5.72	2.42	(y direction)
3	–	–	2.80	(z direction)
4	15.61	21.69	16.87	(x direction)
5	16.78	40.07	17.11	(y direction)

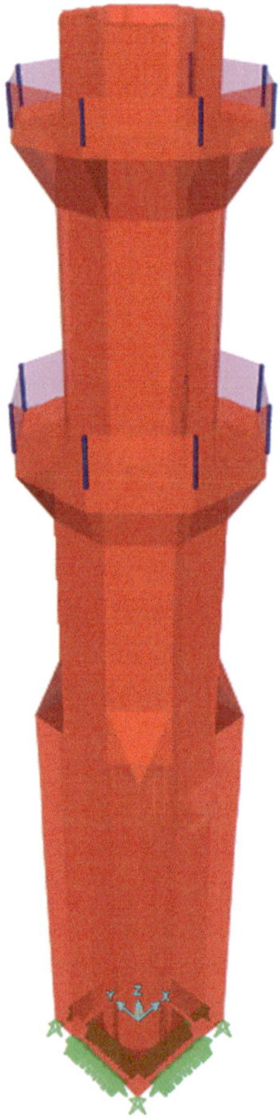

Fig. 11 Detailed finite element model of the minaret. Red colour represents the solid and shell elements, blue colour represents frame elements, and green colour represents the soil springs

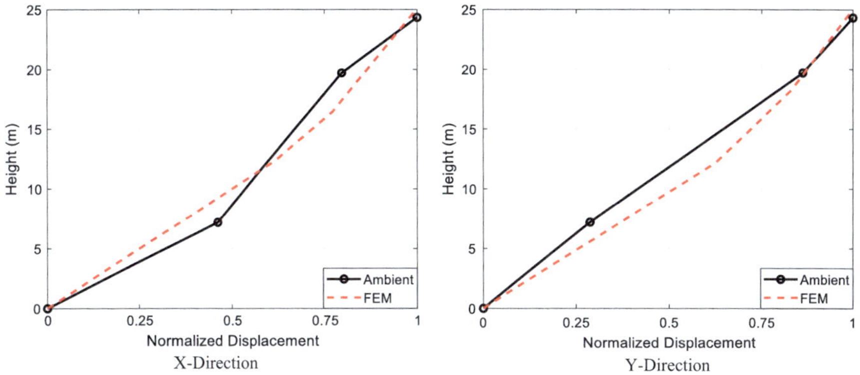

Fig. 12 The normalized horizontal displacement of the FE model modal analysis and the ambient vibration measurement points in X-Direction (mode 1) and Y-Direction (mode 2). The sensors' locations are denoted with circles

mode description of Mode 1 is the motion in the x-direction, while Mode 2 is in the y-direction. Good agreement between the normalized displacement of the FE model and ambient measurements is observed at the measurement points. Relatively less agreement took place at transition zones above the baseand the balcony slabs due to lack of measurements at these points due to the ismpracticality of installing sensors at these locations without damaging the minaret's body.

Moreover, the FE model results gave a reasonably good estimate of the minaret lateral deformation. It adapts the precise geometry, sections, openings, and stiffness variation along with the minaret height, while the ambient vibration plot represents the linear piecewise connection between the limited available three measurement points. Also, the disagreement between the results may be attributed to the numerical model deficiency in simulating the minor deterioration in the actual limestone bricks or fill material. Discarding the staircase modelling has a negligible effect on the minaret global response (Sezen et al. 2008). The previous results exhibit the field measurements' significance for such structures on tuning and evaluating the FE model's response. The FE model's tuning represents the model's mesh sensitivity analysis to determine the model's fine discretization to produce its dynamic characteristics.

The limestone selected for the numerical model has a compressive and tensile strength of −27.47 and 5.40 MPa, respectively, as previously mentioned. The principal positive (tensile) and negative (compressive) stresses are investigated, noting that the principal stresses are oriented by definition so that the associated shearing stress is zero. Using the principal stresses as a failure criterion is allowed and restricted to brittle materials, such as limestone. The peak principal stresses analysis due to the gravity loading cases is illustrated in Fig. 13. The results show that the peak principal compressive stresses are -1.02 MPa, and the peak tensile stresses are 0.23 MPa, for the three orthogonal principal stress tensor σ_{11}, σ_{22}, andσ_{33}. In detail, the peak principal compressive and tensile stresses are −0.21, −0.21, −1.02, and 0.23, 0.18,

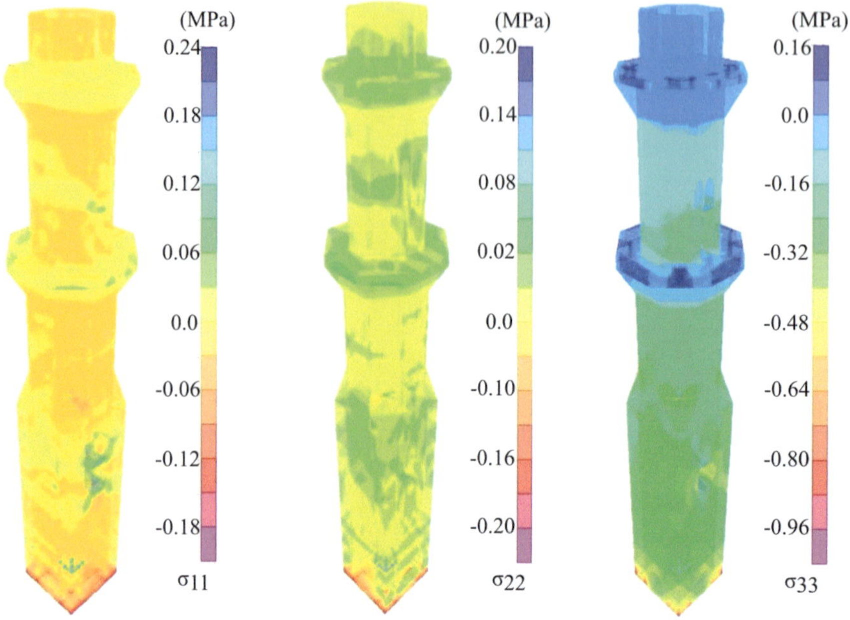

Fig. 13 Principal stresses on the minaret body due to gravity loading

0.13 MPa for stress σ_{11}, σ_{22}, and σ_{33}, respectively. The principal stresses are acceptable within the minaret construction's material stress limits defined by Imam (2001), and they indicate that the minaret can withstand its weight with no cracking.

4.3 Earthquake Response Spectrum Analysis

In this section, the linear dynamic response spectrum analysis is conducted to draw the minaret body's maximum stresses, considering the gravity load due to the minaret's self-weight. Two response spectrum analysis cases are investigated: the 1992 Cairo earthquake and the C-MCSI response spectra.

4.3.1 1992 Cairo Earthquake Spectrum

The 1992 Cairo earthquake case is selected to determine the numerical model's accuracy in detecting the minaret response since the minaret survived the Cairo earthquake in reality, with neither damage nor cracks observed after the earthquake shaking. Moreover, the 1992 Cairo earthquake's acceleration response spectrum obtained from the MCSI analysis is investigated since there are no known ground motion records for the event. Therefore, the 50th percentile response spectrum of the 1992 Cairo earthquake is used to conduct the response spectrum analysis for the minaret model in a bidirectional horizontal direction. The principal stress results are plotted in Fig. 14. The colour limit was set to the allowable stress range, and thedark blue colour represents regions exceeding the allowable tensile stress. The results show that the peak compressive stresses are always under the allowable compressive strength with peak values of −0.21, −0.21, −1.00 MPa for σ_{11}, σ_{22}, andσ_{33}, respectively. On the other hand, the tensile stresses are high, but not enough to exceed the tensile strength, as the peak recorded principle tensile stress tensor was 2.1(0.38σ_t), 1.3(0.24σ_t), 5.1(0.94σ_t) MPa in the minaret body for σ_{11}, σ_{22}, and σ_{33}, respectively, of the material tensile strength *(σ_t)*. As expected, the numerical model introduced an acceptable behaviour of the minaret under the Cairo earthquake response spectrum through producing undamaging peak tensile and compressive stresses within the minaret, especially at the transition zone or near the openings.

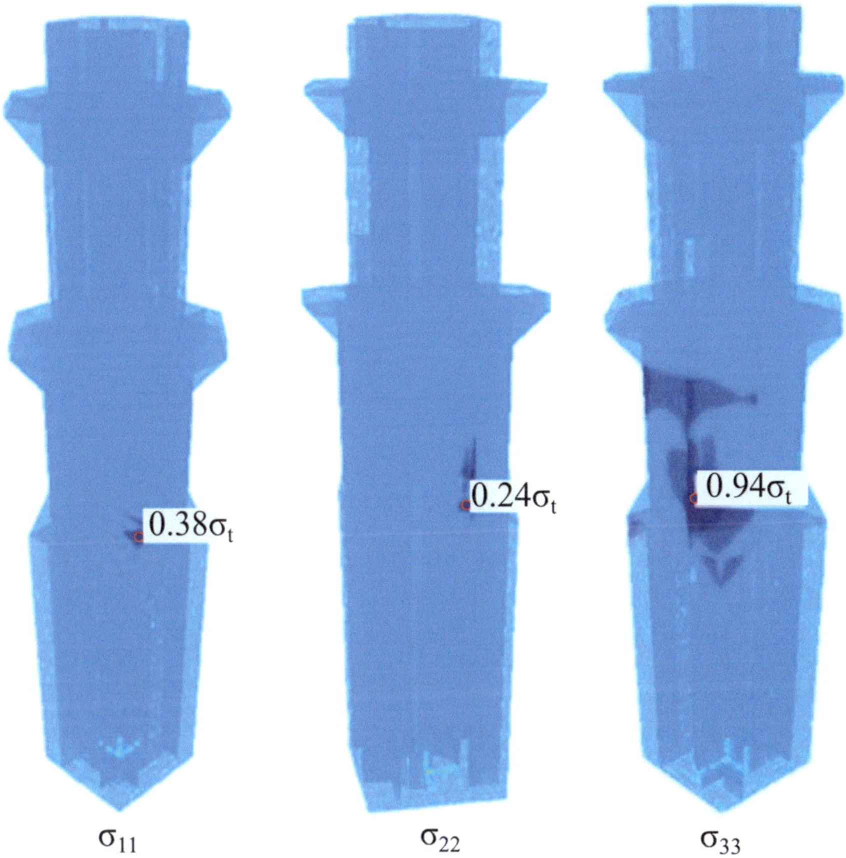

Fig. 14 Peak tensile principal stresses as a factor of the material tensile strength (σ_t) on the minaret body under the 1992 Cairo earthquake response spectrum

4.3.2 C-MCSI Spectrum

The calculated C-MCSI at the 50th percentile spectrum is applied at the minaret base in both horizontal directions. The principal stress results are depicted in Fig. 15, and they show that the minaret would suffer significant tensile stresses exceeding the limestone tensile strength, which denotes enormous cracking and even crushing in the minaret, particularly at the base and at the transition zone right above the base. The peak tensile and compressive stresses are $4.6(0.85\sigma_t)$, $4.9(0.91\sigma_t)$, $8.9(1.64\sigma_t)$ MPa and −0.20, −0.19, and −0.93 MPa. This level of tensile stresses predicts severe damage or even the minaret's collapse under the C-MCSI response spectrum. The previous results demonstrated that the tensile stress concentrations took place at the end of the transition zone between the square base and the hexagon shaft, where the reduction of the minaret cross-sectional area occurs.

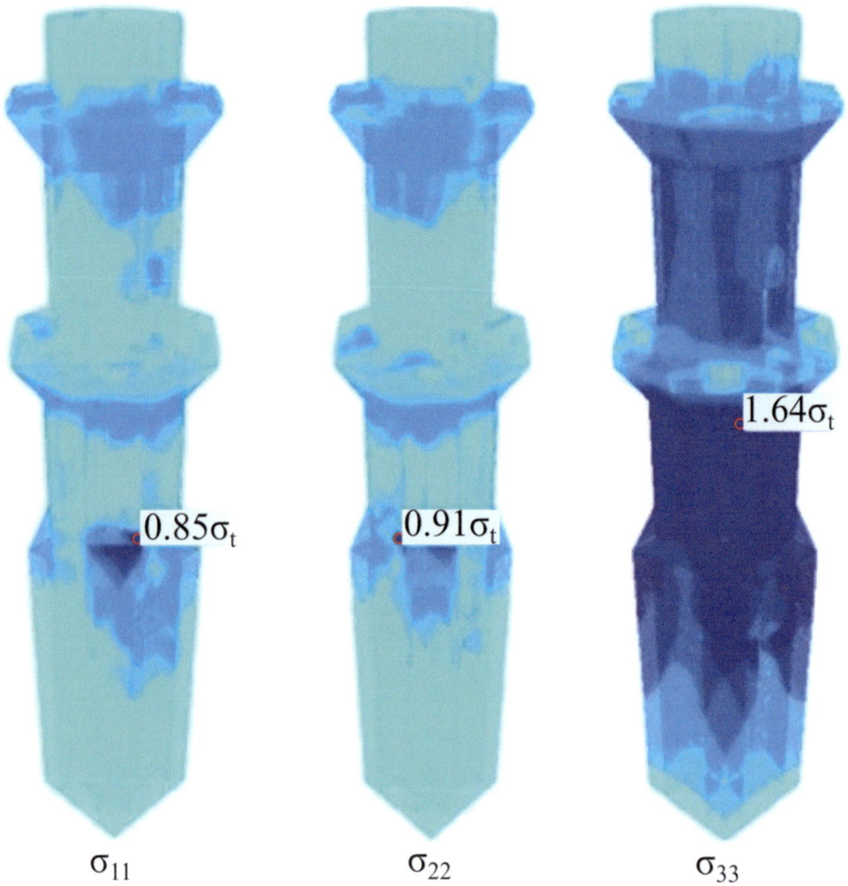

Fig. 15 Peak tensile principal stresses as a factor of the material tensile strength (σ_t) on the minaret body under the C-MCSI response spectrum

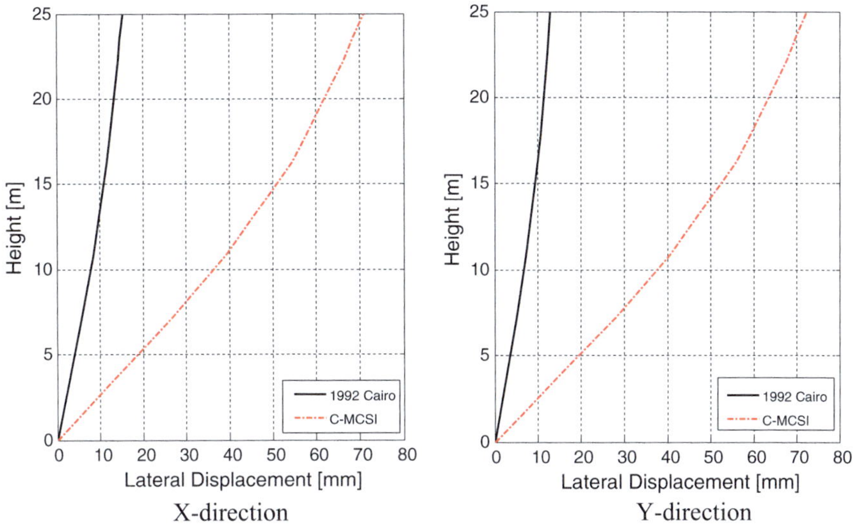

Fig. 16 Maximum horizontal displacement in X-direction (left panel) and Y-direction (right panel) under the 1992 Cairo and C-MCSI response spectra

The peak horizontal displacement along the minaret height in both horizontal directions under the 1992 Cairo and C-MCSI response spectra action is illustrated in Fig. 16. The results show that the C-MCSI analysis case developed higher lateral displacement than the 1992 Cairo case in both directions. Furthermore, the peak lateral displacement under the C-MCSI spectrum is about 70.8 and 72.8 mm in x and y-direction, respectively, compared with 15.3 and 13.0 mm under the 1992 Cairo response spectrum in x and y-direction.

4.4 Time History Analysis

The seven pairs of acceleration time histories selected in Sect. 5 are considered for conducting the minaret's dynamic time history analysis. The peak lateral displacement at the transition zone above the base and the top of the minaret in x- and y-directions are illustrated in Table 2; their fluctuation between the two orthogonal directions is mainly due to the difference between the ground motions components (NS or EW) assigned for each direction. However, it is worth mentioning that a slight difference exists in the peak displacement between the x- and y-direction due to the geometry irregularity (i.e., openings, imperfect symmetry, etc.) as seen in the C-MCSI response spectrum results in Table 2, although an identical response spectrum is applied in the x and y-direction at the minaret base.

The floor response spectra for 5% damping have been computed at the top of the minaret using time history analysis as shown in Fig. 17, where the mean of the

Table 2 Maximum horizontal displacement at the transition zone and top of the minaret subject to time history analysis

Measured point	Maximum horizontal displacement [mm]			
	Top		Transition zone	
Direction	X-dir	Y-dir	X-dir	Y-dir
Seven ground motion pairs	51.41	39.80	31.57	24.76
	63.49	54.65	39.04	33.93
	59.72	54.91	36.82	34.31
	60.20	44.37	37.17	27.75
	64.10	49.53	39.46	30.85
	59.07	31.14	36.35	19.38
	57.51	52.00	35.47	32.39
Mean	59.36	46.63	36.55	29.05
C-MCSI RS	70.80	72.80	43.70	45.20

seven response spectra is also depicted for both orthogonal directions. The spectral acceleration at the top of the minaret (base of the mabkhara or cap) is also investigated. The minaret's top describes a specific shape of the minaret finial, and it is always considered the weakest part of the minaret, i.e., more likely to collapse during ground shaking. The minaret cap (currently are not in place) already experienced damage and complete failure earlier, as noticed comparing Fig. 1b and c (Mayer and Speiser 2007). Thus, the time history analysis could be needed in the future restoration and put in place of the mabkhara. Substantial amplification occurs in the response spectra with peak ground acceleration (PGA) of 1.37 and 1.28 g, in x- and y-direction, respectively. While the input motions have an average response spectrum with a PGA of 0.4 g, as shown previously in Fig. 5.

5 Conclusions

Four earthquake scenarios consisting of one recent and two historical earthquakes are adopted to model the ground motion and compute the minaret site's seismic input using the NDSHA approach. The 1992 and 950 scenarios were used as seismic sources to model NS cross-sections under the minaret, while the 857 earthquake scenario was used as the earthquake scenario for the modelling of the EW cross-section.

We selected the seismic input (response spectra and time histories) from the database computed for the historical Cairo by Hassan et al. (2020) for the evaluation of the dynamic performance of the minaret of the Madrasa the Princess Tatar al-Higaziya, which will help in proposing a seismic conservation strategy for this valuable structure. We provide the MCSI and C-MCSI acceleration response spectra

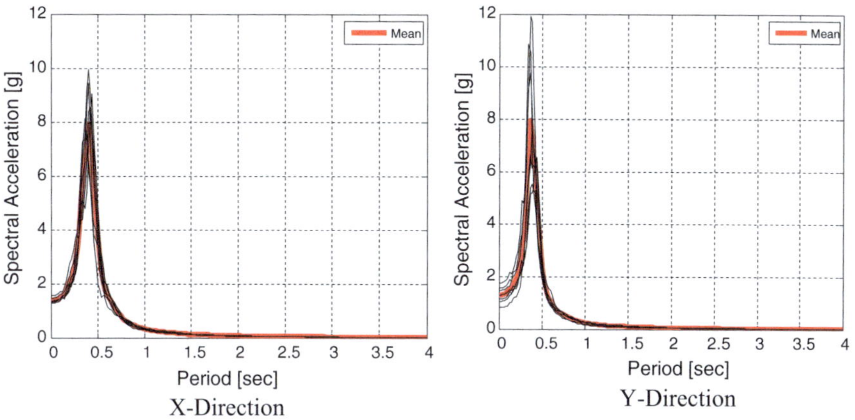

Fig. 17 Acceleration response spectra at the top of the minaret for 5% damping, in X-direction (left panel) and Y-direction (right panel)

at the minaret site to assess the minaret's dynamic behaviour by joining both seismological and engineering knowledge to understand the performance of this monument under earthquake action. The results could help recommend improved mitigation measures for the minaret while also representing a step towards the risk reduction and management of this historic structure.

A detailed numerical model for the minaret was established, accompanied by seismic sensors and ambient vibrations to evaluate the numerical model. A careful visual inspection was performed to summarize the minaret body's construction materials and elements and investigate the minaret cracking status. Two seismic analysis types were conducted for the minaret numerical model, namely the response spectrum analysis and the time history analysis. The response spectrum analysis was selected to replicate the 1992 Cairo earthquake excitation scenario since no earthquake records were preserved near the minaret's site. The numerical model adequately captured the minaret response without indicating any signs of cracking or damage, which matches the minaret's actual behaviour during the 1992 earthquake. Furthermore, the response spectrum analysis was also applied for the proposed C-MCSI$_{SS}$ spectrum; the computed C-MCSI$_{SS}$ response spectrum and time history analyses predict severe damage to the minaret. The analysis under the C-MCSISS predicts significant lateral displacements at the top of the minaret and excessive tensile stress concentration, particularly at the geometric transition zone between the squared base and the hexagon shaft.

Itis worth mentioning that the small vertical separation gap between the minaret and the wall of the Madrasa may pose a pounding potential between the adjoining structures due to the minaret's horizontal vibration. This threat will be covered in the future work of the Madrasa.

Finally, since the minaret is expected to suffer severe damage against the anticipated earthquake shaking scenario, a vital protection plan is recommended to avoid any future damage or collapse to the structure. Stitching the walls with pre-stressed rebar or reinforcement of the walls' inner side with incorporated steel are pervasive ways for retrofitting and protecting historical monuments and structures. Besides, skins of reinforced concrete coating or fibre-reinforced plastic (FRP) on the walls' outer side may enhance the walls' tensile strength, hence improving the minaret's performance against the anticipated intense earthquake scenario.

Acknowledgements This work was partially supported by the Egyptian Ministry of Higher Education (Cultural Affairs and Missions Sector, Cairo) and by the (MAE-MHE) bilateral Egyptian-Italian project "Advanced seismic hazard assessment in the Nile Delta, including the site effects from distant earthquakes." under the Science and Technology Development Fund (STDF), Egypt, Grant No 25991 and 25553. Also, we would like to thank NRIAG for providing the noise sensors and the field logistics. Special thanks to the Ministry of Antiquities for facilitating the work and their fruitful cooperation. The authors are debited to Prof. Abbas M. Abbas, Franco Vaccari, and Dr. Nagy Naguib for their valuable contribution. The first author wants to express his thanks and appreciation for Prof. Giuliano Panza for the continuing support and inspiration.

References

Baker JW (2011) Conditional mean spectrum: tool for ground motion selection. J Struct Eng 137:322–331

Baker JW, Cornell AC (2006) Spectral shape, epsilon and record selection. Earthquake Eng Struct Dynam 35:1077–1095. https://doi.org/10.1002/eqe.571

ECP-201 (2011) Egyptian Code of Practice No. 201 for calculating loads and forces in structural work and masonry. National Research Center for Housing and Building, Ministry Housing, Utilities and Urban Planning, Cairo

El-Attar AG, Saleh AM, Zaghw AH (2005) Conservation of a slender historical Mamluk-style minaret by passive control techniques. Struct Control Health Monit 12(2):157–177

ElGabry M, Hassan HM (2021) Updated seismic input for next generation of the Egyptian building code. In: Sustainable issues in infrastructure engineering. Springer, Cham, pp 55–79

Fasan M (2017) Advanced seismological and engineering analysis for structural seismic design. PhD thesis, Trieste University, Italy

Gaber H, El-Hadidy M, Badawy A (2018) Up-to-date probabilistic earthquake hazard maps for Egypt. Pure Appl Geophys 1–28

Gorshkov AI, Hassan HM, Novikova OV (2019) Seismogenic nodes (M≥5.0) in Northeast Egypt and implications for seismic hazard assessment. Pure Appl Geophys 176(2):593–610

Hassan HM, Panza GF, Romanelli F, ElGabry MN (2017a) Insight on seismic hazard studies for Egypt. Eng Geol 220:99–109

Hassan HM, Romanelli F, Panza GF, ElGabry MN, Magrin A (2017b) Update and sensitivity analysis of the neo-deterministic seismic hazard assessment for Egypt. Eng Geol 218:77–89

Hassan HM, Fasan M, Sayed MA, Romanelli F, ElGabry MN, Vaccari F, Hamed A (2020) Site-specific ground motion modeling for a historical Cairo site as a step towards computation of seismic input at cultural heritage sites. Eng Geol 268:105524

Imam HF (2001) Technical study on the structural condition of the Tatar al-Higaziya Madrasa and minaret (monument#36). Historic Cairo Project, Stage 3, Group 3. Cairo, Egypt

Mayer W, Speiser V (2007) A future for the past: restorations in Islamic Cairo; 1973–2004. German Archaeological Institute, Cairo

Rinaldin G, Fasan M, Noé S, Amadio C (2019) The influence of earthquake vertical component on the seismicresponse of masonry structures. Eng Struct 185:184–193

Sawires R, Peláez JA, Fat-Helbary RE, Ibrahim HA (2016) Updated probabilistic seismic-hazard values for Egypt. Bull Seismol Soc Am 106(4):1788–1801
Sayed MA, Go S, Cho SG, Kim D (2015) Seismic responses of base-isolated nuclear power plant considering spatially varying ground motions. Struct Eng Mech 54(1):169–188. https://doi.org/10.12989/sem.2015.54.1.169
Sezen H, Acar R, Dogangun A, Livaoglu R (2008) Dynamic analysis and seismic performance of reinforced concrete minarets. Eng Struct 30(8):2253–2264
Sykora D, Look D, Croci G, Karaesmen E (1993) Reconnaissance report of damage to historic monuments in Cairo, Egypt following the October 12, 1992 Dahshur earthquake. Army Engineering Waterways Experiment Station Vicksburg MS Geotechnical Lab
Toni M (2012) Site response ad seismic hazard assessment for the southern part of Cairo city, Egypt. Unpublished PhD thesis, Faculty of Science, Assiut University, p 151
Vesic AB (1961) Beams on elastic subgrade and Winkler's hypothesis. In: Proc. 5 th. Int. Conf. on Soil Mech. Found. Engrg. Paris, pp 845–50
Williams C (2008) Islamic monuments in Cairo: the practical guide. American Univ in Cairo Press

Remote Sensing

Notes on Artificial Intelligence and Big Earth Observation Data for the Study of the Human Past

Rosa Lasaponara, Xinyuan Wang, and Nicola Masini

Abstract In the recent decades, the availability of Earth Observation technologies (from satellite, aerial and ground) for the study and preservation of the human past is stepping into a golden age. The currently available data, tools and services (including open cloud resources to process big satellite data) opened up a new frontier of possibilities and applications which at the same time also pose several challenges to be faced. All of the diverse aspects ranging from the research and innovation to the application issues, namely from the science to services, must to be tackled by the scientific community in conjunction with the "end uses needs" to ensure an effective and reliable applicability. The use of space technologies, big data, artificial intelligence (AI) for the study of the human past highlights the multi-trans and inter-disciplinary scientific and technical aspects not only for the novel concepts and approaches proposed, but also for the development between and across diverse disciplines In this paper, the state of the art in the field of EO Big data and artificial intelligence for the study of human past is brief summarized.

Keywords Cultural heritage · Archaeology from space · Copernicus big data · Sentinel data · Optical images · Radar data · Environmental risk · Discovering · Artificial intelligence

R. Lasaponara
National Research Council of Italy, CNR-IMAA, C.da S. Loja 85050, Tito Scalo, PZ, Italy
e-mail: rosa.lasaponara@imaa.cnr.it

X. Wang
Chinese Academy of Sciences -CAS-RADI, No. 9 Dengzhuang South Road, Beijing 100094, China
e-mail: wangxy@radi.ac.cn

N. Masini (✉)
National Research Council of Italy, CNR-ISPC, C.da S. Loja 85050, Tito ScaloItaly, PZ, Italy
e-mail: nicola.masini@cnr.it

G. M. El-Qady and C. Margottini (eds.), *Sustainable Conservation of UNESCO and Other Heritage Sites Through Proactive Geosciences*, Springer Geology,
https://doi.org/10.1007/978-3-031-13810-2_29

1 Introduction

In the recent decades, the availability of Earth Observation technologies (from satellite, aerial and ground) for Cultural Heritage (CH) is stepping into a golden age characterized by an increasing growth of both classical and emerging multidisciplinary methodologies, addressed to the study and documentation of the human past (Lasaponara and Masini.2008; Lasaponara and Masini 2011; Luo et al. 2019; Opitz and Herrmann 2018). The main critical, challenging aspect of the use of EO in archaeology is a lack of correspondence between the great amount of data and information from diverse technologies (satellite, aerial, ground RS) and effective methods to extract information linked to the study of the human past (Lasaponara and Masini 2012). So that, today, archaeologists have the possibility and opportunity to use an ensemble of diverse technologies, which however require, as preliminary steps, the setting up of ad hoc data processing and integration methodologies for archaeological investigations, analysis and interpretation.. All of the diverse aspects ranging from the research and innovation to the application issues, namely from the science to services, must to be tackled by the scientific community in conjunction with the "end uses needs" to ensure an effective and reliable applicability also exploiting the EO data, information and products today also available free of charge, as in the case of Copernicus initiative (https://www.copernicus.eu/en). In particular, since 2014 satellite Copernicus data provided from the diverse Sentinel platforms can be free downloaded from the ESA web site (Copernicus Open Data Hub) along with open software tools as SNAP developed for the data processing (https://step.esa.int/main/download/snap-download/).

The diverse Sentinel platforms provide data (see Fig. 1) for multipurpose applications mainly focusing security and monitoring issues. Data from active and passive Sentinel sensors can be usefully integrated each other's and suitably support new operational applications which at the same time pose several critical issues, as those linked with the processing and interpretation issues, for transforming data into useful information. Nevertheless, it must be considered that all the Sentinel platforms provide satellite big data which require strong hardware infrastructures or cloud facilities [today also available as open and free tools as Google Earth Engine© (GGE)]. To cope with the need of Big data storage, processing and management, in 2017, EC funded the set up of cloud and edge computing to make available an open European cloud to deploy applications and data processing centralized with no need to transfer and duplicate petabytes of data. So that EC and ESA supported the developments of the so-called Data and Information Access Service (DIAS), in service since 2018 (see Fig. 2).

DIAS platforms provide the "foundations" on which build applications exploiting Copernicus data, along with other tools developed ad hoc to enable a data manipulation easier also for user with limited technical expertise of the earth observation

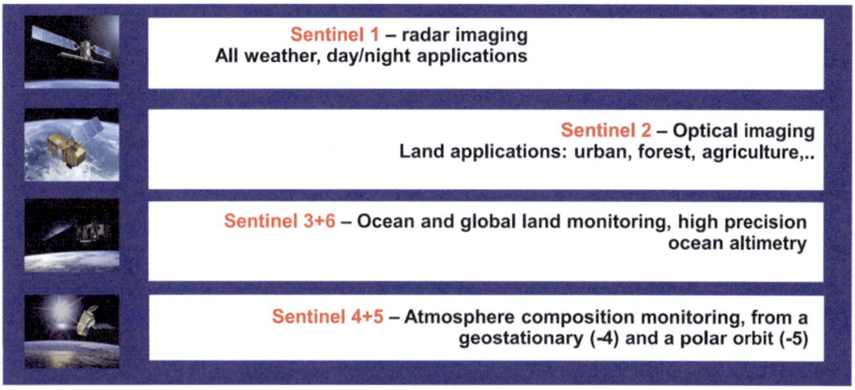

Fig. 1 Overview of Sentinel sensors, data typologies, and the principal applications

domain Sentinel Hub initiative (https://www.sentinel-hub.com/), Data Cube Facility Service (Fig. 3) which provides fast access to a considerable amount of EO information in order to establish a "bridge from Space to Applications". (https://eo4society.esa.int/2019/05/21/european-data-cube-facility-service-an-eo-resource-factory/).

Big data are generally characterized using the so-called Vs to capture their complex nature, that is generally not an absolute "categorization" but strongly linked and depending on the a given application:

- Volume—amount of data
- Velocity—generation, or analyzed
- Variety—differences in data sources
- Veracity—uncertainty of data
- Validity—the suitability
- Volatility—temporal validity
- Value—usefulness of the information
- Visualization—displaying and showcasing
- Vulnerability—security and privacy
- Variability—the changing meaning of data

On the other side, the main technological challenges related to Big data are:

- Heterogeneity—differences in structure
- Uncertainty—data reliability
- Scalability—sizing the workflow and infrastructure Timeliness—real-time requirements
- Fault tolerance—sensitivity to errors
- Data security—privacy issues, data leaks
- Visualization—displaying of information

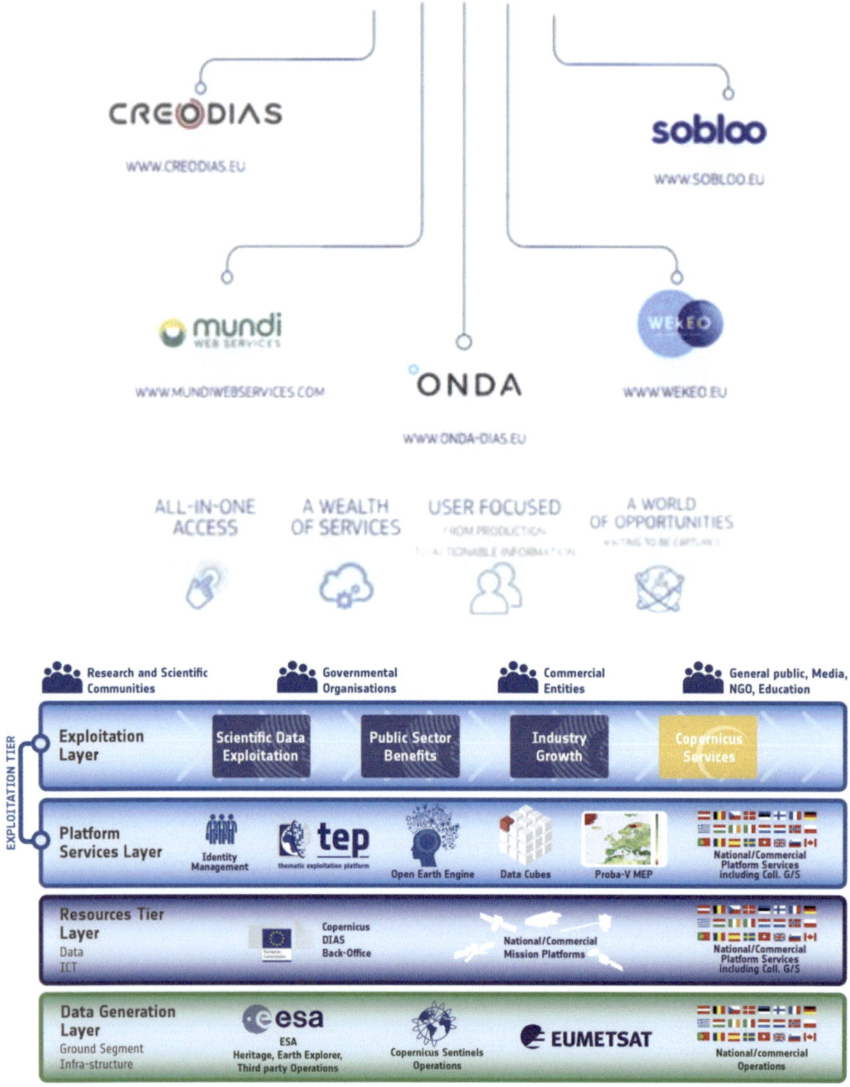

Fig. 2 Up: Data and Information Access Services. They are five cloud-based platforms funded by EU to provide centralised access to Copernicus data and information, as well as to processing tools (https://www.copernicus.eu/en/access-data/dias). Bottom: the EO platforms eco-system (https://eo4society.esa.int/2019/05/21/european-data-cube-facility-service-an-eo-resource-factory/

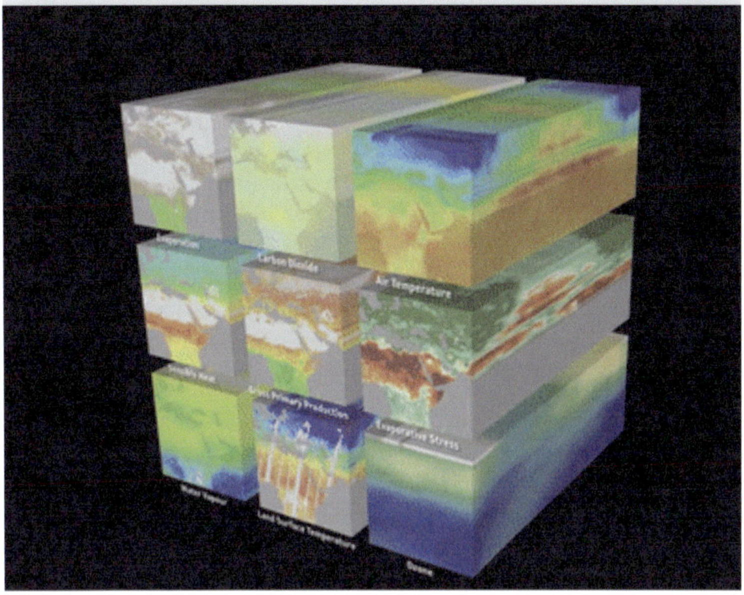

Fig. 3 European Data Cube Facility service ESA courtesy

2 State of the Art: BIG EO Data for the Study of the Human Past

In the last two decades the use of EO in CH has been strongly increasing thanks to the technological improvements of the sensors ever more useful for the study of the human past and ancient landscapes for a large variety of environs including desert, tropical and Mediterranean ecosystems. In particular, the latest generation of satellite sensors opened up new frontiers and possibilities offering ever-closer and more comprehensive look at the earth's surface; providing a detailed imaging today also available at 0.30 cm from both satellite optical as WorldView-3 satellite and SAR data (COSMO-SkyMed and TerraSAR-X).

The recognition of archaeological features from above is based on the use of the so-called archaeological proxy indicators, adopted for landscape archaeology and for detecting buried archaeological structures through their visual interpretation (Lasaponara and Masini 2012) (Fig. 5). The most common archaeological proxy indicators are generally known as crop, soil, shadow, and damp marks and are caused by the presence of buried remains and traces of ancient environs still fossilized in the modern landscape. These features induce spatial anomalies (in vegetation growth and/or status, surface moisture content, micro-reliefs) that are generally not visible in situ but only evident from above even if are very subtle not permanent signals (Masini et al. 2017). In the last decades, significant improvements have been obtained in the identification of archaeological proxy indicators from VHR satellite optical and SAR data (Kalayci et al. 2019; Masini and Lasaponara 2017; Jiang et al. 2016). Starting from VHR satellite based studies and applications (see, for example Luo et al. 2019), new applications and developments are expected particularly from open data such as satellite Sentinel-1 (S-1) and Sentinel-2 (S-2), whose use poses intriguing scientific challenges, as the following:

(i) Is the spatial resolution of S-1 and S-2 a critical limitation being that, except for a few examples, the majority of archaeological investigations have been performed using very high-resolution satellite data?
(ii) Is it possible to capture the spectral signatures of archaeological proxy indicators as soil and crop marks considering that the identification of archaeological features is a very complex issue: (a) due to their subtle nature, and, moreover, (b) are not permanent signals but only visible in specific observational conditions?

Recently, a few studies assessed the potentiality of S-1 and S-2 data in cultural heritage domain, including archaeological heritage monitoring, landscape archaeology, and the detection of archaeological proxy indicators in diverse regions of Europe, Asia and Africa, and opportunities to use them for the automated detection of archaeological features (Agapiou et al. 2014; 2019; Elfadaly et al. 2020; Khalaf and Insoll 2019; Orengo et al. 2020; Tapete and Cigna 2017; Zanni and De Rosa 2019).

Undoubtedly, today the availability of big and open satellite data offers big opportunities and big challenges, relevant for multi- and inter-disciplinary studies as "space archaeology" (Fig. 4) including the site discovery, monitoring and preservation, today considered a priority at European and international level with important cultural, social, and economic repercussions.

This is clearly highlighted by the increasing number of papers (Kvamme 2013; Schneider et al. 2015; Orengo et al. 2020; Trier et al. 2018) and dedicated conferences and workshops on archaeological object detection in remote sensing data and from the fact that in 2017, the European Commission (EC) organized a specific workshop "to assess the potential of Copernicus in support of cultural heritage preservation and management".

As an example, Abate et al. (2020) and Elfadaly et al. (2020) focused on the use of satellite Sentinel-2 and Sentinel-1 data, respectively for assessing their capability in the identification of archaeological buried remains in the Mediterannean ecosystems of the "Tavoliere delle Puglie" (Foggia, Italy). The investigations were performed using multi-temporal Sentinel-2 and Sentinel-1 data. In detail, Abate et al. (2020) used a multitemporal analysis of 2016–2018 Sentinel-2 data set to capture the spectral signatures of soil and crop marks and characterize their temporal behavior using Time Series Analysis and Spectral Un-mixing. Tasseled Cap Transformation and Principal Component Analysis were also adopted to enhance archaeological features. Results from investigations were compared with independent data sources and enabled us to (i) characterize the spectral signatures of soil and crop marks, (ii) assess the performance of the diverse spectral channels and indices, and (iii) identify the best period of the year to capture the archaeological proxy indicators.

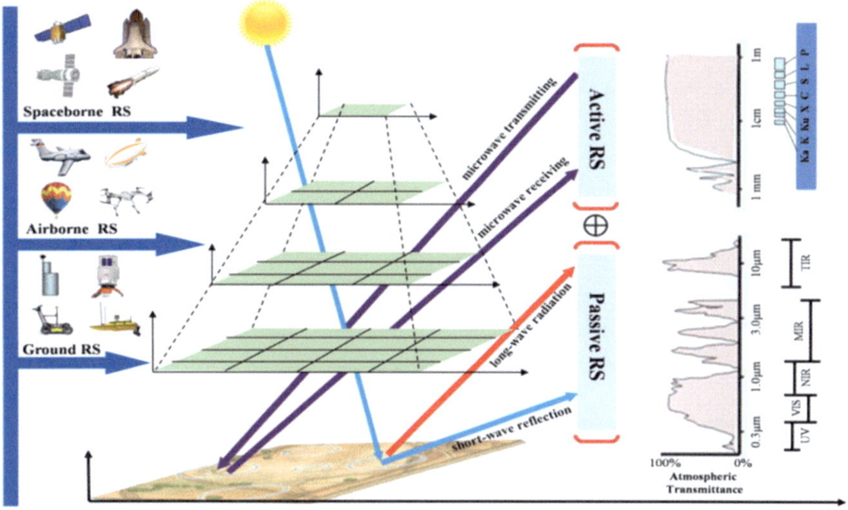

Fig. 4 Multiscale and multi sensor Remote Sensing based approach to archaeology and cultural heritage management (from Lei Luo et al. 2019)

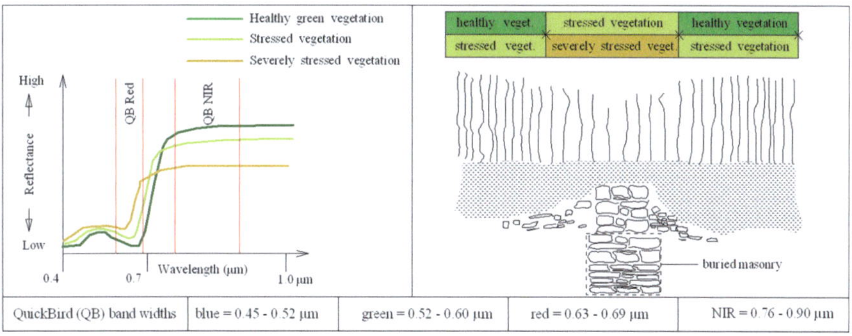

Fig. 5 Example of archaeological proxy indicators: Crop marks

Additional very important results of these investigations were (i) the discovery of unknown archaeological areas and (ii) the setup of a database of archaeological features devised ad hoc to characterize and categorize the diverse typologies of archaeological remains detected using Sentinel-2 Data.

Elfadaly et al. 2020 assessed the capability of SAR Sentinel 1 in the imaging and detection of palaeo-landscape features in the same area as in Abate et al. 2020. The results from the Sentinel 1 (S-1) data were successfully compared with independent data sets, and the comparison clearly showed an excellent match between the S-1 based outputs and ancient anthropogenic transformations and landscape features (see Fig. 7).

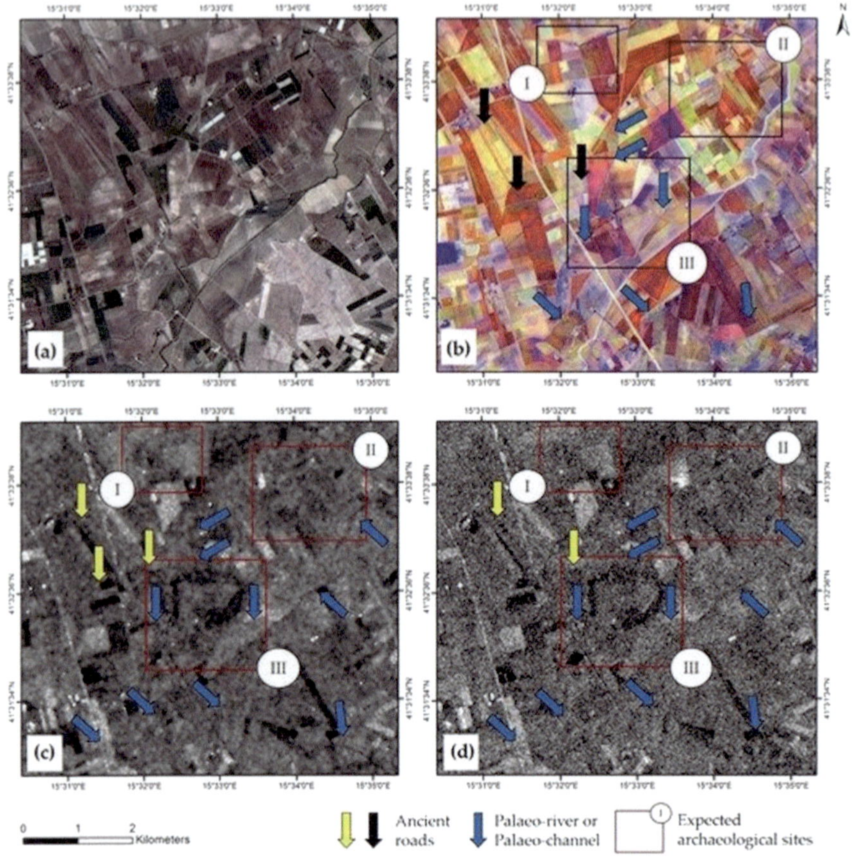

Fig. 6 Satellite based study of palaeoenvironment: identification of palaeorivers in Norther Apulia (South Italy) using Sentinel 1 and 2

Agapiou et al. (2019) evaluated the potential of the synergic use of multitemporal Sentinel-1 and Sentinel-2 dataset to image and semi-automatically extract archaeological features of Neolithic landscape in Thessalian plain.

Khalaf and Insoll (2019) explored open source satellite imagery, including Sentinel-2 for the monitoring and protection of Islamic archaeological landscapes in Ethiopia.

Finally, Zanni and De Rosa (2019) identified and reconstructed the ancient viability between the Roman cities of Aquileia (Aquileia, Italy) and Singidunum (Serbia), by means vegetation indices derived from Sentinel-2 data.

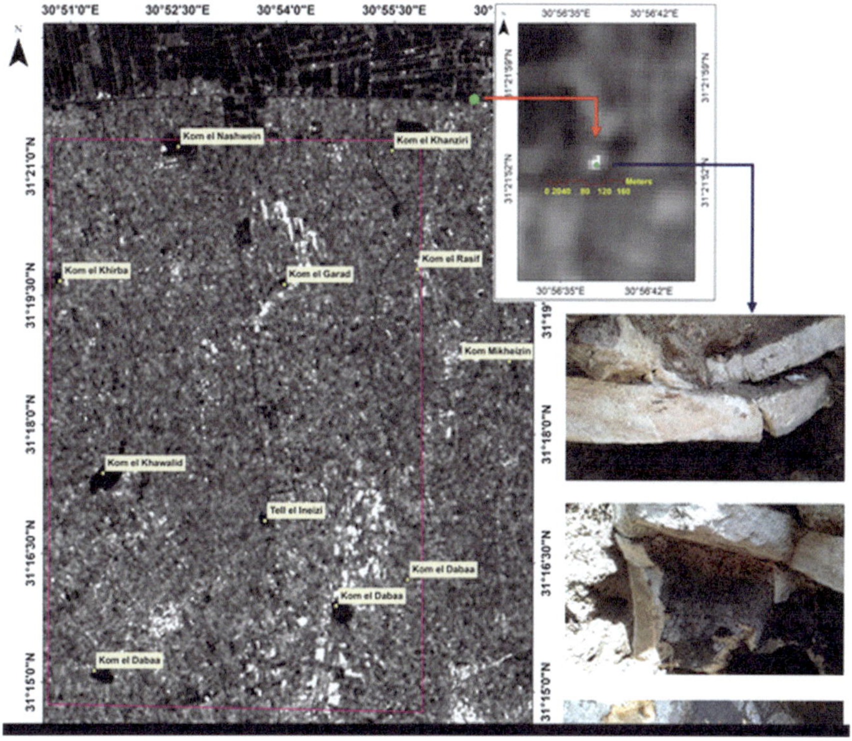

Fig. 7 SAR based identification of archaeological features

3 State of the Art: LiDAR Based Archaeology

The difficulties in the documentation, survey, and material collection increase in areas characterized by the presence of dense vegetation (Lasaponara and Masini 2009). In these cases, the LiDAR-based analysis requires special attention both to: (i) pre-processing -point clouds processing and classification- to avoid that the removal of low vegetation can also determine loss of archaeological information, and (ii) enhancement of LiDAR-Derived Models (LDMs), based on topographical modeling parameters (slope, convexity) and relief visualization techniques (Local Relief model, Sky View Factor, Openness) (Hesse 2010; Zaksek et al. 2011; Doneus 2013).

Over the past two decades, LiDAR has found increasing popularity in archaeology, and many archaeological landscape projects across Europe, America and Asia (Chase et al. 2011; Evans et al. 2013; Masini and Lasaponara 2020) focused on/or incorporated LiDAR.

The available data set types (from point cloud to DTM and orthophoto) allow to use more approaches to extract features, also in automated way, linked referable to

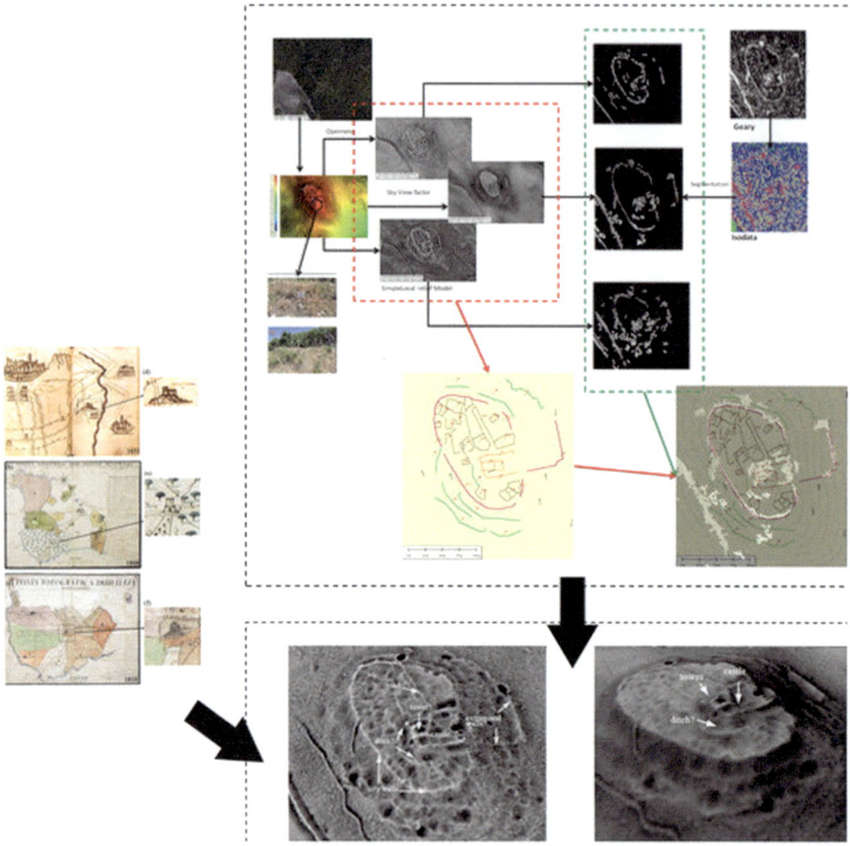

Fig. 8 LiDAR based semiautomatic extraction and interpretation of archaeological features. Case study related to a medieval village including a castle in Basilicata (South of Italy) (Masini et al. 2018)

diverse archaeological proxy indicators, including those linked to changes in micro-topography and vegetation growth (Masini et al. 2018; Trier et al. 2021) and multiple archaeological feature classes (Van Der Vaart and Lambers 2019) (see Fig. 8).

4 State of the Art: UAV Archaeology

Close range RS based on the use of UAV based imagery, including RGB, multispectral and thermal infrared is today widely used in archaeology also because available at low cost. The large amounts of data, sometimes more than we can handle, make necessary the application of AI for their effective in operational scenarios of CH.

The most recent UAV systems offer advanced opportunity to exploit active and passive data traditionally acquired from aerial and ground platforms such as hyperspectral imagery, LIDAR and Ground Penetrating Radar (GPR) data, with an expected economic saving and reduction of data acquisition times. Nevertheless, all **of these technologies are rarely used today in archaeological applications**, considered still quite expensive and complex.

5 State of the Art: Ground Remote Sensing for Archaeology

Ground Penetrating Radar (GPR), Magnetic surveys, along with other geophysical prospection have been (since long time) and are today widely used in archaeology; but except a few examples (Agapiou et al. 2017; Masini et al. 2017) rarely fully integrated with satellite and aerial data sets. This may be due to the fact that geophysics as well as aerial and satellite remote sensing belong to diverse scientific communities (experts and specialists) and there is an urgent need of sharing experience and background to facilitate the development between and across the diverse disciplines.

However, when geophysics is functionally integrated with archaeological research and excavation, it can become a valid predictive tool, even in a diachronic key vision (Fig. 9).

6 State of the Art of the Data Integration for the Study of the Human Past

As a whole, respect to the past, the archaeologists and CH decision makers are more aware of the benefits of EO technologies in terms of reduction of costs and time of archaeological investigations, and opportunities to support strategies addressed to conservation and preservation of cultural assets thanks to the:

(i) improvement in spectral and spatial resolution of sensors, evermore useful for site discovery and preventive archaeology;
(ii) availability of multiscale, multi-temporal and multi sensor data useful to investigate changes due to human activity in areas of cultural interest and monitoring risk in archaeological sites.

In particular, Copernicus “facilities” can suitably support the risk estimation and monitoring also exploiting the combination of satellite data and environmental measurements obtained from diverse sources including airborne and ground-based measurements (Fig. 10).

An example of data integration was specifically devised for the monitoring and knowledge improvement of Nasca Lines in Peru (see Fig. 11). This was achieved by the integration and reuse of diverse remote sensing datasets, including multispectral passive satellite data, UAV based L-bad SAR imaging and close range

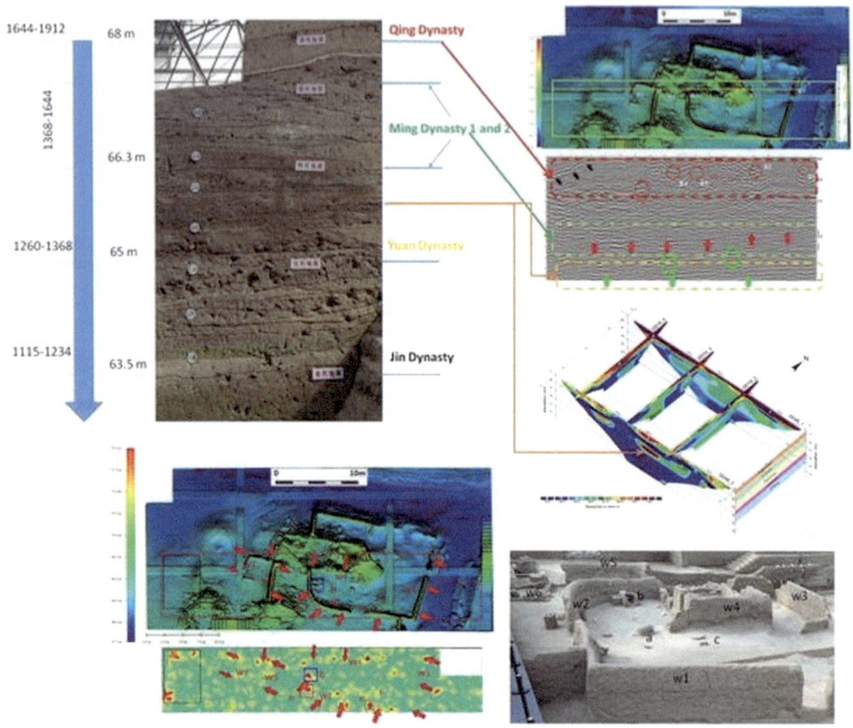

Fig. 9 Geophysical data integration for the diachronic detection of anthrophogenic layers in Kaifeng (Henan, China) [Masini et al. 2017]

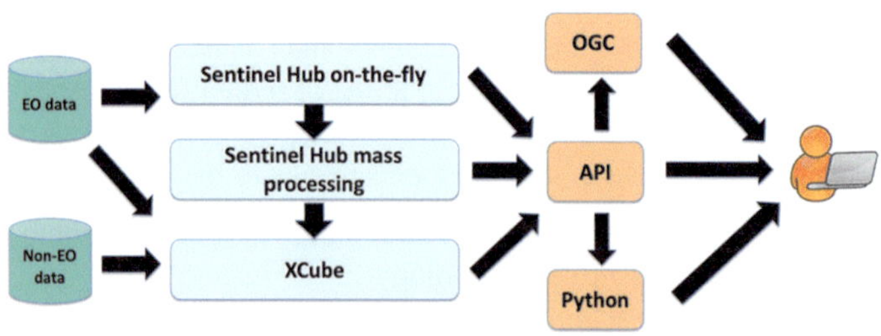

Fig. 10 Scheme of the EO and non –EO data (ESA courtesy)

photogrammetry. An automatic data processing was set up to detect disturbance features.

In particular, machine learning based approach was devised for the automatic extraction of geoglyphs and disturbance features mainly linked to vandalism and impact of anthropogenic activity. The multiscale and multisensor dataset along with in situ analyses allowed to understand the relative chronology of the geoglyphs.

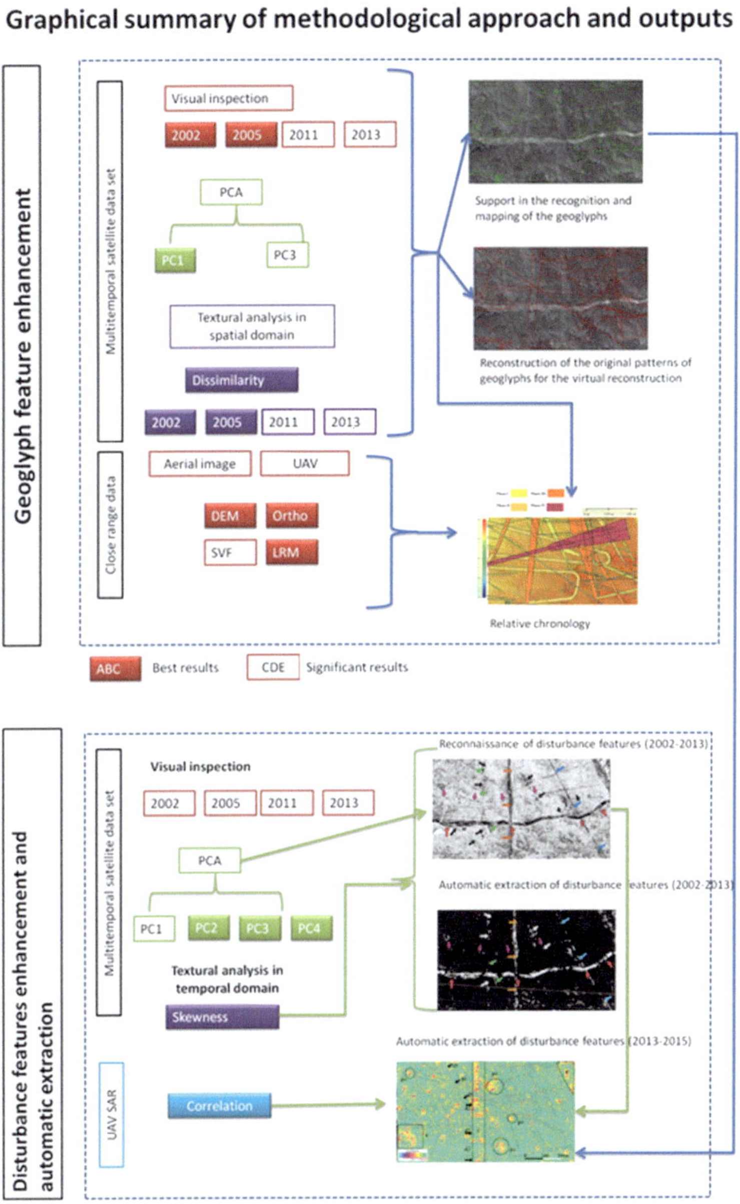

Fig. 11 Nasca geoglyphs: EO and Machine learning based approach for the extraction of geoglyphs and disturbance features (Masini and Lasaponara 2020)

Undoubtedly, the integration of data from diverse sources and complementary sensors as, for example, satellite-based imaging and information with manned and unmanned data have significant potential to contribute into decision making for supporting the preservation of unique or un-renewable resources as cultural heritage and archaeological landscape.

7 Artificial Intelligence and Big Earth Observation Data for the Study of the Human Past

EO Big data pose big challenges related to the analyses and information extraction. To this aim, several methodologies for automatic data processing based on Machine and Deep learning algorithms are available today in open and commercial software tools.

Archaeological geophysics have been already benefited from AI approaches used as a powerful tool used to maximize the extraction of the information content for large areas surveyed using multi-sensor arrays. Machine and Deep learning methods have been successfully used for processing large volume of geophysical data (Küçükdemirci & Sarris 2020) collected over huge areas.

Relating the study of the human past from above, up to now we can distinguish two are the categories of methods used (for additional information see Lambers et al. 2019).

(i) One is based on prior knowledge of the properties of the expected archaeological objects It includes: (i) template matching based methods (searching for simple geometric shapes such as circles, quadrangular etc..), (ii) knowledge based algorithms which require detailed knowledge about the expected objects, (iii) (GE) OBIA-based approaches using image segmentation.
(ii) The second family of methods, known as machine learning, seems to be more promising also.in archaeological domain, and most commonly based on Randon Forest and Convolutional Neural Networks (CNNs). The latter is a deep learning algorithm and is an image-classifier recently implemented for: (i) archaeological object detection using LiDAR data (Lambers et al. 2019); and (ii) site detection (mounds) using open-access optical satellite data (Caspari and Crespo 2019).

Nevertheless, it is important to consider that up to now AI in satellite, and, therefore also in satellite based archaeology, is under-exploited. This due to the pre-processing issues, as (but not only) the need to have a highly refined cloud detection, a significant reduction of noise particularly critical for the study of human past being that archaeological features are very subtle and not permanent signals.

However, recently, a dedicated workshop was focused on Big Data and Artificial Intelligence for Earth Observation to highlight the advances achieved in this field (https://ec.europa.eu/info/events/workshop-big-data-and-artificial-intelligence-earth-observation-2020-nov-19_en). In particular, the Copernicus Access Platform Intermediate Layers Small Scale Demonstrator-Candela—devised and set 4 analytics tools deployed in CREODIAS,

- Earth Observation data mining for classification, allowing users to refine their query by iteratively specifying a set of relevant and non-relevant images.
- Deep Learning for Change Detection on time series for optical and radar Earth observation data. The tools provide generic change detection maps for every couple of respectively optical and SAR images.
- Semantic search and indexation on the output of the Earth observation library and non-image data, to allow users to make requests using multi-criteria.
- Data fusion techniques to merge data that came from various sources, enabling to combine multiple image sources for classification.

8 Conclusion

Big data has emerged in the past few years providing opportunities to improve and/or enable research and decision support applications with unprecedented value for digital CH and archaeology. The possibility to fast analyses relatively large, varied and rapidly changing huge quantities of data has sped up the work during the diverse phases of application ranging from survey, mapping, documentation, exploitation and monitoring at diverse scales of interest, moving from small artefacts to architectural structures and landscape scale. There is, therefore, no doubt that EO big data will significantly change the scientific approach, data analysis and methodologies as well as discoveries of unknown archaeological sites. Moreover, these technologies are non- invasive survey methods very reliable not only for the discovery of lost archaeological remains/sites, but also useful to investigate cultural landscapes, assess the condition of archaeological features that is a mandatory step for the preservation and management of Cultural properties and historic sites.

References

Abate N, Elfadaly MN, Lasaponara R (2020) Multitemporal 2016–2018 Sentinel-2 data enhancement for landscape archaeology: the case study of the Foggia Province, Southern Italy. Remote Sens 12:1309

Agapiou A, Alexakis DD, Sarris A, Hadjimitsis DG (2014) Evaluating the potentials of sentinel-2 for archaeological perspective. Remote Sens 4:2176–2194

Agapiou A, Alexakis D, Hadjimitsis DG (2019) Potential of virtual earth observation constellations in archaeological research. Sensors 19:4066

Agapiou A, Lysandrou V, Sarris A, Papadopoulos N, Hadjimitsis D (2017) Fusion of satellite multispectral images based on ground-penetrating radar (GPR) data for the investigation of

buried concealed archaeological remains. Geosciences 7(2):40. https://doi.org/10.3390/geosciences7020040
Caspari G, Crespo P (2019) Convolutional neural networks for archaeological site detection—finding "princely" tombs. J Archaeol Sci 110:104998
Chase AF et al (2011) Airborne LiDAR, archaeology, and the ancient Maya landscape at Caracol, Belize. J Archaeol Sci 38:387–398. https://doi.org/10.1016/j.jas.2010.09.018
Copernicus Open Access Hub. Available online: https://scihub.copernicus.eu/dhus/#/home. Accessed 11 March 2020
Copernicus services in support to Cultural Heritage (2019) https://www.copernicus.eu/sites/default/files/2019-06/Copernicus_services_in_support_to_Cultural_heritage.pdf. Accessed 20 Aug 2020
Doneus M (2013) Openness as visualization technique for interpretative mapping of airborne LiDAR derived digital terrain models. Remote Sens 5(12):6427–6442. https://doi.org/10.3390/rs5126427
Elfadaly A, Abate N, Masini N, Lasaponara R (2020) SAR sentinel 1 imaging and detection of Palaeo-landscape features in the mediterranean area. Remote Sens 12:2611
Evans DH et al (2013) Uncovering archaeological landscapes at Angkor using LiDAR. Proc Natl Acad Sci USA 110:12595–12600. https://doi.org/10.1073/pnas.1306539110
Hesse R (2010) LiDAR-derived local relief models a new tool for archaeological prospection. Archaeol Prospect 17(2):67–72. https://doi.org/10.1002/arp.374
https://step.esa.int/main/download/snap-download/
https://www.copernicus.eu/en/access-data/dias
https://eo4society.esa.int/2019/05/21/european-data-cube-facility-service-an-eo-resource-factory/
Jiang A, Chen F, Masini N, et al (2016) Archeological crop marks identified from Cosmo-SkyMed time series: the case of Han-Wei capital city, Luoyang, China. Int J Dig Earth https://doi.org/10.1080/17538947.2016.1254686
Kalayci T, Lasaponara R, Wainwright J, Masini N (2019) Multispectral contrast of archaeological features: a quantitative evaluation. Remote Sens 11:913. 1309 23 of 27 12
Khalaf N, Insoll T (2019) Monitoring Islamic archaeological landscapes in ethiopia using open source satellite imagery. J Field Archaeol 44:401–419. https://doi.org/10.1080/00934690.2019.1629256
Küçükdemirci M, Sarris A (2020) Deep learning based automated analysis of archaeo-geophysical images. Archhaeol Prospect. https://doi.org/10.1002/arp.1763
Kvamme KL (2013) An examination of automated archaeological feature recognition in remotely sensed imagery. In: Bevan A, Lake M (eds) Computational approaches to archaeological spaces. Left Coast Press, Walnut Creek, CA, USA, pp 53–68, ISBN 9781611323467
Lambers K, Verschoof-van der Vaart WB, Bourgeois QPJ (2019) Integrating remote sensing, machine learning, and citizen science in Dutch archaeological prospection. Remote Sens 11:794
Lasaponara R, Masini N (2008) Advances in remote sensing for archaeology and cultural heritage management. In: Proceedings of international EARSeL workshop "advances in remote sensing for archaeology and culturale heritage management", Rome 30 September–4 October, 2008. Aracne, Roma, ISBN: 978–88–548–2030–2
Lasaponara R, Masini N (2009) Full-waveform airborne laser scanning for the detection of medieval archaeological microtopographic relief. J Cult Herit 10S:e78–e82. https://doi.org/10.1016/j.culher.2009.10.004
Lasaponara R, Masini N (2011) Satellite remote sensing in archaeology : past, present and future. J Archaeol Sci 38(9):1995–2002. https://doi.org/10.1016/j.jas.2011.02.002
Lasaponara R, Masini N (2012) Remote sensing in archaeology: from visual data interpretation to digital data manipulation. In: Lasaponara R, Masini N, (eds) Satellite remote sensing. A new tool for archaeology. Springer, Dordrecht, The Netherlands, pp 3–16
Luo L, Wang X et al (2019), Airborne and spaceborne remote sensing for archaeological and cultural heritage applications: a review of the century (1907–2017. Remote Sens Environ 232:111280

Masini N, Capozzoli L, et al (2017) Towards an operational use of geophysics for Archaeology in Henan (China): archaeogeophysical investigations, approach and results in Kaifeng. Remote Sens 9(8):809. https://doi.org/10.3390/rs9080809

Masini N, et al (2018) Medieval archaeology under the canopy with LiDAR. The (Re)discovery of a medieval fortified settlement in Southern Italy. Remote Sens 10:1598. https://doi.org/10.3390/rs10101598

Masini N, Lasaponara R (2017) Sensing the past from space: approaches to site detection. In: Masini N, Soldovieri F (eds). Sensing the past. From artifact to historical site. Springer International Publishing, pp. 23–60. https://doi.org/10.1007/978-3-319-50518-3_2

Masini N, Lasaponara R (2020) On the reuse of multiscale LiDAR data to investigate the resilience in the late medieval time: the case study of Basilicata in South of Italy. J Archaeol Method Theory. https://doi.org/10.1007/s10816-020-09495-2

Opitz R, Herrmann J (2018) Recent trends and long-standing problems in archaeological remote sensing. J Comput Appl Archaeol 1(1):19–41. https://doi.org/10.5334/jcaa.1

Orengo HA, Conesa FC, et al (2020) Automated detection of archaeological mounds using machine-learning classification of multisensor and multitemporal satellite data. PNAS 117(31):18240–18250. https://doi.org/10.1073/pnas.2005583117

Schneider A, Takla M, Nicolay A et al (2015) A template-matching approach combining morphometric variables for automated mapping of charcoal kiln sites. Archaeol Prospect 22:45–62

Sentinel Online—ESA (2020) https://sentinel.esa.int/web/sentinel/home. Accessed 11 March 2020

SNAP Tutorials—STEP. https://step.esa.int/main/doc/tutorials/snap-tutorials/. Accessed 11 March 2020

Tapete D, Cigna F (2017) Trends and perspectives of space-borne SAR remote sensing for archaeological landscape and cultural heritage applications. J Archaeol Sci: Rep 14:716–726. https://doi.org/10.1016/j.jasrep.2016.07.017

Trier ØD, Cowley DC, Waldeland AU (2018) Using deep neural networks on airborne laser scanning data: results from a case study of semi-automatic mapping of archaeological topography on Arran, Scotland. Archaeol Prospect 2018:1–11

Trier OD, Reksten JH, Løseth K (2021) Automated mapping of cultural heritage in Norway from airborne lidar data using faster R-CNN. Int J Appl Earth Obs Geoinf. https://doi.org/10.1016/j.jag.2020.102241,95,102241)

Verschoof-van der Vaart WB, Lambers K (2019) Learning to look at LiDAR: the use of R-CNN in the automated detection of archaeological objects in LiDAR data from the Netherlands. J Comput Appl Archaeol 2(1):31–40. https://doi.org/10.5334/jcaa.32

Zakšek K, Oštir K, Kokalj Ž (2011) Sky-view factor as a relief visualization technique. Remote Sens 3:398–415. https://doi.org/10.3390/rs3020398

Zanni S, De Rosa A (2019) Remote sensing analyses on sentinel-2 images: looking for Roman roads in Srem Region (Serbia). Geosciences 9:25

The Potential of Satellite Interferometry for Geohazard Assessment in Cultural Heritage Sites

Federico Raspini, Silvia Bianchini, Davide Festa, Matteo Del Soldato, Pierluigi Confuorto, Pablo Ezquerro, and Nicola Casagli

Abstract A continuous monitoring system of ground deformation, based on radar images acquired by ESA (European Space Agency) Sentinel-1 constellation, is active over the Tuscany Region (Central Italy). The potential of repeat-pass satellite SAR (Synthetic Aperture Radar) interferometry has been exploited to investigate spatial patterns and temporal evolution of regional and local ground deformation that affect cultural heritage sites. With millions of measurement points, ground deformation maps for Tuscany Region provide information that can be exploited to scan wide areas and to flag ground instabilities. These areas become targets for detailed analysis with high resolution sensors (*e.g.*, COSMO-SkyMed satellites of the Italian Space Agency) to create a virtual constellation, in which different satellite data sources are synergically used to create a more effective and robust Earth Observation system. The potential of a virtual constellation is presented and discussed through the case study of Pistoia, a city whose origins date back to the Etruscan civilization.

Keywords Geological hazard · Satellite InSAR · Monitoring · Mapping · Cultural heritage

1 Introduction

According to the UNESCO (United Nations Educational, Scientific and Cultural Organization), the World Heritage Sites (WHS) must be safeguarded with an adequate protection system, in order to guarantee their conservation, integrity and authenticity.

F. Raspini (✉) · S. Bianchini · D. Festa · M. Del Soldato · P. Confuorto · N. Casagli
Earth Sciences Department, University of Firenze, Via La Pira, 4, 50121 Firenze, Italy
e-mail: federico.raspini@unifi.it

Present Address:
P. Ezquerro
Geohazards InSAR Laboratory and Modeling Group (InSARlab), Geological Survey of Spain (IGME), 28003 Madrid, Spain

N. Casagli
National Institute of Oceanography and Applied Geophysics, OGS, Sgonico, Trieste, Italy

G. M. El-Qady and C. Margottini (eds.), *Sustainable Conservation of UNESCO and Other Heritage Sites Through Proactive Geosciences*, Springer Geology,
https://doi.org/10.1007/978-3-031-13810-2_30

Currently, many UNESCO sites are threatened by geohazards, but the safeguard of these sites does not seem to be wide-ranging and only sudden geological events draw the attention of WHS managers and are considered as factors that undermine the protection of the properties (Valagussa et al. 2020). Unfortunately, not only these geological factors can threaten cultural and natural heritage, but also slow-kinematic phenomena, such as slow-moving landslides and ground-subsidence (Pastonchi et al. 2018).

According to Pavlova et al. (2017) about 60% of the WHL (World Heritage List) sites are exposed to at least one geological hazard, with earthquakes and landslides among the most frequent. Nevertheless, the level of attention placed on geohazards appears to be inadequate, especially if compared with other potential threatening factors, such as wars or uncontrolled urban development (Canuti et al. 2009). In Italy protection and safeguard of WHS is a critical problem, being the country with the highest number of UNESCO sites in the world, and with 45% of them affected by landslides.

In this framework, a greater knowledge of the potential impact of such phenomena on WHS and, more in general, on all cultural heritage sites, could be obtained with the geoinformation retrieved through the satellite InSAR (Interferometric Synthetic Aperture Radar) techniques (Tapete and Cigna 2017). In the last decades, satellite radar data have been recognized as a valuable tool for deformation analyses, not only for geological and environmental applications (Casagli et al. 2017) but also for the management and conservation of built heritage in cultural and rural sites (Alberti et al. 2017; Tapete and Cigna 2017).

Early applications of satellite SAR in archaeology date back to 1980s (Adams et al. 1981) and enabled relevant discoveries, revealing hidden features and paleo-landscapes in subtropical territories and arid environments. Following these pioneering studies, the potential of spaceborne SAR data has been fully exploited and many applications have been performed, using SAR acquisitions for mapping and monitoring ground deformation affecting cultural heritage sites. This flourishing of applications has been fostered by the development of single and multi-interferogram SAR techniques (A-DInSAR, Advanced Differential Interferometry SAR), a family of approaches which allows to retrieve very accurate measurements (with a millimetric accuracy) of terrain and structures displacements and which provides information to support the management of risks due to any deformation phenomena (*e.g.*, landslide, subsidence).

In this framework, a major role is played by the European satellites Sentinel-1 and by the Italian constellation COSMO-SkyMed. Thanks to the systematic images acquisition (with an effective repeat pass of 6 days), the Sentinel-1 mission is capable to scan wide areas and to flag ground instabilities. The very high resolution of COSMO-SkyMed acquisitions (up to 3 m when acquiring in StripMap mode) is ideally suited to refine pattern of deformation.

In this work, we exploited C-band Sentinel-1 and X-band Cosmo-SkyMed InSAR data to detect anthropogenic-induced local subsidence that affect the city of Pistoia, a cultural heritage site in the Tuscany Region, Central Italy, the first region in the world to have a continuous ground deformation monitoring using Sentinel-1 constellation

(Raspini et al. 2018). The aim of this work is to present and exploit the potential of a virtual constellation, in which different satellite data sources are synergically used to create a more effective and robust Earth Observation system for geohazard assessment in cultural heritage sites.

2 The Tuscany Region

Located on the Tyrrhenian side of the Italian peninsula (Fig. 1), Tuscany, with an area of 22.985 km^2, extends between the regions of Liguria, Emilia-Romagna, Marche, Umbria and Lazio. Tuscan territory includes also a little islands archipelago composed by six main islands and other small rocks. Tuscany is administratively subdivided into 10 provinces with the regional capital, Firenze, located in the northern part of the Region.

The main reliefs rise in the northern and eastern parts of the region, and they belong to the Northern Apennine system, an arcuate orogenic belt that marks the border of the region from NW to SE. The Northern Apennines is a NE-verging fold and thrust belt, consisting in a thick sedimentary package of carbonate and silicoclast deposits.

Tuscany has been the first region in the world to adopt, in October 2016, a continuous ground deformation monitoring using Sentinel-1 images. Subsidence phenomena, as well as landslides and floods, are among the main geohazards affecting the Tuscany Region. The monitoring of related ground deformations plays a key role in their management to avoid long-term effects on buildings, infrastructures and cultural heritage.

At the end of 2019, Italy, with a total of 55 proprieties, is the nation with the highest number of sites inscribed in the WHL, sharing the first position with China. Seven of these 55 proprieties, both cultural and natural sites, are located in the Tuscany Region and they have been inscribed in the WHL in different years. The UNESCO cultural heritage sites in Tuscany are the Historic Center of Florence; Piazza del Duomo in Pisa; the Historic Center of San Gimignano; the Historic Center of Siena; the Historic Center of Pienza; Medici Villas and Gardens in Tuscany (14 proprieties), and Val d'Orcia. Besides the aforementioned cultural sites, the whole Tuscan territory is studded by dozens of other cultural heritage sites, symbolizing the flourishing economic, cultural, and social development that has involved the Tuscan territory over several centuries of history.

Some of these sites are affected by land subsidence, some other threatened by landslides. Considering that we can neither instrument all the sites nor prevent all the risks posed by ground deformation and that it is not feasible to monitor with conventional techniques the whole heritage, a different strategy for risk mitigation urged to be conceived.

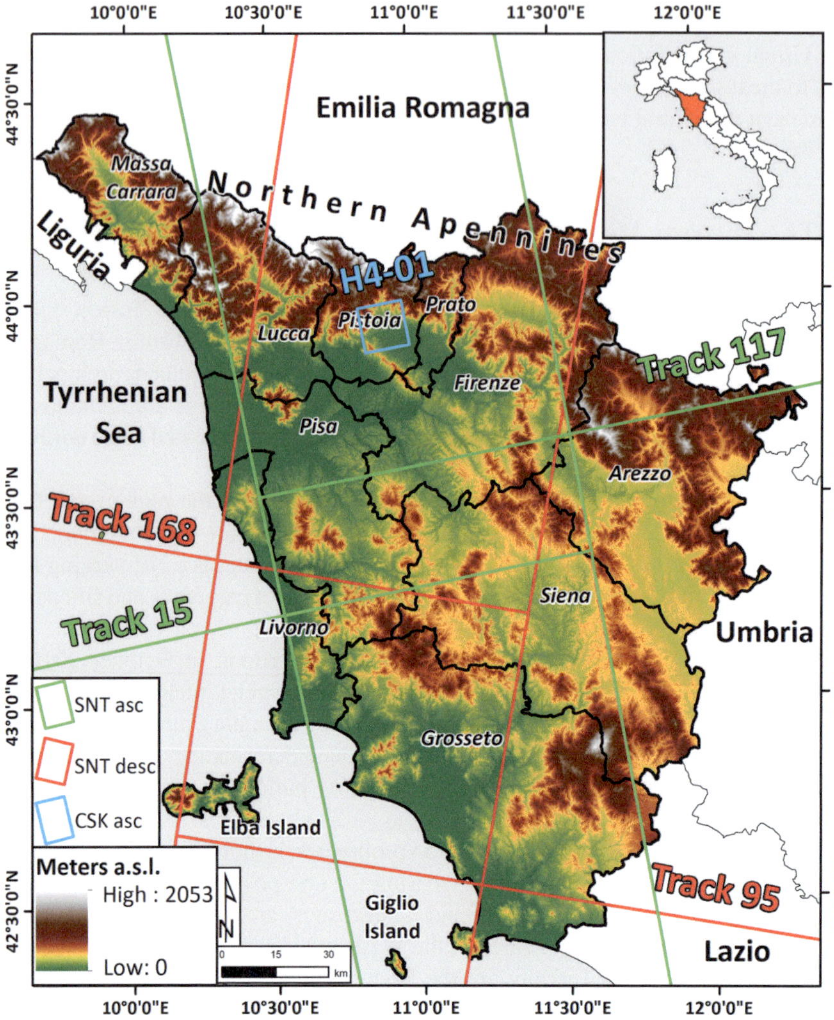

Fig. 1 Physical setting of the Tuscany Region shown through a Digital Elevation Model. Tracks of Sentinel-1 and COSMO-SkyMed satellites used in this work are also included

3 The Virtual Constellation

The ground deformation maps for the Tuscany Region have been generated by analysing two large stacks of C-band Sentinel-1 acquisitions by means of the SqueeSAR algorithm (Ferretti et al. 2011), an algorithm which represents the evolution of PSInSAR (Ferretti et al. 2001), the first technique belonging to the PSI (Persistent Scatterer Interferometry) family specifically implemented for the processing of multi-temporal radar imagery. The novelty of PSInSAR algorithm is the capability of identifying a network of coherent radar targets, *i.e.*, point-like-targets (PS, Permanent Scatterers widely present over urban fabric) within the radar image. Such pixels correspond to manmade objects, outcropping rocks, debris areas or buildings that register a stable radar signal over the whole observation time period.

The development of SqueeSAR technique contributed to extend the field of application of PSInSAR (Ferretti et al. 2011) to natural terrain, overcoming the main limitation of this technique. SqueeSAR estimates deformation rates not only from point-like-targets but also from partially coherent pixels, called Distributed Scatterers (DS). DS points correspond to homogeneous ground surfaces (*i.e.*, uncultivated areas, deserts, debris covered slopes and scattered outcrops) and not to single objects.

LOS (Line of Sight) deformation rate can be estimated with an accuracy theoretically lower than 0.1 mm/yr (Colesanti et al. 2003). PSI analysis is designed to generate time-series of ground deformations for individual reflectors. The accuracy of the single measurement in correspondence of each SAR acquisition ranges from 1 to 3 mm (Colesanti et al. 2003). Each measurement is referred temporally and spatially to a unique reference image and to a stable reference point.

All the provided ground displacement measurements fulfil a set of basic requirements from the regional authorities: (*i*) regional coverage, (*ii*) <1 cm accuracy, (*iii*) continuous delivery of information, (*iv*) timely delivery of updated products, (*v*) low cost (considering the number of measurement points for the spatial extent of Tuscany).

Sentinel-1 data allows to scan wide areas to detect hotspots (Bianchini et al. 2012), namely narrow unstable zones characterized by high deformation. To these hotspots, priority must be established when planning field surveys and *in situ* validation campaigns, so that field work time and efforts can be optimized and significantly reduced.

To refine the analysis of on-going deformation patterns at local scale, some areas became targets for detailed analysis with higher resolution COSMO-SkyMed images, to create a virtual constellation, where different satellite data sources are used in synergy to create a more effective and robust Earth observation system. In a virtual constellation different space segments overlap, in order to meet different requirements.

The use of different sensors may help to mitigate intrinsic limitations of any particular sensor, providing a common set of Earth Observation data. For the target areas medium resolution, wide area (Sentinel-1) and high resolution—detailed scale (COSMO-SkyMed) data are merged and integrated to derive information to contribute to a quantitative analysis as most precise as possible.

3.1 Sentinel-1 Results at Regional Scale

In Fig. 2 velocity maps obtained from the processing of all the Sentinel-1 images, from December 2014 to December 2019 are reported. Each velocity map includes more than 900,000 points and provides a synoptic view of the regional displacement

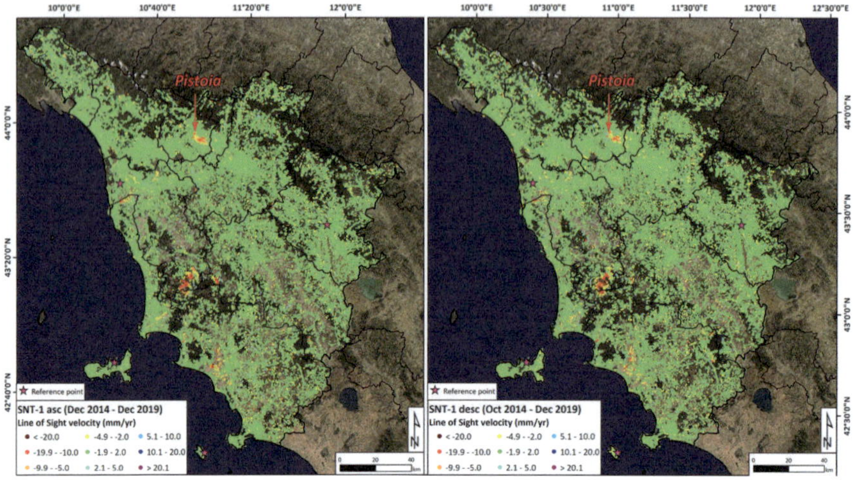

Fig. 2 Overview of the results obtained by SqueeSAR processing of C-band Sentinel-1 data in the two acquisition geometries (ascending on the left and descending on the right). Measurement point are represented by dots coloured according to their mean velocity along the line of sight (LOS) of the satellite, expressed in millimetres per year, with positive values of mean velocity representing displacements towards the satellite, and negative values representing displacements away from the satellite

field. Moreover, through the analysis of displacement time series, it is possible to highlight "anomalies of movement", *i.e.*, changes in ground deformation temporal patterns (Raspini et al. 2018).

These maps include a wealth of quantitative information on a range of phenomena, including slow-moving landslides (with the possibility to highlight the acceleration of movements that typically precede collapse and to progressively refine the forecasting of the failure time), subsidence (related to aquifer depletion or load imposition), uplift, consequences of geothermal resources exploitation and on the impact of these phenomena on man-mad infrastructures, including cultural heritage sites. The information contained in Fig. 2 are particularly valuable, as they allow, scanning wide areas, to spot unstable zones, supporting the definition of priorities, starting from the situation deemed to be most urgent. Significant deformation is revealed in the zone of Pistoia (central part of the Region), a place with a long history of land subsidence related to overexploitation of several confined aquifers present at different depths.

3.2 COSMO-SkyMed Results at Local Scale

The area target of a detailed COSMO-SkyMed analysis is the urban fabric of Pistoia, a typical medieval city with a population of approximately 91,000 inhabitants and which hosts important cultural heritages, such as the Duomo and the San Giovanni Baptistery that attract several tourists every year. Indeed, the city was appointed as the Italian Capital of Culture in 2017 by the Italian Ministry of Cultural Heritage. Pistoia is a centre of Etruscan settlements before becoming a Roman colony in the sixth century BC, along the important road *Via Cassia*. From the fifth century the city was a bishopric, and during the Lombardic kingdom it was a royal city and had several privileges.

Pistoia's most splendid age began in 1177 when it proclaimed itself a free commune: in the following years it became an important political centre, erecting walls and several public and religious buildings. A network of underground channels (called "*gore*" in Italian) runs below the historical city center. The "*gore*" channels, built in Medieval Age, were used for water supply and are today considered as a touristic attraction. In addition to the historical buildings and the cultural heritage, Pistoia is famous for its plant and flower nursery activities mainly widespread in a south-eastern area of the city, called Bottegone, close to the Prato province to the East.

A set of 59 COSMO-SkyMed images acquired in Stripmap mode (3 × 3 m of ground resolution) and covering the period from January 2016 to March 2018, were processed using the Coherence Pixel Technique (CPT), developed by the Remote Sensing Laboratory (RSLab) at Universitat Politecnica de Catalunya (Mora et al. 2003; Blanco-Sánchez et al. 2008; Confuorto et al. 2017) (Fig. 3).

CPT algorithm is based on the exploitation of spatial coherence, increasing the number of measurement points especially in rural areas through the use of Distributed Scatterers (DS). CPT results confirmed that the city of Pistoia is affected by a large subsidence with a mean value of 1 cm/yr and peaks of 2.5 cm/yr in the historical center and along the highway located south of the urban area. Groundwater depletion and compaction of fine-grained deposits are likely the causes of the measured subsidence. Almost 50,000 measurement points (MP) have been identified in the urbanized area of Pistoia, with 7,500 MP within the Middle Age (built during the XIV century A.C.) city walls. Such density of points ensures an almost spatially continuous coverage of information on surface deformation and related hazards. This makes COSMO-SkyMed InSAR ideally suited to map the extension of threatened areas, improving confidence on spatial pattern of the examined phenomenon. A further benefit of multi-images techniques is the generation of time series of the relative position for each MP in correspondence of each SAR acquisition, allowing analysis of temporal evolution of displacement and calculation of cumulated displacement for the investigated period of time.

4 Discussions and Conclusions

In the last two decades, satellite interferometry has experienced major developments, gaining increasing attention within the communities of scientists, practitioners, stakeholders and end users. Despite the unique capability to observe very large areas, this

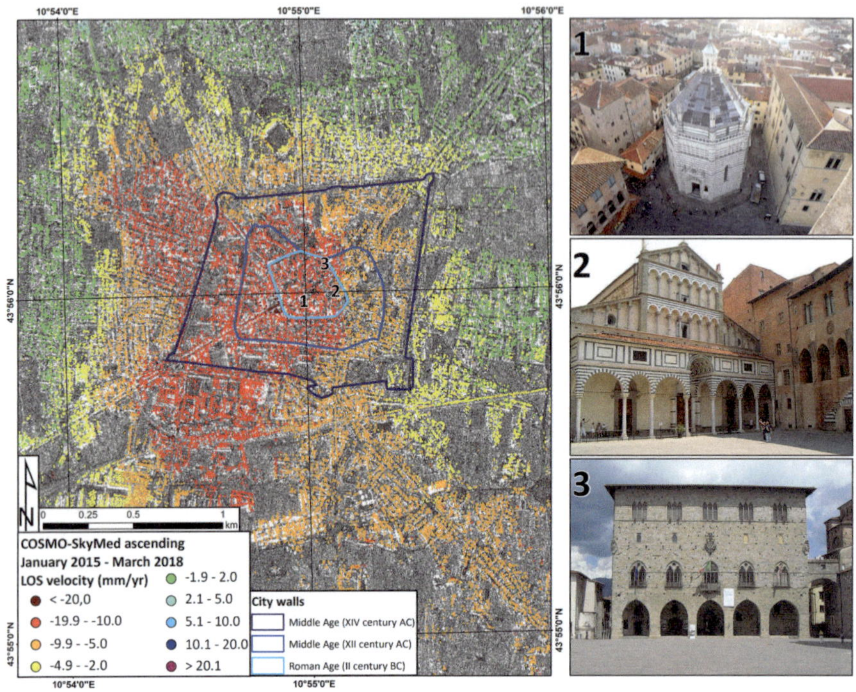

Fig. 3 COSMO-SkyMed-based ground deformation map of the urban area of the city of Pistoia. Traces of historical city walls and location of some of the most important cultural heritage sites are reported

technology has been typically applied to the analysis of relatively small sites or for post-event analysis, concentrating on the spatial investigation of ground movements, and possibly look retrospectively to capture precursory signals of an already set up motion.

The Tuscany Region adopted in 2016 a monitoring operative service based on Sentinel-1 InSAR technology aimed at providing a streamline of surface deformation measurements covering wide areas, ideally suited to spot unstable areas. Thanks to the regional-scale mapping capability the systematic and regular SAR observations and rapid product delivery, Sentinel-1 mission provides accurate updates of surface movement on a continuous basis. This constellation profitably contributes to the design and implementation of a monitoring system capable of monitoring and measuring ground deformation affecting cultural heritage sites at an appropriate temporal scale. Considering the abundance of cultural heritage sites in Italy, very often interacting with natural hazards, the urge of a continuous and systematic monitoring, as the one represented here, is of utmost priority.

The use of COSMO-SkyMed constellation, which proved its potential for detailed analysis, represents a value-added to refine pattern of deformation at local scale and helps to overcome the necessary trade-offs that govern the design and implementation of a monitoring system at regional scale, improving confidence on spatial pattern of the examined phenomenon.

Acknowledgements The ground deformation monitoring system presented in this paper has been requested, founded and supported by the Regional government of Tuscany, under the agreement "Monitoring ground deformation in the Tuscany Region with satellite radar data". Authors are grateful to TRE ALTAMIRA for having processed Sentinel-1 data for the Tuscany Region.

References

Adams RE, Brown WE, Culbert TP (1981) Radar mapping, archeology, and ancient Maya land use. Science 213(4515):1457–1468

Alberti S, Ferretti A, Leoni G, Margottini C, Spizzichino D (2017) Surface deformation data in the archaeological site of Petra from medium-resolution satellite radar images and SqueeSAR™ algorithm. J Cult Herit 25:10–20

Bianchini S, Cigna F, Righini G, Proietti C, Casagli N (2012) Landslide hotspot mapping by means of persistent scatterer interferometry. Environ Earth Sci 67(4):1155–1172

Blanco-Sanchez P, Mallorquí JJ, Duque S, Monells D (2008) The coherent pixels technique (CPT): An advanced DInSAR technique for nonlinear deformation monitoring. In: Earth sciences and mathematics. Birkhäuser, Basel, pp 1167–1193

Canuti P, Margottini C, Fanti R, Bromhead EN (2009) Cultural heritage and landslides: research for risk prevention and conservation. In: Sassa K, Canuti P, (eds) Landslides—disaster risk reduction. Springer, Berlin/Heidelberg, Germany, pp 401–433

Casagli N, Frodella W, Morelli S, Tofani V, Ciampalini A, Intrieri E, Lu P (2017) Spaceborne, UAV and ground-based remote sensing techniques for landslide mapping, monitoring and early warning. Geoenviron Disast 4(1):9

Colesanti C, Ferretti A, Prati C, Rocca F (2003) Monitoring landslides and tectonic motions with the permanent scatterers technique. Eng Geol 68(1):3–14

Confuorto P, Di Martire D, Centolanza G, Iglesias R, Mallorqui JJ, Novellino A, Calcaterra D (2017) Post-failure evolution analysis of a rainfall-triggered landslide by multi-temporal interferometry SAR approaches integrated with geotechnical analysis. Remote Sens Environ 188:51–72

Ferretti A, Prati C, Rocca F (2001) Permanent scatterers in SAR interferometry. Geosci Remote Sens IEEE Trans 39(1):8–20

Ferretti A, Fumagalli A, Novali F, Prati C, Rocca F, Rucci A (2011) A new algorithm for processing interferometric data-stacks: SqueeSAR. Geosci Remote Sens IEEE Trans 49(9):3460–3470

Mora O, Mallorqui JJ, Broquetas A (2003) Linear and nonlinear terrain deformation maps from a reduced set of interferometric SAR images. IEEE Trans Geosci Remote Sens 41:2243–2253

Pastonchi L, Barra A, Monserrat O, Luzi G, Solari L, Tofani V (2018) Satellite data to improve the knowledge of geohazards in world heritage sites. Remote Sensing 10(7):992

Pavlova I, Makarigakis A, Depret T, Jomelli V (2017) Global overview of the geological hazard exposure and disaster risk awareness at world heritage sites. J Cult Heritage 28:151–157

Raspini F, Bianchini S, Ciampalini A, Del Soldato M, Solari L, Novali F, Del Conte S, Rucci A, Ferretti A, Casagli N (2018) Continuous, semi-automatic monitoring of ground deformation using Sentinel-1 satellites. Sci Rep 8(1):1–11

Tapete D, Cigna F (2017) Trends and perspectives of space-borne SAR remote sensing for archaeological landscape and cultural heritage applications. J Archaeol Sci Rep 14:716–726

Valagussa A, Frattini P, Crosta GB, Spizzichino D, Leoni G, Margottini C (2020) Hazard ranking of the UNESCO world heritage sites (WHSs) in Europe by multicriteria analysis. J Cult Herit Manag and Sustain Dev 10(4):359–374

Instability Processes and SAR Data Analysis in the Pompeii Archeological Park

Carla Iadanza, Gabriele Leoni, Daniele Spizzichino, Alessandro Trigila, Claudio Margottini, Massimo Osanna, Bruno de Nigris, Alberta Martellone, Mario Costantini, Elena Francioni, Francesco Trillo, and Federico Minati

Abstract The Italian Institute for Environment Protection and Research (ISPRA) signed in 2015 a three-year agreement with the Pompeii Archaeological Park aimed at analyzing and interpreting ground motion measures obtained by satellite monitoring as well as the geomorphological processes affecting the unexcavated areas. COSMO-SkyMed SAR data were processed by e-GEOS Company in the framework of an agreement between the Italian Ministry of Cultural Heritage and Activities and Finmeccanica Group (nowadays Leonardo Group). A collection of information about recent instability processes in the Archaeological Park was carried out. Limits and constraints of satellite monitoring applied to cultural heritage protection and conservation have been investigated. The results highlighted the potential of the satellite interferometric technique, above all concerning the monitoring of the archaeological protective shelters, the effectiveness of the shoring measures and the identification of those areas where the instrumental monitoring should be implemented.

Keywords UNESCO · Interferometric satellite monitoring · Archaeological Park · Instability · Cultural Heritage

C. Iadanza · G. Leoni · D. Spizzichino · A. Trigila
ISPRA, Geological Survey of Italy, Via V. Brancati 48, 00144 Rome, Italy

C. Margottini (✉)
Former Scientific Attaché, Embassy of Italy in Egypt, 15, Abd El-Rahman Fahmy Str., Garden City, Il Cairo, Egypt
e-mail: claudio.margottini@gmail.com

M. Osanna · B. de Nigris · A. Martellone
Pompeii Archaeological Park, Pompei, Italy

M. Costantini · E. Francioni · F. Trillo · F. Minati
e-GEOS–Italian Space Agency/Telespazio, Rome, Italy

G. M. El-Qady and C. Margottini (eds.), *Sustainable Conservation of UNESCO and Other Heritage Sites Through Proactive Geosciences*, Springer Geology,
https://doi.org/10.1007/978-3-031-13810-2_31

1 Introduction

The ancient city of Pompeii (Naples, Italy) is a Roman settlement destroyed by the eruption of the Mount Vesuvius in 79 AD. Volcanic ashes, pumices and debris that poured down during the volcanic eruption sealed the city, favouring the optimal preservation during centuries. Excavation of the archaeological site of Pompeii started in 1728. The historical, artistic and cultural value of this Roman city was universally acknowledged in 1997, when it was endorsed to the World Heritage List of the United Nations Educational, Scientific and Cultural Organization (UNESCO) (Veneranda et al. 2018). Since then, the number of tourists has been steadily increasing and reached 3.8 million visitors in 2019. Pompeii is the most visited cultural site of Italy after the Colosseum of Rome.

The ancient city of Pompeii with its 66 hectares, 44 of which are excavated, is divided into 9 *Regiones* (district), and contains almost 1500 domus (houses).

This research is carried out in the framework of two different agreements. The first agreement, between Finmeccanica Group (nowadays Leonardo Group) and Ministry of Cultural Heritage and Activities, is aimed at processing Synthetic Aperture Radar (SAR) data from COSMO-SkyMed satellite images by e-GEOS Company. The second one is a three-year agreement (2015 2018) between the Pompeii Archaeological Park and the Italian Institute for Environment.

Protection and Research (ISPRA, Geological Survey of Italy) aimed at analyzing and interpreting ground motion measures from SAR data and at evaluating the geomorphologic processes affecting the unexcavated areas. The targets are:

- to test strengths and weaknesses of SAR interferometry technique and of satellite monitoring data interpretation to detect ground deformations in the archaeological area;
- to identify a standard methodology for the use of SAR data, in order to highlight potential damage conditions and to analyze their evolution to support Archaeological Park's management and maintenance activities.

With respect to the previous applications of SAR interferometry (Iadanza et al. 2013; Spizzichino et al. 2017) the innovation concerns with the monthly reprocessing of satellite radar images that makes this technique act as a monitoring system.

The methodology followed these operational steps:

1. collection of information on past damage events in the Archaeological Park and creation of events' database mainly regarding the same period of available satellite data;
2. preliminary geomorphological zoning of unexcavated slopes and instability processes classification;
3. SAR interferometry monitoring data interpretation,

    3.1. back analysis of some recent years' collapse events, to assess the predictive capability of the technique by identifying displacement trends before the collapse;

3.2. definition of a semi-automatic methodology to identify anomalies and accelerations of the displacements and to compare them by in situ monitoring surveys.

2 Collection of Information on Past Damage Events and Geomorphological Zoning

ISPRA and Archaeological Park's staff performed a systematic collection of information (news, reports, etc.) about recent instability processes for the whole area occurred since year 2005.

This survey led to the creation of a geodatabase of instability and collapse events recording 35 events and related information (time, location, description, triggering, data source, reliability, photos). The map of past instability events points out some critical areas: most events of the 2005–2017 period occurred in Regio V, in Regio IX and along the north-eastern side of Via *dell'Abbondanza* (Fig. 1).

Through the dates of occurrence of the instability processes and collapse phenomena it was also possible to define both the number of collapses per months (Fig. 2) and the number of collapses per year (Fig. 3).

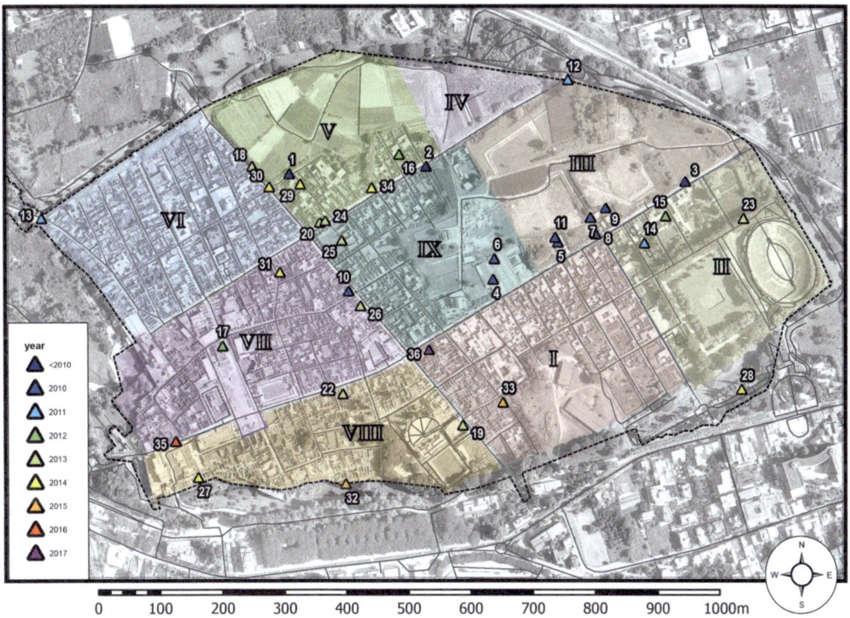

Fig. 1 Map of instability phenomena occurred in the Pompeii archaeological area (classified by year; time interval 2005–2017)

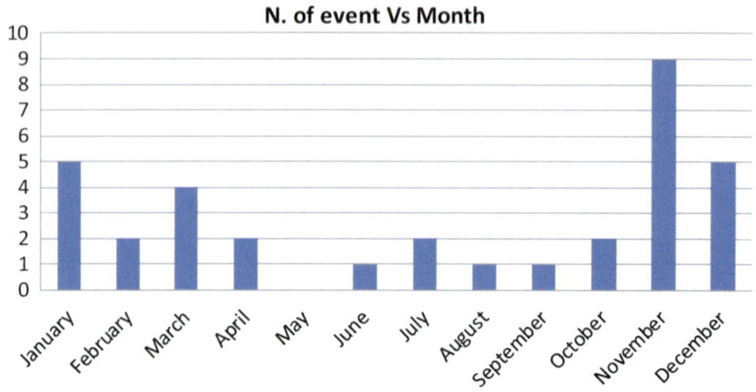

Fig. 2 Number of collected events sorted by months (time interval 2005–2017)

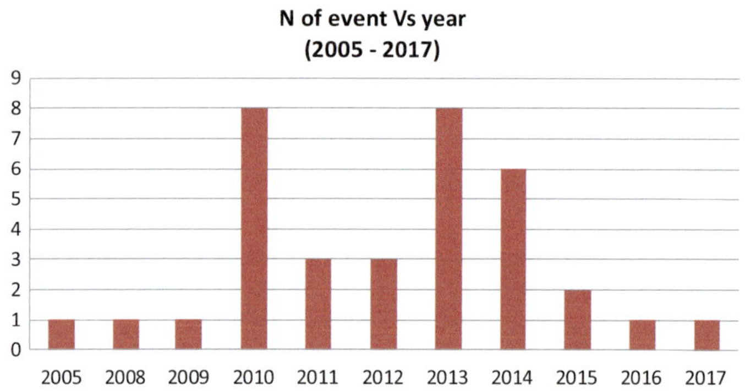

Fig. 3 Number of collected events sorted by years (time interval 2005–2017)

Most of the events occurred in years 2010, 2013 and 2014, particularly during winter season (since November to January), most likely due to heavy rainfall (Fig. 3). These correlations will be useful for future analyses relating to the definition of the rainfall trigger thresholds.

Together with the past events data collection a geomorphological survey on the whole archaeological area has been carried on during year 2016, paying more attention to slopes not yet excavated.

Main morphological processes of the area (small slides, earth-flows, rock falls, etc.) have been mapped to obtain an overall layout of most active areas, on which further investigation and in situ monitoring are needed, to assess new excavations and manage future exploiting. An example is reported in Fig. 4.

Most active slopes are located in Regio II and V, and along the northern side of Via *dell'Abbondanza* (Figs. 1 and 5).

Fig. 4 Instability process affecting small portion of the unexcavated slope in the northern sector (Regio V)

3 Satellite Radar Interferometry in the Pompeii Archaeological Park

Satellite radar interferometry technique (see Rocca et al. 2000) allows millimetric measurement of ground displacements on wide areas, processing a huge amount of measuring points (called Persistent Scatterer, PS) radiometrically stable over time, such as buildings, stones, metal frames, etc. Satellites with radar sensors onboard follow polar orbits, acquiring earth images by two different geometries: Ascending when the satellite move from south to north and Descending from north to south. Interferometry analysis on a stack of images acquired in the same area over a certain period provides measurement point's position (coordinates and elevation), mean annual velocity and time series of displacements. Data are acquired along the Line Of Sight (LOS), i.e. the line connecting the sensor with the target.

In the Pompeii area radar interferometry technique shows a few constraints related to some peculiar conditions:

- vertical thin walls (thickness range 10–60 cm) and absence of roof;
- small distance between adjacent walls (1–2 m) resulting in shadowing effects that affect radar beams;
- areas with low radar reflection, e.g. green cultivated areas or bare soil on the slopes bounding unexcavated areas.

Fig. 5 Geomorphological features map

Interferometry analysis herein presented has been performed by e-GEOS Company using a recent technique called Persistent Scatterer Pairs (PSP; Costantini et al. 2014). The PSP technique is a proven SAR interferometry method characterized by the fact of exploiting in the processing only the relative properties between close points (pairs) in order to overcome atmospheric artefacts (which are one of the main problems of SAR interferometry).

By PSP technique were analyzed satellite radar images in radar X-band (3.1 cm wavelength) coming from COSMO SkyMed constellation, with mean revisiting time of 16 days, acquired in HIMAGE mode (Stripmap) and characterized by pixel spatial resolution of 3 m, both in Ascending and Descending geometry. More in detail the historic analysis considered only Ascending images in a first period (May 2010 to March 2012), then both geometries have been analyzed, Ascending since October 2012 and Descending since May 2011 until November 2016, processed on a monthly base (Fig. 6). PS point positioning is characterized by a sub-meter standard deviation in both East, North and Vertical, while the deformation measurement has millimeter accuracy.

4 Back Analysis on Some Instability Events

With the aim of interpreting satellite monitoring data and of assessing their application, a back analysis has been conducted on some case studies in the Pompeii archaeological area. SAR radar interferometry technique is useful to predict wall collapse that is affected by slow deformations (e.g. bulges, leaning walls). On the other hand, when the wall collapse is instantaneous and precursory elements are missing or occurring in a time shorter than the satellite revisit time, this technique cannot be helpful.

The methodology was applied by analyzing mean annual velocities of displacements recorded by measure points close to the collapse and their time history until the event. Back analysis has been conducted on 7 events, occurred between November 2013 and January 2017, which locations were well known from photos and description.

These events involved 4 domus' walls, the Porta Nocera necropolis, one retaining wall (not excavated slope) and the southern slope of Regio VIII.

Case study analysis points out a few constraints to use satellite radar interferometry as a monitoring system. In the case of the wall in Fig. 7 some PSs seem to be stable during the months before the collapse, thus suggesting a brittle break of the wall. But in many cases the time interval between the last acquisition and the rupture is quite long (spanning from 8 until 77 days) so that it is not possible to detect an imminent structure's deformation. In addition, PSs that are stable even after an event could belong to a part of the wall adjacent to the collapsed one. Some case misses' measurements points at all, maybe due to shadowing effects (Fig. 8).

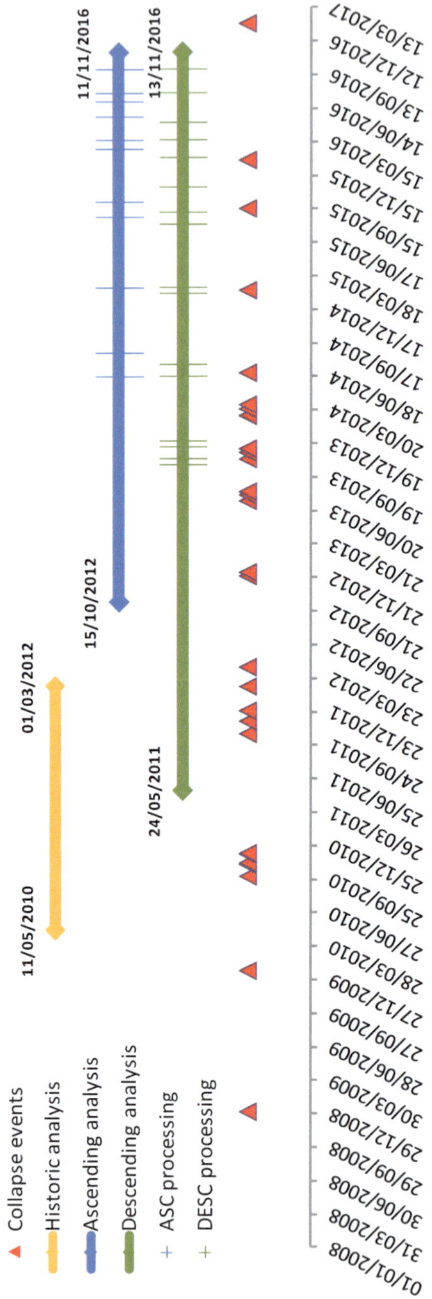

Fig. 6 Historic analysis, Ascending analysis, Descending analysis and date of occurrence of collapse events

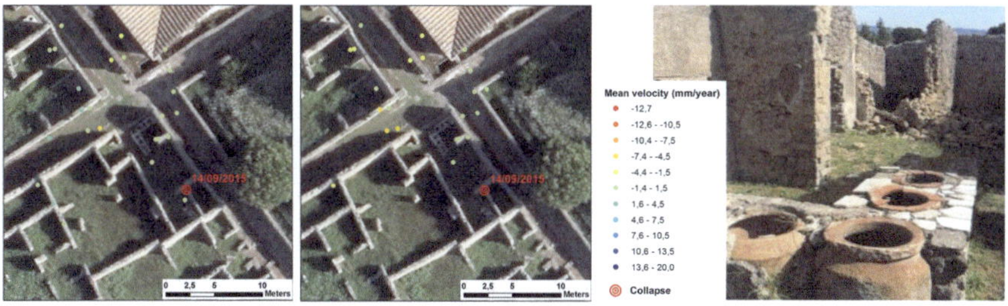

Fig. 7 Event #33, September 14th, 2015: CSK Ascending image processing until August 21st, 2015 (pre-event, left). CSK Ascending image processing until October 1st, 2015 (post-event, centre). The pre-event image shows a PS at about 1-m distance from the collapse that was stable until 24 days before, then it disappears in the post-event processing, apparently due to the loss of coherence caused by the rupture. Photo of the event (right). This event should be interpreted as a brittle break of the wall

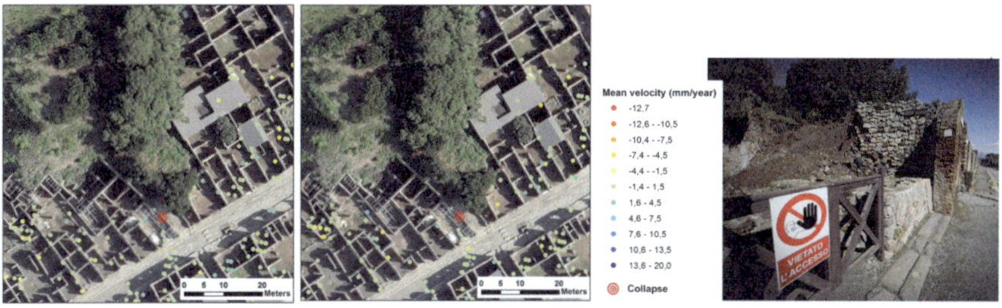

Fig. 8 Event #34 March 2nd, 2014: CSK Descending image processing until December 23rd, 2013 (pre-event, left). CSK Descending image processing until June 17th, 2014 (post-event, centre). There is no PS on the perimeter wall, probably located in a shadow area due to presence of vegetation. Photo of the event (right)

5 Semi-automatic Detection of Anomalies or Accelerations

Semi-automatic detection of possible anomalies or accelerations of the measurement points' deformation trend, using empirical thresholds on displacement, velocity and acceleration, aims to highlighting the most critical conditions over the entire archaeological site, for prevention purposes.

This task support site managers to plan surveys to check the real degree of stability. This prevention activity is strategic, because of the great amount of measurement points all around the study area (about 15,000 PSs in Descending geometry and 24,000 PSs in Ascending geometry—SAR images processing March 2016, Figs. 9 and 10).

As a first hypothesis an empirical displacement velocity threshold of ±3 mm/yr. has been considered, that led to drop out more stable PSs, thus obtaining a smaller dataset, 271 PSs in Descending geometry and 737 PSs in Ascending geometry. The threshold value corresponds to approximately the 1st and the 99th percentile of the

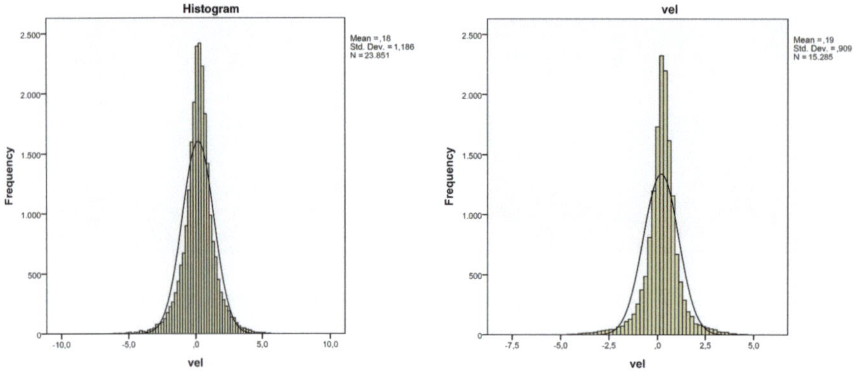

Fig. 9 Frequency histogram of PS velocity in Ascending geometry (left) and descending geometry (right)

PS velocity in DESC geometry and between the 1st-2nd percentile and the 98th-99th percentile in ASC geometry (Fig. 9).

Concerning with the mean annual velocities sign, where positive means a movement towards the satellite and negative means a movement away from the satellite, it is noteworthy that lowering and settlement of roof and cover frameworks cause the presence of PSs with negative velocities.

On the contrary wall damages causing displacements with horizontal component, such as bulges and leaning walls, may cause positive or negative velocities, depending on the angle between the movement direction and the LOS.

A cumulative displacement threshold of ± 10 mm has been considered (Fig. 11) and compared with the previous one.

The latter seems a better option to highlight time series characterized by a long nearly stable phase and a deformation phase occurring in a short time. This displacement could exceed 10 mm, even if the mean annual velocity that is considered over the entire measuring time is still smaller than the velocity threshold (Fig. 12).

The last threshold considered the displacement's acceleration that highlight movements occurred during last measurements. The ratio between the velocity during the last 4 acquisitions (about 2 months for Descending geometry and about 3 months for Ascending geometry) and the mean annual velocity during the entire period (about 2.5 years) has been calculated (Fig. 13).

It is worth mentioning that the more are the images and longest is the time coverage of acquisition the more accurate is the evaluation of displacement.

Combining last two thresholds (i.e. more than 10 mm of cumulate displacement and acceleration ratio between last 4 acquisitions and the entire period greater than 3) led to select as critical about 1% of measurement points (160 MPs in Ascending geometry and 167 MPs in Descending geometry).

In order to check the presence of actual deformation effects on archaeological remains (damages, fractures, bulges, leaning walls, etc.) and their accordance with

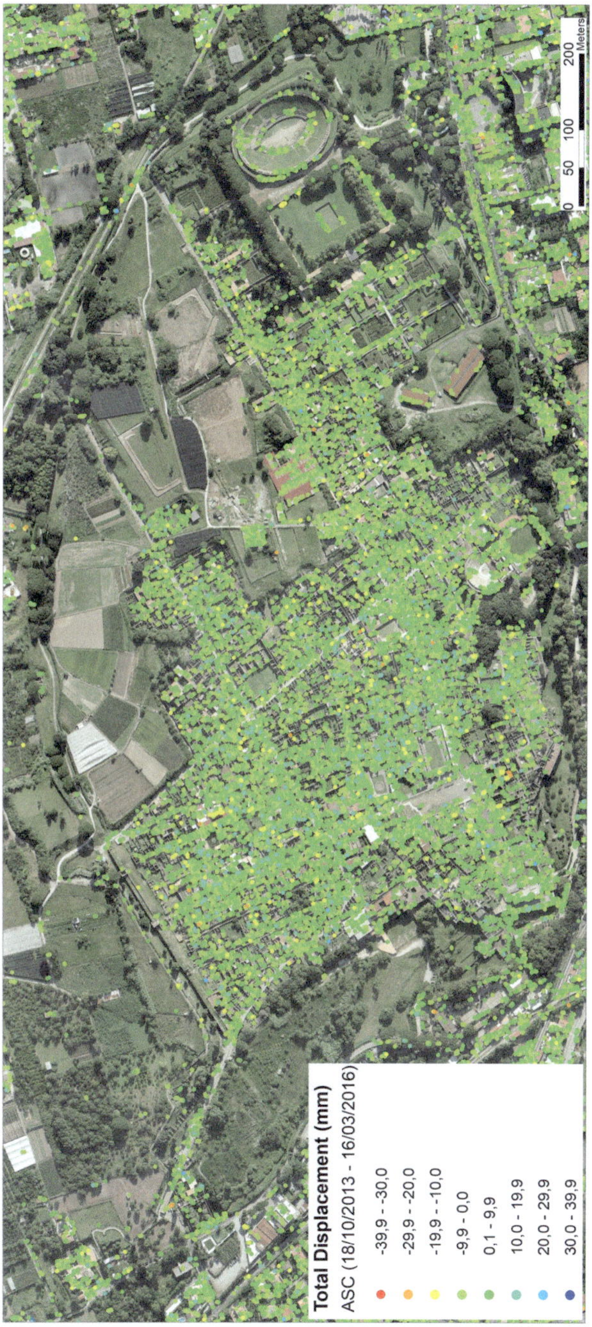

Fig. 10 PSs in Ascending geometry (measuring time October 18th, 2013 to March 16th, 2016)

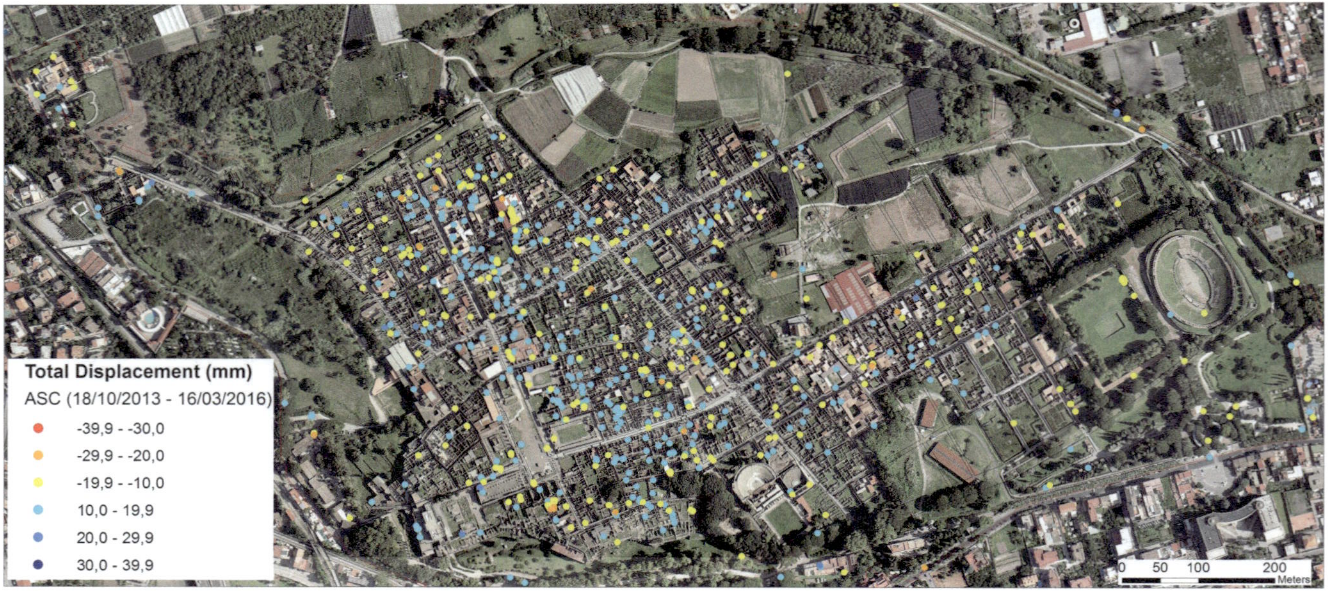

Fig. 11 Filtered PSs with cumulative displacement greater than ± 10 mm in Ascending geometry (measuring time October 18th, 2013 to March 16th, 2016)

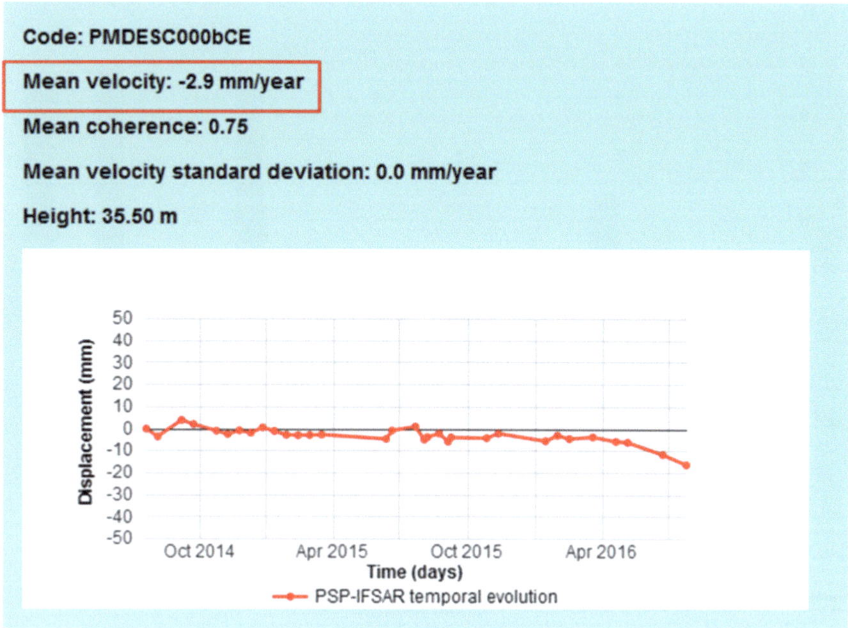

Fig. 12 Comparison of velocity and displacement threshold for one measurement point

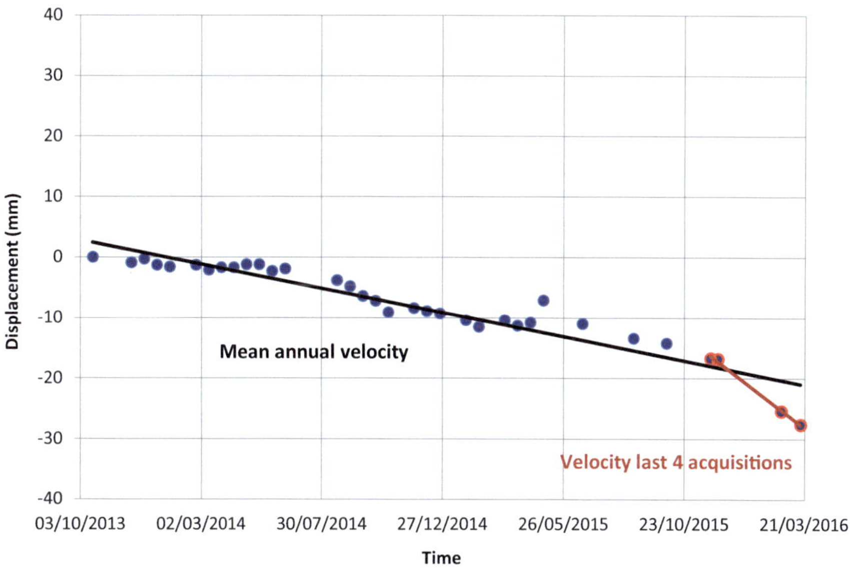

Fig. 13 Comparison of mean annual velocity during the entire period and velocity during the last 4 acquisitions

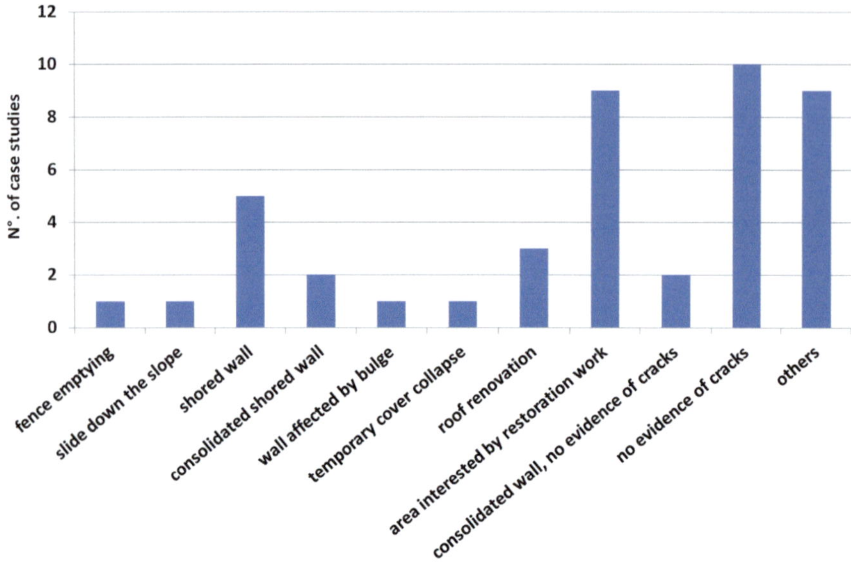

Fig. 14 Distribution of case studies inspected on the June 2016 survey

satellite monitoring, in June 2016 a field survey has been conducted, particularly referring to areas with cumulative settlements exceeding 10 mm.

Out of 44 case studies analyzed, 20% were referred as to consolidation and restoration work in progress; 23% were referred as to walls affected by bulges, damages, shored up and/or consolidated with or without evidence of ruptures; 9% were related to the renovation of the roof coverage or to their settlement; 23% of sites do not show evidences of deformation/rupture by visual inspection (Figs. 14, 15 and 16).

At *Casa dei Vettii* site satellite monitoring points out that several PS are affected by a cumulative negative displacement (lowering) by Ascending geometry ranging between 10 and 20 mm, since October 18th, 2013 until March 3rd, 2016 (Fig. 15).

Pompeii Archaeological Park's managers visually verified the deterioration of a wooden beam temporary supporting the roof coverage, in accordance with satellite data.

Preliminary results highlighted that the satellite interferometry technique is capable to identify in the archaeological area mainly the roof coverage firmness over times and the effectiveness of wall shoring systems. Particularly the roof analysis is more affordable because of the presence of several PSs on the same element.

Moreover, this technique is useful to check the effectiveness of shoring structures. About this issue it is very important to exchange information with site managers in order to combine monitoring data with site interventions and their temporal evolution.

Future developments of this research will be addressed to: (i) improve future acquisition and processing of satellite interferometry data, also with the installation of passive corner reflector; (ii) analysis of pluviometric data and identification

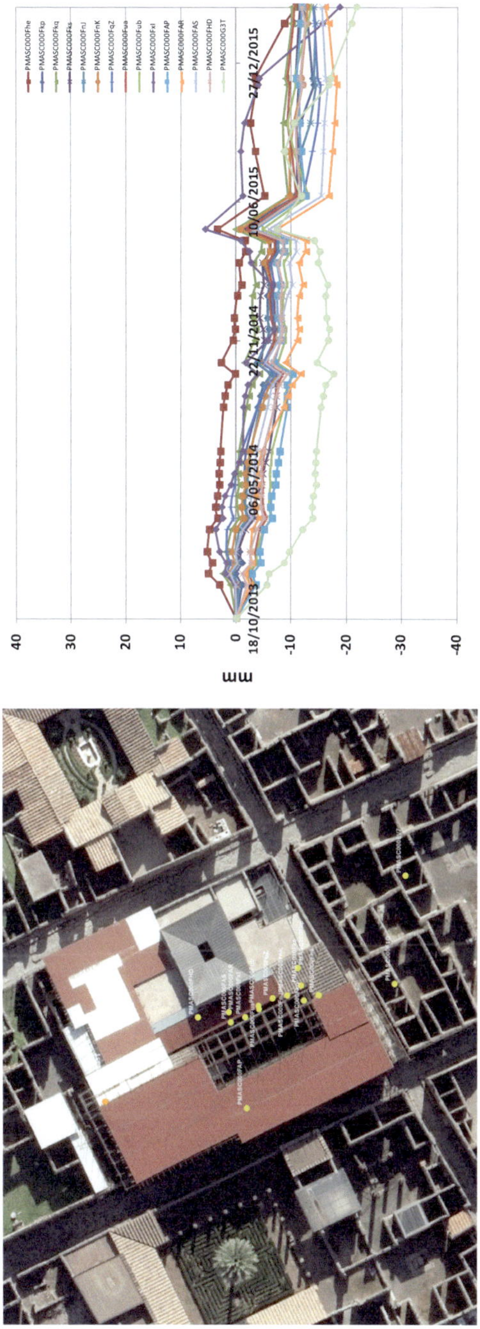

Fig. 15 (left) Measurement Point at Casa dei Vettii, Ascending geometry, October 18th, 2013 to March 16th, 2016. (right) PSs displacements' time history

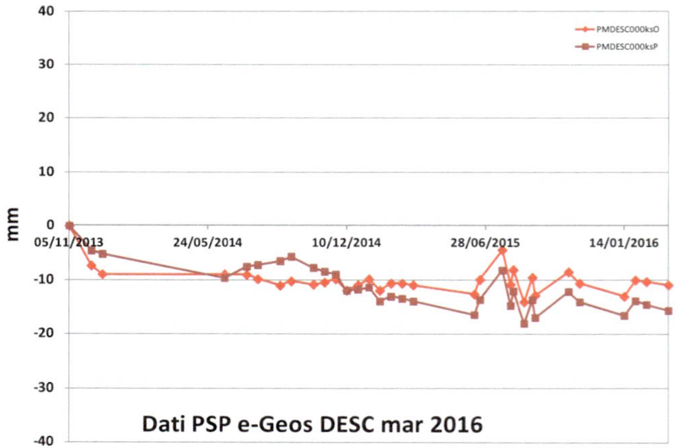

Fig. 16 (left) Shored wall with a recently installed metallic structure—June 2016 field survey; (right) displacements' time history of two measurement points near the wall in Descending geometry, November 5th, 2013 to March 18th, 2016

of possible rainfall triggering thresholds, (iii) support for the analysis of in situ monitoring data and possible implementation and integration of the current system.

Acknowledgements The authors are very grateful to all the Pompei Archaeological park workers and technicians for their help and support during all the survey activities.

References

Costantini M, Falco S, Malvarosa F, Minati F, Trillo F, Vecchioli F (2014) Persistent scatterer pair interferometry: approach and application to COSMO-SkyMed SAR data. IEEE J Sel Top Appl Earth Obs Remote Sens 7:2869–2879

Iadanza C, Cacace C, Del Conte S, Spizzichino D, Cespa S, Trigila A (2013) Cultural heritage, landslide risk and remote sensing in Italy. In: Sassa K, Canuti P, Margottini C (eds) Landslide science and practice: vol 6: risk assessment, management and mitigation. Springer, pp 491–500

Rocca F, Prati C, Monti Guarnieri A, Ferretti A (2000) Sar interferometry and its applications. Surv Geophys 21:159. https://doi.org/10.1023/A:1006710731155

Spizzichino D, Margottini C, Brustia E, Cigna F, Comerci V, Dessì B, Guerrieri L, Iadanza C, Leoni G, Tapete D, Trigila A, Vittori E (2017) Satellite monitoring applied to natural hazards and cultural heritage: the PROTHEGO project. In: Conference proceedings 11° workshop tematico di telerilevamento-AIT Bologna, 27–28 giugno 2017

Veneranda M, Fdez-ortiz De Vallejuelo S, Prieto-Taboada N, Maguregui M, Marcaida I, Morillas H, Martellone A, De Nigris B, Osanna M, Castro K, Madariaga JM (2018) In-situ multi-analytical characterization of original and decay materials from unique wall mirrors in the House of Gilded Cupids, Pompeii. Herit Sci 6:40. https://doi.org/10.1186/s40494-018-0205-2

Remote Sensing Applications for Cultural Heritage Sites Sustainability: Case Studies from Egypt

Abdelaziz Elfadaly, Mohamed A. R. Abouarab, Ayaat Shams Eldein, and Rosa Lasaponara

Abstract Egypt has a long history extended to the prehistory, Neolithic, predynastic, Pharaonic, Greek and Roman, Coptic, and Islamic eras. Because of the climate changes, the land use/land cover changes, and the land-reclamation activities, the groundwater level under the reconstruction and the monuments walls became high and present big risk on these heritage sites. This study aims to show some studies applied on some archaeological sites in Egypt using Satellite data and GIS tools. In this study, Optical (Landsat, Quickbird, Orbview, Spot, and Sentinel2) have been analysed by RS and GIS software (ArcMap, Snap, and Envi) used to detect the geo-environmental problems around the archaeological in Egypt. In order to detect the land use/land cover changes close to the heritage sites, the spatial distribution analysing, spatial autocorrelation, Ripley's *K* function, Hot Spot Analysis, Spatial Cluster Analysis (Ripleys K Function), and the supervised classification methods were applied. The results of this study proved that the urban sprawling close to the studied areas was enormous and caused big risk on the archaeological areas. Such studies can support the decision makers with the required information about the environmental status around the archaeological sites for keeping these sites in safe condition.

Keywords Optical images · Spatial analysis · Environmental risk

---

A. Elfadaly
NARSS, National Authority for Remote Sensing and Space Sciences, Cairo, Egypt
e-mail: abdelaziz.elfadaly@narss.sci.eg

M. A. R. Abouarab · A. S. Eldein
Faculty of Arts, KFS, Department of Archaeology, University of Kafrelsheikh, Kafrelsheikh 1501, Egypt
e-mail: mabouarab@art.kfs.edu.eg

A. S. Eldein
e-mail: ayat.shams@art.kfs.edu.eg

R. Lasaponara (✉)
CNR-IMAA, C.da S. Loja, 85050 Tito Scalo (PZ), Italy
e-mail: rosa.lasaponara@imaa.cnr.it

G. M. El-Qady and C. Margottini (eds.), *Sustainable Conservation of UNESCO and Other Heritage Sites Through Proactive Geosciences*, Springer Geology,
https://doi.org/10.1007/978-3-031-13810-2_32

1 Introduction

Since along time; Egypt incudes rich heritage sites, artefacts, objects, as well as the immovable archaeological sites, such as sites, historic, and rock art sites. In Egypt, due to the increasing rates of inhabitants in both the Nile Valley and the Nile Delta, urban and agricultural sprawling became a general trend in the landscapes. Recently, this expansion represents aggressions against the archaeological and historical sites (Elfadaly et al. 2020).

In addition, the daily and seasonal cycles of dry and wet weather, acting aggressive deterioration on the surface and inside the stone block foundation. Also, the much closed groundwater to the land surface may cause deterioration in the heritage sites. Recently, many organizations that interest in the heritage sites (e.g. UNOSAT, UNITAR, and UNESCO) are now able to monitor damage to cultural heritage via satellite imagery (Elfadaly et al. 2019a). During the last decades, Remote sensing applications produced the geo-information data can preserve the heritage sites that could assess the geo-hazard around the archaeological sites (Elfadaly and Lasaponara 2019b; Elfadaly and Lasaponara 2019a; Abate et al. 2020). Remote sensing and GIS techniques can create a historical database such as the hazard maps may be generated, indicating which sites are potentially in dangerous situation (Elfadaly et al. 2019d). In this study, multi-satellite data were analysed to detect the land use/land cover changes around some archaeological sites in North and South of Egypt. Also, some innovation models were created to decrease the environmental risk around these heritage sites.

1.1 Study Area

The study includes three heritage sites in North of Egypt (Catacombs of Mustafa Kamel at Alexandria, Qayitbay Citadel at Rosetta, and Mosque of Sultan Al-Zahir Baybars at Cairo), besides two archaeological areas in South of Egypt (Seti I, Ramesseum temples, and Medinet Habu at West Luxor and Horus Temple at Edfu) (Fig. 1).

1.2 Risk Definition

Unfortunately, most of the Egyptian monuments are suffering from the bad geo-environmental status, additional to the continually increasing population. Many of the environmental problems around the archaeological areas come as a result of urban and agriculture sprawling that cause the uprising of the groundwater level (Fig. 2) (Elfadaly et al. 2018a).

Fig. 1 Shows the study areas in Egypt includes Alexandria, Rosetta, Cairo, Luxor, and Edfu by Google Earth plus Engine

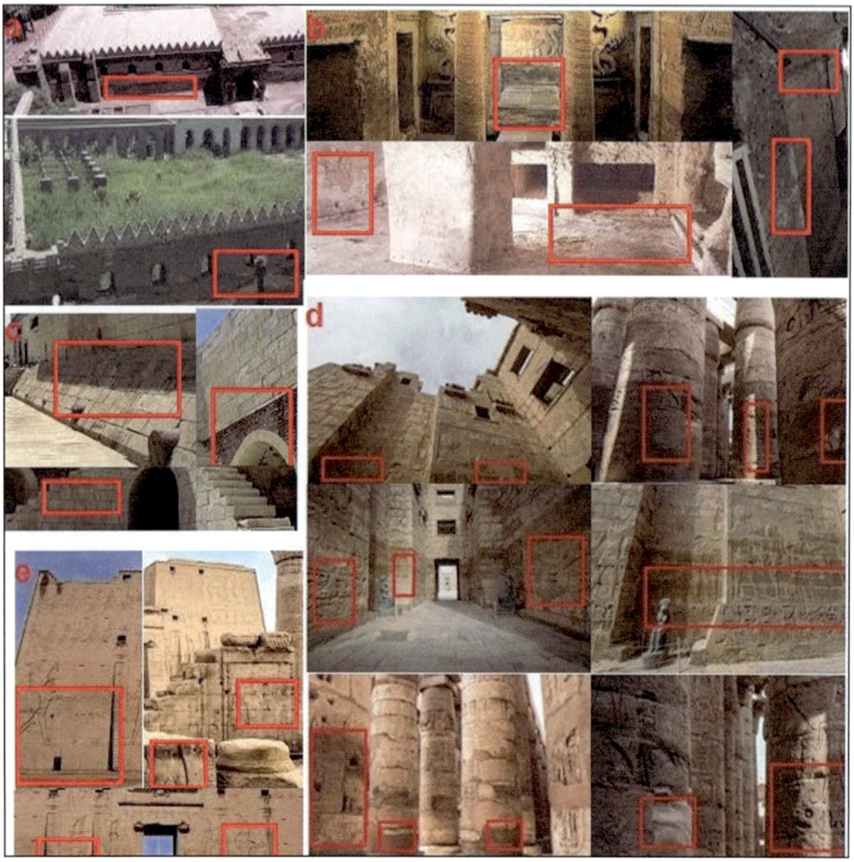

Fig. 2 Shows the threatened archaeological walls and decorations in Al-Zaher Mosque (**a**), Catacombe of Mostafa Kamel (**b**), Qaytbay Castle (**c**), Medinet Habu (**d**), and Horus Temple (**e**)

2 Data and Methods

2.1 Data

The required satellite imagery for the study areas were downloaded from the USGS Earth Explorer (Landsat TM, Coronaj-3, GeoEye's Orbview) (http://edcsns17.cr.usgs.gov/EarthExplorer/), and Sentinel-2A and Sentinel-2B data were downloaded from ESA web site (https://scihub.copernicus.eu/dhus/#/home). The National Authority for Remote Sensing and Space Science (NARSS) provided the study with the high resolution data (Spot3 1993, Spot4 2011 and Quickbird) (Table 1). The images interpretation are done with ENVI 5.1 ArcGIS10.5, SNAP 7.0.0 software. The obtained images are studied and analyzed to detect the changes in the layers based on past and present satellite data.

Table 1 Data collection of the study including the data resolution, acquisition date, and the source

Satellite	Sensor	Resolution (m)	Acquisition date	Source
Corona	KH-4A	1.8 m	Multi-dates	USGS
Landsat	MSS, TM, ETM, ETM+, and L8	60 m MSS 30 m Others	Multi-dates	USGS, GLCF
GeoEye's Orbview-3	(Pan)	1.00 m	Multi-dates	USGS
SPOT 3	HRV	20 m	November 1993	NARSS
Spot4	2 × HRVIR	10 m	April 2011	NARSS
Quickbird2	XS/P	0.6 m	November 2005	NARSS
Sentinel	2A,B	10 m after resampling	Multi-dates	USGS, ESA

2.2 Methods

This study aims to assess the current status of Al-Zaher Mosque at Cairo, Catacombe of Mostafa Kamel at Alexandria, Qaytbay Castle at Rosetta, Medinet Habu, Ramesseum and Seti I temples at Luxor, and Horus temple at Edfu using Remote sensing and GIS techniques to identify and map areas that are affected by uncontrolled land use that cause critical threats for these cultural sites. The integrated analysis was done through the integrated GIS and Remote sensing techniques.

2.3 Unsupervised and Supervised Classification

At the beginning, satellite data were layer stacking, dark subtract, geometric correction, unsupervised classification, and supervised classification using ArcMap and Envi software. The classification process was done through calculate the Euclidean distance between every two multivariate variables (vectors) depends on the sum of squared differences of their components. Therefore, given two vectors x and y, the Euclidean distance between them is computed as (Eq. 1) (Elfadaly et al. 2018b);

$$d_E(x, y) = \sqrt{(x - y)^T (x - y)} \tag{1}$$

2.4 Spatial Distribution Analysing (Getis-Ord and Hot Spot)

On the other hand, the spatial distribution analysing by Getis-Ord and Hot Spot was conducted using ArcMap software. Gi* statistic works has done by looking at each feature (vector) within the context of neighboring features. A high z-score and small p-value for a feature indicates a spatial clustering of high values. While, a low z-score value and minimum p-value indicates a spatial clustering of low values. While, a z-score near zero indicates no apparent spatial clustering. To be a statistically significant hot spot, a feature will have a high value and must be surrounded by other features with high values as well. Getis-Ord Gi* statistic is presented as (Eqs. 2–4) (Elfadaly et al. 2017a);

$$\text{Getis} - \text{Ord}G_i^* = \frac{\sum_{j=1}^{n} w_{i.j}x_j - \overline{X}\sum_{j=1}^{n} w_{i.j}}{S\sqrt{\frac{\left[n\sum_{j=1}^{n} w_{i,j}^2 - \left(\sum_{j=1}^{n} w_{i.j}\right)^2\right]}{n-1}}} \tag{2}$$

where;

$$\overline{\mathrm{X}} = \frac{\sum_{\mathrm{j}=1}^{\mathrm{n}} \mathrm{x_i}}{\mathrm{n}} \tag{3}$$

And;

$$\mathrm{S} = \sqrt{\frac{\sum_{\mathrm{j}=1}^{\mathrm{n}} \mathrm{x_i^2}}{\mathrm{n}} - (\overline{\mathrm{X}})^2} \tag{4}$$

2.5 *Spatial Autocorrelation (Global Moran's I)*

One of most used tools in the spatial distribution analysis is the spatial autocorrelation (Global Moran's I), Generally, spatial autocorrelation tool is measured with Moran's I ranged within values between +1 as high correlation and −1 as perfect dispersion. It identifies statistically significant hot spots and cold spots within a set of weighted features within a context of neighboring features. The Moran's I Statistic for spatial autocorrelation is given as (Eq. 5) (Elfadaly et al. 2017b);

$$I = \frac{n}{S_o}\frac{\sum_{i=1}^{n}\sum_{j=1}^{n} w_{i,j}z_i z_j}{\sum_{i=1}^{n} z_i^2} \tag{5}$$

2.6 *Band Indices*

The computation of the MNDWI (Modified Normalised Difference Water Index) has been used to show the water feature that appear in greater positive values. The greater enhancement of water in the MNDWI-image will result in more accurate extraction of open water features. Also, the Built-Up Index (NDBI) has been used because it is able to enhance the new built-up areas. Urban land classifications are largely based on the use of this specific index, i.e. a mathematical combination of diverse spectral channels. Also, The identification of the built-up and water features in the study areas were calculated according the (Eqs. 6 and 7) (Lasaponara et al. 2016);

$$\mathrm{MNDWI} = \frac{\mathrm{Green} - \mathrm{SWIR}}{\mathrm{Green} + \mathrm{SWIR}} \tag{6}$$

$$\mathrm{NDBI} = \frac{\mathrm{SWIR1} - \mathrm{NIR}}{\mathrm{SWIR1} + \mathrm{NIR}} \tag{7}$$

2.7 *Ripley's* K *Function*

Ripley's *K* function definition is the tool which characterize the spatial structure of point or polygon patterns by graph. K Function Graphic Usage tool requires projected data to accurately measure distances between the total extracted features includes the points and polygons. Respectively, the tool output is a table with fields: Expected K and Observed K containing the expected and observed K values. When the L (d) transformation is applied, the Expected K values will always match the Distance value (Lasaponara et al. 2017).

3 Results and Discussion

3.1 *Detection of Urban Changes*

The change detection was mainly focused on the detection of urban feature changes using images acquired in the different investigated years (see Table 2). In more detail, the analysis of Corona, Landsat, Orbview, Quick bird, Spot, and Sentinel-2A imagery

Table 2 Spatial Autocorrelation Values for the built-up area around Medinet habu, Ramesseum and Seti I temples between 1967 and 2000

Spatial Autocorrelation Value	1999	2008	2018
Moran's Index	0.133702	0.131808	0.114997
Expected Index	−0.000322	−0.000334	−0.000349
Variance	0.000007	0.000007	0.000004
z-score	50.899377	49.497299	56.709045

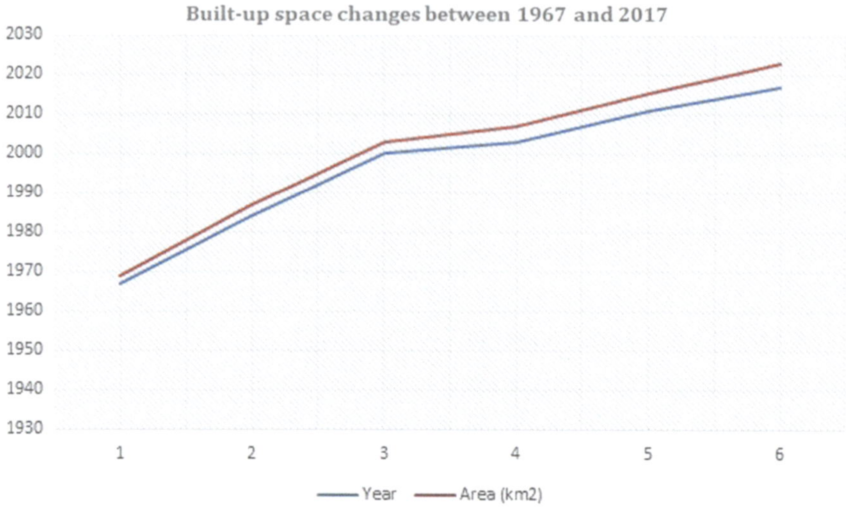

Fig. 3 Shows the built-up space changes by graph around the Ramesseum Temple, Medinet Habu, and Seti I temple at Luxor between 1967 and 2017

revealed that the urban area increased during the long-term period. These changes highlighted that the urban sprawling increased exhibiting the main direction in the changes chronology. The main phenomenon in the encroachment is the increment in built-up area between 1967 and 2018 in all the study areas (Figs. 3, 4, 5 and 6).

3.2 Hot Spot Analysis (Getis-Ord Gi*)

Getis-Ord and Hot Spot results revealed that the urban sprawling around Medinet Habu and the Ramesseum and Seti I temples was chronology focused around the heritage sites that appeared in the red color. This phenomenon defined the geo-environmental risk increment around these sites in the last decades (Figs. 7, 8, 9, 10, 11 and 12).

3.3 Spatial Cluster Analysis (Ripleys K Function)

The result of the spatial Cluster Analysis (Ripleys K Function) tools revealed the spaces between the urban feature near and around Medinet habu, Ramesseum and Seti I temples between 1967 and 2000 became closer that appear in the Moran's Index, Expected Index, Variance, z-score values (Table 2). The distance between the Expected K and Observed K in the Function Graphic showed that they became very

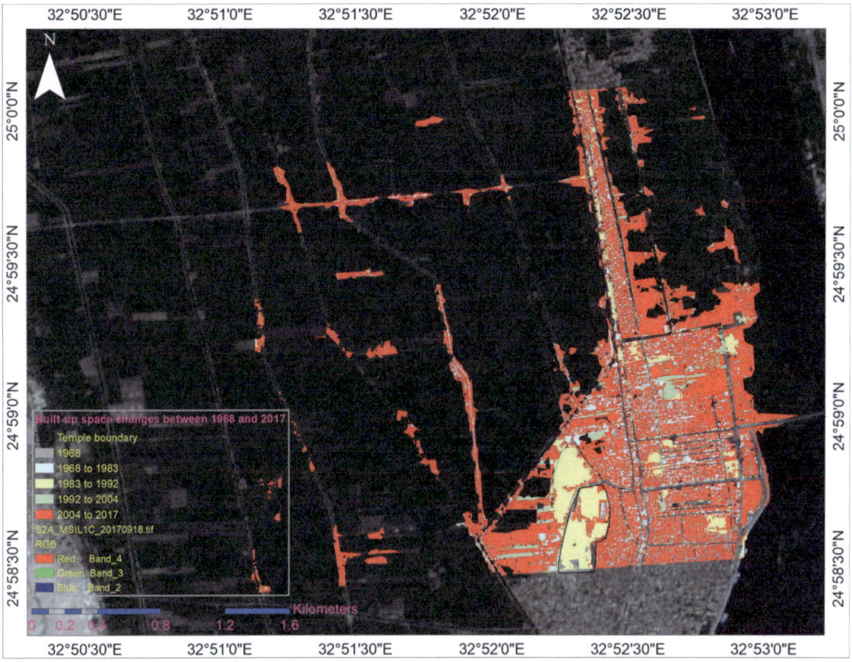

Fig. 4 Shows the built-up space changes around Horus Temple at Edfu between 1968 and 2017

close which means that the urban sprawling was the main phenomenon in the studied period (Figs. 13, 14 and 15).

3.4 *Spatial Autocorrelation Moran's I*

The result of the Spatial Autocorrelation Moran's I Given the z-score of 50.8993770335, that means that there is a less than 1% likelihood that this clustered pattern could be the result of random chance for the built-up area around the Catacomb of Mostafa Kamel at Alexandria in 1999. But, given the z-score of 49.4972994962, there is a less than 1% likelihood that this clustered pattern could be the result of random chance for the built-up area in 2009. On the other hand, given the z-score of 56.7090451833, there is a less than 1% likelihood that this clustered pattern could be the result of random chance for the built-up area in 2018 (Figs. 16, 17 and 18). These values reflect that the urban encroachment was the main phenomenon around the Catacomb of Mostafa Kamel between 1999 and 2018.

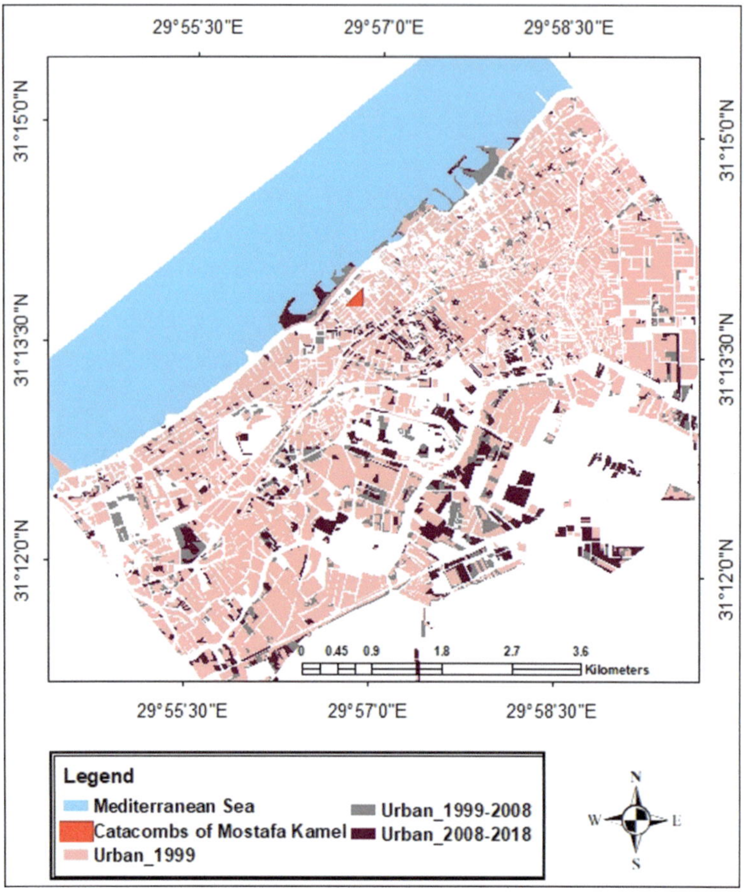

Fig. 5 Shows the built-up space changes around the Catacomb of Mostafa Kamel at Alexandria between 1999 and 2018

3.5 Band Indices

The result of the MNDWI and NDBI indices pointed out that the changes occurred in Qayitbay Citadel at Rosetta between 1984 and 2017 were actually enormous. Significant changes in the built-up area focused around the Citadel area in the West and South sides. The encroachment in the built-up area appeared in black color in the MNDWI analysed satellite image, but showed in white color in the NDBI analysed satellite image (Figs. 19, 20, 21 and 22).

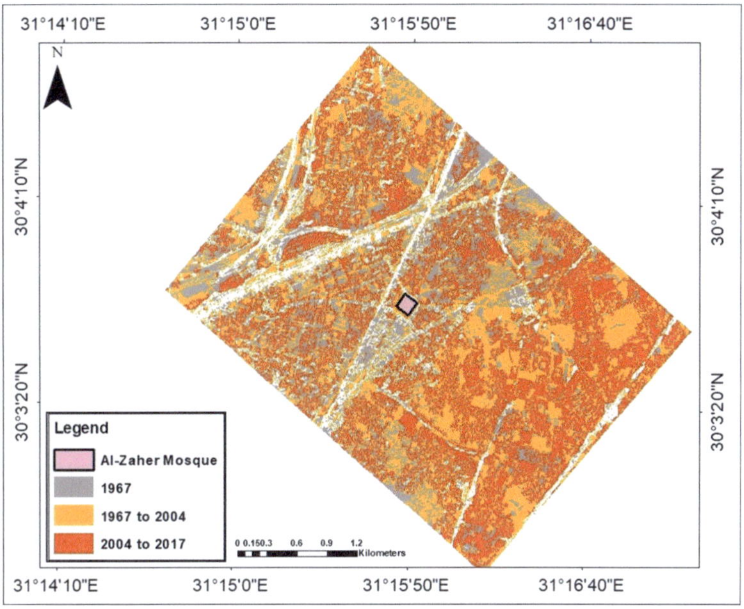

Fig. 6 Shows the built-up space changes around Al-Zaher Mosque at Cairo between 1967 and 2017

4 Recommendation

Remote sensing and GIS techniques can be met by a Zonation System in Alexandria (Lasaponara et al. 2018), Cairo, Rosetta, Luxor, and Edfu archaeological sites that applies different management policies to different boundary zones (according to UNESCO) (http://whc.unesco.org/en/soc/3597). The distances have been chosen as a result of the environmental situation in the study area. The archaeological area must be surrounded by three areas. The first boundary is between the heritage and the core area (monitoring 50 M). The second boundary is between the core and the buffer zone (education, training-human settlements and research station or experiment 50 M). The last is between the buffer zone and transition zone (tourism and recreation 50 M) (Fig. 23). Depending on the some suggested points, drainage system has been proposed to withdrawal the groundwater. The suggested model is used to withdraw the wastewater slowly. The wastewater will be transferred with tubes to water recycle station (Fig. 24).

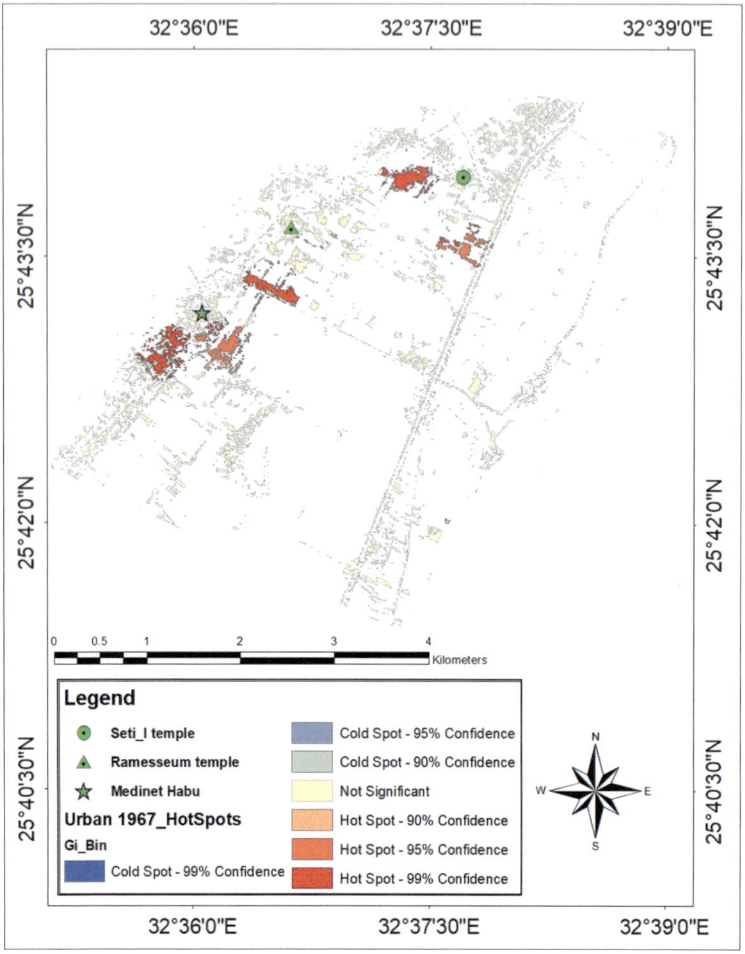

Fig. 7 Shows the hot spotted regions around Medinet Habu, Seti I temple, and Ramesseum temple in 1967

5 Conclusion

Most of the Egyptian monuments are suffers from groundwater. The problem is increased by the poor drainage system, urban sprawling, agriculture encroachment, and seawater rising around these heritage sites. Change detection analysis showed the differences between images of the same area at different times. Suggestions and Solutions established a suitable engineering system groundwater lowering in the study area. The cases studied presented in this paper show the feasibility of

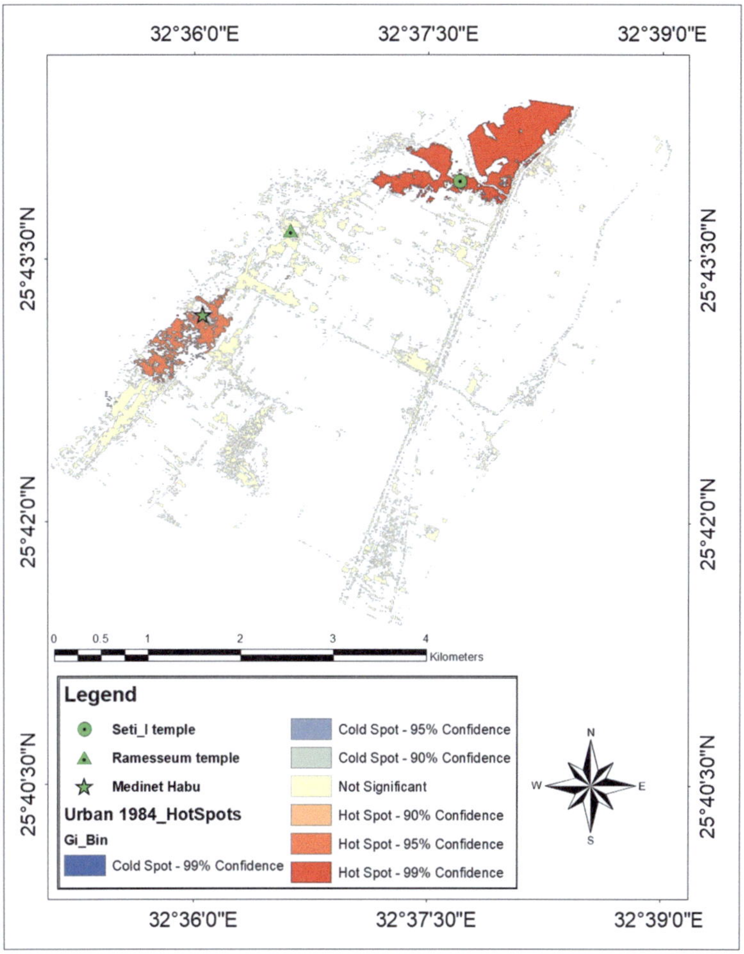

Fig. 8 Shows the hot spotted regions around Medinet Habu, Seti I temple, and Ramesseum temple in 1984

integrating aerial images analysis in the area, which can be used as the basis for helping in detect and create a suitable solution to keep the archaeological area. The results of this study highlights the importance of using new scientific tools and techniques in the archaeological geo-prospection.

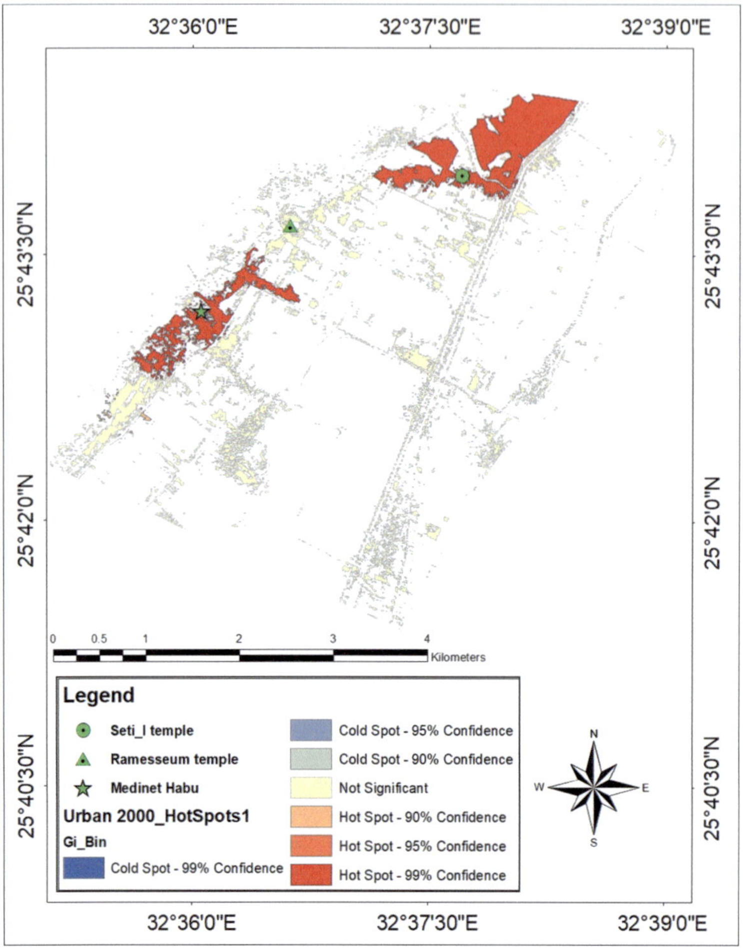

Fig. 9 Shows the hot spotted regions around Medinet Habu, Seti I temple, and Ramesseum temple in 2000

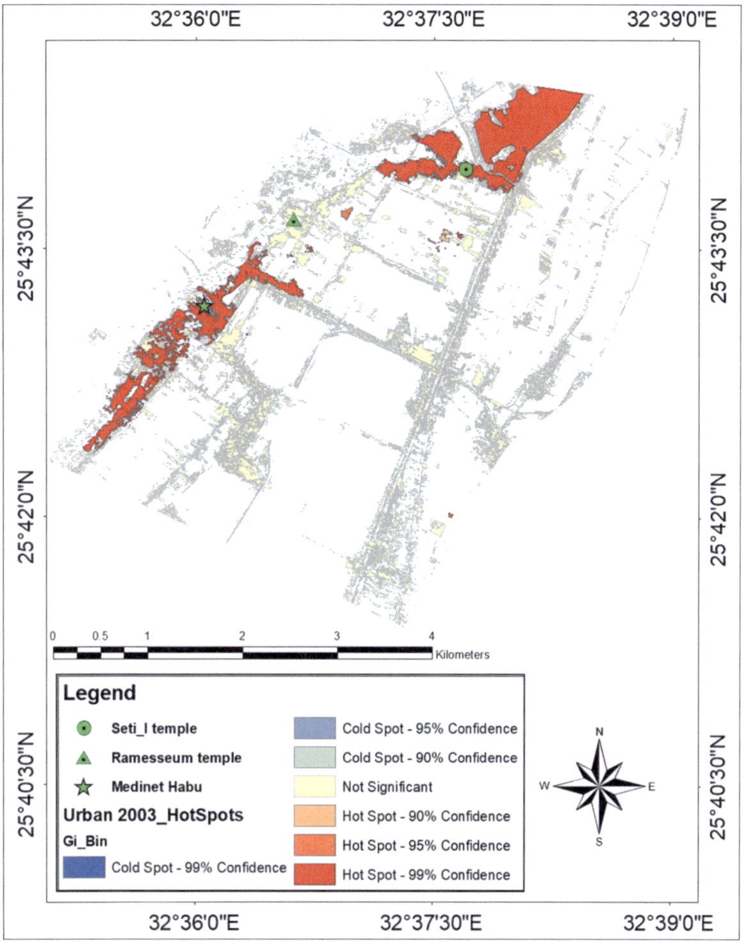

Fig. 10 Shows the hot spotted regions around Medinet Habu, Seti I temple, and Ramesseum temple in 2003

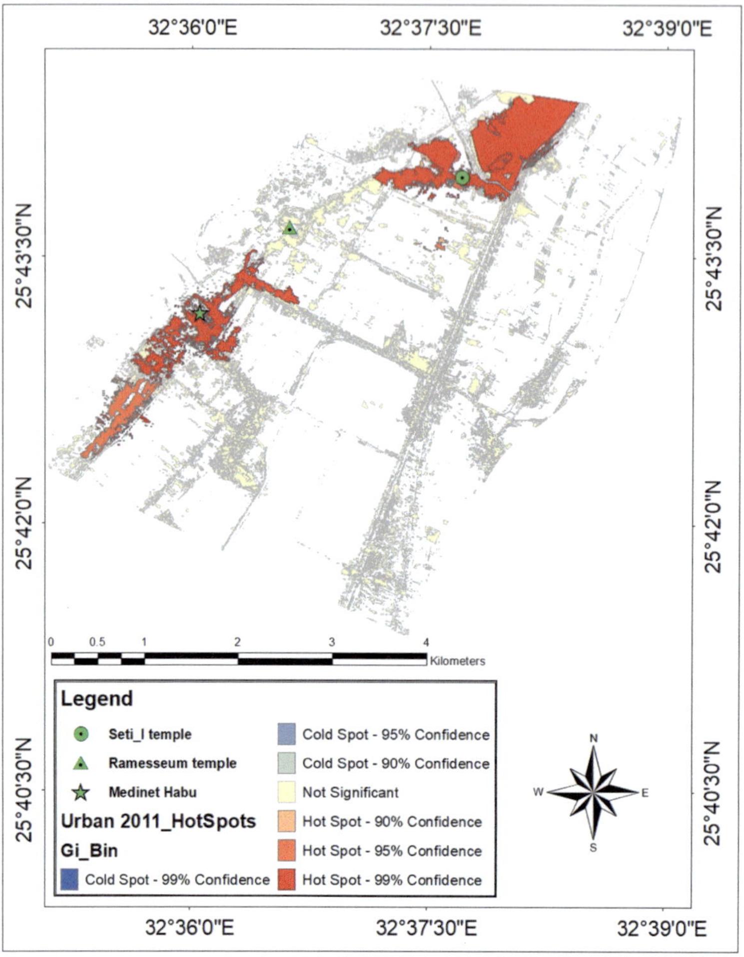

Fig. 11 Shows the hot spotted regions around Medinet Habu, Seti I temple, and Ramesseum temple in 2011

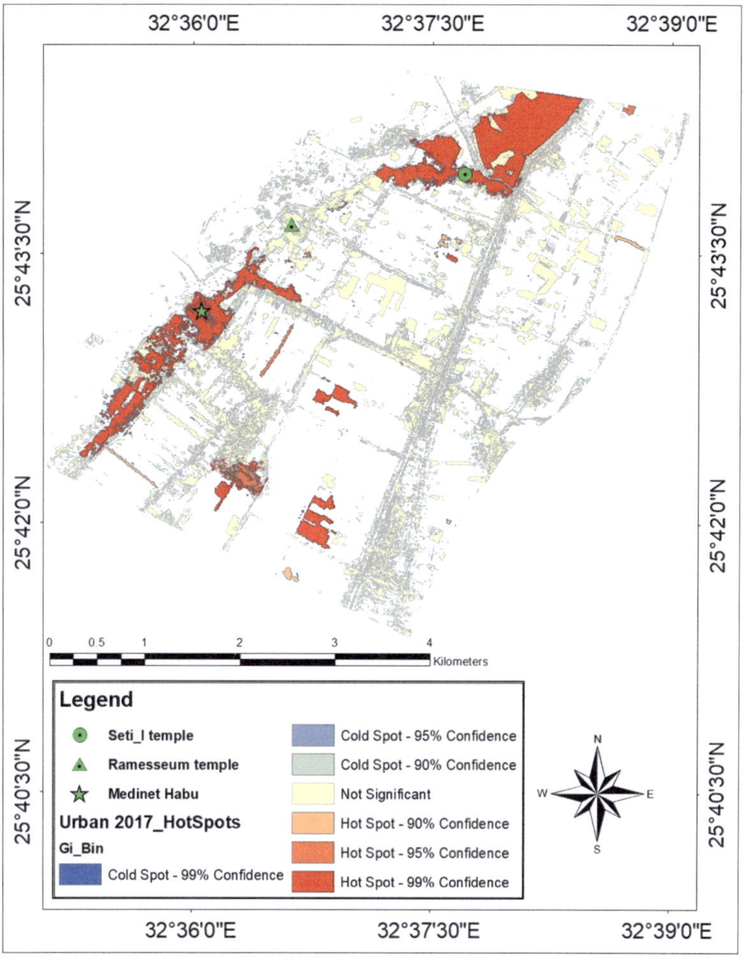

Fig. 12 Shows the hot spotted regions around Medinet Habu, Seti I temple, and Ramesseum temple in 2017

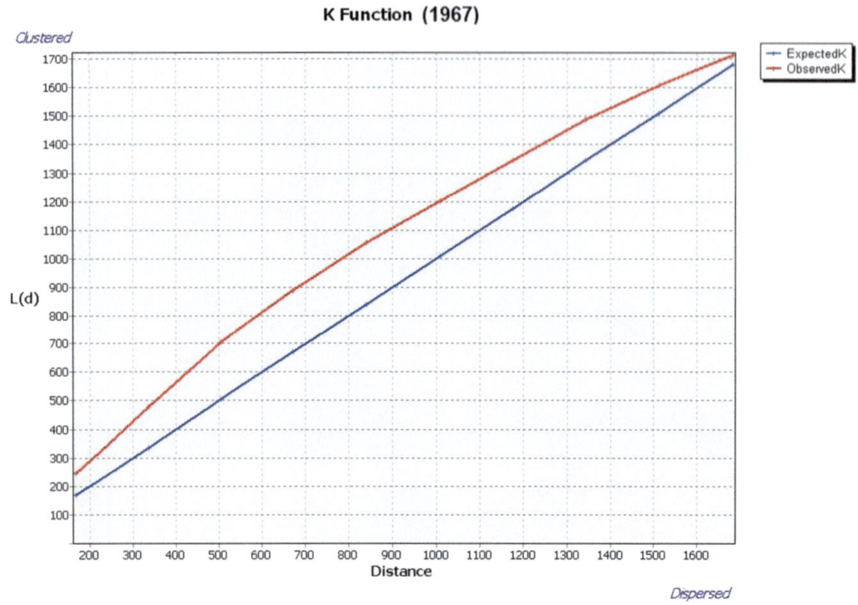

Fig. 13 Shows the spatial cluster value (K Function) by graph around Medinet Habu, Seti I temple, and Ramesseum temple in 1967

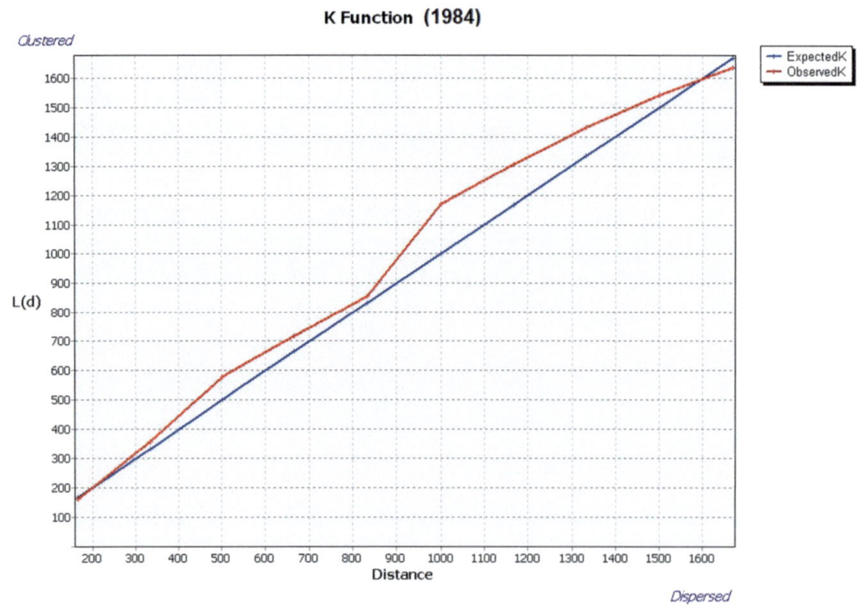

Fig. 14 Shows the spatial cluster value (K Function) by graph around Medinet Habu, Seti I temple, and Ramesseum temple in 1984

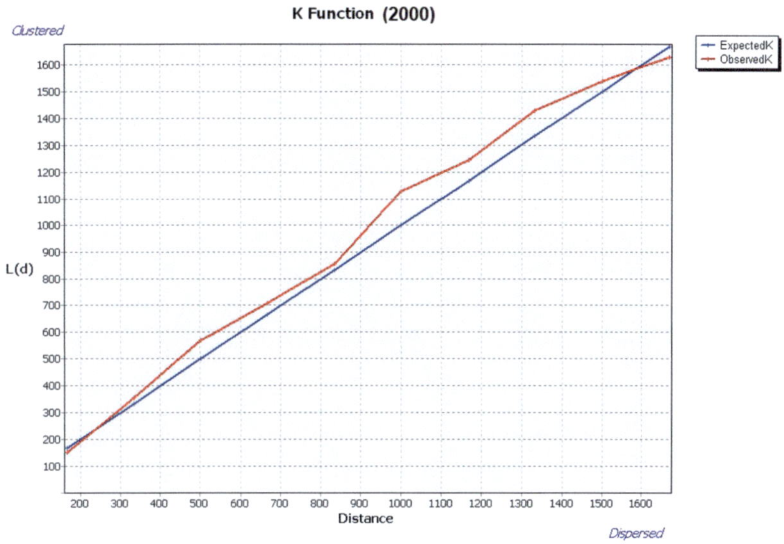

Fig. 15 Shows the spatial cluster value (K Function) by graph around Medinet Habu, Seti I temple, and Ramesseum temple in 2000

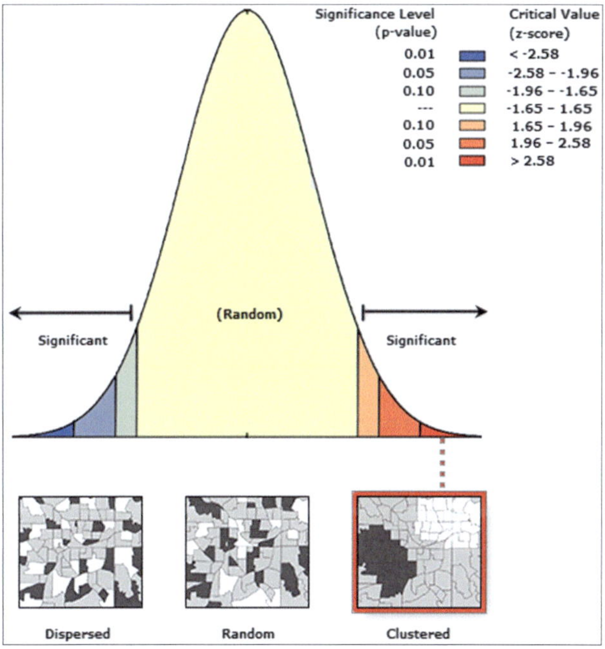

Fig. 16 Shows the z-score value by graph for the built-up area around the Catacomb of Mostafa Kamel at Alexandria in 1999

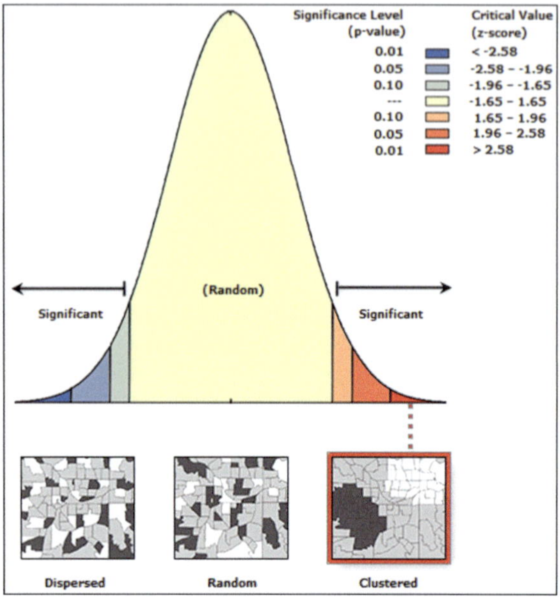

Fig. 17 Shows the z-score value by graph for the built-up area around the Catacomb of Mostafa Kamel at Alexandria in 2008

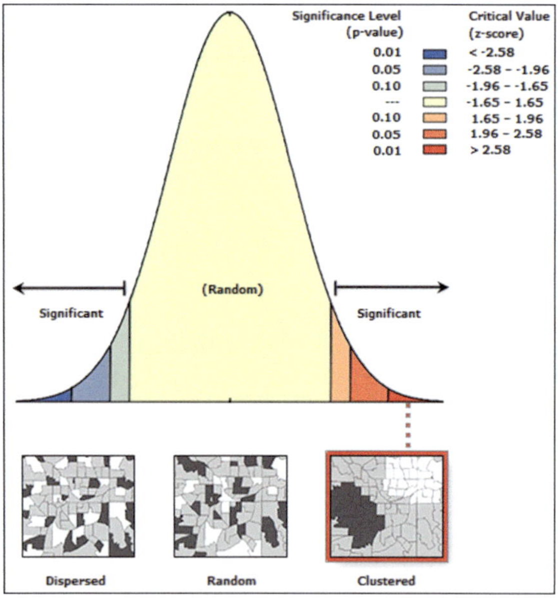

Fig. 18 Shows the z-score value by graph for the built-up area around the Catacomb of Mostafa Kamel at Alexandria in 2018

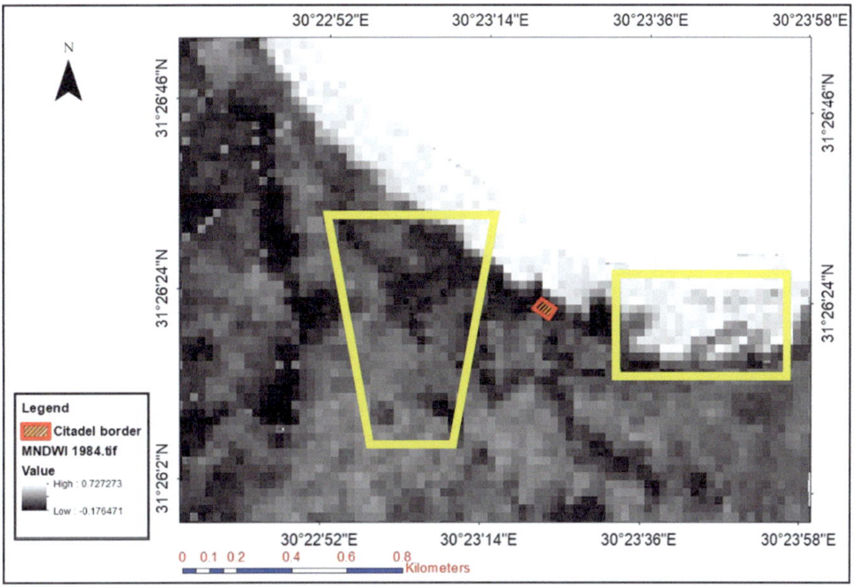

Fig. 19 Shows the result of the MNDWI index bordered by the yellow line for study area of Qayitbay Citadel at Rosetta by Landsat image in 1984

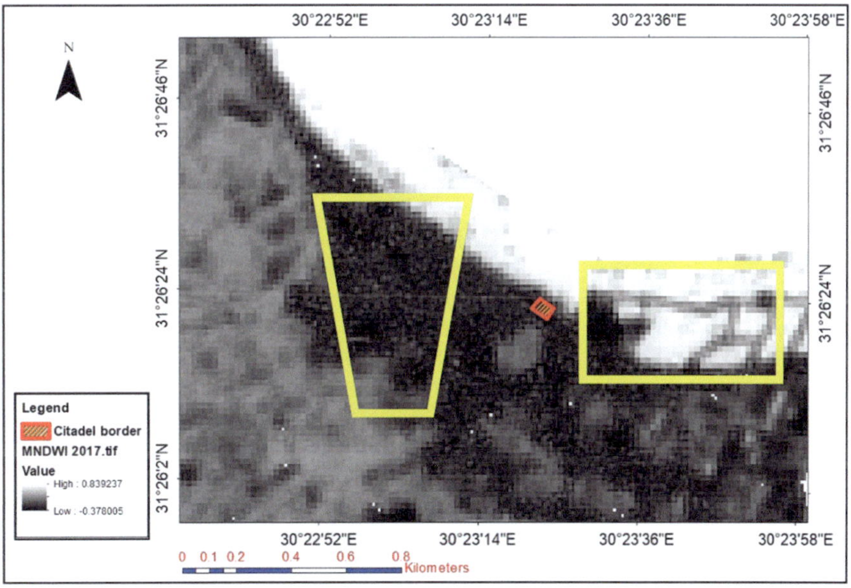

Fig. 20 Shows the result of the MNDWI index bordered by the yellow line for study area of Qayitbay Citadel at Rosetta by Sentinel2 image in 2017

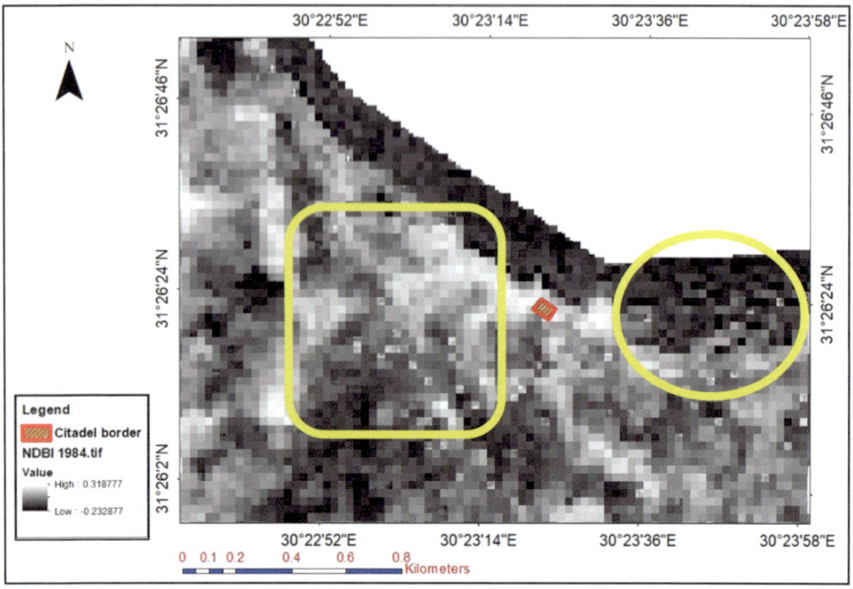

Fig. 21 Shows the result of the NDBI index bordered by the yellow line for study area of Qayitbay Citadel at Rosetta by Landsat image in 1984

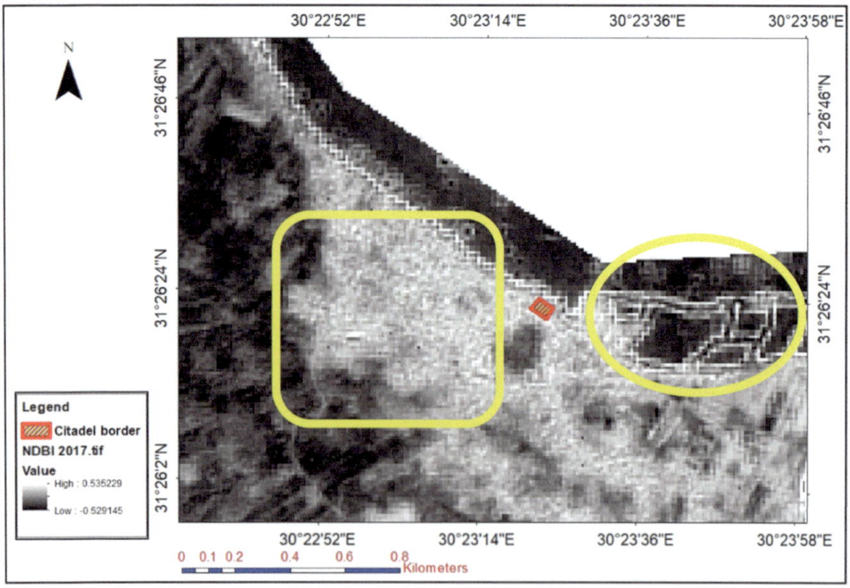

Fig. 22 Shows the result of the NDBI index bordered by the yellow line for study area of Qayitbay Citadel at Rosetta by Sentinel2 image in 2017

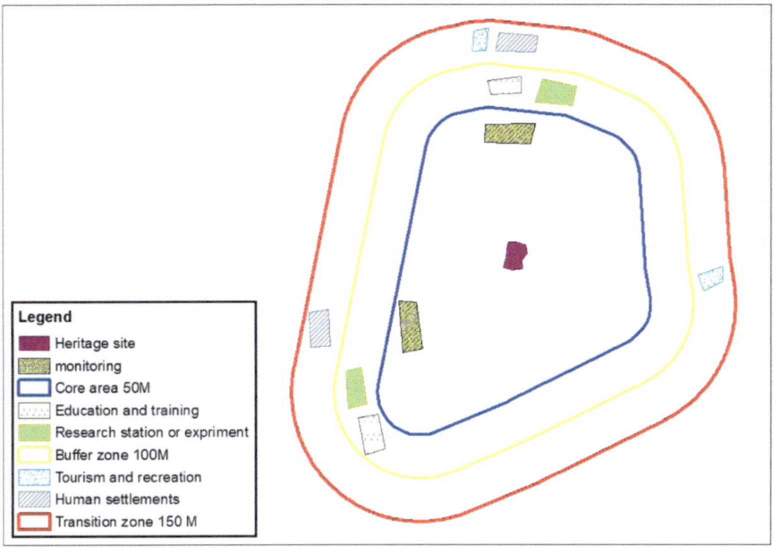

Fig. 23 Shows imagination plan for the recommended activities around the heritage site according to UNESCO suggestion

Fig. 24 Shows suggested model for decreasing the groundwater level around the heritage site

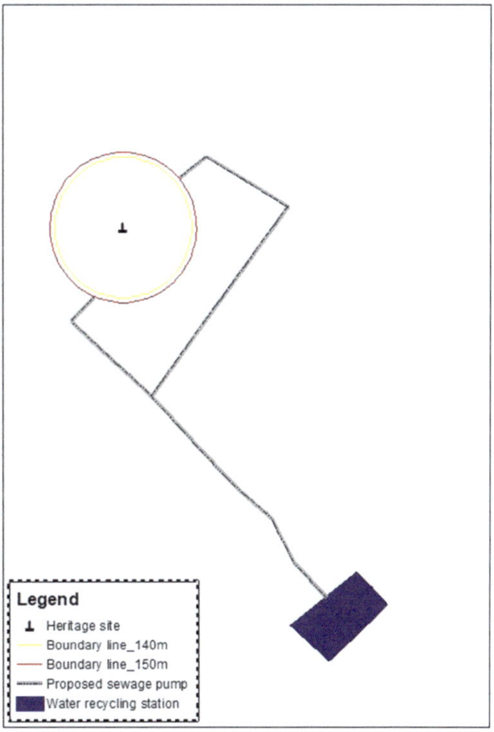

Acknowledgements The authors would like to express them appreciation to National Authority for Remote Sensing and Space Sciences (NARSS), Faculty of Arts, University of Kafrelsheikh (KFS), and National Research Centre, Italy (CNR) for supporting and funding the publication.

References

Abate N, Elfadaly A, Masini N, Lasaponara R (2020) Multitemporal 2016–2018 sentinel-2 data enhancement for landscape archaeology: the case study of the Foggia province, Southern Italy. Remote Sens 12(8):1309

Elfadaly A, Wafa O, Abouarab MA, Guida A, Spanu PG, Lasaponara R (2017a) Geo-environmental estimation of land use changes and its effects on Egyptian Temples at Luxor City. ISPRS Int J Geo Inf 6(11):378

Elfadaly A, Lasaponara R, Murgante B, Qelichi MM (2017b) Cultural heritage management using analysis of satellite images and advanced GIS techniques at East Luxor, Egypt and Kangavar, Iran (a comparison case study). In: International conference on computational science and its applications. Springer, Cham, pp 152–168

Elfadaly A, Attia W, Qelichi MM, Murgante B, Lasaponara R (2018a) Management of cultural heritage sites using remote sensing indices and spatial analysis techniques. Surv Geophys 39(6):1347–1377

Elfadaly A, Attia W, Lasaponara R (2018b) Monitoring the environmental risks around Medinet Habu and Ramesseum Temple at West Luxor, Egypt, using remote sensing and GIS techniques. J Archaeol Method Theory 25(2):587–610

Elfadaly A, Lasaponara R (2019a) Cultural heritage management using remote sensing data and GIS techniques around the archaeological area of ancient Jeddah in Jeddah city, Saudi Arabia. Sustainability 12(1):1–15

Elfadaly A, Lasaponara R (2019b) On the use of satellite imagery and GIS tools to detect and characterize the urbanization around heritage sites: the case studies of the catacombs of Mustafa Kamel in Alexandria, Egypt and the Aragonese Castle in Baia, Italy. Sustainability 11(7):2110

Elfadaly A, Abouarab MAR, El Shabrawy RRM, Mostafa W, Wilson P, Morhange C, Silverstein J, Lasaponara R (2019a) Discovering potential settlement areas around archaeological tells using the integration between historic topographic maps, optical, and radar data in the northern Nile delta, Egypt. Remote Sens 11(24):3039

Elfadaly A, Murgante B, Qelichi MM, Lasaponara R, Hosseini A (2019b) A comparative analysis of temporal changes in urban land use resorting to advanced remote sensing and Gis in Karaj, Iran and Luxor, Egypt. In: International conference on computational science and its applications. Springer, Cham, pp 689–703

Elfadaly A, Abate N, Masini N, Lasaponara R (2020) SAR sentinel 1 imaging and detection of palaeo-landscape features in the Mediterranean area. Remote Sens 12(16):2611

Lasaponara R, Elfadaly A, Attia W (2016) Low cost space technologies for operational change detection monitoring around the archaeological area of Esna-Egypt. In: International conference on computational science and its applications. Springer, Cham, pp 611–621

Lasaponara R, Murgante B, Elfadaly A, Qelichi MM, Shahraki SZ, Wafa O, Attia W (2017) Spatial open data for monitoring risks and preserving archaeological areas and landscape: Case studies at Kom el Shoqafa, Egypt and Shush, Iran. Sustainability 9(4):572

Lasaponara R, Elfadaly A, Attia W (2018) Using remote sensing and GIS techniques for monitoring the environmental status the problems and the solutions around Esna temple at Luxor, Egypt, p 1983

Optical downloading data (Landsat, Orbview, and Corona), USGS (2020). http://edcsns17.cr.usgs.gov/EarthExplorer/. Accessed 30 Oct 2020

Sentinel2 downloading data. https://scihub.copernicus.eu/dhus/#/home. Accessed 30 Oct 2020

UNESCO (2020) http://whc.unesco.org/en/soc/3597. Accessed 30 Oct 2020

Rome Walls Satellite Monitoring and Protection by Prothego Methodology

Gabriele Leoni, Daniele Spizzichino, Marina Marcelli, and Cristina Carta

Abstract The Aurelian Wall, one of the monuments belonging to the historic centre of Rome (within the UNESCO world heritage list since 1980) was built at the end of the third century (271-275 AD). It ran over 19 km, with 18 main gates and 383 towers. In addition, the Janiculum Wall was erected in 1641-1644 AD to protect the Vatican and the *Trastevere* district. The present paper describes the activities carried out jointly between Geological Survey and *Sovrintendenza Capitolina*, including the implementation of dedicated Geographic Information System (GIS) platform as a data repository and spatial data elaboration. The whole research was also implemented within the general framework of PROTHEGO project (www.prothego.eu). The damage information layer has been compared with detailed urban hazard maps (*e.g.* seismic, landslide, subsidence) and with the PS (InSAR measured points). The satellite monitoring data and analysis is based on the processing of COSMO-SkyMed image data (time interval since April 2011 to March 2014). These results, made possible thanks to a multidisciplinary approach, give an important contribution to define priorities of conservation intervention.

Keywords UNESCO · Satellite monitoring · Rome · Cultural Heritage

---

G. Leoni · D. Spizzichino (✉)
ISPRA–Dipartimento per il Servizio Geologico d'Italia, Rome, Italy
e-mail: daniele.spizzichino@isprambiente.it

G. Leoni
e-mail: gabriele.leoni@isprambiente.it

M. Marcelli · C. Carta
Sovrintendenza Capitolina – Direzione Interventi su Edilizia Monumentale, Rome, Italy

G. M. El-Qady and C. Margottini (eds.), *Sustainable Conservation of UNESCO and Other Heritage Sites Through Proactive Geosciences*, Springer Geology,
https://doi.org/10.1007/978-3-031-13810-2_33

1 Introduction

The Historic Centre of Rome has been added to the UNESCO World Heritage List in 1980; ten years later, following the request made by the Holy See to expand the site to the extraterritorial properties inside the site of Rome and the Basilica of St. Paul Outside the Walls, the International Council of Monuments and Sites stated the new registration in the World Heritage List under the name:"*The Historic Centre of Rome, the properties of the Holy See in that City enjoying Extraterritorial Rights and S. Paolo fuori le Mura*". The property includes a series of monuments of incomparable artistic value, produced over almost three millennia of history and encompasses the whole historic center of Rome, included within the urban walls, which can be considered a real symbol of the UNESCO Site.

The source of the municipal properties comes from a law of 1847, the *motu proprio* of pope Pius IX, which defined the powers of the new Municipality with regard to public services, in particular the maintenance of aqueducts, buildings and roads. With this law the office "Public education, ancient and modern monuments ", the future Superintendence, was born. Today the Cultural Heritage owned by the Capital is maybe the most consistent and variegated in the world: it consists of ancient and modern museums, monuments, fountains, historical gardens and archaeological sites and monuments. The mission of the *Sovrintendenza Capitolina* is mainly related to the conservation, the management and the enhancement of all historical, artistic, and archaeological properties of the Municipality of Rome.

The Aurelian Wall, built at the end of the third century (271-275 AD) when Rome was a city with a population of more than 1 million people, ran for over 19 km, with 18 main gates and 383 towers. It has been used as a significant military defence of the city of Rome during the Middle Age and after, until 1870, when the *Bersaglieri* Corp of the Kingdom of Italy breached the wall near Porta Pia and captured Rome.

The wall also defined the boundary of the city of Rome up until the nineteenth century, with the built-up area being confined within the walled area, that is today the UNESCO Site (Spizzichino et alii, 2016). In 1641-1644 a stretch of defensive wall was erected by Pope Urban VIII, as a completion of Leonin Wall, to defense the Vatican Hill and the *Trastevere* district: the Janiculum Wall (*Mura Gianicolensi*).

The uninterrupted life of the monument for over 1700 years, the continuous restoration works, the cuts and demolitions for urban development had weakened the structure. Recently, after the collapse of a part of the northern sector, a newspaper called the *Mura Aureliane* "the giant with feet of clay", because of its fragility. Now the forces of the *Sovrintendenza* are mainly focused on counteracting the deterioration of the monument, facing damages caused by water (rain and damp), air pollution, stress due to traffic vibrations, vegetation and biological agents besides unpredictable events such as earthquakes.

The *Progetto Osservatori* (Observers Project) is a GIS created in 2010 as a support tool for operational decisions of the *Sovrintendenza*. It consists of a database and procedures, aimed at programming maintenance interventions for the conservation of monuments under observation, in particular city walls and ancient aqueducts. The

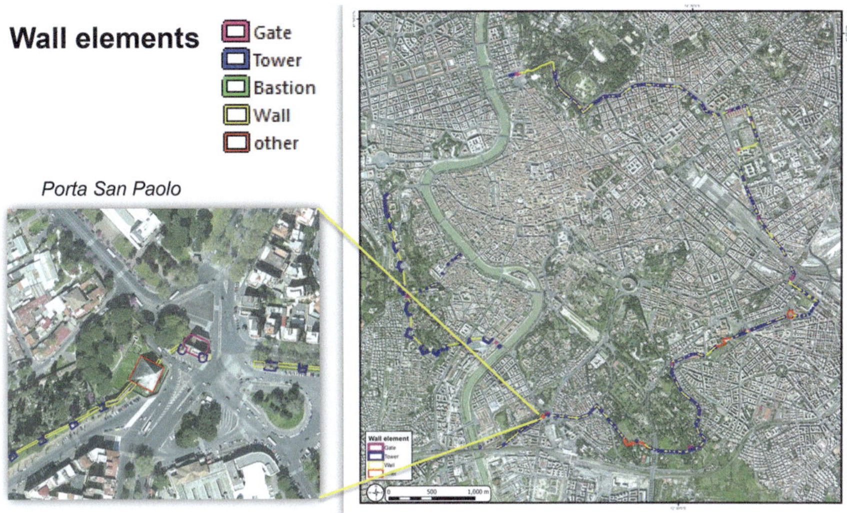

Fig. 1 Mura Aureliane and Gianicolensi overview

working team has been composed by an interdisciplinary group of technicians of different origin: archaeologists, architects, restorers, engineers. After having mapped the singular elements of the walls and proceeded with field survey, an overall analysis of hazard and vulnerability has been realized and a preliminary risk map of the structures has been implemented (see Fig. 1).

So far more than 700 surveys have been recorded and 13.5 km of roman walls and 3.6 km of papal walls are monitored. Currently the purpose of the Project is to identify priority areas of intervention and to carry out a cost analysis, in order to have an effective tool to plan restorations in a short time and to prevent catastrophic events.

2 Geological and Geohazard Setting

Rome has been established on the alluvial plain of the Tiber River, between the western margin of the Apennine chain and the Tyrrhenian Sea. Current geological setting derived from the Plio Pleistocene extensional phase due to the opening of the Tyrrhenian sea, that caused the subsidence of the chain margin and the volcanism along the coast. The result was the widening of the Tiber valley, and the subsequent alternating depositional and erosive phases related to the fluctuations of the sea level due to Quaternary climate oscillations. During the depositional phases the valley was filled by marine and continental deposits, and by volcanic deposits, mainly pyroclastic. (Mattei et al. 2008) (see Fig. 2).

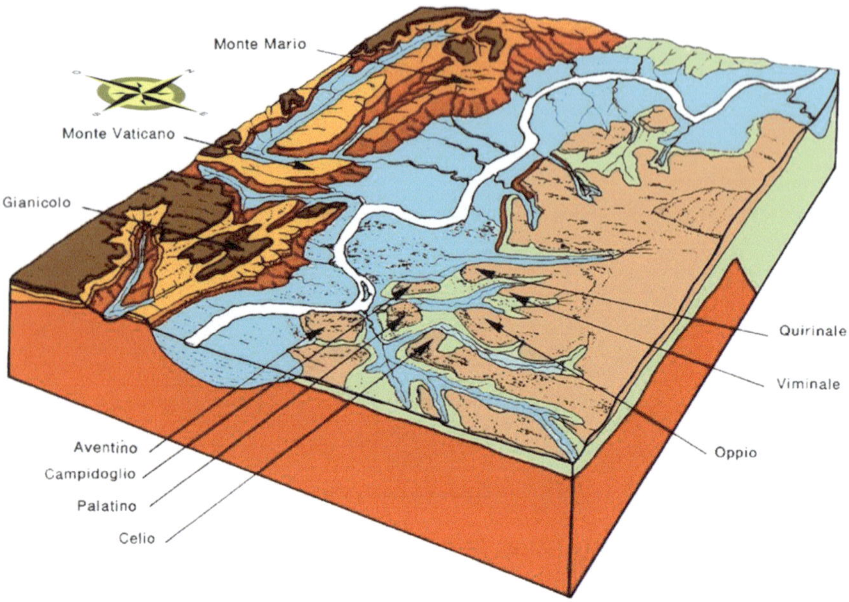

Fig. 2 Geological setting (Faccenna et al. 1995)

The city was founded where the wide valley narrows among the foothills made of pyroclastic deposits from Colli Albani (southeast), that from Sabatini Mountains (northwest) and the structural high made of Neogenic marine deposits of Monte Mario and Vatican Hill (west).The geological setting of Rome, along with the most recent geomorphological evolution, driven by climate fluctuations and shaped by Tiber River and its tributaries streams, strongly influenced ancient Rome history and still affects the typology and the spatial distribution of geohazard. Ancient Romans chose to settle on the wide and flat top of the Seven Hills, safe from the frequent floods that, now as then, are the main hazard of the area. During the first urban development the left bank of Tiber River has been reclaimed, by filling the marshes and by building a complex drainage system made of a network buried channels, flowing into the *Cloaca Maxima.*

Since Rome became the Capital of Italy (1870 AD), flood prone areas along Tiber River, upstream and downstream of the city centre, were densely urbanized and a great protection system was implemented, building embankments and reclaiming wide areas. The urban sprawl occupied recent soft alluvial sediments, increasing the geohazard due to natural subsidence and compression of loose sediments and made ground (Carta et al. 2017).

The hill slopes are mostly covered at the top by hard pyroclastic layers that are frequently affected by rockfall, due to erosion of underlying soft deposits. Although landslide hazard is not widespread, locally it can produce severe effects because of the dense urbanization of the area, as at Monteverde Hill.

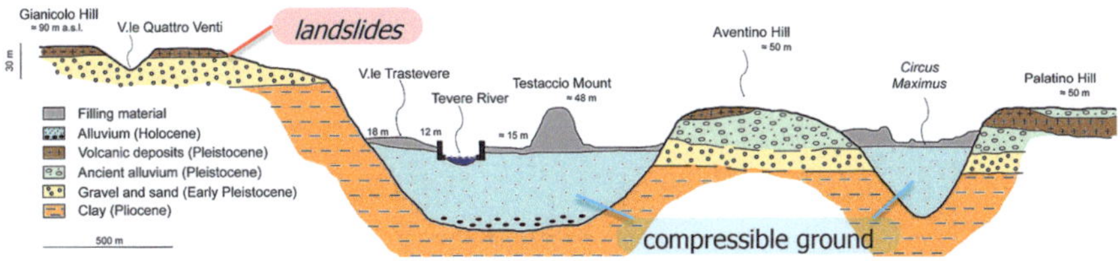

Fig. 3 Geohazard setting (modified, from: Del Monte et al. (2013))

The seismicity of the city is mainly related to activity of the Alban Hills, that triggers small, shallow and local earthquakes, and to the tectonic activity of the Apennine range, with strong (up to Mw = 7) but far earthquakes. Indeed the seismic amplification can produce severe effects in the areas filled with more recent alluvial deposits and made ground.

Another type of geohazard is due to the network of caves that are often shallow: most of them are old mines of soft volcanic rock (tuffs and pozzolana) used as building material. These caves, that are still not totally discovered, can produce sinkholes due to the fall of the vault.

Finally it is worth mentioning the geohazard due to the late volcanic activity of Alban Hills, that results in weak seismicity, ground uplifts, thermal anomalies and soil gas emissions. (Comerci et al. 2015) (see Figs. 3 and 4).

3 Monitoring

The analysis of Aurelian and Janiculum Walls by InSAR data has been carried out on the Permanent Scatterer (PS) dataset provided by the Italian Ministry for the Environment, Land and Sea (MATTM), obtained in the framework of the *Piano Straordinario di Telerilevamento Ambientale* (PST-A). This dataset consists of PS ground motion databases from 3 different satellite imagery missions: ERS 1/2 (yrs 1992 2000), ENVISAT (yrs 2002 2010) and COSMO SkyMed (yrs 2011 2014). Because of the differences in time span and ground resolution of PS data, each dataset has been analysed separately, looking at major evidences of ground motions.

Because of the resolution and the update, the COSMO-SkyMed dataset provides information better than the previous (Table 1). It reaches a density of PS of nearly one point per hundred square meters, so the analysis focused on this dataset.

Unfortunately the comparison of Ascending and Descending geometry highlights that they seem to lead to a different behaviour of the overall vertical movement (Fig. 5): the Ascending geometry shows a considerable stability of eastern and western sides, where recent sedimentary and volcanic deposits lay over the bedrock, while central and southern zones, characterized by more thick alluvial deposits from the filling of the Tiber river valley, seem to be affected by lowering, *i.e.* subsidence due to compaction of soft deposits. For Ascending geometry, the average displacement

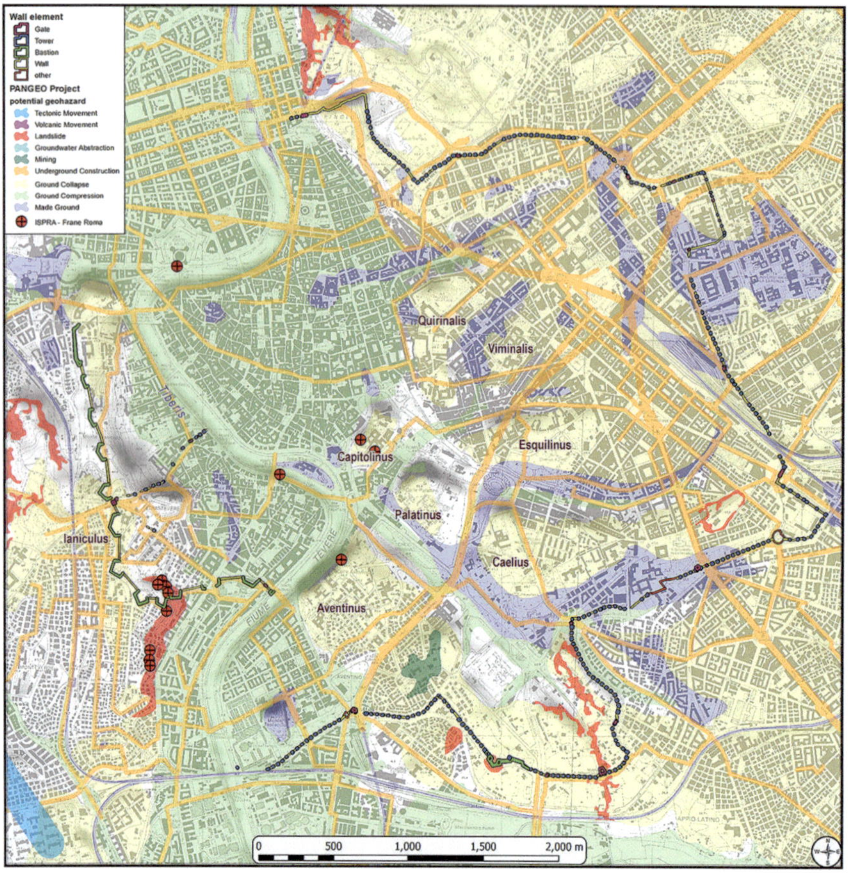

Fig. 4 Geohazard assessment map of Rome Walls

Table 1 PS density per km^2 for Rome City center area

PS dataset	Data range	ASCending	DESCending
ERS-1/2	04/1992–12/2000	208	341
ENVISAT	11/2002–07/2010	716	747
COSMO-SkyMed (CSK)	04/2011–03/2014	9864	8475

rate along the Line Of Sight (LOS) is −0.75 mm/yr in the total study area, according with the areal prevalence of the subsiding zones. The Descending geometry reflects the same trend in the different zones, but the average displacement rate along the Line Of Sight (LOS) is +0.36 mm/yr, i.e. 1.1 mm/yr higher than the Ascending, thus pointing out an overall uplift that is not consistent with the geologic setting.

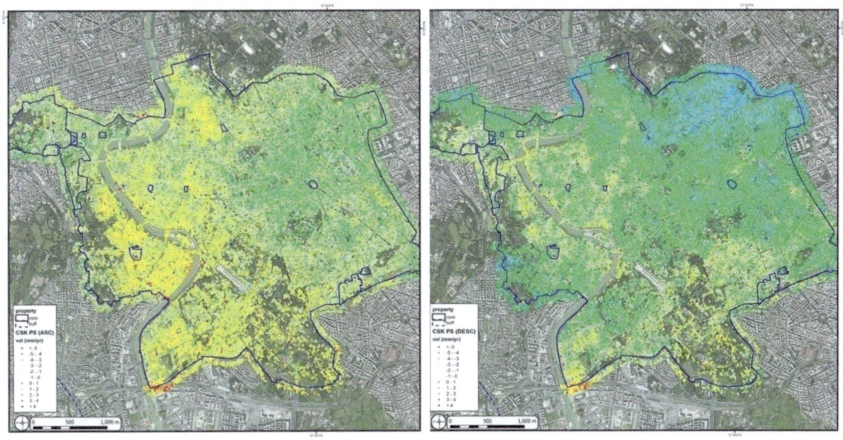

Fig. 5 COSMO-SkyMed PS displacement velocity: ASCending (left), DESCending (rigth). Colour legend: Green = stable; warm colours = lowering; cold colours = uplift

For the Descending geometry, in the northeastern sector, the general uplift hit the anomalous value of 1.5 mm/yr, i.e. nearly 2 mm greater than the Ascending geometry. Because of this uncertainty for the Descending geometry, only the Ascending geometry has been used to investigate the relationship between ground movement from PS and wall's damages.

The planar georeferencing has been analyzed, finding out that it seems to be less accurate than the spatial resolution: despite the difficult identification of PS with respect to their actual ground element, a random analysis has been performed in different zones, looking at alignments of cluster of points depicting geometries rather than single points.

This analysis pointed out an overall planar shift between PS and their actual position of about 2 m toward Southeast for the Ascending geometry and about 3 m toward North for the Descending geometry.

Anyway, the construction technique of PS implies that each point results from the analysis of targets in a 3 m wide pixel, thus the high detail of positioning doesn't agree very well with the georeferencing accuracy, which depends on the grid spacing. The vertical ground motion has been evaluated by statistics on the displacement rate of PS and by the analysis of time series of target points in the whole study area: the result is that in this area the PS displacements between following acquisitions are usually affected by small fluctuations, like a background noise, reasonably due to seasonal variation (e.g.: thermal expansion, water table fluctuations, etc.). These minor variations have been considered negligible under a threshold of **±2 mm**. Moreover, it is usual that measurement points consist of couples of PSs, very close each other and similar in displacement trend. Because of all these constraints on planar and vertical accuracy it seems more meaningful, for the displacement trend, to consider the anomalies of cluster of points, rather than single points. For this reason, each feature of the walls with presence of anomalous displacement from PS

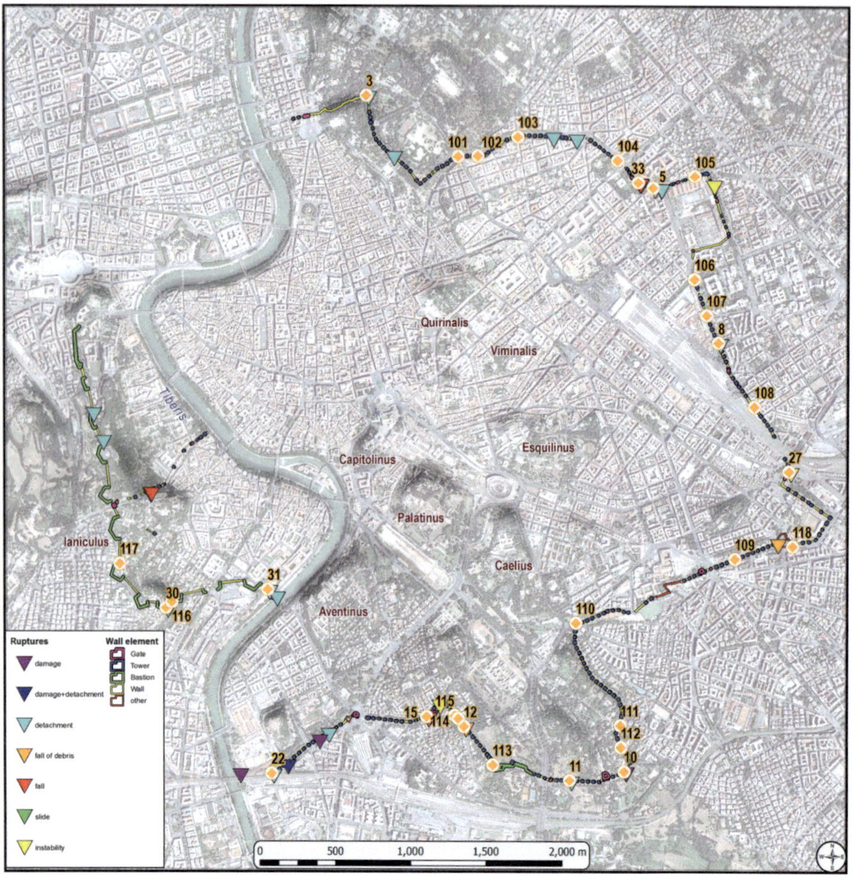

Fig. 6 Map of elements of *Mura Aureliane and Gianicolensi*, ruptures and sites selected for detail survey

and/or hit by one of the known damages has been analyzed in a buffer area of 50 m, in order to highlight potential differential movements (Fig. 6).

4 Exposed Elements Description

Each selected feature described below is marked by an ID, a Name and an Address and it has been analyzed in terms of: i) structural typology of the wall element and eventual damage description, ii) geo-hazard affecting the area, iii) evidences from PS analysis. The latter considers, as mentioned before, only the Cosmo Sky-Med Ascending PS dataset, spanning from 22/04/2011 to 13/03/2014.

All the east portion of the Aurelian Wall, i.e. sectors A-G (Muro Torto—Porta S. Giovanni) and sectors J L (Porta Metronia—Porta S. Paolo), lay over the ignimbrite plateau resulting by the overlap of pyroclastic deposits from Sabatini and Albani volcanoes. The geologic setting causes a potential rockfall hazard in steep slopes and a moderate hazard of collapse of underground cavities.

5 Synthesis of the Analysis

The analysis of the 26 sites highlights different combination of wall ruptures and ground motions. Time history's analysis of PS for the measurement leads to classify ground motion in 3 main displacement styles, positive (i.e. uplift) or negative (i.e. lowering):

1. stable for displacement rate between -2 and $+2$ mm/yr, that is practically the back noise of measures in the whole study area;
2. weak displacement rate is between ± 2 e ± 3 mm/yr, and, although it's quite small, usually points out a significant warning for early ruptures or widespread instability;
3. significant displacement rate usually exceeds ± 3 mm/yr and it is coupled with clear evidences of presence or absence of ruptures.

With respect to the given measurement period (22/04/2011 to 13/03/2014) recorded ruptures may have occurred before (previous), during (in), after (post) or during a longer period (along); but ruptures may be also not clear in a fragile area (susceptible) or clearly absent (none).

Following table analytically shows rupture period and displacement rate measures for each site (see Fig. 7).

6 Some Examples of the PS Application

Some of the anomalies and deformation trends found during the study and summarized in Table 2, are better described below in order to highlighting the limits and potentiality of the technique (see Table 3).

6.1 ID 113-Bastione Del Sangallo

The Bastione is part of an about 400 m long modern (XVI century) fortification that replaced the ancient wall.

This part of the walls was in fact entirely rebuilt by the architect Antonio da Sangallo at the behest of Pope Paul III who feared a new attack to Rome by the

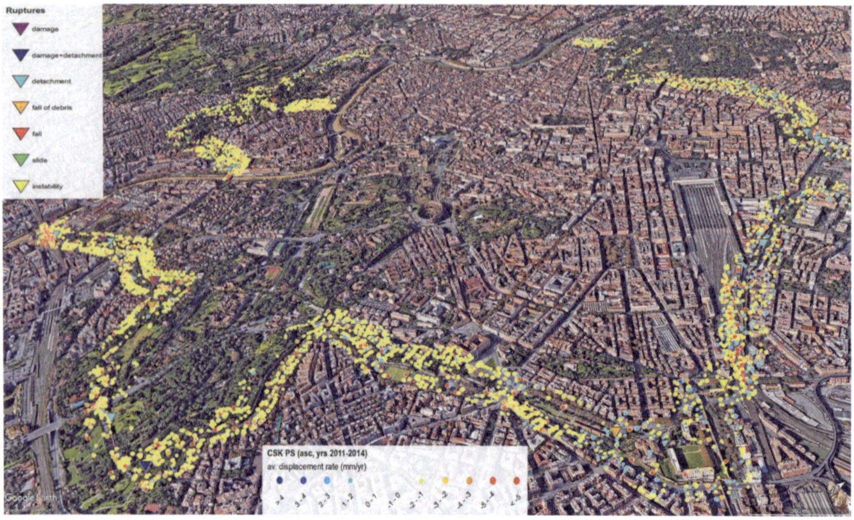

Fig. 7 PS analysis in a buffer area of 50 m

Turks, after the one suffered in 1527 by the mercenary troops of Charles V. Already in 2001, this portion had important problems of instability due to poorly regulated infiltrations and promoted by the uncontrolled roots development.

On western external side can be found, at least since 2008, two vertical cracks that run nearly the total height of the wall.

The study carried out using the satellite monitoring technique has in fact highlighted the presence of a clear deformation anomaly (verified directly in situ) and triggered by the presence of a pair of vertical fractures that cross the entire facing from the top to the base (see Fig. 8). Right on both cracks one PS each points out a 3 mm/yr lowering, a bit greater than the surroundings that is 1.9 mm/yr.

These fractures are probably due to local infiltration phenomena and lack of maintenance of the trees at the top of the wall.

6.2 *ID 114-Part L21-L23*

The wall between Towers L21 and L23 is about 75 m long; it includes a first part, adjacent to Tower L21 with many reconstructions through centuries at different heights and a system of metal rods at the top (Fig. 9).

At the northern edge of this part, adjacent to a recent vehicle arched pass through the walls (Largo Giovanni Chiarini), a couple of PSs records a 3.0 mm/yr lowering, while the surroundings are lowering only by 1.2 mm/yr.

In this specific section, we are faced with a problem of conservative restoration and consolidation that has never been completely solved.

Table 2 Ruptures and displacements data of described features

ID	Feature name	Rupture		Displacement		
		Date	Timing	Site average	Peak value	Differential
3	Muro Torto	04/10/2006	Previous	−0.57	N.a	N.a
101	Part A26-A27	None	No	−0.68	−8.55	−7.87
102	Tower B04	None	No	−0.39	−4.70	−4.31
103	Tower B11	None	No	−0.63	−4.60	−3.97
104	Tower C08	None	No	−0.51	3.50	4.01
33	Part D02-D03	2008–2012, 27/01/2017, 16/05/2017	Along	−0.41	N.a	N.a
5	Tower D05	20/11/2011	In	−0.20	−3.20	−3.00
105	Tower CP11	None	No	−0.03	−7.00	−6.97
8	Part E14-E15	01/11/2007	Previous	0.05	4.75	4.70
108	Tower F07	None	No	−0.18	−5.80	−5.62
27	Part F19	Before 2007	Along	−0.26	−5.80	−5.54
110	Tower H17	None, but fragile	Susceptible	−1.10	−3.20	−2.10
111	Porta Latina	None	No	−1.33	−6.65	−5.32
112	Part K04-K05	None	No	−1.30	−3.25	−1.95
10	Tower K10	Since 2009	Along	−1.62	N.a	N.a
11	Part L07-L08	15/04/2001	Previous	−1.25	N.a	N.a
113	Bastioni Sangallo	Since 2008	Along	−1.91	−3.10	−1.19
12	Part L20-L21	Since 2006	Along	−2.29	−4.60	−2.31
114	Part L21-L23	None, but fragile	Susceptible	−1.20	−3.10	−1.90
115	Tower L23	None	No	−1.77	−4.65	−2.88
15	Tower L31	31/01/2014	In	−1.42	−4.30	−2.88
22	Part M16-M17	March 2011	In	−4.60	−11.04	−6.44
31	Porta Portese	November 2016	Post	−0.93	−2.20	−1.27
30	Bastione05	None, but fragile	Susceptible	N.a	N.a	N.a

(continued)

Table 2 (continued)

ID	Feature name	Rupture		Displacement		
		Date	Timing	Site average	Peak value	Differential
116	Bastione04	From Nov 2011 to Jun 2014	In	−1.00	−5.30	−4.30
117	Part 07-08	Since 2008	Along	−0.80	−6.10	−5.30
3	Muro Torto	04/10/2006	Previous	N.a	N.a	N.a

Table 3 Synthesis of PS displacement versus rupture period

			Rupture period					
Displacement from PS			Previous	In	Post	Along	Susceptible	None
Style	mm/yr	Class	A	B	C	D	E	F
Uplift	>3	I	1					1
Weak uplift	$2 \div 3$	II						
Stable	$-2 \div 2$	III	2		1	3	2	1
Weak lowering	$-2 \div -3$	IV		2		1	1	1
Lowering	<−3	V		2		2		6

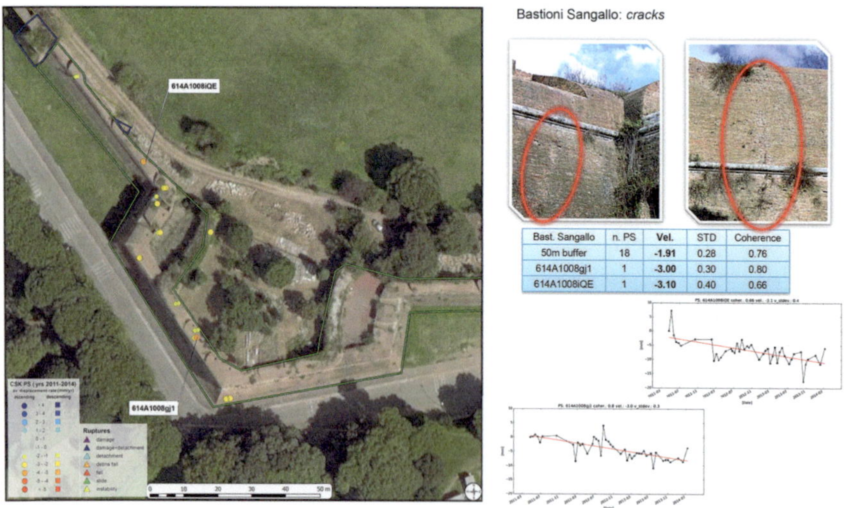

Fig. 8 Cracks affecting the Bastione Sangallo and their temporal evolution

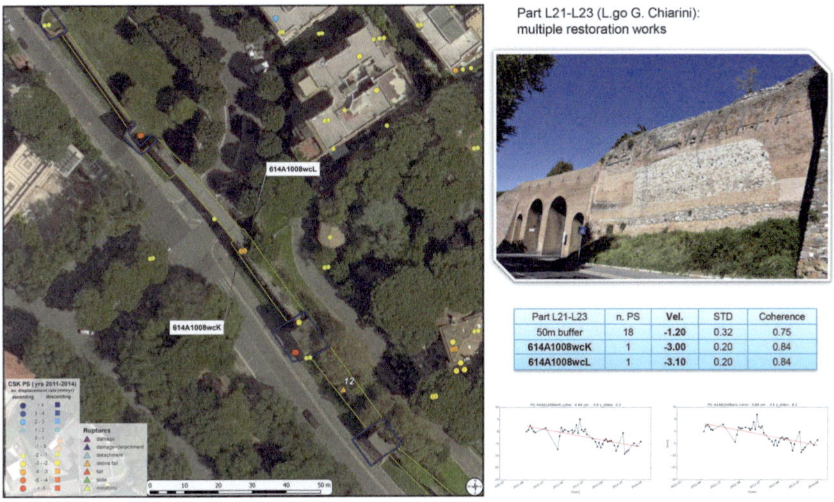

Part L21-L23	n. PS	Vel.	STD	Coherence
50m buffer	18	-1.20	0.32	0.75
614A1008wcK	1	-3.00	0.20	0.84
614A1008wcL	1	-3.10	0.20	0.84

Fig. 9 Displacement due to the presence of unstable multiple restoration works

In this situation, monitoring has in fact highlighted how the various mitigation and safety interventions, that have taken place over the years, have never definitively solved the stability problems connected to the driveway opening of this stretch of wall (see Fig. 9). This specific example demonstrates how the interferometric technique is also very useful to verify the effectiveness over time of consolidation and mitigation interventions.

6.3 *ID 22-Part M16-M17*

During March 2011 curtain wall fell from the Part M16 M17, between the ancient Mattatoio and Viale di Campo Boario buildings inside the walls and the rail bed outside, close to the S. Paolo rail bridge.

In this section, we are in the presence of one of the main interactions between the ancient walls and the modern city transformation. The section was in fact interrupted to allow the passage of one of the main railway lines of the city. In this case we are not faced with a localized problem but with a generalized interaction between soil and the whole structure. The southern rail bed, about 5 m higher than the alluvial plain, was raised during years 1907–1910 for the construction of the rail bridge and it cut the wall causing the loss of about 200 m of Aurelian Walls. The river bank is made of Holocene alluvial clay, interbedded with peat levels, so it is heavily affected by subsidence, due to the presence of compressible ground.

Natural settlement is worsened by stress induced by continuous train transit, even because it is still the only rail bridge of the city. The abundance of metal parts, such as rails, ensure a high PSs density, all of them recording high subsidence rate. The average lowering of about 60 PS around Part M16 M17 is of 3.6 mm/yr, but the distribution is inhomogeneous: inside the wall, PS show an average lowering rate of about 1.7 mm/yr, while outside the wall the average lowering is 4.6 mm/yr. (see Fig. 10). The only 12 PSs, of the entire dataset around the walls, that exceed the lowering rate of 10 mm/yr are right in this area.

It is certainly one of the areas of greatest attention and risk, given the complex interaction between natural hazard, vulnerability and the multiple presence of various exposed elements.

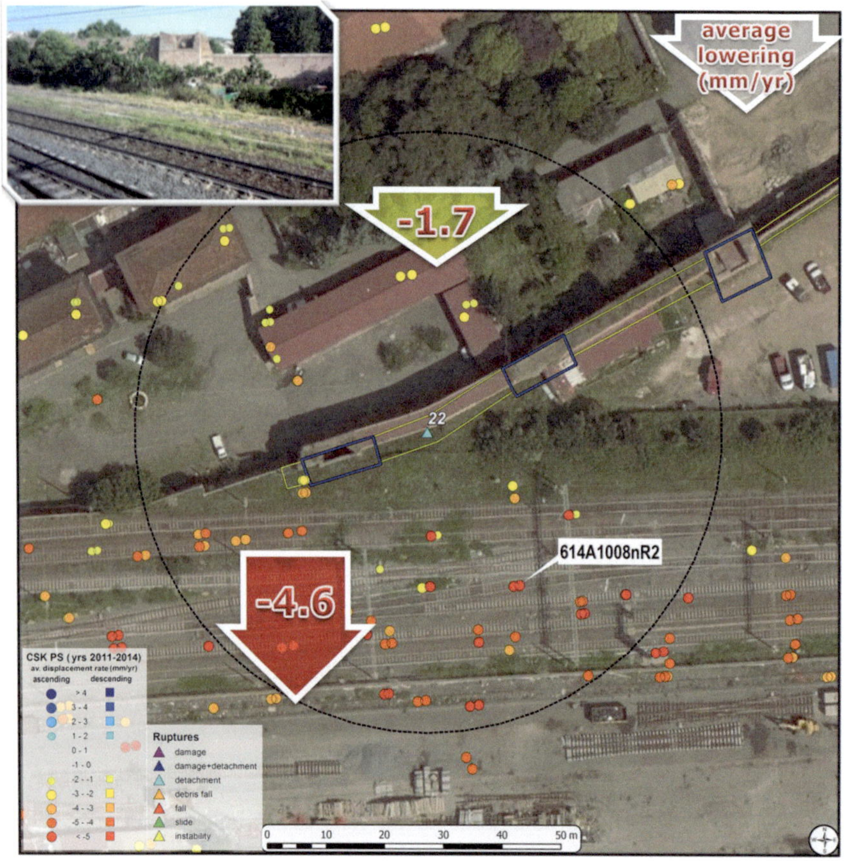

Fig. 10 General subsidence affecting the interaction between walls and railway

7 Conclusion

The present research describes and summarizes the research activities carried out along the Aurelian Wall by using the satellite monitoring techniques.

PS techniques provide a large number of site-wide deformation measurements with small cost and without visual impact. At the same time, PS are proxy data useful to define warning thresholds and to build site scale early-warning systems. The good results foster the improvement of multidisciplinary approach. Ongoing and future high-resolution monitoring will allow stakeholders to plan interventions.

Preliminary results highlighted that the satellite interferometry technique is capable to identify both displacements and deformation trend along the archaeological linear infrastructure belonging to urban area such as walls or aqueduct.

Moreover, this technique is useful to check the effectiveness of restoration and stability works, as well as the effects of maintenance absence along the walls. About this issue it is very important to exchange information with site managers in order to combine monitoring data with site interventions and their temporal evolution.

In situ validation is mandatory to verify each anomaly and it is necessary the ordinary maintenance of the structures, especially concerning weed control.

Damp and water infiltration on the walls must be controlled and prevented, as well as the effectiveness of past interventions.

Future developments of this research will be addressed to: (i) improve future acquisition and processing of VHR satellite interferometry data, also with the possible installation of passive corner reflector; (ii) a combined infrared thermal analysis must be implemented to avoid infiltration an walls rapid collapse, (iii) support for the analysis of in situ monitoring data and possible implementation and integration of the current system.

Acknowledgements The authors are very grateful to all the Sovrintendenza Capitolina ai Beni Culturali staff and technicians for their help and support during all the survey activities.

References

Alberti S, Ferretti A, Leoni G, Margottini C, Spizzichino D (2017) Surface deformation data in the archaeological site of Petra from medium-resolution satellite radar images and SqueeSAR™ algorithm. J Cult Heritage 2510-20. Article no S1296207417300432. https://doi.org/10.1016/j.culher.2017.01.005

Carta C, Cimino MG, Leoni G, Marcelli M, Margottini C, Spizzichino D (2017) SAR interferometry monitoring along the ancient Rome CityWalls–the PROTHEGO project case study. Geophysical Research Abstracts vol 19, EGU2017-13462-1, 2017 EGU General Assembly 2017 © Author(s) 2017. CC Attribution 3.0 License

Cecchi R, Gasparoli P (2010) Prevenzione e manutenzione per i Beni culturali edificati. Procedimenti scientifici per lo sviluppo delle attività ispettive. Alinea, Firenze

Comerci V, Vittori E, Cipolloni C, Di Manna P, Guerrieri L, Nisio S, Succhiarelli C, Ciuffreda M, Bertoletti E (2015) Geohazards monitoring in Roma from InSAR and in situ data: outcomes of the PanGeo project. Pure Appl Geophys 172

Del Monte M, Fredi P, Pica A, Vergari F (2013) Geosites within Rome city center (Italy): a mixture of cultural and geomorphological heritage. Geogr Fis Dinam Quat 36

Esposito D et al (eds) (2017) Le Mura Aureliane nella storia di Roma, 1. Da Aureliano a Onorio. Atti del convegno; Roma, 26 marzo 2015, Roma TrE-Press

Faccenna C, Funiciello R, Marra F (1995) Inquadramento geologico stutturale dell'area romana. Memorie Descrittive della Carta Geologica d'Italia 50

Marcelli M, Carta C, Baranello C (2019) Un sistema informativo geografico per la gestione e il monitoraggio dei grandi complessi monumentali lineari di Roma (mura e acquedotti). Un focus sul settore M delle Mura aureliane. In: Dalla mappa al GIS. Laboratori in rete: ricerca, didattica, progettualità. Atti del XI Seminario di studi storico-cartografici, Roma

Marra F, Rosa C (1995) Stratigrafia e assetto geologico dell'area romana. Memorie Descrittive della Carta Geologica d'Italia 50

Mattei M, Funiciello R, Parotto M (2008) Roma e contesto geodinamico recente dell'Italia Centrale. Memorie Descrittive della Carta Geologica d'Italia 80

Spizzichino D, Margottini C, Brustia E, Comerci V, Dessì B, Guerrieri L, Iadanza C, Trigila A, Vittori E (2016) UNESCO Cultural heritage Vs. Natural hazards at European scale. The PROTHEGO project. In: Proceedings of the 35th IGC 27 August-4 September 2016 Cape Town, South Africa. https://www.americangeosciences.org/geotimes/35th-igc-full-text-abstracts-available

Valagussa A, Frattini P, Crosta GB, Spizzichino D, Leoni G, Margottini C (2020) Hazard ranking of the UNESCO world heritage sites (WHSs) in Europe by multicriteria analysis. J Cult Heritage Manag Sustain Dev 10(4):359–374. https://doi.org/10.1108/JCHMSD-03-2019-0023

Valagussa A, Frattini P, Crosta G, Spizzichino D, Leoni G, Margottini C (2021) Multi-risk analysis on European cultural and natural UNESCO heritage sites. Nat Hazards 105(3):2659–2676. https://doi.org/10.1007/s11069-020-04417-7

New Policies for Management and Conservation of Heritage Sites

The Project of Parco Archeologico Del Colosseo and the Italian Network of Archaeological Parks: From Satellite Monitoring to Conservation and Preventive Maintenance Policies

Alfonsina Russo, Irma Della Giovampaola, Daniele Spizzichino, Gabriele Leoni, Alessandro Coletta, and Maria Virelli

Abstract The monitoring projectof the *Parco archeologico del Colosseo* was inspired by the need to build up a system of protection and conservation, based on the principles of a sustainable exploitation. Within this framework, the Park has launched a static and dynamic monitoring project consisting of five fundamental levels of activities: (i) a database of all the historical data of the monuments, together with the existing graphic and photographic documentation (namely digital documentation archive); (ii) visual monitoring carried out by teams of technicians dedicated to the inspection and control of monuments, also thanks to dedicated app that will allow to send data to the central system; (iii) satellite monitoring (historical analysis of the satellite data) going directly into the system and analysed in order to monitor possible ground deformation; (iv) in situ monitoring from traditional geotechnical instruments; (v) experimental activities, such as the use of H-BIM in the archaeological field, aimed at monitoring by diagnostic instruments. Basically, the project involves the creation of a multi-parameter system of permanent control of the entire archaeological area, with the associated indicators of the level of risk, based on the combined use of innovative technologies. The proposed new paradigm for the monitoring (combining satellite monitoring, downstream services and in situ calibration data), will be proposed also at national level through the implementation of the Italian Network of Archaeological Parks (INAP). Some examples of satellite monitoring application will be presented and illustrated in order to stress the roles of new EO technologies in the field of conservation and maintenance polices.

Keywords UNESCO · Satellite monitoring · Parco archeologico del Colosseo · Cultural heritage

A. Russo · I. D. Giovampaola (✉)
Parco Archeologico del Colosseo, Piazza S. Maria Nova, 53, 00186 Rome, Italy
e-mail: irma.dellagiovampaola@cultura.gov.it

D. Spizzichino · G. Leoni
ISPRA, Geological Survey of Italy, Rome, Italy

A. Coletta · M. Virelli
ASI – Italian Space Agency, Rome, Italy

G. M. El-Qady and C. Margottini (eds.), *Sustainable Conservation of UNESCO and Other Heritage Sites Through Proactive Geosciences*, Springer Geology,
https://doi.org/10.1007/978-3-031-13810-2_34

1 Introduction: The Monitoring Project of the Parco Archeologico Del Colosseo

All archaeological sites are affected by changes due to a natural decay related to the ageing. If it compromises the functionality of the cultural property it becomes pathological and results in degradation. The monitoring, carried out with the use of innovative technologies, is a preliminary tool to an effective planned maintenance activity and therefore preventive conservation. Regarding these aspects the Parco archeologico del Colosseo (see Figs. 1 and 2) took a strategic direction of a gradual transition from a plan of monitoring to a constant and planned conservation activity. The monitoring project of the Parco archeologico del Colosseo (that started in a systematic way only in 2018) was inspired by the desire to build a sustainable system of protection and conservation, then allowing a proper tourism valorisation. With these objectives in mind, Irma Della Giovampaola, responsible for the monitoring of the Parco archeologico del Colosseo, has developed a static and dynamic monitoring project consisting of five fundamental activities:

- database of the historical data of the monuments, together with the existing graphic and photographic documentation
- visual monitoring for the inspection and control of monuments carried out by teams of technicians, also thanks to an app that will allow them to send data to the central system
- satellite monitoring: the historical analysis of the satellite data will flow into the system and will be analysed in order to monitor ground changes in relation to the monuments

Fig. 1 Parco archeologico del Colosseo: view of the Roman Forum, in the background the Flavian Amphitheatre

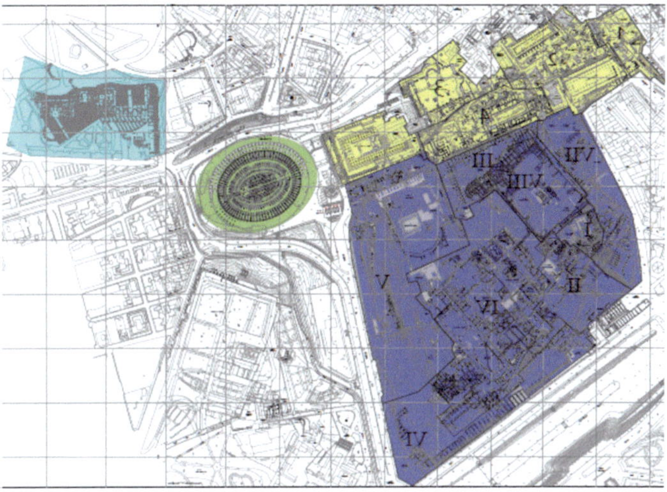

Fig. 2 The zoning of the Parco archeologico del Colosseo for the management of monitoring and maintenance activities (Russo and Della Giovampaola 2020a)

- monitoring from ground diagnostic instruments, for which specialist support will be essential
- experimental activities, such as the use of H-BIM in archaeology, aimed at monitoring by diagnostic instruments.

Basically, the project involves the creation of a multi-parameter system of permanent control of the entire archaeological area, with the associated indicators of the level of risk, for which it is necessary to combine innovative technologies.

In this way, the project will allow to plan, in an effective and timely manner, the necessary interventions for both ordinary and extraordinary maintenance, thus providing not only an operational tool, but also a management system for the Park with a better use of its financial resources.

As far as the first aspect is concerned, the web-GIS platform (by Irma Della Giovampaola and Alessandra Petretto), the core of the monitoring system currently being set up (Fig. 3), will be able to interrelate data according to defined parameters in order to indicate the state of conservation of the archaeological heritage: depending on the level of importance of the damage identified, a notification or an alert will be generated, the purpose of which is to schedule maintenance or on-demand interventions (Della Giovampaola 2021a, b).

With regard to the inspection activities, among those already started by the Parco within the project, the one related to the vegetation presence (conducted by Gabriella Strano) is particularly important for an archaeological area.

The monitoring of the latter is in fact preparatory to the mechanical and biological interventions of weed control which is, if not properly managed, the most widespread cause of damage and degradation to the monumental heritage. The vegetation with

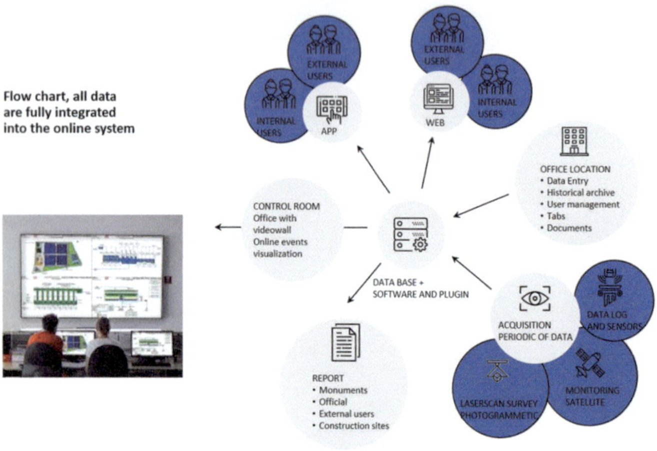

Fig. 3 The monitoring system of the Parco: diagram of the expected flows and of the implementation of all the data (Russo and Della Giovampaola 2020a)

woody root systems, abundantly present in the Parco, is in fact the highest degree of Dangerousness Index of the species (Fig. 4).

The mechanical and biological interference is caused by the weight on the walls, by the extension of the radical apparatus able to solubilise the mortars and the different types of building materials, creating structural damages to the archaeological artefacts.

The work that the Parco is carrying out is to identify the typology of the vegetal presences, their degree of dangerousness, the damage already caused on the materials that constitute the artefacts and the risks of future damage, putting them in reciprocal relation with the other data considered essential for the planning of conservation and maintenance interventions of cultural heritage.

Another fundamental aspect taken into consideration by the Parco for the development of the parameters useful to define the state of conservation of the property, is the detailed geomorphologic and geologic map (by Carlo Rosa) (Fig. 5).

The maps are currently being drawn up with the data obtained:

- from the survey of the existing geological outcrops, accompanied by adequate photographic documentation,
- from geologic and archaeological literature and archival documents,
- by the examination of the stratigraphy of the numerous boreholes, known from the literature and/or directly performed.

The geologic data collected in this way, including the relevant bibliography, have been included in the web-GIS platform of the Parco's monitoring working group. This is linked to the inventory of underground caves conducted by the Parco, most of which have already been identified and mapped in their underground shape, in some cases with the support of laser scanner. This database is currently being

Fig. 4 Palatine Hill, southern slope—Weed monitoring in the archaeological context (Russo and Della Giovampaola, 2020a)

completed. The results of the cave inventory have been made available to ISPRA (Daniele Spizzichino, Gabriele Leoni, Paolo Guarino) for the evaluation of the risks related to their presence and the management of transport and visitor flows within the Parco.

Another fundamental aspect is the diagnostic instrumentation, for which the Parco has carried out a census of what has already been installed in the archaeological area over time, in order to define a data processing system managed by the web-GIS platform of the monitoring project. The instrumental diagnostic tools are accompanied by satellite monitoring, already tested in the past for a short period, to obtain information on ground displacements, structures, and buildings. The use of satellite SAR interferometry technique applied to Cosmo-SkyMed images is combined with the advantage of being able to use the archives of radar images that allow us to deduce, in an extensive manner, the evolution in time of more than twenty years of deformation processes. For these reasons, the Parco also considered fundamental the satellite historical analysis of the archaeological area, carried out since 2010 until 2019. The satellite images, provided by ASI (Alessandro Coletta and Maria Virelli), were processed on commission by e-GEOS with interferometric technique. The data thus processed fed the web-GIS platform of the Parco's monitoring project

Fig. 5 Geological map: geomorphology of the Roman Forum before and after roman presence (Russo and Della Giovampaola 2020a)

and were made available to ISPRA for the interpretation of the ground effects. In addition, in order to identify a protocol for the use of satellite data (whether they are the result of current monitoring or historical analysis) on cultural heritage in the field of conservation and protection, the Parco's monitoring project has designed a test, that has already started in the period up to this publication: due to its complexity, the Flavian Amphitheatre was chosen as the research area, in particular the South side, in order to implement the sensor system already existing on the North side, taking into account the information provided by the PS (Fig. 6). The project includes the development of a system that allows not only the ground validation of satellite data, but also the most suitable type of instrumentation for this purpose, in order to define a replicable protocol, at least in some aspects, on a larger scale, facilitating further activities expected in the Parco monitoring project.

As part of the activity carried out by the monitoring working group, in 2019 the Parco has promoted an international conference on this topic, whose proceedings, pending the publication of this special issue, have already been published (Russo and Della Giovampaola 2020b).

The International Conference, which was awarded with the medal of the President of the Republic, is a milestone in the world of cultural heritage protection: for the first time, in fact, by placing multidisciplinarity as a fundamental resource, the Parco archeologico del Colosseo brought together representatives of institutions active in

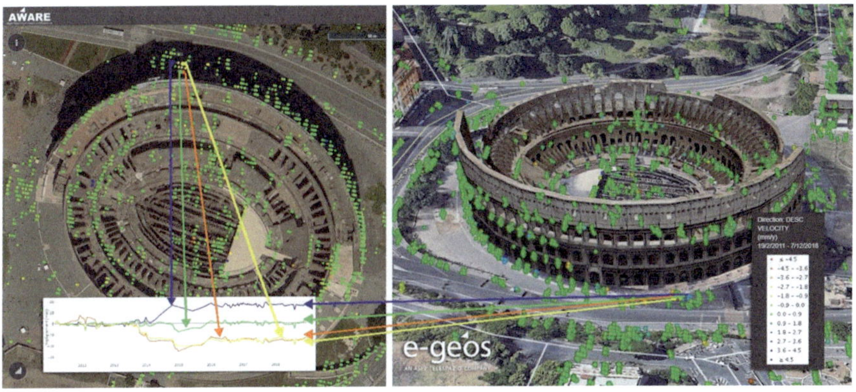

Fig. 6 The satellite historical analysis of the Flavian Amphitheatre (Russo and Della Giovampaola 2020a)

the experimentation of new methods of conservation, to discuss the challenges of the new millennium: reducing vulnerability and increasing resilience through the use of low-impact technologies.

In particular, the proceedings illustrate the relevance of monitoring and scheduled maintenance in archaeological areas, with the aim of defining a protocol to face the consequences of climate change impact and increasing effects of extreme weather events. Through the comparison among apparently distant research areas, this volume outline new monitoring methods suitable to be applied to cultural heritage, without neglecting the ordinary but complex management of archaeological areas. Given the need to use innovative technologies, the Parco has established a Technical Advisory Board, whose members, are: *Italian Space Agency* (Alessandro Coletta and Maria Virelli), the *National Institut of Geophysics and Volcanology* (Arrigo Caserta and Fawzi Doumaz), the *Italian Institute for Environment Protection and Research* (Daniele Spizzichino and Gabriele Leoni), the *Interdepartmental research Centre of Sapienza, University of Rome* (Franco Gugliermetti), the *Carabinieri for the protection of Cultural Heritage (TPC)* and also individual experts such as Dr. Vito D'Adamo, Gen. Carlo Magrassi and Prof. Pietro Valentino. The board will allow a confrontation with different specialist sectors for a better assessment of the static and dynamic risk of the whole archaeological area.

The Technical Advisory Board is composed by those public research institutions, with which the Parco has established agreements for research and investigations in the field of monitoring cultural heritage; the contribution of each institution is reported below.

2 The Activities of the Technical Advisory Board of Parco

2.1 Italian Space Agency (ASI)

The monitoring activities implemented (low impact and high technological content) in the whole Parco area applied satellite monitoring techniques with SAR Interferometry technology, for the identification of deformation phenomena (past or still in progress) of the ground, in order to identify local anomalies for future risk assessment.

To carry out this analysis, Italian Space Agency provided a total of 273 images acquired by the COSMO-SkyMed Constellation (150 acquired in ascending geometry and 123 acquired in descending geometry), with mean revisiting time of approximately 16 days, acquired in Stripmap-HIMAGE mode, characterized by spatial resolution of 3 × 3 m. The analysis was implemented in two phases: the first involved the analysis of all historical data and images available from 2011 to 2018, useful for obtaining displacement and velocity maps for the investigated time window. The second phase involved a specific study on the images acquired only for 2019. These analyses are useful for monitoring the effects of the excavation carried out near the park to build the new section of the underground line.

In 2003 the Italian Space Agency (ASI) and Italian Ministry of Defence (MoD) commissioned and funded the COSMO-SkyMed (COnstellation of small Satellites for Mediterranean basin Observation, Fig. 7) dual earth observation system (civil and military), which currently represents the largest Italian investment in the field of Earth observation (Caltagirone 2014). Initially, the system consists of a constellation of four Low Earth Orbit mid-sized satellites, each equipped with a multi-mode high-resolution Synthetic Aperture Radar (SAR) operating at X-band (3.1 cm wavelength) and fitted with particularly flexible and innovative data acquisition and transmission equipment. Following, to the 4 First Generation satellites a Second Generetion of COSMO-SkyMed satellites was added, also consisting of 4 identical satellites always equipped with an X-band SAR, positioned on the same orbital plane. On December 18, 2019, the first Second Generation satellite was launched, the second satellite were launched in February, 1st 2022 and the launches of the third and fourth satellites are expected in the next years.

The satellites of the COSMO-SkyMed constellation (Fig. 8) will follow a heliosynchronous low orbit (about 700 km above the Earth surface) around the Earth passing close to the poles; the satellites will be phased at 90° from one another and each of them will run 14.8125 orbits a day, that is 137 orbits every 16 days. Hence each satellite will repeat the same ground track every 16 days, but the same ground track will be repeated by a satellite every four days. Thanks to the characteristics of high spatial resolution (up to 1 m for civil applications), short revisit time (up to 1 day in the tandem configuration) and different types of polarization available, the four satellites of the COSMO-SkyMed constellation will allow frequent repeated observations of a same site. The active SAR technology guarantees that very high-resolution images will be taken in any condition of illumination and weather. The COSMO-SkyMed is

Fig. 7 COSMO-SkyMed satellite

natively conceived as a Dual-Use (Civilian and Defence) end-to-end Earth Observation System aimed to establish a global service supplying provision of data, products and services compliant with well-established international standards and relevant to a wide range of applications, such as Defence/Intelligence Applications, such as borders surveillance, Risk Management, Scientific and Commercial Applications. The constellation is also used for cartography and for monitoring the evolution of natural phenomena, such as volcanic eruptions and oil spills in the sea in any part of the world; also, as for providing detailed and timely data on regions hit by an earthquake or a tsunami, where traditional data collection and communication systems are likely to be disrupted.

The set of requirements, imposed at highest level, has brought to the following needed performances (Table 1):

- large amount of daily acquired images
- satellites worldwide accessibility

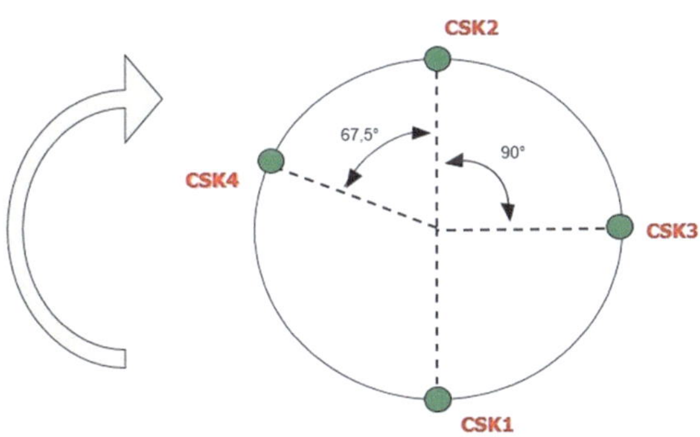

Fig. 8 COSMO-SkyMed First Generation constellation configuration

Table 1 COSMO-SkyMed sensor main characteristics

Radar sensor	X-band
Radar frequency	9.60 GHz
Resolution	1 m, 3 m, 15 m, 30 m
Orbital altitude	619 km
Angle of inclination	97.86

- all weather and day/night acquisition capabilities
- very fine image quality (e.g. spatial and radiometric resolution)
- possibility of image spatial resolution trade-off with size, at most possible extent and including sub-meter resolution

Each satellite can operate in different modes (Spotlight, StripMap HIMAGE and PingPong, and ScanSAR Wide and Huge) providing SAR images of different coverage and spatial resolution (Fig. 9).

- **SpotLight Mode:** the extension of the frame is about 11 km × 11 km, in azimuth and range direction respectively, the ground resolution is about 1 m × 1 m.
- **Stripmap Mode:** it is the most common imaging mode. Two different implementations of this mode are provided: the Himage and the PingPong.

The Himage is characterized by a swath width of about 40 km and, an azimuth extension for the standard product (standard frame) of about 40 km (corresponding to an acquisition of about 6.5 s); ground resolution is about 3 m × 3 m.

In the PingPong mode the acquisition is performed in strip mode alternating the signal polarization between two of possible ones, i.e. VV, HH, HV and VH. This mode is characterized by a swath width value of about 30 km and an azimuth extension for the standard product of about 30 km (standard frame) corresponding to an acquisition of about 5.0 s; ground resolution is about 15 m × 15 m.

- **ScanSAR Mode:** it allows larger swath in range with respect to the Stripmap one, but with a less spatial resolution.

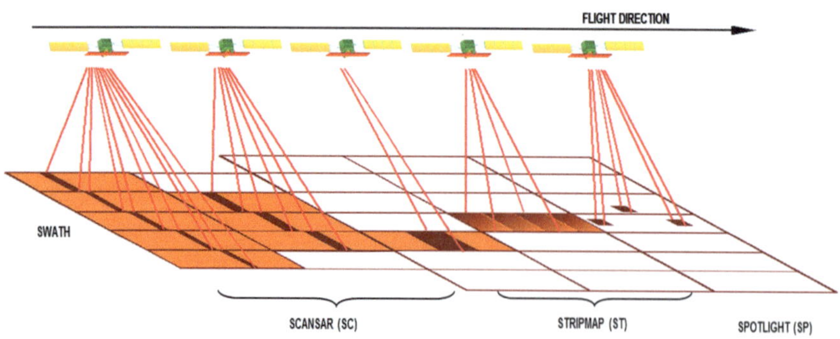

Fig. 9 COSMO SAR acquisition modes

The two different implementations allowed for this acquisition mode are WideRegion and HugeRegion, with a spatial resolution selectable from 30 m × 30 m to 100 m × 100 m.

After the launch of the fourth satellite at the end of 2010 from Vandenberg Launch Facility in California, the COSMO-SkyMed constellation became fully operational, and to date all four satellites have completed their nominal operational life cycle, or 5.25 years (Virelli 2014). Starting from the launch of the first satellite in 2007, COSMO-SkyMed has proved to be a useful and very efficient instrument of monitoring in case of emergencies. It has indeed been used in case of landslides, floods, volcanic eruptions, earthquakes and whatever could be of interest for Civil Protection, such as environment and climate monitoring, disaster and risk management, surveying coastlines and hydrogeological resources. Nevertheless, COSMO-SkyMed represents a fundamental contribution in the monitoring of the National territory, being able to acquire up to 1800 images per day. Since the launch of the first satellite it has been evident the huge importance of getting a comprehensive and updated catalogue of images acquired all over the world.

With the aim of creating a regularly updated interferometric archive, according to the specific needs of the Civil Protection Department, in 2009 the ASI activated the Map Italy project, a full interferometric mapping service of the whole National territory based on every-16 days Stripmap Himage acquisitions, both in Ascending and in Descending orbit direction, using the COSMO-SkyMed system (Figs. 10, 11 and 12).

Thanks to this project, historical series of images are acquired on the Italian territory in order to use them for interferometric analysis of instability phenomenon and

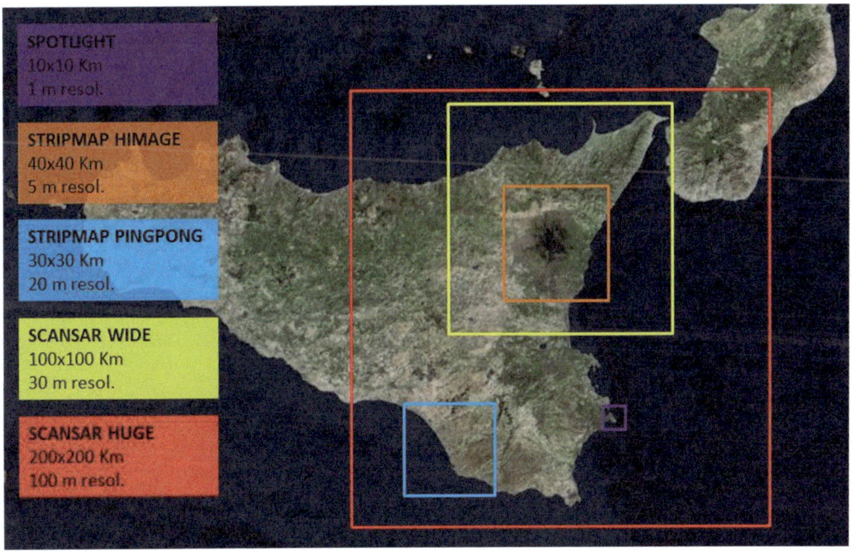

Fig. 10 A pictorial view of the 5 major types of COSMO-SkyMed acquisition modes

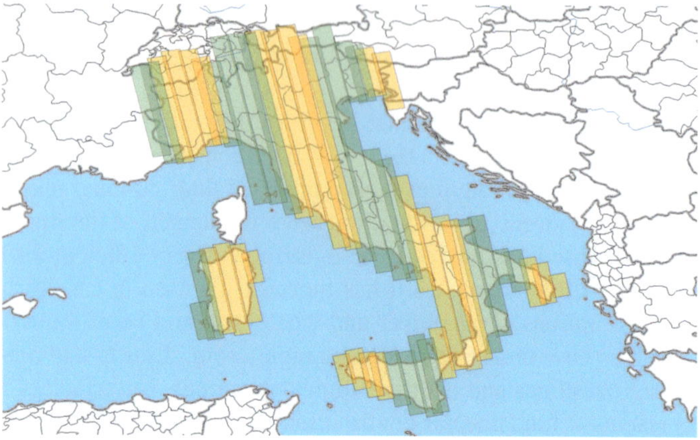

Fig. 11 Map Italy ascending geometry

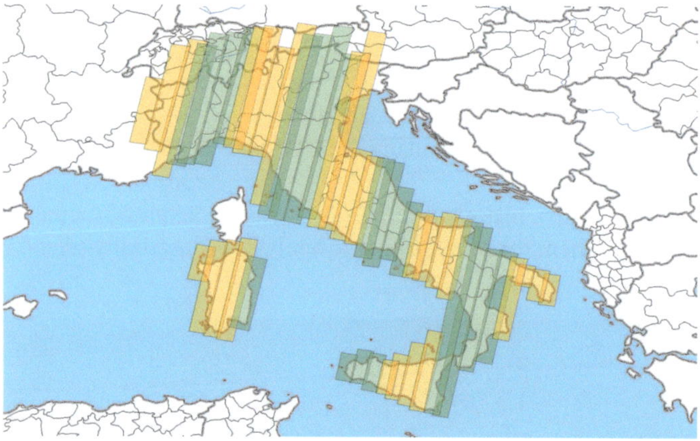

Fig. 12 Map Italy descending geometry

endogenous risk of the same territory (seismic and volcanic phenomena, landslides, subsidence, etc.) and to routinely and intensively populate a specific interferometric historic archive as a National geographic reference. Due to the strategic importance of Map Italy project, it was agreed to increase the priority level of this interferometry mission, which started as a "background mission".

The Map Italy data archive has proven invaluable for tracking ground deformation and surface change. The same availability of long time series acquired regularly on the same site, up to hundreds of images over a decade of observation, is a fundamental requisite for being able to set up ordinary high frequency monitoring activities. In this way, it is possible to verify the state of conservation and, on the contrary, identify events of damage, accidental or intentional, natural or anthropic, which could intervene over time, in certain cases even without any warning or possibility of intervening, except that after the event.

To analyse the COSMO-SkyMed images provided by ASI, e-Geos used one of the satellite radar interferometry techniques called Persistent Scatterer Pairs (PSP; Costantini et al. 2014). The Satellite Radar Interferometry techniques (see Rocca et al. 2000) allow the millimetric measurement of ground displacements over large areas, processing an enormous quantity of measuring points (called Persistent Scatterer, PS) stable from a radiometric and geometric point of view over time, such as buildings, stones, metal frames, etc.

Radar Interferometry analysis on a stack of images acquired in the same area over a certain period provides measurement points position (coordinates and elevation), mean annual velocity and time series of displacements. Considering the limitations of all the interferometric techniques, in particular the fact that the measurement is along the line of sight (LOS) and the geometric distortions, in order to obtain the maximum information from interferometric analysis, both ascending and descending geometry have been used.

3 National Institute of Geophysics and Volcanology (INGV)

The INGV network has an array geometry and has been designed to allow exploiting the seismic noise information relevant for site effect studies. These latter analyse the ability of the site to amplify the soil shaking in the range of frequencies of engineering interest, typically 0.2–10 Hz, ability to focus/defocus soil shaking in different parts of the site.

The monitoring network prototype is realized by an array of seismic stations (Fig. 13) able to record seismic noise, so called ambient vibration. The seismic noise is soil shaking in absence of earthquakes, in other words, it is the background signal. Nowadays it is well known that huge information concerning the features of the soil shaking is hidden in the seismic noise.

The soil shaking features represent the input to the selected monument and so by monitoring also the monument, we can investigate the complex dynamics that drive the soil-building interaction. In such a way, we will achieve and provide useful information for whom is in charge for checking the "health of archaeological buildings" mainly for safety and resilience purposes.

This is the first seismic array experiment, and it will be compliant with the Parco monitoring needs. We plan to use such array at least for one month then, according to the obtained results, other geometries will be adopted in order to refine and better understand the soil-monument interaction and finally quantify the monument response to the soil input.

In order to deploy our geophysical asset, we usually acquire the geometries of the space thanks to photogrammetry techniques, sometime using drones and other times with classical image shooting around the site.

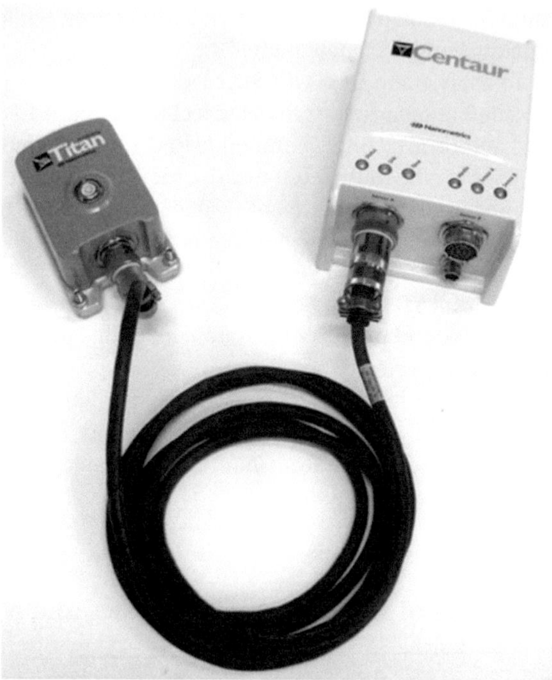

Fig. 13 Seismic station (INGV, A. Caserta, F. Doumaz)

The obtained result is an up-to-date planimetry and 3d model if needed. In case of Colosseum, as low-cost experimental technique, we made a first 3D model using a 6 m pole mounting a GoPro stabilized by a gimbal, walking and shooting every 2 s Geotagged and high-resolution pictures.

Covering the most possible parts of the monument, the model will help for the positioning of the seismic stations for the array taking advantage of a 3D immersive view since we need to position on a same vertical several stations in the complex architecture of Colosseum monument (see Fig. 14).

4 Italian Institute for Environment Protection and Research (ISPRA)

The ISPRA signed in 2019, a scientific agreement with the Parco aimed at analysing and interpreting ground motion measures obtained by satellite monitoring as well as supporting the geomorphologic processes study affecting the archaeological area. During the last year ISPRA tested the application of satellite monitoring techniques with SAR Interferometry technology, for the identification of ground deformation phenomena (past or still in progress), in order to identify local anomalies for future risk assessment (Spizzichino & Margottini 2021; Alberti et al. 2017).

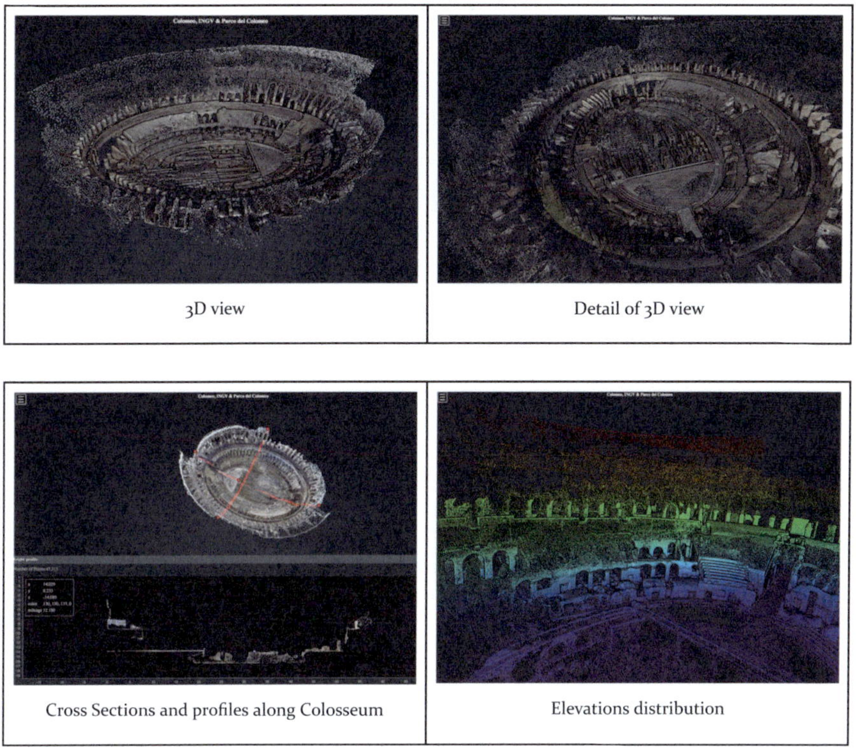

Fig. 14 INGV server when the points cloud can be explored in 3D http://kharita.rm.ingv.it/3dpointclouds/Colosseo_Livello1ply/ (A. Caserta-F. Doumaz)

The analysis was implemented in two phases: the first involved the analysis of all the historical data and images available (2011–2018) useful to obtain displacement and velocity maps for the investigated time window (Fig. 15).

The second phase involved a specific study only for the 2019, in order to verify the observations of the on-site monitoring system during the excavation of the TBM in the park area for the construction of a new underground line. In the following sections are reported the details of the implemented analysis and some specific application examples.

One of the main monitoring activities implemented (low impact and high technological content) for the park area, was the application of satellite monitoring techniques with SAR Interferometry technique, to detect past or still in progress ground deformation phenomena.

The first example comes from the analysis of Descending PS time history on the East sector of the Flavian Amphitheatre (Fig. 16).

During the last month of 2015 and first months of 2016 a specific maintenance work was implemented exactly in the same area where a cluster of PS highlights an abrupt lowering of the wall of about 15 mm.

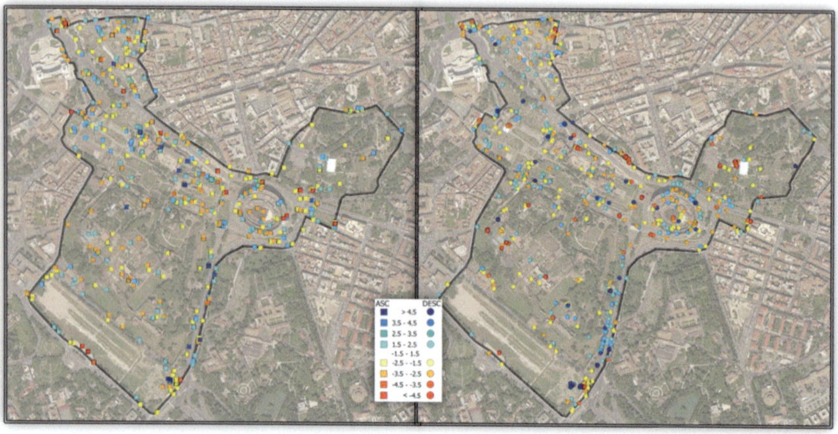

Fig. 15 PSP-IFSAR e-Geos elaboration March 2011–2018 (Parco archeologico del Colosseo)

The mitigation measures consisted in the protective wrapping of some pillars, with the aim of preventing flaking and deformation. The detail of the intervention is visible in Fig. 17.

The present study represents a prototype model potentially exportable to other cases from a theoretical and methodological point of view. The study promotes the extensive use of satellite monitoring in archaeological area as fundamental tool for maintenance activities. Preliminary results highlighted that the satellite SAR interferometry technique is capable to identify in the archaeological area both past and present displacement trend. Moreover, this technique is useful to check the effectiveness of shoring structures. About this issue it is very important to exchange information with site managers in order to combine monitoring data with site interventions and to follow their actual effectiveness. Future developments of this research will be addressed to: (i) improve future processing of SAR interferometry data, also with the installation of passive corner reflector; (ii) support for the analysis of in situ

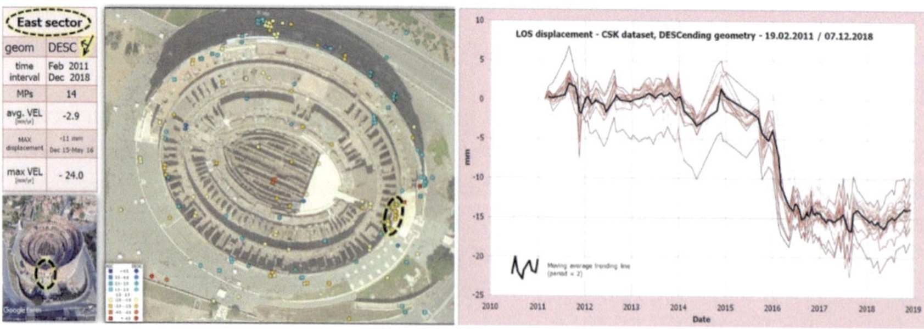

Fig. 16 Data sheet and time series displacement for the Descending PS cluster in the East sector (Parco archeologico del Colosseo)

Fig. 17 Detail of pillar wrapping in the monitored area in 2020 (Parco archeologico del Colosseo, photo of D. Spizzichino)

monitoring data and possible implementation and integration of the current system. The job described is part of a collaboration agreement between Technical Monitoring Board and Parco.

5 Interdepartmental Research Centre of Sapienza, University of Rome (CITERA)

In recent years, monitoring techniques have begun to make use of new integrated operational paths for the processing, sharing and representation of information based on range imaging, computer vision, crowd sourcing and satellite assets. In this reference framework, CITERA has recently developed, together with a public–private consortium, two biennial projects financed by ARTES Demonstration Project calls of ESA (European Space Agency). The first, RECIPE (Resilience in art CIties: Planning for the Emergencies) completed at the end of 2019, had the goal of developing a SaaS platform capable of providing, using photogrammetric methods, low-cost 3D models by acquiring photos taken by tourists (with their smartphones phone and/or tablet) with crowdsourcing techniques based on a dedicated app; spatial earth observation (EO) georeferencing (GNSS), communication (SAT) had been used to scale the models, to insert them into GIS products and to cope with the absence of a cellular network for natural disasters, war events, difficult accessibility.

The comparison among 3D models, produced in different times, was able both to highlight measurable variations in the dimension (with level of details, LoD, depending on the quality of the photos, in the range 10–2 mm/m), and to appreciate colour changes of the surface. The second project, still at its beginning, called VADUS

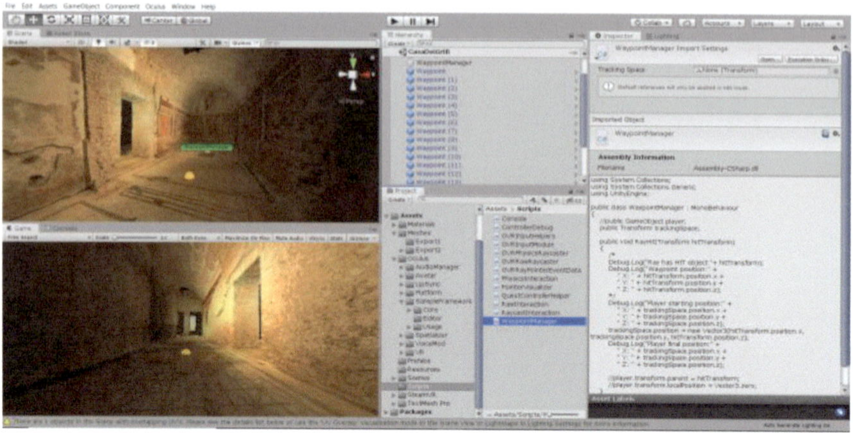

Fig. 18 View in the Unity 19 C3 Interface of Casa dei Grifi (Parco archeologico del Colosseo) during VR production phase (CITERA, F. Gugliermetti)

(Virtual Access and Digitalization for Unreachable Sites), will produce high resolution 3D model, through the integration between the range-based and image based approaches; the first will use a database of close-range photos taken by trained staff, the second a LIF (Laser Induced Fluorescence) and a RGB-ITR- RGB (Imaging Topological Radar; RGB Red, Green, Blue and IR InfraRed) for the characterization of colorimetric and chemical aspects of the surface.

This approach will overcome the limits of VADUS related both to crowdsourcing, in terms of quality, completeness and LoD of the 3D products (about 1 mm/m), and of materials characterization, but with much higher costs. Also, in this case, the georeferencing and the correct insertion of the artefact within the archaeological sites will be obtained through EO and GNSS satellite techniques. A further processing chain will allow to produce VR (virtual reality) and AR (augmented reality) in high resolution and 360° views, also on request in hyperspectral mode, giving quantitative information on state of conservation by comparing different, in time, 3D.

Another expected result is linked to the possibility both of using VR or AR, even at high resolution, for touristic visits that can be managed by the 5G network (Figs. 18 and 19), and of supporting the management of cultural heritage through data sharing techniques such as HBIM (Historical Building Information Modelling).

6 The Parco Archeologico Del Colosseo and the Italian Archaeological Parks Network

Finally, among the activities carried out in the field of cultural heritage conservation by the Park there is the realization of a *"multisystemic monitoring system also through the satellite analysis of the national archaeological and monumental heritage"* as

Fig. 19 Parco archeologico del Colosseo: images from the VR of Casa dei Grifi (CITERA, F. Gugliermetti)

part of the *"national project of monitoring and programmed maintenance of the artistic and monumental heritage"*, of which the Parco archeologico del Colosseo is the leader and which involves the archaeological parks of Ostia Antica, Paestum, Pompeii, Herculaneum, Campi Flegrei as partners with whom to set up a control room for the satellite monitoring of archaeological heritage. The main objective is the realization of a multisystemic monitoring system through satellite analysis of the national archaeological and monumental heritage as part of the "National Project of monitoring and programmed maintenance of the artistic and monumental heritage".

7 Conclusion

In summary, the method implemented by the Parco consists of:

(1) Map the needs of the Parks in the field of Earth Observation.
(2) Identify the Copernicus services already operational and those to be activated on demand (e.g. specific COSMO-SkyMed services and, in the near future, Prisma).
(3) Identify specific needs for which satellite services need to be combined with ground-based tests by diagnostic instrumentation, linked to an ad-hoc platform that will be responsible for processing ground/space data.
(4) Identify possible improvements and customisations within the already operational central services to meet more requirements and to a greater extent.

The aim pursued by the Parco archeologico del Colosseo is to create a model of approach to issues related to monitoring and planned maintenance, important for other archaeological sites and national and international monuments. We believe that this is an essential condition to leave such a great heritage to future generations.

References

Caltagirone F, Capuzi A, Coletta A, De Luca GF, Scorzafava E, Leonardi R, Rivola S, Fagioli S, Angino G, L'Abbate M, Piemontese M, Zampolini FE, Torre A, De Libero C, Esposito PG (2014) The COSMO-SkyMed dual use earth observation program: development, qualification, and results of the commissioning of the overall constellation. IEEE J Select Top Appl Earth Observ Remote Sens 7(7):2754–2762. https://doi.org/10.1109/JSTARS.2014.2317287

Costantini M, Falco S, Malvarosa F, Minati F, Trillo F, Vecchioli F (2014) Persistent scatterer pair interferometry: approach and application to COSMO-SkyMed SAR data. IEEE J Sel Top Appl Earth Obs Remote Sens 2014(7):2869–2879

Della Giovampaola I (2021a) SyPEAH: The WebAPP System for Protection and Education to Archaeological Heritage in the Parco Archeologico del Colosseo. Geosciences 11(6):246. https://doi.org/10.3390/geosciences11060246

Della Giovampaola I (2021b) Piano sostenibile di tutela e valorizzazione del patrimonio archeologico e di educazione continua al patrimonio culturale: SyPEAH (a platform System for the Protection and Education of Archaeological Heritage), in Bullettino Comunale della Commissione Archeologica di Roma CXII, 2021, pp 61–75. https://doi.org/10.48255/J.BCAR.CXXII.2021.04

Alberti S, Ferretti A, Leoni G, Margottini C, Spizzichino D (2017) Surface deformation data in the archaeological site of Petra from medium-resolution satellite radar images and SqueeSARTM algorithm, Journal of Cultural Heritage (2017), https://doi.org/10.1016/j.culher.2017.01.005

Iadanza C, Cacace C, Del Conte S, Spizzichino D, Cespa S, Trigila A (2013) Cultural heritage, landslide risk and remote sensing in Italy. In: K. Sassa P, Canuti C, Margottini (eds) Landslide science and practice, vol 6 Risk Assessment, Management and Mitigation. Springer, pp 491–500

Russo A, Giovampaola ID (eds) (2020b) Monitoraggio e manutenzione delle aree archeologiche. Cambiamenti climatici, dissesto idrogeologico, degrado chimico-ambientale, atti del convegno internazionale di studi, Roma Curia Iulia, 20–21 marzo 2019, Bibliotheca Archaeologica, 65, Rome

Rocca F, Prati C, Monti Guarnieri A, Ferretti A (2000) Sar interferometry and its applications. Surv Geophys 21:159. https://doi.org/10.1023/A:1006710731155

Russo A, Della Giovampaola I (2020a) Il monitoraggio e la manutenzione delle aree archeologiche. Il Piano per il futuro del Parco archeologico del Colosseo, in Monitoraggio e manutenzione delle aree archeologiche. Cambiamenti climatici, dissesto idrogeologico, degrado chimico-ambientale, atti del convegno internazionale di studi, Roma Curia Iulia, 20–21 marzo 2019, a cura di A. Russo, I. Della Giovampaola, Roma 2020a (Bibliotheca Archaeologica, 65), pp 13–31

Spizzichino D, Margottini C, Brustia E, Cigna F, Comerci V, Dessì B, Guerrieri L, Iadanza C, Leoni G, Tapete D, Trigila A, Vittori E (2017) Satellite monitoring applied to natural hazards and cultural heritage: the PROTHEGO project. In: Conference proceedings 11° workshop tematico di Telerilevamento - AIT Bologna, 27–28 giugno 2017

Spizzichino D, Margottini C (2021) Chapter 11 Satellite monitoring of geo-hazards affecting cultural heritage. In A Research Agenda for Heritage Planning Edited by Eva Stegmeijer and Loes Veldpaus Monograph Book Published. 256. 07 Sep 2021 Print ISBN: 9781788974622 eISBN: 9781788974639. https://doi.org/10.4337/9781788974639. Collection: Geography, Planning and Tourism

Virelli M, Battagliere ML, Coletta A (2014) ASI COSMO-SkyMed Mission overview and data exploitation. IEEE Geosci Remote Sens Mag 2(2):64–66. https://doi.org/10.1109/MGRS.2014.2317837

Adaptation, Traditions and Conservation: The Case of the Asante Traditional Buildings World Heritage Site in Ghana

Christopher Wetcher

Abstract Integrating traditional and modern conservation systems provides an idiosyncratic approach for the management of heritage sites, whose traditional custodians rights were disposed in favor of the State Based Management Systems (SBMS). Using the Asante Traditional Buildings World Heritage site (Ghana) as a case study, this paper which is emanating from my Ph.D. studies seeks to examine the relationship between traditional and current management approaches, as well as the associated mechanisms applied in furthering the conservation of the site. This analysis is buttressed against how legal stewardship of the site by Ghana Museums and Monuments Board (GMMB) and its international appropriation as a World Heritage site, have collectively and inadvertently resulted in its progressive deterioration. The integration of traditional management perspectives into the State Based Management Systems has the potential of developing an integrated management model building on the resilience and mechanism of Traditional Management Systems. Such integration is the kind of adaptation required in ensuring World Heritage sites have sustainable conservation approaches, including advancing debates on the localness of cultural heritage and its impact on globalization of conservation through instruments such as the 1972 World Heritage Convention.

Keywords Conservation · Integrated management · Asante traditional buildings · State-based management · World Heritage

1 Introduction

The Asante Traditional Buildings (ATB) date back to the late 17th Century, when the Asante Kingdom was established and reached its peak in the eighteenth century.

C. Wetcher (✉)
Department of Archaeology and Heritage Studies, University of Ghana, P. O. Box LG 3, Accra, Ghana
e-mail: cwetcher@st.ug.edu.gh

Ministry of Education, Ghana National Commission for UNESCO, Accra, Ghana

G. M. El-Qady and C. Margottini (eds.), *Sustainable Conservation of UNESCO and Other Heritage Sites Through Proactive Geosciences*, Springer Geology,
https://doi.org/10.1007/978-3-031-13810-2_35

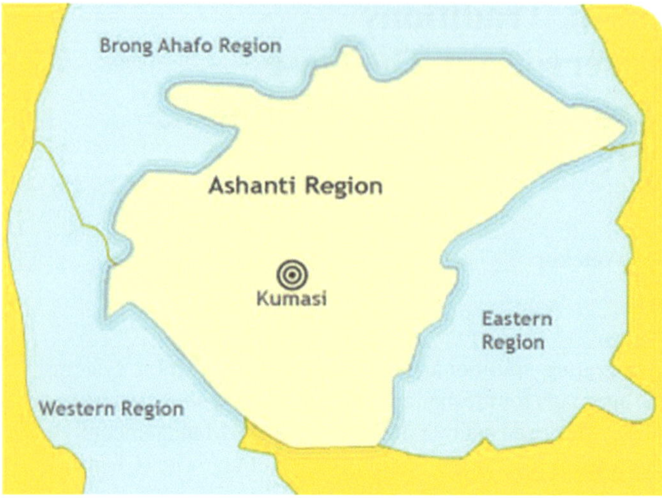

Fig. 1 Map of the Asante Region

Kumasi became the new capital of this Kingdom. The Asante Traditional Buildings are the last remaining testimony of the great Asante civilization and these include the ten shrines/fetish houses namely Abirim, Asawase, Asenemaso, Bodwease, Ejisu Besease, Adarko Jachie, Edwenase, Kentinkrono, Patakro and Saaman which are now World Heritage Sites. Most of the buildings are to the north-east of Kumasi, with only Patakro, to the south of Kumasi – in the Asante Region of Ghana (Fig. 1).

Meticulously arranged around courtyards, the ATB's are constructed of timber, bamboo and mud plaster and originally had thatched roofs (Prussin 1980). The unique decorative bas-reliefs that adorn the walls are bold and depict a wide variety of motifs (Fig. 2).

Common forms of designs on the ATB's include spiral and arabesque details with representations of animals, birds and plants, linked to traditional "Adinkra"[1] symbols (Prussin 1980; Joffroy et al. 1998). As with other traditional art forms of the Asante kingdom, these designs are not merely ornamental, they also have symbolic meanings, associated with the ideas and beliefs of the Asante people, and have been handed down from generation to generation. Their rich color, and the skill and diversity of their decorations are the last surviving examples of a significant traditional style of architecture that epitomized the influential, powerful and wealthy Asante Kingdom of the late 18th to late nineteenth centuries. The ATB reflect and reinforce a complex and intricate technical, religious and spiritual heritage (Prussin 1980; Joffroy et al. 1998). The traditional religion that was practiced in the ATB, took the form of consulting with the deities to seek advice on specific situations, or before an important initiative (Prussin 1980).

[1] Asante symbols that represent concepts and aphorisms.

Fig. 2 Different Motifs and Designs on the ATB

The ATB's are rectangular in shape with usually more than one courtyard for various purposes. The buildings were primarily constructed of mud plastered onto a timber frame work with steeply pitched roofs covered with thatch. The upper section of the buildings is painted with white clay, whilst the plinths and the lower section are painted with a red laterite and polished to a dull shine (Fig. 3).

These buildings are associated with the institution of chieftaincy and/or the indigenous animist religion (Prussin 1980); they served as palaces, shrine for the powerful deities who protected the Asante Kingdom and as homes for the affluent. Also known as "Obosomfie" (house of a god), the ATB are mostly made up of four buildings

Fig. 3 The ATB Polished to a dull shine

enclosing a central courtyard. Three of the buildings are open to the courtyard with raised floors called "dampons". One of these is reserved for the drummers, with its three plain walls giving effective resonance to their playing. Another room, usually the one opposite, is used by the singers who accompany the drumming during religious ceremonies. The third open room is used as a cooking area where ceremonial meals were regularly prepared to be partaken of by the gods. The fourth building houses the shrine and it is closed by decorated walls or intricate open-work screen walls which allows for ventilation and lighting, yet it provides an unusual and mysterious atmosphere. The shrine itself is placed on a raised and often ornamented platform or "dais" and this room can only have entered by the priest and his attendants known as "nsumankwaafo". An important feature in the courtyard is the "nyame dua" or altar for the sky god. This is in the form of a tree or a forked post between whose branches, a calabash, pot or brass basin is wedged, and into which sacrificial offerings are deposited. It is placed to the right of the shrine room entrance. Such altars were formerly also found in every Asante compound and the oldest member of the home would not eat before putting some of the food into it for "Onyame"[2] (Prussin 1980).

2 Precolonial and Post-colonial Conservation Approaches in Ghana

As a locally driven phenomenon, heritage is intricately interwoven and embedded in the indigenous knowledge systems of the Indigenous and Descendant Communities (IDC) of Africa (Taruvinga 2019). These systems/practices serve as the protective and management substratum for heritage properties, sites and landscapes which are considered sacred (Mahachi and Kamuhangire 2008; Murimbika 2006). Indigenous communities have a rich history of being creators and users of heritage, including controlling the exploitation of any other resources in their locality from time immemorial. This is enshrined in the Traditional Management Systems (TMS) governing IDCs, now loosely referred to as Local Communities (Taruvinga 2019). Most communities in Ghana in the precolonial period safeguarded, preserved and sustainably used their heritage resources through social and religious sanctions.

Considering the importance of ancestral connections at heritage sites in Africa (Pikirayi 2016; Chirikure et al. 2015), and the fact that the conservation of African heritage resources is closely intertwined with communal worldview, ontological and eschatological beliefs, and social formations (Sheridan 2008: 16), it will be improper not to involve the local communities in the management process as such an act will lead to the betrayal of their pioneering spirits in conservation (Taruvinga 2019). With the coming of the colonial masters and the introduction of their alien concepts of conservation and heritage, the traditional management practices were negatively positioned (Taruvinga 2019) and to some extend demonized (Gblerkopr

[2] Lesser god.

2005) without recourse to the fact that the cultural and symbolic significance of historic built environment is a key part in the understanding of Africa's heritage (Eboreime 2009, 2008:1).

In Ghana, "ethnic" groups have practices that are connected to their heritage sites and resources. These practices are enshrined in taboos, rituals and sacrifices (Kanpeyeng et al. 2011; Anquandah 1997, 2003). The Asante people of Ghana have rich cultural and religious practices. The several manifestations of these cultural and religious practices include the ATB's. The remaining ATB which primarily serves as shrine houses are deeply entrenched in the indigenous knowledge practices of the Asante communities. Conservation of these buildings which were carried out by certain groups of people were also interwoven in the traditional knowledge of the Asante people. These management practices persisted until the 1960's when the buildings were declared national monuments and were placed under the custody of the then newly established Ghana Museums and Monuments Board. Considered unscientific, these knowledge practices were gradually being dissociated from the ATB's by SBMS. The seemingly disconnection of the communities and their knowledge system from the ATB's and the inability of GMMB to effectively conserve them has left the buildings at the mercies of the unfavorable weather conditions. Coupled with this is the challenge of the introduction of both eastern and western religions which have vehemently demonized the traditional religion thereby leaving almost all the ATB's virtually empty with no priest to mediate between the communities and their gods. This shift in the management approach to heritage in Africa during the colonial period resulted in the overriding of TMS by the State Based Management Systems (SBMS). These SBMS emanated from the colonial laws which primarily focuses on the monumentality and concepts of protection and identification borrowed from outside Africa without acknowledging the historical and cultural values that the communities attached to their heritage (Negri 2008).

The colonial laws empowered government institutios resulting in local communities being marginalised in the management of their heritage. These approaches have been intensified with the ratification of the World Heritage Convention in Ghana, which has further empowered the SBMS. This is so because the World Heritage Convention has not been domesticated and localized to allow for community engagement and involvement at World Heritage sites. The heritage laws of Ghana are still based on the colonial concepts of the protection of cultural heritage. These concepts are alien to the local communities who are not inspired to implement these concepts at the ATB sites. Negri (2008), notes that the scope of these colonial laws only covers modalities for cataloguing, recording, listing, and declaring heritage items; the rights and obligations of the owner, holder and public agencies towards the protected items; and the control of the trading of these items and the regulation of their export. He further notes that these laws do not state the modalities according to which the knowledge of this heritage can be returned to the populations or in which form it will participate in the social life and development of society. In the context of Ghana, the GMMB's rules and regulations are spelt out in National Liberation Council Decree (NLCD) 387 of 1969 (now known as Act 387 of 1969), and Executive Instruments (EI) 29 of 1973 and 42 of 1972 which are derived from the colonial concepts of heritage protection. The rules and regulations of GMMB only addresses the following; Act 387 of 1969 addresses: control of antiquities; establishment and functions of Board meetings; proceedings and remuneration; members of the Board; officers and employees funds; accounts and audit regulations. EI 29 of 1973 mainly addresses: the export of antiquities; the sale of antiquities; national monuments. EI 42 of 1972 addresses: the list of national monuments. These laws which undergirds the SBMS does not include TMS in its operation. The colonial laws consider TMS as unscientific and to a large extent devilish. It is however imperative that for an inclusive and effective management of the ATB's and other heritage properties in Ghana, the heritage laws are reviewed to integrate TMS. This will go a long way to remove the top-down management approach of the ATB's and further give the communities some space and voice in the management process.

In the mid 1970's Ghana through an Act of Parliament ratified the World Heritage Convention. Subsequent to the ratification Ghana was able to serially list the ATB's and the Forts and Castles on the World Heritage List. Munjeri (2009:15), sums up the World Heritage Convention as 'a unique legal instrument which based on the idea that some cultural and natural heritage sites are of universal and exceptional importance and therefore need to be protected as part of the common heritage of humanity. After the ratification of the World Heritage Convention, there has been no efforts by the parliament of Ghana to integrate the Convention and its operational guidelines into the country's heritage laws. In addition to this there have not been any public education and sensitization on the Convention and its operational guidelines. The few workshops that have been organized have also not made provisions for the people staying in the communities where the World Heritage Sites are located to participate. The communities do not know about the Convention. It does also appear that the World Heritage Convention has not properly been integrated in the SBMS and development plan of the country. This has been demonstrated in instances where the country has taken certain initiatives at some World Heritage Sites without recourse to the Operational guidelines. To domesticate the World Heritage Convention, it is imperative that the country aligns the Convention with the existing laws as well as the TMS of the local communities.

3 The Asante Traditional Buildings and World Heritage Perspectives

The ATB was listed on the World Heritage List (WHL) at the Fourth Session of the World Heritage Committee Meeting held in Paris from 1–5 September 1980. Even though there appear to be some form of arrangement between the local communities and GMMB on the management of the site, this is not institutionalized in SBMS. While the World Heritage Convention has one of its pillars as communities, their involvement in the process has been indirectly through the State Party. Additionally, modernity and introduction foreign religions demonize traditional religion and practices in Africa.

Since the inscription of the ATB on the WHL, there have been a plethora of conservation works by GMMB, UNESCO and other partner organizations but this has not properly addressed challenges at the site. The cataclysmic nature of these factors has led to the loss of some parts of the buildings while others have been restored using concrete by the communities in the absence of proactive approach by GMMB. Several attempts were made by local communities to get GMMB and the government (District Assembly) to come to the rescue of the buildings but this consistently failed. These non-complaint interventions were meant to help the buildings resist the destructive nature of the weather.

From 1996 to the present, these identified factors have been recurring and it appears that GMMB is not taking any concrete steps to address them, including developing a strategy for the management of the ATB. This has led to the increasing encroachment of the buffer zones of most of the ATB, while—some of them are in a state of abandonment with only care takers who just sweep around and then lock the buildings up.

The emerging problem is that the site is not effectively managed and is increasingly becoming vulnerable to both human and natural induced threats, while the local community wisdom and traditional approaches have not been harnessed to address the situation. This situation compromises the standing of the site as a World Heritage site. Specifically, issues of authenticity and integrity have been compromised due to the weak protection and ineffective management systems being applied by GMMB. Below is a table with the summary of the identified threats and management issues of the 1996, 1997 and 1999 monitoring mission by UNESCO to the ATB. The mission was intended to ascertain the State of Conservation (SOC) of the sites.

1996 (SOC)	Climatic issues; intensification of agricultural development making the traditional building materials—thatching materials, bamboo, specific timber species—less easily obtainable
1997 (SOC)	Insufficient resources available for the conservation of fragile Ashanti buildings; Tropical climatic conditions; Proliferation of insects; Intensification of agriculture
1999 (SOC)	Insufficient resources available for the conservation of fragile ATB; Management plan; Financial resources; Need for conservation and management plans for the entire site; Important restoration work carried out without any serious scientific research

4 Adopting an Integrated Approach for Managing the Asante Traditional Buildings

Considering the conservation challenges bedeviling the ATB, it is necessary to adopt an integrated management approach in order to retain their significance. Such a management approach would draw from the existing TMS, SBMS, and international Conventions. This approach is intended to help ensure that all the values of sites, including their Outstanding Universal Values, national, regional and local values are respected and maintained not just as an international obligation but as part of a national and local community responsibility on caring for heritage and the well-being of society.

The involvement of communities (local people) in the management process has been considered as an important approach to effective management of heritage sites by a plethora of scholars (Chirikure et al. 2015; Jopela 2011 and 2010; Fowler 2002; Maradze 2003; Ndoro and Pwiti 2001; Blanchard and Trotter 2001), and it is about time the wisdom of local communities is considered as indispensable in the management process of the ATB and heritage sites in Africa at large. With regards to the management of the ATB, their historical, spiritual and socio-cultural relevance within the Asante Communities is an indication that an integrated approach is required. This approach will help to involve and engage the local communities who are indispensable in the management process of the ATB. Actively involving the communities in the decision-making framework by the government institutions (GMMB) mandated to manage the ATB's will inspire the communities to significantly contribute to the management process. This will further help to remove the top-down approach by the SBMS which has disempowered the communities who are the primary owners of the ATB. In addition to the SBMS rendering the local people impotent in the management of the ATB, the SBMS has further caused the loss of the TMS which served as the protective substratum of the site. As indicated in the 2003 Convention for the Safeguarding of Intangible Cultural Heritage (ICH) that the continuous practice of the ICH of a site makes the site viable—hence the discard of the associated TMS of the ATB and objectification of the same clearly demonstrates the betrayal of the pioneering spirits of the ATB (Taruvinga 2019).

5 Discussion

5.1 Towards an Integrated Management Approach

The role of traditional management practices and knowledge in the formation the Asante Kingdom of which the ATB (gods and shrines) was instrumental is an indication that an integrated approach should be adopted by GMMB. To develop an integrated approach for the management of the ATB, the following needs to urgently be addressed: documentation of the TMS of the ATB; community empowerment;

review of the antiquated laws on heritage and domestication of the World Heritage Convention.

5.2 Documentation of the TMS of the ATB

While the ATB have survived the conditions of the weather from the precolonial period through the resiliency of the TMS of the Asante people, these systems have not been documented. Their survival till this date is an indication that some efforts were made by the local communities to preserve the site. There is need to undertake a detail research and documentation of the ATB and the associated TMS. A proper and detailed documentation of the associated TMS of the ATB will help to furnish the site managers and GMMB at large with the right/pioneering methods and approaches to the conservation of the ATB's.

5.3 Community Empowerment

Communities are the primary owners of heritage. They are often regarded as source communities (Crosby 2015). The understanding of the source communities about heritage is totally different from the researchers or so-called experts who go into the communities to undertake research. They are the custodian of the best practices and management tools for ensuring the preservation of their heritage. To ensure that the ATB are properly preserved and managed, it is very important that communities where the ATB are located are empowered to protect the buildings. As a way of empowering the communities, GMMB should understand that since the local community live with the ATB and to some extent have oversight responsibilities over the buildings, they should be actively involved in all decision-making process regarding the management of the sites. Furthermore, GMMB through the palaces/chiefs should constantly engage the communities by educating them on their roles as primary owners of the buildings in conserving them. If possible, the GMMB should also champion an annual program at each of the ATB sites to help revive the spirits of the ATB and make them viable. It must be noted that as a shrine, the ATB were used and conserved simultaneously. In this regard initiating programs and activities at the sites will help in the management of the ATB. This will help democratize the management and conservation of the building and eventually get rid of the SBMS which do not give room for TMS. In order to ensure that the source communities feel respected and part of the management process—the local communities and the mandated state institution (GMMB) must co-control, co-own and co-produce knowledge/ideas or policies for the management of the ATB. With reference to Crosby's (2015) work at Fombori in Mali—The Culture Bank, a similar community framework or model constituting all people in the communities can be put in place for the management of the ATB. This will enforce multivocality, inclusivity and people centered approach towards the preservation of the ATB.

5.4 *Review of the Antiquated Laws on Heritage and Domestication of the World Heritage Convention in Ghana*

To effectively manage heritage sites in Ghana including the two world heritage sites i.e. the forts and castles and the ATB, there is a need for a review of the antiquated and colonial laws that for a long time has relegated the traditional management systems to the backstage. In 2013, there was an attempt by the Government of Ghana to review the Heritage laws through a proposed "Draft Ghana Heritage Resources Authority Bill". The proposed bill makes provisions for the proclamation and protection of National Monuments and Sites, repairs and alterations, control of monuments and sites, control of commercial use, pre-development impact assessment, conservation areas, site management, historic shipwrecks and change of ownership, among many other aspects. The Bill also included changing from GMMB to Ghana Heritage Resources Authority and the appointment of a Council representing cross-cutting scientific interests among them the Forestry, Archaeology and Heritage, Architecture and Environmental experts. The Bill envisioned clearly defined functions around museum services, monuments and sites services, and conservation. It is recommended that work on this proposed bill is resumed to ensure that the Heritage laws of Ghana is well positioned to meet international standards and good practices. With the view of having an integrated approach for the management of the heritage sites and World Heritage Sites in Ghana, particularly the ATB, it is also necessary for the Proposed Bill to be elaborated through broad stakeholder consultations to intricately interweave the wisdom and knowledge practices of communities who possess the pioneering spirits for the management of heritage sites in the Bill. Furthermore, the Bill should also be made to be consistent with the Operational Guidelines for the implementation of the World Heritage Convention with the view of domesticating the Convention for the effective implementation of same.

Acknowledgements I would like to express my special thanks of gratitude to the organizers/the committee of this Symposium, with the theme: "Sustainable Conservation of UNESCO and other heritage sites through proactive geosciences", for offering me a full grant to participate in this important symposium.

I would also like to thank Dr. Pascal Taruvinga for his love and immense support in guiding me in putting this paper together.

References

Anquandah J (1997) Cape Coast Castle and Fort Saint Jago, Elmina Ghana: archaeological reconnaissance survey phase two. Report for Midwest Universities Consortium for International Activities (MUCIA) Ghana

Anquandah, J. (2003). The arts of Koma-Bulsa. In Ghana yesterday and today, 135–49. Paris: Muse´e Dapper.

Blanchard L, Trotter RT (2001) What is community? An evidence-based definition for participatory public health. Am J Pub Health 91:1929–1938

Chirikure S, Mukwende T, Taruvinga P (2015) Post-colonial heritage conservation in Africa: perspectives from drystone wall restorations at Khami World Heritage site. Zimbabwe, Int J Her Stud. https://doi.org/10.1080/13527258.2015.1103300

Crosby VT (2015) The culture bank: micro-credit living objects and community development in West Africa. In: Silverman AR (eds) Museum as process: translating local and global knowledges. Routledge, London and New York, pp 189–207

Eboreime OJ (2008) Understanding cultural sites and the preparation of Nomination Dossier for World Heritage Listing. In: Biu ZA (ed) Conservation and management of heritage sites in Nigeria. Abuja, UNESCO, pp 22– 27

Eboreime J (2009) Challenges of heritage management in Africa in Ndoro W, Mumma A, Abungu G (eds) Cultural heritage and the law protecting immovable heritage in English-speaking countries of Sub-Saharan Africa (Rome: ICCROM), pp 1–6

Fowler P (2002) World Heritage cultural landscapes, 1992–2002: a review and prospect. In: UNESC (eds) Cultural landscapes: the challenges of conservation, World Heritage 2002 Shared Legacy, Common Responsibility Associated Workshops; World Heritage Papers, Ferrara, Italy, pp 16–33

Gblerkpor WN (2005) An archaeological investigation of the Krobo Mountain dry-stone terraces. Unpublished M.Phil. Thesis, University Ghana. Legon

Joffroy T, Moriset S, Misse A (1998) Asante traditional buildings - asante traditional buildings. Accra, Ghana: UNESCO, ICCROM, CRATerre-EAG. P 36

Jopela A (2010) Traditional custodianship of rock art sites in Central Mozambique: a case study from Manica district. Stud African past 8:161–177

Jopela A (2011) Traditional custodianship: a useful framework for heritage management in Southern Africa. Conserv Manag Archaeol Sites 3(2–3):103–122

Kankpeyeng BW, Nkumbaan SN, Insoll T (2011) Indigeous cosmology, art forms and past medicinal practices: towards an interpretation of ancient Koma Land sites in northern Ghana. Anthropol Med 18(2):205–216

Mahachi G, Kamuhangeri E (2008) Administrative arrangements for heritage resources management in Sub-Saharan Africa. In: Ndoro W, Mumma A, Abungu G (eds) Cultural heritage and the law. Protecting immovable heritage in English speaking countries of Southern Africa. ICCROM conservation studies 8. Rome: ICCROM, pp 43–51

Maradze J (2003) Back to the old school? Revival of traditional management systems in Zimbabwe. Papers from the 14th general assembly and scientific symposium of ICOMOS, Victoria Falls, October 2003, accessed in February 2005. www.internationalicomos.org/victoriafalls 2003, Unpublished

MunJeri D (2009) Introduction to international conventions and charters on immovable cultural heritage in Ndoro W, Mumma A, Abungu G (eds) Cultural heritage and the law protecting immovable heritage in English-speaking countries of Sub-Saharan Africa (Rome: ICCROM), pp 13–24

Murimbika MT (2006) Sacred powers and rituals of transformation: an ethno-archaeological study of rainmaking rituals and agricultural productivity during the evolution of the Mapungubwe state, AD 1000 to AD 1300. PhD thesis, University of the Witwatersrand

Ndoro W, Pwiti G (2001) Heritage management in southern Africa: local, national and international discourse. Pub Archaeol 2(1):21–34

Negri V (2008) Introduction to heritage law in Africa in Ndoro W, Mumma A, Abungu G (eds) Cultural heritage and the law protecting immovable heritage in English-speaking countries of Sub-Saharan Africa (Rome: ICCROM), pp 7–12
Pikirayi I (2016) Archaeology, local knowledge, and tradition: the quest for relevant approaches to the study and use of the past in Southern Africa. Research Gate, pp 112–135
Prussin L (1980) Traditional Asante architecture. The MIT Press Journal. 13(2):57–87
Sheridan MJ (2009) The environmental and social history of African sacred groves: a Tanzanian case study. Afr Stud Rev 52(1):73–74
Sheridan MJ (2008) The dynamics of African sacred groves: ecological, social and symbolic processes. In: Sheridan MJ, Nyamweru C (eds) African sacred groves: ecological dynamics and social change. Ohio University Press, Athens
Taruvinga P (2019) Betrayal of the pioneering spirit and approaches: domesticating international commitments on sustainable development at World Heritage sites in Zimbabwe. Unpublished

Management of Cultural Assets in Sudan from the Perspective of Sustainable Development Goals

Naoyo H. Sekihiro

Abstract Nubian heritage sites are situated between the Arab Republic of Egypt (hereinafter Egypt) and the Republic of the Sudan (hereinafter Sudan). But, the recent tourism and general interests are focused only on the Nubian heritage sites in Egypt. Also, the positive image of Sudan has not yet fully recovered in Japan, although Sudan has been making progress in the post-conflict reconstruction process. This paper aims to make a strong suggestion, from the perspective of the Sustainable Development Goals (SDGs), to treat the Nubian heritage sites both in Egypt and Sudan comprehensively, for which the international campaign to safeguarding monuments in Nubia was launched in the 1960s. This international campaign also influenced the development and adoption of the World Heritage Convention of UNESCO in 1972. The Nubian heritage sites which were moved and reconstructed in the Sudan National Museum in Khartoum should be taken care of equally as those in Egypt in the coming years. SDG 11 "Sustainable cities and communities" has a specific target with the aim to "strengthen efforts to protect and safeguard the world's cultural and natural heritage". However, Culture has a transversal role to play in achieving all the SDGs. The comprehensive re-presentation of the historical significance of the Nubian heritage sites will motivate the people in Sudan to achieve sustainable development by capitalizing on its rich cultural heritage. The case study presented in this paper may also provide inspirations for other countries and communities in post-conflict situations.

Keywords Nubian heritage sites · World heritage · The Sudan National Museum · SDGs

---

N. H. Sekihiro (✉)
Kyoto City Archaeological Research Institute, Kyoto, Japan
e-mail: naoyo@pearl.ocn.ne.jp

G. M. El-Qady and C. Margottini (eds.), *Sustainable Conservation of UNESCO and Other Heritage Sites Through Proactive Geosciences*, Springer Geology,
https://doi.org/10.1007/978-3-031-13810-2_36

Fig. 1 Books about Nubia in Japan

1 Introduction

Author's first visit to the Republic of the Sudan (hereinafter Sudan) was in 2007. That was for the study of Predynastic Egypt, the chance to change the recognition to Sudan and Nubia and the first step to construction the relationship with the cultural assets in Sudan. This paper focus on Nubian heritage sites (hereinafter Nubian sites) between Sudan and the Arab Republic of Egypt (hereinafter Egypt) based on my experience and purpose to clear the adequate preservation and management style of cultural assets in Sudan which influenced the establishment of World Heritage Convention from the perspective of Sustainable Development Goals. In this paper Nubian sites mean the sites became the subject to the salvage campaign by UNESCO in the 60s and not the original area from 1st to 4th cataract. Figures 1, 3, 4, 5, 6, 7, 8, 9, 10, 11, 12, 13, 14, 15, 16, 17, 18, 19, 20, 21, 22 are photographed by the author.

2 Recognition to Nubian Sites in Japan

The aspects that Nubian sites influenced to form the idea of 'World Heritage' and World Heritage Convention is finally shared in Japan recently. But after World Heritage become popular its origin continued to be ignored and even more the sites situated both in Egypt and Sudan. This tendency is occurred not only in the public but also in the most specialists related to World Heritage in Japan. Unfortunately, this information is not common still now. NHK (Japan Broadcasting Cooperation) broadcasted the program about World Heritage in November 2019 it only treated Abu Simbel but not treat the other reconstruction in Egypt and in Sudan. This is the reality in Japan.

Professor Hachishi SUZUKI who was a Japanese archaeologist and temporary research employee of Japan Embassy in Egypt surveyed in Nubian sites but Japanese government did not send the official archeological expedition because of the technical reason. After his survey both in Sudan and Egypt he introduced the Nubian area in Japanese language and his book is not too much to say still a Bible of Nubian study in Japan (Fig. 1).

Author encountered Nubian sites as the relation to preservation of cultural assets in Japan. The dismantling and stripping of its wall painting of ancient tumulus – *Takamatsuzuka* tumulus and *Kitora* tumulus were under discussing at Nara National Research Institute to which I belonged. Nubian sites were considered as the practical example for the preservation method by dismantling and stripping at that time. The direct visit is only way to know it at that time because there is no information about the circumstances and management condition after the reconstruction. For this reason, author visited three preservation areas by boats between Aswan to Abu Simbel in Egypt and the museum garden at Sudan National Museum for researching the condition and circumstances of reconstruction and the state of visitor during a week in ten years ago. Author continues the lecture and poster exhibition for the public based on the information above until now but further efforts will need to broad the information of location, people who lived there, the original feature of World Heritage and the existence and significance of the reconstruction at the Sudan National Museum in the future.

3 Nubian Sites Except for Abu Simbel and Philae

This section is purpose for describing the outline and current condition of reconstructed Nubian sites in three preservation area in Egypt Nubia and Sudan Nubia.

3.1 Egypt Nubia

The current condition of Nubian sites in Egypt was reported in Japanese language (Sekihiro 2008, 2010). In Japan, only Abu Simbel is introduced as Nubian sites to the public as usual because the expense of its reconstruction was partially covered by Japan and included as the main place of sightseeing. Abu Simbel is also always mentioned as its influence to the establishment of World Heritage or World Heritage Convention but not others. Actually many other reconstructions in the three preservation area exist between Aswan and Abu Simbel as below.

3.1.1 New Kalabsha

New Kalabsha is situated about 750 m south from Aswan High Dam and constitute from Kiosk of Qertassi, Temple of Beit al-Wali, Temple of Kalabsha, Temple of Gerf Hussein and the rock drawings (Fig. 2).

Kiosk of Qertassi

The original place of Kiosk situated about 40 km south from New Kalabsha. It was built in Roman era and measures ca. 10 m north to south and ca.7.5 m east to west. Kiosk of Emperor Trajan in Philae has the similar style because both have the sandstone roof. The other temples are in the Qertassi but only this kiosk was moved and reconstructed by Egyptian Antiquity in 1962 (Fig. 3).

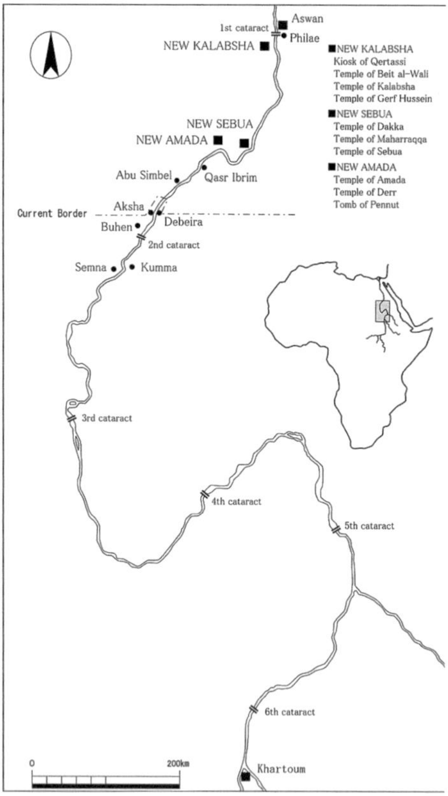

Fig. 2 Three preservation areas and sites position

Fig. 3 Kiosk of Qertassi in New Kalabsha

Fig. 4 Temple of Beit al-Wali in New Kalabsha

Temple of Beit Al-Wali

The original place of temple situated about 300 m north from Temple of Kalabsha. The positional relationship with Temple of Kalabsha was kept after the reconstruction. This temple was built by Ramesses II and consisted from the pylon, the forecourt, the vestibule and the sanctuary with half-cave style. Four God statues in the sanctuary were destructed in Christian era and plastered. For this reason, the original decoration was well preserved. This temple was moved and reconstructed by Egyptian Antiquity with the financial supports from America in 1962–65 (Fig. 4).

Fig. 5 Temple of Kalabsha in New Kalabsha

Temple of Kalabsha

The original place of temple situated about 50 km south from Aswan. This temple was constructed by Thutmose III and Amenhotep II and dedicated to the local god Mandulis. In the end of Ptolemaic period and Roman period it was reconstructed and most of the current structure attribute to Roman period. It consists of the approach from the Nile to west, the pylon, the forecourt, the hypostyle, two vestibules and the sanctuary. This is the biggest and independent structure only used sandstone in Nubia. Former West Germany expedition moved and reconstructed in 1961–1963 (Fig. 5) and also found the small another sanctuary in Ptolemy IX at the sanctuary and the basement. It was moved to Elephantine at Aswan. Huge Ptolemaic pylon was transferred as the reward.

Temple of Gerf Hussein

This temple is the half-cave style and dedicated to Ptah by Ramesses II. It is not big scale but has the similar structure with Abu Simbel. The columns in the hypostyle were figured the statue of Ramesses II but these are represented fleshier and more calm face than the Egyptian statue of same period. The temple itself was too weak to move somewhere so Egyptian Antiquity moved and reconstructed a part of it in 1964. Many rock drawings put around this temple as original place (Fig. 6).

3.1.2 New Sebua

New Sebua is situated about 150 km south from Aswan High Dam and constitute from Temple of Dakka, Temple of Maharraqqa and Temple of Sebua.

Fig. 6 Temple of Gerf Hussein and Rock Drawings

Fig. 7 Temple of Dakka in New Sebua

Temple of Dakka

The original place of temple situated about 40 km north from New Sebua. This temple is dedicated to Thoth and former structure had existed from Middle Kingdom or New Kingdom. But the full-scale construction had been started by King Arkamani at Meroe around 220 BC. Ptolemy IV continued the construction and extended the pylon and the sanctuary in Roman period. For this reason, the Roman sanctuary attaches the original sanctuary of Armkamani's period. This temple was moved and reconstructed by Egyptian Antiquity in 1962–65 (Figs. 7, 8). Unpaved road is constructed between Temple of Sebua and Dakka.

Fig. 8 Temple of Dakka (left) and Maharraqqa (lower right)

Temple of Maharraqqa

The original place of temple situated about 50 km north from New Sebua. This small Roman temple is dedicated to Serapis but not completed decoration on the wall and column. It is not unusual to attach the spiral staircase in the temple. Egyptian Antiquity moved and reconstructed in 1962–65 (Fig. 8).

Temple of Sebua

The original place of temple situated about 4 km east from New Sebua. Two temples, Amenhotep III and Ramsess II were situated in Sebua but only the latter moved to New Sebua. This temple was dedicated to Amen Re and Re-Horakhty with half-cave style constituted from the two forecourts, the pylon, the inner court, the vestibule, the antechamber and the sanctuary. St. Petro was drawn on the plaster which is covered the wall after destroying the statue of Ancient Egyptian god in the sanctuary because this temple was utilized as Christian church. Egyptian Antiquity moved and reconstructed by the financial support from America in 1962–65 (Fig. 9). The Christian painting was moved to Cairo. Tourists can see the beautiful Pharaonic painting in the temple. The cultural assets of both periods have been preserved until now.

3.1.3 New Amada

New Amada is situated about 180 km south from Aswan High Dam and constitute from Temple of Amada, Temple of Derr and Tomb of Pennut.

Fig. 9 Temple of Sebua in New Sebua

Fig. 10 Temple of Amada in New Amada

Temple of Amada

The original place of temple situated about 2.6 km south from New Amada. This temple started to construct in the reign of Tuthmosis III and completed in the reign of Amenhotep II. But it rebuilt in the reign of Tuthmosis IV at the hypostyle. The temple was constituted from the pylon which made by mud brick, the hypostyle, the vestibule, the sanctuary and dedicated to Amen Re and Re-Horakhty. The problem was too fragile to reconstruct to the other place. French expedition and the Egyptian Antiquity moved all structure of temple and put on the rail and moved in 1964–65. Amada is the shortest structure from the original place because whole temple moved after its jack up (Fig. 10).

Fig. 11 Temple of Derr in New Amada

Temple of Derr

The original place of temple situated about 11 km southwest from New Amada. It situated on the eastern bank of the Nile. Ramsess II constructed and constituted from the two hypostyles, the sanctuary. This temple was dedicated to Ptah, Amen Re, deified Ramsess II and Re-Horakhty. As Temple of Gerf Hussein, this temple style is as same as Abu Simbel. The god statue at the sanctuary was destructed in Christian period. Though the sandstone column was hardly damaged Egyptian Antiquity dismantled in 1964 and reconstructed in 1971–1973 (Fig. 11).

Tomb of Pennut

The original place of tomb situated at Aniba about 40 km west from New Amada. Pennut was the proxy viceroy of Nubia in the reign of Ramesses VI. This rock tomb was constituted the vestibule, the niche for the god statue and shaft. Last judgement in the book of dead, the personality of Pennut himself and life history of him were inscribed on the wall of vestibule. In Aniba there are many tombs for the governors and their families from 18 to 20th dynasty but tomb of Pennut was only moved and reconstructed by Egyptian Antiquity in 1962–65 by the financial support from America. The wall decoration before robbing can find the report by G. Steindorff in 1937 (Fig. 12).

Qasir Ibrim

Qasir Ibrim is also Nubian sites which keeps its original place but not opened to the public (Fig. 13). It situates about 240 km south from Aswan and became a small island on the Nasser Lake. The sites are situated originally on the summit of hill

Fig. 12 Tomb of Pennut in New Amada

Fig. 13 Qasir Ibrim

about 50 m in its height. Taharqa who was a king of Kush, constructed the temple which was used as the fortress in Ptolemaic period. In addition, temples, sanctuaries and tombs were also constructed after New Kingdom. The stela which shows 4th year of the reign Amenhotep I is the oldest artifact that we can know the exact date.

Tourists are allowed to visit above sites by boats as three- or four-nights cruise from Aswan or Abu Simbel (Fig. 14). This means that the people who can visit three preservation areas are limited to the tourist, tourist guide and the site manager. Most tourists are from foreign countries and local people are much fewer than in Abu Simbel and Philae because of their access.

Fig. 14 Temple of Kalabsha from the Cruise Boat's Upper Deck

3.2 *Sudan Nubia*

The current condition of Nubian sites in Sudan was also reported in Japanese language (Sekihiro 2008, 2010). The Sudan National Museum completed before the reconstruction of Nubian sites at Khartoum (Fig. 15). German architect F. Hinkel contributed to that project. He made the plan to layout the canal in the center of garden as 'Nile River' and the structures put on both side of this canal as the original geographical relationships. The visitor can see them right hand after through the entrance gate. The structures moved from the national border were Temple of Aksha, Tomb of Djehutihotep, Temple of Buhen, Temple of Semna and Temple of Kumma. The relief of Gebel Shaikh Suleiman and the column from Faras are arranged around the canal, too. Each structure has the ingenious devise under the ground for taking consideration into the humid climate in Khartoum. Lead foil was spread under the structure floor and it defends the structures from dampness. Unfortunately, no point can see this system directly but the good condition of each structure represents the effectiveness of this method.

3.2.1 Temple of Aksha

This temple was situated western side of Nile River and dedicated to the deified Ramsess II. It consists of the pylon, the forecourt, vestibule and sanctuary with the independent style (Porter and Moss 1995:120, 127). The reconstruction was operated by the financial support from France with the work of Buhen temple simultaneously in 1962- 63 (Säve-Söderbergh 1987: 145). It was the hard work because of the salt damage. South side of south wall at vestibule was reconstructed under the open facilities (Fig. 16).

Fig. 15 The Sudan National Museum (SNM)

Fig. 16 Temple of Aksha in SNM

Fig. 17 Tomb of Djehutihotep in SNM

3.2.2 Tomb of Djehutihotep

This rock tomb was situated eastern side of Nile River at Debeira (Säve-Söderbergh 1987: 145). Its reconstruction was operated simultaneously with Aksha and Buhen temple in 1963 (Säve-Söderbergh 1987: 147). Djehutihotep himself was the Egyptianized Nubian governor. The slates from al-Markhiyat mountain which situates north Omdruman were utilized to this reconstruction. The tomb reconstructed west–east axis as original. A Small room for the instruction situates in southern part of reconstruction with the photo about moving and reconstruction and brief explanation about the original place in English and Arabic. The vestibule measures ca.6.2 m west–east and ca.4.5 m north–south. Wall painting on north side of west, west side of north and east side of north remain their color still now (Fig. 17). The fade of color and peeling is progressing though all entrances of this tomb are always closed and turn off the lights.

3.2.3 Temple of Buhen

This temple was situated at western side of Nile River and constructed the fortified city which measures ca.150 m east–west and ca.180 m north–south. Temple for Isis and Min by Amenhotep II and Temple for Horus of Buhen by Hatshepsut and Tuthmosis III were constructed in this city but only Horus temple was moved to Khartoum. It consists from the pylon, the forecourt, the vestibule, the sanctuary and three side rooms. The reconstruction was operated by the financial supports from England and America in 1963–67 (Säve-Söderbergh 1987: 145).

Shellac was used to strengthen the sandstone during the reconstruction. The temple was covered by sand at first, because there are no adequate cranes at that time in Sudan. The blocks removed from upper to lower layer gradually like ancient method. Removed block was transported by the slide which used the logs of mahogany

Fig. 18 Temple of Buhen in SNM

because they are not expensive. The 6 m big cloth for covering the blocks was woven by the native ladies. A part of forecourt, vestibule and the sanctuary were moved and reconstructed. Blocks that original place was unknown were displayed on the west wall of facilities. Slate of al-Markhiyat Mountain were utilized to the floor as same as in Tomb of Djehutihotep. The reconstruction is covered by the movable and vault facility which have the window for sun light in the ceiling, three parts divided and putted on the rail (Fig. 18). This facility is the nest style moving from the entrance to opposite side but not moved like that in two decades. The full open style after all the facility moved to one side can see on the figure of report by Caminos (1974: Pl. 7, 8). The visitor can compare the good conditioned wall paintings with the current ones because the explanation panel was prepared in front of Buhen.

3.2.4 Temple of Semna

This temple was situated at western side of Nile River and constructed the fortified city in Middle Kingdom which measures ca.95 m east–west and ca.120 m north–south. The temple for Dedwen and Tuthmosis I were constructed in the fort city and former temple moved and reconstructed. The cemeteries in Middle Kingdom and New Kingdom and Rock drawings were also founded in the site but they submerged unfortunately. Temple of Semna was dedicated to Dedwen of Nubian god and deified Senusert III. Tuthmosis III was also concerned its construction (Caminos 1998a: 12–15). The reconstruction was operated by the financial support from Belgium in 1963–67(Säve-Söderbergh 1987: 145). The method of reconstruction was as same as Buhen and 146 blocks was transported to Khartoum. East and west portico and the sanctuary were reconstructed and the blocks unknown places were presented at the western part of west portico. Slates were spread as the floor. The big difference from Buhen facility is the birds preventing net covering the temple (Fig. 19). A part of roof was recovered and this is also the different aspects from Buhen. The facility

Fig. 19 Temple of Semna in SNM

of Buhen is triples structure but Semna is the single structure and it can move to west side by the rail.

3.2.5 Temple of Kumma

This temple was situated at opposite side of Semna, and eastern side of Nile River. It was the fortress constructed by Senusert III in Middle Kingdom. This temple was dedicated to Khnum and Tuthmosis II started to construct and it was succeeded by Tuthmosis III. Amenhotep II was repaired extensively (Caminos 1998b: 1–4). The reconstruction was supported financially by Holland in 1963–67(Säve-Söderbergh 1987: 145). The method of reconstruction was as same as Buhen and Semna. 480 blocks were transported to Khartoum.

The temple reconstructed at the opposite side of the psudo Nile as the original geographical relation. The single facility was utilized as in Semna. Slates were spread as the floor as same as Buhen and Semna. Second forecourt and the sanctuary which situates at north-western part of temple were reconstructed and the part of roof was recovered. Birds preventing net covered the ceiling as same as Semna. The facility was not moved in these two decades (Fig. 20).

4 Current Condition of Sudan

The main purpose of this paper is present the significance of reconstructed Nubian sites in Sudan as mentioned above. The issue at modern Sudan and the current condition of preservation and management of cultural assets are related to this purpose.

Fig. 20 Temple of Kumma in SNM

4.1 Background

The people started the full effort for Sudan revolt on December in 2018 after raising the price of bread and 30 years government come to end finally. Although Sudan has been making progress in the post-conflict reconstruction process the positive image of Sudan has not yet fully recovered in Japan. The positive image was wiped out under the economic sanctions and State Sponsors of terrorism in quarter of century. This affection is deeper than expected it seems Sudanese themselves are sometimes in the circumstances which cannot avoid to be not in active for the presentation of their history and culture. These situation prevent the true recovery of Sudan and it should be changed immediately.

4.2 Feature and Current Situation of Cultural Assets in Sudan

Nubian sites in Sudan were moved to the Sudan National Museum. The open-air exhibition was not able to reappearance the periphery of original sites in Egypt but reflected the original geographical relation clearly in the garden. But not many specialists related to World Heritage know this fact now. There are three reasons of this situation.

First is that most of researchers were Egyptologist. Sudan has been studied only as a part of ancient Egypt in a long time.

Second is the affection of back ground as above. The archeological research, study and the preservation of archaeological and historical sites had not been recognized its importance by the former government. There are no circumstances to focus their own history before Islamic era. Therefore, the ancient sites and their history sometimes accepted inadequately in the public of Sudan (not mean in academic).

Third is the original diversity. As mentioned, Sudan had the tendency to have been studied as a part of ancient Egypt. It caused the concentration to the history and culture along the Nile. But ancient Sudan was influenced not only from Egypt but many areas like Darfur, South Sudan, eastern desert, Ethiopia and Red Sea because the trade was so active. This diverse culture is not easy to explain simply for its own feature hence difficult to understand easily. Next section is to show the practical way to solve the above discussions.

5 Cultural Assets in Sudan from the Perspective of SDGs

Recently cultural assets started to focus on new international standard, Sustainable Development Goals (SDGs). Japan Consortium for the International Cooperation in Cultural Heritage (JCIC) held the Symposium in 2019 but this standard has not been accepted broadly to the specializes for the cultural assets in Japan yet.

My first relation to SDGs was in 2016. The public photo exhibition which was arranged by Ministry of Foreign Affairs in Tokyo at Global Festa- Hope the world 'leaving no one left behind'. The framework of Sudan National Museum Exhibition was under construction at that time and some photos were presented at the exhibition to broad the activities and idea of exhibition. The concept of 'leaving no one left behind' was the good encouragement to the work because only the negative information about Sudan were shared or completely neglected in Japan society.

SDGs is purpose for the social change and especially for the unsustainable current world change to the sustainable world. This is not only required in Sudan but Japan on the sociality of archaeology and the publicity of academic results and so on. We can find the same situation at the other countries or 'Africa considered as one country'. Some Japanese researchers don't accept SDGs as new armchair theory because they believe the theory from UN doesn't cover each local field.

Actually speaking, this idea is not completely new. In Japan there is similar way of thinking 'Sanpou-yoshi (benefit for all three sides)' in the last eighteenth century. It shows that the good business should take into consideration the benefits the three sides, the customer, society and vendor. Most big companies in Japan keep this mind as their management philosophy. Management philosophy is not as completely same as the academic research standard but they seem to have the same basic direction to the society.

JCIC symposium in 2019 defined SDGs has four fields, society field, economy field, environment field and cross-sectional filed to former three, and unfortunately not included 'culture'. But this is not unnatural and no need to make another field of Culture because the SDGs is originally for the social change. If anything, culture is the most important and related to all fields, goals and targets. This is shown 'Culture as driver and enabler for sustainable development' by UNSECO in 2017 (Fig. 21). By this, we can understand that culture contains all comprehensively.

Fig. 21 Culture in SDGS by UNESCO moving forward the 2030 Agenda for Sustainable Development (UNESCO moving forward the 2030 Agenda for Sustainable Development. https://en.unesco.org/creativity/sites/creativity/files/247785en.pdf; p.16; 2021, 225 final accesses)

Goal 11, Target4 describes about the cultural assets clearly but the cultural assets in Sudan seem to be related with other Goals and Targets. Five goals and targets can be pointed out the relation on this paper as below;

Goal 1, Target 5: By 2030, build the resilience of the poor and those in vulnerable situations and reduce their exposure and vulnerability to climate-related extreme events and other economic, social and environmental shocks and disasters.

During the long economic sanctions and manipulation of information the circumstances of economy, society and environment were forced to be vulnerable and it is one of the reasons to become poverty. The basic research of the artifacts and preservation and management of cultural assets like Nubian heritage sites in Sudan National Museum will contribute the recovery of positive image and broad the correct information of history and culture which lost in these two or three decades. Furthermore the resilience of will be reconstructed.

Goal 4, Target 7: By 2030, ensure that all learners acquire the knowledge and skills needed to promote sustainable development, including, among others, through education for sustainable development and sustainable lifestyles, human rights, gender equality, promotion of a culture of peace and non-violence, global citizenship and appreciation of cultural diversity and of culture's contribution to sustainable development.

History education about pre-Islamic period is uncommon in Sudan. That is reason why the result of ancient history study by Sudanese historians and archaeologists is not easy to broad to the public and to construct the neutral or comprehensive identity. Sudanese is not uninterested in their pre-Islamic history but just limited to access to the related information. The direct access to the research and the study of Nubian heritage sites in Sudan will contribute to history education at school, the establishment of identity, the promotion of cultural diversity, the preservation of cultural assets and the succession of traditional culture. Then, they will provide the recognition not only Nubian heritage in Egypt but in Sudan influenced to World Heritage convention, too.

Goal 8, Target 9: By 2030, devise and implement policies to promote sustainable tourism that creates jobs and promotes local culture and products.

Sudan Government recognizes the importance of tourism but only limited people how to do for it. The information of comprehensive and local history is not common enough to understand the significance and value of historical heritage correctly. For example, the iron making was experimented and opened to the public in Meroe but it will need more time to broad its information to all area in Sudan. The basic study of artifacts and Nubian heritage sites in the museum and the publication on Books and SNS will also contribute to broad the information correctly and promptly, and construct the foundation of sustainable tourism in Sudan.

Goal 10, Target 2: By 2030, empower and promote the social, economic and political inclusion of all, irrespective of age, sex, disability, race, ethnicity, origin, religion or economic or other status.

Misunderstanding and prejudice impede above fundamentally. Sudan history is introduced partially and the government is not afforded to solve this issue because of its high multi-ethnicity. Only the conflict in Sudan has been focused until now but the foreign mission has continued the archaeological research between the Campaign of UNESCO in 60's and the end of North–South conflict. Many artifacts from those research in the Sudan National Museum are stored and contribute to understand Sudan history neutrally and comprehensively without the stereotype like Gold, Black Pharaoh and 'Nuba wrestling' in case of Japanese recognition.

Goal 12, Target b: Develop and implement tools to monitor sustainable development impacts for sustainable tourism that creates jobs and promotes local culture and products.

The current situation is the previous step to consider the impacts after activating the tourism and the few people are engaged in it. The objective grasp of actual condition will be essential to develop and implement tools for the field of cultural assets to monitor sustainable development impacts. The correct understanding to the artifacts and Nubian heritage sties in the Museum, the grasp and improvement of preservation circumstance are important. This information should be shared not only archaeologists and experts of the cultural assets but also the other related institution for the international cooperation.

6 Conclusion

The recent tourism and general interests are focused only on the Nubian sites in Egypt. Strictly speaking too much focus only Abu Simbel. As mentioned above, there are three preservation area of reconstruction, New Kalabsha, New Sebua and New Amada in Egypt. There are the reconstructions in Sudan, too. All of them were from one cultural area by the same rescue campaign. Many engineers and specialists were collaborating under the concept for Saving Nubia. There is no difference of the attitude to the reconstruction between Sudan and Egypt.

Another reason of this suggestion is for the positive image of Sudan. Although Sudan has been making progress in the post-conflict reconstruction process it has not yet fully recovered in Japan and maybe not in the other countries. Author pointed out 5 goals and targets of SDGs to discuss this issue on this paper. These goals and targets are not absolute but a kind of factor for starting discussion.

Many students visit the Sudan National Museum (Fig. 22) and its open exhibition in a daytime and sometimes held the cultural event or concert in the garden. If all these people understand why the temples and tombs moved to this museum their visit and event will be more effective and empower the tourism and the construction of identity on them.

The comprehensive re-presentation of the historical significance of the Nubian sites will also motivate the people in Sudan to achieve sustainable development by capitalizing on its rich cultural heritage and good opportunity to remind the public about the inherent meaning of World Heritage and World Heritage Convention again.

Fig. 22 Students at 2nd floor in SNM

References

Caminos RA (1974) The New Kingdom Temples of Buhen. 2 vols. Egyptian Exploration Society, London

Caminos RA (1998a) Semna and Kumma I. The Temple of Semna. Egyptian Exploration Society, London

Caminos RA (1998b) Semna and Kumma II. The Temple of Semna. Egyptian Exploration Society, London

Säve-Söderbergh T (1987) Temples and tombs of Ancient Nubia. Thames and Hudson, London

Sekihiro N (2008) The Nubian sites now: the first step to the World heritage convention. Quart of Archaeol Stud 218:12–17 [In Japanese]

Sekihiro N (2010) Current conditions of the Nubian sites and maintenance issues of cultural heritage in Sudan. J West Asian Archaeol 11:139–148 [In Japanese]

Improving the Georgian National Programme for the Preservation of the Rock-Cut Sites Through Interdisciplinary Geosciences

Nikoloz Antidze, Claudio Margottini, Tamar Meliva, Daniele Spizzichino, and Manana Vardzelashvili

Abstract Present paper summarizes the activities that National Agency for Cultural Heritage Preservation of Georgia has put in place over the last ten years. More specifically, all the phases that led to the mapping of the sites to be preserved and the actions taken to protect them are reported and summarized. Among a huge number of tasks, the Agency's main missions were to lead, coordinate and implement state program for the identification, conservation and rehabilitation, protection and promotion of tangible and intangible cultural heritage, monitor and inventory of tangible and intangible heritage assets throughout the country and provide cultural heritage expertise. This programme is the result of the long process, started in Vardzia Rock-cut Complex and not yet completely developed.

Keywords Rock-cut sites · Vardzia · Geosciences

1 Introduction

Georgia is one of the oldest countries and is special with its unique heritage. During the very long history, it always has lived on the crossroads where the large civilizations met. Located in between Europe and Asia, the country always has been of a great interest for different civilizations, thus culture obviously has an influence of the East as well as the West. This little country has very rich nature as well as the culture.

N. Antidze · T. Meliva · M. Vardzelashvili
National Agency for Cultural Heritage Preservation of Georgia, Tbilisi, Georgia

C. Margottini (✉)
Former Scientific Attaché, Embassy of Italy in Egypt, 15, Abd El-Rahman Fahmy Str., Garden City, Il Cairo, Egypt
e-mail: claudio.margottini@gmail.com

C. Margottini · D. Spizzichino
UNESCO Chair On Prevention and Sustainable Management of Geo-Hydrological Hazards, University of Florence, Florence, Italy

ISPRA – Dpt. For the Geological Survey of Italy, Rome, Italy

G. M. El-Qady and C. Margottini (eds.), *Sustainable Conservation of UNESCO and Other Heritage Sites Through Proactive Geosciences*, Springer Geology,
https://doi.org/10.1007/978-3-031-13810-2_37

The architectural and urban heritage of Georgia is rich with diverse examples of sacral and secular buildings, fortifications, historic settlements and towns. Cultural landscapes, populated from ancient times, represent testimony of harmonious co-existence of a man with nature. The country is full of heritage, more than ten thousand places has been designated as the Cultural Heritage Monument or Object. Therefore, preservation, rehabilitation and popularization of Georgian Cultural heritage, are placed as one of the main state priority of the country.

The difficult socio-political conditions in the country at the end of XX century dropped it isolated from contemporary worldwide processes and it was not able to follow the innovations and to introduce to new approaches in the field. Only for last decades, it became possible to improve arisen situation: In a very short timeframe, there were improved a lot of tasks regarding cultural heritage. Being aware, that Georgian cultural heritage required transferring on the different, new model of management that would be adequate to the worldwide standards, became the main background for founding National Agency for Cultural Heritage Preservation of Georgia, which is an operational institution responsible for the implementation of national cultural heritage policy.

Among a huge number of tasks, the Agency's main missions are to lead, coordinate and implement state program for the identification, conservation and rehabilitation, protection and promotion of tangible and intangible cultural heritage, monitor and inventory of tangible and intangible heritage assets throughout the country and provide cultural heritage expertise.

The National Agency is based on eighteen museums and museum-reserves museums, placed in the different regions of Georgia. During almost a decade the special attention is paid to the long-term protection of the rock-cut cultural heritage in Georgia, even more, the national program for the multidisciplinary research and preservation of this type of heritage is being established, to include all rock-sites and ensure their proper protection, both for the heritage and the visitors. This programme is the result of the long process, started in Vardzia Rock-cut Complex and not yet completely developed (Margottini et al. 2015).

The sites and remains, considering the energy relief, climatic setting and the low strength of the parent material, are not in equilibrium with the current changing environment. They exhibit an intrinsic fragility, especially with respect to environmental threats, during the time. In fact, they are continuously impacted and weathered by several internal and external factors, both natural and human-induced, with rapid and/or slow onset. These include major sudden natural hazards, such as earthquakes or extreme meteorological events, but also slow, cumulative processes such the erosion of rocks, weathering, compounded by the effect of climate change, without disregarding the role of humans, especially in conflict situations.

A general approach, implemented in the conservation activities, may include a very detailed interdisciplinary study, to understand rock degradation processes and causative factors and followed by field conservation work. Both phases are in parallel with monitoring. The mitigation and conservation phase are mainly related to the re-discovering of traditional knowledge and sustainable practices and is based on the application of local conservation techniques.

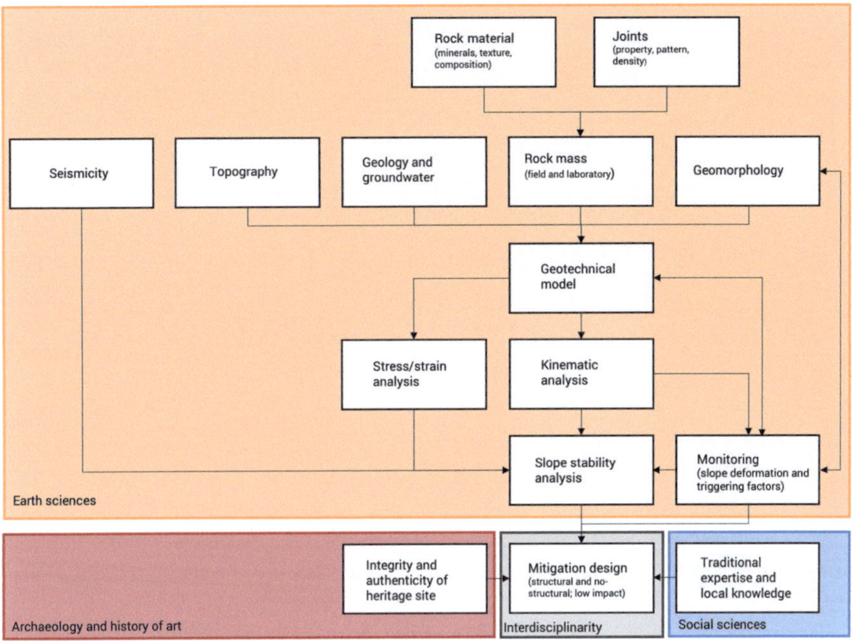

Fig. 1 Site engineering approach for the mitigation of rock slope instabilities in rupestrian sites

The methodology used in the present approach is primarily aimed at the protection of the heritage sites but, at the same time and when possible, at the empowerment of local communities, to be trained to safeguard and to manage the site. Protecting heritage from natural threats is, in fact, not a simple engineering work but also a fundamental step to be given priority together with other humanitarian concerns. This is especially relevant at a time when traditional knowledge and sustainable practices, which ensured a certain level of protection from the worst effects of natural hazards or human-made disasters, are being progressively abandoned and/or forgotten.

A general approach for the mitigation of rock slope instabilities in Georgian Cultural Heritages is reported in the following Fig. 1.

2 The Vardzia Showcase

Vardzia Rock Cut City, as an outstanding sample of harmonious confluence of manmade and natural monument, is most distinguished heritage of Georgian medieval, represented in the World Heritage Tentative List from 1993 (Fig. 2).

Unfortunately, located in the most active seismic region of Georgia and same time being on one of the routes of major invasions it has suffered several dramatic destructions.

Fig. 2 Vardzia landscape frontal view

Ongoing natural processes of rock weathering and erosion, combined with seismic activity and thus frequent landsides and rock falls have been harming integrity of the site during the centuries and till recent time, causing the direct threat to its existence. Vardzia rock cut city destruction has reached the most dramatic state in the first half on twentieth century, when several collapses have occurred on almost abandoned site.

The extremely endangered condition of the monument is described in the reports of that period, and then also several major interactions were made to stop the ongoing process of destruction. Interventions of that period were mostly point targeted, based on the technologies of that period (Fig. 3).

This way a massive concrete support wall in the facade of the monument was constructed; simple rain water avoiding channels were laid out of stones on top of the monument, etc.

Fig. 3 Past mitigation measures, reinforced concrete wall and wooden scaffolding (1960–1973)

These urgent measures were local/partial solutions not resolving the major problem of the monument – its ongoing process of natural destruction.

The processes of deterioration become more and more dangerous for humans visiting the site. Even during the middle of twentieth century site had average of 15,000 visitors per year, nowadays (as of 2019) this number has increased more than ten times, about 180,000 visitors. Worth mentioning that World Heritage Committee in 1999, while admitting its outstanding universal value, highlighted the need for improved protection management and legal framework of the nominee site. For this reason, the Committee referred the dossier back to the State authorities for implementing above mentioned improvements.

During the twentieth century and first decade of twenty-first century, the State tried its best to provide the stabilization measures of particular portions of damaged areas. During these years the main actions had been focused on controlled demolishment of the weaken rocks (so called cleaning) by groups of climbers to avoid the uncontrolled processes of possible collapses equally dangerous for humans as well as for cultural heritage. No systematic multi-disciplinary survey had been implemented to elaborate the general plan for the safeguarding of the site.

Complexity of the dual, manmade and natural monument represents a major challenge for the protection and sustainable functioning of Vardzia. Therefore, implementation of proactive safeguarding strategy to ensure the structural stability and proper state of conservation of the Complex was an urgent issue to maintain integrity and authenticity of Vardzia Rock Cut Complex.

Due to the lack of experience and knowledge in this specific field of monitoring and conservation of Rupestrian heritage sites, where the approach should be based mostly on the interdisciplinary earth sciences and advanced non-intrusive technologies, internationally well-known expert, Prof. Claudio Margottini, represented ISPRA (Geological Survey of Italy), together with his strong scientific team was invited for methodological assistance by National Agency for Cultural Heritage Preservation of Georgia (Fig. 4).

On the other hand the National Agency mobilized the all scientific-educational institutions in Georgia for collaborating with Italian colleagues within the project. The leading methodological counterparty from Georgian site became Cultural Heritage and Environment Research Center of Ilia State University, providing active involvement of researchers and students.

While the first introductory activities have been done in 2012, the implementation time frame of the project was 2014–2019 and considered the elaboration and implementation of the urgent measures and long-term strategy to mitigate the rock fall and landslide hazards in Vardzia (Margottini et al. 2016).

All the intervention were focusing on low impact and/or recovering traditional techniques.

Based on the interdisciplinary Italian-Georgian studies, established sustainable monitoring system and field missions, the Conservation Plan for urgent interventions was elaborated in 2015 that has been completed in 2019.

Most relevant was the ground based radar, surveying the site with an accuracy of "mm", from 800 m away (Fig. 5).

Fig. 4 Field survey of the international expert team during the first International Workshop on "Landslide hazard assessment and sustainable monitoring techniques for the safeguard of Vardzia Monastery site Vardzia (Georgia)". 16–19, October 2012

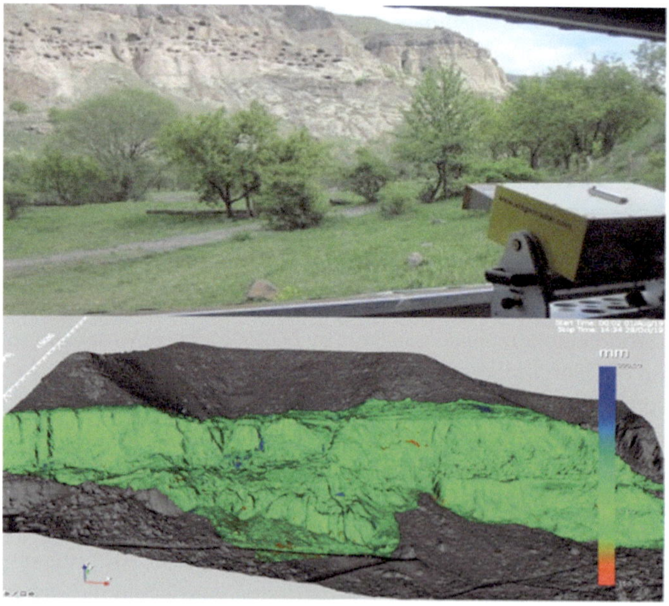

Fig. 5 Ground Base Radar installed in 2012 to monitoring the site

The analysing of monitoring data proves that the enormous efforts done jointly by Italian and Georgian experts were successfully for stabilization of rock movements and further deterioration of the Monument. Another most important achievement of the project was the establishment of the complex monitoring system (Elashvili et al. 2016), that considers gathering a various data (movement, deformation, humidity, precipitation, temperature, seismic tremors, etc.) near real-time automatic processing followed by the further analyses by scientific team and as a result elaborating and implementation of particular hazards mitigation proposals that should ensure the long-term structural stability of Vardzia (Boldini et al. 2017).

Due to the successfully implemented pilot project, and based on the capacity built in Vardzia, since 2018 on the decision of Georgian Government, it was expanded to the "Rock Cut Heritage Safeguard State Program" (Frodella et al. 2020) and now involves several major Rupestrian complexes in Georgia: Vanis Kvabebi, Uplistsikhe and David Gareja (Figs. 6 and 7).

The later one since 2018 is in the List of 7 most endangered heritage sites of Europa Nostra. Due to its interdisciplinary, modern approaches and technical advances,

Fig. 6 Rock cut city of Uplistsikhe

Fig. 7 Lavra monastery in David Gareja complex

sustained monitoring approach, the project is also considered as a one of the kinds, also on international level.

The last, but not least achievement of the project is the established collaboration platform between Georgian and Italian scientific schools. This exclusive possibility of sharing of experience and knowledge jointly focusing on solving the stability problems in rock cut complexes in Georgia, has paramount impact on development of professional capacity and approaches of Georgian institutions. Therefore, the project demonstrates successful results of collaboration between partner countries in the field of cultural heritage and enhanced the institutional partnership of different scientific and educational institutions of Georgia and Italy.

3 Advanced New Technology Applications

The new advanced technologies, such as the above mentioned Ground Based Radar, climate-monitoring and seismic-monitoring devices have been installed to study the tendencies of deterioration of the surfaces, as well as to provide warning alarm system in order to implement the appropriate measures in case of increasing danger of collapse (Elashvili et al. 2016).

Because of the multidisciplinary studies (hydrological, geological, mechanical, physical, chemical, kinematical etc.) and the data received from the monitoring system, the plan of the first action was elaborated continuously in a dynamic way.

In line with studies, the first physical works aiming at consolidation of the most sensitive areas started in 2015. The medieval tunnel that was almost deteriorated from the main rock was anchored by Georgian engineers in collaboration with international experts, who provided the methodological instructions and supervised the field works to ensure the best quality of the conservation.

This was followed by the scheduled interventions (Boldini et al. 2017) on the most urgent areas and systematic actions to complete the conservation/consolidation of the urgent areas of Vardzia Rock cut complex to ensure its stabilization and maintenance (Fig. 8).

4 Conclusions

In conclusion, the Conservation of Vardzia Rock Cut Complex is considered as one of the most ambitious and knowhow projects being implemented in Georgia for the last decades. The multidisciplinary and multi-institutional participation supported by the best international expertise, together with the actions carried out on basis of investigations and studies became the best model of the conservation management and a guideline for similar sites in Georgia.

Since 2017 the projects on other rock-cut sites has been launched, which finally led to the establishment of the National Programme of the Preservation of the Rock-cut

Fig. 8 Conservation and consolidation of urgent areas in Vardzia

Sites (David Gareji, Uplistsikhe, Katskhi Column etc.). One of the most important results of the project is increased professional skills of Georgian specialists in the most comprehensive and sensitive issues of mix heritage conservation according to the international standards and best practices.

Acknowledgements The research and works described in this paper were realized by the support of the Georgian government and the great effort of the National Agency for Cultural Heritage Preservation of Georgia experts' team. The authors are very grateful to all the site managers, workers and technicians, as well as to all the national and international Universities and Research Institutes involved since the beginning in the conservation program.

References

Boldini D, Guido GL, Margottini C, Spizzichino D (2017) Stability analysis of a large-volume block in the historical rock-cut city of Vardzia (Georgia). Rock Mech Rock Eng (2017). https://doi.org/10.1007/s00603-017-1299-7

Elashvili M, Vacheishvili N, Margottini C, Basilaia G, Chkhaidze D, Kvavadze D, Spizzichino D, Boscagli F, Kirkitadze G, Adikashvili L, Navrozashvili L (2016) Concept of complex environmental monitoring network – Vardzia rock cut city case study. Geophysical research abstracts, vol 18, EGU2016–13460–2, 2016 EGU General Assembly 2016 © Author(s) 2016

Frodella W, Elashvili M, Spizzichino D, Gigli G, Adikashvili L, Vacheishvili N, Kirkitadze G, Nadaraia A, Margottini C, Casagli N (2020) Combining infrared thermography and UAV digital photogrammetry for the protection and conservation of rupestrian cultural heritage sites in Georgia: a methodological application. Remote Sens 12(5):892. https://doi.org/10.3390/rs1205 0892

Margottini C, Antidze N, Corominas J, Crosta GB, Frattini P, Gigli G, Giordan D, Iwasaky I, Lollino G, Manconi A, Marinos P, Scavia C, Sonnessa A, Spizzichino D, Vacheishvili N (2015) Landslide

hazard, monitoring and conservation strategy for the safeguard of Vardzia Byzantine monastery complex, Georgia. Landslides 12:193–204
Margottini C, Spizzichino D, Crosta GB, Frattini P, Mazzanti P, Scarascia G, Beninati L (2016) Rock fall instabilities and safety of visitors in the historic rock cut monastery of Vardzia (Georgia). Volcanic Rocks and Soils – Rotonda et al. (eds) Taylor & Francis Group, London, pp 372–378. ISBN 978–1–138–02886–9.Volcanic Rock and Soil

Aswan Declaration

The Aswan Declaration on Sustainable Conservation of UNESCO and Other Heritages Sites Through Proactive Geosciences

Claudio Margottini and Gad Mohamed El-Qady

Abstract Cultural Heritage sites are considered the cornerstone for most of the nations' civilizations. Consequently, UNESCO put the sustainable protection and preservation of cultural heritage sites on the top of its priorities in the 2030 agenda. Recognizing that UNESCO World Heritage Sites, and other heritage sites, are located in various geographical settings and their territories may be partly or entirely exposed to various natural hazards and extreme weather events, a proper proactive policy is mandatory. The international community represented by the participants of the symposium on "sustainable conservation of UNESCO and other heritages sites through proactive geosciences" has agreed to promote the preventive protection of cultural heritage against geohazards, by use an interdisciplinary and transdisciplinary approaches to conservation and protection of heritage sites, which shall bring together Arts, History, Sciences, Engineering, Traditions, Management and Socio-economy disciplines, among others, driven by the sustainability principles. Among earth sciences, engineering geology and geotechnical engineering, information communication and technology (ICT), remote sensing and imaging, structural engineering and health monitoring, and early warning systems, play an essential role in conservation and management of cultural properties. We have recommended to the international community to continue providing the needed support to build resilience of heritage sites, including through the optimization of ecosystems, aligned with the Sendai Framework for Disaster Risk Reduction, the 2030 Agenda for Sustainable Development, the Paris Agreement, and the New Urban Agenda.

Keywords UNESCO · Heritage sites · Geohazards · Proactive geosciences · Sustainability

---

C. Margottini (✉)
Embassy of Italy in Egypt, Cairo, Egypt
e-mail: claudio.margottini@gmail.com

G. M. El-Qady
National Research Institute for Astronomy and Geophysics (NRIAG), Cairo, Egypt
e-mail: gadosan@nriag.sci.eg

G. M. El-Qady and C. Margottini (eds.), *Sustainable Conservation of UNESCO and Other Heritage Sites Through Proactive Geosciences*, Springer Geology,
https://doi.org/10.1007/978-3-031-13810-2_38

1 Introduction

Cultural heritage represents the legacy of humankind on planet Earth. It witnesses millennia of people adaptation to their environment, as demonstrated in many monuments, sites and cultural landscapes. Such historical landmarks are subjects to continuous changes and to the influence of modern growth and development. The impacts and alterations result from several internal and external conditions stemming from both natural and anthropogenic factors, such as rapid (i.e. earthquakes, landslides, floods, debris flows, slope movements) and slow onset (i.e. geological and geotechnical subsidence, soil and coastal erosion, sinkholes, hydro-geological conditions) including climate change. Nowadays, cultural heritage shows evidence of the impact of geohazards and weathering and call for the need to rethink 'sites' conservation and management plans. Consequently, geoscience discipline and affiliated empirical research studies and innovations in technology may need to bring new paradigm for the preservation of cultural properties providing a resourceful platform for learning.

In the past decades, the shift in disciplines from working inward to opening up to inter-disciplinary ways of thinking draws special attention to the added value of merging Arts with Sciences among other disciplines for better management and preservation of cultural heritage. The 'New School' of thoughts is manifested by many showcases and projects on the mitigation of risks of geohazards which aim is to maintain the integrity and authenticity of cultural and historical sites heritage. The advanced scientific investigations involve a better understanding of the natural processes, coupled with conservation approaches mainly based on sustainable practices including the use of traditional knowledge in the recovery techniques and building local capacities to ensure effective conservation works with time.

2 The Context

The Aswan declaration was prepared during the 1st Symposium on "sustainable conservation of UNESCO and other heritages sites through proactive geosciences", jointly organized in December 2019 by the Embassy of Italy in Egypt, the National Research Institute for Astronomy and Geophysics of Egypt, the UNESCO Office in Cairo (Egypt), and the Egyptian Academy of Scientific Research and Technology.

In this context the Symposium brought together researchers and practitioners from various disciplines to share experiences and lessons learned on the management of heritage sites, facing the risks of geohazards and other geo-environmental threats around the world. Investigated threats include slope instabilities in soil and rock material, weathering, bearing capacity, hydrogeology and seismicity. All topics falling in the disciplines of engineering geology, geotechnical engineering and rock mechanic were considered.

Various hazards were interlocked with innovative technology in scientific investigation which are now providing a strong support to sustainable conservation of heritages. Major achievements are presently coming from geophysics, laser scanning and virtual reality, structural and environmental monitoring and remote sensing.

Finally, a proposal of new policies for management and conservation of heritage sites were established, as background for the Aswan Declaration. Considering that most of the heritages sites in the word are suffering for geohazards and other geo-environmental threats, it is evident that a specific policy aimed at the sustainable conservation of such sites must be implemented. Such new policy has to considered the shared contribution of earth sciences, archaeology and history of art and, finally, social sciences. That is because if earth sciences are essential in understanding the impacting threat, archaeology and history of art are of basic importance in understanding the integrity and authenticity of the site. In the meantime, any conservation projected dealing with a cultural heritage should enhance, when possible, traditional knowledge and local expertise, to guarantee the long-term maintenance of the work from local population, then to ensure sustainability (Fig. 1).

The involved scientific community was representative of 19 Countries from all over the world, as reported in Fig. 2. This is ensuring a representativeness from various school of thought and approaches, which reflect mixed social and scientific conditions.

Many case studies were presented and discussed among participants. Such cultural heritage sites reflect a variety of endangered case histories, methodological approaches, involved sciences and expertise from all over the world (Fig. 3). They deeply contributed to the synthesis reported in the declaration, which hopefully reflects most of the variability of exposed heritage elements and threats.

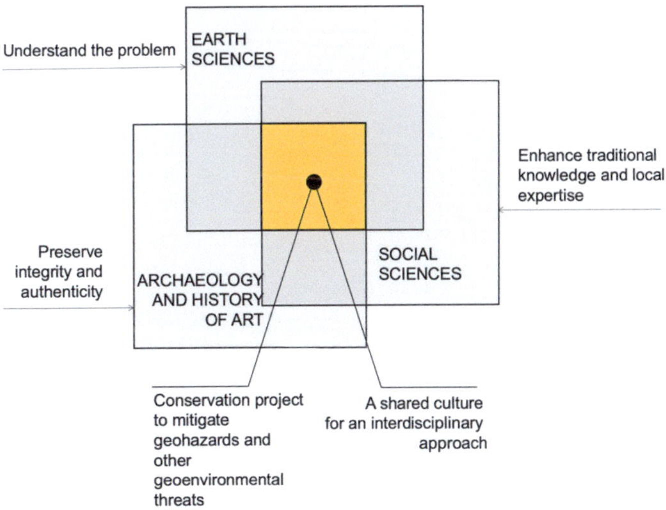

Fig. 1 The methodological approach proposed in the Aswan Declaration

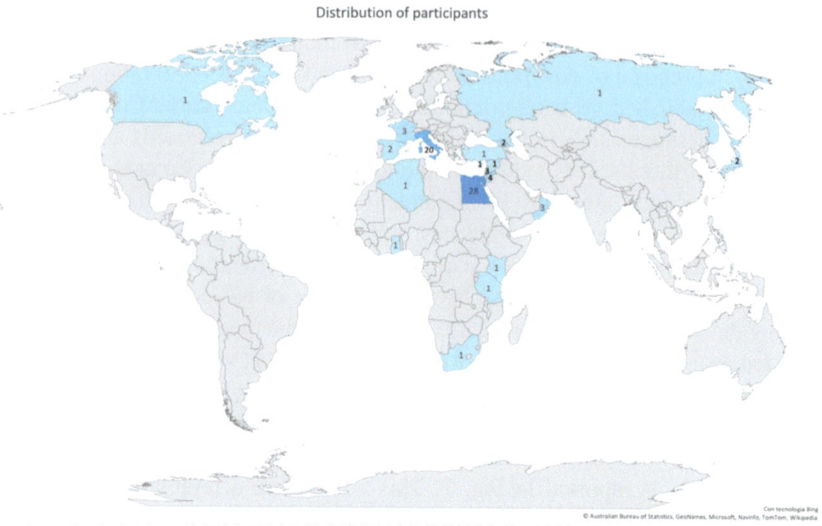

Fig. 2 Distribution of participants to the 2019 Luxor-Aswan Symposium on "sustainable conservation of UNESCO and other heritages sites through proactive geosciences"

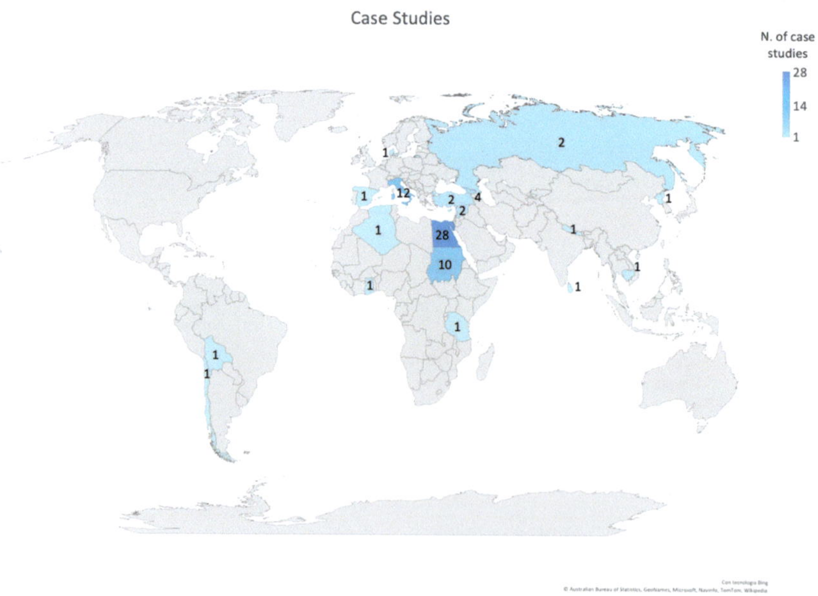

Fig. 3 Discussed case studies

3 The Aswan Declaration

At the closing session of the symposium, the participants (Table 1) actively discussed about a common framework for the sustainable conservation of heritage sites through proactive geosciences and a final joint statement was signed: the Aswan declaration. The text of the declaration is following, while the signed version is reported in Annex 1.

Aswan Declaration

12 December 2019

The international scientific community operating in the field of Geosciences and new technologies for Cultural Heritage protection and conservation, meeting at Aswan in December 2019,

Recognizing that UNESCO World Heritage Sites, and other heritage sites, are located in various geographical settings and their territories may be partly or entirely exposed to various natural hazards and extreme weather events.

Acknowledging that in recent decades, many sites of significant outstanding cultural value have suffered damage, occasionally irreversible, from natural and human-induced processes, with rapid and slow onset, including climate change, resulting in the destruction of countless historical properties, museums and archives that hold the history of humanity within their walls and structures.

Acknowledging that, cultural landscapes, hold natural heritages and provide ecological services valuable to human kind, are being threatened and destroyed.

Recalling that it is urgent and critical to anticipate, plan for and reduce disaster risk in order to more effectively protect cultural heritage, and thus strengthen their resilience, as stated in the Sendai Framework for Disaster Risk Reduction, and the need to rethink 'sites' conservation and management plans.

Recalling that the 2030 Agenda for Sustainable Development calls to strengthen efforts to protect and safeguard the world's cultural and natural heritage (Goal 11.4).

Recognizing cultural heritage role in stimulating participation and responsibility as a key element to reach the targets and goals of the New Urban Agenda, and that the international community has agreed to promote the protection of cultural heritage including through the use of new technologies and techniques.

Acknowledging that geoscience disciplines and affiliated empirical research studies and innovations in technology may provide new paradigms for the preservation of cultural properties providing rich and solid platforms for learning.

Recalling that in the past decades, the shift in disciplines from working in silos to interdisciplinarity and trandisciplinarity draws special attention to the benefit brought from these inter-sectoral initiatives for better management and preservation of cultural and natural heritage. Nowadays, the advance in methodology and techniques offers innovations at all level. It gives birth to a better understanding of the natural processes, coupled with conservation approaches and sustainable practices integrating traditional knowledge in the recovery techniques and building local capacities to ensure effective conservation through time.

Recognizing that earth sciences, and particularly engineering geology and geotechnical engineering, information, communication and technology (ICT), remote sensing and imaging, structural engineering and health monitoring, and early warning systems, play an essential role in conservation and management of cultural properties.

Recalling that the protection of Cultural Heritages from geotechnical and geological hazards is positioned at borderline of Earth Science discipline and Cultural Heritages Conservation.

We the attending participants recommend to countries, end-users and stakeholders:

- to strongly acknowledge the need for interdisciplinary and transdisciplinary approaches to documentation, conservation, protection and valorization of heritage sites, which shall bring together Archaeology, Arts, History, Sciences, Engineering, Traditions, Management and Socio-economic disciplines, among others, driven by sustainability principles.
- to develop a strategy where local knowledge and traditional conservation techniques, as well as new technologies and methodologies, are implemented on the basis of deep knowledge of natural hazards and their impacts.
- to the international community to continue supporting knowledge sharing in the area of sustainable conservation of UNESCO and other cultural and natural heritage sites through proactive geosciences, and other related activities such as the development of the UNESCO Atlas on Natural Hazards.
- to the international community to continue providing the needed support to build resilience of heritage sites, including through the optimization of ecosystems, aligned with the Sendai Framework for Disaster Risk Reduction, the 2030 Agenda for Sustainable Development, the Paris Agreement, and the New Urban Agenda.
- to develop interdisciplinary and transdisciplinary multilingual educational programs in science and technology for knowledge transfer and capacity building.

Recommend to the international community to continue providing the needed support to build resilience of heritage sites, including through the optimization of ecosystems, aligned with the Sendai Framework for Disaster Risk Reduction, the 2030 Agenda for Sustainable Development, the Paris Agreement, and the New Urban Agenda.

Table 1 Participants to the symposium

Name	Surname	Nationality
Hamidatou	Mouloud	Alger
Andreas	Wolter	Canada
Kyriacos	Themistocleus	Cyprus

(continued)

Table 1 (continued)

Name	Surname	Nationality
Abbas	Mohamed Abbas	Egypt
Abdel-Monem sayed	Mohamed	Egypt
Abdel-Hamid	Mohammed Elbshbeshy	Egypt
Ahmed	Elhadi Sherif	Egypt
Mohamed	ElGabry	Egypt
Mostafa	Hussein	Egypt
Mariam	Sallam	Egypt
Samar	Ali Ahmed Ghareeb	Egypt
Hany	ElSayed	Egypt
Mohamed	Marzouk	Egypt
Ahmed	Hamed	Egypt
Mohamed	Ibrahim	Egypt
Amin	Ibrahim Mohammed Ahmed	Egypt
Fathi Abdelaziz Elfadaly	Abdelazi	Egypt
Yasser	El Shayeb	Egypt
Nevine	Aly	Egypt
Waad	Abu ElEla	Egypt
Ghareeb	Sonbol	Egypt
Ayman	El-Ashmawi	Egypt
Adel	Okasha	Egypt
Sabri	Abdel Kareem	Egypt
Said	Soliman	Egypt
Mohamed	Elqady	Egypt
Hesham	ElKaliouby	Egypt
Mohamed Mohy	El Karmoty	Egypt
Hesham	Eid	Egypt
Hany	Helal	Egypt
Amr	Farouk	Egypt
Gad Mohamed	El-Qady	Egypt
Mustapha	Meghraoui	France
Catherine	Meghraoui	France
Jair	Torres	France
Tamara	Meliva	Georgia
Kirkitadze	Giorgi	Georgia
Christopher	Wetcher	Ghana
Yoshinori	Iwasaki	Japan
Sekihiro	Naoyo	Japan
Bushra	Nabass	Jordan
Tahani	Salhi	Jordan

(continued)

Table 1 (continued)

Name	Surname	Nationality
Khaled	Hani Mufleh Al-Amrien	Jordan
Adwan	Abdel Aziz	Jordan
Marco	Fasan	Italy
Rosa	Lasaponara	Italy
Nicola	Masini	Italy
Franco	Vaccari	Italy
Massimiliano	Pieraccini	Italy
Irma	Della Giovampaola	Italy
Guido	Gottardi	Italy
Pietro	Laureano	Italy
Luigi	Petti	Italy
Marco	Camorani	Italy
Giovanni	Fiorini	Italy
Giuseppe	Spilotro	Italy
Maria Dolores	Fidelibus	Italy
Daniele	Spizzichino	Italy
Francesca	Schiavello	Italy
Daniela	Boldini	Italy
Federico	Raspini	Italy
William	Frodella	Italy
Claudio	Margottini	Italy
Luisa	Russo	Italy
George	Eshiamwata Were	Kenia
Elsa	Sattout	Lebanon
Gergian	Raffi	Lebanon
Tania	Zaven	Lebanon
Ahmed	Al Alawi	Oman
Safa	Al Abdali	Oman
Majid	Alabdali	Oman
Yuri	Manstein	Russia
Reda	Sbeinati	Syria
José Antonio	Merodo	Spain
Jose Ignacio	Gallego Revilla	Spain
Heinz	Ruther	South Africa
Zuberi Salum	Mabie	Tanzania
Mahmut	Göktuğ Drahor	Turkey

Annex 1

Aswan Declaration

12 December 2019

The international scientific community operating in the field of Geosciences and new technologies for Cultural Heritage protection and conservation, meeting at Aswan in December 2019,

Recognizing that UNESCO World Heritage Sites, and other heritage sites, are located in various geographical settings and that their territories may be partly or entirely exposed to various natural hazards and extreme weather events.

Acknowledging that in recent decades, many sites of outstanding cultural value have suffered damage, occasionally irreversible, from natural and human-induced processes, with rapid and slow onset, including climate change, resulting in the destruction of countless historical properties, museums and archives that held the history of humanity within their walls and structures.

Acknowledging that, cultural landscapes, contain natural heritage and provide ecological services valuable to human kind, are being threatened and destroyed.

Recalling that it is urgent and critical to anticipate, plan for and reduce disaster risk in order to more effectively protect cultural heritage, and thus strengthen its resilience, as stated in the Sendai Framework for Disaster Risk Reduction, and the need to rethink 'heritage sites' conservation and management plans.

Recalling that the 2030 Agenda for Sustainable Development calls to strengthen efforts to protect and safeguard the world's cultural and natural heritage (Goal 11.4).

Recognizing the role of cultural heritage in stimulating participation and responsibility as key elements to reach the targets and goals of the New Urban Agenda, and that the international community has agreed to promote the protection of cultural heritage including through the use of new technologies and techniques.

Acknowledging that geoscience disciplines and affiliated research studies and innovations in technology may provide new paradigms for the preservation of cultural properties providing rich and solid platforms for learning.

Recalling that in the past decades, the shift in disciplines from working in silos to interdisciplinarity and trandisciplinarity draws special attention to the benefit brought from these intersectoral initiatives for better management and preservation of cultural and natural heritage. Nowadays, the advance in methodology and techniques offers innovations at all level. It gives birth to a better understanding of the natural processes, coupled with conservation approaches and sustainable practices integrating traditional knowledge in recovery techniques and building local capacities to ensure effective conservation through time.

Recognizing that earth sciences, and particularly engineering geology and geotechnical engineering, geophysics, information, communication and technology (ICT), remote sensing and imaging, structural engineering and health monitoring, and early warning systems, play an essential role in conservation and management of cultural properties.

Recalling that the protection of Cultural Heritages from geotechnical and geological hazards is positioned at the borderline of Earth Science disciplines and Cultural Heritages Conservation.

We the attending participants recommend to countries, end-users and stakeholders:

to strongly acknowledge the need for interdisciplinary and transdisciplinary approaches to documentation, conservation, protection and valorization of heritage sites, which shall bring together Archaeology, Arts, History, Sciences, Engineering, Traditions, Management and Socio-economic disciplines, among others, driven by sustainability principles.

to develop a strategy where local knowledge and traditional conservation techniques, as well as new technologies and methodologies, are implemented on the basis of deep knowledge of natural hazards and their impacts.

to the international community to continue supporting knowledge sharing in the area of sustainable conservation of UNESCO and other cultural and natural heritage sites through proactive geosciences, and other related activities such as the development of the UNESCO Atlas on Natural Hazards.

to the international community to continue providing the needed support to build resilience of heritage sites, including through the optimization of ecosystems, aligned with the Sendai Framework for Disaster Risk Reduction, the 2030 Agenda for Sustainable Development, the Paris Agreement, and the New Urban Agenda.

to develop interdisciplinary and transdisciplinary multilingual educational programs in science and technology for knowledge transfer and capacity building.

Aswan, 12 December 2019

Yuri A. Manshtein (Russia) Adel Okasha

(ITALIA)

Saied Abdel Hamid

MIX
Papier aus verantwortungsvollen Quellen
Paper from responsible sources
FSC® C105338

If you have any concerns about our products,
you can contact us on
ProductSafety@springernature.com

In case Publisher is established outside the EU,
the EU authorized representative is:
Springer Nature Customer Service Center GmbH
Europaplatz 3, 69115 Heidelberg, Germany

Printed by Libri Plureos GmbH
in Hamburg, Germany